Lecture Notes on Data Engineering and Communications Technologies

Volume 68

Series Editor

Fatos Xhafa, Technical University of Catalonia, Barcelona, Spain

The aim of the book series is to present cutting edge engineering approaches to data technologies and communications. It will publish latest advances on the engineering task of building and deploying distributed, scalable and reliable data infrastructures and communication systems.

The series will have a prominent applied focus on data technologies and communications with aim to promote the bridging from fundamental research on data science and networking to data engineering and communications that lead to industry products, business knowledge and standardisation.

Indexed by SCOPUS, INSPEC, EI Compendex.

All books published in the series are submitted for consideration in Web of Science.

More information about this series at http://www.springer.com/series/15362

Subarna Shakya · Robert Bestak ·
Ram Palanisamy · Khaled A. Kamel
Editors

Mobile Computing and Sustainable Informatics

Proceedings of ICMCSI 2021

Springer

Editors
Subarna Shakya
Institute of Engineering
Tribhuvan University
Kirtipur, Nepal

Ram Palanisamy
Gerald Schwartz School of Business
St. Francis Xavier University
Antigonish, NS, Canada

Robert Bestak
Czech Technical University in Prague
Praha, Czech Republic

Khaled A. Kamel
Department of Computer Science
Texas Southern University
Houston, TX, USA

ISSN 2367-4512 **ISSN 2367-4520 (electronic)**
Lecture Notes on Data Engineering and Communications Technologies
ISBN 978-981-16-1865-9 **ISBN 978-981-16-1866-6 (eBook)**
https://doi.org/10.1007/978-981-16-1866-6

This Springer imprint is published by the registered company Springer Nature Singapore Pte Ltd.
The registered company address is: 152 Beach Road, #21-01/04 Gateway East, Singapore 189721, Singapore

We are privileged to dedicate the proceedings of ICMCSI 2021 to all the participants, committee members, and editors of ICMCSI 2021.

Preface

It is with deep pleasure that I write this preface to the proceedings of the International Conference on Mobile Computing and Sustainable Informatics (ICMCSI 2021) held in, Tribhuvan University, Nepal, on January 29–30, 2021.

This international conference serves a forum for researchers to address mobile networks, computing models, algorithms, sustainable models, and advanced informatics that supports the symbiosis of mobile computing and sustainable informatics.

The conference covers almost all the areas of mobile networks, computing, and informatics. This conference has significantly contributed to address the unprecedented research advances in the areas of mobile computing and sustainable informatics. ICMCSI 2021 is dedicated to explore the cutting-edge applications of mobile computing and sustainable informatics to enhance the future of mobile applications. The papers contributed the most recent scientific knowledge known in the field of mobile computing, cloud computing, and sustainable expert systems. Their contributions helped to make the conference as outstanding as it has been. The local organizing committee members and their helpers have put much effort ensuring the success of the day-to-day operation of the meeting.

We hope that this program will further stimulate research in mobile communication, sustainable informatics, Internet of Things, big data, wireless communication and pervasive computing and also provide practitioners with better techniques, algorithms, and tools for deployment. We feel honored and privileged to serve the best recent developments to you through this exciting program.

We thank all authors and participants for their contributions.

Kirtipur, Nepal — Prof. Dr. Subarna Shakya
Praha, Czech Republic — Robert Bestak
Antigonish, Canada — Ram Palanisamy
Houston, USA — Khaled A. Kamel

Acknowledgements

International Conference on Mobile Computing and Sustainable Informatics (ICMCSI 2021) would like to acknowledge the excellent work of our conference organizing committee and keynote speakers for their presentation on January 29–30, 2021. The conference organizers also wish to acknowledge publicly the valuable services provided by the reviewers.

On behalf of the editors, organizers, authors, and readers of this conference, we would like to thank the keynote speakers (Dr. R. Kanthavel and Dr. Joy Iong Zong Chen) and the reviewers for their time, hard work, and dedication to this conference. The conference organizers would like to acknowledge all the technical program committee members for the discussion, suggestion, and cooperation to organize the keynote speakers of this conference. The conference organizers also wish to acknowledge the speakers and participants who attended this conference.

Many thanks are given to all the persons who helped and supported this conference. ICMCSI 2021 wishes to acknowledge the contribution made to the organization by its many volunteers. The members have contributed their time, energy, and knowledge at a local, regional, and international levels.

We also thank all the chairpersons and conference committee members for their support.

Contents

About the Editors

Prof. Dr. Subarna Shakya is currently Professor of Computer Engineering, Department of Electronics and Computer Engineering, Central Campus, Institute of Engineering, Pulchowk, Tribhuvan University, and Coordinator (IOE), LEADER Project (Links in Europe and Asia for engineering, eDucation, Enterprise and Research exchanges), Erasmus Mundus. He received M.Sc. and Ph.D. degrees in Computer Engineering from the Lviv Polytechnic National University, Ukraine, 1996 and 2000, respectively. His research area includes government system, computer systems & simulation, distributed & cloud computing, software engineering & information system, computer architecture, information security for e-government and multimedia systems.

Robert Bestak received the Ph.D. degree in Computer Science from ENST Paris, France (2003), and M.Sc. degree in Telecommunications from Czech Technical University in Prague, CTU, Czech Republic (1999). Since 2004, he has been Assistant Professor at the Department of Telecommunication Engineering, Faculty of Electrical Engineering, CTU. He took part in several national, EU and third-party research projects. He is Czech Representative in the IFIP TC6 organization and Vice-Chair of working group TC6 WG6.8. He serves as a Steering and Technical Program Committee member of many IEEE/IFIP conferences (Networking, WMNC, NGMAST, etc.), and he is a member of the editorial board of several international journals (*Computers and Electrical Engineering, Electronic Commerce Research Journal*, etc.). His research interests include 5G networks, spectrum management and big data in mobile networks.

Prof. Ram Palanisamy is Professor of Enterprise Systems at the Gerald Schwartz School of Business, St. Francis Xavier University, Canada. He obtained his Ph.D. in information systems management from Indian Institute of Technology (IIT), New Delhi, India. He had academic positions at Wayne State University, Detroit, USA; University Telekom Malaysia; and National Institute of Technology, Tiruchirappalli, India. Palanisamy's research has appeared in several peer-reviewed articles in several journals, edited books and conference proceedings.

Dr. Khaled A. Kamel is currently Chairman and Professor at Texas Southern University, College of Science and Technology, Department of Computer Science, Houston, TX. He has published many research articles in refereed journals and IEEE conferences. He has more than 30 years of teaching and research experience. He has been General Chair, Session Chair, TPC Chair and Panelist in several conferences and acted as Reviewer and Guest Editor in referred journals. His research interest includes networks, computing and communication systems.

Mitigating the Latency Induced Delay in IP Telephony Through an Enhanced De-Jitter Buffer

Asif Karim, Eshtiak Ahmed, Sami Azam, Bharanidharan Shanmugam, and Pronab Ghosh

Abstract IP telephony or voice over IP (VoIP) at present is promising a shining future for voice services. There are several technical aspects which make the technology attractive; on the other hand, few technical loopholes and shortcomings make user's experience less than optimal and also bring forth significant security issues. This paper offers a technical dissection of the quality of service (QoS) of VoIP. "Signaling" part of VoIP has been discussed based on the Session Initiation Protocol (SIP) along with propositions to tackle problem like jitter that often causes latency in communication. To address the issue of jittering, an alteration in the working mechanism of de-jitter buffer has been put forward where it is shown that addition of few extra variables within the de-jitter buffer to synchronize the packet arrival and release timing can certainly improve the user experience. Reducing the latency is of prime importance to voice data services as it directly affects the acceptance trend of VoIP among mass consumers. The scale of improvement has also been compared to that of a normal jitter buffer as well as a detailed illustration has been provided on Session Initiation Protocol (SIP), a key component of the overall system that makes thing happen. The proposed modification in the de-jitter buffer has been illustrated along with positive results. It shows a one-third improvement in the average latency, resulting into twice as better performance and nearly halved latency.

A. Karim · S. Azam (✉) · B. Shanmugam
College of Engineering and IT, Charles Darwin University, Darwin, Northern Territory, Australia
e-mail: sami.azam@cdu.edu.au

A. Karim
e-mail: asif.karim@cdu.edu.au

B. Shanmugam
e-mail: bharanidharan.shanmugam@cdu.edu.au

E. Ahmed
Faculty of Information Technology and Communication Sciences, Tampere University, Tampere, Finland
e-mail: eshtiak.ahmed@tuni.fi

P. Ghosh
Department of Computer Science and Engineering, Daffodil International University, Dhaka, Bangladesh

S. Shakya et al. (eds.), *Mobile Computing and Sustainable Informatics*, Lecture Notes on Data Engineering and Communications Technologies 68,
https://doi.org/10.1007/978-981-16-1866-6_1

Keywords Signaling · VoIP · SIP · Jitter · De-jitter buffer · IP telephony

1 Introduction

Voice over Internet Protocol (VoIP), which in other words, can be called IP telephony, is a technology which serves the purpose of both voice and multimedia communications using the Internet Protocol (IP) networks. More specifically, it refers to the services such as voice communication, SMS, and multimedia via the public Internet, rather than using the public switched telephone network (PSTN). VoIP with a bit enhancements and modifications is indisputably destined to challenge the prevalent PSTN-based phone system. VoIP means the transmissions of voice data traffic in packets over a data network primarily using protocols such as Session Initiation Protocol (SIP), and this data network can be Internet or an intranet.

There are several synonyms in the market of this technology which is based on various Internet Protocols (IP), such as IP telephone and packetized voice. Usually, the user first initiates call on one side of Internet, and the analog voice signal is then digitized and compressed. Now, this compressed signal is packetized into SIP packets and just simply delivered to the transmission media. At the receiving end, this process is just reversed [1]. The adoption of this technology is on the rise with each year. Some of the advantages for its widespread adoption include easy and inexpensive installation as well as maintenance. Plus, VoIP scales up or down really efficiently, which is a real convenience for the companies as they do not always know how many employees it will have next year, it may have more, or employee number can also be slashed. So, phone line restructuring becomes a painless issue which hardly costs anything significant [2]. A VoIP phoning system supports all the traditional features of PSTN-based lines, and novel features are constantly being provided by the vendors as value-added service. Further, setting up conference calls, an extremely important need for any businesses, is really easy in IP-based telephony and offers commendable quality of service. Besides, VoIP can be used for fax services as the traditional fax machines cost a lot over long distances, and quality attenuation over long distance is also an issue [3]. In VoIP-based services, these bottlenecks are non-existent. Simultaneous functionality is also an important feature of modern IP telephones, such as working with emails from the IP phone set at the same time carrying on conference calls using that very phone set. Thus, the usage of VoIP is only set to increase.

However, as with many technologies in their forming stages, VoIP also has few performance issues to be considered. This paper not only addresses some of these design issues, but also brings forth few propositions to improve the shortcomings. Although VoIP is likely to take over the traditional telephone system, this mechanism will not be of much convenience to business organizations as they will have to risk their business due to some exacting loopholes in the VoIP's working model. For several reasons, the experience of using it can be hampered such as high amount of traffic passing through same bandwidth or delay in the process of compressing and then decompressing the information from sender to receiver [4]. Another major concern of this technology is the security.

There are number of loopholes in this aspect [5]. Among myriad of security concerns on a VoIP network, Address Resolution Protocol (ARP) spoofing has been one of the trickiest to deal with. In ARP attacks, the attacker broadcasts spoofed announcement of the MAC address forcing subsequent IP packets to flow through the attacker's host. This consents the eavesdropping of communications between two users [6]. The attacker can easily exploit the clear and plain format of a SIP packet to his advantage and can incur damage. There are lot of mechanisms for encrypting textual data in a network such as public–private key, ciphers, hash functions, and deploying image-keys. [7], but using cryptosystems for voice data often compromises the quality and latency even further. Cryptographic algorithms itself may show performance issues while processing [8]. While the bandwidth and its sharing is a problem that is user and condition specific, the delay factor and security concerns can be considered as problems on a technical level.

A VoIP system works by converting analogue voice signals into digital, then transmitting those as data over the broadband\Internet communication channel. It is a very efficient way of making calls—for a start, once it is set up, it is considerably cheaper than normal phone lines.

1.1 Problem Formulation

Delay is generated in a VoIP network due to multifarious reasons. One of the significant reasons for this is Firewall inspection of SIP packets. Delay is the amount of time it takes to reach a packet to its destination. Traditionally in PSTN, a delay of 150 ms is acceptable. Therefore, in a VoIP model, if it exceeds 150 ms, the users at both ends can become perplexed to who should continue speaking and who should be listening, as they experience an incessant gap between words or fractions of a word [9]. The limitations caused by this problem have been illustrated in a later section of this paper.

Even though there are multiple concerns to be addressed in terms of improving the VoIP services, this paper will look more into the quality of services (QoS) issues rather than security threats. The objective of this paper is to pinpoint few of the snags and propose a model to address the delay issue.

2 Relevant Works

Jitter is the congestion that occurs when a large number of Internet connections attempt to compete with each other at the same time, resulting in many tiny packets of information vying to use the same IP network; this can germinate significant delays in voice communication. There have been a good number of studies where this problem has been taken into consideration. One way includes making two different virtual switches out of one switch. However, it has the problem of non-uniformity among switches in that not all the switch supports this kind of functionality. Further, there has been some development in side areas that aids in helping the router to prioritize VoIP packets over normal data packets to process accordingly. One such technology is DiffServ. It is a quality of service (QoS) protocol that prioritizes IP voice and data traffic to help preserve voice quality, even when network traffic is heavy. It uses the type of service (ToS) bits in the IP header to indicate traffic importance and drop preference [10]. One other technique that has been developed to fasten the speed at which the packets flow is header compression. Header compression tries to leverage the repetitive bytes found in the header of a VoIP packet [11]. This attempt has somewhat been successful due to the significant size of the packet header to the payload. This technology is already integrated into many routers. However, if this applied too rapidly, the offshoot will be a negative one as it will, in fact, compound the jitter and delay factor. So, the best practice is to balance with care.

A number of studies have been done where the researchers proposed solutions of delay buffers to be implemented for controlling delays of all kinds including jitter [12, 13]. The idea here is that the voice packets will be stored in a buffer temporarily and will be sent out in a consistent manner, and thus, the packets will reach to destination in a certain interval of time. The users will then have a fairly good expectant time of hearing the voice from the other end and can adjust accordingly. This is a popular implementation and thus found its way into many hardware vendors doorstep. Many routers include this facility also. Another study highlights the importance of de-jitter buffer and analyzes behavior of jitter buffers with and without packet reordering capability and measures the additional packet loss incurred by packets dropped in buffer on top of the measured network packet loss [14]. However, it does have problems such as a larger de-jitter buffer, which causes higher latency inflicted on the system, while a smaller jitter buffer causes higher packet loss. A de-jitter buffer is architected to eradicate the effects of jitter from the decoded voice stream.

De-jitter buffer, as shown in Fig. 1, buffers each arriving packet for a short span before sending it out. This surrogates additional delay and packet loss for jitter. There are two types of de-jitter buffers. A **fixed de-jitter buffer** maintains a constant size, whereas an **adaptive jitter buffer** has the faculty of adjusting its size dynamically in order to optimize the delay/discard trade-off [9].

Both fixed and adaptive de-jitter buffers are equipped with the facility of automatically adjusting to changes in delay. For example, if a step change in delay of 15 ms transpires, then some short-term packet discarding can occur there, resulting from the change; however, the de-jitter buffer would be instantly realigned.

The current implementation of the de-jitter buffer still has some limitations which affects its performance. As example, if there is too many packets in the buffer at a certain point of time, a lot of the packets have to be discarded as their TTL had already crossed the accepted margin, caused by the buffer overrun problem. Making the buffer larger to tackle the problem could result in much higher level of latency. As there is no specific way to say for sure when which packet will arrive, it could happen that packets will arrive in a different order than the actual or expected one.

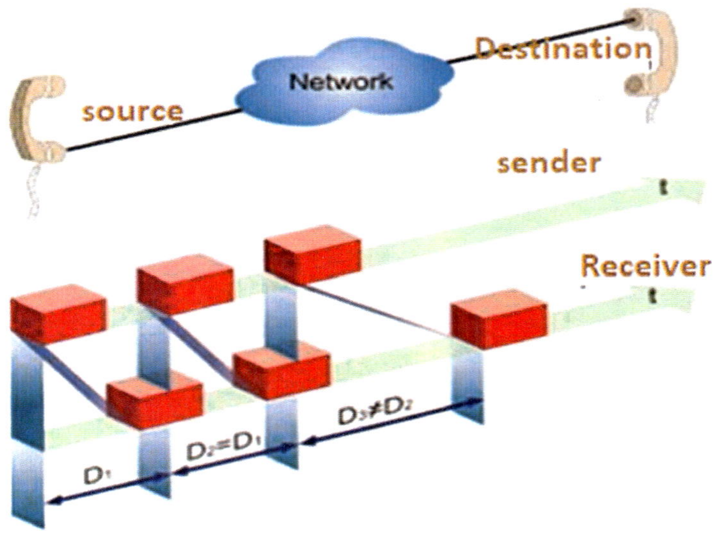

Fig. 1 De-jitter buffer

At this point, setting a fixed delay can create a latency or break in the conversation. Also, larger packets tend to incur network congestion, introducing more and more delay.

Number of studies and research have already been accomplished on this topic, but the problems still remain intact. Many of the previous works actually cite a workaround instead of furnishing a complete solution. A recent work highlighted the issue of *jitters* in VoIP and proposed a solution that uses optimal packet call flow routing (CFR) model in real voice optimization to reduce jitter in VoIP systems [15]. Another study focused on presenting how to achieve maximum VoIP connection quality with optimum de-jitter buffer delay [16]. The results show that five-sixths of the connections were of either high or medium quality having a relatively small delay with the de-jitter buffer. In [17], the parameter selection of de-jitter buffer has been considered as the key to improve the service performance. The simulation results indicate that both the transmission delay and bit error rate can be reduced with packet size and buffer length being optimized. Table 1 summaries the above-discussed related works.

All of these works have proposed good solutions using de-jitter buffers; however, some of the existing problems with de-jitter buffer such as the buffer overrun problem and latency induced by bigger buffer size have not been addressed. This research aims to propose a potential solution in this regard.

Table 1 Summary of the above-discussed systems

Authors	Proposition	Shortcomings
Salem et al. [12]	Use large memory buffer to hold the packets to transmit later in a consistent manner	Does not address the associated higher Latency due to large buffers
Voznak et al. [13]	Also deploy similar buffer with extended memory usage	Suffers from the latency issue that can be seen with large memory buffers
Rodbro and Jensen [14]	Uses header compression	Unless balanced properly, header compressions can increase jitters
Adebusuyi et al. [15]	Uses techniques such as real voice optimization	Does not perform well in 4G-based VoIP services
Lebl et al. [16]	Takes an approach to address a number of different parameters related to QoS	The outcome is not consistent and may vary according to the buffer size

3 VOIP Technique and Network Structure

After the connection is established, the voice data is converted into digitized form. However, as digital data requires a lot of bits, it is compressed. Now, the sample of voice is packetized to be sent over the Internet. These packets are wrapped with Real Time Protocol (RTP). The packets are synchronized by means of identification number, inserted in the RTP header for each packet. Each packet is then further wrapped around User Datagram Protocol (UDP). UDP carries RTP as payloads. As the packets reach the other end, the whole process is reversed, that is the packets are disassembled and put into the proper order; digital bits are extracted out and decompressed, converted to analog signal, and finally delivered to the user's handset. The role of Session Initiation Protocol (SIP) here is that it encompasses all the aspects from call initiation to call termination; and to do that, it includes number of related protocols such as RSVP, RTP RTSP, SAP, and SDP.

VoIP can be configured and settled into different topologies and configurations. But before discussion of how the VoIP is configured, some related terms such as media gateways and call managers or soft switch need to be understood.

3.1 Media Gateways and Call Managers

Media gateways (servers) act as a translation unit between disparate telecommunications networks [1]. VoIP media gateway (MG) performs conversion between TDM voice to Voice over Internet Protocol. Media gateways are controlled by a media gateway controller known as call managers or soft switches, providing the call control and signaling functionality [1]. Communications via these two entities are done through protocols such as SIP (which will be discussed later), MGCP, or H.248.

AMG has to perform several operations in a VoIP network. Primarily,

- Carries out A/D conversion of the analog voice channel.
- Transforms a DS0 or E0 to a binary signal compatible with IP or ATM.
- Multi-vendor interoperability.
- Transport of voice mainly using IP-based RTP/RTCP.

3.2 *Topologies*

There are number of ways in which a VoIP network can be structured [18]. One of which is the use of PC and the employment of a router. An easy and inexpensive way to use VoIP is the 1:1 VoIP gateway architecture.

.

3.3 *Session Initiation Protocol (SIP)*

Session Initiation Protocol (developed by IETF) is a protocol which is termed as an application layer signaling protocol. SIP primarily deals with interactive communication using multimedia elements. Signaling means to initiate, modify parameters, and terminate sessions between end users. SIP does exactly this. SIP calls may be terminal-to-terminal, or they may require a server to intervene [19]. Note that SIP shares a close affinity with IP. A SIP message looks very much like a HTTP message and thus in plain text. SIP addresses are also very much like emails such as sip:name@a_domain.com.

The user can usually set up call in two modes, called redirect and proxy, and servers are designed to handle these modes.

Besides signaling, SIP can also be used to invite other people in a conversation or can simply be employed to start a new session. SIP is completely independent of the type of content and the mechanism used to convey the content [20]. SIP is there only for signaling purposes, not to worry about any other aspects of media flow or type. SIP simply works as glue inside the whole process among different layers and protocols.

To support session services, there are five facets of multimedia session management for SIP:

User Locality: Identifying which end system will be used for communication.

User Potentiality: Determining the media and media parameters to be used for this communication.

User Availability: Determining whether or not the called party is enthusiastic to engage in communications.

Session Setup: Setting up the session parameters at both the called and calling parties.
Session Management: Including the transfer and termination of sessions, the modifying of session parameters, and the invoking of session services.

3.4 SIP Elements

Before going into the discussion of how SIP works, some terms of the world of IP telephone using SIP need to be understood.

(1) ***User Agents (UA)***: An endpoint in a VoIP system, normally IP phones or media gateways. It usually has an interface toward the users. When user A wants to call user B, he fires up the appropriate program containing a SIP UA. The user interacts with user B through the interface. Naturally, user B at the other end also has a similar kind of program, and he can either accept or reject the invitation from user A. The UA also incorporates various media tools to actually handle the media content. Normally, the UA establishes the session, and the media tool it contains handles the content of the session. In many applications, these two come under one program. For example, a video conferencing software [21].
(2) ***Redirect Server***: Redirect servers aid the SIP UAs by furnishing location information where the user can be sensed. From Fig. 5, it can be seen how redirect servers perform their action. For example, say user A wants to call user B (located at domain.com) and thus sends an invitation. But at domain.com, a redirect server is positioned to handle the incoming calls, so, the UA of user A, instead of contacting with the intended address, contacts either SIP:A@192.168.1.1 or SIP:A@192.168.1.2. These addresses are different from the intended address. The redirect server may also propose which of the address is more likely. It then simply sends back messages that contain this information, and the UA of A tries to establish session with these addresses. However, it can also return address of other server which may have more information, and in other words, this process can be chained [21].
(3) ***Proxy Server***: The aim of the proxy is same, but in this case, instead of returning the possible address information, the proxy tries those addresses by itself in order to establish the session. Note that in that case too, there may be intermediary proxies and location servers. Oftentimes, a single server can be configured as both redirect and proxy, and these two can also coexist in a single system [21]. In practice, a SIP-based proxy server facilitates the establishment and maintenance of communication channel between two SIP addresses. Any SIP device can communicate to another SIP device, but in order to achieve that, they deploy a go-between, called a SIP proxy, to begin the communication, which then drops out, allowing point-to-point communication.
(4) ***SIP Register***: An application typically running on a server that aids the UAs to register themselves so that they are able to receive calls. The registrar can

or cannot be colocated with the proxy or redirect servers. Generally, the user provides information about his location [21].

4 Limitations

This section will have a detail discussion on the limitations of VoIP mentioned earlier.

Among various and diverse kinds of performance issues relating to VoIP, "Delay" is the most crippling one. The delay in a VoIP network can be for many reasons. A VoIP network must be able to deliver packets within a standard max of 150 ms. This puts a major consideration on how much security overhead can be included. Also, it leaves very little margin for error recovery in case of an error in packet delivery. One type of delay is called **fixed delay** and is mostly occurred within the signal processing systems [22], such as the processing delays within the voice coders/decoders (codecs) that make the A/D signal conversions and are also detected within the physical transmission systems, such as the copper pairs. The **variable delays** come from queuing times at packet processing points, such as routers and switches, plus transmission variables, such as the path that a particular packet or series of packets takes within the network [22]. Firewall routers such as double socket routers re-establish an IP packet flow on the inner side of the router after it has been disconnected on the outer side. This aids in a regulating IP packets in a consistent diction but also introduces delays. Sometimes, there is a delay when the firewalls update the iptables.

Also, larger packets tend to incur network congestion, introducing an additional delay on top of the delay incurred in each hop the packet travels through. This type of non-uniform, variable delay is known as **jitter**. Jitters transpire because not all the packet will take the same route, and this actually is more of a problem as this can introduce the arrival and processing of packets out of order. RTP is based on UDP, so there are no reassembles at the protocol level. However, rearranging can be done in application layer, but it tends to be slow, which, needless to say, compounds the delay malefactor. Figure 2 details how this jitter happens.

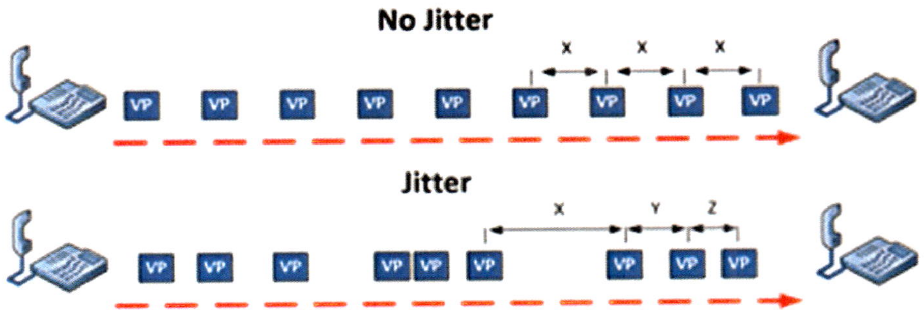

Fig. 2 VoIP jitters [23]

In Fig. 5, the amount of time (say, z_p) it requires for packets *A* and *B* to send and receive is equal when there is no jitter. But when the packets confront delay in the network, this uniformity of z_p is affected, and the expected packet may be received *after* it is expected. This is why a *jitter buffer*, which hides inter-arrival packet delay variation, is essential [22]. Voice packets in IP networks have highly variable packet-inter-arrival intervals. Common practice is to count the number of packets that arrive late and create a ratio of these packets to the number of packets that are successfully processed. Then, the ratio can be utilized to adjust the jitter buffer to target a predetermined, allowable late-packet ratio; this, in most cases, compensates for delays. Note that these buffers can be either dynamic or static. Quality of service (QoS) is a make or break for the acceptance of VoIP. The ever exponential growth of Internet user and the development of multitude of multimedia-oriented services exert increasing demands on Web servers and network links, which can cause overloaded end-servers and congested network links [24]. These issues almost make it indispensable that the design of VoIP, or for that matter any Internet-based application, should strive to achieve every bit of possible design efficiency.

5 The Proposed Solution

Several solutions have been introduced to get an upper hand on obstacles like delay and jitter. In this section, this study will present a proposed way around for the issue of jitter.

5.1 Modification in De-Jitter Buffer

To tackle the problem of jitter and fixed delay, the implementation of de-jitter buffer is common. However, as has been said before, if there are too many packets in the buffer, many of the packets having their TTL already crossed the accepted margin must be discarded due to the buffer overrun problem. Also, the next problem is if the size of the buffer is large then it causes too much latency. Because there is no guarantee that after exactly what span of time, the packets will arrive. Thus, one packet that has been formed later may appear earlier in the buffer than a packet that has been formed and sent before the previous one.

Thus, if the delay length of the buffer is set to 200 ms, all the packets will be played out after 200 ms, and thus, it may create latency or break in the conversation for a packet that has been reached earlier. So, the proposed solution will bring some modification in the working procedure to the de-jitter buffer. The modified buffer, as shown in Fig. 3, will have two extra variables, say $\text{Syn}_{\text{f\_pk}}$ and $\text{Syn}_{\text{n\_pk}}$; $\text{Syn}_{\text{f\_pk}}$ will record the first packet's synchronization number, (that is, 1), i.e., the packet serial number, inscribed in the header of the packet itself. It will then, instead of saving the

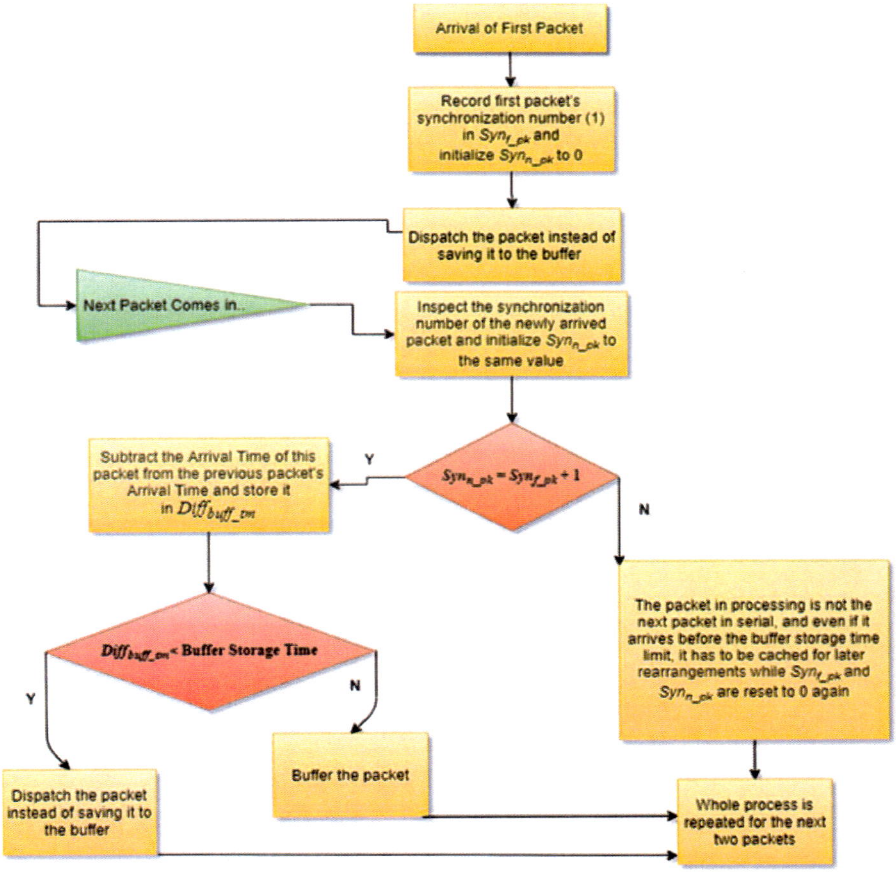

Fig. 3 Flow diagram of the proposed modification

packet, dispatch it. Now, the next packet comes, at this moment, $\text{Syn}_{\text{f\_pk}}$ equals to 1, and $\text{Syn}_{\text{n\_pk}}$ equals to 0.

The buffer will inspect the synchronization number of this packet, and if this number is one greater than Syn_{f_pk}, i.e., $Syn_{n\_pk} = Syn_{f\_pk} + 1$, then it will subtract the arrival time from the previous packet's arrival time and store it in another variable, say $Diff_{buff\_tm}$. Now, if $Diff_{buff\_tm}$ is less than the buffer storage time, it will again, instead of saving the packet, forward it.

However, if $Syn_{n\_pk} <> Syn_{f\_pk} + 1$ (unequal), then the packet in processing is not the next packet in serial, and even if it arrives before the buffer storage time limit, it has to be cached for later rearrangements, while Syn_{f_pk} and Syn_{n_pk} are reset to 0 again. For the next two packets, the same mechanism is repeated. This way, cached packets that arrive before the expected delay margin will be propagated instead of probable drop-out after a certain period, and it will not create latency, in-effect improving the overall performance. Synchronization is also maintained using the same logic. However, there will still be some codec-related delays that need to be accounted for.

6 Results and Analysis

The impact of reducing the delays in jitter can have significantly strong effect in the reduction of latency. In this section, this dramatic effect will be explained through facts and figures. The observable latency in a VoIP service is often the combination of parameters such as *average jitter* and *average latency* into one single metric known as "Effective Latency". It is calculated as

$$\text{effective\_latency} = \text{avg.\_latency} + (2 * \text{ avg.\_jitter}) + 10.0 \tag{1}$$

In Eq. 1, effect of jitter is usually being doubled as its impact is relatively high on the overall performance, and a general constant of 10 ms (milliseconds) is added to account for the delay caused by the respective codec. *Average latency* can be measured using usual concepts [25], for instance, if the sum of measured latency is 960 ms and the number of latency sample is 30, then the average latency is 32 ms (960/30). *Average jitter* is calculated by comparing the interval when RTP packets were sent to the interval at which they were received [26]. For instance, if the first and second packets leave 75 ms apart and arrive 95 ms apart, then the jitter is 20 ms (95–75). Now, if we sample such differences, for example, 5 times, and receive series of jitters such as 20, 45, 36, 15, and 14, then the average jitter is 26 ms; it basically follows the Eq. 2 (***P*** denotes the specific packets departure and arrival time in ms).

$$\text{Avg.\_jitter} = \frac{\sum ((P_2 - P_1) + (P_4 - P_3) + \cdots + (P_n - P_{n-1}))}{N} \tag{2}$$

Now, before implementing the improvements in a normal de-jitter buffer, the average latency for five consecutive minutes had been recorded as 25, 32, 33, 28, and 50 ms and the corresponding average jitters as 35, 48, 62, 42, and 65 ms (Using Eq. 2). These two parameters have now been used to calculate the effective latency using Eq. 1, and the result was found to be a sharp increase in effective latency that the user will experience.

The effective latency in milliseconds for five consecutive minutes has been calculated as 105, 138, 167, 122, and 190, respectively, using Eq. 1. Figure 4 illustrates the rise of latency, or in other words, declination of quality of service (QoS) in the conversation graphically.

It can clearly be seen from Fig. 4 that even though there are some improvements around fourth minute (T4), the general trend shows an increase in effective latency, which causes disruption in service to heighten sharply as well. Once a delay associated with jitters starts to materialize, the quality oftentimes may deteriorate rapidly in subsequent seconds.

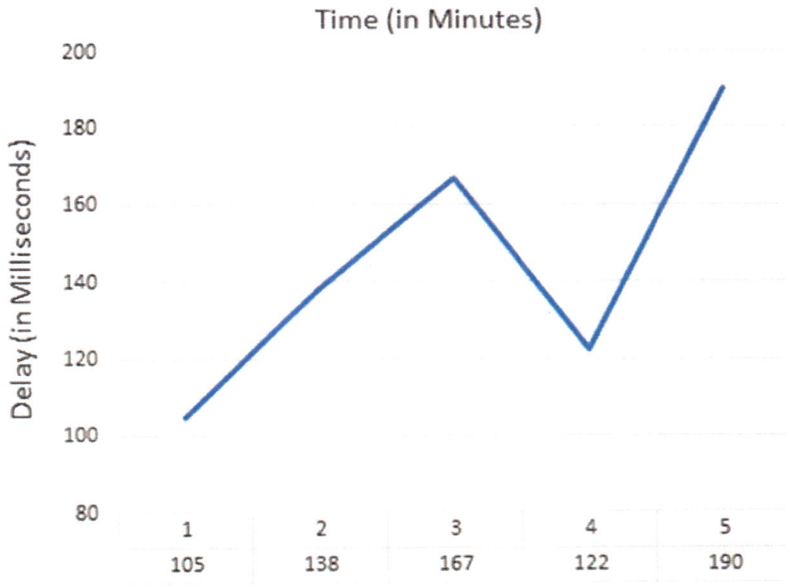

Fig. 4 Effective latency in normal jitter buffer

Though sometimes the curve will have opposite direction, that is quality may get better with time, however, it can be the side-effect of other network factors, such as throughput, bandwidth or congestion getting improved.

As can be seen from Fig. 5, a one-third improvement in the average latency due to the improved design in the jitter buffer may double the performance in the quality of the conversation. It basically halves the effective latency. Such is the importance and effect of improvement in the jitter buffer, which can clearly affect considerable positive impression in the overall quality of service. Thus, an efficient design and construction of de-jitter buffer indeed is a crux part of a VoIP system. Additionally, the proposed study presents better result in terms of comparison with earlier discussed systems that include application of large memory blocks.

7 Conclusion

As data traffic continues to increase and cross that of voice traffic, the convergence and integration of these technologies will not only continue to rise, but also will streamline the way for a truly unified and seamless avenue of communication. VoIP can provide significant benefits and cost savings. Since voice and data traffic can be integrated, the necessary infrastructure to provide both services is reduced. Also, bandwidth will be utilized properly as bandwidth on a network is rarely completely flustered with data traffic, and circuit switching calls usually waste a lot of bandwidth

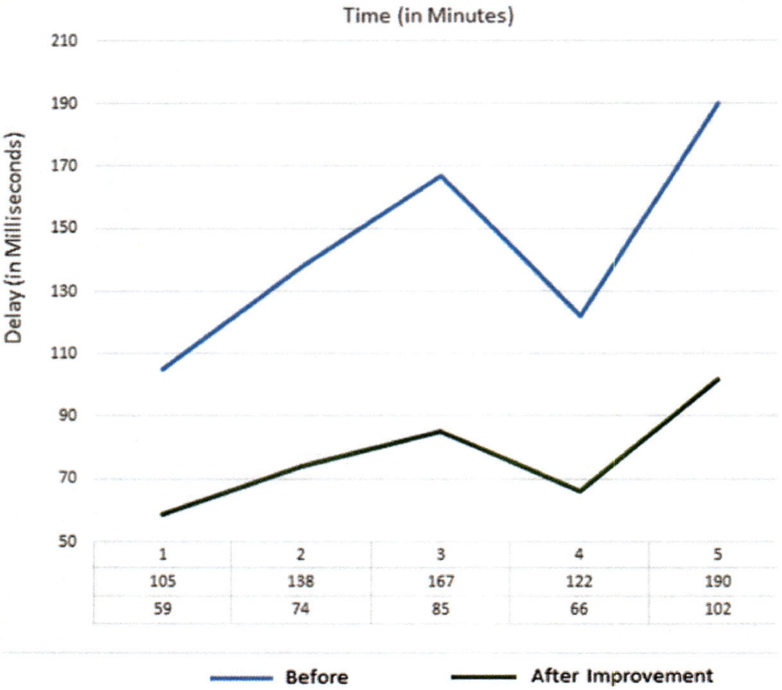

Fig. 5 Comparison of effective latency experienced before and after the implementation of the recommended improvements in jitter buffer

in general. VoIP is still in its embryonic condition in many areas of its development. It truly is a revolutionary technology that will pose many challenges to circuit switched infrastructure.

This paper discusses the issues of latency caused by jitter and also proposes an improved de-jitter buffer which can certainly have a positive effect on the technology. The implementation results indicate many positives. It shows a one-third improvement in the average latency due to the improved design in the jitter buffer, resulting into potentially doubling the performance in the quality of the conversation. Additionally, the effective latency factor has been observed to be nearly halved. The detailed working procedure of SIP and the importance of quality of service (QoS) in a VoIP network have also been discussed. Although the results are very promising and could potentially improve the quality QoS of the VoIP service, more improvements can be done. This solution needs to be tested on a larger scale to identify possible improvement suggestions. Traditional noise issues such as additive and subtractive noises may affect the performance, and thus, further work is needed to factor against such issues for a better quality of service.

References

1. R. Kuhn, T. Walsh, S. Fries, *Security Considerations for Voice over IP Systems* (National Institute of Standards and Technology, Department of Commerce, U.S, 2005).
2. L. Uys, Voice over internet protocol (VoIP) as a communications tool in South African business. Afr. J. Bus. Manage. **3**(3), 089–094 (2009)
3. *Adding Reliable Fax Capability to VoIP Networks*. Dialogic White Paper (2010)
4. N.L. Shen, N.A. Aziz, T. Herawan, *End-to-End Delay Performance for VoIP on LTE System in Access Network*, in Lecture Notes in Computer Science, vol. 7804 (Springer, Berlin, Heidelberg, 2013)
5. E.B. Fernandez, J. Palaez, M. Larrondo-Petrie, Security patterns for voice over IP networks. J. Softw. **2**(2), 19–29 (2011)
6. Z. Trabelsi, W. El-Hajj, ARP Spoofing, in *Information Security Curriculum Development Conference*, Sept 2009
7. A. Karim, Multi-layer masking of character data with a visual image key. Int. J. Comput. Netw. Inf. Secur. **9**(10), 41–49 (2017)
8. M.E. Haque, S.M. Zobaed, M.U. Islam, F.M. Areef, Performance analysis of cryptographic algorithms for selecting better utilization on resource constraint devices, in *1st International Conference of Computer and Information Technology (ICCIT)* (2018)
9. C.D. Nocito, M.S. Scordilis, Monitoring jitter and packet loss in VoIP networks using speech quality features, in *IEEE Consumer Communications and Networking Conference (CCNC)*, Jan 2011
10. J. Li, Q. Cui, The QoS research of VoIP over WLAN, in *International Conference on Communications, Circuits and Systems* (2006)
11. Y. Wu, J. Zheng, K. Huang, The performance analysis and simulation of header compression in VoIP services over cellular links, in *9th Asia-Pacific Conference on Communications*, Sept 2003
12. A.M. Abdel Salam, W. Elkilani, K.M. Amin, An automated approach for preventing ARP spoofing attack using static ARP entries. Int. J. Adv. Comput. Sci. Appl. **5**(1) (2014)
13. M. Voznak, A. Kovac, M. Halas, *Effective Packet Loss Estimation on VOIP Jitter Buffer* (Springer, Lecture Notes in Computer Science, LNCS-7291, 2017), pp. 157–162
14. C. Rodbro, S.H. Jensen, Time-scaling of sinusoids for intelligent jitter buffer in packet based telephony, in *Speech Coding, IEEE Workshop Proceedings* (2002)
15. K. Adebusuyi, E. Hilary, G.K. Ijemaru, A research of VoIP jitter on packet loss in GSM voice over IP systems. Int. J. Innov. Technol. Explor. Eng. (IJITEE) **6**(12), 5–12 (2017)
16. A. Lebl, M. Mileusnić, D. Mitić, Ž. Markov, B. Pavić, De-jitter buffer role in improving VOIP connection quality—examples from practice, in *International Scientific Conference On Information Technology and Data Related Research* (2017)
17. H. Gao, H. Cheng,Simulation of de-jitter buffer parameter optimization of in power PTN. Stud. Opt. Commun. **1** (2018)
18. W. Wang, K. Bourg, Brian lane and James farmer, in *VoIP Architecture, FTTx Networks*, pp. 307–322, Dec 2017
19. D. Noworatzky, How to troubleshoot voice quality problems in VoIP phone systems. https://info.teledynamics.com/. Accessed 8.9.20 (2018)
20. C. Davids, V.K. Gurbani, S. Poretsky, *Terminology for Benchmarking Session Initiation Protocol (SIP) Devices: Basic Session Setup and Registration*, RFC 7501, Apr 2015
21. X. Yang, R. Dantu, D. Wijesekera, Security issues in VoIP telecommunication networks, in *Handbook on Securing Cyber-Physical Critical Infrastructure*, pp. 763–789 (2012)
22. L. Zheng, L. Zhang, D. Xu, Characteristics of network delay and delay jitter and its effect on voice over IP (VoIP), in *IEEE International Conference on Communications*, Aug 2002
23. A.S. Amin, H.M. El-Sheikh, Scalable VoIP gateway, in *The 9th International Conference on Advanced Communication Technology*, Feb 2017
24. J.K. Lee, K.G. Shin, NetDraino: saving network resources via selective packet drops. J. Comput. Sci. Eng. **1**(1), 31–55 (2007)

25. P.C. Saxena, J. Sanjay, R.C. Sharma, Impact of VoIP and QoS on open and distance learning. Turkish Online J. Distance Educ. **7**(3) (2006)
26. M. Amin, Voip Performance measurement using QoS parameters, in *The 2nd International Conference on Innovations in Information Technology* (2005)

A Cyber-Safety IoT-Enabled Wearable Microstrip Antenna for X-Band Applications

R. Praveen Kumar, S. Smys, Jennifer S. Raj, and M. Kamarajan

Abstract In recent days, wearable antennas are getting adequate attention in many areas. The antennas should be lightweight and its should be easily integrated with other systems. To match with the system requirements, conduction mode of textiles and conductive threads are used in many places. A specified shape (badge) microstrip wearable antenna for X-band applications is proposed. The specified shape (badge) antenna is simulated using CST MW Studio. Two different parameter antennas were designed and proposed with one slot in and the size is 1 mm. The specified badge-shaped antenna is structured with rectangular substrate and rectangular ground, whereas the radiating element will be in badge shaped. The specified antenna consists of four conductive threads in truncated patch and a circular ring patch antenna. The antennas are placed in indoor and outdoor positioning. This antenna was embedded on the top of the military beret. The proposed specified shape antenna is embedded for vehicle movement detection. This proposed antenna is often used in speed gun which is used to detect the real-time speed of the vehicle. The different parameters like frequency, voltage standing wave ratio, E-Field, S11 parameter and H-Field were analyzed using simulation tool. The working frequency will be 10.483 GHz for both antennas. The standing wave ratio for antenna with slot will be 1.37 and without slot will be around 1.34.

Keywords Wearable antenna · CST MW Studio · Vehicle speed detection · VNA · Directivity · Badge-shaped antenna · Cotton substrate

R. Praveen Kumar (✉)
ECE, Easwari Engineering College, Chennai, India

S. Smys
CSE, RVS Technical Campus, Coimbatore, India

J. S. Raj
ECE, Ganamani College of Engineering and Technology, Namakkal, India

M. Kamarajan
ECE, Mohamed Sathak A.J. College of Engineering, Chennai, India

S. Shakya et al. (eds.), *Mobile Computing and Sustainable Informatics*, Lecture Notes on Data Engineering and Communications Technologies 68,
https://doi.org/10.1007/978-981-16-1866-6_2

1 Introduction

Wearable technology is being prevalent since more than a decade. Wearable antennas play an integral part in the design of wearable electronics. These antennas are mainly used to detect and analyze some physiological parameters like temperature and heart rate [1, 2]. These antennas are integrated into wearable devices like smartwatches and Bluetooth devices [3] that can be hidden by any fashion accessory for many applications like location tracking [4–6], audio recording and many more. Wearable technology came into picture during 1995–1997, wherein Jennifer Healey and Rosalind Picard designed prototypes of wearable devices and integrated them with fashion accessories and collected physiological data from the person who was wearing it. Later in 2004, a fashion design label called "cute circuit" introduced the concept of Bluetooth connected electronics [7]. Wearable antennas are preferred for the fact that they have the flexibility of substrate [8–10] and ease of integration [11] into commonly used fabrics. Although they have some disadvantages like low efficiency because of their contact with the human body [5, 12, 13] which acts as a noisy source while capturing desired electromagnetic radiation, and the placement of antenna adds a decrement in the performance characteristics of wearable antennas since the geometry of the surface matters for the antenna to work effectively. The importance of wearable antenna technology is now rapidly growing in par with the interest of consumers on wearable electronics. Wearable antennas can a dipole, a microstrip patch or any other geometry. Conductive textiles and threads are incorporated to integrate dielectric substrate and radiating substance. This article proposes two microstrip wearable antennas, one with a rectangular slot of 1 mm width in the center and the other without any slot. These two antennas are of badge-shaped geometry and are desired to work for X-band (8–12 GHz) applications. The operating frequency is around 10.5 GHz and this antenna may be used for speed detection of vehicles. The dielectric substrate is chosen as cotton fabric whose dielectric constant $\varepsilon = 1.35$ and loss tangent value of 0.0009. Both the badge-shaped antennas are designed to work for 10.48 GHz. This antenna may be incorporated on the uniform sleeve worn by traffic policemen to detect speed of the vehicles. The cotton substrate provides the advantage of flexibility when worn on the sleeve. The antenna is simulated using CST MW Studio then fabricated and tested using vector network analyzer. Figure 1 shows the basic structure of a microstrip antenna; Fig. 2 shows the geometry of a general textile antenna.

This article contains the following sections: Sect. 2 provides an idea about the existing approaches of wearable antennas. Section 3 contains the detailed design of the badge-shaped microstrip antenna and the performance results of both simulated and fabricated antenna. Finally, Sect. 4 contains conclusion and future work.

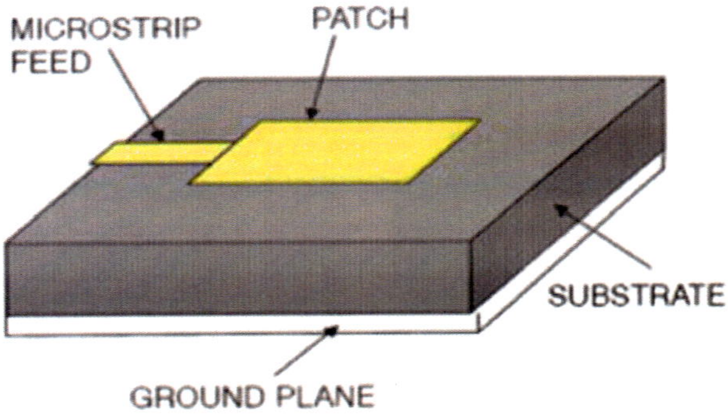

Fig. 1 Basic structure of microstrip antenna

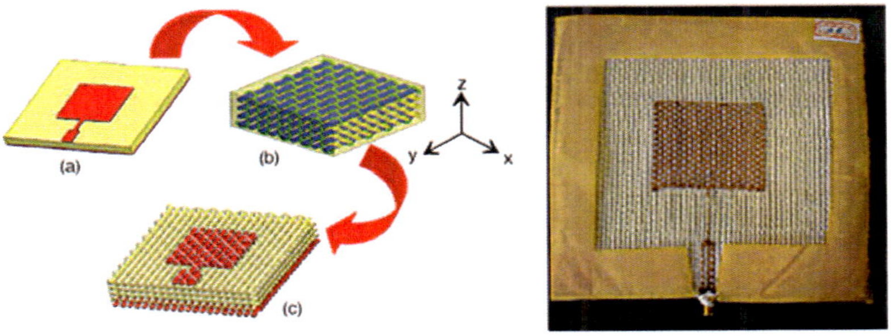

Fig. 2 Geometry of textile antenna

2 Existing Approach

Over the past few years, there have been many developments in the wearable antenna. Lee and Tak [6] reported a truncated patch antenna integrated into the beret of the army officers. It was designed to give the live location of the officers at any time. The circular ring patch that was developed was found to be having an operational frequency of 915 MHz with simulated return loss of 10 dB. In another report given by Zhi Xu, Thomas Kaufmann and Christophe Fumeaux [7, 14, 15] show the use of textile fabric in the design and fabrication of wearable and flexible antenna. On studying the antenna properties and characteristics of the antenna, it was seen the antenna's functionality did not change even under extreme bending conditions from 90° to 180°. The study showed that the functionality did not change even under extreme stress.

3 Antenna Design

3.1 Antenna Configuration

The proposed wearable microstrip antenna is designed with and without a rectangular slot. The antenna without slot is shown in Fig. 3a and with slot is shown in Fig. 3b. The microstrip antenna consists of three components: ground plane, dielectric substrate and radiating element. The ground plane and radiating element are fabricated using a thin copper sheet of thickness 0.035 mm. The radiating element is designed in the shape of a typical badge which is mounted on the dielectric substrate of thickness 1.2 mm with dielectric constant of 1.35 and loss tangent $\delta = 0.0009$. The type of feed used here is microstrip line feed. A rectangular slot is of width 1 mm and it is mounted vertically at the center of the radiating element.

The transmission line model method is used to analyze the microstrip antenna. The transmission line model provides a set of mathematical expressions in order to analyze the essential parameters of the antenna.

To find width,

$$W = \frac{C_0}{2f_r}\sqrt{\frac{2}{\varepsilon_r + 1}} \tag{1}$$

To find length,

$$L\frac{C_0}{2fr\sqrt{\varepsilon_{\text{reff}}}} - 2\Delta L \tag{2}$$

To find the fringing length,

$$\frac{\Delta L}{h} = 0.412\frac{(\varepsilon_{\text{reff}} + 0.3)\left(\frac{W}{h} + 0.264\right)}{(\varepsilon_{\text{reff}} - 0.258)\left(\frac{W}{h} + 0.8\right)} \tag{3}$$

To find the effective dielectric constant,

$$\varepsilon_{\text{reff}} = \frac{\varepsilon_r + 1}{2} + \frac{\varepsilon_r - 1}{2}\left[1 + 12\frac{h}{W}\right]^{-1/2}, W/h > 1 \tag{4}$$

To find the length and width of the ground plane,

$$\begin{aligned} L_g &= 6h + L \\ W_g &= 6h + w \end{aligned} \tag{5}$$

The ground plane and substrate have the same dimensions and ε_r of cotton is 1.35. The optimized dimensions of the antenna are shown in Table 1. The optimal antenna

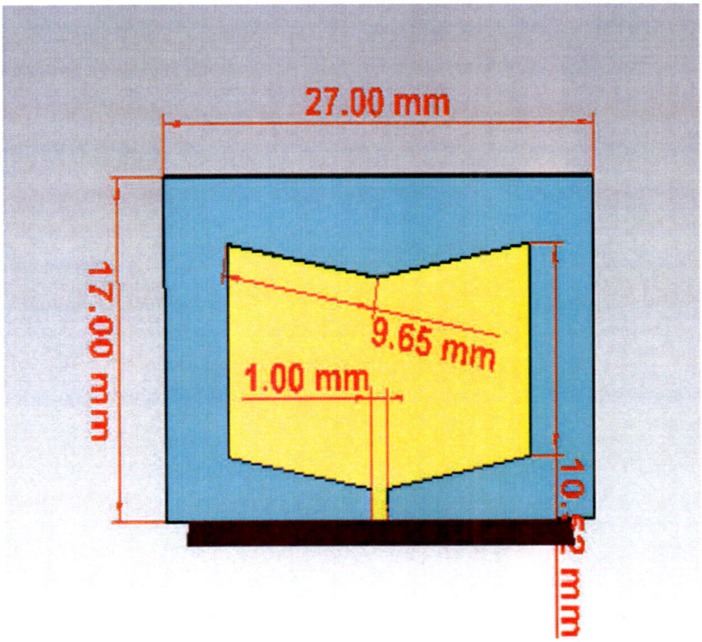

(a)

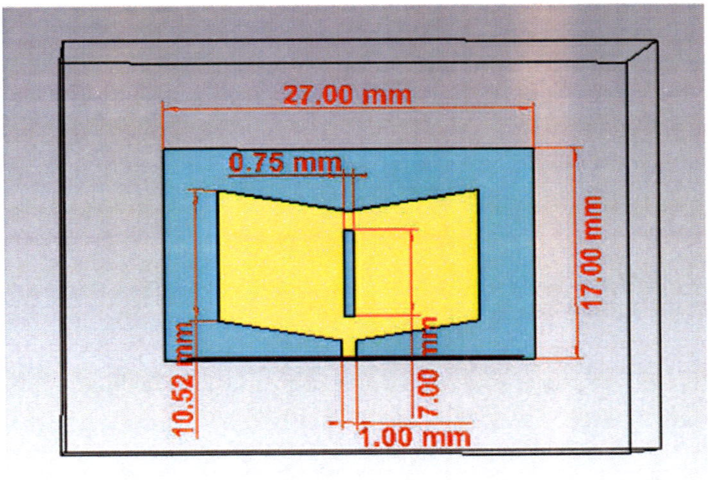

(b)

Fig. 3 a Antenna without slot, **b** antenna with slot

Table 1 Parameters and dimensions of the antenna

Parameter	Dimensions (mm)	Description
H	1.2	Substrate (cotton) height
L	17	Length of the substrate and ground plane
W	27	Width of the substrate and ground plane
F_L, S_L	18.417, 7	Feed length and slot length
F_W	1	Feed width and slot width
T	0.035	Patch and ground plane thickness
L_P	10.521	Patch length
W_P	19	Patch width
ε_r	1.35	Dielectric constant of substrate

parameters such as S_{11}, VSWR and gain are obtained using the simulation software CST MW Studio.

3.2 Antenna Performance

From the figure, it is observed that the measured S_{11} is slightly different from the simulated value and this may be due to losses because of hand fabrication. The front and back view of the fabricated microstrip antenna without slot is shown in Fig. 4a, b, respectively, and with slot is shown in Fig. 5a, b, respectively. The comparison of the antenna parameters without and with slot is shown in Table 2. The comparison between simulated and measured results is shown in Table 3.

3.3 Simulated Results

The results are obtained for various antenna parameters such as return loss S_{11}, VSWR, radiation pattern for both without and with slot. Figure 6 shows the return loss graph for the antenna without slot at a resonant frequency of 10.48 GHz, and the return loss S_{11} is observed to be − 16.72 dB. Figure 7 shows the VSWR graph for the antenna without slot at a resonant frequency of 10.48 GHz, and the VSWR is observed to be 1.341. Figure 8 shows the radiation pattern in polar plot for the antenna without slot at a resonant frequency of 10.5 GHz. It has a main lobe magnitude of 9.45 dBi with direction 7.0°. The angular width is 64.8° at 3 dB and it has a side lobe level of − 10.5 dB. Figure 9 shows the radiation pattern in 3D plot for the antenna

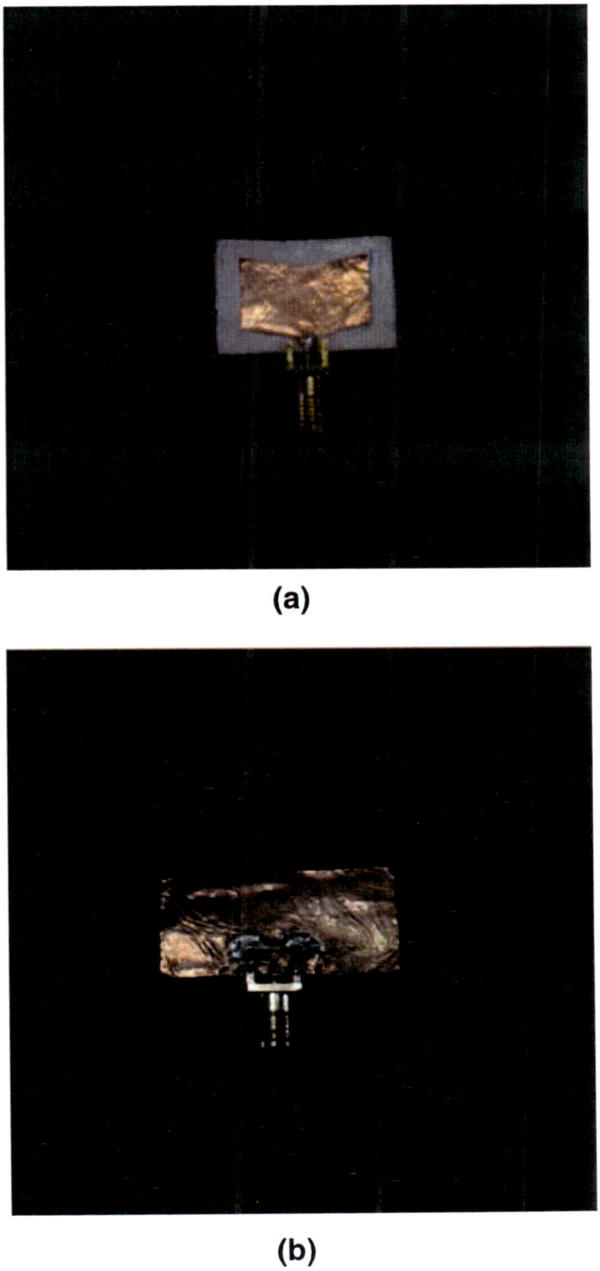

Fig. 4 **a** Badge Shaped antenna without slot (Front view), **b** Badge Shaped antenna without slot (Back view)

Fig. 5 **a** Badge shaped antenna with slot (front view), **b** Badge shaped antenna with slot (Back view)

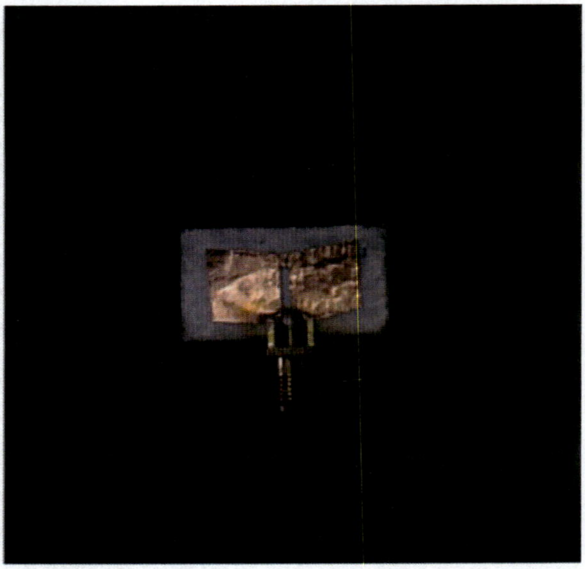

(a)

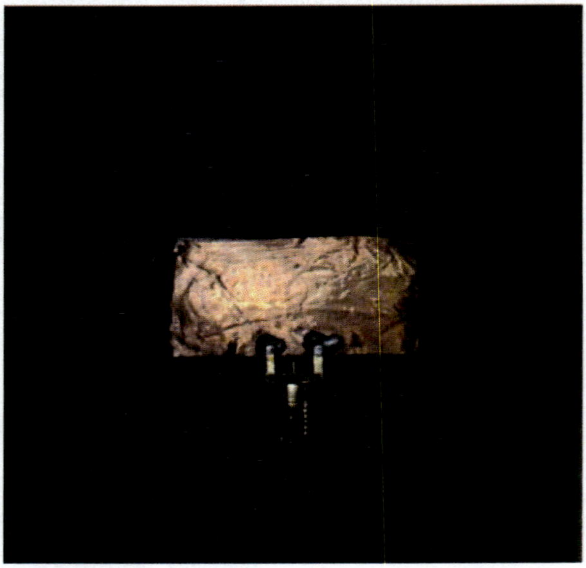

(b)

Table 2 Comparison between parameters of without and with slot

S. No.	Parameters	Without slot	With slot
1	S_{11}	− 16.72	− 16.13
2	VSWR	1.341	1.369
3	Main lobe magnitude	9.45 dBi	9.46 dBi
4	Angular width	64.8°	64.8°
5	Side lobe level	− 10.5 dB	− 10.5 dB
6	Radiating efficiency	− 2.049 dB	− 2.001 dB
7	Total efficiency	− 2.144 dB	− 2.108 dB
8	Directivity	9.437 dBi	9.442 dBi
9	E-Field	49.64e + 03	49.49e + 03
10	H-Field	107.2	106

Table 3 Comparison between simulated and measured results

S. No.	Parameters		Simulated value	Measured value
1	S_{11} (dB)	Without slot	− 16.72	− 19.10
		With slot	− 15.96	− 12.35
2	VSWR	Without slot	1.341	1.25
		With slot	1.378	1.63

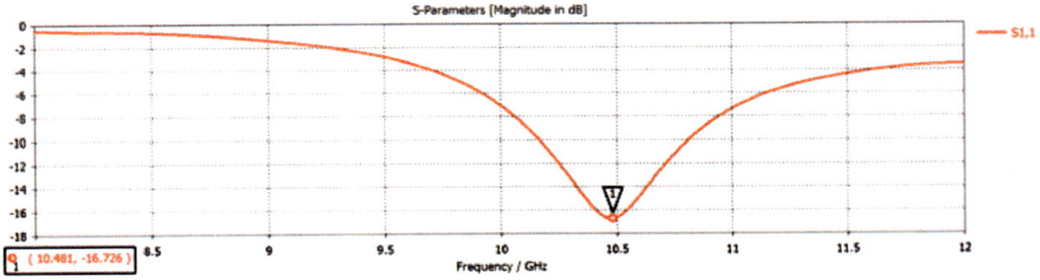

Fig. 6 Return loss (S_{11})

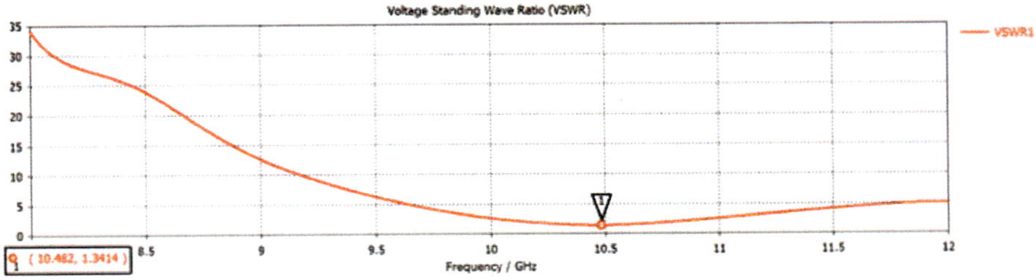

Fig. 7 VSWR

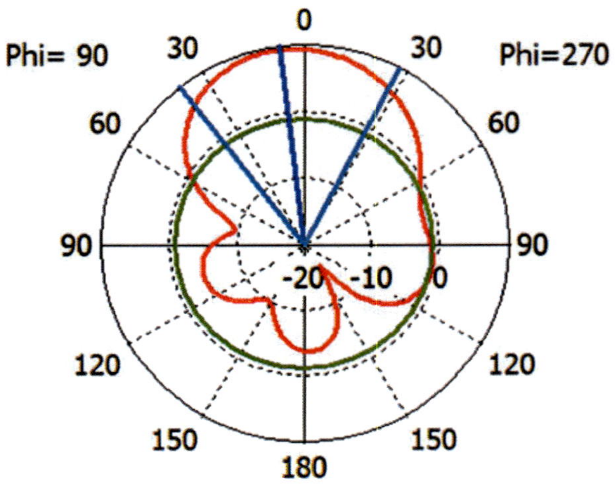

Fig. 8 Radiation pattern (polar)

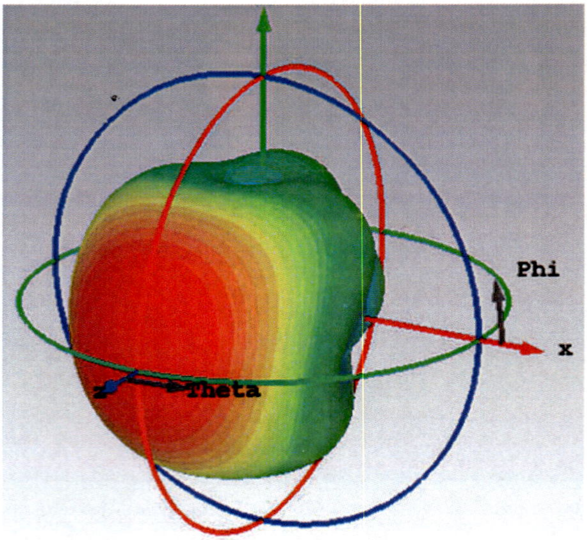

Fig. 9 Radiation pattern (3D)

without slot at a resonant frequency of 10.5 GHz. It has a radiating efficiency of − 2.049 dB and total efficiency of − 2.144 dB. The directivity is 9.437 dBi.

Figure 10 shows the return loss graph for the antenna with slot at a resonant frequency of 10.5 GHz, and the return loss S_{11} is observed to be − 16.13 dB. Figure 11 shows the VSWR graph for the antenna with slot at a resonant frequency of 10.5 GHz, and the VSWR is observed to be 1.369. Figure 12 shows the radiation pattern in polar plot for the antenna with slot at a resonant frequency of 10.5 GHz. It has a main lobe magnitude of 9.46 dBi with direction 7.0°. The angular width is 64.8° at 3 dB and it has a side lobe level of − 10.5 dB. Figure 13 shows the radiation pattern in 3D plot

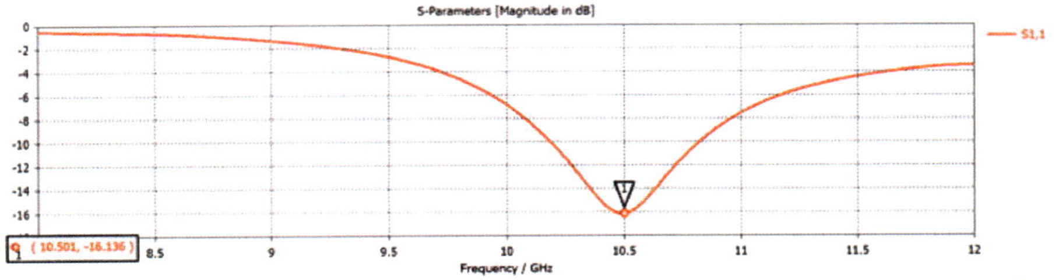

Fig. 10 Return loss (S_{11})

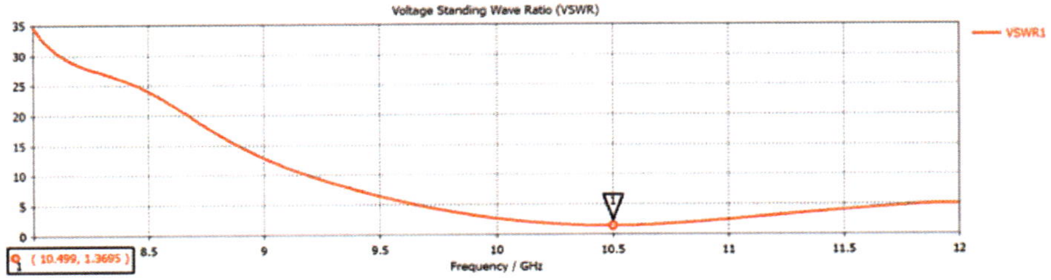

Fig. 11 VSWR

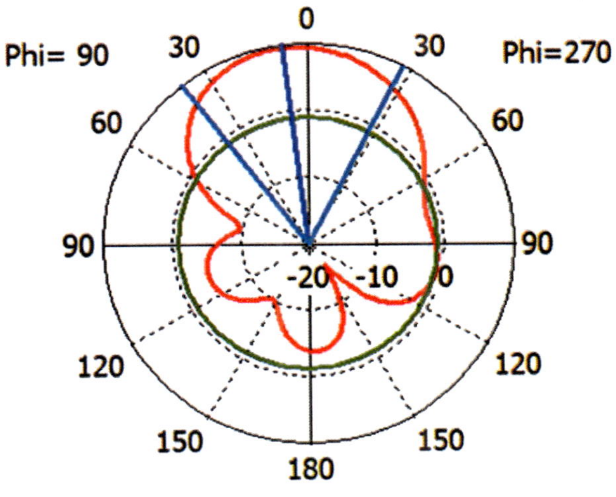

Fig. 12 Radiation pattern (polar)

for the antenna without slot at a resonant frequency of 10.5 GHz. It has a radiating efficiency of − 2.001 dB and total efficiency of − 2.108 dB. The directivity is 9.446 dBi.

Fig. 13 Radiation pattern (3D)

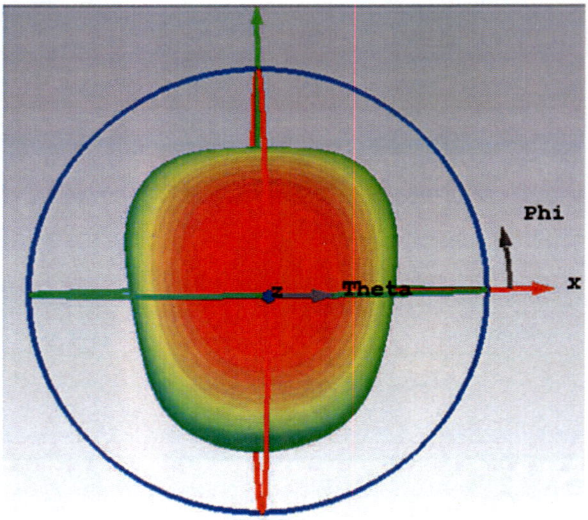

3.4 Measured Results

The antenna is tested with the help of vector network analyzer (VNA). The antenna parameters including return loss, VSWR and gain are measured. The vector network analyzer with the antenna connected to it is shown in Fig. 14. Figure 15 shows the return loss graph for the antenna without slot at a resonant frequency of 10.1 GHz, and the return loss S_{11} is observed to be − 19.10 dB. Figure 16 shows the VSWR graph for the antenna at a resonant frequency of 10.1 GHz, and the VSWR is observed to be 1.25. Figure 17 shows the return loss graph for the antenna with slot at a resonant frequency of 10.47 GHz, and the return loss S_{11} is observed to be − 12.35 dB. Figure 18 shows the VSWR graph for the antenna at a resonant frequency of 10.47 GHz, and the VSWR is observed to be 1.63. Figure 19 shows the testing of antenna with VNA.

4 Conclusion and Future Work

Two badge-shaped wearable antennas were designed and fabricated. Due to higher frequency of operation, the dimensions of the antennas were small in the order or 27 × 17 × 0.035 mm^3. The antenna that is designed with a rectangular slot of 1 mm width has s_{11} value of − 15.96 as simulated result and − 12.35 when measured using a VNA. The one without any slot has s_{11} value of − 16.72 as simulated result and − 19.10 when measured using VNA. The peak gain for antenna without slot was found to be 9.437 dBi and for that of antenna with 1 mm slot was found to be

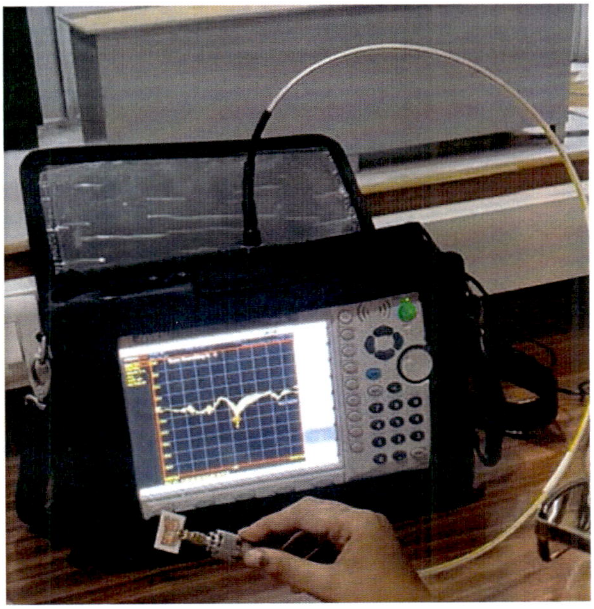

Fig. 14 Vector network analyzer

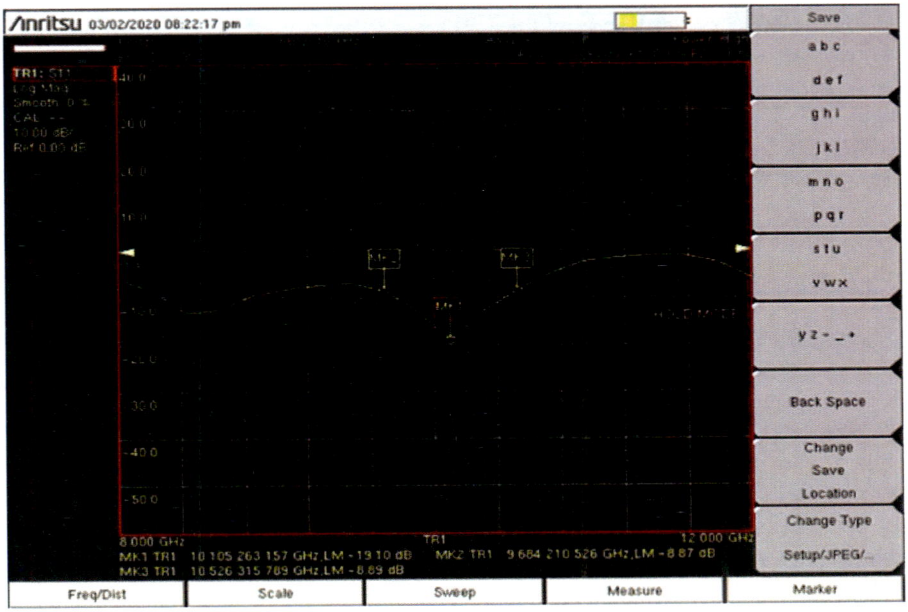

Fig. 15 Return loss (S_{11})

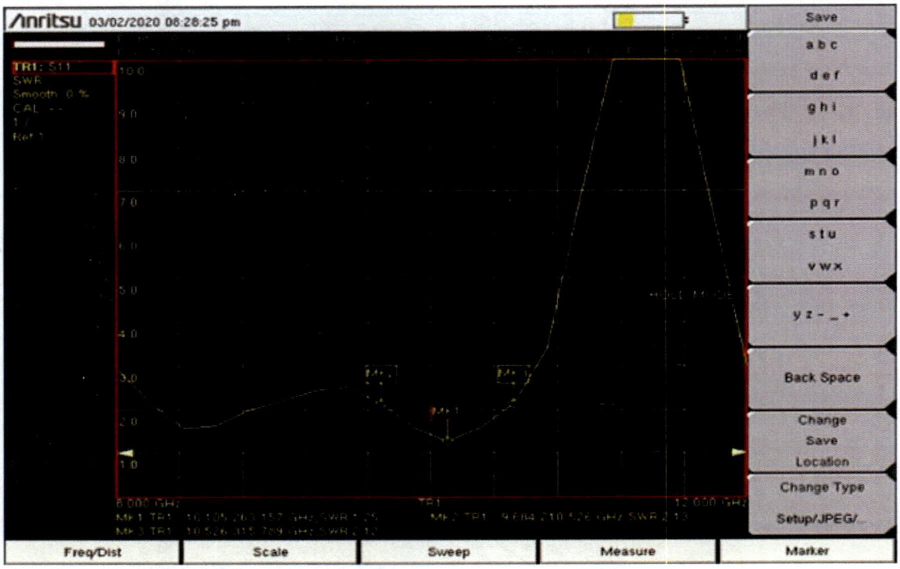

Fig. 16 VSWR

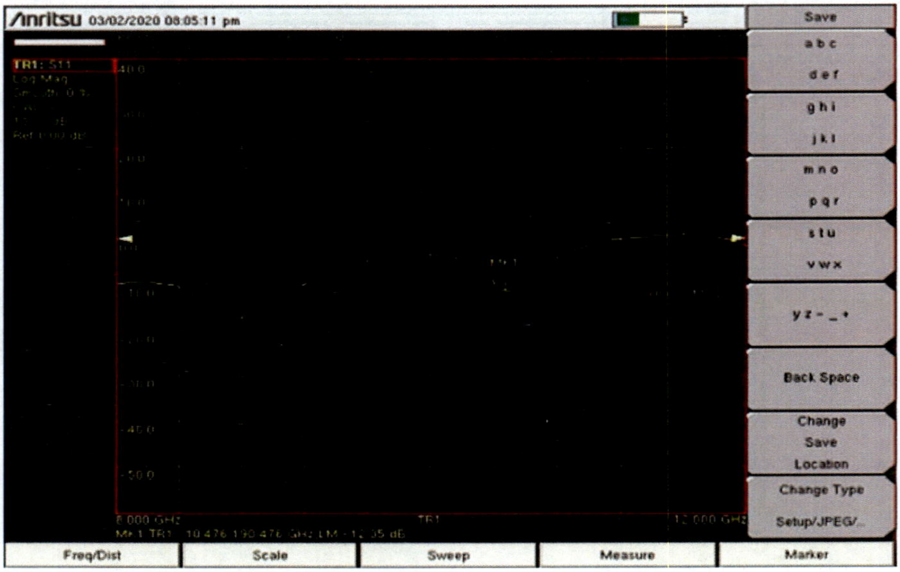

Fig. 17 Return loss (S_{11})

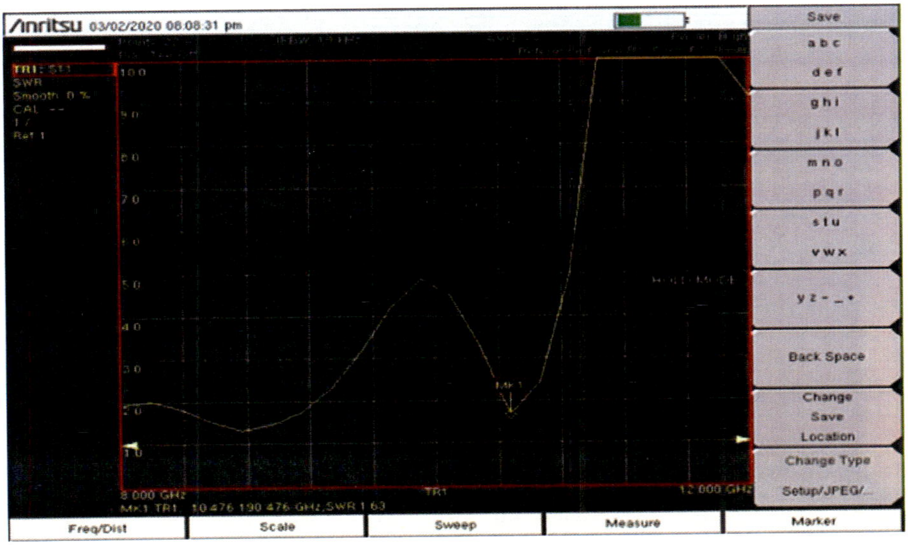

Fig. 18 VSWR

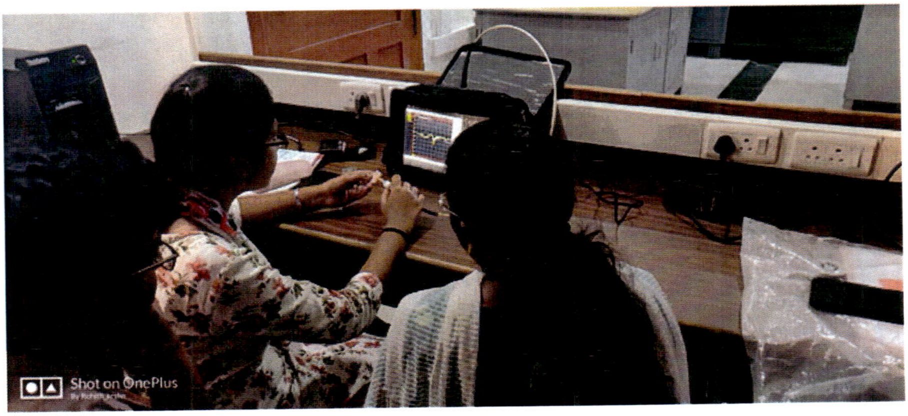

Fig. 19 Testing of antenna with VNA

9.446 dBi. Upon comparing among these two fabricated antennas, it is found that the badge-shaped antenna without slot was more efficient. The antenna with slot was found to accept 65–70% of the given input power/voltage. The antenna without slot accepted 85–90% of the given input power. The VSWR values of both antennas show an average of 87% agreement between simulated and measured results. Also, the s_{11} values of these antennas show an average of 81% agreement between simulated and measured results. The future work may involve incorporating concepts of energy harvesting and re-configurability in order to implement in real time. The substrate can also be altered to obtain different radiation pattern or even improve efficiency.

This antenna may also be integrated with any embedded system module for real-time implementation of the product. The shape of the radiating patch can also be changed and even modifying the dimensions of the antenna components may result in different outcomes like frequency shift, wider bandwidth, higher directivity, etc.

References

1. C.A. Winterhalter et al., Development of electronic textiles to support networks, communications, and medical applaictions in future U.S. military protective clothing systems. IEEE Trans. Inf. Technol. Biomed. **9**(3), 402–406 (2005)
2. K. Shafique, B.A. Khawaja, M.A. Tarar, B.M. Khan, M. Mustaqim, A. Raza, A wearable ultra-wideband antenna for wireless body area networks. Microw. Opt. Technol. Lett. **58**(7) (2016)
3. A. Baroni, P. Nepa, H. Rogier, Wearable self–tuning antenna for emergency rescue operations. IET Microw. Antennas Propag. 1–11 (2015)
4. E.K. Kaivanto, M. Berg, E. Salonen, P. de Maagt, Wearable circularly polarized antenna for personal satellite communication and navigation. IEEE Trans. Antennas Propag. **59**(12), 4490–4496 (2011)
5. P.S. Hall et al., Antennas and propagation for on-body communication systems. IEEE Antennas Propag. Mag. **49**(3), 41–58 (2007)
6. H. Lee, J. Tak, Wearable antenna integrated into military berets for indoor/outdoor positioning systems. IEEE Antennas Wireless Propag. Lett. **16** (2017)
7. Z. Xu, T. Kaufmann, C. Fumeaux, Wearable textile shielded stripline for broadband operation. IEEE Microw. Wirel. Components Lett. **24**(8) (2014)
8. S.J. Chen, T. Kaufmann, D.C. Ranasinghe, C. Fumeaux, A modular textile antenna design using snap-on buttons for wearable applications. IEEE Trans. Antennas Propag. **64**(3), 894–903 (2016)
9. I. Locher, M. Klemm, T. Kirstein, G. Troster, Design and characterization of purely textile patch antennas. IEEE Trans. Adv. Packag. **29**(4), 777–788 (2006)
10. G.A. Cavalcante, D.R. Minervino, A.G. D'Assunc¸ao, Valdemir P. Silva Neto and Adaildo G. D'Assunção, A compact multiband reject inverted double-e microstrip filter on textile substrate. Microw. Opt. Technol. Lett. **57**(11) (2015)
11. Y. Bayram et al., E-textile conductors and polymer composites for conformal lightweight antennas. IEEE Trans. Antennas Propag. **58**(8), 2732–2736 (2010)
12. R.P. Kumar, S. Smys, A novel report on architecture, protocols and applications in Internet of Things (IoT), in*2018 2nd International Conference on Inventive Systems and Control (ICISC)*, Coimbatore, 2018, pp. 1156–1161. https://doi.org/10.1109/ICISC.2018.8398986
13. D.K. Anguraj, S. Smys, Trust-based intrusion detection and clustering approach for wireless body area networks. Wireless Pers Commun. **104**, 1–20 (2019). https://doi.org/10.1007/s11277-018-6005-x
14. P.C. Sharma, K.C. Gupta, Analysis and optimized design of single feed circularly polarized microstrip antennas. IEEE Trans. Antennas Propag. AP **31**(6), 949–955 (1983)
15. P. Jyothirmani, J.S Raj, S. Smys, Secured self organizing network architecture in wireless personal networks. Wireless Pers. Commun. **96**, 5603–5620 (2017)

Keyword Recognition from EEG Signals on Smart Devices a Novel Approach

Sushil Pandharinath Bedre, Subodh Kumar Jha, Chandrakant Patil, Mukta Dhopeshwarkar, Ashok Gaikwad, and Pravin Yannawar

Abstract Technological advancement in the field of electroencephalography (EEG) based on brain activity classification extends a variety of significant applications, namely, emotion recognition, muscular moment analysis, neurological disorders identification, the prediction of the intensions, machine controlling in smart devices, and healthcare devices. In this article, a novel approach is introduced for EEG-based digit and keyword recognition for smart devices like mobile, tablets, etc. EEG signals recordings of 10 subjects (i.e., 7 male and 3 female) were acquired from the age group 20–25 years, and volunteered to imagine digits and keywords. An multiple feature extraction algorithms were employed such as short-time Fourier transform (STFT), discrete cosine transform (DCT), and discrete wavelet transform (DWT) to extract the feature from EEG data. The dimension of the feature space was reduced by employing linear discriminant analysis (LDA). The normalized features were passed through diverse nature of multiple classifiers, namely, support vector machine (SVM), k-nearest neighbor (KNN), random forest (RF), Naïve Bayes (NB), multi-layer perceptron (MLP), and convolution neural network (CNN) to perform classification analysis. By analysis and comparison of the classifiers, the MLP outperformed to claim over the rest of the classifiers in both digit and keyword classification with 96.43% and 92.36% recognition accuracy, respectively.

Keywords BCI · EEG signals · Keyword recognition · Emotive · Smart devices

1 Introduction

The advancement in the field of brain-computer interface (BCI) leads to several application developments through non-invasive EEG. There is plenty of research work proposed by various researchers across the world. Among all BCI development techniques, EEG has been the most accepted method due to its high resolution of temporal signal data, safety, convenience, and usability. EEG data can be collected from 10 to 20 electrodes placed on the scalp.

S. P. Bedre · S. K. Jha · C. Patil (✉) · M. Dhopeshwarkar · A. Gaikwad · P. Yannawar
Dr. Babasaheb Ambedkar University, Aurangabad, Maharashtra 431004, India

S. Shakya et al. (eds.), *Mobile Computing and Sustainable Informatics*, Lecture Notes on Data Engineering and Communications Technologies 68,
https://doi.org/10.1007/978-981-16-1866-6_3

An electroencephalogram (EEG) measures activities within the human brain from neural currents known as electrical activity due to the changes in electric activities in the human brain influence EEG patterns, and these EEG patterns can be used to operate BCI systems. BCI systems can be divided into two categories such as invasive and non-invasive. And this research is focused on non-invasive BCI. As EEG signals are influenced by different body movements such as hand movements, eye movements, and other muscular activities, there is another chance of getting noisy data due to power line obstruction. Hence, signals acquired from electric activities must be preprocessed. The process of EEG data preprocessing is also known as a signal enhancement. Many researchers introduced artifact removal techniques from EEG data such as independent component analysis (ICA), principal component analysis (PCA), common spatial patterns (CSP), spatio-spatial patterns (CSSP), common average referencing (CAR), and adaptive filtering. It is observed that most of the researchers have adopted ICA, PCA, adaptive filtering, and CAR methods for noise removal.

Processing of EEG signals has been widely accepted in a wide range of applications, in particular, emotion classification, seizure detection or prediction, motor imagery classification, sleep state classification, drug effects diagnosis, fatigue identification, hand movement detection, mental task classification, and motor imagery classification.

There exist many non-invasive EEG-based research studies, which are discussed in successive section.

Rosado et al. [1] proposed a system that automatically detects epileptiform discharges in the EEG signals through a multi-stage detection algorithm. Wavelet transform and mimetic analysis techniques were used for EEG signal analysis and classification was done with the help of fuzzy logic. This study may contribute to the diagnosis of epilepsy.

Phothisonothai et al. [2] used spontaneous EEG singles to develop a brain-computer interface. Leaf and right-hand imaginary moments were captured and features were extracted through the autoregressive (AR) model and the band power estimation (BPE) techniques. The classification was carried out using a three-layer feed-forward neural network and compared with linear discriminant analysis (LDA). A similar approach based on hand movements was investigated by Robinson et al. to develop BCI through EEG signal data. Slow and fast hand movements were captures through EEG signals. Temporal-spatial-spectral resolution of EEG signal was computed from wavelet-common spatial pattern (W-CSP) algorithm and Fisher linear discriminant (FLD) classifier was used to perform classification. Alomari et al. used imaginary fist movements in their research to develop BCI. Root mean square (RMS) and the mean absolute value (MAV) were adopted to extract EEG-based fits movement features. Then, support vector machines (SVMs) and neural networks (NNs) classifiers were employed to accomplish the classification task.

Zhang et al. [3] contributed to solving challenges faced in brain activity recognition using analysis of electroencephalography (EEG). The proposed system can cope with problems like low recognition accuracy occurred by the massive noises and the low signal-to-noise ratio in EEG signals. They used an autoencoder layer to refine raw EEG signals collected from the multi-brain multi-class activity of the person. The psychological data recorded from stimulating subjects by exposing emotional pictures. Further, the classifier was trained with three classes as negative–positive, and neutral conditions. KNN algorithm was used to perform classification.

Jatupaiboon et al. [4] proposed a system that can classify happy and unhappy emotions evoked by classical music and pictures. The SVM classifier was used to classify power spectral density (PSD) features.

Jirayucharoensak et al. [5] employed deep learning network (DLN) to recognize emotions from nonstationary EEG signals. A stacked autoencoder with a hierarchical feature learning method was used to implement DLN. The work presented by Ackermann et al. [6] demonstrates techniques, methods for feature selection, and classification for emotion recognition from EEG data.

Alotaiby et al. [7] reviewed on channel selection and classification methods from EEG signals. This study helps to reduce computational complexity (noise and high dimensional data) and reduces the amount of overfitting of data caused by the selection of unnecessary channels. The study suggests wavelet transform is effective for feature extraction. Also, Lotte et al. [8] presented a review on classification algorithms for EEG-based BCI. Feature extraction, feature selection, channel selection, classifiers, and performance measures are briefly discussed in this work. The study suggests that band power is wildly used for feature extraction and SVM, LDA has been used in most of the existing work based on EEG-based BCI. In their previous work, Lotte [9] provided an overview and tutorial for EEG-based signal processing development of the BCI system toward mental state recognition.

Amarasinghe et al. [10] employed self-organizing maps and artificial neural networks for the classification of thought patterns. EEG signals were acquired from the emotive EPOC headset.

Chen et al. [11] used EEG signals for fatigue assessment caused by the excessive duration of watching DTV. Gravity frequency of power spectrum and power spectral entropy was studied to analyze the level of fatigue.

Dong et al. [12] proposed a system that uses EEG signals to recognize two types of mental states such as "agreement" and "disagreement". Self-relevant sentences were used to acquire EEG data.

Huang et al. [13] developed a system that uses behavior measures of the fluctuations in the EEG signals for the detection of the arousal states of humans. Hidden Markov models (HMM) were used to estimate the probability of change in the EEG signals.

Horlings et al. [14] presented a system for emotion recognition using EEG data. Valence and arousal emotions were collected from EEG signals and classified into five different classes. A threefold cross-validation technique was used for train and test data.

Fabiani et al. [15] developed brain-computer interface (BCI) for controlling cursor movements. Mu or beta frequency bands over the sensorimotor cortex were extracted from the EEG signals.

Amin et al. [16] proposed a method for feature extraction and classification using EEG signals. The study suggests Fisher's discriminant ratio (FDR) and principal component analysis (PCA) methods are useful for feature normalization. Traditional classifiers for instance and Naïve Bayes (NB), K-nearest neighbors (KNN), multi-layer perceptron (MLP), and support vector machine (SVM) were employed for the classification of the EEG signals.

Hadjidimitriou et al. [17] proposed a system music assessment using time–frequency (TF) feature analysis. Nine subjects have participated during the data collection, subjects were asked to listen to music based on self-reported rating and familiarity. KNN and SVM were employed for categorization data between "like" and "dislike" categories.

Hashimoto et al. [18] make an effort toward the development of a BCI system based upon imaginary foot movements recorded through EEG signals. Topographies and time–frequency map was used for the analysis of the EEG signal data. A similar approach was adopted by Herman et al. [19] in their study motor imagery (MI) patterns are analyzed, relevant frequency patterns have been identified and quantified. In their experiments, the imaginary hand movements were recorded and used for classification. SVM and RDF classifiers were used to quantify and identification EEG of imaginary hand moments.

Hosseinifard et al. [20] employed nonlinear analysis of EEG signals for the classification of depressed patients and normal persons. There are 45 normal subjects and 45 depressed patients who contributed to data collection, correlation dimension, Higuchi fractal, detrended fluctuation analysis (DFA), and Lyapunov exponent are computed over EEG signals and used as features. Feature classification was carried out using KNN, LDA, and logistic regression (LR).

Iacoviello et al. [21] designed a brain-computer interface for classification of the emotions, for this a real-time classification method was proposed. EEG signals which were collected over self-induced emotions have been considered. The low amplitude signals from EEG signals were classified by adopting SVM, PCA, and wavelet transform.

Erfanian et al. [22] exploited independent component analysis (ICA) for the development of the BCI system. Multi-channel EEG signals were used to study the effect of performance in BCI.

Valente et al. [23] studied EEG patterns to evaluate the sensitivity of the EEG signals of Angelman syndrome (AS) patients. Seventy EEG signals are analyzed from 26 patients. Classification of EEG patterns was done in three categories, namely theta patterns (TP), delta patterns (DP), and posterior discharges (PDs).

Jaiswal et al. [24] proposed a technique for the classification of epileptic EEG signals. Local neighbor descriptive pattern (LNDP) and one-dimensional local gradient pattern (1D-LGP) techniques are used for feature extraction from EEG signals. EEG features have been classified into two categories: epileptic seizure and non-seizure. Several conventional classifiers are compared and validated such as SVM, nearest neighbor (NN), ANN, and decision tree (DT) using tenfold cross-validation.

Jayarathne et al. [25] developed an authentication system similar to biometric authentication systems. While EEG data collection, subjects are asked to visualize the number of their ATM PIN. Subsequently, the mental conditions of the subjects are recorded through EEG signals. The common spatial patterns (CSP) method was employed for feature extraction and LDA was used for classification purposes.

Jrad et al. [26] developed a BCI system based on EEG signals and sensor weighting support vector machines (SW-SVM) was employed. In the proposed BCI system, two datasets, namely an error-related potential (ErrP) dataset and P300 dataset were used.

Another work on EEG-based emotion recognition has proposed by Khosrowabadi et al. [27]. During EEG data collection, 26 subjects were participated and asked to monitor four emotional states. Self-organizing map (SOM) is used for detecting the boundaries between scores of separable regions. The performance was evaluated through fivefold cross-validation using the KNN classifier.

Kunze et al. [28] proposed a novel approach to discriminate habits and reading activities of the users while reading different genres of the document. In this research, the emotive EGG system was used for acquiring EEG data and a band pass filter was used to remove the noise from the signal.

Lahane et al. [29] proposed an approach for emotion recognition using EEG signals. Kernel density estimation (KDE) technique was used to extract features from EEG data. Artificial neural network (ANN) was used as the classifier.

Li et al. [30] presented work for evaluating driver fatigue using EEG signal data. Data collected from 16 channels have been divided into three bands, i.e., theta, alpha, and beta. Gray relational analysis (GRA) was employed for an indication of driver fatigue. Kernel principal component analysis (KPCA) was used to optimize the electrodes from EEG. Similar work based on real-world fatigue proposed by Lin et al. [31] for analysis of alertness level using EEG-based longitudinal study. Liu et al. employed KPCA–HMM algorithm to extract features and Kolmogorov complexity (Kc), approximate entropy (ApEn) used as a complexity parameter of EEG signals to analyze mental fatigue.

Liu et al. proposed a system for human emotion recognition and visualization through EEG. This study presents a fractal dimension algorithm for emotion recognition and a 3D virtual environment for visualizing emotions.

McBride et al. developed a system for the detection of early Alzheimer's disease using EEG signals. This work was proposed by Nicolaou et al. utilized the entropy feature extracted from EEG signals for the early detection of epilepsy. The classification was carried out by employing an SVM classifier. Nie et al. also adopted an SVM classifier to develop a system for emotion recognition from EEG signals that are acquired during watching movies, delta (1–4 Hz), theta (4–8 Hz), alpha (8–13 Hz), beta (13–30 Hz), and gamma (36–40 Hz) bands are used and passed through fast Fourier transform (FFT) to constitute a feature set. Classified emotions are categorized into two categories such as positive emotions and negative emotions.

Reference [16] proposed a novel method for the analysis of EEG signals. The second-order linear time-varying autoregressive (TVAR) process was used for the parametric representation of EEG signals.

Rebsamen et al. presented a novel approach to compute cognitive workload using EEG signals. In this research, 16 subjects were participated, all subjects asked to perform the mental arithmetic task with different difficulty levels. Three classes, viz.

relaxed, low workload, and the high workload was used to the categorized cognitive workload.

Shi et al. employed SVM and extreme learning machines (ELM) to estimate the vigilance of operators using the EEG-based approach.

This research focused on a novel approach for digit and keywords recognition from EEG data. Therefore, EEG signal data in the experiment were exploited. This literature study indicates that there is no work done on this particular area of research. The proposed system will utilize EEG signal data to classify digits and keywords which is described in subsections as follow.

2 Materials and Methods

To accomplish the proposed keyword recognition system on smart devices, a five-stage methodology is adopted such as EEG data acquisition, preprocessing, feature extraction, normalization, and classification model. The accuracy of the system is a critical aspect of consideration of any EEG-based BCI system. Thus, normalization algorithms are applied over raw EEG data to remove artifacts from EEG data. This section provides an overview of the proposed system and ANTARANG framework for the interpretation of EEG data.

Figures 1a and 2b depict the system flow involved in the classification of keywords. The methods and materials used in this research are discussed below.

2.1 Overview

The subject is required to gear with emotive EEG set at the time of data acquisition as well as during testing samples. The electrode placed on the scalp or subset of electrodes in an EEG device may dislocate during data acquisition and this may lead to bad contact with the scalp, and therefore an artifact or poor quality signal may be obtained. However, there might be an electrodes placed on the scalp may also have mechanical faults, for instance, frayed wiring, which can completely or partially reduce the signal obtained. Such electrodes can induce artifacts into the signals. So in a preprocessing step, FAST independent component analysis (ICA) was applied on EEG sample data to removing artifacts, and resulting ICs were passed for feature extraction. Basically, in biomedicine, extraction, and separation of statistically independent sources underlying multiple measurements of biomedical signals extracted and analyzed with the help of ICA.

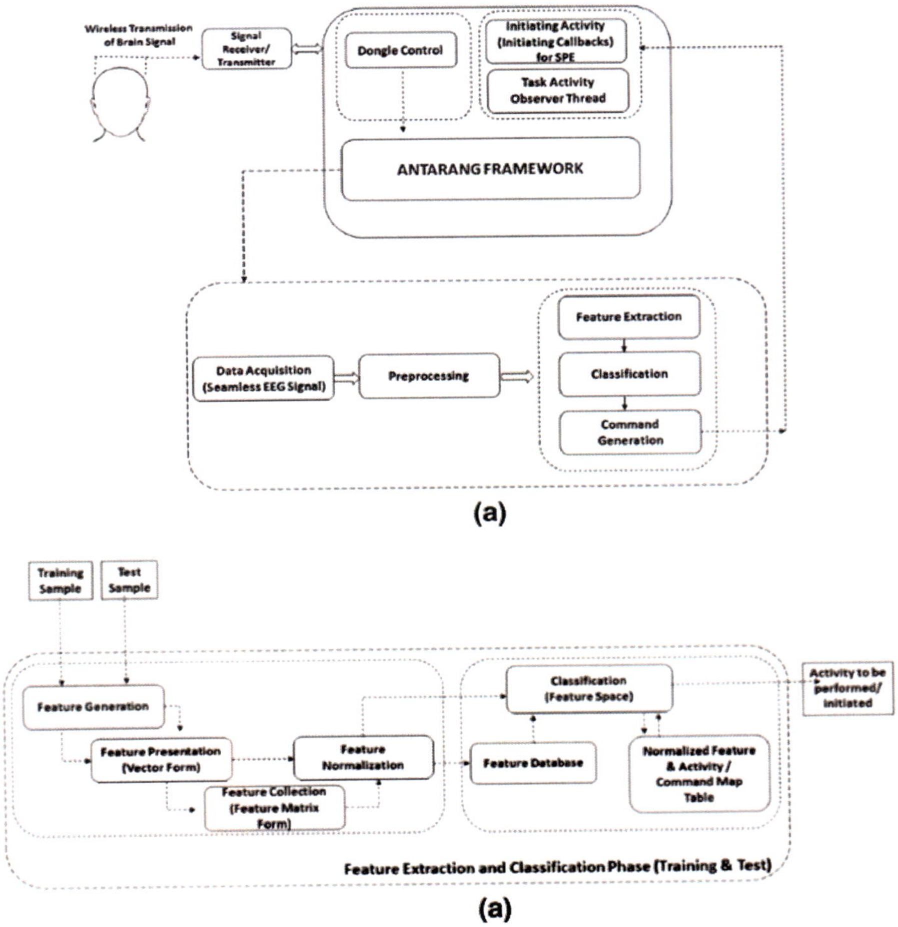

Fig. 1 **a** and **b** Block diagram of thought processing system 'ANTARANG'

Fig. 2 **a** Brain lobes and **b** emotive EPOC device for brain wave data acquisition

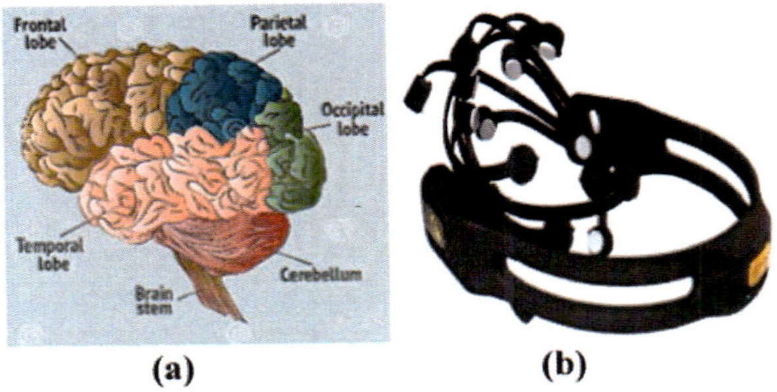

2.2 Data Acquisition

The subject is required to place emotive EPOC headgear to acquire EEG signals corresponding to the activity. The data is acquired by the emotive headset is transferred to the wireless dongle connected to the device seamlessly via Bluetooth mechanism. The dongle control mechanism connected to the smart device acts as a receiver. This will store the test sample primarily in the storage of the smart device and be handed over to the ANTARANG framework.

For the implementation of most of the system, researchers have developed their own datasets according to the requirements that meet to research objectives. In the same manner, EEG signal dataset is developed using the EmoEngine device. Figure 2a, b depicts the brain lobes and emotive EPOC device, respectively.

The data was acquired from the emotive EPOC headset and saved in an output file for analysis. The emotive headset which sends the data about the activity performed by the subject to the remote smart device through the available communication mechanism. The data is stored on the smart device and further be used for training and testing of the samples over smart devices. Traditionally, the data received from the subject is seen for five broad spectral sub-bands, namely, delta (0–4 Hz), theta (4–8 Hz), alpha (8–16 Hz), beta (16–32 Hz), and gamma waves (32–64 Hz), and these sub-bands have been frequently used EEG signals which are generally adopted in clinical interest. The accurate information regarding neuronal activities can be extracted from these five sub-bands and, consequently, there could be some changes in the EEG signal, which are not so obvious in the original full-spectrum signal, can be amplified when each sub-band is considered independently. Each EEG segment was considered as a separate EEG signal resulting in a total of 125 EEG data segments.

The brain is formed using five lobes, these lobes perform all the critical neurological activities such as the frontal lobe controls the activity of speech, thought, emotions, problem-solving, skilled movements, parietal lobe identifies and interprets sensations such as touch, pain, etc. The occipital lobe collects and interprets visual images that are sight, and the temporal lobe controls the activities related to hearing and storing memory and the cerebellum controls the coordinate's familiar movements. Similarly, the relationship between brain lobes that the excreted (energy) frequency of signal is given in Table 1.

The dataset is developed and contains 12 keywords {'Close', 'Copy', 'Cut', 'Delete', 'New', 'Ok', 'Open', 'Paste', 'Pause', 'Play', 'Start', 'Stop'} and EEG

Table 1 Signal type, frequency and its origin

Type	Frequency range (Hz)	Origin
Delta	0–4	Cortex
Theta	4–8	Parietal and temporal
Alpha	8–13	Occipital
Beta	13–20	Parietal and frontal
Gamma	20–40	Parietal and frontal

signal recordings of 10 subjects (i.e., 7 male and 3 female) from the range of age group (20–25) were considered. The total volume of the database is 1200 (12 * 10 * 10) samples. The subject who is familiar with these keywords was chosen for data collection, all selected subjects are healthy with no mental and physical disorders. The setup for data collection was developed at Vision and Intelligent System Laboratory of the Department of Computer Science and Information Technology, Dr. Babasaheb Ambedkar Marathwada University, Aurangabad. All subjects were asked to sit comfortably on an armchair and a screen was placed next to them in an electromagnetically protected room. The written consent was taken from each subject for the recording of EEG signals before collecting EEG data. All subjects were instructed that this experiment has been designed to be used for brain-computer interface applications. A simple display screen in PowerPoint was arranged for the data collection under the proposed research work. This system generates a keyword signal with an interval of 5 s. After every 5 s, the next number was shown on the screen. To collect proper signals from EEG, all subjects were demonstrated the display keyword screen, so that they become more familiar with the task. The same process was repeated five times. So the total volume of the keyword dataset is 1200 samples. After the data collection process was done, the collected dataset was passed to a subsequent stage as per the proposed methodology that is data preprocessing.

2.3 Data Preprocessing

While data acquisition, the system receives a large number of EEG signals as a vector of each EEG signal. These signals were saved in a formed matrix. Since EEG signals may carry a vast amount of artifacts in them, these artifacts and noise affect the performance of the classification system. To build an efficient keyword classification system, apply the FAST independent component analysis (ICA) on EEG data. Noisy EEG signals contain independent original signals that occurred due to various conditions while data acquisition. Fast ICA can solve the problem by which each EEG data vector separates from noisy data through decomposition. It resulting in independent components (ICs), these ICs were passed for feature extraction. In this study, fast ICA symmetric approach is used for preprocessing of EEG signals. This approach estimates all components (all weight vectors) simultaneously. The following Eq. (1) was used to orthogonalized the steps.

$$w = (ww^T)^{\frac{-1}{2}} w \tag{1}$$

where w is the weights vector matrix $(w_1 w_2 \ldots w_n)^T$ and $ww^T = QDQ^T$ is an eigenvalue decomposition.

The square root of ww^T is obtained through the following equation.

$$(ww^T)^{\frac{-1}{2}} = QD^{\frac{-1}{2}} Q^T \tag{2}$$

where Q represents the matrix of eigenvectors and the diagonal matrixes of eigenvalues are calculated in matrix D. This process is repeated till the stop condition fulfills. The stop condition is described by Eq. (3) follow.

$$1 - \min\left(\mathrm{abs}\left(\mathrm{diag}\left(w^T w_{\mathrm{old}}\right)\right)\right) < \varepsilon \tag{3}$$

ε is the constant.

3 Feature Extraction

3.1 DWT Feature Extraction

In the keyword classification, a new technique for a nonparametric method is used for feature extraction that is based on multiresolution analysis of wavelet transform. Since short-time Fourier transforms and FFT are not able to obtain joint-time frequency resolution that is used for approximation and detail extraction of EEG signal. Since the DWT is a good candidate for expanding nonstationary EEG signals onto basic function, by shifting and contracting a single prototype function (the mother wavelet—$\psi_{x,y}$) to select signal under consideration. Equation 4 shows mother wavelet function $\psi_{x,y}(t)$ as follows.

$$\psi_{x,y}(t) = \frac{1}{\sqrt{x}}\psi\left(\frac{t-y}{x}\right) \tag{4}$$

where $x, y \in R, x > 0$ and R represents wavelet space.

X and y parameters denoted to shifting and scaling factor. Since mother wavelet has limitation is to satisfy the acceptability condition.

$$C_\psi = \int_{-\infty}^{\infty} \frac{|\psi(\omega)|^2}{\omega} \mathrm{d}\omega < \infty \tag{5}$$

where $\psi_{(\omega)}$ is the Fourier transform (FT) of $\psi_{x,y}(t)$,

To obtain time–frequency representation a pair of filters that cut the frequency domain in the half of the signal repeatedly performed. The DWT [25] breaks down the signal into approximation coefficient (AC) and detailed coefficients (CD). Further, the CA can be divided into new detailed coefficients and new approximations. The set of detailed coefficients and approximation was produced by iteratively performing the steps above at various scales or levels. The AC and CD are applied for EEG signals allow compressing useful data to the first few coefficients. Therefore, only these coefficients can be used for classification using machine learning algorithms. This kind of data compression may dramatically reduce input vector size and decrease the time required for training and classification. These features were calculated for all the samples of the 'Keyword set'.

3.2 FFT Feature Extraction

In order to develop any robust BCI system using EEG signals, the efficiency depends upon the prominent features extracted from the EEG signal. These signals are nonlinear and highly complex. In this study, the spectral centroid (SC) and spectral entropy (SE) features are computed through fast Fourier transform (FFT); therefore, the EEG signals are framed into five seconds of the frame and 50% of overlap. The beta channel is considered, since this channel has a high-frequency range beta 1 (13–20 Hz), beta 2 (21–30 Hz), which is responsible for focusing the action selection network mechanism of activation of frontal lobes of the human brain. Butterworth's fourth-order filter is adopted to extract the beta frequency band from the EEG signal. Following Eq. (6) is used to extract FFT coefficients [26] from the beta channel.

$$X_k = \sum_{n=0}^{N-1} x_n e^{-i2\pi k \frac{n}{N}} k = 0, 1, 2, 3, \ldots N - 1 \tag{6}$$

where X_k are coefficient of FFT, N is the number of EEG data samples, and n is a total number of points in FFT.

Spectral entropy and spectral centroid from EEG signal is calculated from following Eqs. (7) and (8).

$$H(x) = \sum_{x \subset X} x_i . \log_2 x_i \tag{7}$$

$$\mathrm{SC} = \frac{\sum_{k=1}^{N} kF[K]}{\sum_{k=1}^{N} F[K]} \tag{8}$$

3.3 Feature Fusion

Approximation coefficient (AC), detailed coefficients (CD), spectral entropy, and spectral centroid features are fused to feed to different classifiers, and the feature fusion was accomplished by following Eq. (9).

$$F = \sum_{i=1}^{i} \mathrm{AC}_i, \mathrm{CD}_i, \mathrm{SE}_i, \mathrm{SC}_i \tag{9}$$

where i is the length of the feature vector.

3.4 Dimensionality Reduction

The linear discriminant analysis (LDA) [27] is the most frequently used method for dimensionality reduction in machine learning. It is assumed that the classification of the data is normally performed by different means of the data, but for each class, the same variance matrix is used. To retain the original dimensions of the data, the optimal orthogonal linear combinations have been calculated by the means of LDA. New base vectors are used to perform the separability of the classes rather than the variance of the entire dataset. The optimization problem can be described in the following Eq. (13).

$$\max_{v} \frac{v^T Q_b v}{v^T Q_v v} \tag{10}$$

$$Q_b = \sum_{i=1}^{C} (\mu_i - \mu)(\mu_i - \mu)^T \tag{11}$$

where Q_b represents the scatter matrix of between classes. μ is the mean of dataset, μ_i denotes the mean of *ith* class.

$$Q_v = \sum_{i=1}^{C} \sum_{j=1}^{n_i} (\mu_i - d_{i,j})(\mu_i - d_{i,j})^T \tag{12}$$

where Q_v represents the scatter matrix of within class, $d_{i,j}$ vector of *ith* class, n_i is used for several samples in ith class, and C is the number of classes. Total scatter matrix Q_t can be obtained from summation of the Q_v and Q_b. From Eqs. (10)–(12) optimization criteria, a generalized eigenvalue–eigenvector is obtained. More reduction can be obtained through obtaining the eigenvectors that correspond to the large eigenvalues.

$$Q_t = Q_b + Q_v \quad \text{where} \quad Q_t = \sum_{i=1}^{n} (\mu_i - \mu)(\mu_i - \mu)^T \tag{13}$$

Q_t is the feature matrix with minimized dimensions is then passed to the classifier for keyword classification.

3.5 Command Map Table and Task Observer Thread

The command map table contains information about the mapped callback corresponding to the idea. The task observer thread observers the activity and invoke/dispatch the task for execution on the smart devices.

3.6 Tools and Software

As part of this work, the preprocessing and feature extraction were implemented in the *SciPy* and *Numpy* library of Python language. Convolutional neural network models were designed using the *Keras library* and run using *TensorFlow* in an attempt to classify the time–frequency representations. The *matplotlib* library was used to create plots of the figures and data visualization.

4 Classifiers Used

4.1 SVM Classifier

Initially, support vector machines (SVM) [28] were used for finding a hyperplane in N (number of features)-dimensional space that separate data points as per the class. In this research, the one-vs-all classification strategy for SVM is employed. It trains the single classifier of each class for the selected class treat as a positive label and the rest of the class labels consider as a negative label, and so on until, it constructs N different binary classifiers. For the *ith* classifier, let the positive examples be all the points in class *i*, and let the negative examples be all the points not in class *i*. Let fi be the *ith* classifier. Classify with Eq. (14).

$$f(x) = \arg\max(i) f_i(x) \tag{14}$$

where f_i is the i_{th} classifier.

4.2 KNN Classification

The KNN [29] classifier used distance-based measures to classify M-classes into N-labels. It is a primitive type of technique it used a different type of distance measure, viz., Euclidian distance, Hamming distance, City Block distance, etc., and the distance function differs with respect to the value of K. Following Eq. (15) has been employed for classification purposes.

$$d(f_1, f_2) = \sqrt{\sum_d (f_1^d - f_2^d)^2} \tag{15}$$

where *d* is the distance between feature *f1* and *f2*.

4.3 CNN Classifier

For the proposed work, CNN model was designed, where EEG keyword dataset data first convolution layer takes this one-dimensional array as input and the convolution operation uses 10 initial convolution filters and a convolutional kernel of size 11. Where, as the activation function the first convolution layer uses *'relu'* (*'ReLu'* or *Rectilinear units* as Activation for Arousal model). In these proposed experiments, the choice of activation functions for this first layer is of cardinal importance, as some functions like *sigmoid or softmax* might be failed to activate neurons of later layers consistently. This improper activation function contributed to develop a defective model. The successive layer is another convolutional neural layer, which again with 100 filters and a 3 * 3 size kernel. This layer adopts *'relu'* as the activation function for both valence and arousal classification. Thus, max pooling layer produces a flat one-dimensional layer with dropout on outputs with a probability of 0 and 5. The final dense layer utilizes the *'softmax'* layer as its activation function. Categorical cross entropy is used as a loss function in this model and *'rmsprop'* used as the optimizer. The experiment was carried out up to 500 epochs and trains the model using batches of 32 experiments each.

4.4 Random Forest

Random forest (RF) [30, 31] classifier utilizes an ensemble learning technique with multiple decision trees through the training process for classification. The output of the RF is individual trees with average prediction. It produces a forest of random numbers of trees. Compare to the normal decision tree method that is based on a rule-based algorithm that predicts the nodes using some set of rules, the RF algorithm uses Gini index [RF2] or information gain to compute root node, and it calculates the root node and twists the features randomly. Majority voting is considered for the selection of next-level nodes that are selected from different classes. The result of the classifier will be handed over to the native command translation mechanism which initiates the activity in the smart processing elements (smart devices).

4.5 *Performance Measure*

The confusion matrix evaluation is used to evaluate the performance of each classifier. A confusion matrix is an $N * N$ table that contains outcomes generated by the classifier. The confusion matrix encompasses four metrics, namely true positive (TP), true negative (TN), false positive (FP), and false-negative (FN).

4.5.1 Training of the Classifiers

In this system, the performance of five classifiers, namely SVM, KNN, RF, MLP, and CNN were compared for EEG-based keyword classification. These classifiers utilize the EEG keyword features mentioned in the data acquisition section as input. In classification experiments, EEG data from 10 subjects is collected for each subject which is given as 10 samples per 12 keywords. For training, 70% of the data is used from the total volume of the dataset and the remaining 30% is utilized for testing the data.

5 Results and Discussion

The feature extraction and selection were carried out using DWT and FFT, respectively. The feature matrix was constituted from all the EEG samples and saved on the disk to pass to different types of classifiers. Data from the feature matrix was divided in the ratio of 70:30, i.e., 70% for training and 30% for testing purposes. Consequently, training and testing features are passed to five different classifiers, namely KNN, SVM, random forest, MLP, and CNN. The result of each classifier is evaluated through a confusion matrix performance measure.

According to the results, the proposed EEG-based classification method can be concluded that there is a difference in human behavior, thinking, and thought process, and significantly there may differ conditions in data acquisition our classifiers achieved higher than 90% accuracy except for one classifier, i.e., random forest. The results are shown in the following section. The detailed result analysis for each classifier is described below. Using the feature fusion of DWT and FFT features as inputs, the accuracy of KNN, SVM, MLP, CNN, and RF are 93.06%, 92.36%, 90.97%, 92.36, and 74.31%, respectively. The highest accuracy among all the classifiers is achieved with KNN followed by SVM, MLP, CNN, and lastly random forest (RF). While classification accuracy differs between the classes, KNN outperforms other classifiers for most of the keywords according to the performance of all classifiers.

In Fig. 4a, a confusion matrix of the KNN classifier has been shown, for feature fusion of Fast ICA and FFT a classification rate of 93.06% has been observed. Considering, 1 sample from class *close*, *copy*, *cut*, *new*, *ok*, *paste,* and *play* are wrongly predicted as class *stop*, *close*, *copy*, *stop*, *copy*, *open*, and *copy*, respectively. Additionally, three samples from the class *stop* are misclassified in a class *play* and *start*. From the confusion matrix, the KNN classifier is used only with 10 samples out of 144 samples were misclassified. The rest of all 134 samples were correctly classified and placed diagonally.

Similarly, from Fig. 4b, out of 144 test samples, 11 were not correctly categorized by the SVM classifier. Three keywords from the class play are wrongly classified into class *start* and *copy*. Two samples from *stop* and *close* keywords were misclassified into class *play*, *start*, and *ok,* respectively. Finally, 1 sample from class *new*, *ok*, *paste,* and *start* are misclassified under the class *play*, *paste*, *close*, and *stop*, respectively. The SVM classifier was able to classify 133 keywords correctly with an overall recognition accuracy of 92.36%.

The confusion matrix for CNN classifier as shown in Fig. 4c, there were a total 144 samples have been tested against true labels. It was seen that out of 17 samples of class '*close*', 15 samples were classified correctly and 2 samples were misclassified so that they found in a class of *ok* and *stop*. Similarly, all test samples of *copy*, *cut*, *pause*, and *start* are correctly classified. Out of these all 144 test samples, the 10 test samples of class *delete*, 09 samples are classified correctly, and 01 sample are misclassified into the class *copy*. The average classification resulting in 90.97% classification accuracy that is out of 144 samples, 131 samples were classified into correct classes, and only 13 samples were misclassified.

The confusion matrix for the MLP classifier is shown in Fig. 4d, from the confusion matrix, it can be seen that 11 samples were misclassified and all rest of the 133 samples out of 144 have been correctly classified. For the class *close* and *ok*, 1 sample from each class is wrongly classified under the class *start* and *paste*, respectively. Two samples from class *open*, *paste*, and *play* are misclassified into the class *cut*, *start*, *close*, *open*, *pause,* and *start*. All test samples of class *copy*, *cut*, *delete*, *new,* and *start* are correctly classified into their respective classes. The overall accuracy reported from the MLP classifier is 92.36%.

In Fig. 4e, a confusion matrix of the random forest (RF) classifier has been shown, for feature fusion of fast ICA and FFT a classification rate of 74.31% has been observed. It is seen that the RF classifier recorded the highest confusion, for the class close, 10 samples out of 17 wrongly predicted. Similarly, out of 13 samples, 6 samples of the class *copy* are misclassified into a class *open, pause*, and *play*. Other confused samples can be seen off-diagonally. Overall 37 samples out of 144 wrongly predicted and 107 samples are correctly classified by the RF classifier. Another observation is that class cut, delete, and copy are rarely misclassified from all the classifier results.

The main objective of this study is to devise a BCI system that can classify EEG-based keywords, i.e., non-invasive. DWT and FFT features were adopted to fuse them for better classification accuracy. Fused features were passed to multiple machine

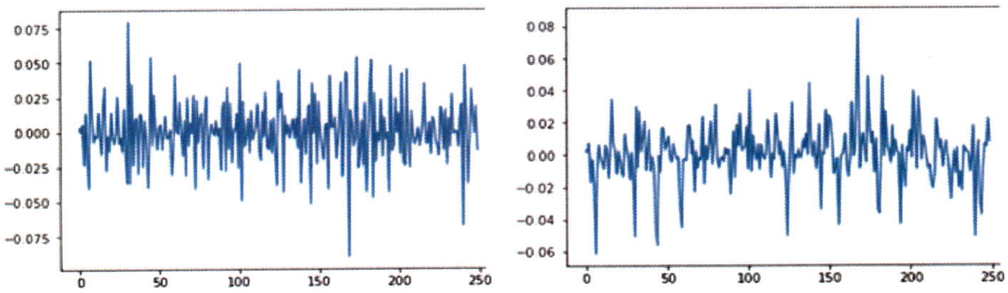

Fig. 3 **a** Actual EEG signal and **b** filtered signal using FICA

learning classifiers, viz. RF, MLP, CNN SVM, and KNN. Figures 1, 2, 3, 4, and 5 where classification outcomes from different classifiers have been reviewed. In this study, LDA algorithms have been deployed to reduce the dimensions of the EEG feature vectors and which is found successful according to the performance of the classifiers. The classification revealed the applicability of the EEG-based keyword classification. A graph that depicts the summary of the classification rates is given in Fig. 5.

6 Conclusion

In this work, a novel approach is proposed for the EEG-based keywords classification based on brain-computer interface (BCI) and smart devices. A total of 12 keywords were considered to constitute 1200 samples of the dataset. DWT and FFT features were exploited from the EEG signal dataset. The extracted features were fused with the assistance of the sum rule. These fused features were passed to the multiple machine learning classifiers. There are five types of classifiers were used over the EEG dataset. The feature vectors were preprocessed with the fast ICA algorithm and dimensionality was reduced through the LDA. The performance of each classifier was evaluated through the confusion matrix system performance measure. The deep classifiers like MLP and CNN were analyzed along with SVM, KNN, and RF for classification purposes. It is observed that the KNN classifier outperformed the overall classifiers with an accuracy of 93.05% recognition rate.

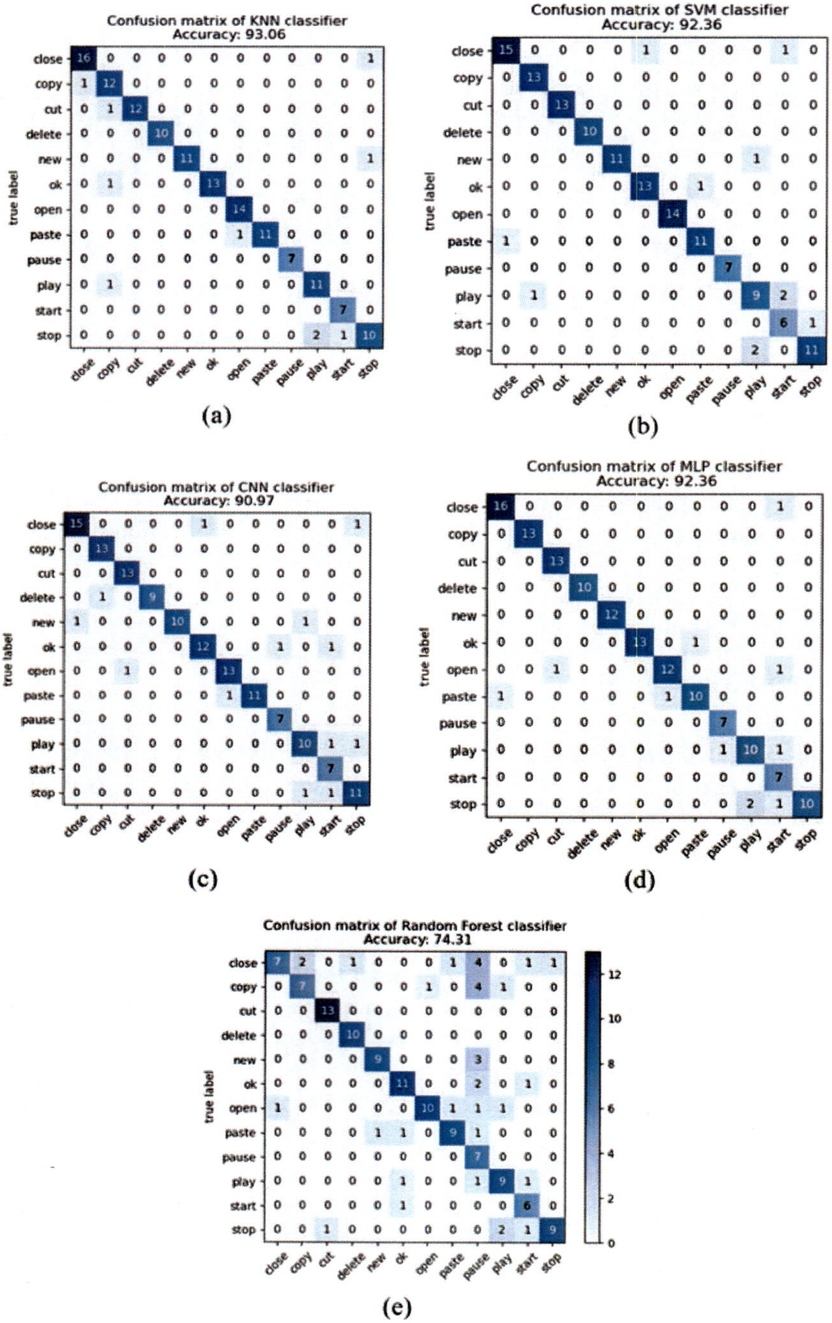

Fig. 4 Confusion matrix of various classifiers for feature fusion, the diagonal figures depict accurately identified samples

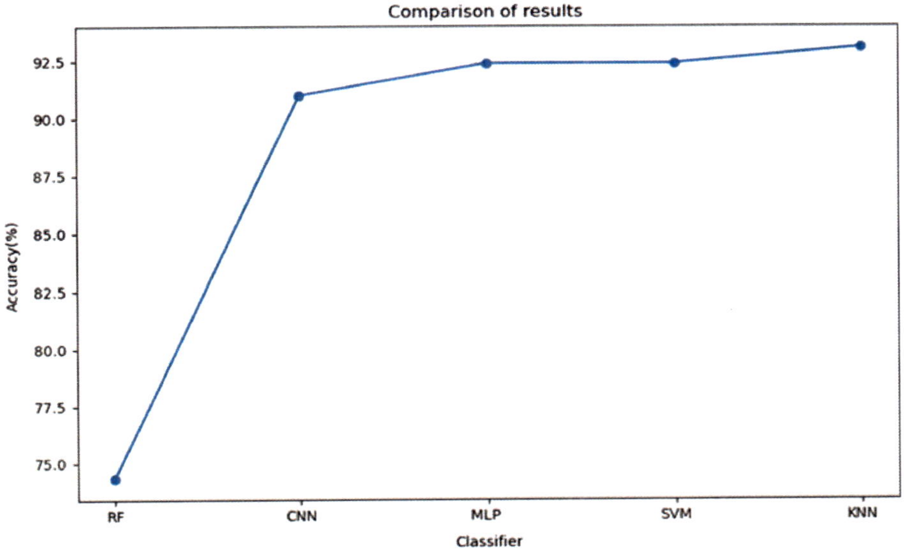

Fig. 5 Curve showing classification rates under different classification schemes

References

1. A. Rosado, A.C. Rosa, *Automatic Detection of Epileptiform Discharges in the EEG*. arXiv: 605.06708 (2016)
2. M. Phothisonothai, M. Nakagawa, EEG-based classification of motor imagery tasks using fractal dimension and neural network for brain-computer interface. IEICE Trans. Inf. Syst. **91**(1), 44–53 (2008)
3. X. Zhang, L. Yao, D. Zhang, X. Wang, Q.Z. Sheng, T. Gu, Multi-person brain activity recognition via comprehensive EEG signal analysis, in *Proceedings of the 14th EAI International Conference on Mobile and Ubiquitous Systems: Computing, Networking and Services*, pp. 28–37, 2017
4. N. Jatupaiboon, S. Pan-ngum, P. Israsena, Real-time EEG-based happiness detection system. Sci. World J. **2013** (2013)
5. S. Jirayucharoensak, S. Pan-Ngum, P. Israsena, EEG-based emotion recognition using deep learning network with principal component based covariate shift adaptation. Sci. World J. **2014** (2014)
6. P. Ackermann, C. Kohlschein, J.Á. Bitsch, K. Wehrle, S. Jeschke, EEG-based automatic emotion recognition: feature extraction, selection and classification methods, in *2016 IEEE 18th International Conference on E-Health Networking, Applications and Services (Healthcom)* (IEEE, 2016), pp. 1–6
7. T. Alotaiby, F.E. Abd El-Samie, S.A. Alshebeili, I. Ahmad, A review of channel selection algorithms for EEG signal processing. EURASIP J. Adv. Signal Process. **2015**(1), 66 (2015)
8. F. Lotte, L. Bougrain, A. Cichocki, M. Clerc, M. Congedo, A. Rakotomamonjy, F. Yger, A review of classification algorithms for EEG-based brain–computer interfaces: a 10 year update. J. Neural Eng. **15**(3), 031005 (2018)
9. F. Lotte, A tutorial on EEG signal-processing techniques for mental-state recognition in brain–computer interfaces, in *Guide to Brain-Computer Music Interfacing* (Springer, London, 2014), pp. 133–161

10. K. Amarasinghe, D. Wijayasekara, M. Manic, EEG based brain activity monitoring using artificial neural networks, in *2014 7th International Conference on Human System Interactions (HSI)* (IEEE, 2014), pp. 61–66
11. C. Chen, J. Wang, K. Li, W. Qiuyi, H. Wang, Z. Qian, G. Ning, Assessment visual fatigue of watching 3DTV using EEG power spectral parameters. Displays **35**(5), 266–272 (2014)
12. S.-Y. Dong, B.-K. Kim, S.-Y. Lee, EEG-based classification of implicit intention during self-relevant sentence reading. IEEE Trans. Cybern. **46**(11), 2535–2542 (2015)
13. R.S. Huang, C.J. Kuo, L.-L. Tsai, O.T.C. Chen, EEG pattern recognition-arousal states detection and classification, in *Proceedings of International Conference on Neural Networks (ICNN'96)*, vol. 2 (IEEE, 1996), pp. 641–646
14. R. Horlings, D. Datcu, L.J.M. Rothkrantz, Emotion recognition using brain activity, in *Proceedings of the 9th International Conference on Computer Systems and Technologies and Workshop for PhD Students in Computing*, pp. II–1, 2008
15. G.E. Fabiani, D.J. McFarland, J.R. Wolpaw, G. Pfurtscheller, Conversion of EEG activity into cursor movement by a brain-computer interface (BCI). IEEE Trans. Neural Syst. Rehabil. Eng. **12**(3), 331–338 (2004)
16. H.U. Amin, W. Mumtaz, A.R. Subhani, M.N.M. Saad, A.S. Malik, Classification of EEG signals based on pattern recognition approach. Front. Comput. Neurosci. **11**, 103 (2017)
17. S.K. Hadjidimitriou, L.J. Hadjileontiadis, EEG-based classification of music appraisal responses using time-frequency analysis and familiarity ratings. IEEE Trans. Affect. Comput. **4**(2), 161–172 (2013)
18. Y. Hashimoto, J. Ushiba, EEG-based classification of imaginary left and right foot movements using beta rebound. Clin. Neurophysiol. **124**(11), 2153–2160 (2013)
19. P. Herman, G. Prasad, T.M. McGinnity, D. Coyle, Comparative analysis of spectral approaches to feature extraction for EEG-based motor imagery classification. IEEE Trans. Neural Syst. Rehabil. Eng. **16**(4), 317–326 (2008)
20. B. Hosseinifard, M.H. Moradi, R. Rostami, Classifying depression patients and normal subjects using machine learning techniques and nonlinear features from EEG signal. Comput. Methods Progr. Biomed. **109**(3), 339–345 (2013)
21. D. Iacoviello, A. Petracca, M. Spezialetti, G. Placidi, A real-time classification algorithm for EEG-based BCI driven by self-induced emotions. Comput. Methods Programs Biomed. **122**(3), 293–303 (2015)
22. A. Erfanian, A. Erfani, ICA-based classification scheme for EEG-based brain-computer interface: the role of mental practice and concentration skills, in *The 26th Annual International Conference of the IEEE Engineering in Medicine and Biology Society*, vol. 1 (IEEE, 2004), pp. 235–238
23. K.D. Valente, J.Q. Andrade, R.M. Grossmann, F. Kok, C. Fridman, C.P. Koiffmann, M.J. Marques-Dias, Angelman syndrome: difficulties in EEG pattern recognition and possible misinterpretations. Epilepsia **44**(8), 1051–1063 (2003)
24. A.K. Jaiswal, H. Banka, Local pattern transformation based feature extraction techniques for classification of epileptic EEG signals. Biomed. Sig. Process. Control **34**, 81–92 (2017)
25. I. Jayarathne, M. Cohen, S. Amarakeerthi, BrainID: Development of an EEG-based biometric authentication system, in *2016 IEEE 7th Annual Information Technology, Electronics and Mobile Communication Conference (IEMCON)* (IEEE, 2016), pp. 1–6
26. N. Jrad, M. Congedo, R. Phlypo, S. Rousseau, R. Flamary, F. Yger, A. Rakotomamonjy, sw-SVM: sensor weighting support vector machines for EEG-based brain–computer interfaces. J. Neural Eng. **8**(5), 056004 (2011)
27. R. Khosrowabadi, H.C. Quek, A. Wahab, K.K. Ang, EEG-based emotion recognition using self-organizing map for boundary detection, in *2010 20th International Conference on Pattern Recognition* (IEEE, 2010), pp. 4242–4245
28. K. Kunze, Y. Shiga, S. Ishimaru, K. Kise, Reading activity recognition using an off-the-shelf EEG--detecting reading activities and distinguishing genres of documents, in *2013 12th International Conference on Document Analysis and Recognition* (IEEE, 2013), pp. 96–100

29. P. Lahane, A.K. Sangaiah, An approach to EEG based emotion recognition and classification using kernel density estimation. Procedia Comput. Sci. **48**, 574–581 (2015)
30. W. Li, Q.-C. He, X.-M. Fan, Z.-M. Fei, Evaluation of driver fatigue on two channels of EEG data. Neurosci. Lett. **506**(2), 235–239 (2012)
31. C.-T. Lin, M. Nascimben, J.-T. King, Y.-K. Wang, Task-related EEG and HRV entropy factors under different realworld fatigue scenarios. Neurocomputing (2018). https://doi.org/10.1016/j.neucom.2018.05.043

Security Analysis for Sybil Attack in Sensor Network Using Compare and Match-Position Verification Method

B. Barani Sundaram, Tucha Kedir, Manish Kumar Mishra, Seid Hassen Yesuf, Shobhit Mani Tiwari, and P. Karthika

Abstract Remote sensor networks are exceptionally simple to ensure organization security. Particularly, basic assaults of different sorts are recorded in remote sensor networks by numerous analysts. Sybil assault is a gigantic damaging assault the sensor network various veritable characters with manufactured personalities are utilized for illicit section an organization. Observing Sybil assault, sinkhole, with wormhole assault multicasting colossal employment in remote sensor organization. Essentially, a Sybil assault implies a hub imagines its character to different hubs. Communication to a criminal hub takes misfortune from data and becomes hazardous in the company. The current technique random password comparison is a method that checks the hub properties by examining neighbors. The review was a Sybil assault with the goal of settling the issue. The study has projected a consolidated comparison with match-position and verification of method (CAM-PVM) and MAP is used for identifying and preventing the section in the Sybil hubs organization. The main aim is to ensure the protection of the organization of remote sensors and to handle multicasting and unicasting assaults of sorts.

Keywords Wireless sensor network · Security · Sybil attack · CAM-PVM · Message authentication method

B. Barani Sundaram (✉) · T. Kedir
Department of Computer Science, College of Informatics, Bule Hora University, Bule Hora, Ethiopia

M. K. Mishra · S. H. Yesuf
Department of Computer Science, University of Gondar, Gondar, Ethiopia

S. M. Tiwari
Department of Computer Science and Engineering, Babu Banarasidas Engineering College, Lucknow, Uttar Pradesh, India

P. Karthika
Department of Computer Application, Kalasalingam Academy of Research and Education, Krishnankoil, India

S. Shakya et al. (eds.), *Mobile Computing and Sustainable Informatics*, Lecture Notes on Data Engineering and Communications Technologies 68,
https://doi.org/10.1007/978-981-16-1866-6_4

1 Introduction

A far-off sensor network includes characteristic noticing, target following, productivity testing, and diverse maintenance decisions. Execution and geology formation has become an enormous task in today's academic research [1]. The use of a remote sensor network in the setting up of jobs is astoundingly important, complemented by security. In a remote sensor organization, things being as they are, may be high or low, taking everything into consideration, avoidance, and the sector of hazardous attacks [2].

Various attacks can also be included on the connection, such as extra dimensions, sinkhole, Sybil, rest, and also unique association forwarding attacks. Several experts have acknowledged their similar organizations, that have useful contraptions, using decentralized and adaptable methods in different trade organizations. For multi-user Web usage, a portion of the device is ready for synchronization. The use of figurations that enhance precision to locate the precise region. Sybil targeting several centers with incorrect ID display or customers' duplicate ID consider about the centers in distant sensor association.

There is a standard code with genuine correspondence between network center points for protection over social network [3]. The mutual association analysis demonstrates the existence of reasonable features or virtual connections in associations. Coventry is an association based on the Web as the most prominent intent of different associations. The positions of the association depend on a steady ID to sort over and shape associations [4]. Fixed establishment or single hop is completely ignored and multi-hop paves the way for the communications [5]. Non-establishment associations may have a more modest scheme in the distant sensor networks.

2 Sybil Attach and Security

2.1 *WSN on Security Attacks*

Different sorts of harmful activities are patent in inaccessible sensor connections [6–8]. A fragment of these is made concerning focuses while others are advanced in a connection and the information association with application layers. Some are made in the approval state [12]. The assaults are at present assigned are dynamic and inactive. The events are composed by sending unlawful data in the affiliation that can affect. Sybil, sinkhole, and sneak are a section of the dynamic assaults. Withdrawn assaults are required to affect the connection assets, for example, lifetime and connection size [9–11].

2.2 Sybil the Attack

The center position or device has different characters that may not necessarily be established. It does not represent the remarkable devices, but actively recognizes the character of yet another stage, causing layoffs within the coordinating events. Sybil attacks degenerate uprightness, stability, and device utilization of data. Some of the limitations are coordination tools, dissemination of air resources, and disclosure of misconduct can be carried out identically. The correspondence association is organized in a sensor network with multiple sensor center points. The remote correspondence between these sensor centers is experienced by the central station. Such center points articulate to the predetermined center points of a predefined number [17, 18]. There will be numerous encryption methods available to prevent external attacks at the center position, but the center points in the correspondence association may be evenly an assault [19, 20].

These kinds of investigator threats are known as the Sybil attack [13–15] for which the center spoofing is considered for the Sybil center point S, as well as the other center, is the average center point N. Specifically, in an authentic communication method, N center points should communicate to each other. In any case, the center point of S arrives here in a different kind within it as an internally known center or sends an assault on the association. A Sybil center attempts to articulate to the neighboring center points while using the character of its traditional center and, in process, the single center point assigns different characters in the area to the different center points of the association which are unconstitutional. Sybil's center is outlined as another character or appropriating legitimate character. Therefore, as an additional part of a center point get into iniquity. This makes strife in the association and it gets fallen. A defective center point that goes into the association with different IDs is showed up in Fig. 1. Likewise, Sybil attacks are described into two structures dependent on the method of attack on the association. They are according to the accompanying.

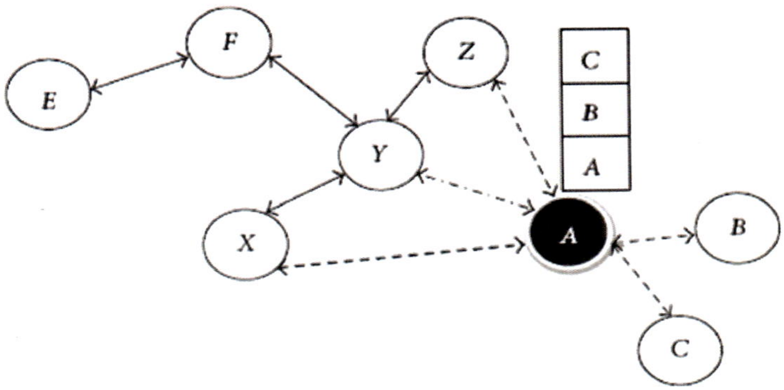

Fig. 1 Sybil attacks with the multiple ID

Table 1 Node $N2$ is the Sybil, acting as $N7$

Node Id	$N3$	$N5$	$N7$ [$N2$]	$N7$	$N8$	$N9$
τ	11:35	11:48	11:07	111:58	11:58	12:07
X	97	164	77	28	64	72
Y	64	94	44	143	155	175
τ	1:06	1:08	1:07	1:07	1:08	1:08

i. Direct Attack and Indirect Attack. In a quick attack, certified centers talk about clearly with Sybil centers, however, is a variant attack of correspondence is through the toxic center.
ii. Fabricated Attack and Stolen Identity Attack. Authentic character centers are used to make unlawful centers. As such, a sensor center point which has an ID of 16-cycle entire numbers makes comparative 16 pieces of ID are made centers. IDs obtained by Sybil center with squashed check the character replication [16].

From Table 1, obviously, the $N7$ data is recreated; it is discovered that the data of the reproduction hub does not coordinate the first $N7$ data in table below. Sybil's movement is related to the use of the CAM-PVM calculation and is recognized in the organization. To provide counteraction to Sybil's action, another MAP calculation is applied alongside CAM-PVM with anticipation of the Sybil movement. The design involves unicast just with multicast correspondence in the organization. The calculation for CAM-PVM and MAP is due to the underneath.

3 Compare with Match-Position and Verification of Method

CAM-PVM calculation has utilized the disclosure with information transfer the organization of the hub's data is check from the Node $N2$ in the Sybil *table*. After the confirmation of the CAM-PVM calculation, the calculation gathers the ID, and current data of hubs, contrasts and beginning data are enlisted. Consequently, the effects of the CAM-PVM calculation can provide accepted hubs in the course to guarantee information transmission. In any case, the specific hubs are treated as obscure hubs, for example, Sybil with information transmission is halted with if an alternate way is chosen.

(1) *Let* $G = \{N1, N2, N3, \ldots, Nm\}$

(2) *Let BS, Admin be the well configured nodes*

(3) *for* $i = 1$ *to* m // Nodes are placed randomly

(4) $Ni \leftarrow$ Location(rand(X), rand(Y))

(5) ID($Ni \leftarrow i$);

(6) *End Loop*

(7) *Let Li be the set of link between pair of nodes in the network*

(8) $Ni \rightarrow BS(\text{Msg}, \tau)$// For every nodes Ni

(9) $U^{*} \rightarrow$ *minimum* distance($Ni, Ni + 1$)

(10) Ni *RES BS*; $RES = \text{ID}, \tau, X, Y$;

(11) $RES \rightarrow$ *iNODEINFO_table*

(12) *End Loop*

(13) $Su\{\text{U}^{*}\} \rightarrow S$

(14) $S \rightarrow RT(S)$

(15) $Ni + 1 \rightarrow Ni(\tau)$

(16) $\tau c = (\text{ID}, X, Y, \tau)$

(17) $U^{*} \rightarrow D$

(18) *for* $i = S$ *to* D //Route Discovery

(19) $RT(S)$ Ni;

(20) Li $\rightarrow (Ni, Ni + 1)$;

(21) $RES \rightarrow$ *iROUTING_table*

(22) *End Loop*

(23) *for* $i = S$ *to* D// Data transmission

(24) *if(current Ni.info ==* iROTINGtable==iNODEINFOtable)then

(25) *if* $\left(\begin{array}{c}\textit{current Ni.info==iROUTINGtable==}\\ \textit{iNODEINFO}_{table}\end{array}\right)$ *then*

(26) Ni *data* $Ni + 1$

(27) *else*

(28) $Ni + 1$ *is blocked as Sybil Node*

(29) *End Loop*

(30) *iROUTINGtable* entries clear

(31) End

4 Message Authentication and Passing (MAP)

The utilization of CAM-PVM is a tedious cycle in practice. The proposed work is to avoid the gadgets in the Sybil action. Every hub ought to allow the verification message. If the source center point assumes the intention dynamically it uses for CAM-PVM, and MAP figuring is used for differentiating the information approval in the present center of the Sybil assault. Where the association G denotes a center N_i that passes the data to center point N_j. And the center N_i sends a message to the center point N_j with its key as msg(N_i), which is delivered by the BS while enrolling in the association G. Center N_j (target center) provides its imperative message msg(N_j), and later, the two keys were confirmed with the base station.

(1) $G = \{N0, N1, N2, \ldots, Nn\}$

(2) *for* $I = 1$ *to* n

(3) rtable = addrec(Ni(id), $Ni(x)$, $Ni(y)$, Ni(msg))

// node id, x, y values of ith node

(4) *End*

(5) *for* $I = 1$ *to* n

(6) *for* $J = 1$ *to* n

(7) $Ni \rightarrow$ *Send(request)* $\rightarrow Nj$

(8) $Nj \rightarrow$ *Send(acceptance)* $\rightarrow Ni$

(9) *If*(msg(Ni), msg(Nj))*exists*(*rtable*)) then

(10) $Ni \rightarrow$ *Send(dataP* $\rightarrow Nj$)

(11) *Else*

(12) *Choose the next neighbor*

(13) *End If*

(14) *End for J*

(15) *End for I*

(16) End

5 Result and Discussion

The entire model is structured using NS2 programming with 25 centers with an association size of 1200 × 1200. Each sensor center point carries on the AODV display. BS of association model, and the center point 18 datasets from center 0 are

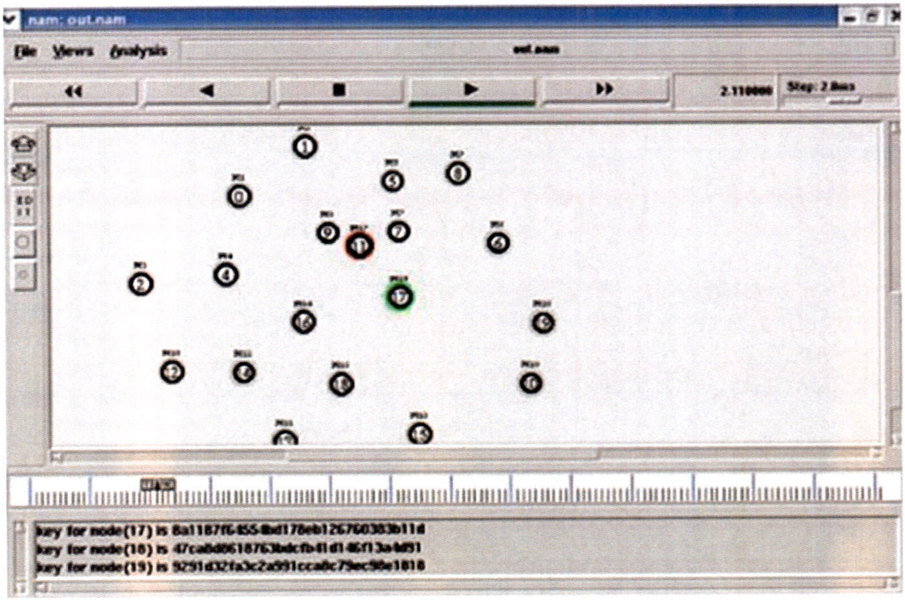

Fig. 2 Simulation to identifying the Sybil node

the two center points represented to progress. Center 0 sends the REQ information to the center point 18, and the center sends a RES message back to center point 0. This can be differentiated by center 12, which sends the RES information with characteristics of center point 18. Later, this could be followed by center 12 and this cannot present its accepted key value, which has a point with center point 18.

The efficiency of the association is controlled by differentiating the throughput when thought the message approval with passing computation in the associated value which is depicted in Fig. 2.

The proposed technique is developed to ensure the recognizable proof of Sybil's attack and the packet can also be destroyed while assessing the normal deferral of packet data movement. Performance of the toxic center, as well as other significant components, decides on the idea of organizing a coordinating appearance. Figure 3 shows the normal deferral of the association data packet before sending the request approval and transmitting the computation. Here, data groups of various sizes are sent at various times. The direct movement of its Sybil centers appears to transfer the data between multiple regions and to agitate the primary centers of the organization. This figure shows that the whole deferral is less likely to occur when the MAP is implemented.

Many centers in the association and the transparency movements of the Sybil center point are reported in the assessment. The partnership centers display various accentuations and distinct quantities as an overview. The test is a comparatively huge to measure the data that are influencing the packet movement. The implementation of

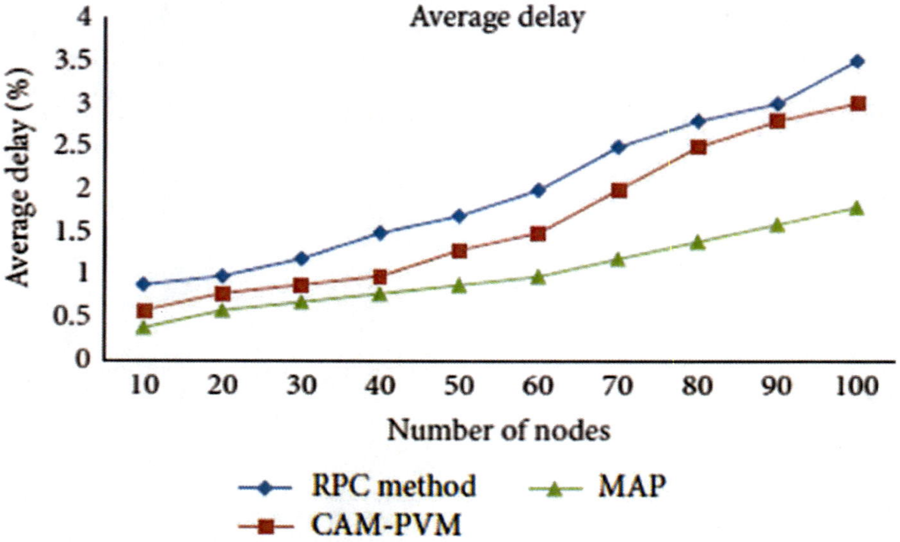

Fig. 3 Examination of normal deferral of information parcel move between existing technique RPC with CAM-PVM and MAP

the proposed method is to resolve when it is submitted with the coordinating events. In the current random password comparison technique, the CAM-PVM possibility happens with the performance of 74% and the MAP exists with 95%. The evaluation of the throughput between the current CAM-PVM and MAP techniques is shown in Fig. 4. It recognizes Sybil's attention and needs to stay away from transmission through aggressor site, and this happens specifically when a territorial structure is passed on during transmission.

6 Conclusion

The message approval with passing technique is applied to check the Sybil center point. The center is Sybil's center point with duplicate ID and the details about the available center points are also explained in this article. The affirmation to use the center point with CAM-PVM is proposed. The CAM-PVM is to verify every center point of the register file with the passing technique and confirmation has been submitted before communication. If a center point does not have approval by the association or base station, it cannot communicate with some other center point in the association. In message affirmation along with the passing technique, the system is extremely effective with some other procedure, and it also reduces time along with the cost.

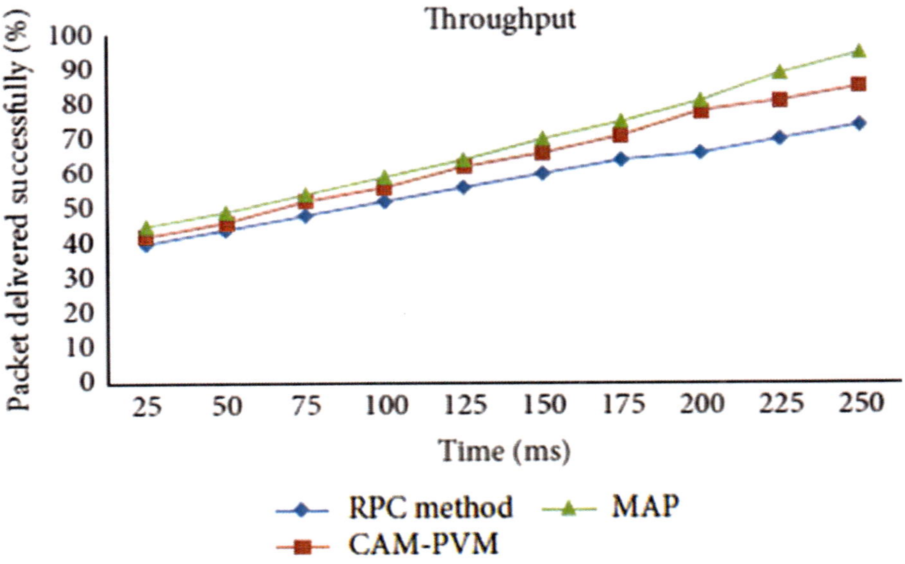

Fig. 4 Correlation of throughput between existing the RPC technique CAM-PVM and MAP

References

1. V. Rathod, M. Mehta, Security in wireless sensor network: a survey. Ganpat Univ. J. Eng. Technol. **1**, 35–44 (2011)
2. A. Modirkhazeni, N. Ithnin, M. Abbasi, Secure hierarchical routing protocols inwireless sensor network; security survey analysis. Int. J. Comput. Commun. Netw. **2**, 6–16 (2012)
3. P. Karthika, P. Vidhya Saraswathi, A survey of content based video copy detection using big data. Int. J. Sci. Res. Sci. Technol. (IJSRST) **3**(5), 114–118 (2017). Online ISSN: 2395-602X, Print ISSN: 2395-6011. https://ijsrst.com/ICASCT2519
4. W. Niu, J. Lei, E. Tong et al., Context-aware service ranking in wireless sensor networks. J. Netw. Syst. Manage. **22**(1), 50–74 (2014)
5. C. Komar, M.Y. Donmez, C. Ersoy, Detection quality of border surveillance wireless sensor networks in the existence of trespassers' favorite paths. Comput. Commun. **35**(10), 1185–1199 (2012)
6. P. Karthika, P. Vidhya Saraswathi, IoT using machine learning security enhancement in video steganography allocation for Raspberry Pi. J. Ambient Intell. Human. Comput. https://doi.org/10.1007/s12652-020-02126-4
7. Z.A. Baig, Pattern recognition for detecting distributed node exhaustion attacks in wireless sensor networks. Comput. Commun. **34**(3), 468–484 (2011)
8. S. Abbas, M. Merabti, D. Llewellyn-Jones, Signal strength based Sybil attack detection in wireless Ad Hoc networks, in *Proceedings of the 2nd InternationalConference onDevelopments in eSystems Engineering (DESE'09)* (Abu Dhabi, UAE, December 2009), pp. 190–195
9. P. Karthika, P. Vidhya Saraswathi, Image security performance analysis for SVM and ANN classification techniques. Int. J. Recent Technol. Eng. (IJRTE) **8**(4S2), 436–442 (2019)
10. S. Sharmila, G. Umamaheswari, Detection of sybil attack in mobile wireless sensor networks. Int. J. Eng. Sci. Adv. Technol. **2**, 256–262 (2012)
11. D.G. Anand, H.G. Chandrakanth, M.N. Giriprasad, Security threats and issues in wireless sensor networks. Int. J. Eng. Res. Appl. 2, 911–916 (2012)

12. P. Karthika, P. VidhyaSaraswathi, Digital video copy detection using steganography frame based fusion techniques, in *International Conference on ISMAC in Computational Vision and Bio-Engineering*, pp. 61–68 (2019). https://doi.org/10.1007/978-3-030-00665-5_7
13. K.-F. Ssu, W.-T. Wang, W.-C. Chang, Detecting sybil attacks in wireless sensor networks using neighboring information. Comput. Netw. **53**(18), 3042–3056 (2009)
14. P. Karthika, P. VidhyaSaraswathi, Content based video copy detection using frame based fusion technique. J. Adv. Res. Dyn. Control Syst. 9, 885–894 (2017)
15. A. Praveena, S. Smys, Efficient cryptographic approach for data security in wireless sensor networks using MES VU, in *2016 10th İnternational Conference on İntelligent Systems and Control (ISCO)* (IEEE, 2016), pp. 1–6
16. M. Adithya, P.G. Scholar, B. Shanthini, Security analysis and preserving block-level data DE-duplication in cloud storage services. J. Trends Comput. Sci. Smart Technol. (TCSST) **2**(02), 120–126 (2020)
17. A. Vasudeva, M. Sood, Sybil attack on lowest id clustering algorithm in the mobile ad hoc network. Int. J. Netw. Secur. Appl. **4**(5), 135–147 (2012)
18. P. Karthika, P. Vidhya Saraswathi, *Raspberry Pi—A Tool for Strategic Machine Learning Security Allocation in IoT*. Apple Academic Press/CRC Press (A Taylor & Francis Group). Proposal has been accepted (provisionally) for the book entitled "Making Machine Intelligent by Artificial Learning", to be published by CRC Press
19. N. Balachandaran, S. Sanyal, A review of techniques to mitigate sybil attacks. Int. J. Adv. Netw. Appl. **4**, 1–6 (2012)
20. G. Jing-Jing, W. Jin-Shuang, Z. Yu-Sen, Z. Tao, Formal threat analysis for ad-hoc routing protocol: modelling and checking the sybil attack. Intell. Autom. Soft Comput. **17**(8), 1035–1047 (2011)

Certain Strategic Study on Machine Learning-Based Graph Anomaly Detection

S. Saranya and M. Rajalakshmi

Abstract "A rotten apple spoils the whole bunch" deciphers the research problem domain. Taking a broader perspective, the existence of anomaly in a graphical community would degrade the global network performance. The anomaly is a hindrance to insight for a better data quality analytics. Though the majority of multidisciplinary contributions in machine learning prevail, a surge towards graph-based learning is gaining a significant importance. Most of the statistical machine learning methodology adopted for outlier identification is inherited in most of the graph-based anomaly detection (GBAD) techniques. This survey aims to establish a broad overview on the state-of-the-art methods for GBAD in a static environment, specifically GBAD techniques by utilizing structural orientation, community-based discovery to seek the quality advancement. To achieve better graph storage on query handling, some graph summary techniques are studied. Intuitive comparative analysis of diverse GBAD algorithm helps to clarify novice researcher in problem solving. Moreover, the survey opens up new research ideas and practical challenges in the realm for a robust futuristic contribution.

Keywords Graph-based anomaly detection (GBAD) · Outlier detection · Big data · Machine learning (ML) · Graph labelling · Structure-based graph anomaly · Community-based graph anomaly · Graph compression

1 Introduction

"A detective finds a culprit in a crime scene". The phrase suites well with the detection activity involved to find an anomaly or a group of anomaly in a big data. The anomaly is hindrance to data preparation for any data analytics process. For different

S. Saranya (✉)
Faculty of Computer Science and Engineering, Coimbatore Institute of Technology, Coimbatore, India
e-mail: saranya.s@cit.edu.in

M. Rajalakshmi
Faculty of Information Technology, Coimbatore Institute of Technology, Coimbatore, India
e-mail: rajalakshmi@cit.edu.in

S. Shakya et al. (eds.), *Mobile Computing and Sustainable Informatics*, Lecture Notes on Data Engineering and Communications Technologies 68,
https://doi.org/10.1007/978-981-16-1866-6_5

applications, it is known popularly as novelty, anomaly, noise, deviation, exception, white crows, indisguise, discordance and abnormality. In the current era, real-world data is dirty, inconsistent, incomplete and noisy that travels with varied sources, in varied sizes and at varied speed.

Data is growing in massive speed which is crucial and challenging to capture, curate, store, view, search, share, transfer, visualize and analyse. Identifying outliers is crucial since data set need to be maintained in some structured format as well as algorithms used for outlier detection in data mining cannot be applied to current real-world data stream. Curse of dimensionality and sparsity in data potentially misleads the performance of outlier analysis. The main characteristics of big data are volume (size), velocity (speed), variety (different formats/complexity), veracity (quality), value (integrity), valence (connectedness), validity (accuracy), variability (evolution), viscosity (data velocity over time), volatility (lifetime), viability (feasible), vocabulary (metadata), venue (location) and vagueness (meaning). These characteristics ensure to overcome the performance degradation is different aspects of real-world scenario and harnesses complex statistical and predictive analytics. The older methods of detecting outlier cannot be suitable for current data stream applications.

The traces left at the crime scene help detective to identify the indisguise, which is always challenging. The anomaly occurs due to man-made errors, device fault, deviation rate, malware, system fault, sampling error, etc. The general challenges in outlier detection are [1, 2]:

a. Imprecise boundary to distinguish normal and abnormal data.
b. Noisy data are often misinterpreted as anomaly and vice versa.
c. Non-universal nature of outlier technique for varying applications.
d. Data availability for training data set should be sufficient.
e. As amount of data sets increases, the outlier detection techniques should evolve as well.
f. Using traditional outlier detection techniques for processing high-dimensional data will lead to high computational costs.
g. In distributed system, the outlier detection algorithms must minimize communication overhead and synchronization overhead between the different sites in which the data resides, as well as the scans over the data.

Zhang categorized outlier into two broad classifications [3]: based on number of data instances and based on type of data where the outlier can be detected as shown in Fig. 1. A point outlier refers to an anomalous individual data with respect to rest of the data. Majority of research area focuses to deal with such outlier. The contextual outlier may be normal at one context and abnormal at other context. Applying detection technique is difficult, if defining context is tedious. A collective outlier is anomalous behaviour by set of related data from entire data set point of view. A point outlier added with context information can be transformed to contextual outlier.

On encountering the outlier, some of the possible measures taken are to correct the errors, check for the distribution assumption, generate the model with or without outlier to evaluate the effect, look for group outliers, sampling and subsampling

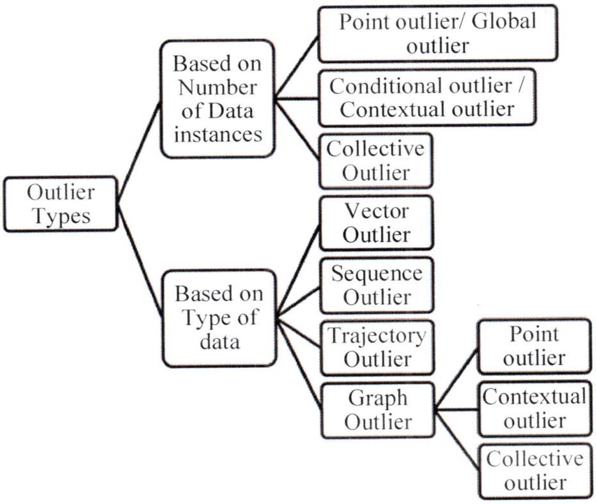

Fig. 1 Generalized classification hierarchy of outlier detection methods

methods. Elimination of outlier removes useful observations for examination. Iterative subsampling is one such elimination strategy that reveals biasing nature in pattern analysis.

The outlier analysis is serviced in several areas such as anti-social behaviour, social networks, road network, telecommunication, banking sector, cyberspace, critical systems, chemical compounds, earth science, sensor networks and image processing. Existence of outlier or anomaly makes the data quality to degrade, thereby making data pre-processing a necessary step in big data analytics [4, 5]. The security mechanism preserves the information for the mobility nature of nodes [6]. Although several data pre-processing techniques exist, most predominant techniques involve retaining data quality, integrity, uniformity and scalability. All these techniques are mutually exclusive to each other. A structured representation always mandates the need to handle such data. Even though majority of data in real world are exhibited relational in nature, still priority is given to graph representation for certain reasons.

The key contributions are listed as follows: (a) Notably different from earlier surveys, state-of-the-art machine learning techniques that are promising to solve every outlier detection issue are summarized. (b) To highlight the influence of graph representation and to explore the need of foundation algorithms in graph anomaly detection strategy. (c) Explore the static GBAD classification methods in terms of structure disorientation, community feature and graph summary strategy for achieving output quality in terms of time and cost. (d) Discuss on some identified issues and challenges in GBAD-based techniques with intent for future research discovery.

2 Generalized Machine Learning Approaches Towards Outlier Detection

"One bad apple spoils the whole bunch". The statement best represents the nature of anomaly's presence in big data analytics. Several machine learning algorithms were proposed by many renowned authors to diagnose the outlier that affects the quality of the data analysis as well ensures to improve processing time. A brief summary of outlier classification approaches in ML: model-based, angle-based, proximity-based and artificial intelligence-based approach is discussed in this section. All the work contributions are listed in chronological fashion to understand the historical follow-up of the methods (Tables 1, 2 and 3).

2.1 Key Challenges Identified in ML Techniques

See Tables 4, 5 and 6.

Some of the challenges handled based on application of interest are [15–17]:

(a) Most existing anomaly detection technique deals with point data rather than a collective data.
(b) Getting a labelled anomalous data instance is quite difficult.
(c) Assumption of most unsupervised learning technique takes the occurrence proposition of normal instances higher than anomalous instance. Assumption bias leads to high false alarm rate.
(d) Most classification results infer linear training time and quite fast testing time inconsiderable to the availability of accurate class label generation.
(e) Missing anomaly, false positive rate occurrence, defining distance measure for graph data is complex are few difficulties observed in nearest neighbour technique with computational complexity O (N2).
(f) Clustering-based approach has quadratic complexity of O (N2d) and for small clusters it is quite fast. Distinction of group anomaly from cluster is a downside by majority of algorithms.
(g) Choosing the right statistical technique with majority of assumptions fails due to varying data sets.
(h) Contextual anomaly techniques quickly detect anomaly at testing phase because of complete view of data analysis, which is not detected by point anomaly technique simply. Definition of the context is necessary to identify anomaly because its presence is hidden using only point anomaly [18].

However, many challenges still remain. Each model-based method is best in its own stake and pinpointing the best method is hardly impractical. Most of the angle-based method works in hybrid versions as unsuitability prevails due to computational complexity of big data applications.

Table 1 Model-based methodology adopted by eminent authors listed in chronological order which approaches to detect outlier

References	Methodology
Kalman (1960)	Kalman filter
Tukey (1977)	Box plot
Rousseeuw and Leroy (1987)	Minimum covariance determinant
Torr and Murray (1993)	Least median squares regression algorithm
Barnett and Lewis (1994)	Grubbs method/Z-value/minimum volume ellipsoid estimation (MVE)/convex peeling/least trimmed squares approach (LTS)
Rousseeuw and Leroy (1996) [7]	
Bishop (1994), Roberts (1998) and Tarassenko (1995)	Gaussian mixture model/extreme value theory
Ruts and Rousseeuw (1996)	ISODEPTH
Parra (1996) and Faloutsos (1997) [8]	Principal component analysis/eigenvalue estimation
Swets and Weng (1996)	Linear discriminant analysis (LDA)
Johnson et al. (1998)	FDC
Baker et al. (1999)	Hierarchical approach
Rousseeuw and van Driessen (1999)	Minimum covariance determinant (MCD)
Tax et al. (1999), Levine (2000), Decoste (2000)	Support vector machine
Laurikkala et al. (2000)	Box plot in univariate and multivariate
SchÄolkopf et al. (2001)	One-class SVM
Shekhar et al. (2001)	Grubb's test for graph structured data
Jordaan et al. (2004)	Robust support vector machines for regression
Liu et al. (2004)	Null space LDA (NLDA)
Maronna et al. (2006)	S-estimators
Filzmoser et al. (2006)	PCOut and sign algorithms (robust principle component strategy)
Nguyen et al. (2010)	Dimensionality reduction/feature extraction for OUTlier detection (DROUT)
Chuang et al. (2011)	Hybrid robust support vector machine for regression
Radovanovi et al. (2015)	Reverse nearest neighbours
Bhattacharya (2015)	Neighbourhood rank difference
Wang et al. (2015)	MST-inspired KNN
Dutta et al. (2016)	Rarity-based outlier detection (RODS)
Pasillas-Diaz (2017)	Feature bagging
Liu et al. (2017)	Homophily
Loperfido (2018)	GARCH model
Proverbio et al. (2018)	Error-domain model falsification (EDMF)

(continued)

Table 1 (continued)

References	Methodology
Cheng et al. (2019)	Entropy-based ensemble anomaly detection (EEAD)

Generally AI is a scrupulous automated machine learning method that can drastically handle scalable applications with better computational efficiency. It is hardly impossible to derive a best method from the set of AI and ML techniques due to its significance of problem associated and model design [19, 20]. However, the future scope lies in advanced learning method such as deep learning.

This section is clearly a comparative study of significant approaches, model-based, proximity-based, angle-based, AI and ML-based, that handles the anomaly detection in broad perspective. Each approach has unique collection of algorithms and supporting methods. Though the challenging factors pertinent to work contribution of eminent authors are studied, still many challenges remain.

3 Graph-Based Anomaly Detection Methods

3.1 Moving Towards Graph Representation

Outliers are extreme values that disguise in different ways and in different dimensions. However, high-dimensional data will have multiple interconnections and interrelation between the entities. Most of the traditional anomaly detection aspect as well as abnormality exhibit relational behaviour. Large interconnected network application such as ecological food web, biological data network, road network, telecom network, social network and space science exhibits interconnections.

Taking into account of interrelation property, graph is proven to be most efficient representation that can handle the interdependencies well. Correlation persistence, adversarial robustness and structural oddity analysis are some highlighting benefits of imparting graph structure. However, it cuts both ways. Due to graph isomorphic nature of graphical approaches, the overall computational performance is degraded. Generating substructure, ranking vertices is highly complex task. In handling real-time data, most detection techniques identify through structural disorientation of streaming data. Memory aspects are limited due to high-dimensional representation of data, which is highly dependable on the nature of the methodology used.

Graph patterns can be retrieved efficiently through graph visualization and incorporation of NoSQL database for graph [21, 22]. Neo4j is a most popular NoSQL database as it models and queries in shorter time than a relational database.

Table 2 Proximity-based/angle-based methodology adopted by eminent authors listed in chronological order which approaches to detect outlier

References	Methodology
Lloyd and Forgy (1965)	K-means
Graham (1972)	Graham's scan
Jarvis (1973)	Jarvis's march (gift wrap algorithm)
Wettschereck (1994)	Majority voting approach
Ng and Han (1994)	CLARANS
Sugihara (1994)	Convex hull
Aha and Bankert (1994), Skalak (1994)	Feature subset selection to optimize KNN
Zhang et al. (1996)	BIRCH
Chan (1996)	Chan's algorithm (shattering)
Knorr and Ng (1998)	DB ($\mathcal{E}$, π)
Allan et al. (1998)	Partitional k-means
Bradley et al. (1999)	K-mediod
Gionis (1999)	Local sensitive hashing (LSH)
Breunig et al. (2000)	Local outlier factor (LOF)
Ramaswamy et al. (2000)	Optimized KNN
Bolton and Hand (2001)	Partition around mediods (PAM) or K-mediods
Tang et al. (2002)	Weighted KNN, connectivity-based outlier factor (COF)
Angiulli and Pizzuti (2002)	Linearization, Hilbert space filling curve
He et al. (2003)	Cluster-based local outlier factor (CBLOF)
Papadimitriou (2003)	Multi-granularity deviation factor (MEDF)
Bay and Schwabacher (2003)	ORCA (randomized nested loop kNN)
Kollios et al. (2003)	Biased sampling OUTlier detection (BSOUT)
Hautamaki et al. (2004)	RKNN (reverse KNN)
Angiulli and Pizzuti (2005)	Hilout algorithms (Hilbert space filling curve + KNN)
Ghoting et al. (2006)	Recursive binning and re-projection (RBRP)
Zhang et al. (2006)	HighDoD
Jin et al. (2006)	Influenced outlierness (INFLO)

(continued)

Table 2 (continued)

References	Methodology
Kriegel et al. (2008)	Angle-based outlier degree (ABOD)
Fan et al. (2008)	Resolution-based outlier factor (ROF)
Kriegel et al. (2009)	Subspace outlier degree (SOD)
Tang et al. (2015)	LINE (embedding)
Kannan et al. (2015)	Mahalanobis distance (MD)-based anomaly detection
Zhang et al. (2015)	Angle-based subspace anomaly detection (ABSAD)
Ou et al. (2016)	HOPE (factorization)
Defferrard et al. (2016)	Laplacian graph spectrum
Cao (2017)	Focal Any-Angle A* (FA-A*)
Li et al. (2017)	Matrix perturbation theory
Qiu et al. (2018)	Matrix factorization + model gram
Pham (2019)	SamDepth *ABOD*
Al-Taei (2019)	Ensemble ABOD
Tripathy et al. (2019)	ABOD + DBSCAN

3.2 *Existing Graph Basis Anomaly Detection Methods (GBAD)*

A graph-type G contains bag of vertices Vi and interconnecting edges Ei, $G = (Vi, Ei)$. With reference to Fig. 1, graph is classified into point outlier, contextual outlier and collective outlier. Point outlier is occurrence of an anomalous vertices or edges. Contextual outlier is a subgraph anomaly occurrence in context to an application. Collective outlier is based on abnormal or suspicious link. A graph is anomalous when following conditions persist: abnormal vertex exists, anomalous edge exists, unexpected presence of vertex label, unexpected presence of edge label, missing vertex and missing edge.

GBAD approaches can be done quantitatively using both static and dynamic approach and qualitatively using visualization concepts as shown in Fig. 2. A graph can be plain (unlabelled), attributed (labelled), directed, undirected, bipartite (equally split into two halves) graph, etc. When a graph structure varies from a graph pattern, then it roots to an anomalous occurrence under the concepts of static graph approach. Dynamic graph looks for the presence of temporal anomalous pattern detection. "A picture is worth a thousand words". The visualization helps to read the structural oddities in the graph. The abnormal values standout as different visualization techniques are applied to suite the application.

Table 3 Artificial intelligence and machine learning approaches adopted by eminent authors to detect outlier is listed in chronological order

References	Methodology
Baum (1972)	Hidden Markov model (HMM)
Dempster (1977) [9]	Expectation maximization (EM)
Hopcroft (1979)	Finite state automatons (FSA)
Vapnik (1982)	Support vector machine (SVM)
Kohonen (1988)	Self-organizing maps (SOM)
Broomhead et al. (1988)	Radial basis function (RBF)
Quinlan (1993)	C4.5
Heckerman (1995)	Bayesian network
Cohen (1995)	RIPPER
Kam (1995)	Random decision forest
Agrawal and Srikant (1995)	Association rule mining
Zhang (1996)	BIRCH
Ester (1996)	DBSCAN
Lane and Brodley (1997)	Similarity-based matching method
Fawcett and Provost (1999)	Rule-based methods
Ghosh et al. (1999)	Artificial neural networks (ANNs)
Warrender et al. (1999)	sliding window method
Billor et al. (2000)	BACON method
Guha et al. (2000)	ROCK
Yu et al. (2002)	FindOut algorithm
Hawkins et al. (2002)	Replicator neural network (RNN)
O'Callaghan et al. (2002)	STREAM
Augusteijn and Folkert (2002)	Multi-layered perceptrons
Pavlov (2003)	Conditional random fields (CRF)
Ye (2004)	Markov model
Keogh et al. (2004)	Window comparison anomaly detection (WCAD)
Wright (2005)	Finite state machines
Aggarwal (2005)	Evolutionary outlier search
Basak (2005)	IHCCUDT
Roth (2004, 2006)	One-class kernel Fisher discriminants
Melin and Castillo (2008)	Fuzzy logic
Liu (2010)	SCiForest
Bengio et al. (2013)	Autoencoder
Xu et al. (2013)	Approximated local outlier factor
Susanti et al. (2014)	Robust regression model

(continued)

Table 3 (continued)

References	Methodology
Perozzi et al. (2014)	DeepWalk
Tang et al. (2015)	Genetic algorithm + FCM
Guha et al. (2016)	Random cut forest on streams
Yang et al. (2016)	Planetoid
Cao et al. (2016)	Deep neural graph representation (DNGR)
Li et al. (2017)	RADAR (residual-based)
Peng et al. (2017)	ANOMALOUS
Bronstein et al. (2017)	Graph convolutional networks (GCN)
Liu (2018)	iForest
Wu et al. (2018)	Susceptible-accepted-recovered model (SAR)
Du et al. (2018)	Dynamic embedding (DNE)
Maya et al. (2019)	Delayed long short-term memory (dLSTM)
Huang et al. (2019)	FeatWalk
Shi et al. (2019)	HERec (heterogenous random network embedded)

The output of outlier detection model is to best identify outlier. Elimination of outlier is not an encouraged methodology, as the existence is necessary for varying application such as fraudulent cases in order to generate robust classification model. The robust classification method can handle large outliers in a data collection rather than non-robust model as it would skew. Different labelling strategy and assigning ranking/outlier score values are incorporated for identification of anomaly.

3.3 *Baseline Link Analysis Ranking Algorithms*

A graph mining incorporate several statistical measures such as eigenvector centrality, betweenness centrality and degree-based. Cosine similarity is not suitable for large graphs as it deficits in time of computation.

Traditional ranking algorithms used in earlier years were random walk (1953), in-degree algorithm (1997), page rank algorithm (1998), hyperlink-induced topic search (HITS) algorithm (1998), SALSA algorithm (2000) is a combination of HITS and PageRank algorithm, PHITS algorithm (2000) is a probabilistic HITS algorithm, randomized HITS (2001) and traffic rank (2002). The majority of algorithm in recent years found to evolve from the traditional base. As a prior focus, the fundamental algorithms such as random walk algorithm, PageRank algorithm and HITS have been considered for discussion.

Table 4 Summary of observed challenging factors of model-based methodology

Methods	Issues observed
Standard deviation (2SD)	Misleading the outliers from mean calculation
Nearest neighbour	False alarm rate
Clustering-based approach	Quadratic complexity. Group anomaly unhandled
CCT [10]	Information loss, eliminate the influential data points
Manhattan distance, linear regression	Better than CCT
Depth-based approach	Assumes border outliers
PCA	Better than depth-based approach
Robust regression	Model fitting
Principle component analysis	Unreliable missing data pre-processing and analysis and better than maximum variance estimation (MDA)
GowPCoA, PairCor [11]	Unreliable
JointM	Considerable
NIPALS, IPCA [12]	IPCA better than NIPALS (high correlation)
RPCA	Better than WLRA (over fitting, low performance on high-dimensionality handling)
DROUT	Better than ERE and APCDA (data preserving inefficiency)
Feature subset selection (bagging)	Model training adopts random subset selection without any strategy
EDMF	Multimodel approach fails insufficient training data parameters and model extrapolation
EEAD	Probabilistic latent space model better than LOF and entropy-based

3.3.1 HITS Algorithm

Most renowned work contribution by Jon Kleinberg based on baseline web search concept using graph representation. HITS algorithm is based on web link that directs to pages relevant to web search induced by the user need. For example, if the user searches for top five mobiles in India, then top page results show ignorance of official mobile phone company sites. This is because resultant page will pop to user with algorithmic appropriateness rather ignoring accurate retrieval. HITS algorithm replaces traditional document matching approach by focusing on weightage of the page through hub weight and good authority score. The good page is a resultant of highest score of the both values. Though this algorithm has slow processing time on query, poor graph construction, slight search deviation, still accuracy of result with algorithm iterative performance is good.

Table 5 Summary of observed challenging factors of proximity-based methodology

Methods	Issues observed
LOF	Suitable only for density-based outliers
LSH	Suitable only for dimensionality reduction purpose
Convex hull	Anomaly inside the hull is left unconsidered
ABOD	Unsuitable for big data application
nn-Hilout	In-memory and disc computation incompatibility
Jarvis's march	Topologically inconsistent
Point-in-polygon (PIP)	Specific to point outlier detection
Tukey depth [13]	Dimensionality reduction not addressed
A*	Inefficient optimal solution, high computational cost
FA-A*	Better than Theta*, A*
MD-based	Requires sample size larger than feature dimensions
LINE	Proximity order based on window length is computationally expensive
Matrix perturbation	Limited to complex problems and accepts only small values
SamDepth ABOD	Requires quadratic computation time but better than proximity-based
Ensemble ABOD	Limited to detecting collective outliers with better time complexity
ABOD + DBSCAN	Storage computational complexity, performance is limited

3.3.2 PageRank Algorithm

Sergey Brin and Larry Page in 1998 proposed this centrality-based algorithm to utilize the link structure of the web page to deliver high precision web page by quality ranking the web page. Different data structures are employed to assist Big Files in organizing and searching results. In short PageRank is an estimation to determine the significance of page distinctive from other such pages as in Eq. 1. A source page receives highest PageRank only when sum of its incoming links are high.

$$\begin{aligned}\text{Page Rank}\,(x) = (1 - \text{damping factor}) + [&\text{damping factor} \\ &* \{\text{Page Rank (page 1)/Total outlink (page 1)} \\ &+ \cdots + \text{Page Rank (page } n\text{)/Total outlink (page } n\text{)}\}]\end{aligned} \tag{1}$$

Generally, damping factor is the probability of a surfer that continues to click randomly. Its advantage is scalability, less query time, less susceptibility to localized link, more efficient and feasibility. Its disadvantage is instance of a ten-year-old research article that has high PageRank than a one-year-old page article because of few citation links in new page. Dead ends, dangling link and spider traps are some issues to be handled.

An intrusion detection strategy on the basis of ordered representation of time delay is studied with comparative study on Google PageRank algorithm [23]. First, the incoming call request sequence, i.e. system execution characteristics, is organized

Table 6 Summary of observed challenging factors of artificial intelligence and machine learning-based methodology

Methods	Issues observed
BACON	Only 50% outliers are handled and certain high residual points are unidentified
Canopy	Better than agglomerative clustering, K-means and EM
MR-AVF [14]	Uncertainty exists to show for diverse application areas
Fuzzy logic	Better than LOF
Sliding window	Memory consumption reduction, complexity of implementation due to dynamic handling
DBSCAN	Optimal parameter selection is iterative
iForest	Problem in setting split point. Computational expensive and visualization are complex. Better than SciForest
IHCCUDT	Better than decision tree. Unbalanced split, search time high
iTree	Restriction limited to tree growth constraint
SciForest	Adopts randomization. Search process is tedious
GCN	Utilize semisupervised learning behaviour with scalability issue. Better than CNN
DeepWalk	Sampling of random walk. Lookup time complexity. Scalability issue
Planetoid	Semisupervised strategy
FeatWalk	Better than DeepWalk, LINE. Achieved scalability and stability with compromises on interoperability
Genetic algorithm	Lack of pertinence and time consuming
Fuzzy logic	Subspace-based better than genetic algorithm. Missing data better than MCR, MPR model-based methods. Inaccurate assumptions

with the help of sliding window concept. The generated pattern records in turn veils path for interrelation graph creation. The prediction of outlier is processed by comparative analysis using PageRank algorithm and error detection method step using hamming distance.

The intention of Wikipedia is to aid best routing in finding appropriate destination website [24]. Some plausible ranking methods follow in-link distance (interrelation), singular-value decomposition (page visit frequency), missing links (rather novel links). Some challenges in hyperlinks are:

a. Every search engine uses the concept of inverted index to best traverse the important links necessary for enhanced retrieval. Determining important links that increases the rank score are often found missing.
b. Sustainability of hyperlink is a complex task.
c. A good web document without any incoming links might get missed out on indexing.
d. Most work contribution focusing on target prediction problem only.

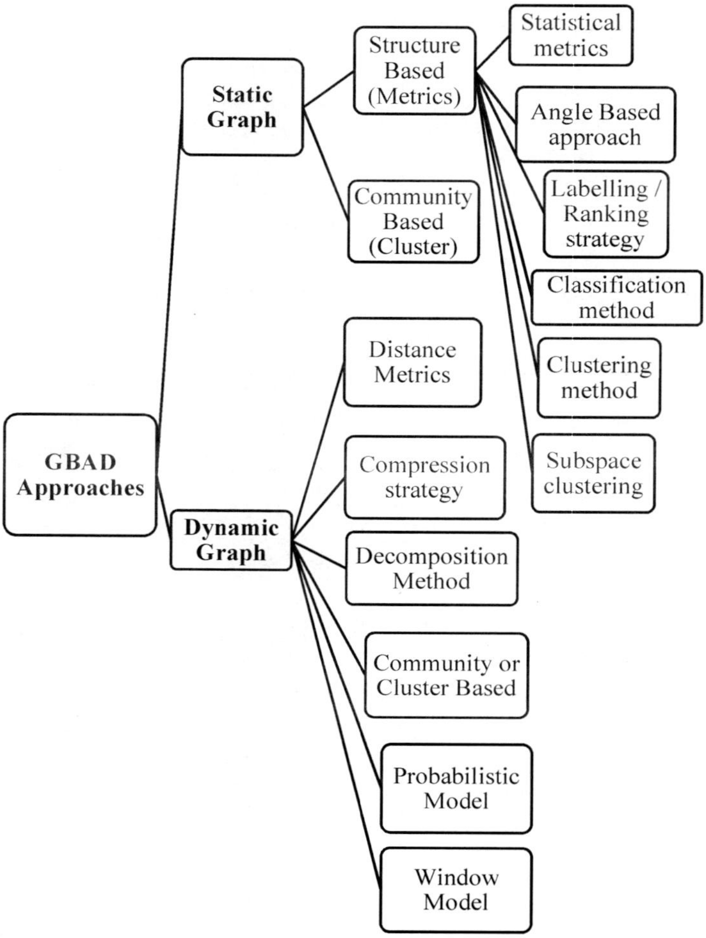

Fig. 2 Graph-based anomaly detection methodology hierarchy

e. Identifying authoritative source node is more complicated than finding appropriate target hub node.
f. Techniques based on concepts such as model training, IR strategy, matrix computation are challenging.
g. Most sources in web page are the root words highly contribute to link the target page. Due to predominant existence of disambiguation and simplicity of the method, this strategy is least exploitable.
h. Cache history logs on every user's web page visit and is often an unexplored analysis area.
i. Raw browsing log collection is hard due to privacy consideration and supervised model prediction is slow and unaffordable.

3.3.3 Random Walk/Blind Walk

Karl Pearson coined the term "random walk" as concept in mathematical statistics in 1905s. Later it gained popularity by use in variety of application domains because of its cascading nature. In web, random walk can be applied to perform probabilistic search, make recommendations for product, friend suggestions and design several other prediction models.

The objective of random walk strategy is to recursively trip on every node in graph [25]. The task assigned to Google Bot is to choose a random page, then move to another through page link encountered. This is purely the exhibition of Markov probabilistic theory. Though the strategy shows quick functionality, high efficient recovery and parallel computation, still shows evidence of worst performance in small graph structures.

Every random walk uses a generalized design pattern to guide the path in its randomness [26]. Some of the techniques designed are walking in grid sections, encountering nodes that fall within the radius, address nodes in random line generation, using step size limit. The state information is utilized for trailing using Metropolis basis strategy with exception to consuming additional space. When compared to computationally expensive density-based approaches such as kernel methods [27], a combination of proximity graph and page rank algorithm is used to detect anomalous data point. Dead links is the stopping criterion for this method which rarely exists.

3.4 Structure-Based Graph Anomaly Score Through Labelling/Ranking Strategy

Structure-based anomaly detection method on a static graph emphasis appropriate selection of ranking metrics is to pave an accurate prediction power. A comparative study on few of the observed methods taking its root of evolution from baseline link analysis algorithm is presented in Table 7. The challenges faced by each method and the techniques adopted to predict the anomaly with labelling strategy in detail are discussed in this section.

Some challenging factors identified in anomaly detection are [28, 29]: (a) Identification of small group anomaly with only traditional density-based technique is proven inefficient. (b) Clusters having equal number of outliers and normal points are indistinguishable. (c) Neighbourhood graph construction is tedious with respect to outlier detection. (d) Most existing techniques require user-defined neighbourhood parameter. (e) Distinguishing point and collective outlier is a issue. To address these issues, Markov chain model is used to assign outlier score to each object suitable to handle static bipartite graph structure and weighted undirected graph structure. Some of the limitations of this proposed method are: (a) expected cover time O (n^3) for some graph, (b) low cardinality with high threshold leads to high false alarm

Table 7 Some comparative study on structure-based graph anomaly score highlighting on the metrics adopted and taking its root from baseline algorithm as discussed in earlier

Algorithm/method	Application	Baseline link analysis algorithm	Existing method of comparison	Metrics used	Evaluation metrics
OutRank	Intrusion detection	Random walk	K-dist (KNN) LOF	Cosine similarity, RBF kernel similarity	Precision, recall, false alarm rate, *F*-measure
GoutRank	Co-purchase network	Random walk	LOF SOF RPLOF SCAN CODA OutRank	Node degree score and eigencentrality score	Area under curve (AUC)
Yagada	Intrusion detection	HITS	GMM LOF K-means	Subdue and KNN	Cumulative frequency distribution
FocusCo	Co-authorship, affiliation network	PageRank	CODA METIS	(Inverse) Mahalanobis distance, Euclidean similarity metric	Normalized mutual information (NMI)
INGC	Co-authorship, affiliation network, co-purchase network	PageRank	K-means LPA	Degree	Precision, recall, *F*-measure, modularity
LPAD	Social network, co-authorship	Random walk	Stranger's random	Abnormality vertex probability, edges probability STDV, sum edge label, mean predicted link label, predicted label STDV, edges probability median, edge count	AUC, TPR, FPR, precision

rate, (c) events are unpredictable, cannot be suitable for finance and stock prediction [30]. Applicable in web search, keyword extraction, text summarization, intrusion detection and PageRanking.

K-nearest neighbours (k-distances) is a widely used representation producing better scalability and accuracy with respect to outlier analysis [31]. In a labelled graph structure, inspecting the presence of anomaly is solved using YAGADA algorithm that utilizes the information of graph structure as well as the characteristic behaviour patterns. In certain security applications such as intrusion inspection, physical Access Control database (ACS), the proposed work detects anomaly two times than any other discretization algorithms. Most detected anomalies exhibit structural abnormality (unusual trail) and numerical outliers (unusual timing data). The computational complexity of the optimized method is logarithmic in nature, whereas in dynamic graph showing worst case complexity O (N^2).

Existing ranking methods [32] focus on either vector-based data (LOF, SOF) or complete graph structures (SCAN algorithm) but not both. Selection of relevant subspaces/subgraphs and assigning the rank values in individual subgraph clusters are challenging. Availability of benchmark data sets is questionable. A graph outlier ranking (GOutRank) and decoupled processing scheme exploiting subspace clusters in an attributed graph are introduced. The rank assignment is based on both graph structural context and attribute set. It generalizes outlier ranking method OutRank. Even though scalability and scoring functionality is improved, still integration of outlier ranking with graph clusters is not to desirable level. The user-preferred clusters and overlapping clusters are unhandled.

User preference to graph mining is incorporated with the so-called focus attributes FocusCO [33]. The approach handles overlapping attributed graph clusters. It uses "chop out" approach to slice the focused candidates to extract good candidates of interest. The main idea of this FocusCO is to identify best structural nodes (BSN) and one that deviates is an outlier. The advantage of this approach is node negligence which supports near-linear time computation complexity.

Outlier identification can be classified into formal and informal methods [34]. The informal methods also called as labelling methods. The three issues with regard to outliers are outlier labelling, outlier accommodation and outlier identification. K-means faces issues with local optimum and lacks in consistency [35]. Label propagation algorithm (LPA) is inconsistent, unstable and random in nature. The technique used to control the influence points on label propagation that is capable to handle static plain graph both directed and undirected edges. The core idea to this work is to identify the graph node with the highest degree as leader node (most influential) value than other. It is time efficient with high accuracy results. But, most algorithms handle only small-size graphs and unsuitable for distributed environment over larger graphs.

Generally identity theft is a serious issue in social network that have more community interconnections than usual. The anomaly is a representation of node with high outgoing link connections [36]. Random forest algorithm is used to generate model framework by handling static plain graph both directed and undirected. To find the irregular vertex, many probabilistic notations are used to rank every graph vertex.

A classical classification model is used to find pattern metadata of link analysis in given graph structure. The classifier predicts accurate results with high performance measure. But, prevalence of bipartite and weighted graphs is unhandled in this technique used.

Table 8 summarizes the related work on set of structural pattern perspective anomaly detection methods grouped by its characteristics. This ventures the users to exploit on the methods depending upon the application problem domain.

3.5 Community-Based Graph Anomaly Score Through Labelling/Ranking Strategy

Community-based anomaly detection aims to divide the graph structure into clusters and diagnoses the existence of outlier. A predominant nature of outlier in community is rarely explicit and mostly implicit. Explicit nature of the outlier is structure distinguishable characteristics that do not belong to any cluster of kind, whereas implicit nature is quite a hidden and conflicting type that disguise themselves in the community. The community-based anomaly detection strategies are depicted in Fig. 3.

Graph clustering is quite a popular research area [37]. A graph may be directed/digraph, undirected, weighted, planar, regular, bipartite, simple acyclic, connected, disconnected, subgraph, clique, isomorphic, isospectral and line graph influences several metrics such as density, betweenness, centrality, adjacency matrix, neighbourhood, degree, cut size, length, path, eigenvalues and normalized Laplacian to venture a definitive objective. Matrix diagonalization is important application in clustering algorithms since computationally efficient to handle linear form of equations. Fuzzy partitioning is not widely accepted approach since vertex-cluster assignment is inappropriate at certain scenarios. The three major problems in graph clustering identified are parameter selection, scalability and evaluation.

A group anomaly detection strategy in overlapped cluster-based evolutionary network is the identified problem definition. The GBAD is broadly classified into two [38]: white crows and indisguise anomaly detection. The popular techniques to detect white crows in nodes, edges and substructures are random walk, relevance search in bipartite graph, KNN density-based method, MDL principle, subdue-based, rarity measurement and tensor-based method, whereas techniques to detect indisguise anomaly are MDL and event-based model [39].

Most clustering-based algorithms (community) focus on aspects of network structure and node attribute properties [40]. A highly robust, scalable algorithm detects communities having multiple intersections with high precision in spite of noise intervention using the node attribute characteristics and edge arrangement. Nodes that have similar property can belong to multiple communities at once. Combining both structure and node attributes is used in many existing works such as single assignment

Table 8 Overview on classic graph properties with respect to observed structure oriented graph anomaly methods

Referred methodology	Graph clustering	Attributed graphs	Attributed subspace	Static bipartite graph	User-preferred clusters	Overlapping clusters	Outlier detection	Scalability	Undirected graph
LOF		x				x	x		x
SCAN	x								
CODA	x	x					x		
OutRank							x		x
GoutRank		x	x	x			x	x	x
Yagada	x	x			x		x	x	
FocusCo	x	x	x	x	x	x	x	x	
RADAR					x	x			
INGC	x						x	x	x
LPAD	x				x	x		x	x

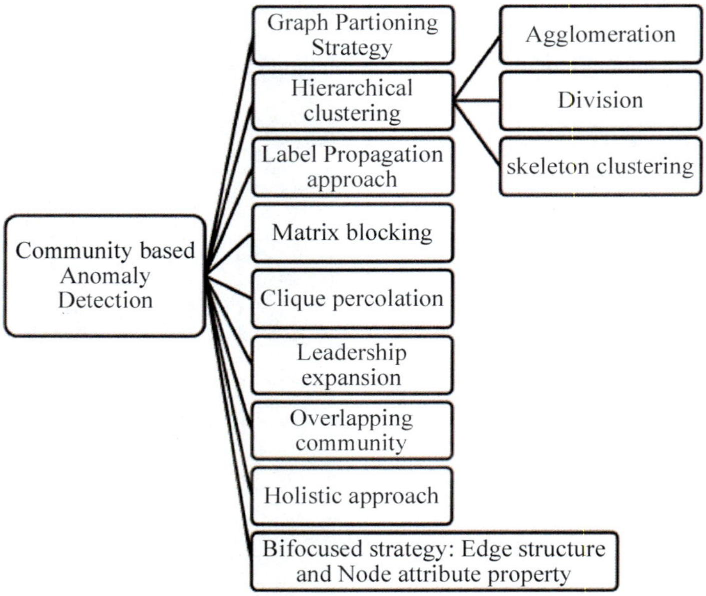

Fig. 3 A broad classification on community-based anomaly detection strategy

clustering, topic-based model, small network graph structure, soft-node community membership aspects.

A universal procedure-oriented benchmark for any community detection algorithm has four modules such as set-up, detection framework, diagnosis and evaluation which is built to improve the efficiency and overall model performance [41]. The popular approaches for community detection are division/hierarchical clustering, partitioning, clique percolation, leadership expansion, LPA, matrix blocking, etc. Propinquity measure for priority of interconnecting elements and revelatory structure for organizing the graph plays key role in this approach. The time cost analysis of different algorithms at each phase helps to provide best summary of algorithms under study.

In graph structure, there are no precise boundary cuts due to multiple overlapping connections [42]. But existing approach focuses only on the structural orientation aspects or the characteristic property of the node but not both. The static community graph structure utilizes neighbourhood quality score, neighbourhood anomaly mining and scalable optimization method. This approach results in high normality score better that existing approach like oddball mishandles the boundary edges. It cannot recover the missing node attributes. In oddball [43], the anomalous behaviour patterns revolve around the rules framed using concepts of egonet and powerlaw.

Community identification is a problem-based classification approach. Apart from many mathematical formulations that precisely classify the community, a broad detection approach irrespective of application domain is classified as follows [44–46]: Cut-based perspective, clustering perspective, stochastic equivalence are based on statistical network model and dynamical perspective. Each technique has specific significance such as cut-based minimizes constraint violation by providing relaxation in optimization problems. Clustering focused to maximize the intranode connection density. The third type focuses to find the regular equivalence that share structural property based on stochastic block model technique. A Markovian-based random walk approach deploys fast time scale to traverse network as a dynamic perspective nature.

Holistic community outlier disguises its identity by sharing the node property of one community at the same time with other [47, 48]. Most existing approaches ignore the structural design of graph resulting in failure to outlier detection. Probabilistic graph model, hidden Markov random field (HMRF) and expectation maximization (EM) algorithm, was proposed to handle static edge-attributed graphs. HMRF model is a parallel consideration of both node information and edge property. EM learns on generated model parameters and infers hidden community labels. But still there exists an unsatisfying contribution towards edge-attributed graphs.

The section focuses to study the performance of the community-based anomaly detection strategy and its ability to scale to big graphs. Most of the community-based methods are evaluated based on community cluster size and run time computation. A broad comparative analysis of each algorithm discussed is reviewed in Table 9.

As each algorithm is distinct by application, sample size and data structure, a unified statistical survey is unrealistic. Hence, methods that adopt area under curve (AUC) evaluation measure on co-authorship data set such as DBLP and co-affiliation data set such as Disney is particularly chosen. Depiction from Fig. 4 shows LPAD is the overall best performing structure-based method with 91% correct predictions and ANOMALOUS tops with 88.35% from observed community-based methods.

3.6 Graph Compression Strategies

Scalability is a necessary criterion for any data analysis. Roughly speaking, big data volume is an important characteristic that eyes on the growing size of the data on real-time highly loading memory capacity. The graph structure can be compressed as it can be stored in several substructures taking memory into account. The condition to perform this strategy is: every compressed data should reproduce the data originality when decompressed. A huge issue lies on the data locality/indexing and lossless data retrieval.

Conditional entropy is a measure of the event occurred with respect to the occurrence of other event similar to a chain reaction. Every event representation is in the binary form, which is likely to vary on its occurrence. For a weighted graph G, most problematic cases occurs when the substructure identified to have size $= 1$ or size $= G$. The proposed strategy is a so-called subdue system [51] that detects the ordered records of duplicate substructure sample. Anomalous substructure detection measure is influenced to identify the subgraph with small substructures where the anomalistic occurrence is high. The challenges is the rarity of the small substructure

Table 9 Some comparative study on community-based graph anomaly score highlighting on the metrics adopted and taking its root from baseline community algorithm as discussed

Method	Application	Baseline link analysis algorithm	Existing methods of comparison	Metrics used	Evaluation metrics
CODA	Co-authorship network	Random walk, holistic community	GLODA DNODA CAN (graph partitioning)	HMRF, EM, Bregman divergence	Precision Sensitivity Runtime
AMEN	Co-authorship, affiliation network, co-purchase network	HITS Direct partitioning	OddBall SODA Conductance	Normality score, anomaly mining, optimization	Mean precision
ConOut (contextual outlier)	Co-authorship, affiliation network, co-purchase network, communication network	Local page rank, hierarchical divisive community	LOF SCAN CODA GoutRank	Baseline scoring, subspace similarity scoring, cluster coverage scoring	F-test t-statistics
ANOMALOUS [49]	Co-authorship, affiliation network, co-purchase network	Page rank, direct partitioning	SCAN LOF Consub + CODA ConOut AMEN RADAR [50]	Matrix regularization, column rank residual decomposition	AUC
HCODA	Co-authorship, affiliation network	Random walk, holistic community	CODA	Hidden Markov random field, expectation maximization	Sensitivity, accuracy
Evolutionary community	Communication network	Random walk, overlapping community, genetic algorithm	Non representative methods	Timeline-based: grown, shrunken, merged, split, born, vanished	Runtime
NIBLPA	Social network	PageRank, clique percolation	LPA KBLPA CNM (hierarchical divisive)	K-shell, node influence	Modularity, NMI, F-measure

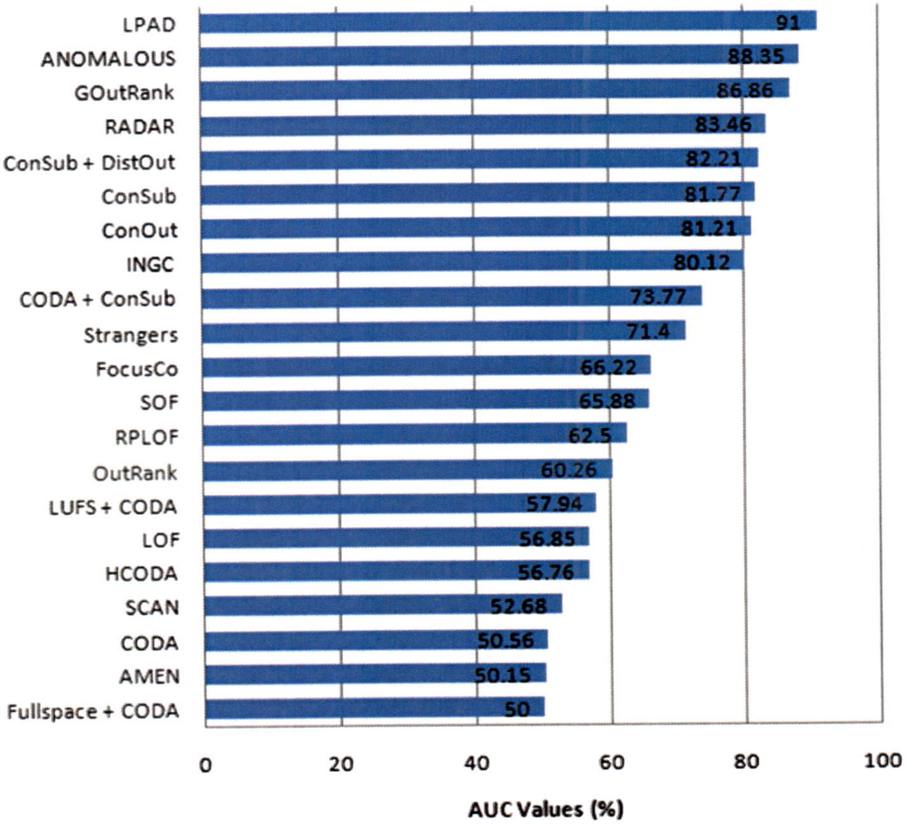

Fig. 4 AUC [%] performance w.r.t structure-based and community-based anomaly detection algorithms for co-authorship relationship data set

occurrence, optimal computation issues based on substructure size selection and also time complexity is exponential in worst-case scenario.

A domain-independent rarity measurement approach discovers unusual links within a graph with contrast to hypothesis that anomaly is a human error. Existing models define the score for interrelationship that handles only bipartite graph and abnormal node without inspecting the substructure abnormality. Information theoretic approach, probabilistic-based and biased substructure detection [19] to perform insertions, modifications and deletions, respectively, is proposed. Abnormality in graph can be discovered through most frequent patterns, probability of pattern identification and maintaining parental relationship between the graph structures are some of the detection ways showing predominantly high performance results.

High dimensionality and volume of graph considerations weaken memory robustness resulting in poor graph visual analysis. Representing the cost cutting nodes together as super-node and similar to edges gives the pattern summary with exception to optimality. This is the work principle of randomized algorithm. Randomized++, upgraded version of randomized algorithm is proposed [58]. Classify each and every node like randomized strategy in sorted manner. In different applications, few nodes are compressed, until the cost value reverses from zero to negative cost with loss of minimum accuracy showing better performance.

Vocabulary-based summarization of graph is an encoding scheme that works on the principle of cost reduction in bit description [59]. It effectively summarizes and understands large closely packed graphs such as cliques. Minimum description length is a popular approach that provides distinct graph summary with minimal cost of computation. This summarization approach generates prioritized sorted list of coinciding subgraphs. VOG involves subgraph generation, subgraph labelling and summary assembly. The highlighting features of this methodology are: (a) simple summary, non-redundant manner, (b) parameter-free, (c) most succinct description, (d) better compression gain, (e) soft clustering and (f) sequentially linear complexity.

Anomalies in bipartite graphs addresses on steps to spot the next community node as well as to figure out the bridging node. However, to detect the three structural anomalies as discussed in GBAD system, if the type of anomaly is not known beforehand, the user may need to run all three GBAD algorithms. Instances-based MDL anomaly detection (IMAD) and instance-based size anomaly detection (ISAD) are methods proposed [60]. Using rule framework and negative weight assignment, the matching anomalous substructure instance can be identified. Memory consumption and time efficiency are better than previous approaches. To run in polynomial time, it is not guaranteed that all of the candidate normative patterns will be generated. It cannot guarantee to discover all anomalies.

To summarize, the graph compression techniques discussed in Table 10 is a categorization of aggregation (super-node, super-edge), attribute (topology, attribute), compression (encode, decode), application (query processing time, hypothetical inference). The contributions of the compression technique with anomaly detection are inadequate, which ventures to be an open research area. Most of the evaluation is carried to measure time efficiency, space cost, accuracy and quality of visualization.

4 Open Challenges in GBAD

a. Identification of anomaly in dynamic graph that evolves over time is a less explored area of research. Anomaly detection has two broad classifications [61]: data streams and evolving graph. Data streams can be handled using statistical method, clustering method, window technique and nearest neighbour technique. Evolving graph is handled using node-based, edge-based and event-based detection methods. Which strategy is best and how can it be utilized?
b. The strategy to choose the time window in a evolving graph model in such a way that to extract useful patterns as well as anomaly discovery exist to be open challenge.
c. Intercommunity interaction conflicts is a futuristic ongoing research area. To mitigate the negative impact created by social conflict on web, how to adopt the

Table 10 Comparative analysis of graph summarization techniques and its contribution towards anomaly discovery

Summarization technique	Existing method	Proposed method	Anomaly detection
Static aggregation-based	Randomized	Randomized++	No
Static attribute-based	Greedy'NForget	VoG	No
	SUBDUE MDL	GBAD-MDL GBAD-P GBAD-MPS	Yes
	SUBDUE	Conditional substructure entropy	Yes
	VoG	StarZIP [52]	No
Compression-based	GBAD	IMAD ISAD Rule coverage	Yes
	Page rank (PR) CC SSSP	Pregel delta encoding [53]	No
	LIGRA	LIGRA+ [54]	No
Application oriented	Random consecIN consecOUT	GrdRandom FlipInOut [55]	No
	PathTree SCARAB	TF-Label [56]	Yes
	RXD	LAD LAD-S [57]	Yes

mobilization strategy to enhance the overall user activity in the community is a problem. Taking into account of every new social inclusion along with dynamic changing social correlations is a challenge to be addressed.

d. Inter- and intracommunity interaction prediction problem unveils the topological scope community in future. Building a scalable interactions prediction model with accurate insight in a dynamic network is highly challenging task.
e. Bioinspired algorithms and swarm intelligence algorithms are suitable to be incorporated in most real-time application areas such as social network, intrusion detection, biological network and road network, e.g. ant colony optimization (ACO) for shortest distance between road network and social members relationship identification. Identifying a suitable relative solution map for every real-time problem with clever algorithms need greater insight.
f. While graph-based representation is gaining importance which is evident through emerging technology such as Neo4j, DEX, VertexDB and et al. Neo4j is an open-source big data analytics tool devised with NoSQL graph database. It is emerging technology that gives accurate predictive insight for better research

analysis [62]. Few application areas, such as social networking, logistics, business management, authorization controls, highly inflict the necessity of such big database environment.

g. How to exploit all available sources in network to accurately detect anomaly focusing on adversarial robustness aspects, as majority of methods focus only on performance and cost.
h. A Clustering or graph-cut algorithm detects multiple communities with implicit ordering by missing the priority criterion. Most discovered communities are not characterized (e.g. clique, star), thus lags in providing user insights.
i. Visualizing the graph is a popular research area to uncover the hidden outliers in big data application. Since graph matching and graph edit distance metrics are impractical and complex, it necessitates the need for distance measures and feature selection methods. Does the optimality, feasibility and outlier prediction discrepancy is the question to test credibility of the adopted strategy.

5 Concluding Remarks

The anatomy of the survey targets to present a research interest in the field of graph anomaly detection for novice learners. The stage-by-stage clarification is met for deeper understanding and clarity in the study of graph anomaly in specific. Eminent work contributions exist in machine learning algorithm renders solution for typical emerging research issues. The need for graph representation with highlighting baseline algorithms defining the blueprint for GBAD is discussed. The core contribution of the work is to unveil the open research problems in the area of GBAD classification for a static network using structural properties and community behaviour. The techniques focus towards algorithms with improved speed of computation and quality of anomalistic detection rate. The graph compression strategy focuses to improve the memory space computation and sub graph retrieval without data loss is also studied.

Identified challenges in machine learning techniques and GBAD techniques render a scope of interest for emerging researchers. Recent trend is utilizing advanced tools and technology to experiment and exploit big data applications without much expense. There is no universally acceptable general-purpose algorithm in anomaly detection due to biased application dependency nature. This necessitates the user to select and play with different algorithms, to achieve optimal and feasible solution at the end.

References

1. A. Jain et al., Big data preprocessing—a survey of existing and latest outlier detection techniques. Int. J. Emerg. Technol. Comput. Sci. Electron. (IJETCSE) **14**(2) (2015)
2. K. Singh et al., Outlier detection: applications and techniques. IJCSI Int. J. Comput. Sci. Issues **9**(1) (2012)

3. J. Zhang, Advancements of outlier detection: a survey. ICST Trans. Scal. Inf. Syst. **13**(01) (2013)
4. A.M.C. Souza et al., An outlier detect algorithm using big data processing and internet of things architecture. Procedia Comput. Sci. **52**, 1010–1015 (2015)
5. X. Xu et al., Recent progress of anomaly detection. Advances in architectures, big data, and machine learning techniques for complex internet of things systems (2019)
6. A. Rajaram, S. Palaniswami, The modified security scheme for data integrity in MANET. Int. J. Eng. Sci. Technol. **2**(7), 3111–3119 (2010)
7. Y. Susanti et al., M estimation, S estimation, and MM estimation in robust regression. Int. J. Pure Appl. Math. IJPAM **91**(3) (2014)
8. S. Dray et al., Principal component analysis with missing values: a comparative survey of methods. Plant Ecol. **216**, 657–667 (2015)
9. A. McCallum, K. Nigam, L.H. Ungar, Efficient clustering of high-dimensional data sets with application to reference matching, in *Proceedings of the Sixth ACM SIGKDD International Conference on Knowledge Discovery and Data Mining; KDD'00* (ACM, New York, NY, USA, 2000), pp. 169–178
10. Z. Abu Bakar et al., A comparative study for outlier detection techniques in data mining, in *CIS* (IEEE, 2006)
11. H.-P. Kriegel et al., Outlier detection techniques, in *16th ACM SIGKDD Conference on Knowledge Discovery and Data Mining* (2010)
12. S. Loisel, Y. Takane, Comparisons among several methods for handling missing data in principal component analysis (PCA). Adv. Data Anal. Classif. (2018)
13. D. Chen, P. Morin, U. Wagner, Absolute approximation of Tukey depth: theory and experiments. Comput. Geom. **46**(5), 566–573 (2013)
14. Koufakou et al., Scalable and efficient outlier detection strategy for categorical data, in *19th IEEE International Conference on Tools with Artificial Intelligence* (2007)
15. A. Koufakov et al., Fast parallel outlier detection for categorical dataset using map reduce, in *IEEE International Joint conference on Neural Networks* (2008), pp. 3297–3303
16. V.J. Hodge, J. Austin, A survey of outlier detection methodologies. Artif. Intell. Rev. **22**(2), 85–126 (2004)
17. V. Chandola et al., Anomaly detection: a survey. ACM Comput. Surv. **09**, 1–72 (2009)
18. A. Patcha, J.-M. Park, An overview of anomaly detection techniques: existing solutions and latest technological trends. Comput. Netw. **51**(12), 3448–3470 (2007). https://doi.org/10.1016/j.comnet.2007.02.001
19. W. Eberle, L. Holder, Anomaly detection in data represented as graphs. Intell. Data Anal. **11**(6), 663–689 (2007). https://doi.org/10.3233/ida-2007-11606
20. P.N. Tan et al., *Introduction to Data Mining* (Pearson Addison Wesley, Boston, 2005)
21. L. Wilkinson, Visualizing big data outliers through distributed aggregation. IEEE Trans. Visual. Comput. Graph. **24**(1) (2018)
22. S. Agrawal, A. Patel, A study on graph storage database of NOSQL. Int. J. Soft Comput. Artif. Intell. Appl. (IJSCAI) **5**(1) (2016)
23. Q. Qian et al., An anomaly intrusion detection method based on PageRank algorithm, in *IEEE International Conference on Green Computing and Communications and IEEE Internet of Things and IEEE Cyber, Physical and Social Computing* (2013), pp. 2226–2230
24. R. West et al., Mining missing hyperlinks from human navigation traces: a case study of Wikipedia, in *ACM International World Wide Web Conference Committee* (2015)
25. A. David et al. Reversible Markov chains and random walks on graphs. Unfinished monograph (2002)
26. S. Vempala, Geometric random walks: a survey, in *Combinatorial and Computational Geometry*, vol. 52 (MSRI Publications, 2005), pp. 573–612
27. Z. Yao et al., Anomaly detection using proximity graph and PageRank algorithm. IEEE Trans. Inform. Forensics Secur. **7**(4) (2012)
28. H.D.K. Moonesinghe et al., OutRank: a graph-based outlier detection framework using random walk. Int. J. Artif. Intell. Tools **17**(1) (2008)

29. P.I. Sánchez, E. Müller, O. Irmler, K. Böhm, Local context selection for outlier ranking in graphs with multiple numeric node attributes, in *SSDBM* (2014)
30. D. Sensarma et al., A survey on different graph based anomaly detection techniques. Indian J. Sci. Technol. **8**(31) (2015)
31. M. Davis et al., Detecting anomalies in graphs with numeric labels, in *ACM CIKM'11*, 24–28 Oct 2011
32. E. Muller, P.I. Sanchez, Y. Mulle, K. Bohm, Ranking outlier nodes in subspaces of attributed graphs, in *IEEE 29th International Conference on Data Engineering Workshops (ICDEW)* (2013)
33. B. Perozzi, L. Akoglu, P. Iglesias Sánchez, E. Müller, Focused clustering and outlier detection in large attributed graphs, in *Proceedings of the 20th ACM SIGKDD International Conference on Knowledge Discovery and Data Mining—KDD'14* (2014)
34. K.S. Kannan, K. Manoj, Outlier detection in multivariate data. Appl. Math. Sci. **9**(47), 2317–2324 (2015)
35. V. Bhatia, B. Saneja, R. Rani, INGC: graph clustering & outlier detection algorithm using label propagation, in *International Conference on Machine Learning and Data Science* (2017)
36. D. Kagan, Y. Elovichi, M. Fire, Generic anomalous vertices detection utilizing a link prediction algorithm. Soc. Netw. Anal. Min. **8**(1) (2018)
37. S.E. Schaeffer, Graph clustering: survey. Comput. Sci. Rev. **1**, 27–64 (2007)
38. Z. Chen, Community-based anomaly detection in evolutionary networks. J. Intell. Inf. Syst. Springer Science+Business Media (2011)
39. R. Jessica et al., A bio-inspired algorithm for searching relationships in social networks, in *Proceedings of the 2011 International Conference on Computational Aspects of Social Networks* (2011)
40. J. Yang, J. McAuley, J. Leskovec, Community detection in networks with node attributes, in *2013 IEEE 13th International Conference on Data Mining* (2013)
41. M. Wang et al., Community detection in social networks: an in-depth benchmarking study with a procedure-oriented framework, in *Proceedings of the VLDB Endowment*, vol. 8, no. 10 (2015)
42. B. Perozzi, L. Akoglu, Scalable anomaly ranking of attributed neighborhoods, in *Proceedings of the 2016 SIAM International Conference on Data Mining* (2016)
43. L. Akoglu, M. McGlohon, C. Faloutsos, Oddball: spotting anomalies in weighted graphs, in *Lecture Notes in Computer Science* (2010), pp. 410–421
44. M. Rosvall, Different approaches to community detection. Extended version of the many facets of community detection in complex networks. Appl. Netw. Sci. **2**, 4 (2017). arXiv:1712.06468v1
45. G. Rossetti, R. Guidotti, I. Miliou, D. Pedreschi, F. Giannotti, A supervised approach for intra-/inter-community interaction prediction in dynamic social networks. Soc. Netw. Anal. Min. **6**(1) (2016)
46. M. Sachan, D. Contractor, T.A. Faruquie, L.V. Subramaniam, Using content and interactions for discovering communities in social networks, in *Proceedings of the 21st International Conference on World Wide Web* (2012)
47. S. Kumar et al., Community interaction and conflict on the web, in *WWW 2018: The 2018 Web Conference*, 23–27 Apr 2018
48. S. Pandhre et al., Community-based outlier detection for edge-attributed graphs. arXiv: 1612.09435v2 [cs.SI] (2017)
49. Z. Peng, M. Luo, J. Li, H. Liu, Q. Zheng, Anomalous: a joint modeling approach for anomaly detection on attributed networks, in *International Joint Conference on Artificial Intelligence* (2018), pp. 3513–3519
50. J. Li, H. Dani, X. Hu, H. Liu, Radar: residual analysis for anomaly detection in attributed networks, in *IJCAI* (2017)
51. C. Noble, D. Cook, Graph-based anomaly detection, in *ACM SIGKDD*, 24–27 Aug 2003
52. D. Batjargal et al., StarZIP: streaming graph compression technique for data archiving. IEEE Access 1 (2019)

53. A. Chavan, An introduction to graph compression techniques for in-memory graph computation (2015)
54. J. Shun, L. Dhulipala, Smaller and faster: parallel processing of compressed graphs with Ligra+ (2015), pp. 403–412
55. O. Goonetilleke, D. Koutra, T. Sellis, K. Liao, Edge labeling schemes for graph data, in *Proceedings of the 29th International Conference on Scientific and Statistical Database Management (SSDBM'17)* (United States of America: Association for Computing Machinery, 2017), pp. 1–12
56. J. Cheng, S. Huang, H. Wu, A. Fu, TF-label: a topological-folding labeling scheme for reachability querying in a large graph, in Proceedings of the ACM SIGMOD International Conference on Management of Data (2013), pp. 193–204
57. F. Verdoja, M. Grangetto, Graph Laplacian for image anomaly detection. Mach. Vis. Appl. **31**, 11 (2020)
58. K.U. Khan et al., An efficient algorithm for MDL based graph summarization for dense graphs. Contemp. Eng. Sci. **7**(16), 791–796 (2014)
59. D. Koutra, U. Kang, J. Vreeken, C. Faloutsos, Summarizing and understanding large graphs. Stat. Anal. Data Min. ASA Data Sci. J. **8**(3), 183–202 (2015)
60. S. Velampalli et al., Novel graph based anomaly detection using background knowledge, in *Proceedings of the Thirtieth International Florida Artificial Intelligence Research Society Conference* (2017)
61. M. Salehi, L. Rashidi, A survey on anomaly detection in evolving data. ACM SIGKDD Explor. Newsl. **20**(1), 13–23 (2018)
62. E. Geepalla, N. Abuhamoud, A. Abouda, Analysis of call detail records for understanding users behavior and anomaly detection using Neo4j, in *5th International Symposium on Data Mining Applications* (2018), pp. 74–83
63. P.I. Gionis, R. Motwani, Similarity search in high dimensions via hashing, in *Proceedings of the 25th International Conference on Very Large Data Bases, VLDB'99* (Morgan Kaufmann Publishers Inc., 1999), pp. 518–529
64. Q. Cheng, Y. Zhou, Y. Feng et al., An unsupervised ensemble framework for node anomaly behavior detection in social network. Soft Comput. (2019)
65. M. Deepa, M. Rajalakshmi, Survey of deep and extreme learning machines for big data classification. Asian J. Res. Soc. Sci. Humanit. Asian Res. Consortium **6**(8), 2502–2512 (2016)
66. F. Angiulli, C. Pizzuti, *Fast Outlier Detection in High Dimensional Spaces*, in *Springer PKDD*. LNAI, vol. 2431 (2002), pp. 15–27
67. F. Angiulli, C. Pizzuti, Outlier mining in large high-dimensional data sets. IEEE Trans. Knowl. Data Eng. **17**(2) (2005)
68. R.L. Graham, An efficient algorithm for determining the convex hull of a finite planar set. Inf. Process. Lett. **1**(4), 132–133 (1972)
69. L. Grandinetti et al., High-performance computing and big data analysis. Commun. Comput. Inf. Sci. (2019)
70. H.V. Nguyen, V. Gopalkrishnan, Feature extraction for outlier detection in high-dimensional spaces, in *Proceedings of the Fourth International Workshop on Feature Selection in Data Mining*. PMLR **10**, 66–75 (2010)
71. J. Gao, F. Liang, W. Fan, C. Wang, Y. Sun, J. Han, On community outliers and their efficient detection in information networks, in *KDD* (2010), pp. 813–822
72. R.A. Jarvis, On the identification of the convex hull of a finite set of points in the plane. Inf. Process. Lett. **2**, 18–21 (1973)
73. J.M. Kleinberg et al., Authoritative sources in a hyperlinked environment. J. ACM **46**(5), 604–632 (1999)
74. K. Senthamarai Kannan et al., Labeling methods for identifying outliers. Int. J. Stat. Syst. **10**(2), 231–238 (2015). ISSN 0973-2675
75. T. Kohonen, Self-organization and associative memory, in *Springer Series in Information Sciences* (1988)

76. K. Sugihara, Robust gift wrapping for the three-dimensional convex hull. J. Comput. Syst. Sci. **49**, 391–407 (1994)
77. L. Xu et al., A hierarchical framework using approximated local outlier factor for efficient anomaly detection. Procedia Comput. Sci. **19**, 1174–1181 (2013)
78. M.M. Breunig, H.-P. Kriegel, R. Ng, J. Sander, LOF: identifying density-based local outliers, in *SIGMOD'00* (2000), pp. 427–438
79. S. Maya, K. Ueno, T. Nishikawa, dLSTM: a new approach for anomaly detection using deep learning with delayed prediction. Int. J. Data Sci. Anal. (2019)
80. N. Billor et al., BACON: blocked adaptive computationally efficient outlier nominators. Comput. Stat. Data Anal. **34**, 279–298 (2000)
81. P. Filzmoser et al., Outlier identification in high dimensions. Preprint submitted to Elsevier Science (2006)
82. P. Cao et al., A focal any-angle path-finding algorithm based on A* on visibility graphs. arXiv preprint arXiv:1706.03144 (2017)
83. Qiu et al., A survey of machine learning for big data processing. EURASIP J. Adv. Signal Process. **2016**, 67 (2016)
84. R.E. Kalman, A new approach to linear filtering and prediction problems. Trans. ASME J. Basic Eng. **183**, 35–45 (1960)
85. S. Brin et al., The anatomy of a large-scale hypertextual web search engine. Comput. Netw. ISDN Syst. **30**(1–7), 107–117 (1998)
86. S. Cateni et al., Outlier detection methods for industrial applications, in *Advances in Robotics, Automation and Control* (2008), p. 472
87. Q. Tan, N. Liu, X. Hu, Deep representation learning for social network analysis. Front. Big Data **2**, 2 (2019)
88. T.M. Chan, Optimal output-sensitive convex hull algorithms in two and three dimensions. Discrete Comput. Geom. **16**, 361–368 (1996)
89. Z. Liu, X. Liu, J. Ma, H. Gao, An optimized computational framework for isolation forest. Math. Probl. Eng. (2018)
90. Z. He et al., Discovering cluster base local outliers. Patten Recogn. Lett. **24**(9–10), 1641–1650 (2003)

Machine Learning Perspective in VLSI Computer-Aided Design at Different Abstraction Levels

Malti Bansal and Priya

Abstract In the past few decades, machine learning, a subset of artificial intelligence (AI), has emerged as a disruptive technology which is now being extensively used and has stretched across various domains. Among the numerous applications, one of the most significant advancements due to Machine Learning is in the field of Very Large Scale Integrated Circuits (VLSI). Further growth and improvements in this field are highly anticipated in the near future. The fabrication of thousands of transistors in VLSI is time consuming and complex which demanded the automation of design process, and hence, computer-aided design (CAD) tools and technologies have started to evolve. The incorporation of machine learning in VLSI involves the application of machine learning algorithms at different abstraction levels of VLSI CAD. In this paper, we summarize several machine learning algorithms that have been developed and are being widely used. We also have briefly discussed about how machine learning methods have transuded the layers of VLSI design process from register transfer level (RTL) assertion generation to static timing analysis (STA) with smart and efficient models and methodologies, further enhancing the quality of chip design with power, performance and area improvements and complexity and turnaround time reduction.

Keywords VLSI · Machine learning · SoC · Integrated circuits · Algorithms · Computer-aided design · SVM · Regression

1 Introduction to Machine Learning and VLSI CAD

Machine learning is an upsurging field which is transforming the world by incorporating the advancement in artificial intelligence (AI) and data science that deals with the systems which can learn from the experience or preexisting data and improve without being specifically and comprehensively programmed for the same. In a nutshell, machine learning enables the computers with human intelligence [1].

M. Bansal (✉) · Priya
Department of Electronics and Communication Engineering, Delhi Technological University (DTU), Delhi 110042, India

S. Shakya et al. (eds.), *Mobile Computing and Sustainable Informatics*, Lecture Notes on Data Engineering and Communications Technologies 68,
https://doi.org/10.1007/978-981-16-1866-6_6

Machine learning centers around developing an effective and efficient learning algorithms that assists the machines to analyze the available data and train itself to accurately predict for the unknown data samples.

In 1950s, a computer program for playing checkers was created by Arthur Samuel who worked with IBM in the AI domain. In this program, Samuel used a scoring mechanism using the locations of the pieces on board and tried calculating the probability of each side winning. Using a minimax technique, the program selects its next move, and this technique eventually got developed into Minimax algorithm. A variety of mechanisms were also developed by Samuel to allow his program become better. His program remembered all the locations on the board that was valued for the reward function. His program had learned from the available data samples and predicted the reward function. Arthur Samuel was the first who coined the term machine learning [2]. Since then, machine learning is gradually evolving in the various domains like quantum computing, robotics, data mining, very large-scale integrated circuits (VLSI), automation, artificial intelligence, Internet of things (IoT), medical, finance, military, etc. As machine learning is mainly running on neural networks and logical and computational algorithms, it makes a system both smarter and beneficial. Due its accuracy, reliability, efficiency and ability to improve, machine learning is excessively effective over the human or natural intelligence [3]. In this paper, primary emphasis has been done on how machine learning is being applied in IC chip design and automating the EDA tools and technologies.

Very large-scale integration (VLSI) deals with the process of integrating hundreds of thousands of transistors on a single chip which is called integrated chips (IC). Due to the integration of large number of transistors, the design of such integrated circuits has turned so complex and time consuming that it becomes a challenging task to achieve it manually by the design engineers. This created the need of a computer-aided design tools for the design, verification and optimization of IC circuits which would run various computers programs that automate the processes involved without much manual calculations. This is often called electronic design automation (EDA). The development of hardware description languages, tools and technologies is helpful in simplifying the design, verification and optimization at various abstraction levels of IC chip fabrication. A variety of EDA tools are used to cut the design cycle and help the engineers to simplify the design process.

Since with the advent of time, technology nodes are dropping in VLSI leading to the lesser chip area, better performance and low power digital circuits. The demand of power, performance and area (PPA) optimization has increased. This is one of the reasons that the complexity of IC is also increasing with time which means that better CAD tools and technologies are required. Machine learning helps VLSI CAD tools to improve and become more efficient. In the further sections, we have discussed about machine learning extending its reach in VLSI at various abstraction levels to solve the design problems dealing with large amount of data, improving on existing algorithms and optimizing the learning methods for automation.

2 The Basic Paradigm of Machine Learning

The basic model of machine learning comprises several steps starting with the collection of data samples from the environment/situations/experiences. This is the preexisting data samples that help the machine to learn a given task. Data is prepared after the collection by removing errors, missing values, repeated values, etc. Next step is the most important, i.e., choosing an algorithm model which differs with the different tasks. There are numerous learning algorithms in machine learning. Now, the model is trained iteratively followed by the evaluation in which the model is tested against unknown data sample. Further, tuning of the model parameter is done to increase the performance of the model proceeded with the final step of making the predictions in the real scenario. The performance in the tasks improves as the machines gain experience while executing the tasks and updating the model each time [4]. The basic model and its process have been represented in the block diagram (Fig. 1).

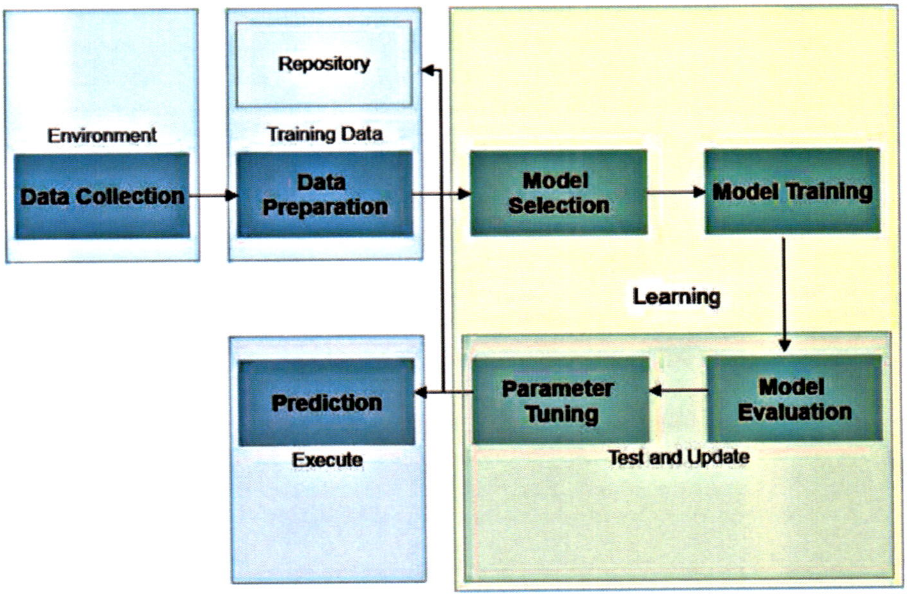

Fig. 1 Basic model of machine learning with process steps

3 Areas of Machine Learning

Machine learning has countless number of applications, and it offers solutions to many real-time issues. Some of the areas where machine learning is being progressively applied are mentioned below [3]:

(i) Face Detection and Recognition, (ii) Visual Perception, (iii) Classification, (iv) Adaptive Systems, (v) Modeling, (vi) Speech and Image processing, (vii) Automation, (viii) Problem Solving, (ix) Genetics, (x) Anomalies Detection, (xi) Games, (xii) Internet of Things (IoT), (xiii) Quantum Computing, (xiv) Medical Diagnosis, (xv) VLSI, (xvi) Stock Market Trading, (xvii) Virtual Personal Assistant, (xviii) Online Fraud Detection, (xix) Speech Recognition, (xx) Traffic Prediction, (xxi) Email spam and Malware Filtering, (xxii) Service Personalization and (xxiii) Computer Vision.

4 Machine Learning Algorithms

The highest level of abstraction in machine learning methods is based on the source of data/information that directs into the learning [4]. It is broadly classified into three categories. These are: (i) unsupervised learning, (ii) supervised learning and (iii) semi-supervised.

In unsupervised learning, only input data is available, and some structure or label needs to be developed to distinguish between the input data samples. In supervised learning, input and corresponding output data samples are available along with the structure/labels. In semi-supervised learning, only some fractions of input data samples have corresponding output pairs, i.e., few of them are labeled or structured [5]. Figure 2 represents the basic classification of machine learning methods and the algorithms used for learning in different methods.

A brief summary of different machine learning algorithms has been represented in the form of table that briefly lists the uses, advantages and its drawbacks which can be useful in the appropriate selection of algorithm and is given in Table 1.

5 Drawbacks of Machine Learning

Although machine learning is very powerful and approved advancement in various domains, it still has its drawbacks which are discussed as follows.

During the training and learning process, a significant volume of data is used. And also the data that we use in the process should be of impartial consistency and high quality which might need the generation of more data, and hence, more time, space and power are required for better quality of results. The other drawback is that some authentic and dependable resources are required in case learning algorithms show time-consuming errors and complexity. It is very important to check that the

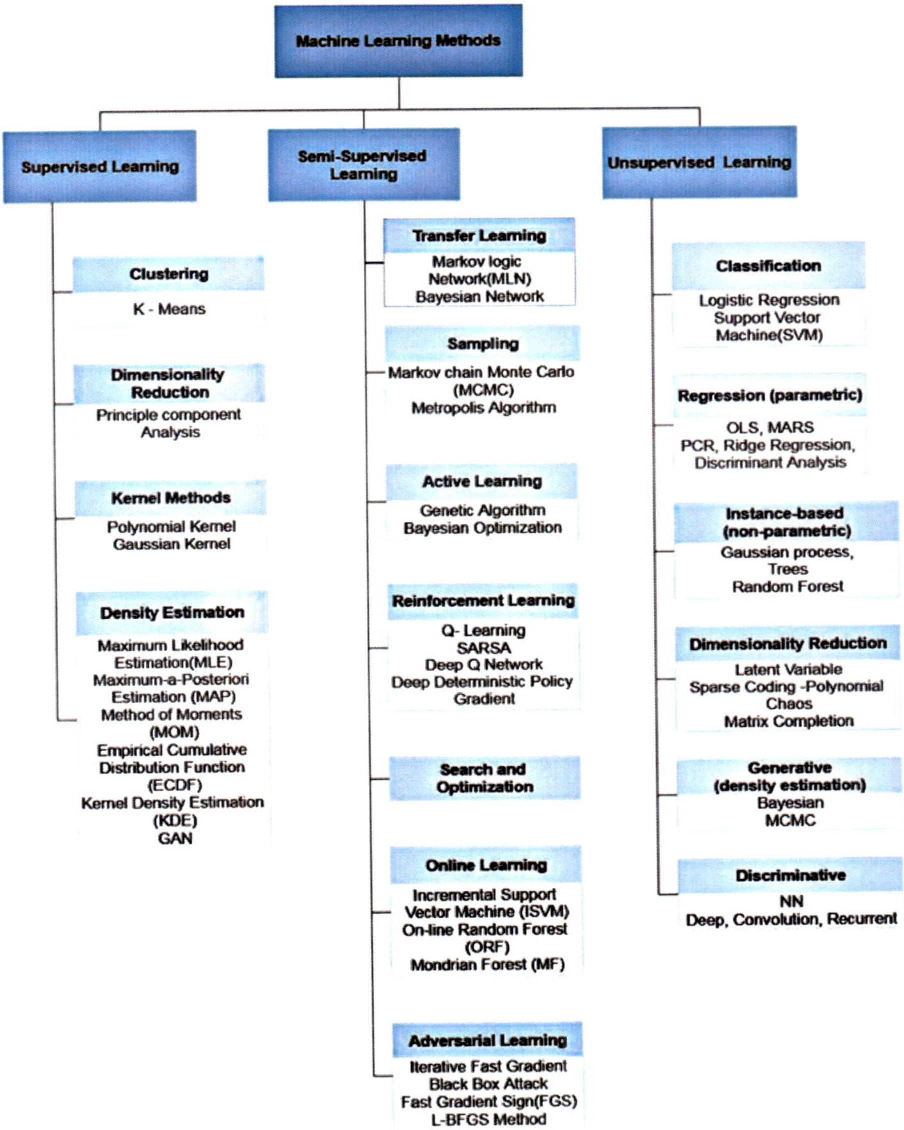

Fig. 2 Classification of machine learning algorithms

algorithms that have been assisted in the process are producing the desired output or not because to get the desired output, we need an accurate learning algorithm with high performance. Selection and availability of such an accurate algorithm is also a challenge. Machine learning still needs a lot of improvements in algorithms and the software that performs the analysis on datasets. Moreover, due to large volume of data, error susceptibility is also high which needs to be taken care of while using a particular

Table 1 Machine learning algorithms and their uses, advantages and its drawbacks

S. No.	Machine learning algorithms					
	Name of algorithm	Author	References	Uses	Advantages	Drawbacks
1.	Gradient Descent	Ray et al.	[6]	To minimize cost function	Efficient, stable error gradient	Never converges for too high and too low learning rate
2.	Linear Regression	Ray et al.	[6]	Models continuous variables, prediction Data analysis process	Easier to understand Easy to avoid over fitting	Not a good fit for nonlinear relationships, cannot handle complex pattern, over simplifies real-word issues
3.	Multi-variate Regression Analysis	Ray et al.	[6]	Used on number of independent variable and single dependent variable	Deeper insight to relationship between variables Models complex real-time issues Realistic and practical	Complex, high knowledge is required for modeling, sample size needs to be high, difficult to analyze
4.	Logistic Regression	Ray et al., Sethi et al., Bhandari et al. and Kohli et al.	[6, 7]	Used on classification problem	Simple to implement Ease of regularization Efficient in computation and training, no scaling required, reliable	Unable to solve nonlinear problem, prone to over fitting, does not work well unless all the variables are identified
5.	Descion Tree	Ray et al.	[6]	Used on regression and classification problem	Suitable for regression Classification problem, easy to interpret and handle, capability to fill missing values, high performance due to efficiency of tree traversal	Unstable, difficult to control size of tree, it may be prone to sampling error, and it gives a locally optimal solution—not optimal solution. Prone to over-fitting

(continued)

Table 1 (continued)

S. No.	Machine learning algorithms					
	Name of algorithm	Author	References	Uses	Advantages	Drawbacks
6.	Support Vector Machine	Ray et al., Choudhary et al. and Gianey et al.	[6, 9]	Used on regression and classification problem	Handles both semi-structured and structured data, can handle complex function, less probability of over fitting, scales up the high-dimensional data, does not get stuck in local optima	Low performance in large datasets, difficult to find appropriate kernel function Does not work in noisy dataset. No probability estimates. Difficult to understand
7.	Bayesian Learning	Ray et al.	[6]	To handle incomplete datasets	Prevents over-fitting, no removal of contradictions required	Prior selection is not easy Distribution can be influenced by prior, wrong predictions possible, complex computation
8.	Naïve Bayes	Ray et al., Sethi et al., Bhandari et al. and Kohli et al.	[6, 7]	Used on binary and multi-class classification problems	Easy to implement, gives good performance, less training data required, scales linearly with predictors and data samples, handles continuous, discrete data. insensitive to irrelevant features	Model often outperforms, too simple, cannot be applied directly, requires retraining, stops scaling when data points are high, more runtime memory required, complex computation for more variables

(continued)

Table 1 (continued)

S. No.	Machine learning algorithms					
	Name of algorithm	Author	References	Uses	Advantages	Drawbacks
9.	K-Nearest Neighbor	Ray et al., Choudhary et al. and Gianey et al.	[6, 9]	Used on classification problems	Simple and easy to implement, cheap and flexible classification, suitable for multi-modal classes	Expensive, computation is distant and intense, less accuracy, no generalization, data large sets

dataset and learning algorithms. Drawbacks can also be related to a specific machine learning algorithm such as nonlinearity, sampling errors, overfitting, noisy datasets, incomprehensible datasets, low performance, complex and expensive computation and insufficient runtime memory.

6 Application of Machine Learning in VLSI CAD Abtraction Levels

There has been substantial advancements in the semiconductors industry over the past five decades mainly due to expeditious technological progress applied in the field of integrated circuits (IC) to improve its performance, reduce the area, power requirements and cost [9]. This technological advancement allowed the incorporation of nearly billions of transistors on a single chip commonly categorized as very large-scale integration (VLSI). Chip designing in VLSI involves number of processing steps at different abstraction levels, concisely called VLSI design flow.

Each abstraction level in VLSI design flow has its customized EDA tool that perfectly covers all the aspects relevant to a given task in the analysis and designing of chips, often referred as computer-aided design (CAD). Figure 3 represents the steps involved in VLSI design flow along with the abstraction levels.

As the chip is designed from system specification to final layout, VLSI CAD requires the addressing of design problems at each level of abstraction. The machine leaning methods that we have listed in the previous section have numerous approaches that can be applied in solving the challenges and problems faced in each step of VLSI design process. Since the main concern behind VLSI CAD was to reduce the increasing complexity at each abstraction level by developing automated simulation or generation/synthesis tools [5], VLSI CAD basically needs methods to model and test the input datasets, and machine learning algorithms serve the same purpose. Many machine learning methods have the capability to deal with the design problems and complexity of VLSI CAD. In further sections, we enumerated the applications of machine learning in various steps of VLSI designing.

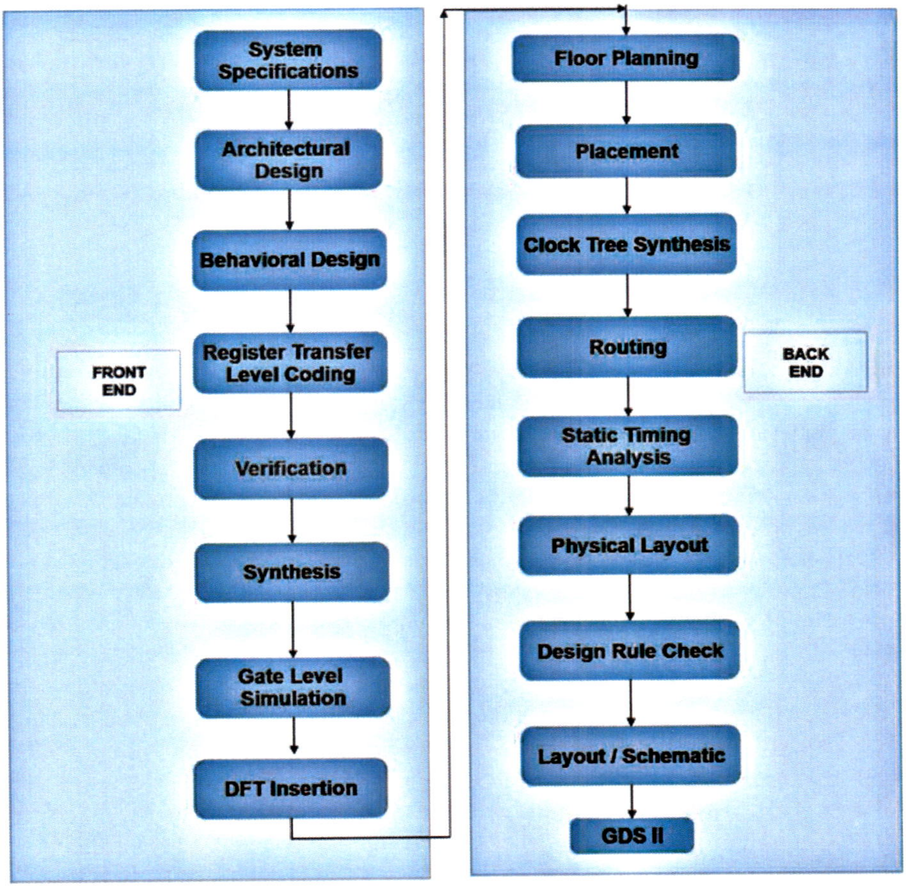

Fig. 3 Process in VLSI design flow

6.1 Machine Learning in Automatic Generation of Assertion in RTL Design

In hardware VLSI design cycle, simulation-based design verification is an important but exhaustive step which takes most of the resources and time. As the design starts getting complex with the integration of more number of hardware components, it becomes challenging for the designer to verify if the design hardware implementation meets the specification or not [5]. The behavior of the design can be monitored using assertions during simulation with the help of properties or temporal logics. Assertions are the instructions that are added in the RTL code to find errors in the design code which even stimulus-based testing cannot find out. Various HDLs have different types of assertions to incorporate in the design code which deals with a specific type of bugs. It is a very challenging task for designers to write effective and specific assertions

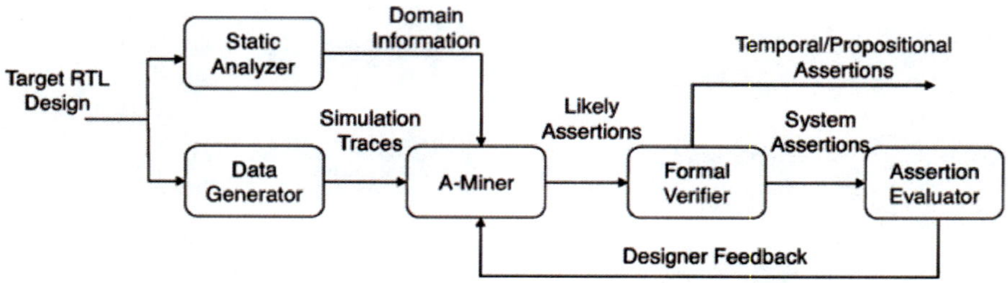

Fig. 4 GoldMine methodology [5]

manually [10]. So, multiple proposals have come up to automatically generate the assertions with the help of machine learning algorithms. The assertions generated by machine are non-uniform and discursive. One of the example is GoldMine, a tool that automatically generates assertions using data science for the analysis and verification of design code [11]. Figure 4 represents the GoldMine methodology in terms of a block diagram.

Data generator is used to simulate the register transfer level (RTL) dcode. A-Miner helps in mining the data obtained from simulation. Static analyzer feeds the domain information from Verilog design code to A-Miner. If the assertions produced by A-Miner are not correct, the verification fails, and assertion evaluator sends a designer feedback to A-Miner to improve the different sets of assertions. This is how the problems with learned assertions are resolved, and it eventually improves the quality of assertions [10]. Basically, A-Miner uses decision tree algorithm to improve the assertion generation process due to its simplicity, preciseness and scalability [12]. Another alternative machine algorithm that can be used is best-gain decision forest (BGDF) algorithm, coverage guided associated mining and PRISM algorithm [5].

6.2 *Machine Learning in Chip Testing*

The major task in chip testing is to decide if a chip should be approved or not which is done by calculating different chip parameters and affixing some predefined criteria. It is presumed that the chip with predefined criteria gets approved satisfying the system design specification [5]. These predefined criteria are operational frequency, chip area, power consumption, timing constraints, performance robustness, etc. These criteria directly affect the manufacturing yield and quality of chip.

Statistical optimization is required to solve the issues related to chip testing. It enhances the manufacturing yield by meeting all the area, timing and power design constraints. On-chip process variation (OCV) is also taken into account while calculating the chip slack which is a parameter that needs to be modeled using random variables [13]. There is additional slack during testing called test margin which is computed using a model for the optimization of chip from the perspective of timing constraints [5].

6.3 Machine Learning for Physical Design Routing

Routing is one of the major steps in back-end of the VLSI design flow that is followed after the placement of pins, logic cells and preplaced cells. It refers to the process of wiring and connecting the pins, logic cells or preplaced cell according to the gate-level netlist produced at the end of front-end of the process [14]. It is a challenge to route billions of flops/logic cells on a chip along with maintaining the design rule checks and constraints.

Since there is an insufficiency of large datasets, physical design routing process faces major issues in the learning using supervised machine learning algorithms [15]. In order to curb the issue of limited dataset, semi-supervised machine learning algorithm can be used. Reinforcement learning also termed as RL approach is one of the semi-supervised learning algorithms which learns from its own experience and hence does not require large datasets [14]. A routing technique based on RL Approach is Alpha RL Router which utilizes the Min–Max game methodology and physical design routing algorithm inspired by an Alpha Go Zero framework which was developed by Google in which software had learned to play the game of Go without any human intervention [16]. The basic methodology of routing issues worked on an assumption where router and cleaner are the two players collaborated as a game. An algorithm such as A-star is employed by the router which helps in performing the initial routing without taking into the consideration of violations in the design rule. Cleaner identifies the breaches in the design rules and chooses the best net that fixes those breaches/violations and rips the old nets and sends them one by one to the router to reroute the nets. Router in the next turn routes the ripped net with the best possible net identified by the cleaner. This process is repeated until all the nets on the chip are routed without any design rule violations. Cleaner focuses on the maximization of its reward score with each step of fixing the design rule violation. It is issued by the router depending upon the quality of prediction made by the cleaner [14]. Figure 5 shows the condensed methodology of router and cleaner.

The optimization in router prediction was done with the help of MCTS algorithm connected in feedback with the neural network (NNET) to predict the next move based on the result obtained from the comparison of probabilities obtained from MCTS and NNET.

6.4 Machine Learning in Physical Design Floor Planning

Due to technological advancements, new additional features and functionalities are being incorporated due to which SoC designs are getting extremely complex since

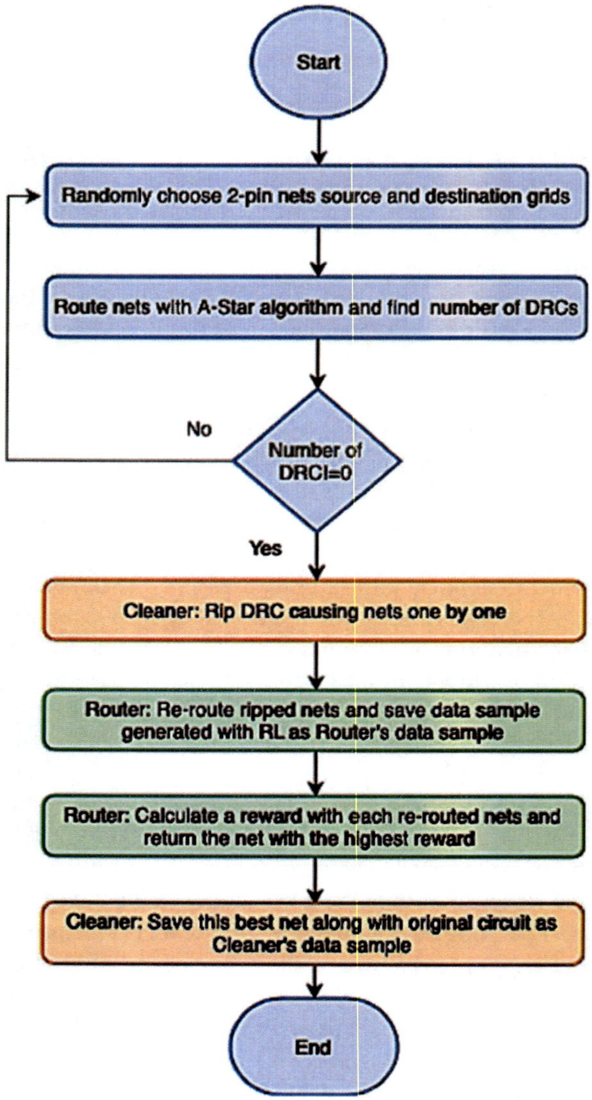

Fig. 5 Alpha-PD-router reinforcement learning-based methodology [14]

they require large number of standard cells, macros and IPs on a single chip. Moreover, it requires high computational speed to which machine learning and neural networks give an excellent assistance. But, there are few issues faced in automating the SoC designs such as the presence of target specific macros, i.e., hard macros, complex connection between different functional blocks and timing constraints associated with blocks or cells for proper functionality of entire chip [17]. Floor planning is the foremost step in physical design flow where most of issues can be dissolved and can be prevented from reoccurring at the later stages where the cost of testing and

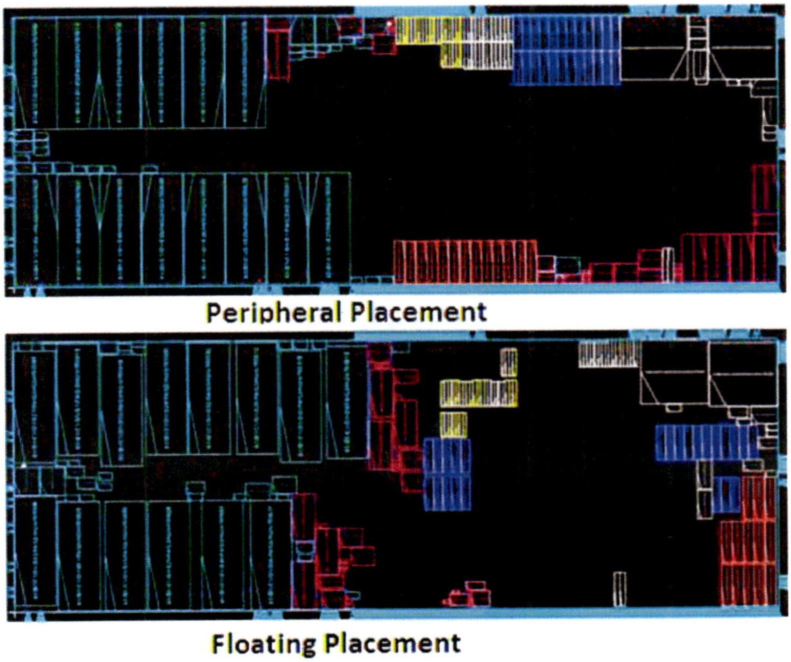

Fig. 6 Different styles of macro placement [17]

removing bugs would be more. This step determines the position of the preplaced and logic cells on the core of the chip, and further, it is followed by the routing and connections of those cells. To perform the placement efficiently, there are several machine learning technologies that have been proposed such as large-scale high-quality macro placement, GPU and mixed-sized global placement [17] (Fig. 6).

Automatic placement techniques segregate the design into sub-blocks depending upon different characteristics, properties and requirements. These sub-blocks have different types of layout styles like peripheral placement and floating placement. These layout styles are chosen by the machine as per the need which is learnt over the experience and available datasets. Power planning is usually done by peripheral placement, and low congestion routing is done by floating placement [17].

6.5 Machine Learning in Static Timimg Analysis (STA)

Static timing analysis contends with the timing performance of the design where it checks the timing constraint of each path and ensures the proper functionality of the design. If any of the paths does not meet the timing constraints, it is said to be a timing violation which eventually affects the performance of the design. The

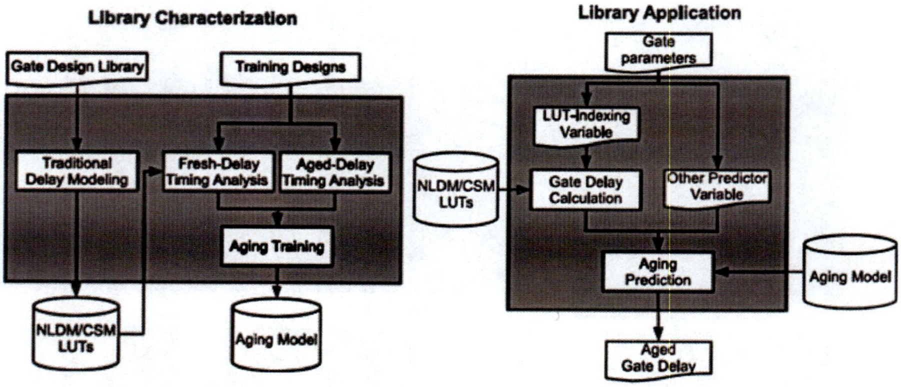

Fig. 7 Characterization and application of libraries in LSTA [18]

challenge associated with STA is high-dimensional correlation. A machine learning-based timing characterization is a potential solution to the above challenge [18]. Support vector machine (SVM) and artificial neural network (ANN) are few of the regression-based machine learning algorithms that can be exploited to correct the path-based timing violations to improve the accuracy [19].

For accurate STA, two main library parameters such as load capacitance and slew are considered to be two dimensions that need to be modeled. These two dimensions get destroyed over a period of time due to dynamic variations like negative bias temperature instability and hot carrier injection also called aging effect. This effect also affects the threshold voltage which ultimately changes timing requirements in STA. Learning-based STA (LSTA) flow can be briefly described in three steps (i) Recognition of challenges related to high dimensions in STA. (ii) Proposing a learning-based model that specifically deals with existing challenges. (iii) Experimenting and evaluating the algorithms on setup libraries included in STA. Figure 7 represents the characterization process of setup libraries in STA during training and application of modified libraries in timing step.

A predefined and preexisting two-dimensional LUT is used as aged cell delays along with a set of training design samples as fresh cell delays in the characterization process of setup libraries. These delays with other parameters included are used as the final training samples for machine learning algorithms. The models learn from these training samples. Finally, this learned model is consolidated to STA library. This modified STA library is applied in the same way, and preexisting ones were applied [18].

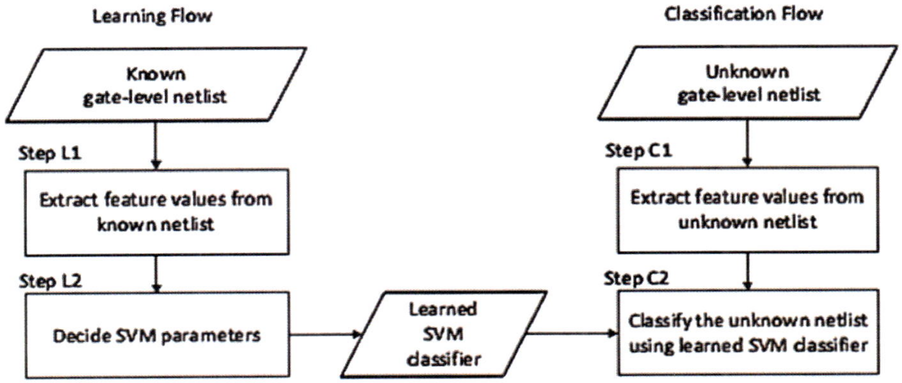

Fig. 8 Flow of learning and classification in gate-level netlist [20]

6.6 *Machine Learning in Gate-Level Netlist*

After the RTL synthesis, the design is implemented in the form of logic gates/cells. Gate-level netlist gives the information about the connectivity of gates and complexity of the circuit design. Designers frequently utilize third-party IC vendors' products which are not attested and dependable since their IC products may contain malwares, also termed as Trojans [20]. Trojans are serious concern as they may lead an IC to fail and may disclose sensitive information. Hence, early detection of Trojans particularly during gate-level netlist is of paramount importance. Machine learning algorithm can be implemented for efficient dynamic and static hardware Trojan detection [21]. Figure 8 shows the flow of learning at gate-level netlist.

Support vector machine (SVM)-based classification can used as machine learning algorithm for the Trojan detection. This classification model classifies the gate-level nets as normal nets and nets with Trojans as 0/1 class. The foremost step while building up the model is to extract the features of Trojan nets based on the samples of nets that are affected with Trojans. In the next step, five features of Trojan nets are considered as five-dimensional vector, and many such vector samples are collected using SVM to rigorously train the classifier. The learned SVM classifier is tested and trained with set of unknown netlists until it starts predicting automatically the correct result with high accuracy.

6.7 *Machine Learning in EDA and IC Design*

Integrated circuit designs and electronic device automation are supported and offered advanced and high value tools and techniques by machine learning.

Figure 9 shows how machine learning and artificial intelligence provide design tools and technologies with newer and improved optimization objectives in the fields

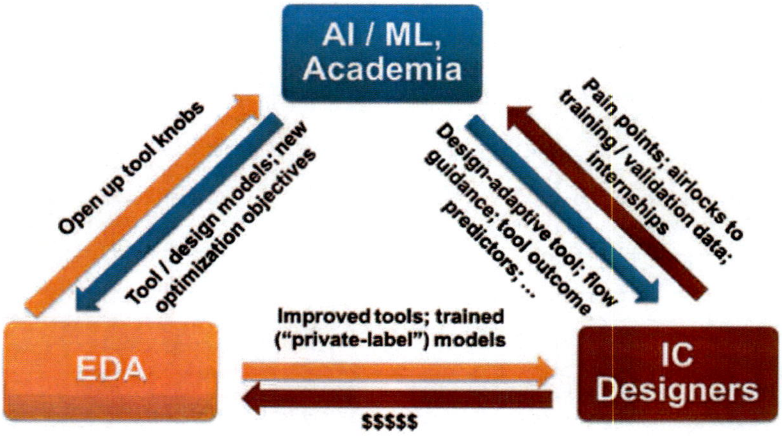

Fig. 9 Machine learning potential in EDA, IC design [22]

of EDA and also foster design adaptive tools, flow guidance, resource management, outcome predictor in the field of IC design.

7 Conclusion and Future Scope

This paper gives an insight on the machine learning methods such as supervised, semi-supervised and unsupervised learning. It also briefly discusses the most frequently and widely used machine learning algorithms that can be specifically used in various domains and focuses on the uses, advantages and drawbacks of these machine learning algorithms. This paper includes a brief description of the most recent research about the machine learning in VLSI computer-aided design. It covers the models and methodologies of machine learning algorithms being applied in the various steps of VLSI Design process such as RTL assertion generation, chip testing, static timing analysis, physical design routing and floor planning, and we learn across the paper that how machine learning automates the complex design, verification and implementation of integrated circuits without much of the human interference. It not only improves the quality of processes but also makes it efficient, accurate and reduces the human efforts.

Machine learning techniques offer great opportunities to EDA and IC design in the near future by opening up the possibilities of improved design convergence and higher levels of optimization. ML models are also expected to implement faster tools and accurate estimations in the respective applications [23, 24]. We aim to make an implementation of these algorithms and perform sets and simulations particularly in one of the steps of VLSI design flow.

References

1. M. Xue, C. Zhu, A study and application on machine learning of artificial intellligence, in *2009 International Joint Conference on Artificial Intelligence*, Hainan Island (2009), pp. 272–274. https://doi.org/10.1109/JCAI.2009.55
2. A. Samuel, Some studies in machine learning using the game of checkers. IBM J. Res. Dev. **3**(3), 210–229 (1959). CiteSeerX 10.1.1.368.2254
3. M. Bansal, Priya, Application layer protocols for Internet of Healthcare Things (IoHT), in *2020 Fourth International Conference on Inventive Systems and Control (ICISC)*, Coimbatore (2020), pp. 369–376. https://doi.org/10.1109/ICISC47916.2020.9171092
4. M. Bansal, Priya, Performance comparison of MQTT and CoAP protocols in different simulation environments, in *Inventive Communication and Computational Technologies*, ed. by G. Ranganathan, J. Chen, A. Rocha. Lecture Notes in Networks and Systems, vol. 145 (Springer, Singapore), pp. 549–560. https://doi.org/10.1007/978-981-15-7345-3_47
5. A. Nayak, K. Dutta, Impacts of machine learning and artificial intelligence on mankind, in *2017 International Conference on Intelligent Computing and Control (I2C2)*, Coimbatore (2017)
6. S. Ray, A quick review of machine learning algorithms, in *2019 International Conference on Machine Learning, Big Data, Cloud and Parallel Computing (COMITCon)*, Faridabad (2019), pp. 35–39
7. P. Sethi, V. Bhandari, B. Kohli, SMS spam detection and comparison of various machine learning algorithms, in *2017 International Conference on Computing and Communication Technologies for Smart Nation (IC3TSN)*, Gurgaon (2017), pp. 28–31
8. I. (Abe) M. Elfadel, D.S. Boning, X. Li (eds.), *Machine Learning in VLSI Computer-Aided Design* (Springer International Publishing, Springer Nature Switzerland AG, 2019)
9. R. Choudhary, H.K. Gianey, Comprehensive review on supervised machine learning algorithms, in *2017 International Conference on Machine Learning and Data Science (MLDS)*, Noida (2017), pp. 37–43
10. A.-N. Du, B.-X. Fang, Comparison of machine learning algorithms in Chinese web filtering, in *Proceedings of 2004 International Conference on Machine Learning and Cybernetics (IEEE Cat. No.04EX826)*, vol. 4, Shanghai (2004), pp. 2526–2531
11. K.H. Yeap, H. Nisar, *Introductory Chapter: VLSI*. https://doi.org/10.5772/intechopen.69188
12. H.D. Foster, A.C. Krolnik, D.J. Lacey, *Assertion-Based Design*, 2nd edn. (Springer Publishing)
13. S. Hertz, D. Sheridan, S. Vasudevan, Mining hardware assertions with guidance from staticanalysis. IEEE Trans. Comput. Aided Des. Integr. Circuits Syst. **32**, 952–965 (2013)
14. J. Han, M. Kamber, *Data Mining: Concepts and Techniques* (Morgan Kaufmann Publishers Inc., San Francisco, 2000).
15. C. Visweswariah, K. Ravindran, K. Kalafala, S.G. Walker, S. Narayan, First-order incremental block-based statistical timing analysis, in *DAC*, San Diego, CA, June 2004, pp. 331–336
16. U. Gandhi, I. Bustany, W. Swartz, L. Behjat, A reinforcement learning-based framework for solving physical design routing problem in the absence of large test sets, in *2019 ACM/IEEE 1st Workshop on Machine Learning for CAD (MLCAD)*, Canmore, AB (2019), pp. 1–6
17. S. Mantik, G. Posser, W.-K. Chow, Y. Ding, W.-H. Liu, "ISPD 2018 initial detailed routing contest and benchmarks, in *Proceedings of the 2018 International Symposium on Physical Design, ISPD'18* (ACM, New York, NY, USA, 2018), pp. 140–143
18. D. Silver, J. Schrittwieser, K. Simonyan, I. Antonoglou, A. Huang, A. Guez, T. Hubert, L. Baker, M. Lai, A. Bolton, Y. Chen, T. Lillicrap, F. Hui, L. Sifre, G. Van Den Driessche, T. Graepel, D. Hassabis, Mastering the game of Go without human knowledge. Nature **550**(7676), 354–359 (2017)
19. T.C. Chen, P.Y. Lee, T.C. Chen, Automatic floorplanning for AI SoCs, in *2020 International Symposium on VLSI Design, Automation and Test (VLSI-DAT)*, Hsinchu (2020), pp. 1–2
20. S. Bian, M. Hiromoto, M. Shintani, T. Sato, LSTA: learning-based static timing analysis for high-dimensional correlated on-chip variations, in *2017 54th ACM/EDAC/IEEE Design Automation Conference (DAC)*, Austin, TX (2017), pp. 1–6

21. A.B. Kahng, M. Luo, S. Nath, SI for free: machine learning of interconnect coupling delay and transition effects, in *Proceedings of SLIP* (2015), pp. 1–8
22. K. Hasegawa, M. Oya, M. Yanagisawa, N. Togawa, Hardware trojans classification for gate-level netlists based on machine learning, in *2016 IEEE 22nd International Symposium on On-Line Testing and Robust System Design (IOLTS)*, Sant Feliu de Guixols (2016), pp. 203–206
23. M. Bansal, M. Nanda, M.N. Husain, Security and privacy aspects for Internet of Things (IoT), in *2021 6th International Conference on Inventive Computation Technologies (ICICT)*, Coimbatore, India, pp. 199–204 (2021). https://doi.org/10.1109/ICICT50816.2021.9358665
24. M. Bansal, S. Garg, Internet of Things (IoT) based assistive devices, in *2021 6th International Conference on Inventive Computation Technologies (ICICT)*, Coimbatore, India, pp. 1006–1009 (2021). https://doi.org/10.1109/ICICT50816.2021.9358662

A Practical Approach to Measure Data Centre Efficiency Usage Effectiveness

Dibyendu Mukherjee, Sandip Roy, Rajesh Bose, and Dolan Ghosh

Abstract If an enterprise develops a data centre application, then it has to afford constantly increasing expenditure for its trade because the price of power consumption always increases and never decreases. If an enterprise wants to handle a power-methodical data centre, then it needs an exceptional observation for all the connections in the transmission series of power which begins from the usefulness grid attachment via the framework of data centre such as the transmission of power, support strategy, cooling, and allotment. Then they are attached to the server. If an enterprise uses conventional procedures to measure the competence of the data centre, then it provides many inappropriate outcomes. Such procedures can comprise in having 'nameplate' or an equipment's estimated power numerals and reckoning the competence of the different power and ingredients of the framework of cooling for assessing the failure of power. This research provides a better straightway reckoning procedure which is named power usage effectiveness or PUE which introduces a more appropriate depiction of the competence of the data centre.

Keywords Data centre · Energy consumption · Green computing · Power usage effectiveness (PUE) · Carbon emission

1 Introduction

The performance related to the power consumption of a data centre is almost operated to squander an extensive volume of power consumption. Nowadays, calculating and modifying the performance of the power consumption of the data centre is beneficial and rational. To some particular users, the expenses contributed on power consumption increase the volume contributed on the accumulation and fitting of IT component. With the particular data being accessible, the procedures for the assessments are completely missing. Procedures of contrasting documented data with devised

D. Mukherjee · S. Roy (✉) · R. Bose · D. Ghosh
Brainware University, Kolkata, India

D. Ghosh
e-mail: dg.cs@brainwareuniversity.ac.in

S. Shakya et al. (eds.), *Mobile Computing and Sustainable Informatics*, Lecture Notes on Data Engineering and Communications Technologies 68,
https://doi.org/10.1007/978-981-16-1866-6_7

specifications of the components are missing. Considering the management of the functioning prices where the upgrading is problematic, the enterprises can better comprehend energy performance in their data centre with the understanding of calculations of its PUE. This PUE occupies the ratio of the entire data centre and the input power to IT load power. The relation between the PUE worth and the performance of the data centre is inversely proportional, as more 'overload' energy gets exerted for the consumption of power of the pressure created by electricity [1].

At present to evaluate data centres, a variety of metrics can be achieved. Nevertheless, a single metric has gradually turned out to be an industry benchmark metric in effect. In the year of 2006, the advent of the power usage effectiveness (PUE) metric was witnessed. Green Grid is a non-profit organization of software experts, launched in the year of 2007 [2]. It has turned out to be the most customary metric to announce the 'energy productiveness' of data centres [3–5]. For the establishment of the amount of energy which in reality is applied for managing the IT component in regards to the entire power pull of a privilege, the PUE is beneficial. This is illustrated in Eq. (1) and Fig. 1. For the evaluation of the consumption of energy of discrete procedures like cooling, in comparison with the IT burden, a partial PUE (pPUE) [6] is applied. At

Fig. 1 Proposed flowchart of PUE in action

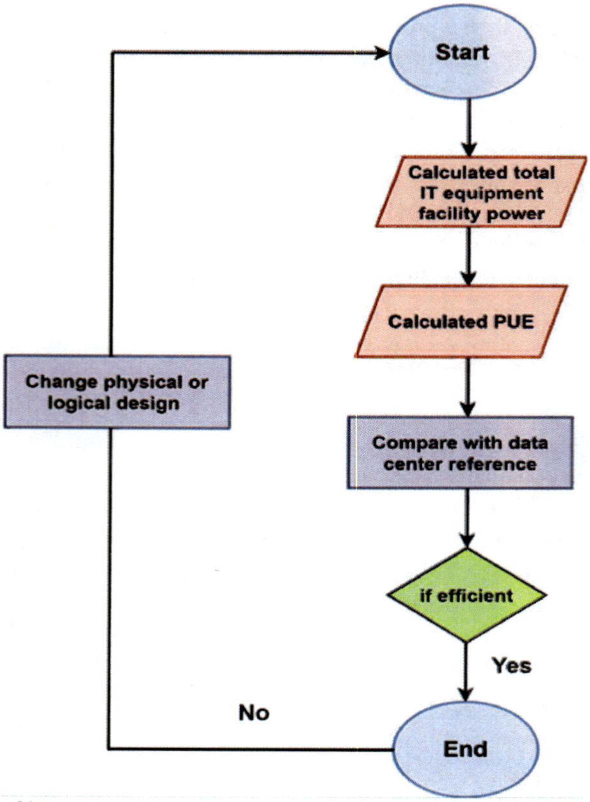

present, the Green Grid explores the idea of the partial PUE (pPUE) in which a value that is akin to PUE, for a subprocedure can be calculated and accounted. A repository is an easy instance of a data centre based on a repository mandatorily comprised of any energy which is applied by the fundamental scopes that assist the repositories and the repository itself. Nevertheless, within this frame of reference, PUE can be a very high-level metric that does not permit any transparency between the retribution of framework which is available to the fundamental framework at variance with the repository itself or between two distinct repositories themselves. It is pPUE that will provide a better procedure to monitor the following levels. Moreover, 'Partial PUE' (pPUE) metrics have been suggested for the assistance to assist to locate the dilemma with the illustration of structural electrical areas in the data centre such as the server floor area, and resolving pPUE for all the areas [4].

pPUE may be applied for rooms in an apartment where versatile works are done, but uncertainly, the total of the pPUE for all areas may not produce the entire PUE of the scope.

$$\text{PUE} = \frac{\text{Total\_Facility\_Energy}}{\text{IT\_Equipment\_Energy}} \tag{1}$$

It can be increased and considered as:

$$\text{PUE} = \frac{(\text{Cooling} + \text{Power} + \text{Lighting} + \text{IT})\text{\_Facility\_Energy}}{\text{IT\_Equipment\_Energy}} \tag{2}$$

Power is the deficit of energy in the allotment procedure of power via line-deficit and different frameworks such as PDU, disorganizations, and lighting. The lighting produces the energy applied for illuminating the data centre and backing rooms. Here, the energy applied with each of the IT components such as servers, networks, storage in the data centre ways [2, 3].

Despite the name of the metric is 'power' usage effectiveness, it evidently estimates the energy exertion of the data centre. Examining the name PUE for energy usage effectiveness or EUE for ignoring the dilemma between power and energy estimations [7] is proposed by Yuventi and Mehdizadeh. The PUE requires being estimated, for more than one year, and then it estimates energy exertion. In the configuration phase of projects for establishing the dynamic 'energy competence' of the data centre, the latter part of the constitution is applied for assisting energy pricing and regulating power consumption. Nevertheless, in the time of configuring phase, PUE values' precise estimation is impossible. Nevertheless, the metric has even now turned out to be a marketing device, with proprietors or constructors applying it for the augmentation of the dynamic 'competence' of their data centre.

It is impossible to have the PUE of 1, as this all the time a few bit of energy exertion for the assistance of the IT components, and a PUE of 1 is always a perfect number. It is evident that at present, for defining the PUE of data centres on an entire globe-covering scale, adequate data is not there [3], but a few types of researches have been made on it. Specific research has shown that 70% of 115 testifiers are acquainted with their PUE. Among them, an all-inclusive average value of 1.69 was accounted for [8]. Next, homogeneous research also shows that for 22 data centres, PUE volumes are discovered between 1.33 and 3 with an average value of 2.04 [9]. Nevertheless, no research among them has given vivid knowledge concerning the behaviour of each scope, the measurement of the control in them, or the procedure of the energy exertion evaluation was executed.

2 Energy Versus Power

The real-time application of the phase PUE was based on the power that was pulled by the IT component, the power that was pulled by the cooling component, and the power deficits in the allocation procedure of electricity. After that, the terminology power usage effectiveness comes. Usually, power gets denoted units in Kilowatts, and it establishes a prompt estimation or ‘snapshot in time’ of the consumption of power. Energy drawn from electricity (kWh) is the consequence of the power (kW) time period which is used in hours. The units of energy will customarily be measured in Kilowatt-hours. When it is said that a power that extracts 20 kW, in reality, extracts that very amount of power for 1 h, it consumes 20 kWh of energy. Power is a prompt estimation; on the contrary, energy is inherent over a period. Usually, though folks are keener in energy, both are of significant roles in the construction and execution of any data centre. In reality, the PUE metric’s clarity offers the mathematics for being well-grounded for one of those two-power and energy. The issues as follows give an impact on the competence of the data centre, related to electricity.

2.1 IT Load

It is a discriminating IT burden which is a very significant issue that very frequently varies. The novel power control attributes to modify the generation of the IT component produces about the IT burden for changing promptly based on requirements. The rate of performance differs severely from the burden. On account of continuously varying requirements, the diagram provides the picture of the change that may happen in the prompt PUE over the period of the day and the IT burden changes over a time in a day. But such prompt PUE value will not be akin due to regular PUE. Homogeneously, the regular PUE may not be the same as the PUE measured in a week or the PUE measured in a period.

2.2 *Effect of Outdoor Condition*

The outside temperature of the air is the most significant issue. If the temperature gets higher, the competence of the data centre gets lesser. The rate of performance diminishes because the thermal-removal procedures contain more energy in the time of the operation of the heat of the data centre that augments on account of percolating outside heat into the data centre.

2.3 *User Configuration*

A massive amount of variance is found in the PUE due to a large number of users' activities. The activities as follows are harming and unusually fluctuating the rate of success. In addition, these depend on the precise configuration of the power and cooling processes such as Temperature Reference Standard Fluctuation, Humidity Reference Standard Fluctuation, Venter Floor Tile Fluctuation, Plenum Fluctuation and Washing Air Strainer Decline. If all of the above operations are carried out, the configuration of the data centre depends on the configuration of the data centre.

2.4 *Product Review Analysis*

The performance is contingent on verifying on account of any one of the situations clarified in this segment. Regular changes bring about changes in the atmosphere of nature. Changes occur on a regular basis on account of the fluctuations in the IT burden, outside temperature and humidity element. IT burdens change on weekends and on weekdays. Inappropriate support of the component is dependent on the data centre activities. Changes on regular basis cause the deterioration of the utilities of one-time estimation of PUE. When comparing the rate of changes and the rate of performance, the estimation of the rate of performance development works based on presentation and there is very little speculation about the electrical request.

3 Related Work

A better in-floor assessment of the exertion of energy was supplied in [10] that evaluates the exertion of the energy of a miniature data centre in Linkoping, Sweden. Unfortunately, the evaluation of the exertion of the energy of such data centre was just directed in the period of the coldest month of the year. A brief duration of evaluation implemented [11] together with the power needed for the IT component just for the span of 1 month. Evidently, this is significant to keep account of the

PUE for an illustrative duration so as to a pragmatic yearly average is able to be achieved, and it is revealed theoretically in [7] and illustrated in [12]. At the time while the pertinent data is accumulated for more than 1 year, it is able to provide an accurate depiction of the entire exertion of energy, for this comprises in any impacts from variation in the atmosphere and IT burden claim. Further research estimated the exertion of the energy of two data centres in Singapore in the interim of 10 min. for the tenure of a week, establishing vivid information about the exertion of the energy of the IT, HVAC, UPS and illuminating procedures [13]. Because of the local climatic condition, gaining experience of comparatively unchangeable temperatures, the data was able to be hypothesized for characterizing an entire year. Despite this potentiality for estimating a shorter duration of time on account of unchangeable temperatures, it may however not be quite lengthy to estimate any kinds of changes in the IT burden.

In spite of the fundamental challenges with such a metric, minimal research continues to severely analyse the PUE semantically. The PUE is evaluated in research along with other typical metrics [14]. In addition, authors in [7] provide a vibrant assessment of the usefulness of PUE, such as metrics of feasibility. The second purpose is to stress that the PUE should be measured over an illustrative span, as it is the metric's illustrated target to date.

From a linguistic analysis, significant computations of exertion energy performed annually in a cyclical manner. In a few forms of analysis, this demonstrates the absence of a vibrant image to measure the rate of energy efficiency. There is an anomaly [15] that provides a vivid image that involves hardware and cooling systems details and assembles data for more than a full year. If the method of estimating the PUE is linguistic, there is no one who will investigate the responsiveness of the PUE to different frameworks or try to reiterate a PUE computation using liberated source data. Because of the responsive function of the industry, all PUE values that exist in the environment depend on stern confidentiality estimates. In this case, the vulnerability of such enterprises to their configuration procedure is considerably uncertain, most of the time; data centres use their energy, which is hidden from public scope [11, 16]. For this reason, the evaluation or validation of the PUE account with reliance is difficult. In cases where energy consumption metrics can be updated, it is crucial that information should be clear in those cases. Such challenges often find it difficult to express recommendations for minimizing energy exertion in data centres.

4 Proposed Work

It is a PUE estimate that provides a high point of the procedure by a well-organized data centre functioned. For better comprehension of the performance in the data centre, it is a must to monitor the procedure of the management for power, transformation and allotment. Such transformation and allotment of power in the data centre, customarily takes place in many phases, and for this, each phase has a lesser amount of appropriate transformation performance. Momentous section of the total facility power gets transformed and allotted. It also produces the consequence of deficit of power in the form of heat. Later, the heat dissolved needs cooling, and at this, the IT component can control within satisfactory atmospheric restrictions that needs supplementary power consumption by the data centre cooling systems that once more appends to the deficit of power from the entirely consumed power at the data centre input [17, 18].

This research allows us to produce a PUE computation vividly on a conventional data centre. The procedures are established for the exploitation of competence. The surveillance is carried on for two months, and display of the PUE varying from month to month applies diverse exploitation. Based on PUE results, the data centre parameters changes accordingly to obtain a best PUE result. Figure 1 shows the flowchart of the actual use of PUE measurement.

5 Result Analysis

In the beginning, this segment allows to reckon the benefit of UPS. Later, the complete volume of UPS is regarded as 40 kVA by the data centre for reckoning the real volume of the UPS. In this research, this is multiplied by power factor 0.9. The tabular representation (Table 1) displays the real and accessible application of such UPS in September 2019 and October 2019. Figures 2 and 3 describe the UPS capacity utilization in September 2019 and October 2019. This article deals with the realistic data from a traditional data centre [4].

Table 1 UPS capacity utilization calculation difference in two months

	UPS capacity utilization calculation month of September '19	UPS capacity utilization calculation month of October '19
Total UPS capacity (kVA)	40	40
Power factor (PF)	0.905	0.905
Total UPS capacity (kW)	36.2	36.2
Current max utilization (kW)	18.6	17.6
Available capacity (kW)	17.6	18.6

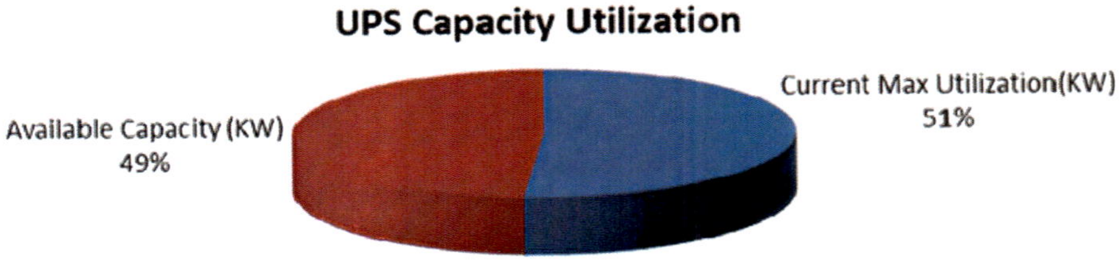

Fig. 2 UPS capacity utilization in the month of September 2019

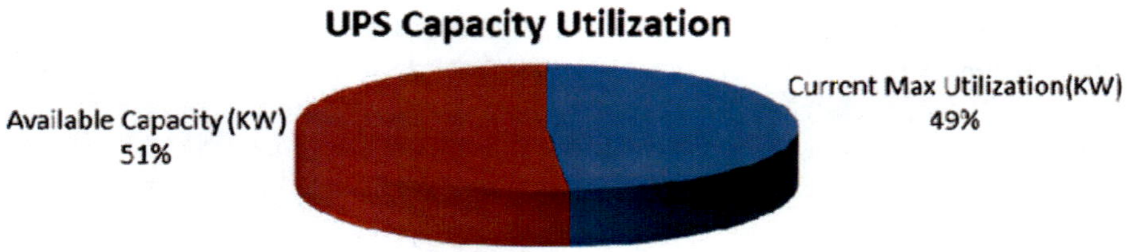

Fig. 3 UPS capacity utilization in the month of October 2019

At this time, measure heat load and CRAC volume exploitation reckoning happened in two distinct months. The tabular representation (Table 2) displays the procedure of the difference of CRAC volume exploitation in two months. Then it supports a benchmark PUE. In such an event, one ton can produce almost 3.5 kW heat load. This study allows to exploit two 10 ton CRAC that can be executed simultaneously. Consequently, the sum of the volume of cooling is almost $10 \times 3.5 \times 2 = 70$ kW heat load. Figures 4 and 5 depict the CRAC capacity utilization in September 2019 and October 2019.

The PUE is measured with the help of the above operation in these two months and obtain the values such as 1.9 and 2.06. Consequently, the measurement of the volume of PUE is lessen that is 2.06 from the upcoming months (Table 3). In such a way, PUE assists the data centre controller for having movement on the exploitation of the power of their data centre. In such an event, the upper limit of IT Load (kW) gets transmitted from UPS output power and average facility load (kW) is conveyed from LT section resource load.

Table 2 Heat load and CRAC capacity utilization calculation

Parameters	Equation	Heat load and CRAC capacity utilization calculation month of September	Heat load and CRAC capacity utilization calculation month of October
IT load (kW)	Same as total IT load power in watts	18.6	17.6
UPS with battery (kW)	($0.04 \times$ Power system rating) + ($0.05 \times$ Total IT load power)	2.4	2.3
Power distribution (kW)	($0.01 \times$ Power system rating) + ($0.02 \times$ Total IT load power)	0.7	0.7
Lighting (kW)	$2.0 \times$ Floor area (ft^2)	1.1	1.1
People (kW)	$100 \times$ Max no. of personnel	0.2	0.2
	Total heat load (kW)	23	22.0
	Available cooling capacity (kW)	47.0	48.0

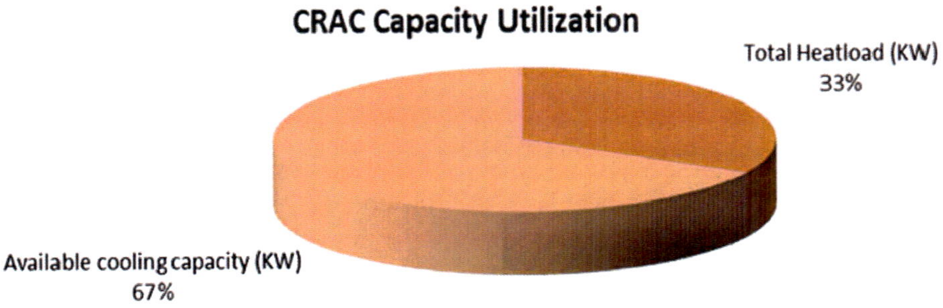

Fig. 4 CRAC capacity utilization in the month of September 2019

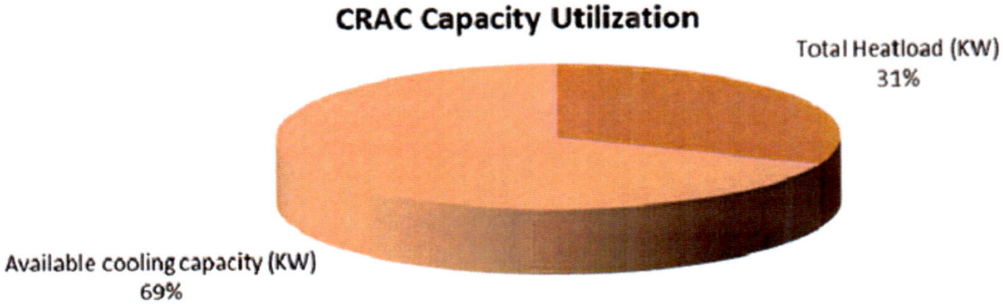

Fig. 5 CRAC capacity utilization in the month of October 2019

Table 3 Power usage effectiveness calculation

	PUE (power usage effectiveness calculation) in the month of September	PUE (power usage effectiveness calculation) in the month of October
Maximum IT load (kW)	18.6	17.6
Average facility load (kW)	35.35	36.23
PUE (power usage effectiveness)	1.9	2.06

6 Conclusion

When controllers are able to achieve optimum outputs, reducing data centre control expenditure, customary data centre configuration procedures would be necessary to replace the operation. The effect of the bulky volume of data centres is regulation of productivity with blemishes. This massive data centre creates an abundant deficit. The reason is every watt of IT burden generated needs a great deal of price for cooling. The data centre controller is provided with a fast and simple protocol to ensure that data centre adeptness is managed by the procedure.

References

1. C. Malone, C. Belady, Metrics to characterize data center & IT equipment energy use, in *2006 Digital Power Forum*, Richardson, TX, USA (2006)
2. The Green Grid, Green Grid Metrics: Describing Datacenter Power Efficiency: Technical Committee White Paper. White Paper #1 (2007)
3. T. Daim, J. Justice, M. Krampits, M. Letts, G. Subramanian, Data center metrics. Manag. Environ. Qual. **20**, 712–731 (2009)
4. R. Bose, S. Roy, H. Mondal, D. Roy Chowdhury, S. Chakraborty, Energy-efficient approach to lower the carbon emissions of data centers. Computing 1–19 (2020)
5. S. Ruth, Reducing ICT-related carbon emissions: an exemplar for global energy policy? IETE Tech. Rev. **28**, 207–211 (2011)
6. T.G. Grid, PUE: A Comprehensive Examination of the Metric. White Paper #49 (2012)
7. J. Yuventi, R. Mehdizadeh, A critical analysis of power usage effectiveness and its use in communicating data center energy consumption. Energy Build. **64**, 90–94 (2013)
8. The Green Grid, Survey Results: Data Center Economizer Use. White Paper #41 (2011)
9. S. Greenberg, E. Mills, B. Tschudi, P. Rumsey, B. Myatt, Best practices for data centers: lessons learned from benchmarking 22 data centers, in *Proceedings of the 2006 ACEEE Summer Study on Energy Efficiency in Buildings*, Pacific Grove, CA (2006)
10. J.F. Karlsson, B. Moshfegh, Investigation of indoor climate and power usage in a data center. Energy Build. **37**, 1075–1083 (2005)
11. V. Avelar, D. Azevedo, A. French, PUE: A Comprehensive Examination of the Metric. White Paper #49 (Green Grid, 2012)
12. M. Patterson, Energy efficiency metrics, in *Energy Efficient Thermal Management of Data Centers*, ed. by Y. Joshi, P. Kumar (Springer Verlag, 2012)
13. K. Kant, Data center evolution: a tutorial on state of the art, issues, and challenges. Comput. Netw. **53**, 2939–2965 (2009)
14. T. Lu, X. Lü, M. Remes, M. Viljanen, Investigation of air management and energy performance in a data center in Finland: case study. Energy Build. **43**, 3360–3372 (2011)
15. D. Mukherjee, S. Chakraborty, I. Sarkar, A. Ghosh, S. Roy, A detailed study on data centre energy efficiency and efficient cooling techniques. IJATCSE **9**, 9222–9242 (2020)
16. D. Azevedo, J.C. Symantec, M.P. Oracle, I.M. Blackburn, Data Center Efficiency-PUE, Partial PUE, ERE, DCcE. The Green Grid (2011), pp. 1–37
17. S. Smys, G. Josemin Bala, Efficient self-organized backbone formation in mobile ad-hoc networks (MANETs). Comput. Electr. Eng. **38**(3), 522–532 (2012)
18. J.S. Raj, Energy efficient sensed data conveyance for sensor network utilizing hybrid algorithms. IRO J. Sustain. Wireless Syst. (04), 235–246 (2012)

Advancing e-Government Using Internet of Things

Malti Bansal, Varun Sirpal, and Mitul Kumar Choudhary

Abstract Internet of Things (IoT) is the most revolutionary and attractive technology of today without which it is nearly impossible to imagine the future due to its applications in numerous fields such as smart cities, home automation, wearable devices, etc., and its ability to make human life much easier via integration with other technologies such as cloud computing and artificial intelligence. In this paper, we have conducted a survey on the ways IoT can be utilized in various sectors of an e-Government such as pollution control, health care and voting. We have also described a six-layer IoT-based model for smart agriculture and studied the benefits of using unmanned aerial vehicles in smart agriculture. This paper further introduces the concept of a "smart" government and its realization via integration of fog computing with IoT leading to a Fog-of-Things (FoT) architecture.

Keywords Internet of Things · e-Government · Social networks · Machine learning · Air pollution · Internet of Health Things · Agriculture · Unmanned aerial vehicles · Voting · Smart government · Fog-of-Things · Smart waste management · Dijkstra's algorithm

1 Introduction

The Internet of Things (IoT) represents an interconnection of various physical objects ("things"), each having a unique identity (IP address). An IP address or logical address is a 32-bit address represented as 4 decimal numbers (such as 173.25.9.1) that is necessary for universal communication independent of underlying physical networks. IoT produces significant insights using a combination of sensors, embedded systems, software and artificial intelligence based on the huge quantity of data generated. The components of IoT are as follows:

M. Bansal (✉) · M. K. Choudhary
Department of Electronics and Communication Engineering, Delhi Technological Univerity (DTU), Delhi 110042, India

V. Sirpal
Department of Electrical Engineering, Delhi Technological Univerity (DTU), Delhi 110042, India

S. Shakya et al. (eds.), *Mobile Computing and Sustainable Informatics*, Lecture Notes on Data Engineering and Communications Technologies 68,
https://doi.org/10.1007/978-981-16-1866-6_8

- **Device**: The physical objects which use sensors to collect data from surroundings. Devices can communicate with other devices via wired or wireless channel.
- **Communication**: The link between devices and the server is made in the data link layer, transport layer, application layer and network layer [1].
- **Radio Frequency Identification (RFID)**: RFID uses radio waves (frequency ~ 125 kHz–2.45 GHz) to authenticate and identify people or objects. An RFID tag consists of a transponder, which, upon getting triggered by an incoming interrogation pulse from an RFID reader, transmits digital data to the reader. RFID uses transmission control protocol (TCP) and HyperText transfer protocol (HTTP) as base protocols. HTTP is used for fetching webpages on the World Wide Web, and TCP is a reliable, connection-oriented protocol in transport layer that also handles flow control to prevent data swamping.
- **ZigBee**: Created by Zigbee Alliance, it is the wireless sensor network technology which uses protocol IEEE 802.15.4. It supports mesh, star and tree topologies [1].

The term **e-Government** signifies a digital model of a traditional government built using various information and communication technologies (ICT) including, but not limited to, IoT, cloud computing and machine learning. An e-Government functions with the aim of enhancing delivery of public services, participation of citizens in the government activities (such as voting) electronically, providing better health and education and improving the agricultural sector. The services of e-Government are classified into three branches, namely government to business (G2B), government to government (G2G) and government to citizen (G2C) [2]:

- **G2B**: Facilitation of the needs and services of the businesses by the government. An example is the e-procurement project in Andhra Pradesh.
- **G2G**: Improving data sharing and communication between government departments/organizations. An example is the National Crime Records Bureau.
- **G2C**: Facilitation of the needs and services of the citizens. Few examples are passport e-seva kendras, IRCTC online booking and online filing of taxes.

The rest of the paper will be organized as follows: In Sect. 2, we will examine air pollution reduction using a combination of social networks, IoT and machine learning. Section 3 studies health care as a key industry in an e-Government architecture and the Internet of Health Things (IoHT) framework. Section 4 examines a six-layer IoT-based agricultural model and further studies the utility of unmanned aerial vehicles (UAVs) in the agricultural sector. Section 5 describes a framework for IoT-based secured voting in an e-Government. Section 6 introduces the concept of "smart" government and examines the feasibility of an IoT-based smart government. Finally, in Sect. 7, we look at a futuristic Fog-of-Things (FoT) framework to counter various problems faced in an IoT-based architecture.

2 Social Networks and Machine Learning

Air pollution is the phenomenon, wherein harmful substances are introduced into the atmosphere like sulphur oxides, carbon monoxide, nitrogen oxides, chlorofluorocarbons, particulate matter and ammonia. Air pollution is a threat to the life of all living beings on the planet, be it humans, animals or plants. It leads to acid rain, reduction of ozone layer thereby contributing to global warming, acute respiratory disorders and even cancer. It is therefore imperative to design a framework for an e-Government service with the aim of reducing air pollution by intelligently educating citizens about its scenario and control tactics.

The framework [2] described here utilizes the power of social networks owing to their high penetration rate in urban and rural areas alike, ability to create online communities easily allowing people to interact with each other, affordability and the unique nature of bringing people of similar interest closer regardless of the physical distance between them. This framework uses a combination of IoT-based pollution sensors, unsupervized machine learning (clustering) algorithm, social networks and smart air pollution education modules to present a smart e-learning platform.

Clustering is the process of grouping similar entities together. A group of similar data points is called a "cluster". The goal is that the points in each cluster have a high similarity with each other and low similarity with a point in another cluster. The difference when compared to regression/classification models is that clustering is an unsupervized algorithm, i.e. only data points (X_i) are provided in the dataset and not the class labels (Y_i) where $1 \leq i \leq n$; n being the number of examples in the dataset.

The framework described here employs ward's agglomerative hierarchical clustering algorithm. It is a bottom-up approach. Each data point is assumed to be a distinct cluster at first. Then, the *similar* clusters are combined together iteratively to form a single cluster. This technique is visualized using a tree-like structure which records the merges, called a dendrogram as shown in Fig. 1.

Fig. 1 Dendrogram

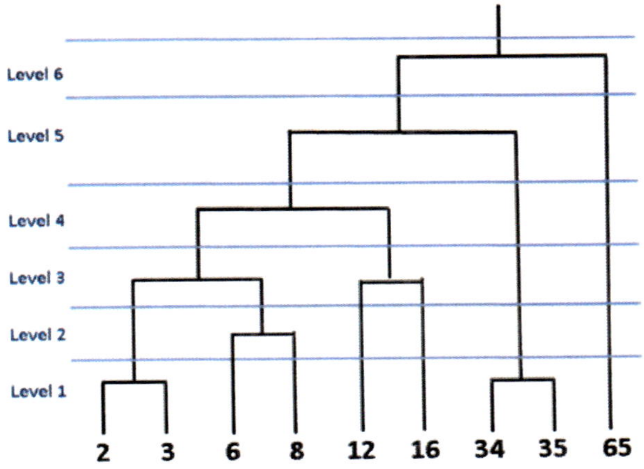

The similarity between two clusters C_1 and C_2 is calculated by using the sum of the squares of distances between points P_i and P_j where $P_i \in C_1$ and $P_j \in C_2$ given by Eq. (1).

$$\text{similarity}(C_1, C_2) = \frac{\sum\left(\text{dist}\left(P_i, P_j\right)\right)^2}{|C_1||C_2|} \tag{1}$$

Ward's method analyses the variance of clusters. To combine two clusters A and B, their merging cost (Δ) is given by Eq. (2).

$$\Delta(A, B) = \frac{n_A n_B}{n_A + n_B}|\vec{m}_A - \vec{m}_B|^2 \tag{2}$$

where n_j is the number of points in cluster j and $\vec{m}_j$ is the centre of cluster j.

The framework consists of the following key steps:

1. Extraction of various user data such as name, native city/town, e-mail ID and current place of living using data from social networks.
2. Clustering on above-extracted data on the basis of native city/town and current place of living. This criterion is used for clustering since awareness can be better spread when related to current place of living or native city/town.
3. IoT pollution sensors are employed to collect data related to air quality for all cities and towns.
4. For a particular city or town, the air quality data and custom designed educational modules are posted via social networks to the cluster of users related to that city.

3 Health Care

The ongoing (at the time of writing this paper) Coronavirus (COVID-19) pandemic has greatly drawn the attention of the entire world towards the importance of robust healthcare infrastructure. The best-in-class healthcare facilities of the world have been overwhelmed by this pandemic. Therefore, it is imperative to utilize the constant technological innovations to improve the same, as a high-priority task. The future of medicine must contain an integrated network of sophisticated medical devices with the aim of enhancing patient care and medical research, with significant reduction in medical chaos. This is a part of G2C activity. IoT-based solutions must be applied to the field of medicine. This is the idea of Internet of Health Things (IoHT).

It is no secret that time is of utmost importance in the case of any medical emergency. IoHT hence works on the principle that the healthcare facilities need to be constantly connected with patients with the mission of prevention of disease complications via health monitoring and improving quality of life for patients with chronic illnesses.

The existence of such a network architecture is of particular importance for patients in rural areas, where there is absence of proper healthcare services, and patients have to travel a long distance (sometimes hundreds of kilometres) to have access to the required services. Usage of IoT-based e-Health systems using biomedical sensors for electrocardiogram (ECG), diabetes, electroencephalogram, oxygen in blood, body temperature, blood pressure, etc., would make the difference between life and death of a patient.

The architecture of an IoHT-based healthcare solution system would have the following major devices/technologies as shown in Fig. 2 [3]. Sensors would be placed to gather data from patients, each sensor being a single node of a wireless body area network (WBAN). This data would be processed at the user terminal (smart phone, smart watch, etc.) by an application. The said user terminal is connected to a gateway by Bluetooth or 6LoWPAN (IPv6 over low-power wireless personal area network). A gateway essentially connects two or more networks together and provides the

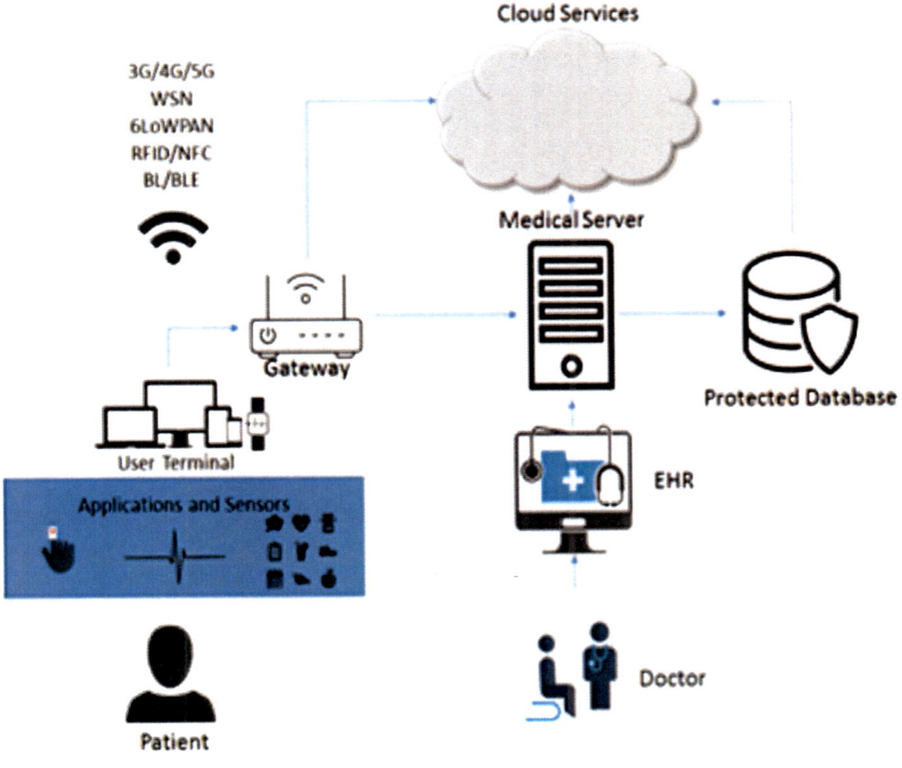

Fig. 2 IoHT-based healthcare system

necessary translation, both in terms of hardware and software. The data gathered can become voluminous. This coupled with the fact that the cloud provides an environment in which fast processing can be done, and machine learning algorithms can be implemented, which can help in medical diagnosis and prognosis, makes cloud storage essential here. The gateway connects to a cloud storage service or a medical server. Also, patient history needs to be stored using electronic health records (EHR) since they play a critical role in medical diagnosis and can be easily accessed by a doctor whom the patient visits. The use cases of the above are varied. Health monitoring for cardiac arrhythmias detection can be done, wherein ECG sensors are used. Patients having Parkinson's disease with symptoms of slowed movement, tremors and balance problems could also be monitored. Stress behaviour over time can be studied for mental health monitoring.

However, the security and privacy of IoHT-based healthcare systems are a key concern given the fact that such devices would be connected to various networks for access anytime, anywhere. Thus, security challenges from the healthcare perspective must be studied. The IoT devices have computational limitations, i.e. low-speed processors with low central processing unit (CPU) speed. Moreover, a device is part of multi-protocol communication, i.e. a proprietary protocol might be used in the local network, whereas communication with IoT service provider might be using Internet protocol (IP). These factors make IoT health devices a target for attackers. Therefore, a fool proof security system for IoHT is needed for its widespread adoption.

4 Agriculture

Today, a major problem faced by any government is that of an exponentially increasing population which in turn leads to a high demand for food. Despite various efforts in the past, the supply rate of food grain has not been able to match the demand rate. This makes the agricultural sector an important area for an e-Government. There is a need of utilizing state-of-the-art technology to develop modern agricultural methods. IoT is leading the way by presenting exciting opportunities in the agricultural sector and has the potential for transforming traditional agriculture to "smart" agriculture.

4.1 Six-Layer IoT-Based Model [4–6] (Shown in Fig. 3)

1. **Physical Layer**: It processes data at root level and transfers to upper layers. It consists of sensors (for sensing parameters in the applied environment), actuators and network devices such as routers, switches and gateways which are managed by microcontrollers present in this bottom most layer.

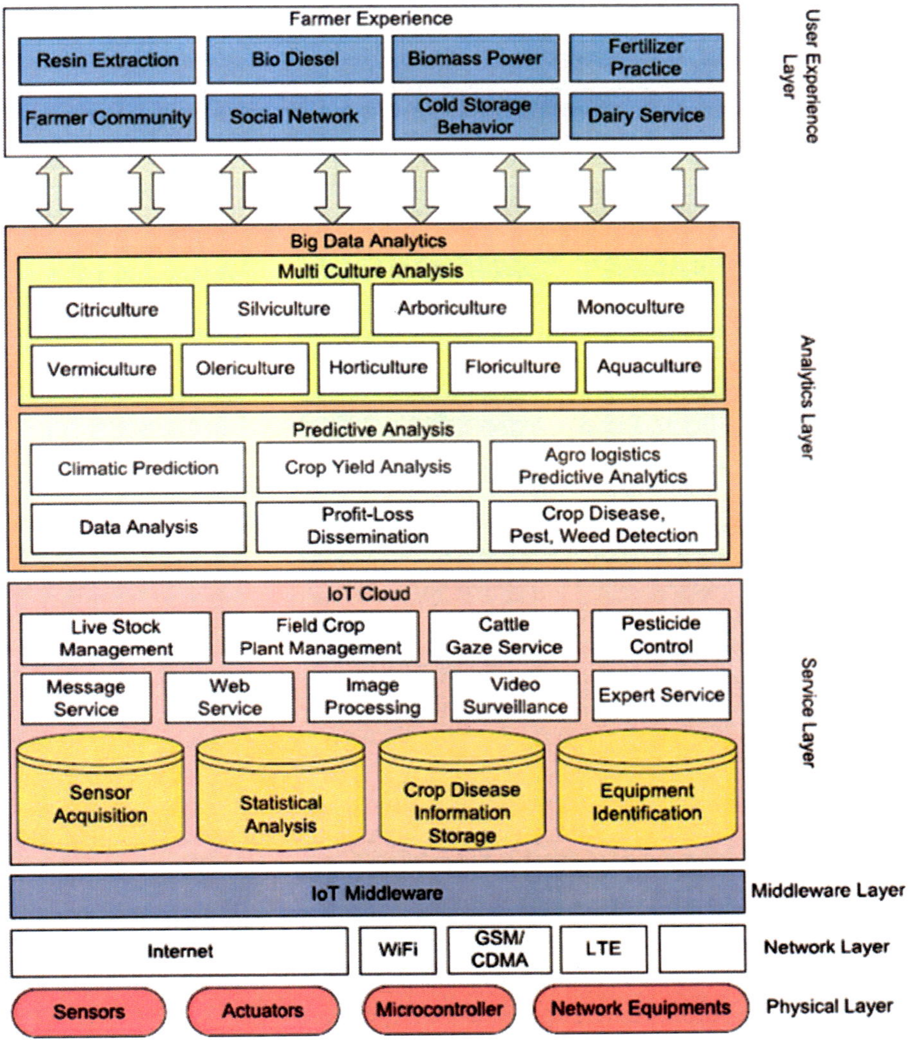

Fig. 3 IoT-based agriculture model

2. **Network Layer**: It deals with communication using Wi-Fi, long-term evolution (LTE, 4G) and relevant protocols like HTTP and simple mail transfer protocol (SMTP).
3. **Middleware Layer**: Middleware such as HYDRA, UBIWARE, SOCRADES, GSN, etc., is responsible for device management, security and interoperation.
4. **Service Layer**: Various services are provided by this layer such as automatic cattle gaze monitoring, pesticide control, insect intrusion, soil condition, etc., in a user-friendly manner by employing the cloud and software-as-a-service (SaaS).

5. **Analytics Layer**: It performs big data analytics with two subparts. First being predictive analytics, such as predicting climate condition in the future, behaviour and pattern of pest attack and weed origination. Second is multi-culture analytics which manages different forms of farming, such as predicting fish breeding in pisciculture, earthworm cultivation in vermiculture used to prepare vermicompost, forest quality in silviculture and prediction of rate of growth of fruits and vegetables in olericulture.
6. **User Experience Layer**: Interaction with farmer takes place through this layer which is the uppermost layer. It includes dissemination of agricultural knowledge via social networks, cold storage analysis, pattern of resin production from trees and dairy services like milk production, cattle disease control, etc., to increase profits.

4.2 Role of Unmanned Aerial Vehicles (UAVs)

A major hurdle when integrating IoT in the agriculture sector is the lack of reliable communication infrastructure in developing countries and rural areas to transmit the data collected using the sensors. In such a scenario, UAVs or drones become a lifesaver, as they can be used to acquire the data for further analysis by communication with the sensors that cover a large area. UAVs offer the advantage of low-cost, real-time and fast surveillance and can be fitted with different kinds of sensors. The water quantity is analyzed by employing thermal sensors as leaves that have water will appear cooler. Weed detection is possible using hyper-spectral sensors that identify the type of plant based on colour of reflected light. Moreover, the green area density can be measured by using near-infrared sensors by noting the normalized difference vegetation index (NDVI) given by Eq. (3).

$$\mathrm{NDVI} = \frac{\mathrm{NIR} - \mathrm{RED}}{\mathrm{NIR} + \mathrm{RED}} \tag{3}$$

where NIR is the near-infrared reflectance, and RED is the visible red reflectance.

5 Voting

In a democratic country, the government is chosen by the citizens through voting in an election. A leader is elected among various candidates who represents the people and works for the development of society. It is imperative to have a fair voting system, for which the most important thing is the security of vote banks. The oldest voting systems were of the form of paper ballot system which prevents hacking but does not account for multiple voting by one citizen or capturing of polling booth by muscle power. A more recent system is that of electronic voting machine (EVM) which saves a lot of trees but is vulnerable to hacking. Hence, there is a need of a more secure voting system which ensures fair elections. Here comes the IoT-based online voting framework [7] described by the author in this section. A person does not have to go to polling booths and wait for an hour to cast their votes. People can even vote from their mobile phones or computers while sitting at home.

5.1 IoT-Based Voting Framework

To cast the vote, firstly identity verification takes place with the help of Aadhar card, and once it is matched with the record already existing in the database, then the fingerprint is matched. To reduce the error in the fingerprint matching, several copies of fingerprints are taken at different time intervals of the voters. In order to ensure fair voting, the system uses the facility of voting status flag which ensures that everyone votes only once. Initially, the flag is FALSE indicating that the person has not casted his vote. Once his identity is verified and the finger print is matched, then he is allowed to cast the vote. Once his voting is done, then flag is made TRUE and his record is locked ensuring that he cannot vote again. So, whenever a person is going to vote via an online server then after all his records are matched, the voting status flag is being checked. If it is false, then he is eligible for voting otherwise his access request is denied. To reduce the traffic over the server network, there are various small local databases to maximize the performance of the system. The general architecture of the framework is shown in Fig. 4.

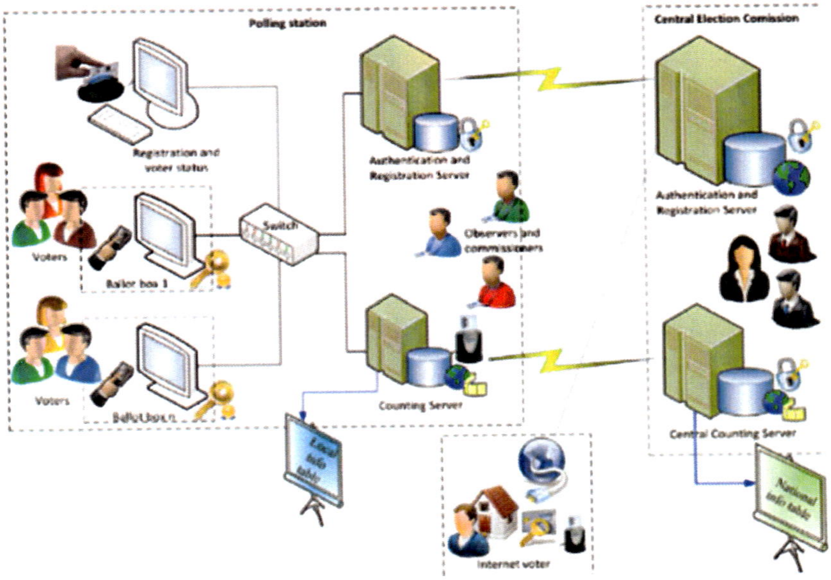

Fig. 4 IoT-based e-voting architecture

5.2 Fingerprint Matching Algorithm

```
Input: MinutiaeF1, template, MinutiaeF2, input
Output: True if matched
          False if not matched
 max_mp=0   //the number of matching points between
two Minutiae.
Make two sets A1', A2' from the endpoints of
      MinutiaeF1 and MinutiaeF2
Make triple sets A and B from A1, A2
  for each triangleA in A do
    for each triangleB in B do
  if (triangleA and triangleB are similar)
     Calculate rotation, deformation, translation
calibration
    Input Minutiae set
     Calculate matching value- val from pre matching
using similar vector triangle
  else
    continue
            end if
     if(val>max_mp) // val is the number of matched
values
     max_mp=val
   end if
   end for
 end for
 if(max_mp>=threshold) //threshold is the threshold
value for matching minutiae sets
 return true
else
return false
end if
```

5.3 Security

All the information stored in a global database or is entered by the voter at the time of voting are kept private, and messages are encrypted with the help of secure hash algorithm (SHA-1) to protect the data from network hacking and data theft. This algorithm works by transforming the data using a hash function. Every time data is entered, SHA-1 converts the data into a key which is unique for each data. Additional security is provided by storing this hash key in the database.

6 "Smart" Government?

Smart governments are essentially an extension of e-Governments, also built using IoT. The framework of a smart government is shown in Fig. 5. Smart government is of two types:

1. **Extension smart governments**, which are a combination of an e-Government and smart cities. It is admin-centric and not transparent. Smart cities are a concept which uses ICT to develop sustainable urban facilities such as electricity, housing and transportation.
2. **Next generation smart governments**, that take features from both smart cities and government 2.0. It is people-centric and transparent. Government 2.0 aims to advance transparency and efficiency by involving government, citizens and companies via integration of Web 2.0.

A major benefit of IoT is the potential it presents for converting e-Governments to smart governments. However, there are certain challenges associated with the same which need to be overcome. Some of these are presented by the author here:

1. **Mind scaping**: It is the process of getting acceptance by the government and the people. Since smart governments benefit citizens by being transparent and community engaged, convincing them is an easy task. However, the major task is of convincing the government. For example, in the Middle East, governments

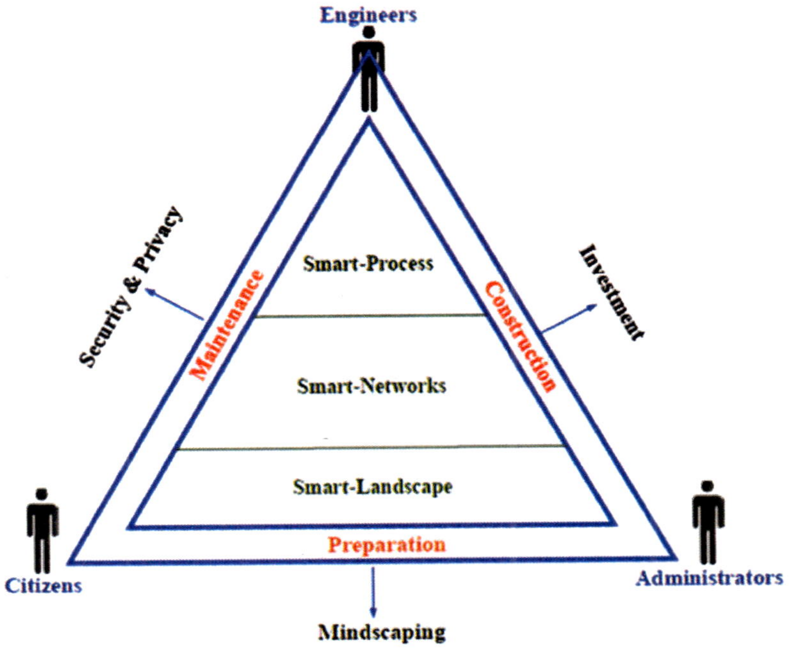

Fig. 5 Smart government framework

are highly hierarchical and centralized. They would not accept the opening of tightly closed administrative systems.

2. **Investment**: In the recent time, many governments have shifted to an e-Government. Now to move towards a smart government, huge investments are required in upgrading to fifth generation (5G) networks, latest sensors, large storage facilities, power supplies, etc.
3. **Security**: Cyber threats are a key concern due to absence of IoT standards. This was brought to attention by the largest distributed denial of service (DDoS) attack till date with a load of 1.2 terabits per second on Oracle Dyn systems by using the Mirai malware to infect a botnet consisting of digital cameras and printers. Another case is the attack on Philips light bulbs to flash an "SOS" message in Morse Code using aerial drones.

7 Future Prospects

In the preceding sections of this paper, the author has examined various IoT frameworks and applications in different areas of an e-Government. But, with such applications comes the increased use of IoT devices and with that comes a large amount of collected data via the sensors [8–16]. There is a need of high computing resources along with large storage space and on-demand and scalable services to process this data. This is where the term Cloud-of-Things (CoT) comes into picture. It couples the power of cloud computing with IoT to provide better storage and processing abilities. However, CoT creates a problem when used in an e-Government scenario. The services of an e-Government are offered to the citizens of a country, which are very large in number and growing every day. Even a small delay can cause problems in critical e-Government services such as e-Health and e-Transport. Since a large amount of data has to be transferred to the cloud, this would lead to congestion (bottlenecks) in the network. This, in addition to other problems such as high latency, security threats and high cost present us with the need of a better and improved framework for the future [17].

The solution to the above problems lies in converting a CoT architecture to a Fog-of-Things (FoT) architecture by the integration of fog computing in the existing framework, in which computation is performed at a data collection edge/source. It is implemented as a layer between the cloud and IoT devices. The FoT architecture [18] for e-Government is shown in Fig. 6.

The IoT layer consists of various sensors which sense temperature, motion, air pressure and moisture, acquire data from the citizens and is sent to the fog layer for processing which then analyses data that needs immediate attention using programs developed in Python or Java and sends the processed information to the government layer and the cloud layer for further action.

A fog node requires the functions of storage, computing and connectivity to a network. Thus, the devices, which act as a fog node, are switches, routers and proxy servers. Insights for the government will be found at the cloud layer. The government

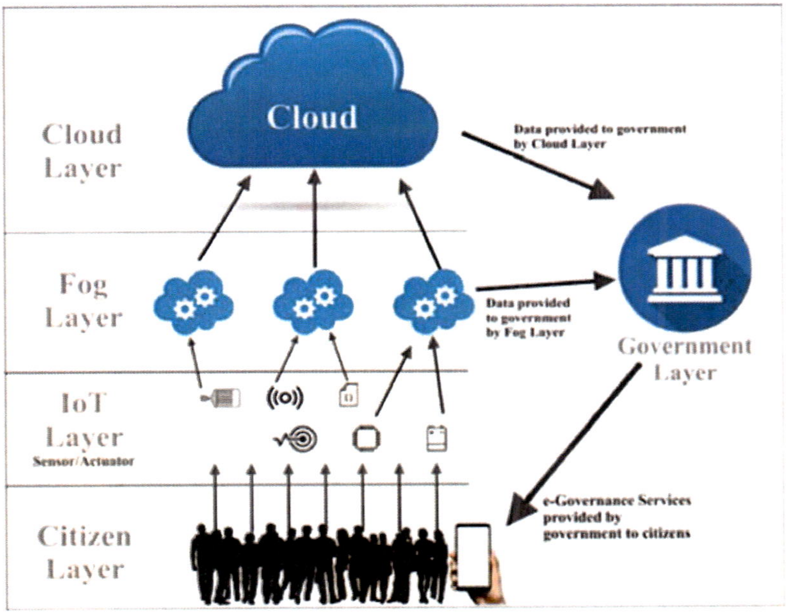

Fig. 6 Fog-of-Things architecture for e-Government

layer provides services to the citizen layer and further acts as communication medium between citizen layer, cloud layer and fog layer. Transformation to a FoT architecture ensures low latency since communication delays are very low and security, as filtering and processing of data take place at fog node and only if needed data is transferred to cloud which ensures confidentiality.

An interesting future use case of the above-FoT architecture is that of smart waste management. An IoT sensor placed in all garbage bins detects the level of garbage. As soon as it crosses a predefined threshold, the concerned authority is alerted. A clustering algorithm such as the one described by the author in Sect. 2 of this paper is used to cluster bins belonging to similar regions. The waste would then be collected following the shortest path formed on the basis of time and cost. The shortest path would be decided on the basis of the Dijkstra's algorithm, which, given any two vertices in a graph finds the length of the shortest path between them. A graph (G) is a data structure comprising of two components, i.e. nodes (vertices, V) and edges (E). On the basis of directional dependencies, a graph is classified as directed or undirected (Fig. 7).

Fig. 7 Graph

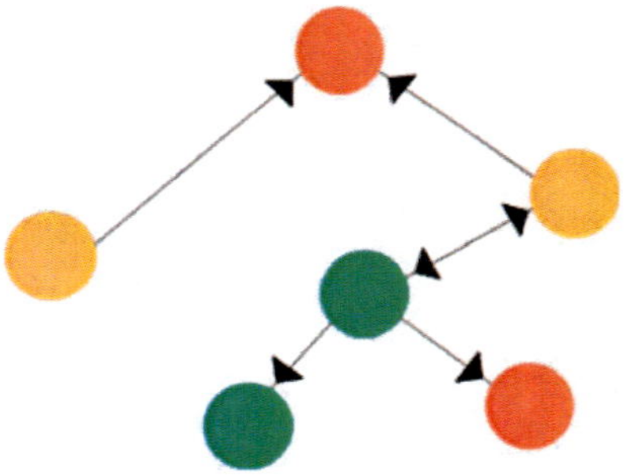

The key steps of the Dijkstra's algorithm are as follows:

1. Assume that all the nodes are unvisited and are at an infinite distance from starting node.
2. Choose the node which is closest to chosen starting node, say *node i*, and mark it as visited.
3. Update the distance of all nodes adjacent to i if the path is shorter than the present distance.

$$G = (V, E)$$

References

1. N. Alam, P. Vats, N. Kashyap, Internet of Things: a literature review, in *2017 Recent Developments in Control, Automation & Power Engineering (RDCAPE)*, Noida (2017), pp. 192–197. https://doi.org/10.1109/RDCAPE.2017.8358265
2. R. Nagarathna, R. Manoranjani, An intelligent step to effective e-Governance in India through E-Learning via social networks, in *2016 IEEE 4th International Conference on MOOCs, Innovation and Technology in Education (MITE)*, Madurai (2016), pp. 29–35. https://doi.org/10.1109/MITE.2016.017
3. J. Rodrigues, D. Segundo, H. Arantes Junqueira, M. Sabino, R. Prince, J. Al-Muhtadi, V. Albuquerque, Enabling technologies for the Internet of Health Things. IEEE Access 1 (2018). https://doi.org/10.1109/ACCESS.2017.2789329
4. P.P. Ray, Internet of things for smart agriculture: technologies, practices and future direction. J. Ambient Intell. Smart Environ. **9**, 395–420 (2017)
5. M. Bansal, Priya, Application layer protocols for Internet of Healthcare Things (IoHT), in *2020 Fourth International Conference on Inventive Systems and Control (ICISC)*, Coimbatore, India (2020), pp. 369–376. https://doi.org/10.1109/ICISC47916.2020.9171092
6. M. Bansal, Priya, Performance comparison of MQTT and CoAP protocols in different simulation environments, in *Inventive Communication and Computational Technologies*, ed. by G. Ranganathan, J. Chen, A. Rocha. Lecture Notes in Networks and Systems, vol. 145 (Springer, Singapore, pp. 549–560). https://doi.org/10.1007/978-981-15-7345-3_47

7. K. Hasta, A. Date, A. Shrivastava, P. Jhade, S.N. Shelke, Fingerprint based secured voting, in *2019 International Conference on Advances in Computing, Communication and Control (ICAC3)*, Mumbai, India (2019), pp. 1–6. https://doi.org/10.1109/ICAC347590.2019.9036777
8. A. AlEnezi, Z. AlMeraj, P. Manuel, Challenges of IoT based smart-government development, in *2018 IEEE Green Technologies Conference (GreenTech)*, Austin, TX (2018), pp. 155–160. https://doi.org/10.1109/GreenTech.2018.00036
9. S. Al-Sarawi, M. Anbar, K. Alieyan, M. Alzubaidi, Internet of Things (IoT) communication protocols: review, in *2017 8th International Conference on Information Technology (ICIT)*, Amman (2017), pp. 685–690. https://doi.org/10.1109/ICITECH.2017.8079928
10. D. Dragomir, L. Gheorghe, S. Costea, A. Radovici, A survey on secure communication protocols for IoT systems, in *2016 International Workshop on Secure Internet of Things (SIoT)*, Heraklion (2016), pp. 47–62. https://doi.org/10.1109/SIoT.2016.012
11. P. Brous, M. Janssen, Advancing e-Government using the Internet of Things: a systematic review of benefits, in *Electronic Government. EGOV 2015*, ed. by E. Tambouris et al. Lecture Notes in Computer Science, vol. 9248 (Springer, Cham, 2015). https://doi.org/10.1007/978-3-319-22479-4_12
12. Z. Engin, P. Treleaven, Algorithmic government: automating public services and supporting civil servants in using data science technologies. Comput. J. **62**(3), 448–460 (2019). https://doi.org/10.1093/comjnl/bxy082
13. S.M.R. Islam, D. Kwak, M.H. Kabir, M. Hossain, K. Kwak, The Internet of Things for health care: a comprehensive survey. IEEE Access **3**, 678–708 (2015). https://doi.org/10.1109/ACCESS.2015.2437951
14. S.B. Baker, W. Xiang, I. Atkinson, Internet of Things for smart healthcare: technologies, challenges, and opportunities. IEEE Access **5**, 26521–26544 (2017). https://doi.org/10.1109/ACCESS.2017.2775180
15. P.V. Garach, R. Thakkar, A survey on FOG computing for smart waste management system, in *2017 International Conference on Intelligent Communication and Computational Techniques (ICCT)*, Jaipur (2017), pp. 272–278. https://doi.org/10.1109/INTELCCT.2017.8324058
16. M. Ayaz, M. Ammad-Uddin, Z. Sharif, A. Mansour, E.M. Aggoune, Internet-of-Things (IoT)-based smart agriculture: toward making the fields talk. IEEE Access **7**, 129551–129583 (2019). https://doi.org/10.1109/ACCESS.2019.2932609
17. M. Bansal, Priya, in *Machine Learning Perspective in VLSI Computer Aided Design* at Different Abstraction Levels, by S. Shakya et al. (eds.), Mobile Computing and Sustainable Informatics, Lecture Notes on Data Engineering and Communications Technologies **68**. https://doi.org/10.1007/978-981-16-1866-6_6
18. N. Tewari, G. Datt, Towards FoT (Fog-of-Things) enabled architecture in governance: transforming e-Governance to smart governance, in *2020 International Conference on Intelligent Engineering and Management (ICIEM)*, London, United Kingdom (2020), pp. 223–227. https://doi.org/10.1109/ICIEM48762.2020.9160037

A New Network Forensic Investigation Process Model

Rachana Yogesh Patil and Manjiri Arun Ranjanikar

Abstract The procedure of the identification and analysis of electronic data is the digital forensics framework. The purpose of the procedure is to retain evidence in its primordial form by gathering, identifying and validating digital information in order to recreate events of the past. For the use of data in a court of law, the context is most appropriate. The evidence aspect of digital forensics requires strict requirements to be followed in court for cross-examination. One major pitfall in the digital forensic analysis is the possible admissibility of collected evidence in the court of law. Digital forensic analysis must comply with the quality of evidence and its admissibility to trial successfully. In this work, we proposed a new network forensic investigation process model. We precisely establish a methodology of investigation for the computer network.

Keywords Digital forensics · Cybercrime · Network forensics · Process model

1 Introduction

Forensics is the method by which evidences are gathered, analyzed and presented with scientific expertise. It focuses on the recovery and examination of latent evidence (forensics means to be placed before the court). This is a reference to latent evidence. There can be several latent facts, from fingerprints left to proof of the DNA retrieved from stains of blood and files on the hard drive. There is no standardization and continuity in the courts and industry since computer forensics is a modern specialty. As a result, as a formal "scientific" discipline, it is not yet recognized [1].

R. Y. Patil (✉) · M. A. Ranjanikar
Pimpri Chinchwad College of Engineering, Savitribai Phule Pune University, Nigdi, Pune, Maharashtra, India

S. Shakya et al. (eds.), *Mobile Computing and Sustainable Informatics*, Lecture Notes on Data Engineering and Communications Technologies 68,
https://doi.org/10.1007/978-981-16-1866-6_9

1.1 Forensic Science

Forensic science is the application of science to criminal and civil laws as governed by the legal standards of admissible evidence and criminal proceedings [2], mainly on the criminal side during criminal investigation. It is the art of collecting evidence for reconstructing the activities of the criminal during or after a crime and providing evidence for prosecution. There are many branches of forensic science; one of them is digital forensics.

1.2 Network Forensics

Network forensics is defined in [3] as "Use of scientifically proven techniques for collecting, combining, recognizing, examining, correlating, analyzing and recording digital evidence from numerous, actively processing and transmitting digital sources for the purpose of uncovering facts associated with the planned objective or measuring the success of unauthorized activities specifically aimed at disrupting, corrupting and/or compromising system components as well as pro-active systems."

1.3 Need of Network Forensics

In today's cyber-world, with increased reliance on computer networks for information and resource sharing, network security has also been a widespread concern for everyone. Three basic principles, confidentiality, integrity and availability, define network security goals. A number of techniques have been developed and applied in security operations to ensure that network systems are safeguarded from threats in all three aspects.

Organizations rely more than ever on their networks, so monitoring and managing these networks are a mission-critical task. But it has become extremely difficult for several reasons to monitor and maintain networks [4].

- The increase in the cyber-security incidents affects variety of institutions and individuals, as well as the increase in cyber-attack sophistication.
- The attacker covers the traces that were used to attempt the attack, making trace back more critical.
- Network security defensive approaches such as firewalls and IDS can only report attacks from the perspective of prevention, detection and reaction.

The alternative approach to network forensics is critical as it has the characteristics of the investigation [5]. Network forensics means that the intruder spends more time and resources shielding his tracks, making the attack expensive.

2 Literature Survey

For forensic investigations, a vast range of digital forensic models have been developed since 2001. Some of these concentrate on either incident response or investigation or highlight a specific investigation process or operation.

Digital forensic investigation model is proposed in [6]. This model has three stages, namely the collection of evidence, the authentication of the evidence and the review of the evidence. This model is concerned with the credibility of the facts and has been developed to adapt to incidents. DFRWS investigative model [7] proposed a framework which consists of the following steps: identification, conservation, collecting, evaluating, examining, presenting and determining. The above framework is improvised, and abstract digital forensics model [8] is proposed containing the stages of planning and approach plan and including the return of facts in order of judgment. Integrated digital investigation process model [9] applied and integrated the usual conventional investigative approach into digital forensic investigation. This was very revolutionary, particularly in the reconstruction phase of scenes of electronics and physical crimes, which is a technique used to identify cyber-criminals. The improved process model for digital investigation [10] is proposed by introducing two new steps to develop the integrated automated investigative process model: trace back and dynamite, which would allow the investigator, from the footprint collected from the secondary crime scene, to locate the primary crime scene for the sole purpose of locating the potential suspect, in the previous iteration, which was a limitation.

The enhanced model for the investigation of cybercrime [11] is proposed which consists of thirteen phases, namely awareness, authorization, preparation, notification, search and detection of evidence, collection of evidence, transport of evidence, proof, evidence storage, evidence analysis, theory, presentation of evidence hypothesis, hypothesis proof/defense and archive storage (used to disseminate about information). A clearer grasp of the investigation is given by the model process and collects much of the knowledge flow for investigating cybercrime. A harmonized process model proposed in [12] is used for readiness of digital forensic testing. This model is holistic in nature and takes due account of readiness, study and the interface between the two forms of activities. Integrated forensic digital process model [13] proposes a structured forensic digital procedure model that allows researchers to adopt a standardized approach to the digital forensic study. A new concept of smart city autonomous vehicle's digital forensics [14] is proposed for smart city automated vehicles which works on investigating autonomous automated vehicle (AAV) cases. The behavioral digital forensics investigation model [15] proposed and evaluated, and it incorporates current best practice with BEA methods for professional analysis of the digital evidence relating to a case. A consumer-oriented cloud forensic process model [16] that directs a customer before during and after an incident is of crucial importance.

The comparative analysis of all the above digital forensic models along with the proposed model is shown in Table 1.

Table 1 Comparative analysis of digital forensic models along with the proposed model

Phases	Models									
	Kruse [6]	Palmer [7]	Saleem [8]	Carrier [9]	Baryamureeba [10]	Ciardhuáin [11]	Valjarevic [12]	Kohn [13]	Al Mutawa [15]	Proposed network forensic investigation process model
Collection	✔	✔	✔	✔	✔	✔	✔	✔	✔	✔
Authentication	✔					✔			✔	✔
Examination	✔	✔	✔	✔	✔	✔	✔	✔	✔	✔
Identification		✔	✔	✔	✔	✔	✔	✔	✔	✔
Analysis		✔	✔	✔	✔	✔	✔		✔	✔
Preservation		✔	✔	✔	✔	✔	✔	✔	✔	✔
Evaluation		✔		✔				✔	✔	✔
Presentation		✔		✔		✔				✔
Determining		✔				✔		✔	✔	✔
Protection		✔	✔					✔	✔	✔
Reporting			✔		✔	✔				✔
Preparation and planning			✔			✔	✔		✔	✔
Archiving and returning			✔			✔				✔
Trace back			✔		✔					✔
Dynamite			✔		✔					

3 Proposed Network Forensic Investigation Process Model

The block diagram shown in Fig. 1 represents the broad level of information about the different components involved in network forensics investigation process.

3.1 Network Forensic Readiness Module

Network forensics only applies to situations where network security technologies are installed on the network at various strategic points, such as IDS, packet analyzers, firewalls and traffic flow measurement applications. The necessary permissions and

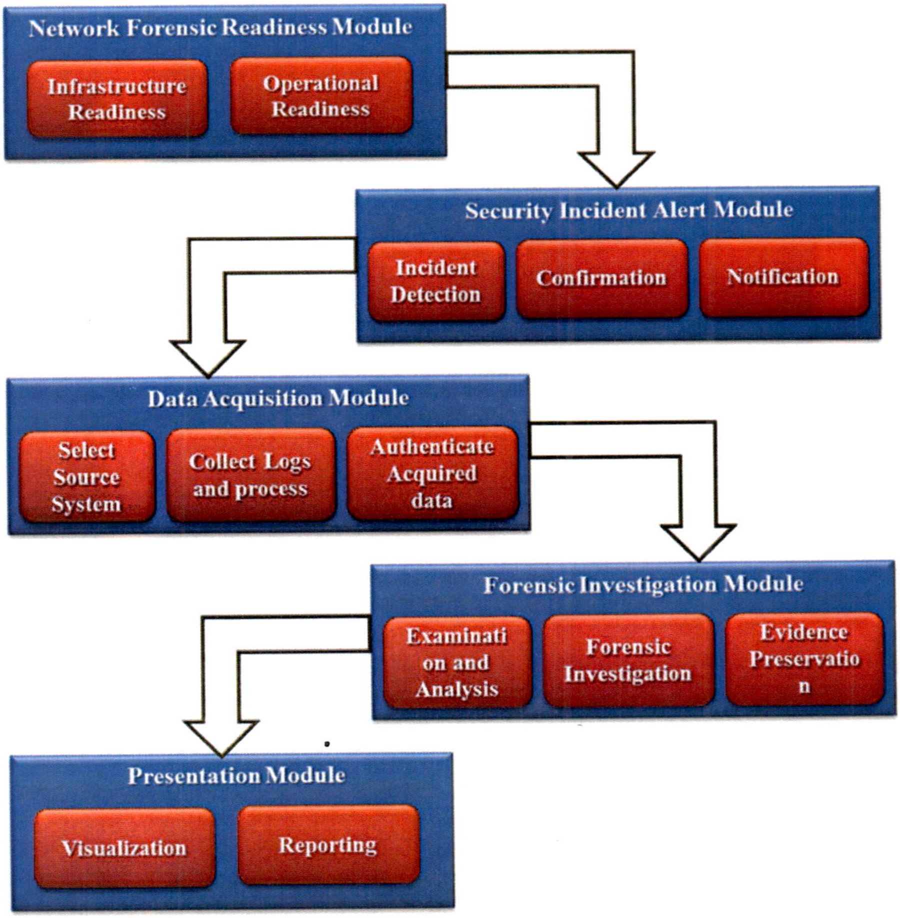

Fig. 1 Proposed network forensic investigation process model

legal warrants are obtained in order not to violate privacy. Forensic readiness of the network addresses the notion of how to maximize the effectiveness of evidence at a minimum operating cost.

The primary intention of infrastructure readiness is to enable the institution to effectively investigate various types of incidents. This will mean that the organization has the right tools and technology to conduct a digital forensic investigation effectively [17].

3.2 Security Incident Alert Module

In order to trigger the rapid response mechanisms, specialized and accurate intrusion alert systems are required for network forensics. This module's main objective is to observe alerts produced by different security mechanisms, describing a breach of security or defilement of policy. It will analyze unauthorized events and noticed anomalies. The attack's presence and nature are determined by different parameters.

3.3 Data Acquisition Module

Sensors used to accumulate traffic data should be used to acquire data. There must be a precise technique that uses consistent hardware and software tools to gather maximum evidence that causes the victim to have minimal impact. This stage is very important as network traffic information changes at a fast-paced speed, and the same trace cannot be generated at a future time. The quantity of data recorded will be extremely large. It takes enormous memory space, and the device requires to be skillful to handle various log data formats properly. The data is collected in the form of logs from different sources for network forensic.

Three types of logs are of interest for network forensics.

- Logs from security software
- Logs from OS
- Logs from various applications
- Network devices logs.

3.4 Forensic Investigation Module

The outcome of this module determines the success of the findings of the investigator that will ultimately be brought before the court of law. The indicators received from the previous phase are classified and correlated using the existing patterns of attack to deduce important observations. Intelligent methods are used to investigate for patterns of data and match attack. The primary aim of this process is to define the route back to the point of origination of the attack from the network or device of a victim through any intermediate networks and communication routes. For the attribution of the attack, the packet captures, and statistics obtained are used. Attribution establishes the attacker's identity and is the hardest part of the forensic network process. The attacker's two simple hiding approaches, IP spoofing and stepping-stone attack, are major issues. The phase of the forensic investigation provides data for the attacker to be prosecuted.

3.5 *Presentation Module*

The observations are presented to legal personnel in an understandable language while explaining the different procedures used to reach the conclusion. Also included is the systematic documentation to fulfill the legal requirements. The findings are often illustrated by visualization, so it can be quickly grasped.

4 Conclusion

The network forensics ensures that the attack is investigated by tracking the attack up to the source and attributing the crime to an attacker, a host or a network. It can forecast potential attacks by creating attack patterns based on current data traces. Authentic proof is often facilitated and can be accepted into a legal framework. To improvise the process of network forensics, we have developed a new network forensic investigation process model which consists of five important phases, namely forensic readiness, incident alert, data acquisition, forensic investigation and reporting. In all instances, it is now important to assess the feasibility and implementation of the model.

References

1. R.Y. Patil, S.R. Devane, Network forensic investigation protocol to identify true origin of cyber crime. J. King Saud Univ. Comput. Inf. Sci. (2019)
2. O. Singh, Network forensics blog (2012)
3. S. Raghavan, Digital forensic research: current state of the art. CSI Trans. ICT **1**(1), 91–114 (2013). F. Author, Article title. Journal **2**(5), 99–110 (2016)
4. R.Y. Patil, S.R. Devane, Unmasking of source identity, a step beyond in cyber forensic, in *Proceedings of the 10th International Conference on Security of Information and Networks*, Oct 2017, pp. 157–164
5. P.R. Yogesh, S.R. Devane, Primordial fingerprinting techniques from the perspective of digital forensic requirements, in *2018 9th International Conference on Computing, Communication and Networking Technologies (ICCCNT)*, July 2018 (IEEE, 2018), pp. 1–6
6. W.J. Kruse, G. Heiser, *Computer Forensics: Incident Response Essentials* (Addison-Wesley, 2002). ISBN 0-201-70719-5

7. G. Palmer, A road map for digital forensic research. Technical report DTR-T001-01, DFRW. Report from the First Digital Forensic Research Workshop, Utica, NY (2001)
8. S. Saleem, O. Popov, I. Bagilli, Extended abstract digital forensics model with preservation and protection as umbrella principles. Procedia Comput. Sci. **35**, 812–821 (2014)
9. B. Carrier, E.H. Spafford, Getting physical with the investigative process. Int. J. Digit. Evid. **2**(2) (Fall) (2003)
10. V. Baryamureeba, F. Tushabe, Enhanced digital investigation process model, in *Digital Forensic Research Workshop*, Baltimore, MD, USA (2004)
11. S.O. Ciardhuáin, An extended model of cybercrime investigations. Int. J. Digit. Evid. **3**(1) (Summer) (2004)
12. A. Valjarevic, H. Venter, A harmonized process model for digital forensic investigation readiness, in *IFIP International Conference on Digital Forensics*, Jan 2013 (Springer, Berlin, Heidelberg, 2013), pp. 67–82
13. M.D. Kohn, M.M. Eloff, J.H. Eloff, Integrated digital forensic process model. Comput. Secur. **38**, 103–115 (2013)
14. X. Feng, E.S. Dawam, S. Amin, A new digital forensics model of smart city automated vehicles, in *2017 IEEE International Conference on Internet of Things (iThings) and IEEE Green Computing and Communications (GreenCom) and IEEE Cyber, Physical and Social Computing (CPSCom) and IEEE Smart Data (SmartData)*, June 2017 (IEEE, 2017), pp. 274–279
15. N. Al Mutawa, J. Bryce, V.N.L. Franqueira, A. Marrington, J.C. Read, Behavioural digital forensics model: embedding behavioural evidence analysis into the investigation of digital crimes. Digit. Investig. (2019)
16. A.N. Moussa, N. Ithnin, N. Almolhis, A. Zainal, A consumer-oriented cloud forensic process model, in *2019 IEEE 10th Control and System Graduate Research Colloquium (ICSGRC)*, Aug 2019 (IEEE, 2019), pp. 219–224
17. P.R. Yogesh, Backtracking tool root-tracker to identify true source of cyber crime. Procedia Comput. Sci. **171**, 1120–1128 (2020)

MDTA: A New Approach of Supervised Machine Learning for Android Malware Detection and Threat Attribution Using Behavioral Reports

Seema Sachin Vanjire and M. Lakshmi

Abstract Android is liable to malware attacks because of its open architecture, massive user base, and easy access to its code. The security investigation depends upon the dynamic analysis for malware detection. In this system, digital samples or system calls were analyzed, and the malicious application created a runtime behavioral profile. The resulting system is further used to detect malware and attribute threat, with selected features analysis. But due to a variety of malware families and execution environments, it is not scalable. Because for every new execution environment, the new feature needs to be engineered manually. MDTA is a portable malware detection framework system. They are also used for detecting different threat acknowledgment using supervised machine learning techniques. MDTA is the best suitable and manageable approach for analyzing behavioral reports using a machine learning algorithm for providing security measures to identify malware without the intervention of the investigator. Additionally, natural language processing (NLP) is used to represent the behavioral report. MDTA is then evaluated on different datasets from diverse platforms and execution environment.

Keywords Android security · Android malware detection · Supervised machine learning

1 Introduction

Malware is a malicious software program with the purpose to wreck computers and servers. Various tools are available for malware detection like static analysis and dynamic analysis of binary code. The investigator or supervisor has to deal with different malicious applications that are intended for different platforms [1–3].

S. S. Vanjire (✉)
Sathyabama Institute of Science and Technology, Jeppiaar Nagar, Rajiv Gandhi Salai, Chennai, India

M. Lakshmi
SRM Institute of Science and Technology, SRM Nagar, Kattangulathur, India
e-mail: laks@icadsindia.com

S. Shakya et al. (eds.), *Mobile Computing and Sustainable Informatics*, Lecture Notes on Data Engineering and Communications Technologies 68,
https://doi.org/10.1007/978-981-16-1866-6_10

The obfuscation of the malware and the payloads created for legitimate uses is even the more challenging part, making it difficult to detect these applications [4–6]. The primary motive is to manipulate the malicious apps' overall behavior at runtime and to mark them as malicious or benign. MDTA will be a portable detection framework for cross-platform applications using supervised machine learning techniques. The system covers simple, avoiding techniques for malware and threats. On the contrary, the dynamic analysis solutions demonstrate more robustness against the strategies of evading, such as clouding and padding [7, 8]. For this reason, dynamic analysis is selected to be the default choice for malware analysis in the system. Static and behavior analysis are sources for security measures used to determine the maleficence of a given system. Here, the emphasis is on dynamic analysis using supervised machine learning techniques to examine the malicious application. The behavioral reports are plugged in, and the activity establishes the classifiers and trains the model to recognize any new challenges and attributes to the unit [9, 10].

The layout of this article is organized as follows. Section 2 describes the literature survey. Section 3 explains the system methodology. Section 4 depicts the system design. Section 5 represents the system algorithm, and Sect. 6 defines the system implementation. Section 7 depicts the result analysis. Section 8 outlines the experimentation, and Sect. 9 presents the discussion. Finally, Sect. 10 concludes the research work.

2 Literature Survey

Sgandurra et al. provide a dynamic analysis approach for classifying ransomware using machine learning [11]. Severi et al. introduce malware sandbox system for tracing malware and execute them [12]. Chen et al. propose a framework to determine and analyze for dynamic API calls of Android using semi-supervised machine learning [13]. Alzaylaee et al. depict that DynaLog is used to work on real malware to detect them from malicious applications [14]. Karbab and Debbabi present an automatic investigation framework to analyze and identify the Android malware [15]. Paul and Filiol contribute a concept of a glass box to dynamically identify malicious applications [16].

3 System Methodology

The aim of the framework is to recognize the malicious application that can come from various platforms such as windows and the Android, which can have diversity. Also, the framework is developed for analysts to perform analysis of diverse, obfuscated applications smoothly and efficiently. In the proposed system, the security investigators rely on dynamic analysis for malware detection in which the digital samples are executed, and reports have been generated based on their runtime behavior. The security investigators use the reports to detect the malware with the manually chosen features. However, the variety of malware and the environment does not use a manual approach. The investigator needs to manually analyze features for new environments. Therefore, this model proposes a portable MDTA that is a new system framework based on supervised machine learning techniques.

4 System Design

In this system context, the user is the security analyst. Figure 1 shows the system architecture to have the smooth working of the system. The user will be analyzing the application to check whether it is malicious or benign. The user will be provided with an interface to collect a behavioral report of an application in a separate environment, so that possibility of infecting real machines is eliminated. The behavioral report is then fed into the feature extraction algorithms by the user. After that, from extracting features, feeding them to malware detection and algorithms of threat attribution, the automated framework can take over. The results of the study are then conveyed in a basic format to the user, which can then be extended to display the results in a different format as well.

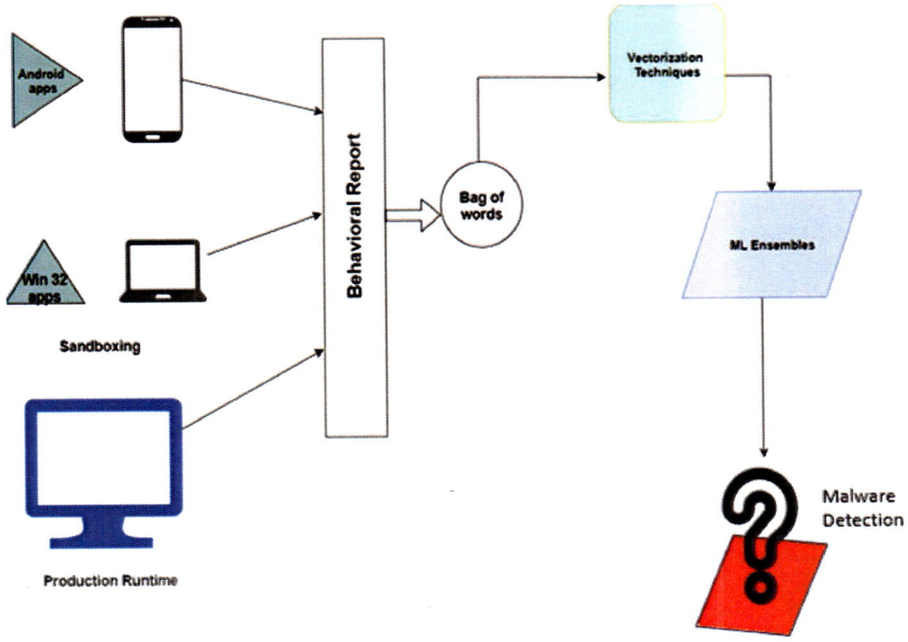

Fig. 1 System architecture

5 System Algorithm

For this system, three algorithms are applied to obtain the final result for malware detection. These algorithms are feature extraction algorithm, build model algorithm, and prediction algorithm. The working of each algorithm is devised as follows.

5.1 Build Model Algorithm

This build model algorithm is used as input to this model. Here, consider the dataset as x-build, and divide it into x-train and x-valid. Here, x-valid should be less as to x-train. Build model algorithm applied on x-build dataset where all machine learning algorithm iteratively called to check threshhold value. The param is the set of data that are unique for all the algorithms. Next, train the model on the x-train and param dataset. To check if the algorithm performs well, use the x-valid set to check the score and threshold. If the score is right, then this generates a tuple that denotes that the model is suitable to be used.

5.2 Feature Extraction Algorithm

In this feature extraction algorithm, the behavior report gathered from the sandbox is the input. The produced performance will be the behavioral report of victories. Use n-gram generators to produce the collective sense of the words. Use the TF-IDF then to extract the features from generated reports which are widely used. To perform analysis on them, algorithms are then applied to the output.

5.3 Prediction Algorithm

This prediction algorithm is used in the system for predicting malware by checking its resultant value from the report. The trained model is then used to classify the application as malicious, and input will be the report of the file. Whether the application is malicious and which family it belongs to will result in the output. The article will again be about achievements. The record of successes is given to the device model. If the value of the detection result is less than zero, then assume that malicious is the application. Then finally check whether this malicious application matches with which malware families and return with that family to which it belongs.

6 System Implementation

The system is designed as per a security analyst's point of view. The first page implements a login implementation where analysts have to log in using the hardwired credentials. At successful login, the application provides a page for a selection of the different machine learning analysis. The analysts can use provided features for training and self-tailoring functionality for different environments. The application implements various machine learning-based algorithms for training and malware detection. The model is trained on the broad set of malicious data provided by online analysis firms. The data used to train the model is based on an analysis of Android and window-based applications. In this system, a total of 13 k sets are used to train the system model for the specific functionality of detecting and attributing family to the application under test. The statistical analytics of this system gives an idea about the efficiency of training with the assistance of pictorial representation such as histogram or heat map. Thus, the analysts can conduct reverse engineering with a considerable amount of applications from either platform windows or Android. They test their findings on online sites to verify the integrity of their results. To affirm the integrity of their results, the online platforms are utilized. For this system, the developer has taken the support of graphical structures for implementation. The graphical structures are data flow diagrams and sequence diagrams. These diagrams are devised below with an explanation.

6.1 Data Flow Diagram

For this system, two different data flow diagrams are used while implementing this system such as DFD level 0 and DFD level 1. Both diagrams are enlightened below.

6.1.1 DFD Level 0

Figure 2 shows data flow diagram level 0. Here, the user (analysts) provides the behavioral report gathered from the framework using a GUI in the raw format. The application implements machine learning (ML) algorithms for feature extraction and the prediction of the maliciousness of the application, respectively. The output will

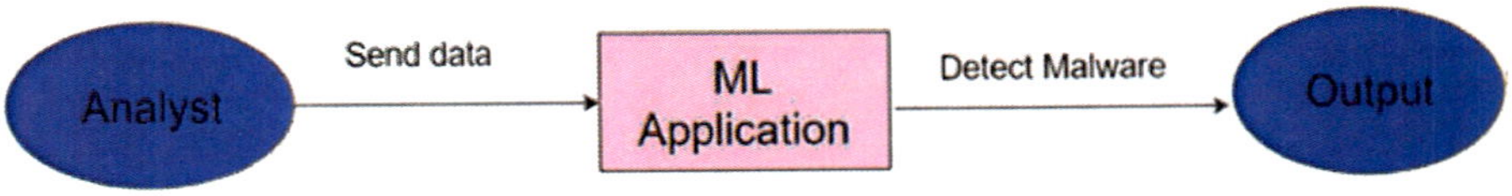

Fig. 2 DFD level 0

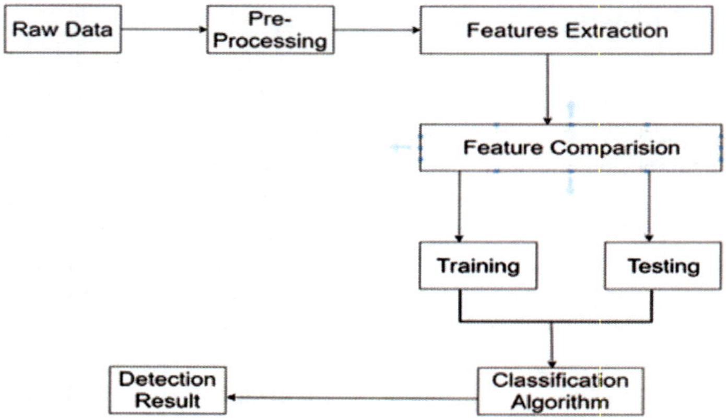

Fig. 3 DFD level 1

be whether the application is malicious or benign and if it is malicious, then to which family it belongs.

6.1.2 DFD Level 1

Figure 3 shows data flow diagram level 1. Here, the user selects the application to be analyzed. Then, the user gives the application to the sandbox environment where the application is executed, and the behavioral report is collected. The user gives the behavioral report of the application after analysis of the application's behavioral study. The report, which is in the raw format, is then victories or features are extracted from the raw report so that the data can be in a suitable format to be used by Machine learning. The data then passes as input to prediction algorithms and algorithms for threat attribution in the required format, providing the output as to whether the applications are malicious or benign.

6.2 Sequence Diagram

Figure 4 is a sequence diagram of the current system. Initially, all the features of the Android or windows systems are collected. Then, the system will generate the behavior report using sandbox in runtime. After this transfer, all collected data are used for report modeling and successes. When this is done, data is transferred to detect malware or maliciousness. The malicious data will be evaluated after this device and the decision to identify malware or threats detection.

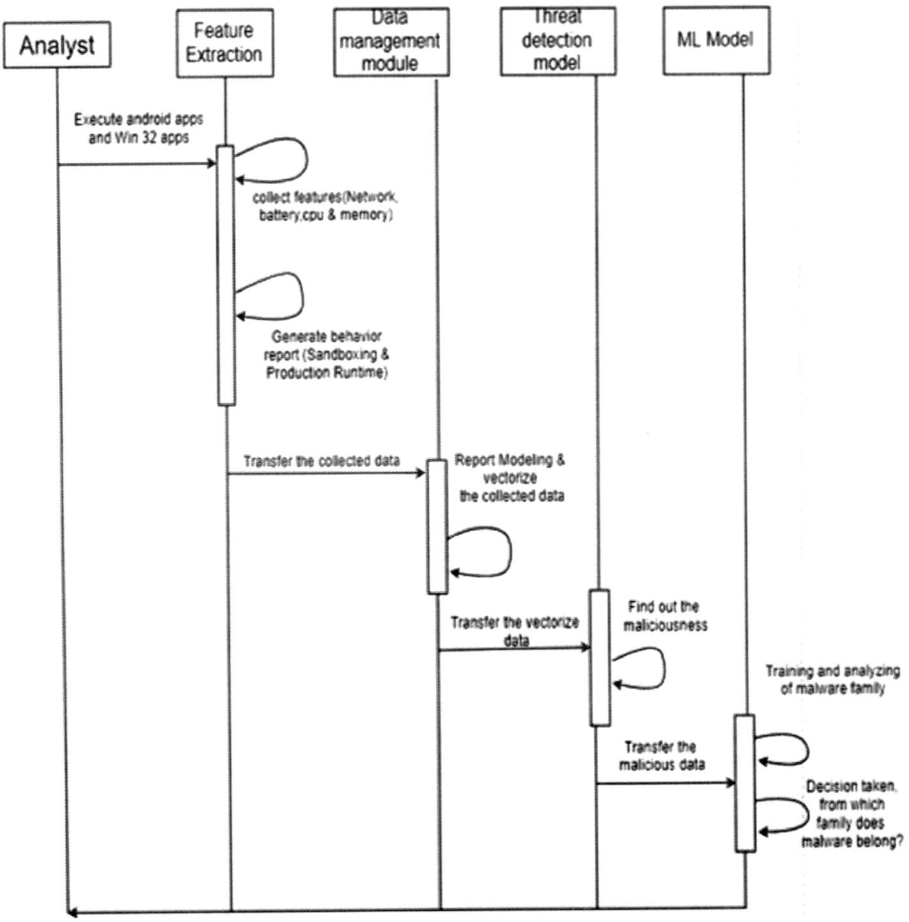

Fig. 4 Sequence diagram

7 Result Analysis

The system is analyzed for two resultant values, one is the accuracy of malware detection, and another is the accuracy of threat attribution. Malware detection accuracy is the percentage to achieve and obtain the percentage value of true positive malware from total malware. Similarly, threat attribution accuracy is to find true threats across all collected threats. It is achieved with given formula as

$$\text{Accuracy} = \text{TP} + \text{TN}/\text{TM} + \text{TB} \times 100 \tag{1}$$

where

TP = True Positive Malwares/Threats
TN = True Negative Malwares/Threats
TM = Total Malwares/Threats
TB = Total Benign.

7.1 Accuracy of Malware Detection

The first accuracy of different supervised machine learning algorithms for malware detection is captured and analyzed. Table 1 shows the accuracy captured for different machine learning algorithms for malware detection.

Table 1 Accuracy of different supervised machine learning algorithms for malware detection

Name of supervised machine learning algorithm	Accuracy of algorithm for malware detection
Decision tree algorithm	88.15
Random forest algorithm	89.44
AdaBoost algorithm	88.67
Gradient boost algorithm	88.85
Gaussian NB algorithm	70.36
Linear regression algorithm	53.93

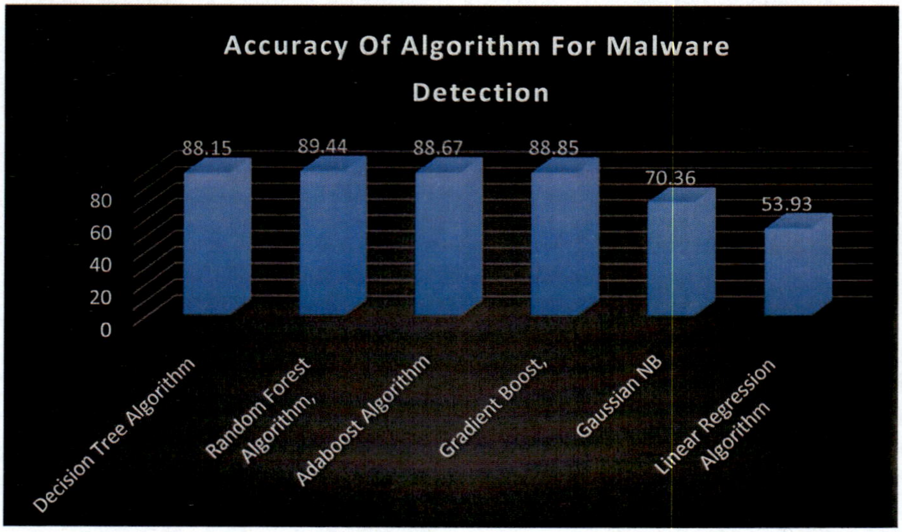

Fig. 5 Accuracy of supervised machine learning algorithms for malware detection

Table 2 Threat attribution accuracy

Threat attribution	Threat attribution accuracy
Threat attribution 1	63.31
Threat attribution 2	66.56
Threat attribution 3	65.15
Threat attribution 4	64.88

Figure 5 shows that various supervised machine learning algorithms are applied for malware detection and threat attribution. Here, decision tree algorithm, random forest algorithm, AdaBoost algorithm, gradient boost algorithm, Gaussian NB algorithm, and linear regression algorithm are applied for malware detection and threat attribution. Figure 5 used various algorithms for building the system model to predict the malware accurately with the highest efficiency. Despite multiple tests performed on the algorithms, resultant observation concluded that the random forest works best always and came on top with the highest accuracy score of about 89–93% and linear regression (LR) developed with the accuracy of only about 50–55%.

7.2 *Accuracy of Threat Attribution*

The accuracy of threat attribution is captured and analyzed. Table 2 shows the accuracy captured for threat attribution.

Here, Fig. 6 shows threat attribution accuracy analysis for different threat attribution performed at a different level. Here, the system uses the API calls data provided

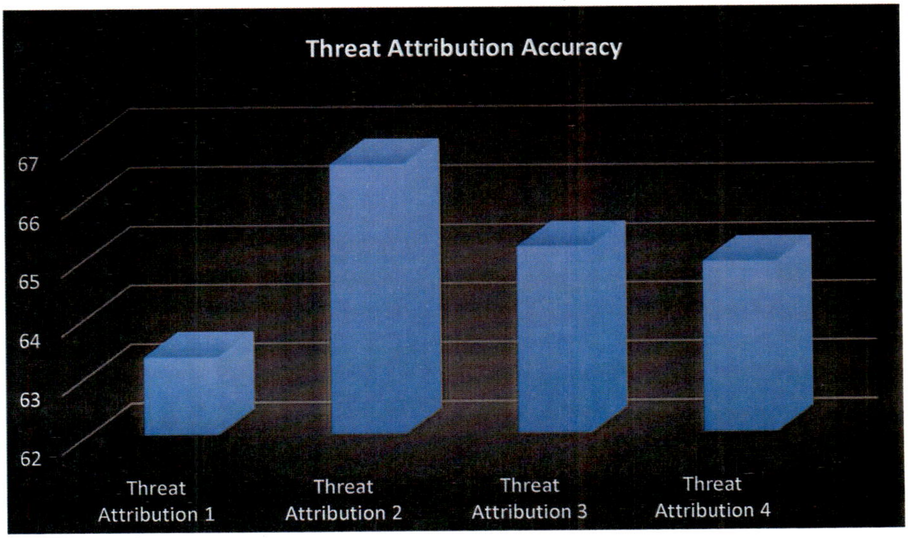

Fig. 6 Threat attribution accuracy analysis

by them at malware for API class accessed from GitHub Web portal. It is a vast dataset of API calls around 2.3 GB of data used in the system for family threat attribution. The samples provided along with the number of samples are as follows. Spyware (832), Backdoor (1001), Virus (1001), Dropper (891), Adware (379), Worms (1001), Downloader (1001), and Trojan (1001). Thus, these are managed to achieve the threat attribution accuracy of 63–67%.

8 Experimentation

The system built was with four major modules such as behavior extraction, behavioral module, static module, and threat classification. The modules are summarized below:

(a) Behavior extraction: In this system module, cuckoo sandbox and Python library are used for the extraction of the behavioral reports and the static information as PE header. Further, the use of the API call is provided for behavioral report extraction at the runtime, and the result is forwarded to the virus detection and classification.
(b) Behavioral module: This module is responsible for the classification of the file based on its behavior at the runtime. The underlying syscalls, memory read/write, and network activity serve best for analyst purpose.
(c) Static module: This module uses a PE header file to extract the header information from the file, and the classification is performed based on the signature of the field.
(d) Threat classification: In this module, exploit all the API calls. The API calls are trained on this module and further extract the API information from the file, and then, the threat is attributed to any given malware environment.

All these above modules are shown in the following screenshot of the system working. It obviously shows static analysis, behavioral analysis, hash analysis, and sandbox analysis. Later, this gives the total count of virus or malware detected in each analysis (Fig. 7).

When malware is found in the Android system or on the Windows system for any file or program, then the current system will retain the model by producing a report. This report is later evaluated to get comprehensive details about malware. Figure 8 illustrates the retraining model's report review. This retraining model report provides details about the accessed EXE file that may or may not contain any benign files.

9 Discussion

Drebin dataset is accessed and used for a given device location to obtain precise results for malware detection and attribution of threats. The Drebin dataset is used as a sandbox for malware identification and attribution of mobile threats to mobile

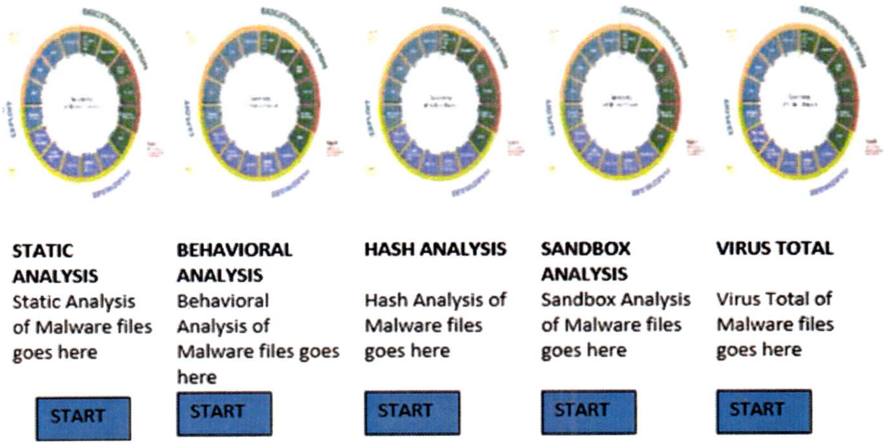

Fig. 7 System malware detection and threat attribution

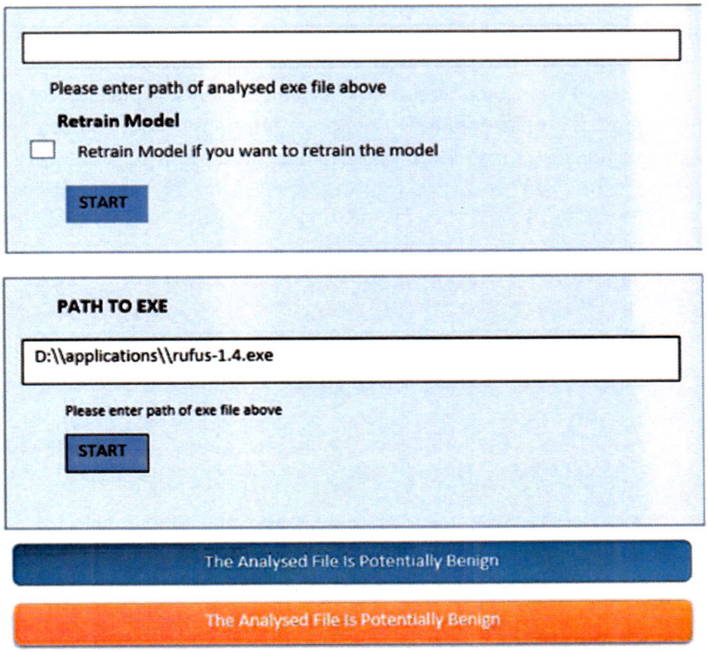

Fig. 8 Retrain model report analysis

devices. Future work for given system is to apply AI (Artificial Intelligence) and deep learning approach for malware detection and threat attribution. This system's dataset could be limited to function in future.

10 Conclusion

Daily users targeting the virtual world of the Internet are growing exponentially, which contributes to security measures. Through exploring new dimensions, the diversity of targeted cyberspace mitigates the problem. Analysis of action is an investigative method for evaluating binary samples and generating reports of actions. A compact and efficient solution is obtained for the retention in this task. The key concept is to model the behavior report to create discriminatory machine learning assemblies, using advanced machine learning techniques. The malware detection task scores above 94% on MALDYOME, Drebin, and MALDOZER datasets. So, 94% accuracy was achieved by the resulting malware identification portability by implementing the system on Android. In the current design, by choosing an appropriate execution environment, MDTA helps to calculate this property for the investigator. To detect malware, the framework effectively uses machine learning techniques. Since the system operates on dynamic analysis, the benefit is that, instead of using manual fingerprints on detection, the system exploits the actions of applications at runtime that tackles the detection process. Since this system architecture works well on a cross-platform framework, the analyst does not need to concern about the system application platform under consideration for the analysis. It is designed to handle Android as well as Windows platform applications.

References

1. D. Moon, H. Im, J.D. Lee, J.H. Park, MLDS: multi-layer defense system for preventing advanced persistent threats. Symmetry **6**(4), 997–1010 (2014)
2. T. Akhtar, B.B. Gupta, S. Yamaguchi, Malware propagation effects on SCADA system and smart power grid, in *Proceedings of IEEE International Conference on Consumer Electronics (ICCE)*, Jan 2018 (SAE, Department of Computer Engineering, 2018), p. 16
3. B. Gupta, D.P. Agrawal, S. Yamaguchi, *Handbook of Research on Modern Cryptographic Solutions for Computer and Cyber Security* (IGI Global, Hershey, PA, USA, 2016)
4. K. Griffin, S. Schneider, X. Hu, T. Chiueh, Automatic generation of string signatures for malware detection, in *SPRINGER RAID 2009: Recent Advances in Intrusion Detection* (Springer-Verlag, Berlin, Heidelberg, 2009), pp. 101–120
5. N. Islam, S. Das, Y. Chen, On-device mobile phone security exploits machine learning. IEEE Pervasive Comput. **16**(2), 92–96 (2017)
6. K. Majeed, Y. Jing, D. Novakovic, K. Ouazzane, Behaviour based anomaly detection for smartphones using machine learning algorithm, in *Proceedings of the International conference on Computer Science and Information Systems (ICSIS2014)*, 2014, pp. 67–73

7. A. Shamili, C. Bauckhage, T. Alpcan, Malware detection on mobile devices using distributed machine learning, in *Proceedings of the 20th International Conference on Pattern Recognition*, 2010, pp. 4348–4351
8. M. Waqar Afridi, T. Ali, T. Alghamdi, T. Ali, M. Yasar, Android application behavioral analysis through intent monitoring, in *Proceedings of the 6th International Symposium on Digital Forensic and Security (ISDFS)*, 2018, pp. 1–8
9. S. Arshad, M. Shah, A. Wahid, A. Mehmood, H. Song, H. Yu, SAMADroid: a novel 3-level hybrid malware detection model for android operating system. IEEE Access **6**, 4321–4339 (2018)
10. A. Saracino, D. Sgandurra, G. Dini, F. Martinelli, MADAM: effective and efficient behavior based android malware detection and prevention. IEEE Trans. Dependable Secure Comput. **15**(1), 83–97 (2018)
11. D. Sgandurra, L. Muñoz-González, R. Mohsen, E.C. Lupu, Automated dynamic analysis of ransomware: benefits, limitations, and use for detection. CoRR (2016)
12. G. Severi, T. Leek, B. Dolan-Gavitt, Malrec: compact full-trace malware recording for deep retrospective analysis, in *Detection of Intrusions and Malware, and Vulnerability Assessment. DIMVA* (2018)
13. L. Chen, M. Zhang, C.-Y. Yang, R. Sahita, Semi-supervised classification for dynamic android malware detection, in *ACM, 2017 Conference on Computer and Communications Security. CCS* (2017)
14. M.K. Alzaylaee, S.Y. Yerima, S. Sezer, DynaLog: an automated dynamic analysis framework for characterizing android applications. CoRR (2016)
15. E.B. Karbab, M. Debbabi, Automatic investigation framework for android malware cyber-infrastructures. CoRR (2018)
16. I. Paul, E. Filiol, Glassbox: dynamic analysis platform for malware android applications on real devices. CoRR (2016)

Investigating the Role of User Experience and Design in Recommender Systems: A Pragmatic Review

Ajay Dhruv and J. W. Bakal

Abstract The world is growing smarter with technology-driven applications in assisting lifestyle of humans. Machine learning has predominantly helped businesses in continuously learning and evaluating human likes and dislikes. Today, deep neural networks have grabbed the attention of many researchers in developing highly complicated yet effective models for prediction and data analytics. Recommender systems being one of the applications of machine learning has caught customer attention especially in the retail and hospitality sector. After a recommendation engine is built, it is evaluated in different tangents like accuracy, recall, etc. However, there is a serious need to shift the research focus to the user experience and design aspect while using such recommender systems. This paper conducts a survey and based on the survey results, a novel conceptualized model labelled 'Cognitive Dynamic Design Engine' (CDDE) is proposed. This model will help businesses to understand customer mindset, perception and usability parameters while presenting recommendations to their customers. This model is proposed to run on top the deep neural network model that is constructed for generating recommendations. This survey is collected from masses of urban and rural parts of a geographical zone. The results of the study show that user experience and aesthetics of the recommendation given to a user inevitably have a direct effect on users buying decisions.

Keywords Recommender systems · User experience · Cognitive dynamic design engine · Usability engineering

A. Dhruv (✉)
Thadomal Shahani Engineering College, Mumbai, India
e-mail: ajay.dhruv@vit.edu.in

J. W. Bakal
Shivajirao Jondhale College of Engineering, Dombivali, India

S. Shakya et al. (eds.), *Mobile Computing and Sustainable Informatics*, Lecture Notes on Data Engineering and Communications Technologies 68,
https://doi.org/10.1007/978-981-16-1866-6_11

1 Introduction

Supervised and unsupervised algorithms have played a significant role in developing prediction-based models. Machine learning models perform better than humans in various applications ranging from banking, advertising to recommend novels or web series to an individual. These models are then applied to traffic controlling systems, e-commerce and e-businesses, hospitality firms and even manufacturing industries. Classification trees, Bayesian networks and regression models have been showing essentially good results in terms of confusion matrices and accuracy bars. However, it is essential to note the bias and variance values while dealing with such algorithms. In machine learning, there could be many bottlenecks affecting the performance of the model and hence one needs to tune either the training or testing set accordingly.

Deep neural networks allow us to create a neural network with multiple hidden layers. But, when we train such deep networks, there arises either the vanishing gradient problem or the exploding gradient problem. There are several techniques in practise to solve these issues, but it is more important to tune the hyperparameters in use like number of hidden units, the learning rate, etc.

Many researchers use transfer learning while developing deep neural nets. Hybrid recommender systems are also adopted by many researchers to understand their behaviour and also have a comparison with the ancient approaches [1]. There are lot of open-source frameworks which allow you to experiment with datasets, analyse them across dimensions of your choice and produce potential results. While developing recommender systems, it is important to pay attention to the user interface aesthetics, the way in which the recommended content is presented and understanding which design can attract customers buying behaviour.

In this paper further, Sect. 2 describes the related work of the diverse recommender engines developed by researchers, their advantages and limitations, the usability factors considered and the evaluation of these models. In Sect. 3, an overview of usability engineering is presented to understand its need in recommender systems. Section 4 presents the experimental setup performed through a survey questionnaire. In Sect. 5, the results of the questionnaire and the observations are projected. Based on the inferences obtained from the survey findings, a recommender model has been proposed in Sect. 6.

2 Related Work

The author proposed a novel GAN recommendation framework that uses auto-encoders as one of the generator, another generator that keeps a track of those items that a user might not be interested in, a Bayesian personalized ranking model as a discriminator, and this framework is tested on real-life data sets on parameters like recall, precision, discounted cumulative gain, mean reciprocal rank and its observed that this framework not only produces higher accuracy rate but also performs better than the ancient GAN-based recommenders [2]. The author proposed a recommender system based on product aspects and user sentiments using LSTM by appropriately tuning the hyper-parameters of the learning algorithm for improved performance [3]. A framework with suitable evaluation metrics was developed to investigate the impact of personalized content recommendations on informal learning from Wikipedia [4].

The Movie Lens dataset is used extensively by many researchers for building recommendation systems [5]. It is used for testing the Multi-Trans Matrix Factorization model with enhanced time weight to acquire changing items and users fuzzy preferences to make a better recommendation with the help of gradient-based optimization algorithm [6]. Recommender systems showcase a tremendous increase in constructing a lot of deep neural networks for prediction of ratings of a product based on the user–item relationships [7]. The author considered temporal information and an implicit feedback mechanism as core paradigms for a music recommender system [8].

It is interesting to understand the effect of unplanned purchase behaviours of customers in e-commerce domain but on the contrary, less importance has been given to the user aesthetics and design of web page navigations [9]. The wide choices of recommender systems like collaborative filtering, content-based, social-based and context-based help in comparing algorithms and their performance on different data sets [10]. A taxonomy of GAN-based models for recommender systems have been presented exhaustively with the research focus aligned at tackling two issues, noise and sparsity in the data [11]. A customer-centric framework helps in evaluating why a customer preferred a recommender system and how was his/her experience by conducting field trials and experiments [12]. Social recommender system is developed by taking a reference of design principles and applying them for a music recommender system [13]. RESNET and other techniques are also becoming popular for feature analysis in deep learning [14].

3 Overview of Usability Engineering

There is a need to migrate from accuracy and other measures to a rich user experience. For instance, while receiving emails on bank and financial statements, the user would like to only know his credit and debit figures at one glance, rather than 3–4 paragraphs of textual content. In today's digital and fast pacing era, the user demands minimal and straightforward, to-the-point content at one view. This has drove attention of researchers towards the different aspects of usability engineering. Usability engineering is concerned with efficient human machine interaction for user comfort and friendliness. The different methods and principles of usability engineering are studied while constructing a model for the same.

Questionnaires are one of the best method for collecting and knowing user requirements, their choices and their cognitive thinking. The section below describes in depth about the various aspects that were covered during the study.

4 Experimental Set-Up

The survey was carried out in online mode by circulating an e-form. The survey captured details of participants across various tangents. These were demographics, user behaviour pattern, product reviews and most importantly UI aesthetics. The questionnaire was disseminated to participants via email, corporate communication and social media platforms, peer chat groups, online forums and communities. The e-form was also disseminated via email to the academia sector comprising of IT and non-IT background students and colleagues. The survey questionnaire is narrowed to understand the role of user interaction design in recommender systems for the people living in Maharashtra. Maharashtra has an approximate population of 11.42 crores and is a state from the Indian subcontinent. The state consists of 6 administrative divisions, further divided into 36 districts, 109 subdivisions and 357 talukas. Although the urban part of Maharashtra is technology advanced, the rural regions are still building infrastructures at a fast pace. Due to Covid-19 pandemic, the e-commerce sector in Maharashtra has expanded their businesses to serve the rural population.

The survey form was anonymous with neither the details of the surveyor nor the personal details of those surveyed were captured ensuring a fairly transparent medium of study. The questionnaire captures a total of 15 choice-based questions along with a choice for descriptive responses. Out of the various e-commerce platforms that are available today, this survey focuses on the online shopping scenario. Since the survey is restricted to the masses of Maharashtra state, the responses are limited to hundreds. The questions are meticulously planned and moderated from industry experts. They aim to understand the users behaviour while operating a recommender system [12]. They majorly aim to understand the role and importance of user interfaces and the design elements when the users are recommended any product on an e-commerce website. The questions are grouped into four different segments, namely—colour and navigation, features, projection of the recommendation and user experience. The questionnaire comprises total 15 questions as below:

1. Do you use e-commerce in day-to-day life?
2. Do you like when you are recommended an item/product by an e-commerce website?
3. How many recommendations do you prefer at a time?
4. Does the user Interface of the recommended item matter to you when you are recommended a product?
5. Which factors play an important role while considering a recommendation?
6. Which mode of communication do you prefer for receiving recommendations of items/products?
7. Which kind of colours tempt you to spend more time on an e-commerce website?
8. How do you like reviews to appear when you are recommended an item?
9. Would you like to know a comparison of similar products when you are recommended a product?

10. Would you like an AI-enabled assistant to solve your queries while you are being recommended a product?
11. Which type of navigation do you prefer on a website?
12. In which form do you prefer recommendations?
13. During cancellation of a product which was bought earlier, would you prefer a recommendation of a better reviewed relevant product of similar cost?
14. Do you prefer to buy those recommended products that have an EMI option for payment?
15. Does discounting of products when bought together encourage you to buy?

5 Findings

Figure 1 depicts the responses of the questionnaire. Majority of the participants were from urban Maharashtra in the age bracket of 22–40. The findings show that there is an equal distribution of responses from male as well female gender groups. It is observed that 97.2% of users would you like to have a comparison of similar products when they are recommended a product/item. 33.9% of the respondents demand for a short video-based user view instead of the traditional text and image-based feedback. 76% of the users prefer light and warm colours for seamless human computer interaction. 91.3% of the respondents agree that the user interface and usability parameters do matter to them when they are recommended an item. Additionally, 55.5% of the users prefer recommendations on social media platforms for their perusal. It is very much interesting to note the fact that 62.6% of the survey participants believe that user interface and aesthetics play a keen role while considering a recommendation in addition to other parameters like brand, cost and reviews. Some users also prefer popups, mouse hovers and comparison friendly user interfaces. It is also observed that the video-based product reviews were the most popular which clearly shows the power of experiential viewership rather than static images. The findings from the survey have helped in understanding the key parameters that users prefer while considering recommendations. The empirical analysis shows that usability is not just about colours and designs, but it is a smooth blend of payment options and offers, navigation, social media analysis, multilingual options, customer retention factors and cross selling—up selling dynamics. In short, for all the 'yes/no' questions that were posed from usability perspective, more than 50% respondents answered the option 'yes'. This shows that there is a serious requirement and demand to understand and prioritize usability factors in a recommender engine rather than simply optimizing the training–testing performance and tuning the hyper-parameters of the algorithm under consideration.

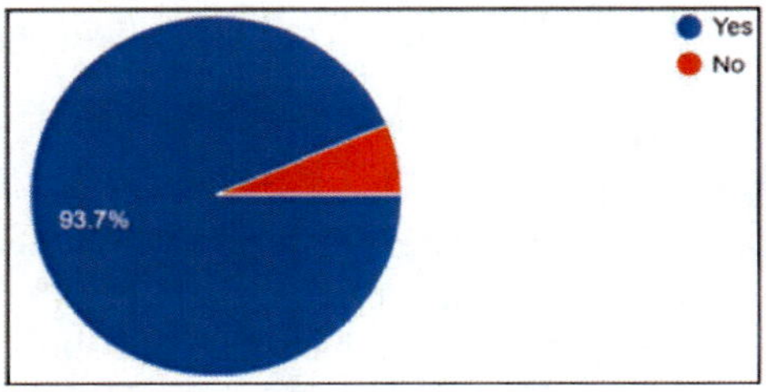

(a) Usage of e-commerce

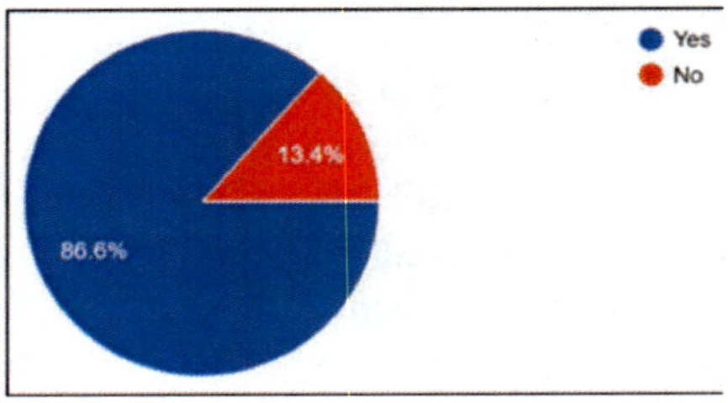

(b) Liking for product recommendation

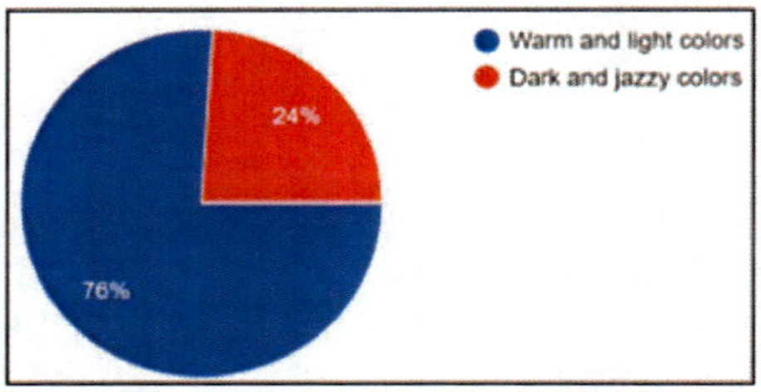

(c) Color preferences

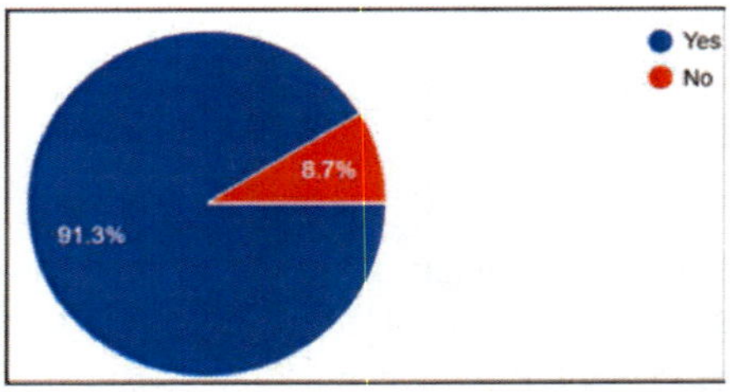

(d) In uence of user interface during recommendation

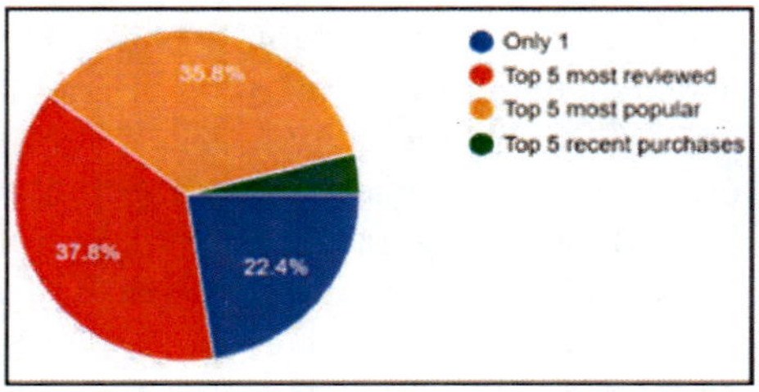

(e) Number of recommendations preferred

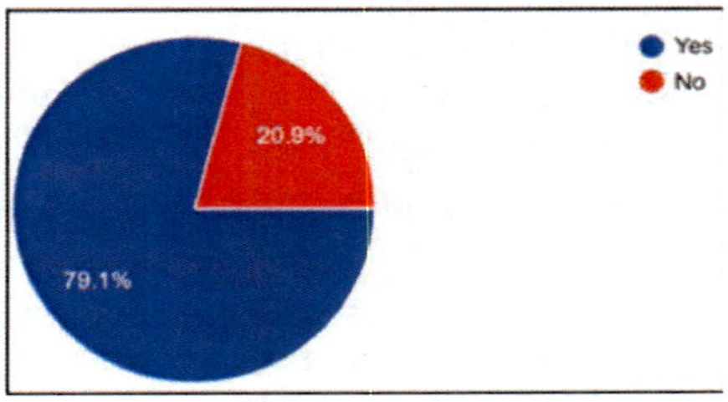

(f) Similar product recommendation during cancellation of a particular product

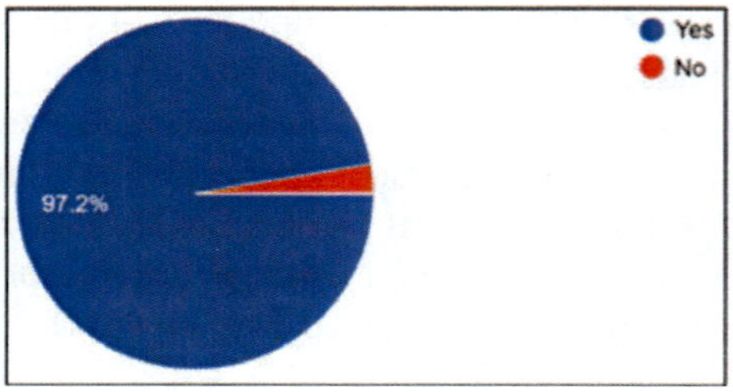

(g) Comparison of similar products during recommendation

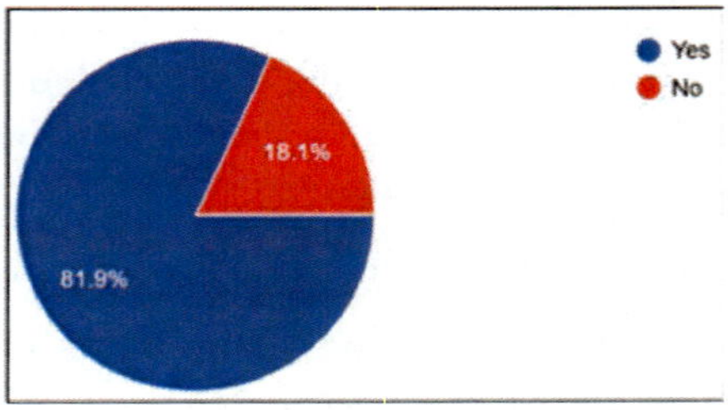

(h) Support for AI enabled assistant

Fig. 1 Survey questionnaire responses in percentage

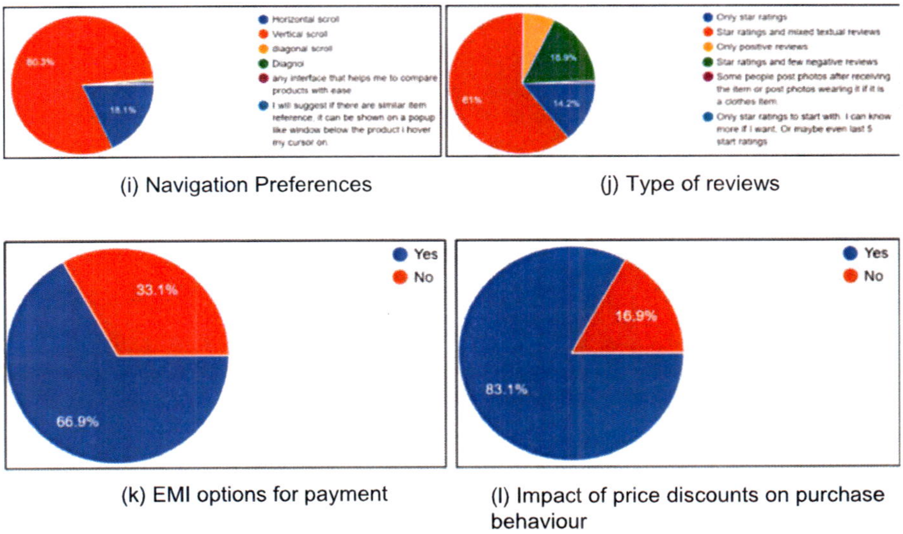

Fig. 1 (continued)

6 Proposed Model

As shown in Fig. 2, the Cognitive Dynamic Design Engine (CDDE) is built after understanding the choices of the respondents from the survey questionnaire. The user–item data pool will serve as a multipurpose input. It will be used to construct

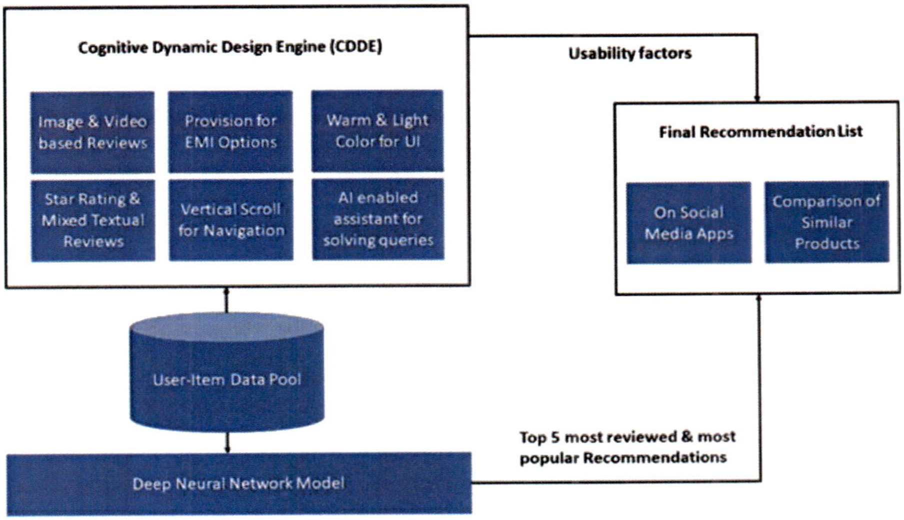

Fig. 2 Proposed model

the deep neural network model for generating recommendations. The data pool will assist the CDDE in establishing various usability factors. The survey results show that more than 50% of the users prefer multimedia-enabled reviews instead of textual ones, provisions for EMI-based flexible payment schemes, warm and light colours, mixed user reviews, vertical scroll and an AI-enabled assistant for solving customer queries. Hence, all these factors form the main components in the CDDE engine. The deep neural network model will generate the top-5 most reviewed and popular recommendations, whereas the CDDE engine will capture other aspects that will impress the users. Lastly, the final recommendation list will be generated and disseminated across the users social media handles and users will also be given a fair amount of options for choosing similar products. Unlike many recommender engines investigated so far, the proposed model pays maximum attention to usability factors followed by the choice of deep learning algorithms.

7 Conclusion

This paper attempts for a paradigm shift in the metrics that are used to evaluate a recommender system. It helps in understanding the mindset of the users when they are on e-commerce retail shopping websites. The questionnaire responses indeed show that user interface and design parameters are utmost important when they are on the web for accepting/rejecting recommendations. The research inevitably shows a strong connect between usability and recommender engines. In addition to a strong and intelligent recommender algorithm, it is shown that the respondents also prefer an equally strong, appealing and out of the box user interfaces. Other factors like precision scores, cumulative gain and accuracy rate will be taken care by the neural network model under consideration, but usability factors and user preferences shall be understood from the design elements of the proposed model. If a recommender engine is built by considering the above model, then it is predicted to have a better customer retention rate for the businesses. The study can also be applied to e-learning platforms, online travel agencies and online medical camps.

8 Future Scope

In the future, the survey can be extended by asking users to rate two e-commerce platforms and answer why they feel the chosen platform is better. We also plan to test the effectiveness of the usability model with the traditional parametric models and understand the tradeoffs and lacunae in their behaviour, if any. More than 90% of the respondents were from urban Maharashtra. So, in the future, we can drill down our study to a rural population specifically. With the help of the questionnaire oated in this study, researchers can explore more questions to capture surveys across a more larger or more smaller geographical zone. The research can also be explored to capture the real time emotions of a user and the influence of their peers on their favourite or not so favourite products. This will help in fine tuning the final recommendation list. A feedback mechanism can be devised to capture why a user really considered a particular recommendation. Although while doing so, lot of care shall be taken to ensure that the user must not get annoyed. Also, the research can be extended for developing a dynamic and an adaptive usability strategy for visually impaired user groups. Finally, one can capture and study the online shopping preferences of the third gender community and propose a model that will recommend items for them by considering design aesthetics as one of the key factors.

Acknowledgements This work is partially supported by 'Vidyalankar Institute of Technology', Wadala, Mumbai, India, under the scheme of 'Higher Studies Sponsorship Policy', 2020.

Conflict of Interest The authors declare that they have no conflict of interest.

References

1. J. Sobecki, Implementations of web-based recommender systems using hybrid methods. Int. J. Comput. Sci. Appl. **3** (2006)
2. D.-K. Chae, J.A. Shin, S.-W. Kim, Collaborative adversarial autoencoders: an effective collaborative filtering model under the GAN framework. IEEE Access **7**, 37650–37663 (2019)
3. A. Da'u, N. Salim, Sentiment-aware deep recommender system with neural attention networks. IEEE Access **7**, 45472–45484 (2019). https://doi.org/10.1109/ACCESS.2019.2907729
4. H.M. Ismail, B. Belkhouche, S. Harous, Framework for personalized content recommendations to support informal learning in massively diverse information wikis. IEEE Access **7**, 172752–172773 (2019). https://doi.org/10.1109/AC-CESS.2019.2956284
5. F. Maxwell Harper, J.A. Konstan, The MovieLens datasets: history and context. ACM Trans. Interact. Intell. Syst. (TiiS) **5**(4), Article 19, 19 pages (2015). https://doi.org/10.1145/2827872
6. J. Zhang, X. Lu, A multi-trans matrix factorization model with improved time weight in temporal recommender systems. IEEE Access **8**, 2408–2416 (2020). https://doi.org/10.1109/ACCESS.2019.2960540
7. M. Almaghrabi, G. Chetty, A deep learning based collaborative neural network framework for recommender system, in *2018 International Conference on Machine Learning and Data Engineering (iCMLDE)*, Sydney, Australia, 2018, pp. 121–127. https://doi.org/10.1109/iCMLDE.2018.00031
8. D. Sanchez-Moreno, Y. Zheng, M.N. Moreno-García, Incorporating time dynamics and implicit feedback into music recommender systems, in *2018 IEEE/WIC/ACM International Conference on Web Intelligence (WI)*, Santiago, 2018, pp. 580–585. https://doi.org/10.1109/WI.2018.00-34
9. Z. Ying, C. Caixia, G. Wen, L. Xiaogang, Impact of recommender systems on unplanned purchase behaviours in e-commerce, in *2018 5th International Conference on Industrial Engineering and Applications (ICIEA)*, Singapore, 2018, pp. 21–30. https://doi.org/10.1109/IEA.2018.8387066
10. M. Jallouli, S. Lajmi, I. Amous, Designing recommender system: conceptual framework and practical implementation, in *International Conference on Knowledge Based and Intelligent Information and Engineering Systems, KES2017*, Marseille, France, 6–8 Sept 2017
11. M. Gao, J. Zhang, J. Yu, J. Li, J. Wen, Q. Xiong, Recommender systems based on generative adversarial networks: a problem-driven perspective. arXiv preprint arXiv:2003.02474

12. B.P. Knijnenburg, M.C. Willemsen, Z. Gantner et al., Explaining the user experience of recommender systems. User Model. User-Adapt. Interact. **22**, 441–504 (2012). https://doi.org/10.1007/s11257-011-9118-4
13. Y. Chen, Interface and interaction design for group and social recommender systems, in *Proceedings of the Fifth ACM Conference on Recommender systems (Rec-Sys'11)* (Association for Computing Machinery, New York, NY, USA, 2011), pp. 363–366. https://doi.org/10.1145/2043932.2044007
14. T. Vijayakumar, R. Vinothkanna, Mellowness detection of dragon fruit using deep learning strategy. J. Innov. Image Process. (JIIP) **2**(01), 35–43 (2020)

A Review on Intrusion Detection Approaches in Resource-Constrained IoT Environment

A. Durga Bhavani and Neha Mangla

Abstract The Internet of things (IoT) has advanced as an innovation to build smart situations such as industrial processes, home automation, smart cities, and environment monitoring. Smart and IoT devices exist in almost every walk of life recently. Since these devices are available in an Internet-connected environment, they are exposed to different vulnerabilities and attacks, typical of any Internet-connected system. Applying conventional intrusion detection and avoidance frameworks on these IoT devices is not practical because of the different asset imperatives in these devices regarding calculation, memory, and energy availability. This survey focuses on different techniques for intrusion detection and avoidance in an IoT environment. The aim of this work is to recognize open difficulties in intrusion detection and avoidance frameworks for the resource-constrained nature of the IoT environment.

Keywords Constrained environment · IoT · Intrusion detection system (IDS) · Anomalies · Attacks

1 Introduction

Internet of things (IoT) is advancing allowing technology with a capacity to identify their condition, associate with each other, and exchange data over the Internet. It is utilized to create a smart environment such as smart homes, health care, and agriculture. The objective of which is to increase human efficiency.

The market for IoT solutions is increasing from 93.5 billion US dollars in 2017 to 225.5 billion US dollars by 2026 according to an outline by Navigant research [1].

IoT has revolutionized many industries including production, manufacturing, finance, health care, and energy. Since IoT devices or objects are connected

A. Durga Bhavani (✉)
Department of Computer Science and Engineering, BMS Institute of Technology and Management, Bangalore, India
e-mail: durga842004@bmsit.in

N. Mangla
Department of Information Science and Engineering, Atria Institute of Technology, Bangalore, India

S. Shakya et al. (eds.), *Mobile Computing and Sustainable Informatics*, Lecture Notes on Data Engineering and Communications Technologies 68,
https://doi.org/10.1007/978-981-16-1866-6_12

Table 1 IoT characteristics

Data gathering	IoT sensors gather large volume of data from their environment for proper functioning. These data has to be protected otherwise there is risk of the data being stolen or compromised
Complex environment	Higher number of devices and the interaction between them is possible due to an advancement in communication technologies. The attack surface area grows as the complexity of the environment increases
Centralized architecture	Data gathered by sensors are sent to a centralized base station for analysis. These centralized base stations become a security pitfall

to the Internet, they are exposed to various vulnerabilities and attacks like any Internet-connected system. Table 1 demonstrates the characteristics which make IoT functional and efficient also as a security weakness.

The attacks in IoT environment can happen through the following three forms

1. Devices
2. Communication channels
3. Software and applications.

An attacker can take advantage of unsecured settings, outdated components, and update mechanisms to launch attacks on devices. By taking advantage of protocol vulnerabilities, spoofing, denial of service, and other such attacks, can be launched on the communication channels.

Vulnerabilities in the software's and applications can prompt different settles on the information and its protection creates. It can also lead to data theft. The attacks like denial of service and distributed denial of service cause considerable damage to IoT services and smart applications to build using the IoT backend. Attack statistics from Identify Theft Resource Center (Figs. 1, 2 and 3) illustrates the severity of attack and losses due to them [2].

Securing the IoT environment from Internet attacks has thus become a major concern. Intrusion detection systems (IDS) are the traditional systems to detect attacks and prevent Internet systems from them. The placement of a typical IDS system in an enterprise is shown in Fig. 4.

Organization	Records	Breach Date	Type	Source	Industry	Estimated Cost
Anthem insurance	78 million	January 2015	Identify theft	Malicious outsider	Healthcare	$100 million
Yahoo	500 million	December 2014	Account access	State sponsored[1]	Business	$350 million
Home depot	109 million	September 2014	Financial access	Malicious outsider	Business	$28 million
JPMorgan chase	83 million	August 2014	Identify theft	Malicious outsider	Financial	$13 billion
Benesse	49 million	July 2014	Identify theft	Malicious insider	Education	$138 million
Korea credit bureau	104 million	January 2014	Identify theft	Malicious insider	Financial	$100 million
Target	110 million	November 2013	Financial access	Malicious outsider	Business	$252 million
Adobe System	152 Million	September 2013	Financial access	Malicious outsider	Business	$714 Million

Fig. 1 World's biggest data breaches

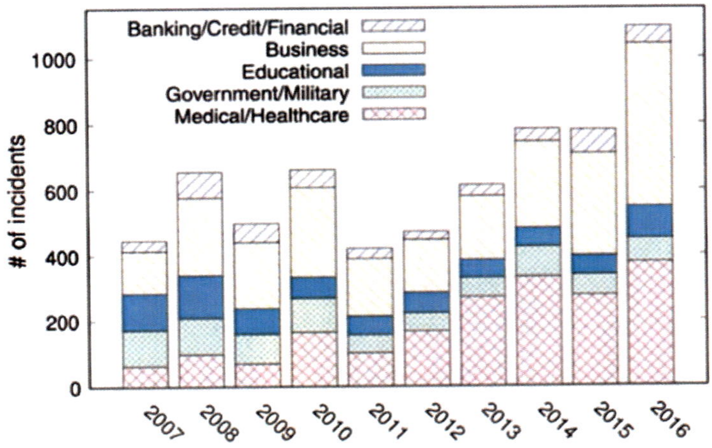

Fig. 2 Sector-wise leakage incidents

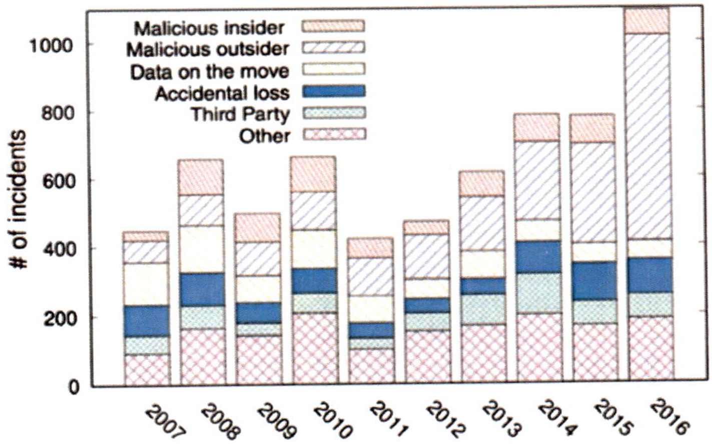

Fig. 3 Incident statistics

Deployed at the network layer, these systems capture the packets and analyze them in real time to detect any attacks. These systems are at the border of the network. Any incoming/outgoing traffic is tapped and provided to the IDS system. The IDS system analyzes the packet and decides to either forward or drop the packet. Analysis of the packets in real time and matching them against large volumes of attack signatures requires higher computational power and larger storage overheads. Conventional IDS are not fully fit for IoT environments due to resource constraints on IoT devices. This could be in terms of computation power, storage, protocol diversity, or energy consumption. This survey analyzes the current intrusion detection and avoidance

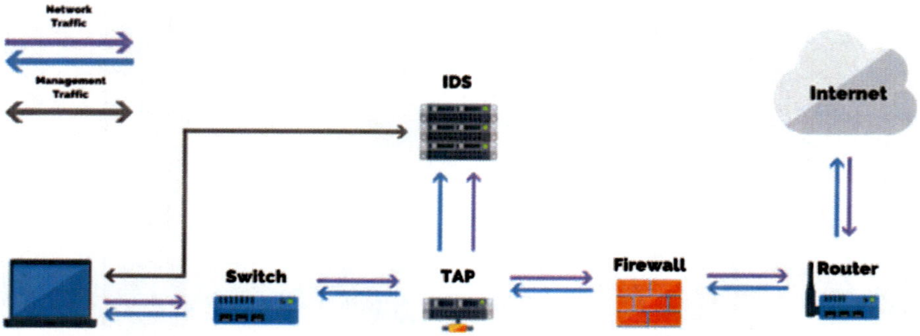

Fig. 4 IDS flow

systems for the IoT environment in terms of indicators like attack detection performance, processing time, and energy consumption. The focus of this review is recognizing the open issues for additional improvement of intrusion detection frameworks for the IoT environment.

2 Review of Intrusion Detection Frameworks for IoT

Event processing IDS structural design for the IoT environment is proposed in [3]. Complex event processing (CEP) is at the core of this solution. CEP helps to detect meaningful complex relationships among the events. The events can be configured by users based on their own interest through transmission events in sensor networks or operating events in enterprise applications. The volume of the frequent occurrence of unauthorized access frequently is detected in real time using CEP. It is a rule-based anomaly detection system, and thus it suffers from the zero-day problem for new attacks.

Authors in [4] proposed a framework to grade the effectiveness of IDS for smart public transport applications using Constrained Application Protocol (CoAP). The work tested three different types of IDS: rule-based, anomaly neural network (NN), and hybrid-based. Following were the inferences from the test

1. Rule-based approach is limited in expressing complex behavior
2. A hybrid approach combining rule-based and anomaly or NN-based has the highest detection efficiency.

The test conducted was limited and the performance impact on the IoT environment was not considered.

Authors in [5] proposed an intrusion detection framework for clustered hierarchical wireless_sensor networks. This solution adopts two ideas of downward and upward projection. The behavior of nodes in each layer is monitored and a sequential probability test is conducted to detect selective forwarding attacks. Due to the statistical method for attack detection, the computation is lightweight, but the approach is limited to selective forwarding attack.

An innovative intrusion detection technique with the constraint-based specification for 6LoWPAN networks was proposed in [6]. The work is established to detect sink_hole attacks with a certain promise on the QoS. Dempster–Shafer evidence theory is applied to evidence from various sources to detect an attack.

A specification-based IDS for detecting topology attacks was proposed in [7]. A specification for a reference node is obtained through simulation and auto profiling. The specification is in form of authentic convention states and advances with relating insights. The IDS monitor node behavior and compare to reference in order to detect attacks. The overhead in the approach is low and can be scaled to larger networks. But the approach is limited to routing topology attacks like sink-hole, rank, local neighbor, and DIS attacks.

Authors in [8] proposed a real-time hybrid intrusion detection_framework for an IoT environment. Sink-hole, wormhole, and selective forwarding assault are recognized by combining anomaly-based discovery with specification-based detection.

The incoming packets are clustered using an unsupervised optimum-path forest algorithm to detect anomalies. To speed up, the clustering is implemented on MapReduce architecture. The suspicious behavior is detected using a voting mechanism. This method was able to detect a sink-hole attack with an accuracy of 76% and a wormhole attack with an accuracy of 96%. The computation requires a MapReduce platform and moving the packets to the MapReduce platform incurs higher network overhead. This could have been solved using distributed clustering with the assistance of fog nodes.

An intrusion detection framework is designed specifically for a smart city environment in [9]. The solution uses two types of detection algorithms. The rule-based detection algorithm is used to detect identified attacks and the anomalies detection system is used to detect unknown attacks. The scheme works on routing attacks.

Authors in [10] proposed an automata model-based intrusion detection system for heterogeneous IoT systems. The solution can identify three types of attacks: jamming, false, and reply attacks. A fine state automata model is constructed for packet transitions and the packets in the network are processed against the automata to detect attacks. Though the approach is easy to carry out, its complexity increases when considered against a higher number of attacks.

A lightweight intrusion detection system combining "fuzzy C means" (FCM) with "principal component analysis" (PCA) is proposed in [11]. Clusters are created using KDDCUP 99 dataset and incoming packets are matched to the closest cluster to detect the attacks. The scheme has high false positives.

An intrusion detection system using a hybrid learning approach is proposed in [12]. The detection has two stages local and global. At the local stage, supervised learning using a decision tree is used to generate correctly classified instances (CCI). At the global stage, CCI is collected from nodes and iterative linear regression is applied to detect maliciousness of a node.

Authors in [13] planned an intrusion detection system relating to the "suppressed fuzzy clustering" (SFC) and "principal component analysis" (PCA) algorithm. The frequency of sampling the data for attack detection is varied dynamically depending on the attack occurrence. In this method, the efficiency of the system is unaffected in a no-attack scenario.

Authors in [14] proposed a component to distinguish the malicious gateway in a clustered IoT environment. The likelihood ratio test is used for detection. The detection is based on the packet drop rate at the gateways. The solution assumes the only the gateway is malicious, but devices are not compromised. Also, attack uplink packets are not considered in the solution.

A lightweight security system based on pattern matching is proposed to detect malicious packets in [15]. The solution practices two novel methods of auxiliary shifting and early decision. The quantity of matching operations and the memory footprint needed for attack detection is reduced due to the two techniques. The method can detect signatures present in a single packet and cannot match against multiple packets spread across the flow.

Authors in [16] proposed a deep packet anomaly detection engine that is ultra-lightweight and suitable for resource-constrained IoT devices. The scheme can identify unusual and typical payloads utilizing proficient bit design coordination requiring bitwise AND activity with contingent counter increment. The packet discrimination is done using a lookup table. The scheme can be realized using hardware. The lookup table increases considering different attack signatures.

A behavioral anomaly-based intrusion detection system is suggested for smart home environment in [17]. Desired behavior and the deviation from the desired behavior are detected using immunity inspired algorithm. Series of IoT sensor data is transformed into event sequences. The event sequences are converted to a numerical form which represents the behavioral identity. Once the identity is established or trained, the sensor data is matched against behavioral identity to detect any deviation. When the deviation crosses the threshold, an alert is issued. There can be multiple profiles and behavioral profile changes over a period of time. The scheme does not work for the temporal changes in the behavioral profile.

Authors in [18] designed and evaluated some intrusion detection mechanisms for the IoT network. The solution is better suited for energy and processing power restraining devices. Intrusion detection is based on a trust management mechanism. Each device calculates the trust score for each neighboring device based on positive and negative experiences. Nodes with trust value lower than a threshold are marked as malicious and singled out in an energy-efficient manner. The scheme can only detect three attacks of selective forwarding, sink-hole, and version number attack.

Authors in [19] proposed an intrusion detection framework called SVELTE. The solution is designed to detect spoofing, selective forwarding, and sink-hole attacks. The attack detection is based on the detection of inconsistency in routing packet against known network topology. But the approach fails in presence of cooperative attacks.

A lightweight distributed IDS system is designed in [20] using the principles of an "artificial immune system" (AIS). The IDS is disseminated across three layers of edge, fog, and cloud. Each layer specializes in the specific role of attack detection. The machine learning intelligence for attack detection is implemented in the cloud layer due to higher complexity. The system works best for botnet attacks.

Authors in [21] proposed an intrusion detection framework utilizing unaided AI. "Density-based spatial clustering of applications with noise" (DBSCAN) is used for unsupervised learning in this work. The solution is able to detect zero-day attacks and unknown anomalies. But the computational complexity is high for deployment in resource-constrained devices.

Ensemble intrusion detection system is planned in [22] to defend against botnet attacks in IoT networks. Decision trees, Naïve Bayes, and artificial neural networks are used for ensembling. Ensemble classifier is found to have higher accuracy than individual machine learning classifiers. The work is based on statistical features learned using past history and cannot identify zero-day attacks.

A threat model is proposed in [23] to analyze the severity of attacks in IoT. A threat assessment model is proposed in this effort to analyze network security in terms of physical threat, information closure, denial of service, spoofing, the elevation of privilege, etc.

Authors in [24] proposed an innovative security model to defend the network against interior and exterior attacks and threats in the IoT environment. The security model utilizing firewall, switches, virtual local area network (VLAN), and authentication, authorization, and accounting (AAA) worker is introduced in this arrangement. Using a firewall with an AAA, server can provide enhanced security. The solution was tested only on a testbed and does not consider the resource-constrained nature of devices.

Authors in [25] proposed a robust approach for malware detection based on the extraction of statistically improbable features. It is a syntactic foot-printing system that determines areas of assessable similarity with well-known malware to discriminate those unknown, zero-day tests.

Authors in [26] proposed a host-based intrusion detection and mitigation framework, called IoT-IDM, to provide network-level protection for smart devices deployed in home environments.

A MapReduce framework for intrusion detection in IoT is proposed in [27]. The solution considers attacks from both sides of WSN and the Internet. Intrusion detection is executed utilizing a directed and unsupervised optimum-path forest model.

Authors in [28] proposed a novel intrusion detection model. The arrangement includes two-layer measurement reduction and two-level grouping. Due to the two-level AI arrangement, the accuracy of detection improves. The solution works only for two Internet attacks.

Deep learning-based intrusion detection using a conditional variation autoencoder is proposed in [29]. Even though the solution has higher accuracy, it is computation-wise complex for the IoT environment. Similar to [29] authors in [30] evaluated the use of deep learning for distributed attack detection.

A novel intrusion_detection technique is proposed in [31]. It relies upon haze processing using "online sequential extreme learning machine" (OSELM) which can admirably interpret the attacks from the IoT traffic. In the proposed system, the existing centralized cloud intelligence in detecting the attack is distributed to local fog nodes to detect the attack at a faster rate for IoT applications. The distributed architecture of fog computing enables distributed intrusion detection mechanisms with scalability, flexibility, and interoperability. But the approach detects attacks only originating from the device side to the cloud side and cannot detect attacks originating from the Internet site and from the other device side.

Authors in [32] proposed an intrusion detection system for the IoT networks using deep learning. A hybrid convolutional neural network model is trained to detect attacks in the IoT network. The approach can work for a wide range of IoT applications. Localization of attack and prediction is not possible in this approach.

Authors in [33] surveyed the role of machine learning algorithms in designing intrusion detection systems for wireless sensor networks. The survey was conducted mainly from the point of view of data theft and compromise (Table 2).

3 Open Issues

The open issues in the existing solutions for data and information security under fog computing environment are listed below

1. Incorporating fog nodes for intrusion detection decision for multifaceted attack detection.
2. Packet handling is not efficient for IoT environment.
3. Lack of multiple operating modes for attack detection.
4. Lack of comprehensive approach that consider attacks on all IoT components.
5. Lack of incorporation of temporal behavioral anomalies.

4 Discussion on Open Issues and Future Research Direction

Issue 1: In most of the solutions, incorporating fog nodes for offloading the computations for intrusion detection at close proximity to devices has not been considered from the viewpoint of detecting attacks and preventing attacks. Also, the multifaceted attack detection from multiple sources like device side, inter-device, and the Internet site is not considered in any of the previous works. Considering fog for multifaceted attack detection would reduce the network overhead and centralized data handling.

Issue 2: All the packets are handled in the same fashion and passed through different stages of attack detection like feature extraction, classification, alert, and prevention. But bypass can be done for certain packets based on type, origination point, and packet signature, etc. By this way of differential treatment, the performance degrades due to the intrusion detection process is reduced and its impacts on other service efficiency are also reduced.

Table 2 Merits and demerits of intrusion detection in IoT environment

References	Approach	Merits	Demerits
[3]	Rule-based anomaly detection system	Real-time processing	Rule-based and lacks dynamic adaptation
[4]	Rule-based, anomaly neural network (NN) based	Lightweight and applicability to CoAP applications	Lacks testing on performance impacts
[5]	Statistical method for attack detection	Low complexity due to statistical test	Limited only to selective forwarding attack
[6]	Rule-based anomaly detection system	Attack detect with QOS guarantee	Limited only to sink-hole attacks
[7]	Specification-based IDS	Low overhead and scalable to larger networks	Limited to routing topology attacks
[8]	Hybrid intrusion detection	Unsupervised clustering enables detecting new anomalies	Higher network overhead
[9]	Rule-based anomaly detection system	Hybrid algorithm for attack detection	Can detect only routing attacks
[10]	Automata model-based intrusion detection	Automata model is easy to implement and has less computational needs	Not scalable for larger number of attacks. Also, the memory footprint increases exponentially and not scalable
[11]	Cluster-based anomaly detection system	Works for larger number of attacks	High false positives
[12]	Hybrid intrusion detection	Two stages of attack detection	False positives cascade and create more damage
[13]	Cluster-based anomaly detection system	Due to dynamic feature extraction frequency, efficiency of system is not affected during no-attack scenarios	The system covers only limited set of attacks
[14]	Cluster-based anomaly detection system	Lightweight detection scheme	Attack on uplink packets not considered
[15]	Pattern matching system	Efficient pattern matching	Limited to single packet
[16]	Deep packet anomaly detection	Ultra-lightweight and suitable for hardware implementation	Can detect only malicious payloads like worms and Trojans
[17]	Behavioral anomaly-based intrusion detection system	Behavioral anomaly detection	Not adaptable to temporal change in behavior

(continued)

Table 2 (continued)

References	Approach	Merits	Demerits
[18]	Distributed intrusion detection	Trust management for devices	Limited to only three attacks
[19]	Svelte-distributed-based	Network topology-based attack detection	Fail to detect cooperative attacks
[20]	Distributed IDS	Distributed attack detection	Role in attack detection for different attacks is not considered in this work
[21]	Anomaly-based detection	Able to solve zero-day attacks and unknown anomalies	Not suitable for resource-constrained devices
[22]	Ensemble intrusion detection system	Ensemble classifiers provide higher accuracy	Not secure against zero-day attacks
[23]	Threat-based analysis	Risk assessment of network	Risk is assessed only for limited attacks
[24]	Network intrusion detection system	Security model involving multiple components	Not tested in resource-constrained environment
[25]	Statistical method for attack detection	Robust statistical feature extraction	Can detect only for malware
[26]	Host-based intrusion detection	On fly detection and open flow architecture is used for framework	Overhead is high
[27]	Hybrid intrusion detection	Multifaceted detection	Works only for routing attacks
[28]	Anomaly-based intrusion detection	Accuracy improvement due to two-tier classification	Limited only for two attacks
[29]	Distributed attack detection	Accuracy is high	Computationally intensive for IoT environment
[30]	Distributed attack detection	Accuracy is high	Even though implementation is distributed, it is still computationally complex in IoT environment
[31]	Distributed intrusion detection	Fog computing	Limited only to device originated attacks
[32]	Hybrid intrusion detection	Deep learning-based intrusion detection	Localization is not supported

Issue 3: The attack detection process can be operating in multiple operating modes with each mode having different performances. The current operation mode can be selected based on multiple parameters like the device's computation needs, residual energy, the current rate of attacks, type of applications, etc. By this way of operating in multiple modes, the performance degrades due to intrusion detection can be reduced in a resource-constrained IoT environment.

Issue 4: There is no comprehensive solution addressing the compromises of all IoT components, right from devices, gateways, communication channels, and base stations. Attacks can happen in any of these components, but most approaches rely only on packet analysis. This problem can be solved by considering a packet, topology, routes, services, and multiple components in the detection approach.

Issue 5: IoT events change over a period of time or depending on the environment it operates, there can be temporal behaviors. The intrusion detection system should not consider this behavior as an anomaly and must in turn adapt itself to temporal and spatial behaviors. This can be done by using semi-supervised machine learning algorithms with continuous feedback-based learning.

5 Conclusion

The Internet of things (IoT) [34] has made a pattern in this mechanically advancing world. Definitely, it has built up its underlying foundations in the field of data innovation, and furthermore in clinical, media transmission, horticulture, and some more. Thus, the framework is more inclined toward attacks and security breaks, and there is an extraordinary test later on ahead to determine these issues.

This paper summarizes the current mechanism going on intrusion detection in the IoT environment. The existing solutions have been detailed, and the problems with each solution are documented. The open areas for further search on designing IDS for IoT environment are discussed with a prospective solution for those issues. Further work will be on the design of efficient solutions to address the identified open issues.

References

1. R. Citron, K. Maxwell, E. Woods, Smart city services market. Technical report. Navigant Research (2017)
2. Identity Theft Resource Center (2017), Available at https://www.idtheftcenter.org/. Accessed 1 Mar 2017
3. C. Jun, C. Chi, Design of complex event-processing IDS in Internet of things, in *2014 Sixth International Conference on Measuring Technology and Mechatronics Automation* (IEEE, Zhangjiajie, 2014), pp. 226–229

4. J. Krimmling, S. Peter, Integration and evaluation of intrusion detection for CoAP in smart city applications, in *2014 IEEE Conference on Communications and Network Security* (IEEE, San Francisco, 2014), pp. 73–78
5. I. Butun, I.-H. Ra, R. Sankar, An intrusion detection system based on multi-level clustering for hierarchical wireless sensor networks. Sensors **15**(11), 28960–28978 (2015)
6. M. Surendar, A. Umamakeswari, InDReS: an intrusion detection and response system for Internet of things with 6LoWPAN, in *2016 International Conference on Wireless Communications, Signal Processing and Networking (WiSPNET)*, Chennai, 2016, pp. 1903–1908
7. A. Le, J. Loo, K.K. Chai, M. Aiash, A specification-based IDS for detecting attacks on RPL-based network topology. Information **7**(2), 1–19 (2016)
8. H. Bostani, M. Sheikhan, Hybrid of anomaly-based and specification-based IDS for Internet of things using unsupervised OPF based on MapReduce approach. Comput. Commun. **98**, 52–71 (2017)
9. V. Garcia-Font, C. Garrigues, H. Rifa-Pous, Attack classification schema for smart city WSNs. Sensors **17**(4), 1–24 (2017)
10. Y. Fu, Z. Yan, J. Cao, K. Ousmane, X. Cao, An automata based intrusion detection method for Internet of things. Mob. Inf. Syst. **2017**, 13 (2017)
11. L. Deng, D. Li, X. Yao, D. Cox, H. Wang, Mobile network intrusion detection for IoT system based on transfer learning algorithm. Clust. Comput. **21**, 1–16 (2018)
12. A. Amouri, V.T. Alaparthy, S.D. Morgera, Cross layer-based intrusion detection based on network behavior for IoT, in *2018 IEEE 19th Wireless and Microwave Technology Conference (WAMICON)* (IEEE, Sand Key, 2018), pp. 1–4
13. L. Liu, B. Xu, X. Zhang, X. Wu, An intrusion detection method for Internet of things based on suppressed fuzzy clustering. EURASIP J. Wirel. Commun. Netw. **2018**(1), 113 (2018)
14. N.V. Abhishek, T.J. Lim, B. Sikdar, A. Tandon, An intrusion detection system for detecting compromised gateways in clustered IoT networks, in *2018 IEEE International Workshop Technical Committee on Communications Quality and Reliability (CQR)* (IEEE, Austin, 2018), pp. 1–6
15. D. Oh, D. Kim, W.W. Ro, A malicious pattern detection engine for embedded security systems in the Internet of things. Sensors **14**(12), 24188–24211 (2014)
16. D.H. Summerville, K.M. Zach, Y. Chen, Ultra-lightweight deep packet anomaly detection for Internet of things devices, in *2015 IEEE 34th International Performance Computing and Communications Conference (IPCCC)* (IEEE, Nanjing, 2015), pp. 1–8
17. B. Arrington, L. Barnett, R. Rufus, A. Esterline, Behavioral modeling intrusion detection system (BMIDS) using Internet of things (IoT) behavior-based anomaly detection via immunity-inspired algorithms, in *2016 25th International Conference on Computer Communication and Networks (ICCCN)*, Waikoloa, 2016, pp. 1–6
18. Z.A. Khan, P. Herrmann, A trust based distributed intrusion detection mechanism for Internet of things, in *2017 IEEE 31st International Conference on Advanced Information Networking and Applications (AINA)* (IEEE, Taipei, 2017), pp. 1169–1176
19. S. Raza, L. Wallgren, T. Voigt, SVELTE: real-time intrusion detection in the Internet of things. Ad Hoc Netw. **11**(8), 2661–2674 (2013)
20. F. Hosseinpour, P. Vahdani Amoli, J. Plosila, T. Hmlinen, H. Tenhunen, An intrusion detection system for fog computing and IoT based logistic systems using a smart data approach. Int. J. Digit. Content Technol. Appl. **10** (2016)
21. F. Hosseinpour, P.V. Amoli, F. Farahnakian, J. Plosila, Artificial immune system based intrusion detection: innate immunity using an unsupervised learning approach. JDCTA Int. J. Digit. Content Technol. Appl. **8**(5), 1–12 (2014)
22. N. Moustafa, B. Turnbull, K.R. Choo, An ensemble intrusion detection technique based on proposed statistical flow features for protecting network traffic of Internet of things. IEEE Internet Things J. 1 (2018)
23. A.W. Atamli, A. Martin, Threat-based security analysis for the Internet of things, in *2014 International Workshop on Secure Internet of Things*, Sept 2014, pp. 35–43

24. S.A. Alabady, F. Al-Turjman, S. Din, A novel security model for cooperative virtual networks in the IoT era. Int. J. Parallel Prog. (2018)
25. P. Faruki, V. Ganmoor, V. Laxmi, M.S. Gaur, A. Bharmal, AndroSimilar: robust statistical feature signature for android malware detection, in *Proceedings of the 6th International Conference on Security of Information and Networks, ser. SIN'13*, Nov 2013 (ACM, New York, NY, USA, 2013), pp. 152–159
26. M. Nobakht, V. Sivaraman, R. Boreli, A host-based intrusion detection and mitigation framework for smart home IoT using OpenFlow, in *2016 11th International Conference on Availability, Reliability and Security (ARES)*, Aug 2016, pp. 147–156
27. M. Sheikhan, H. Bostani, A hybrid intrusion detection architecture for Internet of things, in *2016 8th International Symposium on Telecommunications (IST)*, Sept 2016
28. H.H. Pajouh, R. Javidan, R. Khayami, D. Ali, K.K.R. Choo, A two-layer dimension reduction and two-tier classification model for anomaly-based intrusion detection in IoT backbone networks. IEEE Trans. Emerg. Topics Comput. **PP**(99), 1 (2016)
29. M. Lopez-Martin, B. Carro, A. Sanchez-Esguevillas, J. Lloret, Conditional variational autoencoder for prediction and feature recovery applied to intrusion detection in IoT. Sensors **17**(9), 1967 (2017)
30. A. Diro, N. Chilamkurti, Distributed attack detection scheme using deep learning approach for Internet of things. Future Gener. Comput. Syst. (2017)
31. S. Prabavathy, K. Sundarakantham, S.M. Shalinie, Design of cognitive fog computing for intrusion detection in Internet of things. J. Commun. Netw. **20**(3), 291–298 (2018)
32. S. Smys, A. Basar, H. Wang, Hybrid intrusion detection system for Internet of things (IoT). J. ISMAC **2**(04), 190–199 (2020)
33. E. Baraneetharan, Role of machine learning algorithms intrusion detection in WSNs: a survey. J. Inf. Technol. **2**(03), 161–173 (2020)
34. N. Mangla, A comprehensive review: Internet of things (IoT). IOSR J. Comput. Eng. 62–72. e-ISSN: 2278-0661, p-ISSN: 2278-8727, Aug 2017

Future 5G Mobile Network Performance in Webservices with NDN Technology

M. C. Malini and N. Chandrakala

Abstract 5G appears to be a people pleaser in the infrastructure of future networks. The 5G network technology is an enhanced network of 4G, which is presently implemented in some countries, and many countries are waiting to recognize the performance of 5G. Web services are emerging technologies that will face future Internet architecture to adopt new features though the existing architecture is integrated with a service provider, service registry, and service requestor. The web service operation is combined with find, publish and bind. The service provider is the hosting platform that controls the service. The service requestor is invoking a request to the service provider. The service registry is a collection of service description where you can search your service description. The primary focus of network testing is signal inspection, the detection of network errors and the identification and correction of frequency problems to improve network efficiency. The HTTP live streaming is the highlighted issue in web service, and through two-way 5G network active though put estimation technique utilized for improving the web service. The latest applications are integrated with web services which are very complex to open usage. They are combined with XML and Simple Object Access Protocol (SOAP). Performance testing is not a common platform for test the web service. The efficiency of the real-time network can be calculated by quality of service which can be facilitated by SOAP. This article describes the microflow that will be taking the execution part while the web service are called. SOAP is an open-source testing tool that helps web service testing and analyzes the network response time and throughput of the services. The automated testing aspect may ensure the verification of the network signal and the identification of errors and the implementation of measures to strengthen the network system. The current test setup is still not enough to enhance the web services and cannot apply the solution. The proposed method will solve common issues and improve the performance of the network.

Keywords NDN named data networking · SOAP Simple Object Access Protocol · MPEG-DASH (Dynamic Adaptive Streaming over HTTP) · 5G network · Throughput estimation · Video streaming

M. C. Malini (✉) · N. Chandrakala
Department of Computer Science, SSM College of Arts and Science, Komarapalayam, Namakkal (Dt), Tamil Nadu 638183, India

S. Shakya et al. (eds.), *Mobile Computing and Sustainable Informatics*, Lecture Notes on Data Engineering and Communications Technologies 68,
https://doi.org/10.1007/978-981-16-1866-6_13

1 Introduction

Web services are consistently advancing improvements that are the result of progressive developments of service providers. Step-by-step changes in web services are becoming increasingly complex, yet accessible accessibility and transparency for platforms [1]. It is produced using Extensible Markup Language (XML) along with the Simple Object Access Protocol (SOAP) [2]. Execution testing of web services is not easy to access the testing same as other platform application testing [3]. The demonstration of the most significant time arranges is frequently estimated by quality of service (QoS) estimation that might be encouraged by these devices [4]. This concept depicts one of the topmost open-source testing tools and advanced improvements of web services with respect to quality, and service testing is finished for the quality improvement and quick response time with end-to-end service [5]. In order to maintain the quality of services, it is required to focus on the individual parameters such as the ratio of data rates, the response time of the provider, the efficiency of services [6]. The mentioned parameters are evaluated the quality of web services that can be assured with the quality of web services. The service response time is mainly evaluated for reducing service reply time. The service response time integrated with a user that an interval between request and service response from the web service. The improvement web service analysis is focused on Simple Object Access Protocol (SOAP) web services. The open-source testing tool Simple Object Access Protocol (SOAP) describes the ideology of the protocol-specific which does utilize for understand the service structure through the computer network [7, 8]. The recent web service providers have adopted the latest trend with can enable various new models of end services to users [9]. The user feedback will depend on many character tics such as ongoing identity, multi-device integration, and service throughput and data delivery ratio. Named Data Networking (NDN) has been presented as future system engineering. The start to finish service throughput calculation for versatile piece-rate video gushing on NDN is the most focused testing points [10]. In particular, the start to service throughput estimation on NDN has all the earmarks of being temperamental, though the web service provider of substance is obscure to the user [11]. Moreover, due to throughput estimation, halfway reserving on NDN switch's Content Store could prompt parcel misfortune temporarily. It introduce a working interest adjustment map, the proposed process, which works by proactively assessing the output in a jump-by-bounce format. The customer hub is then helped with the latest accessible start to finish transfer speed. In this way, the video player can instantly adjust to the difference in arrange condition. The web service execution assessment utilizing NDN-JS and DASH-JS on the arrangement organizes an exhibit that the new arrangement gives better normal stream bit-rate and expends less system transmission capacity than the customary framework. The active performance principle is used to implement the performance of the network, which offers the viability of the service structure in various network conditions [12]. Basically, the NDN system doesn't run through IP-based, and it's been working through naming also the messaging protocol exchanging the structured data about the implementation in web services in 5G network. The highlighted points exchanging the naming-based messaging protocol will be added advantage to improve the web service using the 5G network [13]. The suggested approach is used to find the bandwidth available for the congestion route, and an optimal solution is assigned to provide the best service. The future Internet architecture is assigned for sending data packets to any named services without allowing response time delays. Further, the user acknowledged the data packet through the congested data path and rated the service experience on the same. In the proposed active throughput calculation methodology, data quality in video streaming is used to minimize service response time. The normal streaming bit rate data is lower than the proposed active throughput average bit rate in video streaming. The proposed system is measured at a bit rate of 4254 kbps, while the existing system has a lesser bit rate of 4400 kbps value on the average video streaming.

This proposed system presents a strategy for active throughput calculation and interest architecture for video gushing over NDN. Our methodology encourages the user to ascertain the base throughput along the conveyance way more precisely. The investigation utilizing NDN-JS and DASH-JS in the arranged system exhibits the productivity of our proposition. In any case, it is important that the proposed arrangement is able to do effectively responding to changes in the accessible transmission capacity when the throughput of each bounce along the conveyance way is known. In future work, the calculation administering bit rate adjustment is refined by considering extra factors, and the reduced response time in retransmission and various network options will examine the usage of our framework on the different web portals [14].

2 Literature Review

In "Web Service Testing Tools": The performance analysis of quality of service when there are multi-task arrangements and usage of web service. To test the task arrangements in quality of service there are many tools equipped with analyzing the quality of services such as SOAP UI, Apache JMeter, Postman, Sauce Labs, Ranorex Studio, Test Rail, Strom, etc. The above-mentioned tools are utilized for measuring the quality of services in a communication network. The improvements are carried on the web application response with accuracy ratio, throughput, and feasibility.

In "The performance Analysis of Web Service Testing Tools" [1], improvements of the services is a major role where the tools will examine the service quality. Web service plays a vital role in a web application that communicates and improves the service quality. The performance analysis of different network features examining the service quality. The parameter taken into consideration for testing the performance is maximum response time, minimum response time, throughput and calculation of bytes.

In "Web Application Performance analysis Testing Tools: The performance of network feature analysis" [2] determines the application feature performance testing with a parameter such as data usage, throughout, viewable duration, number of hit pages, and CPU utilization.

The performance improvement of web service using SOAP UI testing is described in detail for web application [15]. It has revised existing performance and improved the behavior in multi-functions with heavy traffic load.

The performance improvement of web service can be tested with different networking conditions conducted by using SOAP UI [16]. This provides the convenience of Web Service Description Language (WSDL) for improving the operations and bit transfer based on the request after importing the Web Service Description Language into the tool.

Elbehiery and Elbehiery (2019), 5G technology is the solution for connecting all the requirements with humans. The 5G network system utilized a transport layer with the quickest live video streaming, entertainment content, and social networking. The real-time applications and Internet of Things (IoT) are integrated with a 5G network and reduces the response time with low latency. The existing 4G network cannot connect with public cloud and multi-cloud at the same time, but the proposed 5G network solution enables business tools, compliance standard, SaaS application monitoring, resource management, and automation. The resource management of public cloud and multi-cloud is maintained properly through the architecture. The planned 5G network encouraging infrastructure contrasted with the current system and split the infrastructure for different resources and web services. This is the core concept to improve the performance across the 5G network, and it will easily adapt the future operations [17].

3 Network Testing Considered in Signal Error Detection and Correction Format

The signal error rate process is determined by comparing the transmitted continuous bits with the received bits, and the number of errors is defined. The total ratio of how many bits received in signal error over the number of total bits received is the signal error rate. The accurate signal rate ratio is 10^{-9} is often considered the minimum acceptable signal error rate for 5G applications [13]. The 5G network basically 10 Gbps data rate feasibility with 1 ms latency. Data communications have more stringent requirements based on consumer request at some positions 10^{-13} is often considered a minimum. The signal rate improvement is based on bandwidth availability. Another method that can be adopted to reduce the signal error rate is to reduce the bandwidth in 5G network. In 5G network, it is inbuilt 1000× bandwidth per unit area. The poor noise signals will be obtained, and the signal-to-noise ratio will then increase. Finally, this results in a reduction of the data throughput attainable.

Similar to Apple's HTTP Live Streaming (HLS) solution, MPEG-DASH works by breaking the content into a sequence of small HTTP-based file segments, each segment containing a short interval of playback time of content that is potentially many hours in duration, such as a movie or the live broadcast of a sports event.

In such all switches assigned to some particular attribute format and different data formats such as audio and video streaming transmitted through this stream. Here, HTTP live streaming is essentially broadcast through without streaming server so that the throughput is most relevant due to the different size of data that can be transferred through this method. 5G-based active method is proposed throughput system to improve the web service communication system [13].

The existing HTTP scheme video streaming protocols are transport protocols. The most common TCP transmission control protocol provides a reliable process for every data transmission. The issues on the existing system are defined clearly in the below steps [12]. Some quality and security issues endure in the existing setup. The NDN-based 5G data network is the proposed network that is implemented for video streaming about 100% coverage with network availability. The minimal network energy usage is with long battery life.

4 System Analysis

The adaptive bit rate scheme is enabling with existing network setup for video streaming option for Internet transactions. Over the TCP/IP network, the Internet video transaction application is implemented through HTTP. Over the TCP/IP network might have network packet access issues since the existing setup firewall will intercept the packet transmission. Dynamic Adaptive Streaming over HTTP (MPEG-DASH) creates the environment standard which allows users to dynamically request the best optimal video, but the video streaming cannot be offered due to firewall intercepting [18].

5 Enhanced System Design

The proposed system implies that the implementation of Dynamic Adaptive Streaming Over HTTP (DASH) through Named Data Networking (NDN) seems to be challenging of fifth-generation network since the transmission of available content disseminated without knowing the destination. The transmission rate is calculated through end-to-end delay which is not a regular one [14]. Further exploits of this issue will occur when the transmission taking a partial cache of content. In this case, the requester might be misled by particle cache content to calculate the throughput with an available bandwidth for further content which does not exist on the content cache. This issue will create packet loss while transmitting the further cache content. Here, versatile bit-rate video streaming over Named Data Networking is calculated to solve the highlighted traditional network problem [19]. The 5G network with the so-called data network architecture efficiently distributes the huge amount of content in the multipath forwarding strategy on a large scale. The idea is to measure the throughput and bandwidth of each NDN router to its neighboring requestors and send the transmission of content to an enhanced 5G network. The throughput estimate will be checked at the request of the user, which will include the option to verify the bit rate of the usable bandwidth. The 5G network throughput estimation on named data networking will find the bandwidth availability. The process of active throughput in congested link delivery path in available bandwidth will make sure video quality in available bandwidth using 5G network [20]. The proven experimental results are clearly showing that the 5G NDN network will provide better performance and quality.

5.1 Steps for Performance Improvements and Existing Analysis

Existing Problems:

(a) Data streaming issues in congestion link
(b) Delay in customer response
(c) Data quality Issues
(d) Cache content Issues
(e) Low data rate and throughput.

These are the areas to concentrate in the existing 3G and 4G networks. Still, improvements are required to improve the above-mentioned points. The experimental methods are clearly saying about to overcome the issues.

5.2 Steps for Overcoming the Issues

The data bit rate output for the new end-to-end throughput and the average value were taken from the specific time period between data transmission in the current 4G conventional network. The efficiency of the data rate is calculated from end-to-end throughput and is greater than the average. Our proposed system will deploy the throughput estimation in the hop-to-hop date transmission [21]. The process of active throughput estimation is performed in named data networking. The interest packet is first sent to the network router further then the router immediately records the timestamp, interest in time, and outbound value. Now the transmission time is calculated between two hops. While the data is sent back to the consumer according to the Pending Interest Table from the service provider then active throughput calculation is estimated successfully. The return data transmission from the service provider will be given by the inbound value, timestamp, and interest out time [22, 23]. The provided parameters append by the router and calculate the throughput of a specific hop. The calculated active throughput value appends to the data packet for the hops.

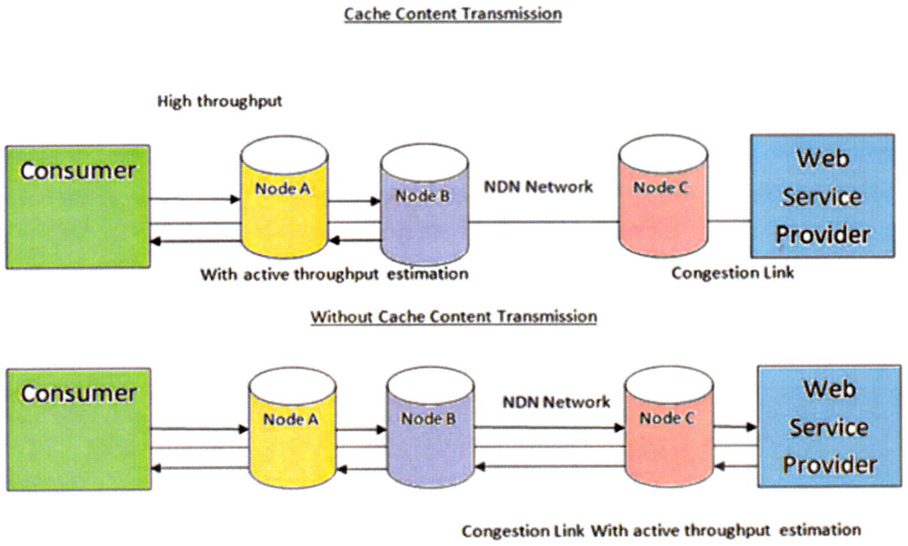

Fig. 1 Active throughput adaption for 5G NDN data streaming

The proposed idea is different from the existing method while processing the interest request from the consumer the active throughput estimation method will check the end-to-end throughput if the value is lower than the average one then immediately the consumer will adapt with the active throughput estimation. Whereas the new 4G system does not have the provision to adjust the hop-to-hop fashion because NDN does not adopt the IP-based transmission and prevents the conversion from application data names to IP addresses, reducing the processing time.

Figure 1 illustrates the cache content transmission while the high-throughput transmission between node B to the consumer, whereas there will be congestion network link between Node C to web service provider (Fig. 2).

In NDN delivery, the web service system novel throughput estimation is derived as the specific time that the allocated medium is used to send data (bits). The active throughput can be estimated by preparing the maximum events that occurred on the assigned medium in the web service slot [24]. Assign Wpi, Wpc, Wps …, Wpn for web service slots which can be utilized for successful data transmission. In such Dc and Ds are the duration slot for the collision and success rate for the data transmission. The average duration is decided for the assigned duration.

5.3 *Algorithm Equations and Operational Methods*

$$\mathrm{Da} = \mathrm{Wpi}\,\mathrm{Di} + \mathrm{Wpc}\,\mathrm{Dc} + \mathrm{Wps}\,\mathrm{Ds} \tag{1}$$

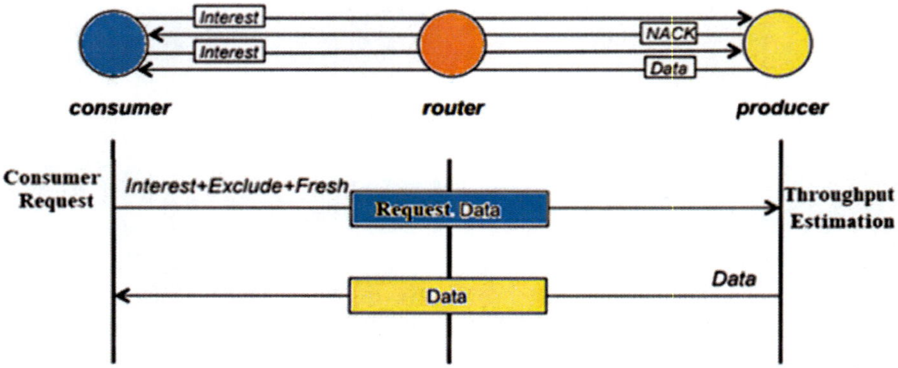

Fig. 2 NDN forwarding plane interest transmission

Now the active throughput At estimated as (1),

$$\begin{aligned}\text{At} &= \text{E1}\big[\text{Web service payload data transmitted in the Assigned time}\big]/\\ &\quad \text{En}\big[\text{Total length of the Assigned time}\big] = \text{Wps} \times \text{E1}\\ &\quad \big[\text{Wservice payload data transmitted in the Assigned time}\big]/\end{aligned}$$

$$\begin{aligned}\text{Da} &= \text{Wps} \times \text{E1}\big[\text{Wservice payload data transmitted in the Assigned time}\big]\\ &\quad /\text{Wpi Di} + \text{Wpc Dc} + \text{Wps Ds}\end{aligned}$$

where E1 [Wservice payload data transmitted in the Assigned time] is the derived average data load size (Units) and Wps × E1 [Wservice payload data transmitted in the Assigned time] is the average size of data successfully transmitted in the assigned slot [25].

By formulating the novel active throughput equation as follows,

$$\begin{aligned}\text{At} &= \text{E1}\big[\text{Web service payload data transmitted in the Assigned time}\big]\\ &\quad /\text{Ds}/\big[1 + \text{Wpc}/\text{Wps} \times \text{Dc}/\text{Ds} + \text{Wpi}/\text{Wps} \times i/\text{Ds}\big]\end{aligned}$$

Figure 3 illustrates the multi-access data communication and solves the congestion link issues and improves the data communication [26]. The above equation and flowchart clearly explain the modification of active throughput estimation and analysis with improvement in data communication [27]. The improved throughput system provides better streaming availability. Based on the throughput, delay time has reduced and the customer response increased. Data access rate has been increased which will impact the delay time and improved the customer service.

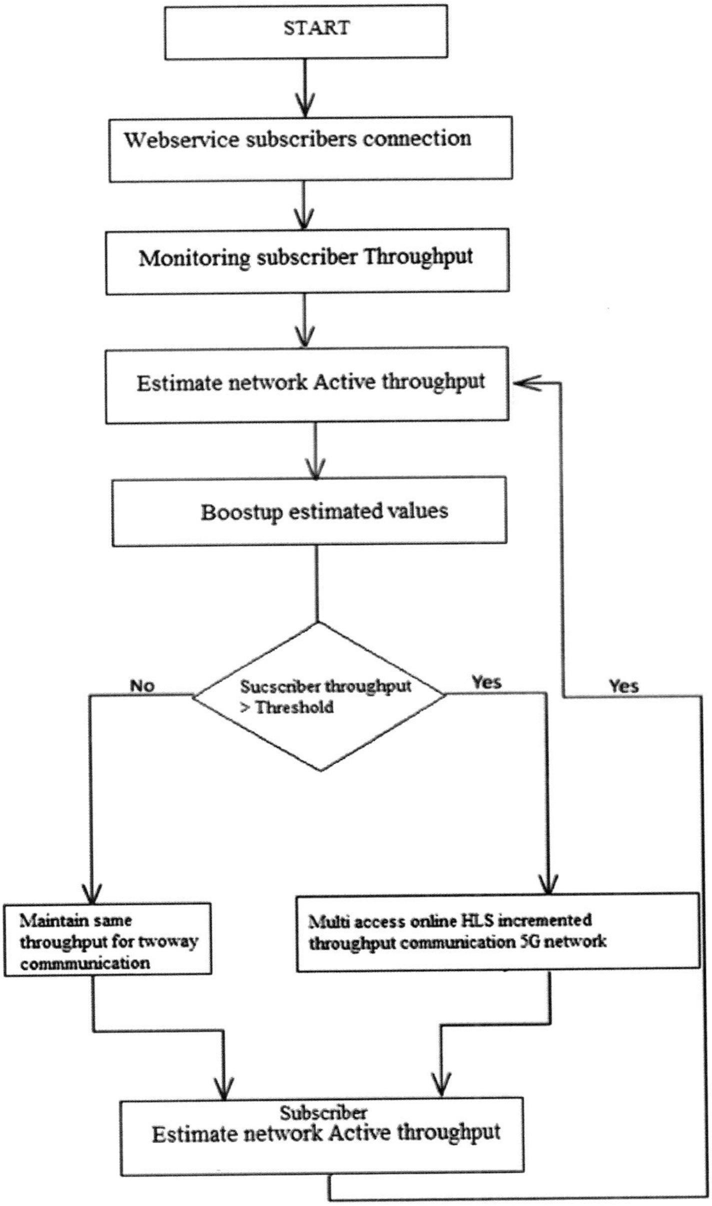

Fig. 3 Active throughput flowchart

Maximum Data Rate = 98 kbps
Number of Source $i = 1$
Data Frame Rate $F = 3$ Packets/min = 3 Packets/60 s = 0.05 Packets/s
Frame Data Size $L = 64$ bytes = 64 * 8 bits = 512 bits
Data Access Rate of the Channel $R = 0.184 * 96{,}000$ bps = 17,664 bps
Transmission Time $T = L/R = 512$ bits/17,664 bps = 17,664 bps
$G = 3 * 0.05$ Packet/s * 0.029 s $= 4.35 * 10^{-3}$
Throughput $S = 4.35 * 10^{-3} * 0.997 = 4.34 * 10^{-3}$.

Consider the bit rate adaption the following terms are followed to maintain the data communication in various congestion aspects. Compared to the previous estimation techniques the ideal and successful events are constant in assigned slots since the static system parameter analysis is necessary to define the values in assigned tasks [28]. The proposed analysis system is assigned for a dynamic system with two-way communication, and the duration for Dc and Ds is mentioned in the duration for the collision, and duration for the success data transmission [29].

Due to the 4G communication, the two-way communication was not enabled with the required throughput level. The 4G throughput range is not enough to broadcast high-range data communication [26, 30] (Table 1).

In Such,

Every Data Communication Bits per symbol: $B_s = \text{Log}_2 M$
Throughput maintains the symbol rate: $R_2 = (1/1 + A)\ B_w$ (Symbol/s)
Congestion Gross Bit Rate: $R_g = B_s R_2 = \text{Log}_2 M(1/1 + A) B_w$ (bps)
Successful Net Data Rate = $R_i = \text{Log}_2 M(1/1 + A) B_w (1 - V_a)$ (bps)

The congestion link has a low throughput. The proposed system will encounter issues in cache content transmission and without cache content transmission. The objective of the proposed work analyzed in terms of data streaming performance improvement in congestion link, customer response time reduced; achieve data quality without cache content transmission [31]. The customer response delay time is reduced due to the proposed algorithm (Table 2).

Compared with the existing system, the current issues not analyzed in future network research should utilize the proposed system to improve the performance [15, 31] (Fig. 4).

6 Conclusion

In this article, several experiments were taken into XML optimizer to find better performance compared with existing analysis and results. The experimental results are identified that SOAP API response time is a competitive one [32]. The maximum size of XML files has reduced 92% which is a better improvement in terms of XML files. SOAP is the most reliable XML optimizer which will provide an easy retry approach in terms of system failure [33]. SOAP has a default web service structure

Table 1 Throughput calculation

Comparison	Throughput calculation				
Algorithm and author details	Video streaming 1 (s)	Video streaming 2 (s)	Started	Minimum level	Maximum level
Active throughput with signal boost	*1.3*	*10.6*	*0.8–0.36*	*0.12*	*0.58*
Existing LR algorithm Author: Sandosh S 2019 (Google Scholar)	1.7	10.5	0.9–0.40	0.13	0.60
Existing Viterbi algorithm Author: Joe Bayley 2019 (arxiv)	1.8	10.6	0.9–0.41	0.13	0.61
Existing safe regression testing Author: Tehreem Masood 2016 (arxiv)	1.9	10.5	0.9–0.42	0.14	0.63
Partitioning algorithm Author Nidhi Arora, Savita Kolhe 2016 (Research gate)	1.9	10.6	0.9–0.39	0.12	0.58

through which WS security features can be enabled. Web service [34] is a crucial task which calculates various parameters and analyses. The proposed system will calculate the throughput and bandwidth from each NDN router and requesters and deliver the content through an improved network. The 5G network active throughput estimation on named data networking will calculate the available bandwidth. The ideal process of active throughput in 5G network congested link delivery path bandwidth will ensure the video quality in the video streaming process. The proposed system experimental results are explicit that 5G NDN network will provide much better quality with less response time.

Table 2 Comparison table task-1 and task-2 restful server and SOAP-WSDL server

	Task	Restful	SOAPWSDL	Restful	SOAPWSDL	Restful	SOAPWSDL	Restful	SOAPWSDL
		ST	ST	ET	ET	TT (min)	TT (min)	MC	MC
Data	*1*	*19.05*	*19.1*	*20.5*	*21.23*	*105*	*133*	*1.009*	*1.641*
	2	21	21.3	22.2	22.58	80	88	0.079	0.087
Data	*1*	*19.1*	*19.15*	*21*	*21.38*	*110*	*143*	*1.122*	*1.787*
	2	21	21.4	22.12	23	72	80	0.072	0.083

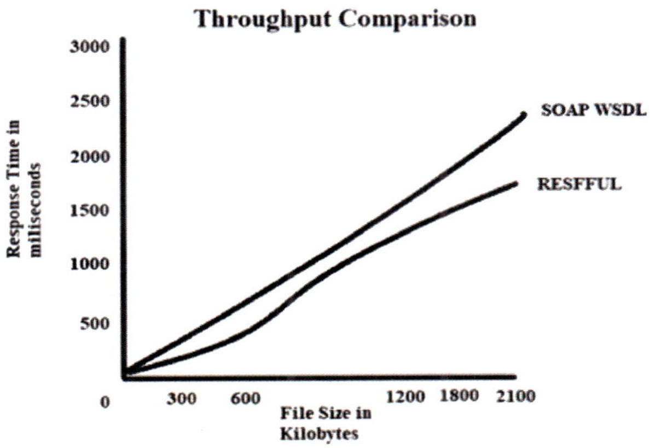

Fig. 4 Throughput comparison

References

1. R. Kumar, A comparative study and analysis of web service testing tools. Int. J. Comput. Sci. Mob. Comput. **4**(1) (2015)
2. S. Kumari, S.K. Rath, Performance comparison of SOAP and REST based web services for enterprise application integration, in *International Conference on Advances in Computing, Communications and Informatics (ICACCI)*, Kochi, 2015, pp. 1656–1660. https://doi.org/10.1109/ICACCI.2015.7275851
3. C. Pautasso, O. Zimmermann, F. Leymann, RESTful web services vs. "big" web services: making the right architectural decision, in *Proceedings of the 17th International Conference on World Wide Web*, Beijing, China, 2008, pp. 805–814
4. I. Ghani, W.M.N. Wan-Kadir, A. Mustafa, Web service testing techniques. Int. J. Adv. Comput. Sci. Appl. (IJACSA) **10**(8) (2019)
5. P. Ahlawat, S. Tyagi, A comparative analysis of load testing tools using optimal response time. Int. J. Adv. Res. Comput. Sci. Softw. Eng. (IJCSE) (2013)
6. J. Leveille, D. Kratz, Building analytic services with SAS® business intelligence. Paper 005-2009
7. Rina, S. Tyagi, Comparative study of performance testing tools. Int. J. Adv. Res. Comput. Sci. Softw. Eng. (2013)
8. R. Coste, L. Miclea, API testing for payment service directive 2 and open banking. Int. J. Model. Optim. **9**(1) (2019)
9. Web Services Conceptual Architecture, https://www.csd.uoc.gr/~hy565/newpage/docs/pdfs/papers/wsca.pdf. Accessed 02 Oct 2014
10. L.L. Iacono, H.V. Nguyen, P.L. Gorski, On the need for a general REST-security framework. Future Internet **11**, 56 (2019). https://doi.org/10.3390/fi11030056
11. S. Hussain, Z. Wang, I. Kalil Toure, A. Diop, Web service testing tools: a comparative study. Int. J. Comput. Sci. (IJCSI) (2013)
12. B. Göschlberger, G. Anderst-Kotsis, *A Web Service Architecture for Social Micro-Learning*, 2–4 Dec 2019 (ACM, 2019). ISBN 978-1-4503-7179-7/19/12, 9
13. J.I.Z. Chen, 5G technology and advancements in connected living-comprehensive survey. J. Electron. **1**(02), 71–79 (2019)
14. A. Soni, V. Ranga, API features individualizing of web services: REST and SOAP. Int. J. Innov. Technol. Explor. Eng. (IJITEE) **8**(9S) (2019). ISSN: 2278-3075

15. R.R. de Oliveira, R.V. Vieira Sanchez, J.C. Estrella, R. Pontin de Mattos Fortes, V. Brusamolin, Comparative evaluation of the maintainability of RESTful and SOAP-WSDL web services, in *2013 IEEE 7th International Symposium on the Maintenance and Evolution of Service-Oriented and Cloud-Based Systems (MESOCA)*, Eindhoven, 2013, pp. 40–49. https://doi.org/10.1109/MESOCA.2013.6632733
16. N. Serrano, J. Hernantes, G. Gallardo, Service-oriented architecture and legacy systems. IEEE Softw. (2014)
17. K. Elbehiery, H. Elbehiery, Int. J. Electron. Commun. Eng. SSRG IJECE J. **6**(8) (2019)
18. Y. Yuan, W. Zhang, X. Zhang, H. Zhai, Dynamic service selection based on adaptive global QoS constraints decomposition. Symmetry **11**, 403 (2019). https://doi.org/10.3390/sym11030403
19. S. Mumbaikar, P. Padiya, Web services based on SOAP and REST principles. Int. J. Sci. Res. Publ. **3**(5) (2013). ISSN 2250-3153
20. N. Negi, S. Chandra, Efficient selection of QoS based web services using modified TOPSIS method. Int. J. Recent Technol. Eng. (IJRTE) **8**(2) (2019). ISSN: 2277-3878
21. H. Haviluddin, E. Budiman, N.F. Hidayat, A database integrated system based on SOAP web service. TEM J. **8**(3), 782–787 (2019). ISSN 2217-8309. https://doi.org/10.18421/TEM83-12
22. Z. Mahmood, Enterprise application integration based on service oriented architecture. Int. J. Comput. **1**(3) (2007)
23. N.M. Ibrahim, M.F. Hassan, A comprehensive comparative study of MOM for adaptive interoperability communications in service oriented architecture. Int. J. Trend Sci. Res. Dev. (IJTSRD) **3**(3). e-ISSN: 2456 – 6470
24. P. Dencker, R. Groenboom et al., Methods for testing web services (2019)
25. C. Fibriani, A. Ashari, M. Riasetiawan, Spatial data gathering architecture for precision farming using web services. Int. Conf. Eng. Technol. Innov. Res. J. Phys. Conf. Ser. **1367**, 012024 (2019). IOP Publishing. https://doi.org/10.1088/1742-6596/1367/1/012024
26. R. Valanarasu, A. Christy, Comprehensive survey of wireless cognitive and 5G networks. J. Ubiquitous Comput. Commun. Technol. (UCCT) 23–32 (2019)
27. P.M. Patil, K. Gohil, R. Madhavi, Web services for mobile computing. Int. J. Adv. Res. Comput. Eng. Technol. **1**(3) (2012). ISSN: 2278 – 1323
28. A.M. Al-Qershi et al., Enhancement of soap web services model performance based on optimized xml files. GSJ **7**(10) (2019). Online: ISSN 2320-9186. Vishawjyoti, R. Agrawal, Comparative study on web service data interchange formats. IIOABJ **10**(2), 27–31 (2019). ISSN: 0976-3104
29. C.S. Keum, S. Kang, I.-Y. Ko, J. Baik, Y.-I. Choi, Generating test cases for web services using extended finite state machine, testcom (2016)
30. A. Arcuri, RESTful API automated test case generation with EvoMaster. ACM Trans. Softw. Eng. Methodol. **28**(1), Article 3 (2019)
31. K. Poobai, S. Awiphan, J. Katto et al., Adaptive bit-rate video streaming on named data networking with active throughput estimation. Association for Computing Machinery. https://doi.org/10.1145/3209914.3209929
32. F. Alshraiedeh, N. Katuk, SOAP and RESTful web service anti-patterns: a scoping review. Int. J. Adv. Trends Comput. Sci. **8**(5) (2019)
33. J.L. Khachane, L.R. Desai et al., Survey paper on web services in IOT. Int. J. Sci. Res. (IJSR). ISSN (Online): 2319-7064
34. K. Kaewbanjong, S. Intakosum et al., QoS attributes of web services: a systematic review and classification. J. Adv. Manag. Sci. **3**(3) (2015)

Survey for Electroencephalography EEG Signal Classification Approaches

Safaa S. Al-Fraiji and Dhiah Al-Shammary

Abstract This paper presents a literature survey for electroencephalogram (EEG) signal classification approaches based on machine learning algorithms. EEG classification plays a vital role in many health applications using machine learning algorithms. Mainly, they group and classify patient signals based on learning and developing specific features and metrics. In this paper, 32 highly reputed research publications are presented focusing on the designed and implemented approach, applied dataset, their obtained results and applied evaluation. Furthermore, a critical analysis and statement are provided for the surveyed papers and an overall analysis in order to have all the papers under an evaluation comparison. SVM, ANN, KNN, CNN, LDA, multi-classifier and more other classification approaches are analyzed and investigated. All classification approaches have shown potential accuracy in classifying EEG signals. Evidently, ANN has shown higher persistency and performance than all other models with 97.6% average accuracy.

Keywords EEG classification · EEG signal · EEG survey

1 Introduction

An electroencephalogram (EEG) is an efficient, cheap-cost, non-invasive test applied to testify the brain electrical activity [1]. EEG is one of the most techniques used to determine an abnormality of the brain functions [1, 2]. EEG signals are computed using electrodes set on the scalp. It is used for monitoring and diagnosing neurological diseases like epilepsy and sleep disorders [3]. Furthermore, EEG signals are utilized for several studies and research such as gaming applications, lie detection, augmented reality, neuromarketing and brain–computer interface (BCI) and others [3, 4].

S. S. Al-Fraiji · D. Al-Shammary (✉)
College of Computer Science and Information Technology, University of Al-Qadisiyah, Dewaniyah, Iraq
e-mail: d.alshammary@qu.edu.iq

S. Shakya et al. (eds.), *Mobile Computing and Sustainable Informatics*, Lecture Notes on Data Engineering and Communications Technologies 68,
https://doi.org/10.1007/978-981-16-1866-6_14

1.1 Motivation

EEG classification plays a vital role in several services and applications based on EEG [5, 6]. It represents an important source for medical activities such as diagnosing people with epilepsy, diagnose sleep disorders, depth of anesthesia, coma, encephalopathies and brain death [7].

- **Time consumption and diagnose availability**: Generally, specialist neurologists analyze the records of EEG visually. This is time-consuming and not always available for remote patients; therefore, machine learning algorithms have been potentially applied for automatic detection and/or prediction of epileptic seizures in EEG form.
- **EEG classification drawbacks in machine learning algorithms**: Machine learning algorithms that have been developed for classification suffer from high stagnation probability, stuck with local optimum, high time requirements, and in persistent results. Technically, it is significantly required to develop a potential classification model that can overcome traditional classification problems and disadvantages.

1.2 Survey Strategy and Evaluation

Potential thirty-two existing approaches are classified into seven groups based on their proposed classification method (Support Vector Machine, Artificial Neural Network, K-Nearest Neighbor, Convolution Neural Network, Linear Discriminant Analysis, Multi-classifier and other classification). Each approach is summarized with their problem statement, proposed solution approach, performance evaluation strategy and results, best achievement. All approaches are analyzed, and critical statement is provided. In this survey paper, performance metrics: accuracy, sensitivity, specificity and processing time are targeted and extracted in order to evaluate previous research. All results of performance measures are depicted and compared. We have concentrated on best achieved results and cross-evaluated inside their groups. Finally, average of best accuracy is calculated for all approaches inside the same group and compared with classification groups. ANN group has shown the highest achieved accuracy. Unfortunately, most studies have ignored the measurement of processing time, and therefore, system performance is not clear for the included studies.

1.3 Paper Organization

The structure of this paper is arranged as follows: Sect. 2 illustrates classification approaches based on Support Vector Machine for EEG Classification. Section 3 describes Artificial Neural Network (ANN) methods for EEG classification. Next,

Sect. 4 describes Convolution Neural Network (CNN) approaches for EEG classification. Then, K-Nearest Neighbor (K-NN) methods for EEG classification are described in Sect. 5. Section 6 illustrates EEG classification methods based on Linear Discriminant Analysis (LDA). Section 7 illustrates Multi-classifier approaches for EEG classification. Section 8 describes a combination of other models for EEG classification. Analysis and evaluation are described in Sect. 9. Finally, Sect. 10 presents conclusion and future work.

2 Support Vector Machine for EEG Classification

Support Vector Machine (SVM) as a successful classification method has been widely applied in machine learning algorithms for EEG signals. Several publications have concentrated on SVM as the core for their proposed solution. Li and Shen [8] have addressed the classification of EEG signals for the mental task, which is one of the main issues of the computer–brain interface (BCI). This approach has been proposed using wavelet packet entropy and Support Vector Machine. This research has applied seven level wavelet packet decomposition to each channel of EEG with db4. Moreover, four spectrum bands (α, β, θ, ς) are extracted and applied by an entropy algorithm that performed on each band. For evaluation, accuracy is the main metric this research used to investigate the success of the proposed model. The obtained results have shown a persistent success for SVM for two-class classification with average accuracy of 87.5–93.0% and for three-class classification with average of 91.4% for subject 1. Colorado State University has been utilized. This approach has provided a high accuracy averaged 93.0%. The provided results and evaluation are considered to be limited which have shown only the accuracy as an evaluation metric.

Shen et al. [9] have addressed the classification of EEG signals. The main goal is to find relationship of stereoscopic acuity and EEG data for the development of 3D technology. A multi-channel selection sparse time window common spatial group (MCS-STWCSG) has been proposed. First, signals based on depth dynamic random-dot stereogram (DRDS) videos are obtained and preprocessed. Second, an improved common spatial pattern (CSP) method applied to select channels. Next, signals segmented by wavelet transform and sliding time window. Then, a common spatial group (CSG) has been applied to extract EEG signal features. Moreover, time–frequency bands and hybrid features have been selected by sparse regression. Finally, Support Vector Machine (SVM) with RBF kernel has been applied for features classification. Accuracy metric has been computed to evaluate the three proposed methods (3C-STWCSG, CCS-STWCSG and MCS-STWCSG) to select different channels. The obtained average accuracy values are 50.96%, 73.13% 87.50%, respectively. EEG signals gathered by international 10–20 systems, embedded 32 lead EEG cap and Neuroscan system. The proposed method has achieved a high performance with accuracy of 94.67%. In fact, this research has failed to show their model performance like processing time. Moreover, accuracy was the only used metric to evaluate the proposed solution.

More potential SVM-based models for EEG diagnose and prediction have been achieved by more researchers and developers like Li et al. [10], Chen et al. [11] and Chisci et al. [12].

3 Artificial Neural Network (ANN) for EEG Classification

George et al. [13] have focused EEG classification problem and discussed how to automatically identify cases of epilepsy and know the seizures patients. The proposed approach has combined several techniques such as tunable-Q wavelet transform (TQWT), entropies, Particle Swarm Optimization (PSO) and Artificial Neural Network (ANN) to classify epileptic seizures and diagnose its types. Technically, the proposed model starts by transforming EEG signals using Tunable-Q Wavelet, extract features, then feature selection with PSO and finally ANN to classify cases. Three different metrics have been used: Accuracy, Sensitivity and Specificity for different experimental cases of EGG datasets. Two types of dataset have been used, first the Karunya Institute of Technology and Sciences (KITS) EEG database and the second Temple University Hospital (TUH). The proposed method has achieved high accuracy of 95.1, 97.4, 96.2 and 88.8% for all the four experimental cases (normal-focal, normal-generalized, normal-focal + generalized and normal-focal-generalized). The authors have not analyzed the different performance of the proposed model for the two deployed datasets.

Another study achieved by Samiee et al. [14] that had classified EEG signals to detect epileptic seizures. One of the challenges is to distinguish regular discharges from nonstationary patterns occurring when seizures. ANN has been proposed as a potential EEG classifier. Discrete Short Time Fourier Transform (DSTFT) has been applied to feature extraction from EEG. ANN represented by a Multilayer Perceptron (MLP) whose performs the classification to distinguish seizure times from seizure-free times. Experimentally, the dataset has been divided into two groups 50 and 50% for training and testing their proposed system. Three different metrics have been used such as Accuracy, Sensitivity and Specificity for different experimental cases of EGG datasets. BONN Dataset University of Bonn, Germany's EEG dataset has been used to examine proposal model. This approach has achieved a high classification accuracy of 99.8%. Although the proposed approach provided high results, when compared with other methods, some earlier studies have achieved higher accuracy. The authors have failed to justify these higher metrics value like (Xie and Krishnan) with 100% accuracy.

More significant ANN-based models for EEG diagnose and prediction have been achieved by more researchers and developers like Guo et al. [15], Luo et al. [5], Satapathy et al. [16] and Bhatti et al. [17].

4 Convolution Neural Network (CNN) for EEG Classification

Raghu et al. [18] have addressed the necessity for recognizing seizure EEG of epileptic patients. In fact, classifying EEG seizure type is a potential requirement for patients' diagnosis and disease control. The main goal is to classify multiple seizure types. Two different approaches have been applied using CNN. First approach transfers learning using a pre-trained network. The second approach tries to extract image features using a pre-trained network and classification using the vector support classifier. In this paper, dataset has been divided into two groups 70 and 30% for training and testing their proposed system. Moreover, they have repeated this methodology for 10 times. Temple University Hospital EEG corpus dataset has been utilized for evaluation. The proposed method has achieved high accuracy of 82.85% and 88.30% by using transfer learning and extract image features approach respectively. Although it is well-known that deep learning model CNN considered to be a time-consuming solution, it is not clear for this paper how good performance can be obtained from the classifier.

Ramakrishnan et al. [19] have focused on detecting patients with epileptic seizure based on Convolutional Neural Network (CNN). This paper has tested both time domain and frequency EEG features and their impact on CNN. CNN has been proposed as a significant deep learning model can be used for EEG classification. The authors experimented both time and frequency domain and found out that time domain features may enhance the proposed system performance. This research has concentrated on the analysis of multiple scenarios with different cases. Three metrics like sensitivity and specificity, classification accuracy are computed to investigate the system capabilities. Two dataset have been used. First, BONN Dataset University of Bonn, Germany's EEG dataset and the second from Boston Children's Hospital CHB-MIT Dataset. This method has provided a high accuracy, up to 98% overall classification. It has provided a good performance as a very short execution time needed.

More significant CNN-based models for EEG diagnose and prediction have been achieved by more researchers and developers like Lian et al. [20] and Mousavi et al. [4].

5 K-Nearest Neighbor (K-NN) for EEG Classification

Awan et al. [21] have discussed the necessity for classification and feature extraction from facial emotions and movements registered using EEG signals. In this paper, the K-Nearest Neighbor Algorithm has been proposed. Initially, Raw EEG signals Segmentation and Selection (SnS) with Root Mean Square (RMS) were applied to extract feature vector of EEG signals. EEG classification was achieved by using a k-nearest neighbor algorithm. Evidentially, accuracy measures were computed in order to verify their results. They have obtained the best result when SRMS was applied. Furthermore, RMS (root mean square), MAV (mean absolute value) and IEMG were applied. KNN was applied on these methods output with accuracy of 80.2%, 80.4% and 71.3%, respectively. EEG signals have been collected from 10 healthy people, aged between 18 and 45 years. This approach achieved a high accuracy of 96.1%. Empirically, accuracy was the only used metric to evaluate the proposed solution. It would be more useful for both academics and developers to have other evaluation metrics obtained like processing time and complexity.

Lahmiri and Shmuel [7] have discussed Classification for Electroencephalogram (EEG) signals to distinguish between seizure intervals and seizure-free intervals in epileptic patients. In this paper, the K-Nearest Neighbor Algorithm has been proposed. The Generalized Hurst Exponent (GHE) estimated at various scales to characterize EEG signal through capturing properties of its multiscale long memory. The moment that computed K-NN has been trained for classification based on GHE estimates. Tenfold cross-validation is utilized and operated 20 times to guarantee more repetitions and randomness. Furthermore, three metrics such as accuracy, sensitivity and specificity have been used as evaluation metrics. The proposed model has outperformed several previous studies models in terms of accuracy. Dataset has been provided by University of Bonn dataset, Germany. The proposed method has achieved a high accuracy of 100%. In fact, this research has failed to show their model performance like processing time.

More significant K-NN-based models for EEG diagnose and prediction have been achieved by more researchers and developers like Oliva and Rosa [22].

6 Linear Discriminant Analysis (LDA) for EEG Classification

You et al. [23] have discussed successful motor imagery electroencephalogram (MI-EEG) classification model through accurate and efficient classification with quick feature extraction. In this paper, a flexible analytic wavelet transform (FAWT) has been proposed as a classification system for motor imagery electroencephalogram (MI-EEG). Technically, (MI-EEG) signals get in band filter as a preprocessing step. Furthermore, feature extraction was applied based on (FAWT). Then, feature selection was implemented, and multidimensional scaling (MDS) was applied to reduce feature dimensions. Finally, classification was achieved by using linear discriminant analysis (LDA). Experimentally, 50–50% (train and test) approach has been performed 10 times for each subject and obtained the resultant average. Accuracy and Maximal MaI have been applied as two evaluation metrics. Accuracy values for all subjects are 85.26% as an average of FAWT + MDS. Two public BCI EEG datasets published by BCI Competition II and III have been used. First dataset, BCI Competition II, 2003 https://www.bbci.de/competition/ii. Second dataset, BCI Competition

III, 2005 https://www.bbci.de/competition/iii/. This method has achieved a high accuracy of 94.29%. Although the proposed method has produced high results compared with other methods, some previous studies have achieved higher accuracy. The authors have failed to justify these higher metrics values like Lee et al. with (97.50% accuracy).

Another study achieved by Hsu [24] that has addressed a feature extraction approach by time-series prediction based on the adaptive neuro-fuzzy inference system (ANFIS) for brain-computer interface (BCI) applications. The main goal is for EEG Motor imagery (MI) classification. In this paper, adaptive neuro-fuzzy inference system (ANFIS) time-series prediction with multiresolution fractal feature vectors (MFFVs) has been applied for motor imagery (MI) EEG classification. Finally, classification has been performed based on linear discriminant analysis (LDA). Accuracy (ACC) and area under curve (AUC) metrics have been computed for model evaluation. MFFV features under ANFIS time-series prediction method have obtained 90.3% and 0.88 for accuracy and AUC, respectively. Evaluation dataset has been obtained from Graz BCI group. This approach has achieved a high accuracy of 93.7%.

7 Multi-classifier Approaches for EEG Classification

Several approaches utilized more than one classifier to recognize EEG signals. Rodrigues et al. [25] have focused on identifying and diagnosing alcohol addicts based on EEG classification through the application of machine learning techniques. In this paper, machine learning techniques have been proposed. Take the EEG data and applied the Wavelet Packet Decomposition (WPD) after that feature extraction has been performed. Finally, the classification performed by several machine learning techniques such as Support Vector Machine (SVM), Optimum-Path Forest (OPF), Naive Bayes, K-Nearest Neighbors (k-NN) and Multi-layer Perceptron (MLP). Experimentally, dataset has been divided into two groups 75 and 25% for training and testing their proposed system. Four metrics like accuracy, sensitivity, specificity, and positive predictive value (precision) are computed to investigate the system capabilities. The best obtained result from Naive Bayes classifier is about 99.78%. On the other hand, the worse result shown from MLP classifier producing 68.62% for specificity, sensitivity and accuracy, and the precision value was 61.85%. KDD dataset that has been originated from an examination of a number of subjects with alcoholic and non-alcoholic cases. This method achieves a high classifier accuracy of 99.78%. Although the proposed method provided high results, when compared with other methods, there are some previous studies that have achieved higher accuracy. The authors have failed to justify these higher metrics values like Upadhyay with (100% accuracy).

Another approach proposed by Piryatinska et al. [26] that has discussed how accurate classification for EEG signal based on efficient feature selection and training. It was found that feature selection represents the most important step in EEG signals classification. The proposed approach was based on the theory of ϵ-complexity of continuous functions. Firstly, ϵ-complexity coefficients of the original signal and its finite differences have been estimated. Secondly, random forest (RF) and support vector machine (SVM) have been evaluated as classifiers. Accuracy metric was used to evaluate the obtained results. Empirically, the proposed model has achieved an accuracy rate of 85.3% with bootstrap confidence interval (CI). This research has discussed previous studies with an accuracy rate of 79.4% as best obtained results. EEG of healthy adolescents and adolescents with schizophrenia symptoms dataset are used that is available on (https://brain.bio.msu.ru/eeg_schizophrenia.htm). This approach has achieved a high accuracy of 88.1%. Although researchers have achieved potential results in comparison with previous methods using the precision metric, their model has failed to achieve better sensitivity like M. Shim (100%) without any justification for this difference.

More significant multi-classifier-based models for EEG diagnose and prediction have been achieved by more researchers and developers like Misiukas Misiūnas et al. [30], Sharmila and Geethanjali [1], Peachap and Tchiotsop [31], Raghu et al. [32], Venkatachalam et al. [2], Diykh et al. [3] and Savadkoohi and Oladduni [33].

8 Other Models for EEG Classification

San-Segundo et al. [28] have addressed the classification of EEG using deep neural network (DNN). This paper has focused mainly on analyzing DNN efficiency in classifying EEG signals for patients with epilepsy. A significant analysis has been achieved on DNN with architecture made up of two layers for feature selection and three layers for EEG classification. Moreover, several EEG signal transforms are implemented and evaluated in order to investigate best transform for potential DNN efficiency. Furthermore, the proposed analysis is achieved with multiple scenarios and cases using two different epileptic datasets. The research has concentrated on the outcome accuracy. The Bern-Barcelona EEG and the Epileptic Seizure Recognition datasets are used to evaluate the overall analysis. Potential results were obtained with different system accuracy based on the applied dataset. The results have shown high accuracy up to 98.9% with the Bern-Barcelona dataset. At the same time, very high accuracy was obtained when classifying non-seizure and seizure recordings. Finally, this research has failed to investigate system performance in order to estimate and assess how practical to be implemented in real-time scenarios.

Hu et al. [27] have addressed the classification of EEG using deep bidirectional long short-term memory (Bi-LSTM) to detect the patients whose suffer from epileptic seizure. In this paper, the (Bi-LSTM) network has been proposed. Raw EEG signals have been initially analyzed by LMD, which suitable for dealing with nonlinear and non-fixed problems. A deep Bi-LSTM model then designed to classify the seizures and non-seizure EEG classification. This research has concentrated on the analysis of multiple scenarios with different cases. Three metrics like sensitivity, specificity and G-mean are computed to investigate the system capabilities. Children's Hospital

Boston CHB-MIT Dataset has been utilized to evaluate the proposed model. The proposed approach has achieved a high mean sensitivity 93.61% and a high mean specificity of 91.85% on the dataset. Although the authors have shown good results when compared with other approaches of EEG classification, they failed to show their model performance like processing time. Moreover, it is clearly noticed some other models have achieved better results without a detailed clarification. Finally, there are more developments to EEG like Dobiáš and St'Astny [29].

9 Analysis and Evaluation

With the aim to have an accurate overview about all the included papers, three main criteria are depicted: accuracy, sensitivity and specificity to be investigated and compared. In fact, most of the research papers concentrated on accuracy metric. Empirically, accuracy metric plays a vital key assessment in classification methods.

We have classified 32 approaches into groups based on their proposed classification method. All results (accuracy, sensitivity and specificity) are depicted in seven tables. Table 1 shows resultant accuracy and sensitivity by Support Vector Machine (SVM) classification for [8–12]. Table 2 shows resultant accuracy by Artificial Neural Network (ANN) classification for [5, 13–17]. Table 3 shows resultant accuracy by Convolution Neural Network (CNN) classification for [4, 18–20]. Table 4 shows resultant accuracy by K-Nearest Neighbor (K-NN) classification for [7, 21, 22]. Table 5 shows resultant accuracy by Linear Discriminant Analysis (LDA) Classification for [23, 24]. Table 6 shows resultant accuracy, sensitivity and specificity by Other Models for EEG Classification for [27–29]. Table 7 shows resultant accuracy by Multi-classifier approaches classification for [1–3, 25, 26, 30–33].

With the aim to have clear visual analysis, all resultant accuracy shown in Tables 1, 2, 3, 4, 5, 6 and 7 is displayed visually by bar chart figures (Figs. 1, 2, 3, 4, 5 and 6). Figure 1 shows resultant accuracy by Support Vector Machine (SVM) classification.

Table 1 Resultant accuracy and sensitivity by Support Vector Machine (SVM) classification

Authors	Dataset	Performance metrics in (%)	
		Accuracy	Sensitivity
Shen et al. [9]	EEG signals gathered by international 10–20 systems	94.67	
Li and Shen [8]	Colorado State University	93	
Li et al. [10]	BONN Dataset University of Bonn	100	
Chen et al. [11]	Bern-Barcelona EEG CHBMIT EEG	94	
Chisci et al. [12]	EEG Freiburg Database		100

Table 2 Resultant accuracy by Artificial Neural Network (ANN) classification

Authors	Dataset	Performance metrics in (%)
		Accuracy
George et al. [13]	(KITS), (TUH)	97.4
Guo et al. [15]	BONN Dataset University of Bonn	99.6
Luo et al. [5]	Multimodal dataset (physiological signals for analyzing emotions) (DEAP) Shanghai Jiao Tong University emotion EEG dataset (SEED)	96.67
Bhatti et al. [17]	BCI dataset EEG signals acquired by Emotiv Epoc	93.05
Satapathy et al. [16]	EEG dataset to detect an epileptic seizure EEG dataset for Eye state prediction	99
Samiee et al. [14]	BONN Dataset University of Bonn	99.8

Table 3 Resultant accuracy by Convolutional Neural Network (CNN) classification

Authors	Dataset	Performance (%)
		Accuracy
Raghu et al. [18]	Temple University Hospital EEG corpus	88.3
Ramakrishnan et al. [19]	*BONN Dataset* University of Bonn Boston Children's Hospital *CHB-MIT*	98
Lian et al. [20]	BONN Dataset University of Bonn	99.3
Mousavi et al. [4]	Sleep-EDF dataset	98.1

Table 4 Resultant accuracy by K-Nearest Neighbor (K-NN) classification

Authors	Dataset	Performance metrics in (%)
		Accuracy
Awan et al. [21]	Collected from 10 healthy people	96.1
Lahmiri and Shmuel [7]	BONN Dataset University of Bonn	100
Oliva and Rosa [22]	BONN Dataset University of Bonn	84

Table 5 Resultant accuracy by Linear Discriminant Analysis (LDA) classification

Authors	Dataset	Performance metrics in (%)
		Accuracy
You et al. [23]	BCI Competition II and III	94.29
Hsu [24]	Graz BCI group	93.7

Table 6 Resultant accuracy, sensitivity and specificity by other models for EEG classification

Authors	Classifier	Dataset	Performance metrics in (%)		
			Accuracy	Sensitivity	Specificity
Hu et al. [27]	Bi-LSTM	Children's Hospital Boston *CHB-MIT*		93.61	91.85
San-Segundo et al. [28]	DNN	The Bern-Barcelona EEG	98.9		
Dobiáš and St'Astny [29]	HMM	Study of Stancák et al.	85.6 ± 0.7		

Table 7 Resultant accuracy by Multi-classifier approach classification

Authors	Classifier	Dataset	Performance metrics in (%)
			Accuracy
Rodrigues et al. [25]	(SVM), (OPF), Naive Bayes, (*k*-NN), (MLP)	KDD dataset	99.78
Piryatinska et al. [26]	SVM, RF	Healthy adolescents and adolescents with symptoms of schizophrenia	88.1
Misiukas Misiūnas et al. [30]	ANN, SVM	Children's Hospital, Affiliate of Vilnius University Hospital Santaros Klinikos	75
Sharmila and Geethanjali [1]	NB, KNN	BONN Dataset University of Bonn	100
Peachap and Tchiotsop [31]	ANN, SVM	BONN Dataset University of Bonn	100
Raghu et al. [32]	SVM, K-NN, MLP	BONN Dataset University of Bonn Ramaiah Medical College and Hospital (RMCH)	99.45
Venkatachalam et al. [2]	PCA and FLD	BCI Competition III	96.54
Diykh et al. [3]	SGSKM	Sleep-EDF database Sleep spindles database	95.93
Savadkoohi and Oladduni [33]	SVM, KNN	BONN Dataset University of Bonn	100

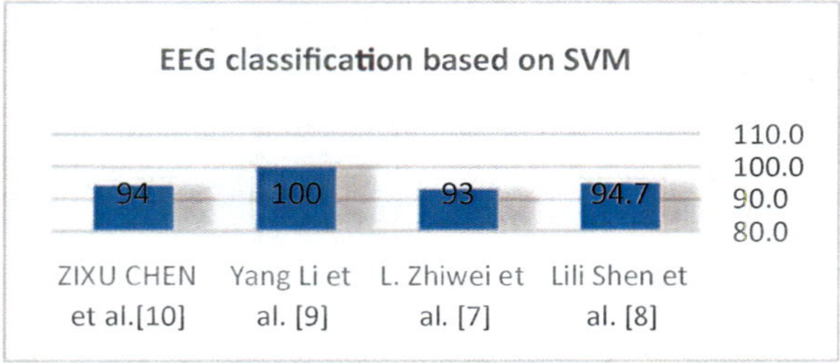

Fig. 1 Resultant accuracy by Support Vector Machine (SVM)

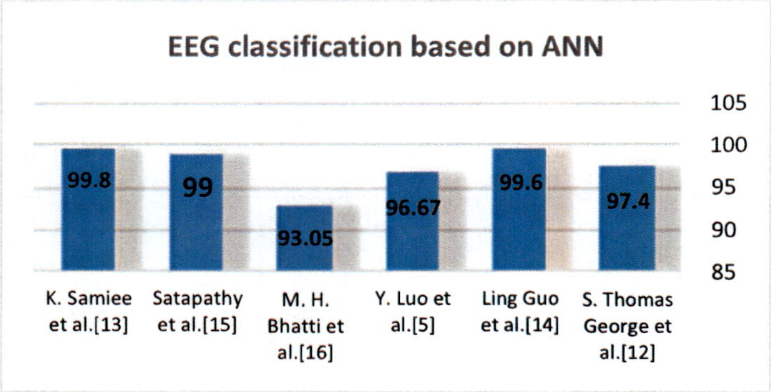

Fig. 2 Resultant accuracy by Artificial Neural Network (ANN)

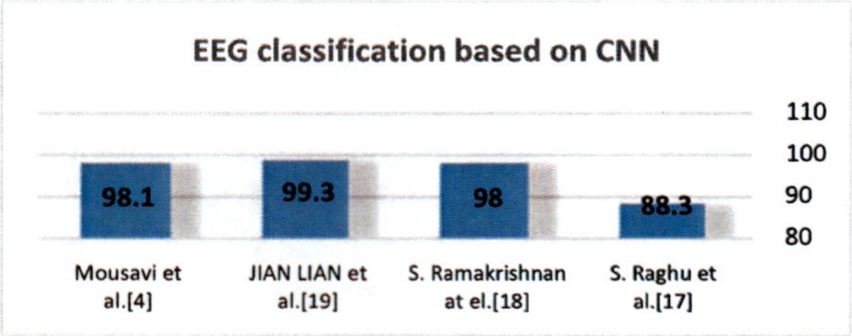

Fig. 3 Resultant accuracy by Convolutional Neural Network (CNN)

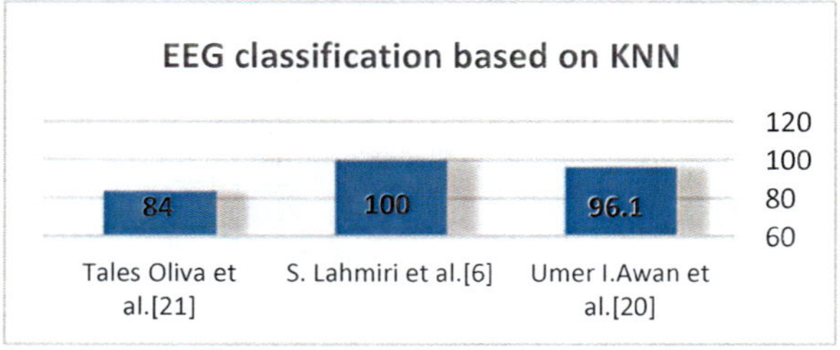

Fig. 4 Resultant accuracy by K-Nearest Neighbor (K-NN)

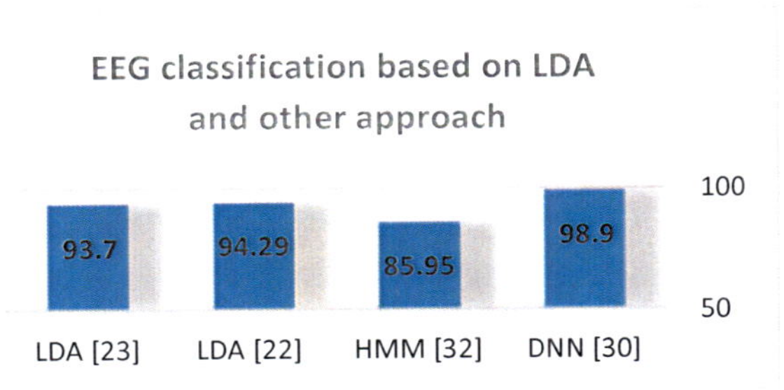

Fig. 5 Resultant accuracy by LDA and other approaches

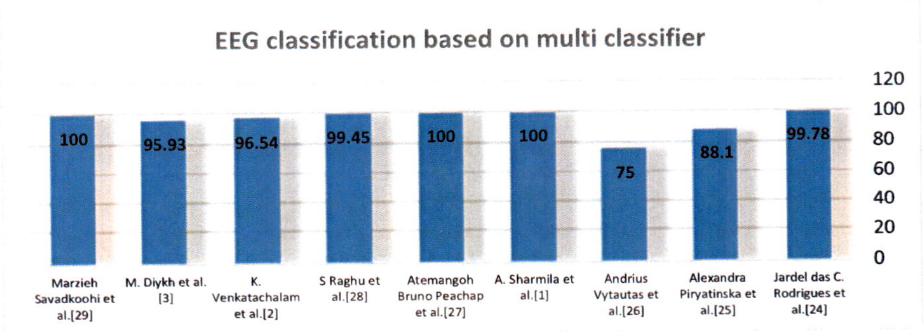

Fig. 6 Resultant accuracy by Multi-classifier approach

Figure 2 shows resultant accuracy by Artificial Neural Network (ANN) classification. Figure 3 shows resultant accuracy by Convolution Neural Network (CNN) classification. Figure 4 shows resultant accuracy by K-Nearest Neighbor (K-NN) classification. Figure 5 shows resultant accuracy by Linear Discriminant Analysis (LDA) and other classification approaches. Figure 6 shows resultant accuracy by Multi-classifier approach classification.

Evidently, Support Vector Machine (SVM) has clearly shown significant accuracy range from 93 to 100%. Artificial Neural Network (ANN) has shown high accuracy results as well start from 93.05 to 99.8%. Moreover, Convolution Neural Network (CNN) classification has shown high accuracy start from 88.3 to 99.3%. On the other hand, K-Nearest Neighbor (K-NN) Classification has shown a slightly variable accuracy start from 84 to 100%. Furthermore, Linear Discriminant Analysis (LDA) Classification has shown almost persistent accuracy start from 93.7 to 94.29%. In order to have precise assessment, we have focused on approaches that include multiple classification methods and investigated as they have shown high accuracy start from 75 to 100%.

In order to have a precise overview and assessment for all classification groups and investigate methods performance persistency, average accuracy is calculated

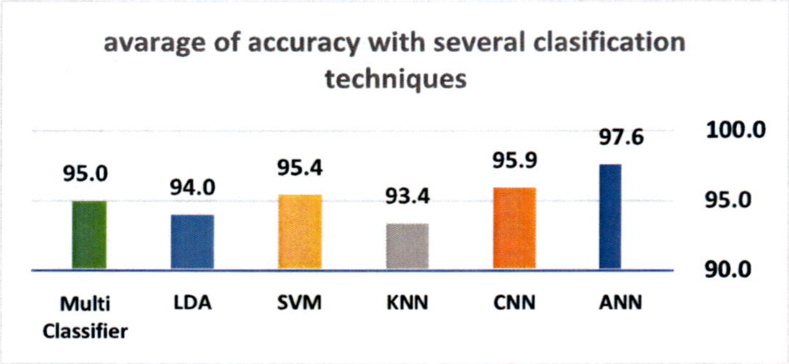

Fig. 7 Resultant average of accuracy with several classification techniques

for all approaches inside the same group. Figure 7 shows the average accuracy for Multi-classifier, LDA, SVM, KNN, CNN, and ANN classification groups. Technically, ANN classification has outperformed all other approaches with 97.6% average accuracy. On the other hand, KNN has shown the lowest performance with 93.4% average accuracy. CNN classification has shown the best second performance with 95.9% average accuracy. SVM classification has shown slight difference that CNN with 95.4% average accuracy.

10 Conclusion

In conclusion, this paper has surveyed 32 main successful approaches for EEG signal classification. All approaches are distributed into seven groups based on the main proposed classification method (SVM, ANN, KNN, CNN, LDA, other, Multi-classifier). The proposed analysis in this survey has targeted the investigation of accuracy, sensitivity and specificity as main criterion. We have looked for processing time as well and they unfortunately lacking of any system performance analysis including the processing time. Generally, all classification groups have shown high capabilities in terms of accuracy metric. Evidently, ANN has outperformed all other models, and KNN has shown the lowest performance.

References

1. A. Sharmila, P. Geethanjali, DWT based detection of epileptic seizure from EEG signals using Naive Bayes and k-NN classifiers. IEEE Access **4**(c), 7716–7727 (2016). https://doi.org/10.1109/ACCESS.2016.2585661
2. K. Venkatachalam, A. Devipriya, J. Maniraj, M. Sivaram, A. Ambikapathy, S.A. Iraj, A novel method of motor imagery classification using EEG signal. Artif. Intell. Med. **103**, 101787 (2020). https://doi.org/10.1016/j.artmed.2019.101787

3. M. Diykh, Y. Li, P. Wen, EEG sleep stages classification based on time domain features and structural graph similarity. IEEE Trans. Neural Syst. Rehabil. Eng. **24**(11), 1159–1168 (2016). https://doi.org/10.1109/TNSRE.2016.2552539
4. Z. Mousavi, T. Yousefi Rezaii, S. Sheykhivand, A. Farzamnia, S.N. Razavi, Deep convolutional neural network for classification of sleep stages from single-channel EEG signals. J. Neurosci. Methods **324** (2019). https://doi.org/10.1016/j.jneumeth.2019.108312
5. Y. Luo et al., EEG-based emotion classification using spiking neural networks. IEEE Access **8**, 46007–46016 (2020). https://doi.org/10.1109/ACCESS.2020.2978163
6. K.K. Al-Nassrawy, D. Al-Shammary, A.K. Idrees, High performance fractal compression for EEG health network traffic. Procedia Comput. Sci. **167**, 1240–1249 (2020). ISSN 1877-0509
7. S. Lahmiri, A. Shmuel, Accurate classification of seizure and seizure-free intervals of intracranial EEG signals from epileptic patients. IEEE Trans. Instrum. Meas. **68**(3), 791–796 (2019). https://doi.org/10.1109/TIM.2018.2855518
8. Z. Li, M. Shen, Classification of mental task EEG signals using wavelet packet entropy and SVM, in *2007 8th International Conference on Electronic Measurement and Instruments, ICEMI*, 2007, pp. 3906–3909. https://doi.org/10.1109/ICEMI.2007.4351064
9. L. Shen, X. Dong, Y. Li, Analysis and classification of hybrid EEG features based on the depth DRDS videos. J. Neurosci. Methods **338**, 108690 (2020). https://doi.org/10.1016/j.jneumeth.2020.108690
10. Y. Li, X.D. Wang, M.L. Luo, K. Li, X.F. Yang, Q. Guo, Epileptic seizure classification of EEGs using time-frequency analysis based multiscale radial basis functions. IEEE J. Biomed. Health Inform. **22**(2), 386–397 (2018). https://doi.org/10.1109/JBHI.2017.2654479
11. Z. Chen, G. Lu, Z. Xie, W. Shang, A unified framework and method for EEG-based early epileptic seizure detection and epilepsy diagnosis. IEEE Access **8**, 20080–20092 (2020). https://doi.org/10.1109/ACCESS.2020.2969055
12. L. Chisci et al., Real-time epileptic seizure prediction using AR models and support vector machines. IEEE Trans. Biomed. Eng. **57**(5), 1124–1132 (2010). https://doi.org/10.1109/TBME.2009.2038990
13. S.T. George, M.S.P. Subathra, N.J. Sairamya, L. Susmitha, M. Joel Premkumar, Classification of epileptic EEG signals using PSO based artificial neural network and tunable-Q wavelet transform. Biocybern. Biomed. Eng. **40**(2), 709–728 (2020). https://doi.org/10.1016/j.bbe.2020.02.001
14. K. Samiee, P. Kovács, M. Gabbouj, Epileptic seizure classification of EEG time-series using rational discrete short-time Fourier transform. IEEE Trans. Biomed. Eng. **62**(2), 541–552 (2015). https://doi.org/10.1109/TBME.2014.2360101
15. L. Guo, D. Rivero, J. Dorado, J.R. Rabuñal, A. Pazos, Automatic epileptic seizure detection in EEGs based on line length feature and artificial neural networks. J. Neurosci. Methods **191**(1), 101–109 (2010). https://doi.org/10.1016/j.jneumeth.2010.05.020
16. S.K. Satapathy, S. Dehuri, A.K. Jagadev, EEG signal classification using PSO trained RBF neural network for epilepsy identification. Inform. Med. Unlocked **6**, 1–11 (2017). https://doi.org/10.1016/j.imu.2016.12.001
17. M.H. Bhatti et al., Soft computing-based EEG classification by optimal feature selection and neural networks. IEEE Trans. Ind. Inf. **15**(10), 5747–5754 (2019). https://doi.org/10.1109/TII.2019.2925624
18. S. Raghu, N. Sriraam, Y. Temel, S.V. Rao, P.L. Kubben, EEG based multi-class seizure type classification using convolutional neural network and transfer learning. Neural Netw. **124**, 202–212 (2020). https://doi.org/10.1016/j.neunet.2020.01.017
19. S. Ramakrishnan, A.S. Muthanantha Murugavel, P. Saravanan, Epileptic EEG signal classification using multi-class convolutional neural network, in *Proceedings of International Conference on Vision towards Emerging Trends in Communication and Networking, ViTECoN 2019*, 2019, pp. 1–5. https://doi.org/10.1109/ViTECoN.2019.8899453
20. J. Lian, Y. Zhang, R. Luo, G. Han, W. Jia, C. Li, Pair-wise matching of EEG signals for epileptic identification via convolutional neural network. IEEE Access **8**, 40008–40017 (2020). https://doi.org/10.1109/ACCESS.2020.2976751

21. U.I. Awan, U.H. Rajput, G. Syed, R. Iqbal, I. Sabat, M. Mansoor, Effective classification of EEG signals using K-nearest neighbor algorithm, in *Proceedings of 14th International Conference on Frontiers of Information Technology FIT 2016*, 2017, pp. 120–124. https://doi.org/10.1109/FIT.2016.030
22. J.T. Oliva, J.L.G. Rosa, Classification for EEG report generation and epilepsy detection. Neurocomputing **335**, 81–95 (2019). https://doi.org/10.1016/j.neucom.2019.01.053
23. Y. You, W. Chen, T. Zhang, Motor imagery EEG classification based on flexible analytic wavelet transform. Biomed. Signal Process. Control **62**, 102069 (2020). https://doi.org/10.1016/j.bspc.2020.102069
24. W.Y. Hsu, EEG-based motor imagery classification using neuro-fuzzy prediction and wavelet fractal features. J. Neurosci. Methods **189**(2), 295–302 (2010). https://doi.org/10.1016/j.jneumeth.2010.03.030
25. J. das C. Rodrigues, P.P.R. Filho, E. Peixoto, N. Arun Kumar, V.H.C. de Albuquerque, Classification of EEG signals to detect alcoholism using machine learning techniques. Pattern Recognit. Lett. **125**, 140–149 (2019). https://doi.org/10.1016/j.patrec.2019.04.019
26. A. Piryatinska, B. Darkhovsky, A. Kaplan, Binary classification of multichannel-EEG records based on the ϵ-complexity of continuous vector functions. Comput. Methods Programs Biomed. **152**, 131–139 (2017). https://doi.org/10.1016/j.cmpb.2017.09.001
27. X. Hu, S. Yuan, F. Xu, Y. Leng, K. Yuan, Q. Yuan, Scalp EEG classification using deep Bi-LSTM network for seizure detection. Comput. Biol. Med. **124**, 103919 (2020). https://doi.org/10.1016/j.compbiomed.2020.103919
28. R. San-Segundo, M. Gil-Martín, L.F. D'Haro-Enríquez, J.M. Pardo, Classification of epileptic EEG recordings using signal transforms and convolutional neural networks. Comput. Biol. Med. **109**, 148–158 (2019). https://doi.org/10.1016/j.compbiomed.2019.04.031
29. M. Dobiáš, J. St'Astny, Movement EEG classification using parallel hidden Markov models, in *International Conference on Applied Electronics*, Sept 2016, pp. 65–68. https://doi.org/10.1109/AE.2016.7577243
30. A.V. Misiukas Misiūnas, T. Meškauskas, R. Samaitienė, Algorithm for automatic EEG classification according to the epilepsy type: Benign focal childhood epilepsy and structural focal epilepsy. Biomed. Signal Process. Control **48**, 118–127 (2019). https://doi.org/10.1016/j.bspc.2018.10.006
31. A.B. Peachap, D. Tchiotsop, Epileptic seizures detection based on some new Laguerre polynomial wavelets, artificial neural networks and support vector machines. Inform. Med. Unlocked **16**, 100209 (2019). https://doi.org/10.1016/j.imu.2019.100209
32. S. Raghu, N. Sriraam, A.S. Hegde, P.L. Kubben, A novel approach for classification of epileptic seizures using matrix determinant. Expert Syst. Appl. **127**, 323–341 (2019). https://doi.org/10.1016/j.eswa.2019.03.021
33. M. Savadkoohi, T. Oladduni, A machine learning approach to epileptic seizure prediction using electroencephalogram (EEG) signal. Biocybern. Biomed. Eng. 1–14 (2020). https://doi.org/10.1016/j.bbe.2020.07.004

Analysis of Road Accidents Using Data Mining Paradigm

Maya John and Hadil Shaiba

Abstract Most road accidents occur due to human errors. The analysis of accidents with respect to different dimensions will help concerned authorities in taking necessary steps to make roads safer. In this paper, we used Apriori algorithm to mine frequent road accident patterns in Dubai during 2017. The features used are the time of accident, accident cause, accident type, age category, road type, day of accident and whether the driver was intoxicated or not. Studies were conducted to identify the accident patterns pertaining to different categories of roads. In E-Route and D-Route roads, less inter-vehicular space resulted in many accidents, whereas in non-major roads, influence of alcohol caused majority of the vehicular collisions. The accidents involving intoxicated drivers were analysed on the basis of time, road type and day of the week. Majority of the intoxication-related accidents were caused by youth. Major chunk of these accidents occured in non-major roads during late night/week hours. Some measures which may help in decreasing accidents are also suggested.

Keywords Road accidents · Apriori algorithm · Frequent itemset mining

1 Introduction

From time immemorial, the dream to move and capture territories was embedded in human minds. In the prehistoric times, walking, running and swimming were the only means of travel. Later on, people started using powerful creatures like horses, camels, donkeys and elephants for transportation [1]. The invention of wheel by 3500 BC was a paradigm shift in the history of human mobility [2]. The discovery of oil marked another era in the history of human transportation. As the number and speed of vehicles increased, accidents also increased.

M. John (✉)
Independent Researcher, Pathanamthitta, Kerala, India

H. Shaiba
College of Computer and Information Sciences, Princess Nourah Bint Abdulrahman University, Riyadh, Kingdom of Saudi Arabia

S. Shakya et al. (eds.), *Mobile Computing and Sustainable Informatics*, Lecture Notes on Data Engineering and Communications Technologies 68,
https://doi.org/10.1007/978-981-16-1866-6_15

According to the World Health Organization (WHO), road accidents account for more than one million deaths every year [3]. The traffic crashes affect both individuals and society in terms of money spent for treating injury and repairing damaged property, increase in commutation time and excessive fuel emissions [4]. The main causes of road accidents include overspeed, influence of alcohol/drugs, distracted driving, poor road infrastructure, unsafe vehicles, etc. On account of the extensive use of mobile phones, there has been an increase in the number of accidents caused due to the distraction of drivers [5]. Safe roads and roadsides, good driving habits, cautious pedestrians and fit vehicles will definitely aid in reducing road accidents.

Data mining techniques are extensively used to analyse road accidents. Kumar and Toshniwal used data mining approaches such as k-means algorithm and association rule mining to analyse road accidents in Deharadun District in India for the years from 2009 to 2014 [6]. Tayeb et al. used association rule mining algorithms such as Apriori algorithm and predictive Apriori algorithm to analyse the relationship between different factors of accidents and severity of accident injuries in Dubai [7]. Ait-Mlouk et al. analysed road accident data of Marrakech Province in Morocco corresponding to the years from 2003 to 2014 [8]. C4.5 algorithm was used to construct the decision tree to extract the rules pertaining to the data. Li et al. used data mining techniques to analyse data corresponding to fatal road accidents obtained from FARS fatal accident dataset [9]. Apriori algorithm-based association rule mining was carried out to analyse the patterns corresponding to fatality rate. K-means-based clustering algorithm was used to identify which states are safe to drive.

Mohamed applied multi-class support vector machine to predict the cause of traffic accidents in United Arab Emirates [10]. Ali and Hamed used Apriori algorithm and EM clustering algorithm to analyse the road accident data pertaining to Alghat Province in Saudi Arabia [11]. Almjewail et al. used K-means and DBSCAN algorithm to analyse road accident data in Riyadh, Saudi Arabia, for the years from 2013 to 2015 [12]. The places with high frequency of accidents (black spots) were identified using DBSCAN algorithm. K-means algorithm was employed to identify the characteristics of accidents. In this paper, road accidents are analysed using Apriori algorithm which identifies the most frequent accident patterns. The analysis is carried out for different types of roads and accidents involving intoxicated drivers.

2 Data and Methods

2.1 Data Description

The dataset was obtained from the official open data portal of United Arab Emirates (https://bayanat.ae). It consisted of 1987 records corresponding to road accidents in Dubai for the year 2017. The original dataset was translated from Arabic to English. It consisted of features such as date of accident, location, type of accident, reason for accident, road status, severity of injury, gender, date of issue of driving licence, occupation, intoxication, status of seat belt, year of vehicle manufacture and insurance company name. The data was pre-processed before experimentation. As far as accidents are concerned, severity of injury is a very important attribute. However, we ignored the injury severity attribute as its value was not mentioned in majority of the cases. All records of features with missing values were removed. This led to a reduction in the number of records to 1750. For the purpose of analysis, new attributes were derived from the existing attributes. The time was categorized into intervals of two hours. In order to perform analysis based on age category, the age group of the person involved in an accident was derived. People above 60 years were categorized as elderly, people below 36 years were considered as youth, and rest were grouped as middle aged. A new feature named "intoxicated" was generated based on the intoxication attribute. The intoxicated attribute was assigned a value "yes" if the value of intoxication attribute was alcohol abuser, drug abuser or drunk. On the basis of the date of accident, a new attribute "day of accident" was derived.

The roads managed by the Roads and Transport Authority (RTA) in Dubai follow a numbering system [13]. The highways connecting Dubai with other emirates in United Arab Emirates (UAE) are designated as Emirates Routes (E-Routes). They are named as letter E followed by two-digit or three-digit number. The major roads connecting the localities in Dubai are designed as D-Routes. Such roads are identified by the letter D followed by two-digit or three-digit number. Hence, roads in the dataset were classified into E-Route, D-Route, Other Roads and Intersection. The attributes which were considered irrelevant for analysis were ignored. The features used for experimentation include time interval, type of accident, cause of accident, location, route type, age group and intoxicated.

2.2 *Apriori Algorithm*

Apriori algorithm is used for generating association rules and mining the frequent itemsets from a database [14]. Support measure is used in identifying the frequent itemsets in a database [15]. Support count is the number of transactions which contain the specified itemset. Levelwise search technique is employed by Apriori algorithm to mine frequent itemset. Here $(k + 1)$ itemsets are analysed based on k-itemsets. Apriori property makes use of antimonotonicity property [16]. According to antimonotonicity property, if an itemset cannot pass a test (support less than minimum support), then the supersets of the itemset also will fail the same test. The advantages of Apriori algorithm include capability to handle large number of transaction data, generation of rules, which are easy to comprehend and discovery of knowledge hidden in a database.

3 Results and Discussion

3.1 Road Type

The percentage of accidents which occurred in different types of roads are shown in Fig. 1. Majority of the accidents occur on less important roads. Nearly 25% of the accidents occur in major highways—D-Route and E-Route. The most common accident patterns in D-Route, E-Route and less important roads are shown in Tables 1, 2 and 3, respectively. In D-Route roads, alcohol influence and less inter-vehicular space attributed to most accidents. The major causes of accidents in E-Route roads are not leaving adequate space between vehicles and sudden deviation. E-Route roads connect major emirates, and hence, it is likely to be used by thousands of people who are not familiar with the roads. This may have led to more accidents due to sudden deviation. In less important roads, most of the accidents are caused due to intoxicated driving. As these roads include many streets which lack designated crossing areas, the chance of pedestrians being run over is more when compared to D-Route and E-Route roads.

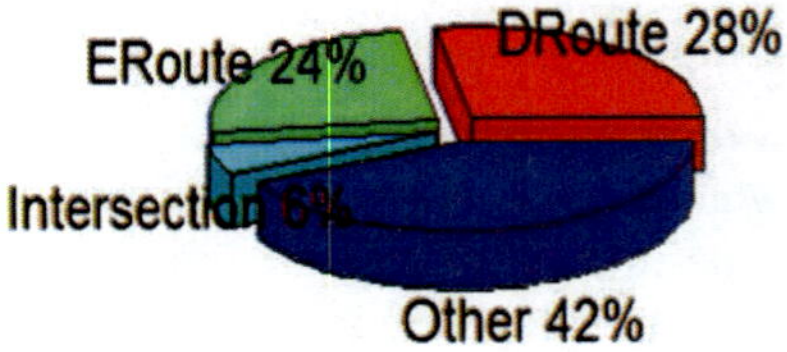

Fig. 1 Analysis of accidents based on road type

Table 1 Most common accident pattern in D-Route roads

Accident type	Accident cause	Age group	Intoxicated
Vehicular collision	Inadequate space	Youth	No
Vehicular collision	Influence of alcohol	Youth	Yes
Vehicular collision	Sudden deviation	Youth	No
Vehicular collision	Inadequate space	Middle aged	No
Vehicular collision	Influence of alcohol	Middle aged	Yes

Table 2 Most common accident pattern in E-Route roads

Accident type	Accident cause	Age group	Intoxicated
Vehicular collision	Inadequate space	Youth	No
Vehicular collision	Inadequate space	Middle aged	No
Vehicular collision	Sudden deviation	Youth	No
Vehicular collision	Influence of alcohol	Youth	Yes
Vehicular collision	Sudden deviation	Middle aged	No

Table 3 Most common accident pattern in less important roads

Accident type	Accident cause	Age group	Intoxicated
Vehicular collision	Influence of alcohol	Youth	Yes
Run over	Not properly estimating road users	Youth	No
Vehicular collision	Sudden deviation	Youth	No
Vehicular collision	Inadequate space	Youth	No
Vehicular collision	Influence of alcohol	Middle aged	Yes

3.2 *Intoxication*

In nearly 30% of accidents, drivers were intoxicated. This stresses the need of increasing inspections to reduce such type of accidents. Figure 2 represents the pie chart showing the intoxication-related accidents with regard to road type and age group. Youth are responsible for nearly 70% of the intoxicated accident cases. As per Fig. 1, there is only 4% difference in the number of accidents in E-Routes and D-Route roads, but according to Fig. 2 there is 11% difference in the percentage of accidents involving intoxicated drivers in these routes. This may be due to the fact that E-Route roads are extensively used by people who travel to Dubai from other emirates for work. The accidents due to alcoholism among these people are likely to be less.

The number of accidents involving intoxicated drivers during different time intervals is pictorially represented in Fig. 3. The major fraction of intoxication-related accidents occur between 10 p.m. and 6 a.m. Being a cosmopolitan city, people in Dubai have a splendid nightlife. The law in Dubai permits pubs, bars, night clubs,

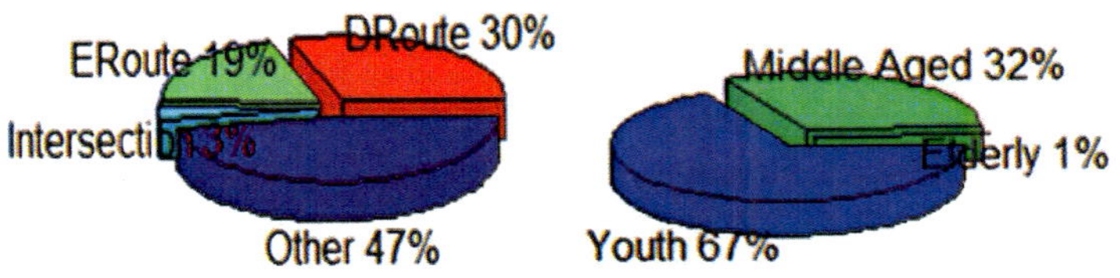

Fig. 2 Analysis of accidents based on road type and age group

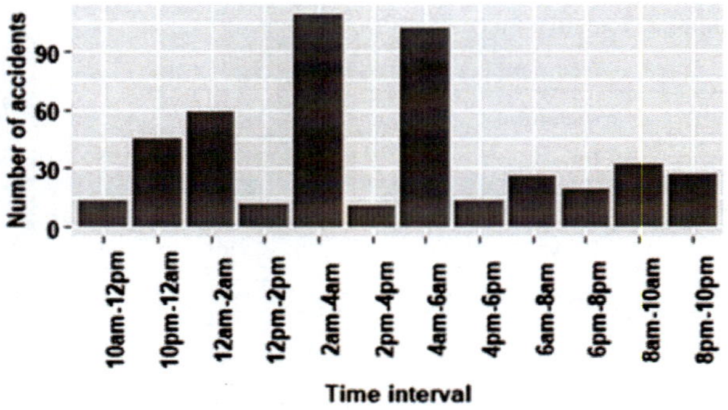

Fig. 3 Analysis based on time

etc., to be open up to 3 a.m. This is likely to contribute to intoxication-related accidents during late night and early morning. Figure 4 represents the plot depicting the number of accident during different days of week during various time interval. On all days of the week, the maximum number of intoxication-related accidents occurs between 2 and 6 a.m. The high number of accidents during Friday may be attributed to the fact that Friday is a holiday and tourists from neighbouring countries visit Dubai on this day [17]. The number of accidents in different types of roads during different time intervals is plotted in Fig. 5. Most intoxication-related accidents occur during night. In all types of roads, the majority of the accidents occur between 12 and 6 a.m.

In the year 2017, nearly 150 types of violations were regarded as punishable traffic offences in Dubai. The penalty may be in the form of fine, black point for drivers or/and confiscation of vehicle. Dubai Police has been using latest technologies to

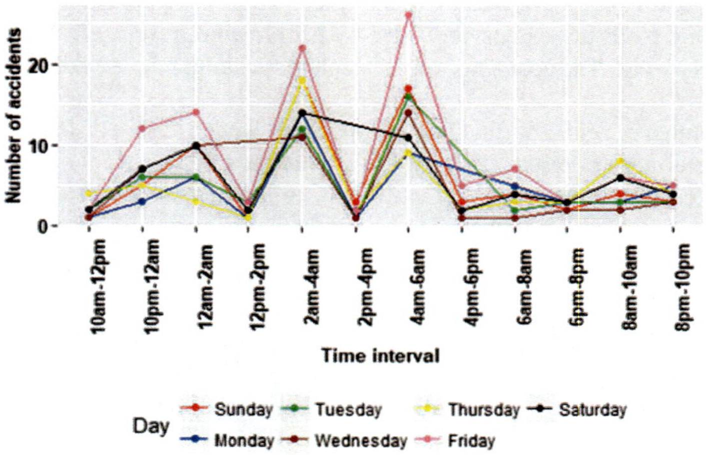

Fig. 4 Analysis based on day and time

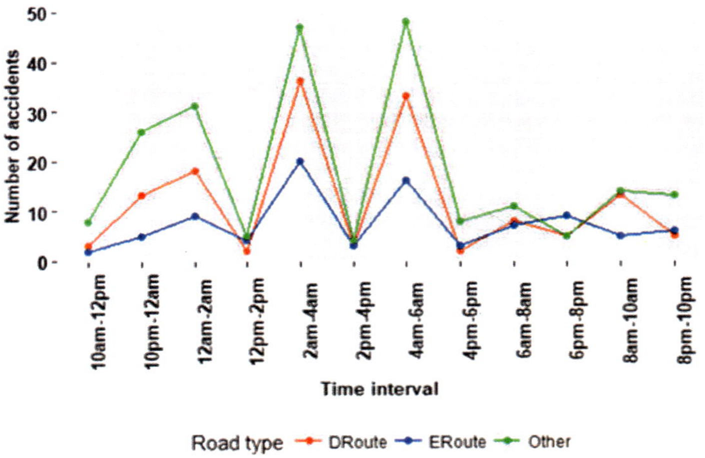

Fig. 5 Analysis based on time and road type

make roads safer. Governmental policies and innovative research have played a great role in drastically reducing the number of accidents in several countries. Authorities in Sweden claimed that their "2 + 1" lane policy played a pivotal role in reducing road accidents. As per this policy, two lanes are devoted for the same direction traffic and the third lane for reverse direction. The direction of traffic in the lanes is changed after every few kilometres. This helps in reducing overspeeding and overtaking. Singapore has introduced advanced warning light which alerts the driver about upcoming traffic signal. Compared to few decades back, roads in Japan are now one of the safest in the world. This is attributed to policies of government, increased infrastructure, stricter enforcement of rules, enhanced public awareness and better traffic systems [18].

The following are certain recommendations suggested on the basis of the study conducted:

- Increasing patrolling on roads especially non-major roads may reduce the accidents due to intoxication as large number of accidents on these types of roads are due to drivers being intoxicated.
- Providing more pedestrian crossing lines in non-major roads with high traffic can help in decreasing the number of pedestrians being run over.
- Compulsory awareness classes may be introduced for drivers who are caught intoxicated. They may also be asked to get involved in socially useful work.
- The 2 + 1 lane policy may be adopted for straight roads which run several kilometres without any intersecting roads [18].
- In order to avoid pedestrians from being knocked down by vehicles, in extremely wide roads, divider may be introduced between lanes. In residential areas, resting areas may be provided for elderly people in the middle of roads. These strategies are currently being used in Singapore [18].

- People should be discouraged to board vehicles being driven by intoxicated drivers. In Switzerland, the passengers travelling with drunk driver are also accountable and might lose their driving license. Japan is very stringent with drivers who are caught drunk. It has adopted the policy of zero blood alcohol level for drivers [18].
- Risky overtaking should be considered as an offence. This policy has been adopted by Norway where overtaking is permitted only on straight roads with high degree of visibility.

4 Conclusions

This study employs Apriori algorithm to unearth frequent accident patterns in Dubai based on data corresponding to the year 2017. The accidents were analysed on the basis of type of road and intoxication. In E-Route and D-Route roads, many accidents were due to less space between moving vehicles. In the case of less important roads, many accidents were caused due to intoxication of the drivers. Youth were involved in majority of accidents where drivers were intoxicated. The major chunk of accidents associated with intoxicated driving occurred between 2 and 6 a.m. Taking into consideration the results obtained, certain measures which may be useful in reducing accidents are also suggested.

References

1. C.C. Northrup, *Encylopedia of World Trade from Ancient Times to the Present* (Routledge, 2015)
2. J.Y. Wong, An introduction to terramechanics. J. Terramech. **21**, 5–17 (1984)
3. X. Xiong, L. Chen, J. Liang, Analysis of roadway traffic accidents based on rough sets and bayesian network. Promet Traffic Transp. **30**, 71–81 (2018)
4. F. Malin, I. Norros, S. Innamaa, Accident risk of road and weather conditions on different road types. Accid. Anal. Prev. **122**, 181–188 (2019)
5. M. Nee, B. Contrand, L. Orriols, C. Gil-Jardine, C.E. Galera, E. Lagarde, Road safety and distraction, results from a responsibility case-control study among a sample of road users interviewed at the emergency room. Accid. Anal. Prev. **12**, 19–24 (2019)
6. S. Kumar, D. Toshniwal, A data mining approach to characterize road accident locations. J. Mod. Transp. **24**, 62–72 (2016)
7. A.E. Tayeb, V. Pareek, A. Araar, Applying association rules mining algorithms for traffic accidents in Dubai. Int. J. Soft Comput. **5** (2015)
8. A. Ait-Mlouk, F. Gharnati, T. Agouti, Application of big data analysis with decision tree for road accident. Indian J. Sci. Technol. **10** (2017)
9. L. Li, S. Shrestha, G. Hu, Analysis of road traffic fatal accidents using data mining techniques, in *15th International Conference on Software Engineering Research, Management and Applications (SERA)* (IEEE, 2017), pp. 363–370
10. E.A. Mohamed, Predicting causes of traffic road accidents using multi-class support vector machines, in *10th International Conference on Data Mining*, Las Vegas, 2014, pp. 37–42
11. F.M.N. Ali, A.A.M. Hamed, Usage apriori and clustering algorithms in WEKA tools to mining dataset of traffic accidents. J. Inf. Telecommun. **2**(3) (2018)

12. A. Almjewail, A. Almjewail, S. Alsenaydi, H. AlSudairy, I. Al-Turaiki, Analysis of traffic accident in Riyadh using clustering algorithms, in *Proceedings of 5th International Symposium on Data Mining Applications* (Springer, 2018), pp. 12–25
13. https://en.wikipedia.org/wiki/Dubai_route_numbering_system
14. B. Lantz, *Finding Patterns—Market Basket Analysis Using Association Rules, Machine Learning with R* (Packt Publishing, 2015)
15. F. Jiang, K.K.R. Yuen, E.W.M. Lee, Analysis of motorcycle accidents using association rule mining-based framework with parameter optimization and GIS technology. J. Saf. Res. (2020)
16. J. Han, M. Kamber, J. Pei, *Data Mining Concepts and Techniques*, 3rd edn. (Morgan Kaufmann, 2015)
17. M. John, H. Shaiba, Apriori-based algorithm for Dubai road accident analysis. Procedia Comput. Sci. **163**, 218–227 (2019)
18. https://sites.ndtv.com/roadsafety/road-safety-india-can-learncountries-1794/

Hybrid Approach to Cross-Platform Mobile Interface Development for IAAS

Yacouba Kyelem, Kisito Kiswendsida Kabore, and Didier Bassole

Abstract Information Access Assistant Service (IAAS) is a collaborative filtered document recommendation system. The validation of the model was carried out by implementing the system on a library management software in a target and closed environment using Web technology. The tests are performed with computers, which limits the use of IAAS. As of today, smartphones occupy a place of choice in population's daily life and moreover, the Web technology used does not promote a good adaptation of interfaces on the mobile. This led us to think about setting up a user interface facilitating human computer interaction on the mobile. The existence of a multitude of mobile platforms, of different sizes and regarding to the context of the use of IAAS, we adopted for cross-platform development. In this paper, we proposed to use the hybrid approach and we used ionic as an implementation tool for our mobile solution. We set up the solution and also performed display tests and evaluation in a closed environment. The test was done by setting up a server, a connection, and three smartphones. The results observed show a good adaptation of the interfaces on the different smartphones but also in relation to the Web interfaces.

Keywords IAAS · Cross-platform · Ionic · Hybrid approach

1 Introduction

In recent years, interaction with information systems has considerably evolved. Advances in technology have led to a profound change in the definition of information systems. Computer is no longer the only representative of consumer systems. Indeed, the miniaturization of devices and the appearance of new means of interaction (smartphone, tablet, PDA, etc.) have made it possible to rethink new means of human–computer interaction. In order to help Web users to access information more quickly, work has led to the implementation of the Information Access Assistance

Y. Kyelem (✉) · K. K. Kabore · D. Bassole
Laboratoire de Mathématiques et d'Informatique (LAMI), Université Joseph Ki-Zerbo, Ouagadougou, Burkina Faso
URL: https://ujkz.net/

S. Shakya et al. (eds.), *Mobile Computing and Sustainable Informatics*, Lecture Notes on Data Engineering and Communications Technologies 68,
https://doi.org/10.1007/978-981-16-1866-6_16

Fig. 1 Illustration of IAAS [2]

System (IAAS) [1]. This system, intended to be used online courses and library systems, has been set up to make recommendations to users.

Putting this notion into practice has led to the implementation and testing of its display on a computer in a closed environment. IAAS interface is conventional, i.e. dedicated to a target device in a known location for a given type of last user.

An example of IAAS display tests on a computer is shown in Fig. 1 [2].

IAAS has been integrated into a library management software (PMB); to launch it just click on the left button at the top of the figure to get the window you see on the right side.

In the context of today's technology that links a very wide variety of interaction devices and a redefinition of the interaction space merits that the design and realization of application interfaces involve new requirements. We particularly refer to mobile devices (smartphone, tablet, PDA, etc.) and the size of their screens, which is highly variable. Faced with such a situation, we are confronted with enormous challenges, in particular the consideration of the interface of IAAS on mobile devices as it was highlighted in the work of [2]. The problem we are addressing is the following: in the framework of the progress in the use of mobile devices, IAAS remains limited in its use.

In order to address this issue, we first adopt a literature review on mobile development to understand the different techniques of mobile programming and IAAS to take stock of those already done. Then, we choose an approach that we will carry out and we will end with display tests and an evaluation of our solution.

2 State of the Art on IAAS

IAAS is a collaborative filtering recommendation system that uses the notice of relevance to make recommendations and adaptations of document lists to users groups. This section summarizes the work of [2].

2.1 *General Mechanism of IAAS*

2.1.1 Notion of Relevance Notice

In IAAS, an user belongs to users group and this one can recommend a document to a group through the relevance notice. In IAAS, the fact that the one who recommends the document does not necessarily belong to the group does not necessarily affect the relevance of his opinion.

In the work of [2], an opinion is a triplet that takes into account the following elements: the documentary unit, the user group, and the relevance weight; noted for example $B = (\mathrm{UD}, \mathrm{Gr}, b)$. In this formula, UD represents the documents or documentary units, Gr is the group and 'b' is the relevance weight and is an integer between 1 and 10, where according to [2]: relevant corresponds to 1, average relevant corresponds to 5 and very relevant to 10. A user who gives his notice of relevance weight 'b' on a documentary unit UDi means that UDi is important for the thematic group Grj. The value of 'b' is called relevance value and one relevance opinion corresponds to one group and to a documentary unit and several relevance notices may exist for a group and a documentary unit.

2.1.2 Collection Mechanism of Relevance Notices for the Benefit of User Groups in IAAS

Any user who consults a document can thank the collection of relevance notices and make an assessment of the document or a part of the document. Thus, by considering a total of n groups and m documents in the information system, each group can receive a maximum of documents recommended to it ranging from 0 to m and each document unit is recommended to the maximum for the different groups ranging from 0 to n. A Grj group is likely to receive several relevance opinions on the same UDi documentary unit: in this case, the recommendation weight is calculated.

The recommendation weight $= pk(\mathrm{UD}i\mathrm{Gr}j)$ is the set of relevance opinions. A group which has not received a relevance notice has a zero recommendation weight, otherwise $Pk\ (\mathrm{UD}i\mathrm{Gr}j) = \mathrm{sum}((\mathrm{UD}i, \mathrm{Gr}j, b))$. The relevance notices are recorded in the group profiles and this therefore constitutes an enrichment for these groups. A relevance notice is registered individually for a recommended document unit. In IAAS, the recommendation weight is used to manage the recommendations. The details of this section can be found in [3]. Algorithm of collection of relevance notices for IAAS has been implemented in [3].

2.1.3 Management of Recommendation in IAAS

The management of recommendations in IAAS is done through the use of the information on the relevance notices of the documentary units sent to groups to make recommendations. Recommendations are given to users as soon as they log on, which is seen by the system as an automatic request to receive recommendations. These are the documentary units recommended to his group that will be recommended to him in turn. Each user receives a list of documents classified in a given order thanks to the recommendation weights. This is obtained by applying the relevance value formula pi, j where the i represents a UD and j a group:

$$\text{Relevance } pi,\ j = \ln(1 + Pk(\text{UD}iGj)) \tag{1}$$

Considering a user group, the pi, j are the different relevance values of each UDi documentary units recommended to it. For a Grj user group, we have a set of pi, j which are the weights noted by the other users. These pi, j are used to sort the UDi recommended to the Grj group by ranking them in descending order and then IAAS delivers the result to the Grj members once they connect to the system. The algorithm for calculating relevance can be found in [3]. The display of the L list resulting from the algorithm does not take into account the inclusion of documentary units [2, 3].

2.1.4 Sorting Documents in IAAS

Sorting documents in IAAS enable to have a list (Lud_0) of document units from the results of queries explicitly launched by users to an external system. A Lud_0 sorted list is obtained using the management of the recommendations described above (Sect. 2.1.3) and the previously enriched user group profiles. We note the Lud_g list of UD to be recommended to the groups and classified according to decreasing pi, j. Concerning the result of the sorting, the user will be interested in having Lud_1 then Lud_2 at the top of the list:

Lud_1 corresponding to the UD documents recommended to its group (Lud_g) and belonging to Lud_0.

Lud_2 is composed of the UD documents from the Lud_0 list not recommended to the user's group. We thus obtain a first list from the sorting: L1 = Lud_1 ||| Lud_2 useful for the user [4, 5]. The algorithm used in this framework can be verified by [3]. The explanation of the functionalities of the algorithm can be found in [2] or [3].

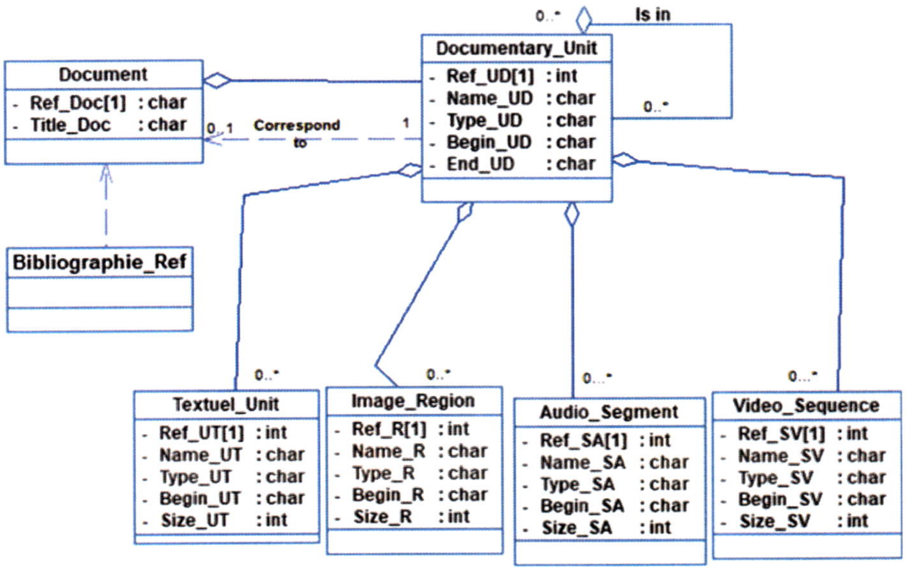

Fig. 2 Model of document [2]

2.2 *Implementation of IAAS*

For the realization of IAAS, [2] set up the document model and that of users model that we present in the following lines.

2.2.1 Document Model

In the case of the IAAS, a piece of audio, video, image, and text is a documentary unit which itself is only part of a document. A piece of sample text is either a section, paragraph, chapter, etc., or any part of a text. The model essentially contains two classes: documents and documentary units, the description of which can be found in [1, 3, 6, 7]. In IAAS, Fig. 2 shows the document unit used.

2.2.2 User Profile Model

IAAS uses the characteristics of users and their groups to manage recommendations and perform filtering. These characteristics are grouped together through the group profile and the user profile [1, 3, 6, 7]. Figure 3 shows the user model used in IAAS.

Two major profiles emerge from Fig. 3:

- All information about users are contained in the User_Profile class;
- The Group_Profile which gathers all the information of the user groups.

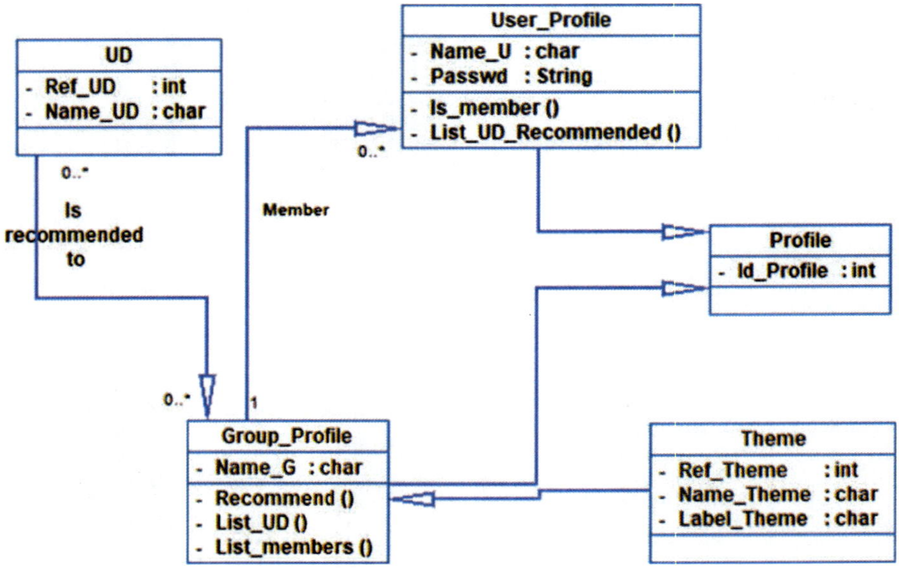

Fig. 3 Model of users' profile [2]

The description of classes, methods, attributes, and cardinalities is the same as in [1, 3].

3 State of the Art on Mobile Development Approaches

Several mobile platforms exist (Android, IOS, Windows Phone, etc.). For each platform, the development of applications follows approaches which we present here, native, Web, and hybrid approaches.

3.1 Native Approach

The native approach is used to develop native applications [8]. Thanks to this approach, developers develop applications for a mobile platform that can only be run on that same mobile platform. This means, for example, that a native Windows Phone application will not be able to run on Android and vice versa. This type of applications enables to get better results and fully exploits all the native features of the mobile platform such as camera access, embedded sensors and contact list, symbol activation or interaction with other applications. If we want to develop an application that needs to run on different mobile platforms (Android, IOS and Windows Phone), we need to develop separate applications with different programming languages such as Java for Android, C# for Windows Phone, and Swift for IOS. This type of application is mainly distributed online through an application store specific to each platform: Apple for IOS, Play Store for Android, Windows Phone Store for Windows Phone, etc. [9–12]. Each platform has a specific application development tool such as Android Studio for Android. The graphical interface of the native applications is a code directly understandable by the device (Fig. 4).

The chart presents the advantages and drawbacks of the native approach (Table 1).

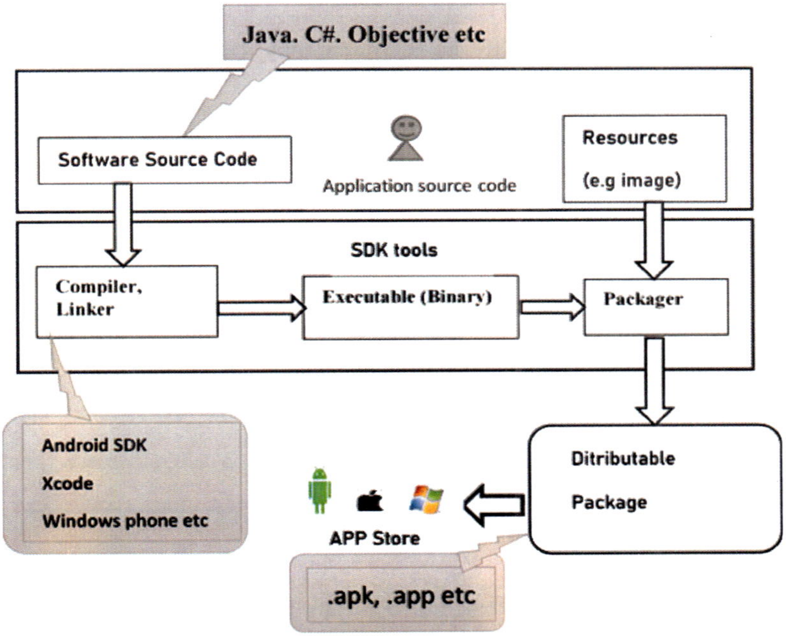

Fig. 4 Native approach architecture [8]

Table 1 Advantages and drawbacks of the native approach [8]

Advantages	Drawbacks
• Multitude of APIs to access all the functionalities of mobile devices such as camera, sensors, network access, file storage, and database • Better performance compared to Web and hybrid applications • Native appearance, fluid interface, and friendly	• Native application development requires good development experience [9] • The same application is developed on several platforms if we want to use them on each platform [13] • Restrictions and costs associated with the development and deployment of certain platforms (e.g., the Apple license development and Apple's approval for distributing the applications to the iTunes Apps store) [11]

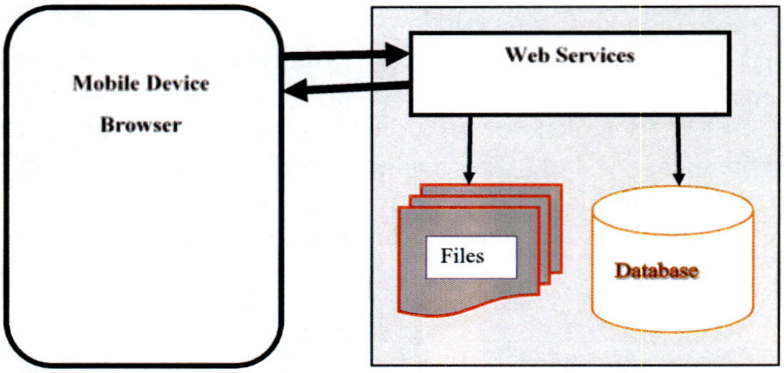

Fig. 5 Web approach architecture [8]

Table 2 Advantages and drawbacks of the Web approach [8]

Advantages	Drawbacks
• Application's development is not too complicated and easy to learn • The processing is done on the server side, and only the interface is brought back to the user • Maintaining the application is simple because application updates and data are performed on the server • The same application is used on multiple platforms using browsers	• The Web applications are not downloadable in online stores • They do not access the device's software or features [10] • They require a connection to run the application [10] • They may experience a malfunction due to connection and network delays • The application developer has less control over how different browsers interpret the content

3.2 *Web Approach*

Mobile applications designed with the Web approach run in the smartphone's browsers and its use requires an Internet connection [8, 12]. In this approach, mobile applications are designed using standard Web technology such as HTML5, CSS3, and JavaScript. These types of applications do not use the functionalities of smartphones because they are sites that are accessed through the smartphone's browser and the associated Web standards. Smartphone Web applications are accessible through a link on a Web site or bookmark in the user's smartphone browser (Fig. 5).

The chart presents the advantages and drawbacks of the Web approach (Table 2).

3.3 *Hybrid Approach*

The hybrid approach is obtained from the native and Web approaches. This approach is used to develop applications that use all or part of the native and Web approaches

[8]. Hybrid applications are downloadable in Web store platforms like native applications. They run in a native container and use the browser engine to display the application interface which is written in HTML and JavaScript. The hybrid application accesses smartphone features that Web applications cannot access such as accelerometer, camera, and local storage on a smartphone. The hybrid approach therefore allows, from a single source code, to generate applications for several platforms and this facilitates source code maintenance and an average development cost. This approach makes it possible to create applications whose graphical interfaces are Web pages (Fig. 6).

The chart presents the advantages and drawbacks of the hybrid approach (Table 3).

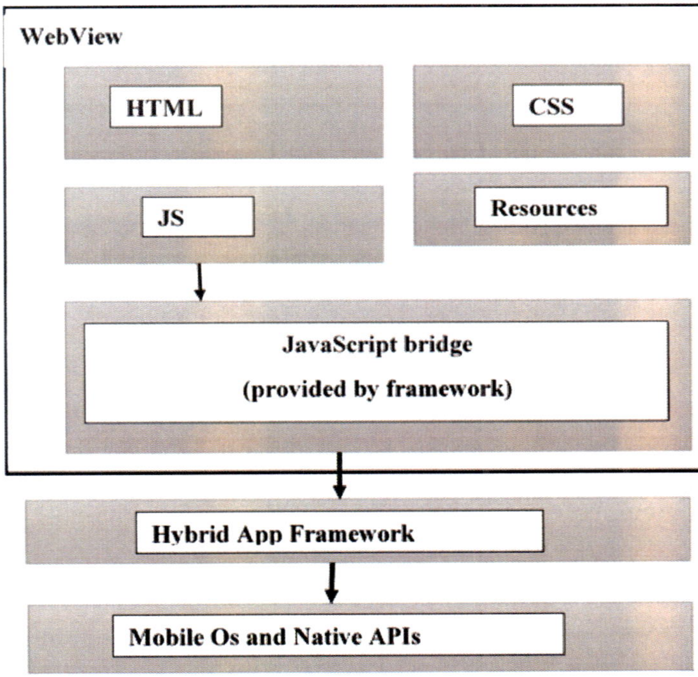

Fig. 6 Hybrid approach architecture [8]

Table 3 Advantages and drawbacks of the hybrid approach [8]

Advantages	Drawbacks
• Hybrid applications is available for download in online stores • The user interface can be reused in different platforms • The application can access the native functionality of the mobile device	• Less efficient than compared to the native application • The interface of the hybrid application will not have the same appearance as the native application

Table 4 Table summarizing the description of the user interface

Context	Justification
Cross-platform	Reach the maximum mobile environment
Accessing the databases	The data are stored in the databases
Access to smartphones features	We want to locate the documents
Using the connection	Can be used anywhere
A good adaptation of the user interface	We want a friendly user interface

4 Comparative Study and Mobile Solution Proposition for IAAS

4.1 Description of the Interface to Be Developed

The idea of our work is to be able to integrate the aspect of interaction with smartphones into IAAS from now on. Since users can use any smartphone (brand, operating system, etc.), we will opt for an interface that can be used by several operating systems (cross-platform). We will experiment our mobile solution with a library management system or an online course management system. Our interface will connect to the network to access the resources of the management system (library or online courses) and then make recommendations (images, text, audio, and video) to all users. In addition, it must have access to the data stored in the database of the management system (library or online course), a better display and above all it must be cross-platform. We envisage having an interface capable of locating documents. We summarize our description in Table 4.

4.2 Proposed Development Approach

According to the state of the art, a distinction is mainly made between the Web approach, the native approach, and the hybrid approach as an approach to the realization of mobile interfaces. According to [9, 14], the choice of the development approach is made according to a certain number of criteria and needs analysis and the objectives of the application to be developed.

Based on the analysis of the strengths and weaknesses of the native approach in Sect. 3.1, it is noted that this approach cannot realize cross-platform interfaces and in addition and requires a high level of experience and cost for the realization of applications. This approach will not help us reach the description of the interface to be developed described in Sect. 4.1. Therefore, we will not use it.

Table 5 Comparative table of approaches according to our context

Context	Native approach	Web approach	Hybrid approach
Cross-platform	No	Yes	Yes
Quality of adaptation	Very good adaptation	Not very good	Good adaptation
Access to smartphones functions	Entirely	Very little	Many of the features

In accordance with the state of the art, the Web approach does not allow the interface to have access to the functionalities of smartphones which, however, is one of our objectives and presents a poor quality of user interface adaptation compared to the hybrid approach.

Although the hybrid approach presents a lower quality interface compared to the native approach, it allows us to do cross-platform development and has access to certain smartphone functionalities. This convinced us to choose the hybrid approach. For the implementation of a cross-platform application using the hybrid approach, several solutions (frameworks) exist [15]. According to the same source, ionic and react native, designed, respectively, by Google and Facebook are the two most used solutions currently. But to implement our IAAS interface on the mobile, we used ionic for reasons of the mastery of this framework. Table 5 summarizes what we said about the choice of the development approach.

5 Implementation and Evaluation

5.1 Functional Needs Model and Analysis

Functional requirements produce a model of the needs focused on the business of the users. We have identified the users of our IAAS system and its use cases. The users are external entities that interact directly with the application. In our case, the users are all people who have the IAAS application on their smartphone. A use case allows the description of the interaction between the users and the system. However, users must log in to IAAS and make recommendations (Fig. 7).

The users must authenticate themselves before reading the recommendations and are free to recommend a document to other users or not. The model implemented is the class diagram model of the user model of [2]. The class diagram allows us to make an interpretation of the instances as well as the methods necessary to implement the interface. It is based on classes and associations.

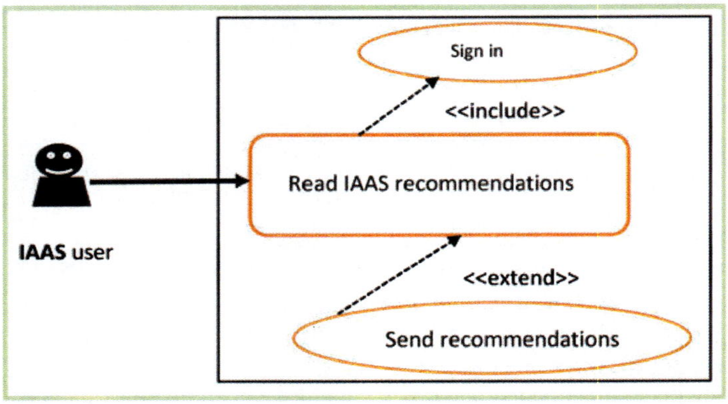

Fig. 7 Description of the interaction between the system and the users

5.2 *Realization of the Proposition*

For the realization of our model on the mobile, we used the PMB library management software and we installed our cross-platform development environment ionic. It has a model view controller architecture.

The model is used to load the data. In particular, it is responsible for creating the databases to send or receive information. We have modified the MySQL database of PMB by adding tables to it. To interact between the database and our interface, we used a PHP file for this purpose.

The view allows us to present the data provided by the model to the controller. In ionic, it is the HTML file that allows us to display the output of the processed data and the CSS files are used to animate the pages.

The controller is the layer that serves as link between the view and the model. It allows, on the one hand, to send commands to the model and to update the view and change the presentation on the other hand. In the case of ionic, these are files with the .ts extension.

We then developed the methods for connecting users, reading recommendations and making recommendations, etc., as described in the user model (Sect. 2.2.2).

5.3 *Display Tests and Evaluation of the Solution*

The display test required a large experimental device and software. These are a server to host all the necessary applications, a Wi-Fi modem to provide communication and smartphones. The mobile devices used are all Android and which composed of a tablet and two other smartphones of different size and characteristics. During our display tests, we noticed a good adaptation of the interface on all three devices compared to the Web version. We also note that on each device we have a very good adaptation of the user interface. Therefore, we can say that hybrid applications have a good interface adaptation compared to Web applications. We have not been able to test our interface with IOS. That is a deficiency in our work.

6 Conclusion

The study of the state of the art allowed us to retain the hybrid approach as a development approach adapted to the design and realization of our mobile interface for IAAS. However, with this approach one develops applications that have a good interface adaptation compared to Web applications. The effectiveness of the implementation of a purely mobile solution has once more strengthened our model. Our approach has enabled us to bring a plus to IAAS, in the sense that mobile users are able to use IAAS efficiently. This is now a major innovation for Internet users. But to further increase the number of IAAS users, it will be even more interesting to be able to generate the interface to test it with IOS, so we have not been able to integrate the localization of documents. As perspectives we intend to implement our interface on IOS while integrating the localization of documents.

References

1. K. Kabore, O. Sié, F. Sèdes, Information Access Assistant Service (IAAS), in *The 8th International Conference for Internet Technology and Secured Transactions (ICITST-2013)*, 9–12 Dec 2013 (IEEE UK/RI Computer Chapter, London, UK, 2013)
2. K. Kabore, A. Peninou, O. Sié, F. Sèdes, Implementing the information access assistant service (IAAS) for an evaluation. Int. J. Internet Technol. Secur. Trans. **x**(x), 2015 (2015)
3. K.K. Kaboré, Système d'aide pour l'accès non supervisé aux unités documentaire. Thèse de doctorat du l'Université de Ouaga 1 Pr Joseph KI-ZERBO, Janvier, 2018
4. M. Tmar, T. Braun, V. Roca, Modèle auto-adaptatif de Filtrage d'Information: Apprentissage incrémental du profil et de la fonction de décision. Thèse de l'Université Paul Sabatier de Toulouse, IRIT, 2002
5. S.E. Robertson, D. Hull, The TREC-9 filtering track final report, in *Proceeding of TREC-9* (2000)
6. A. Jameson, B. Smyth, Recommendation to groups, in *The Adaptive*, ed. by P. Brusilovsky, A. Kosa, W. Nejdl. LNCS, vol. 4321 (Springer-Verlag, Berlin Heidelberg, 2007), pp. 596–627
7. R. Burke, Hybrid web recommender systems, in *The Adaptive Web: Methods and Strategies of Web Personalization*, ed. by P. Brusilovsky, A. Kobsa, W. Nejdl. Lecture Notes in Computer Science, vol. 4321 (Springer-Verlag, Berlin Heidelberg New York, 2007), this volume
8. M. Lachgar: Approche MDA pour automatiser la génération de code natif pour les applications mobiles multiplateformes, 08 July 2017, pp. 15–18
9. S. Xanthopoulos, S. Xinogalos, A comparative analysis of cross-platform development approaches for mobile applications, in *Proceedings of the 6th Balkan Conference in Informatics*, Sept 2014 (ACM, 2013), pp. 213–220
10. C.R. Raj, S.B. Tolety, A study on approaches to build cross-platform mobile applications and criteria to select appropriate approach, in *2012 Annual IEEE India Conference (INDICON)*, Dec 2012 (IEEE, 2012), pp. 625–629

11. P. Smutný, Mobile development tools and cross-platform solutions, in *2012 13th International on Carpathian Control Conference (ICCC)*, May 2012 (IEEE, 2012), pp. 653–656
12. T. Melamed, B. Clayton, A comparative evaluation of HTML5 as a pervasive media platform, in *Mobile Computing, Applications, and Service* (Springer, Berlin Heidelberg, 2010), pp. 307–325
13. D. Sambasivan, N. John, S. Udayakumar, R. Gupta, Generic framework for mobile application development, in *2011 Second Asian Himalayas International Conference on Internet (AH-ICI)*, Nov 2011 (IEEE, 2011), pp. 1–5
14. S. Charkaoui, Zakariaadraoui, E.H. Benlahmar, Etude comparative des outils de développement mobile multiplateforme (2014)
15. https://blog.bam.tech/business-news/react-native-vs-ionic, 17 July 2020

Self-organizing Data Processing for Time Series Using SPARK

Asha Bharambe and Dhananjay Kalbande

Abstract Time series data can be found at many places in our day-to-day life. As the world is moving toward Internet of things (IoT) and sensors, data is obtained in enormous amounts. These data from multiple sources need to be combined as a single unit in order to make inferences or predictions. As the velocity of data generation increases, big data perspective needs to be applied to time series. Big data time series representation should be flexible enough to accommodate different time series. With multiple time series collected from different sources and at different intervals, it poses a challenge to combine these time series into a single one to perform analysis. In this paper, we propose a big data approach to represent time series that provides solution to these challenges and helps in analysis of the data. The proposed approach provides an efficient way to combine multiple time series and perform analysis on it.

Keywords Big data · Time series · Temporal RDD · SPARK

1 Introduction

In principle, time series data (temporal data) is different from a static data with regard to the time of occurrence of an event. A static data could describe only events occurred while temporal data additionally associates all events a time stamp of occurrence. Pervasiveness of time series data across various areas can be observed in day-to-day life from applications in weather data, trend of stock prices, social media trends, heath tracking devices, etc. to various science and engineering applications such as sensor data, satellite trajectory data, and process variable data from sensors.

A. Bharambe (✉)
V.E.S. Institute of Technology, Mumbai, India

D. Kalbande
Sadar Patel Institute of Technology, Mumbai, India
e-mail: drkalbande@spit.ac.in

S. Shakya et al. (eds.), *Mobile Computing and Sustainable Informatics*, Lecture Notes on Data Engineering and Communications Technologies 68,
https://doi.org/10.1007/978-981-16-1866-6_17

Temporal data analysis provides meaningful information, which can be used further for prediction. For instance, health monitoring devices logs various health-parameters every second (or part of it). Such healthcare time series data can be used to infer the heath information, and in critical cases, it can be used to avoid life threatening situation. Often more than one temporal data sets are required in order to make good prediction through their correlation.

With the progress of Internet of things (IoT) and evolution of fast networked sensors vast amount of data is collected and stored [1], while most of them are temporal data sets. Spark [2] is widely used data processing platform for big data sets as it provides better performance as compared to Hadoop MapReduce [3]. Spark comprises large set of operators with cyclic data flow support and it accepts language-integrated queries that provides an ease of big data analysis.

In Spark, typically a data is loaded as text files and converted into custom classes or supporting data type with respect to the programming language chosen. However, a general data model of Spark does not take temporal aspects of the data into account nor it has dedicated data types and operators for temporal data, for instance see [4, 5].

In this article, we propose an architecture that introduces the temporal data in Spark as temporal resilient distributed dataset (Temporal RDD) over Spark's own resilient distributed dataset (RDD) [6]. Such systems will have multilayered architecture for representing the data and performing queries on it.

2 Problem Definition

Representation and storage of continuous stream of data using relational databases is not suitable as vast amount of data flow leads to scaling problem.

This problem can be addressed by the using Hadoop and NoSQL databases [7]. Hadoop utilizes parallelization approach to perform analysis on the data through the use of tools like Spark.

Spark provides scalable and fault tolerant parallel distributed processing with aggressively cached in-memory distributed computing using low latency high-level APIs and stack of high-level tools. It parallelizes the applications across the clusters while hiding the complexities of distributed systems programming. Its primary core abstraction is called as resilient distributed dataset (RDD). Essentially it is just a distributed collection of elements that is parallelized across the cluster. In Spark 1.3, dataframe API that enables the user to write a SQL-like program in a declarative manner was introduced. They can be used to achieve high performance by leveraging advantages of various optimizers. Further, in Spark 1.6 dataset API was introduced that facilitate the users for writing a generic program, such as machine learning in a functional manner. It was also designed to achieve superior performance by reusing the advantages of optimizers. These three sets of APIs, i.e., RDD, dataframe and dataset, provides a way to represent the temporal data.

An objective of this paper is time series analysis of big data that comprises identification of the nature of phenomenon over an interval of time and utilize this information for prediction of phenomenon ahead of time.

As Spark's API is not sufficient for time series data analysis, we propose an architecture, in which we introduce Temporal RDD in addition to Spark's RDD, dataframes and datasets. This approach includes a set of operations such as join, difference, autoregression, and correlation.

3 Related Work

Time series data representation is professionally researched topic for smaller scale. For big data, the research is limited to few libraries been made available with only few models incorporated.

The first approach to implement time series data as an extension of Hadoop MapReduce is Spark-TS [8, 9]. This open-source library provides API for munging, manipulating, and modeling time series data. It provides Autoregressive Integrated Moving Average (ARIMA) model, Exponentially Weighted Moving Average (EWMA) model and Generalized Autoregressive Conditional Heteroskedastic (GARCH) models for forecasting. However, the library is not in active development and the last update made was on 17th March 2017.

The second approach is a distributed time series analysis framework for Spark called Huohua [10]. It is an open-source library which provides a variety of functionality which includes analysis functions using grouping, temporal join, summarizing, aggregating, etc. To enable the temporal join, a time aware data structure- a Time Series RDD was proposed, which uses the temporal locality. Time Series RDD extends the capabilities of RDD to execute on Spark framework. A time range is associated with each partition in Time Series RDD so that it is known that all data between a time-frame lies in the same partition. Group function in the framework allows the rows to be grouped with exactly the same timestamps. This grouping can be done based on data having same time stamp or group data before a time or after time. Huohua enables fast distributed temporal join of large unaligned time series. The temporal join provides two types of joins left Join and future Left Join. In temporal join, a typical matching criteria has two parameters direction (look backward or look forward) and a window (how much to look backward or look forward). The framework although provided significant speedup against the traditional RDD, is not supported and has evolved into the new framework called Flint which incorporated the future suggestions for Huohua framework.

Flint [11, 12] augments Sparks functionality for time series analysis library. This open-source library is based on Time Series RDD, which is a time series data structure and a collection of utility functions. It provides functionality for time series manipulation and time series modeling. It provided a time series dataframe which is an extension of dataframe which has time-awareness. The framework has a component for time series manipulation and one for time series modeling. It focuses on time series manipulation and provides the functions such as AsofJoin.

4 Proposed Work

In this paper, we propose an architecture that represents the temporal data in spark as Temporal RDD as shown in Fig. 1. The system has a multilayered architecture for representing the data and performing queries on it.

The bottom layer over the Hadoop Distributed File System (HDFS) [13] is the Spark-RDD layer, based on that the Temporal RDD is built as a second layer. This layer combines various time series data and read it as a single key-value pair in RDD by performing a join on the time attribute. The third layer is the querying layer, through that one can query the Temporal RDD and gets results. Since all the queries to be executed for analysis/prediction are with regard to time factor, we consider the time as a key while remaining attributes as values.

In join operation, different time series data are joined together with a time key. During the join, if there is any mismatch in the frequency of data representation, meaning they can be represented at various levels like daily data vs weekly data or monthly data. This data needs to be synchronized either by compressing the time series data or expanding it to match the time series for specified duration. Compressing the time series data refers to mapping to higher level of data thus accommodating larger time range. Expanding time series data means to map the data to lower level of data.

Fig. 1 Proposed architecture for temporal RDD

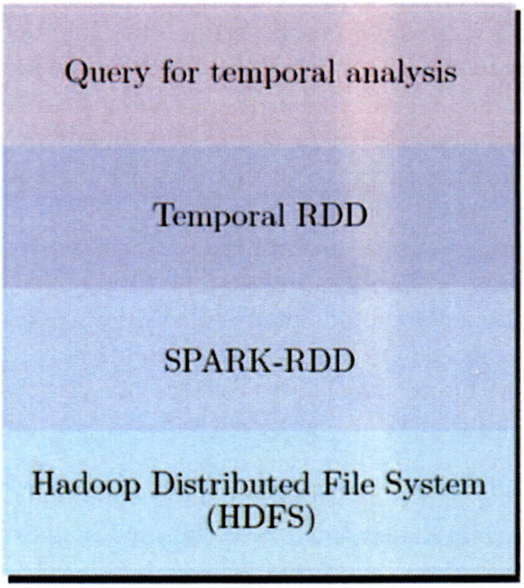

For example, if time series T1 captures data on daily basis and time series T2 captures it on weekly basis, then one can have an option of converting either daily data to weekly by applying compression or converting weekly data to daily by applying expansion. See Tables 1 and 2 for joining of Web site daily visitors count and subscriber's count via expansion. Table 1 shows two datasets, of which one is daily, and the other is weekly dataset. If a join has to be performed on these datasets, they lack the many common dates, and the result would lead to loss of data. Compression the join would map daily data to weekly and then perform a join. This mapping can be either an aggregation or min/max value. Similarly expanding the join would map the weekly data to daily data by repeating the previous/latest data or repeating the nearest value. Example of expansion with repeating previous value is shown is Table 2.

If compression or expansion is not possible by above method, then the joined time series can be created by replacing the missing value by the appropriate value. For instance, see Tables 3 and 4 that illustrates two attributes, namely, temperature and number of disease cases. If the attribute No of cases is repeated, then it gives false data. So, in such cases cumulative frequency can be used for representing the data as shown in Table 4

Operations that can be performed on time series are correlation, autocorrelation, comparison of two time series, pattern matching, clustering, forecasting. For this

Table 1 Two time series data with daily and weekly time

Date (daily)	No. of visitors	Join	Date (weekly)	No. of subscribers
04/10/2019	150		04/10/2019	100
05/10/2019	270		06/10/2019	150
06/10/2019	300		13/10/2019	120
...	...		20/10/2019	140
12/10/2019	250		...	...
14/10/2019	310			
...	...			

Table 2 Two time series data with expansion time intervals

Date	No. of visitors	No. of subscribers
04/10/2019	150	100
05/10/2019	270	100
06/10/2019	300	150
...	...	...
12/10/2019	250	150
13/10/2019	250	120
14/10/2019	310	120
...	...	

Table 3 Two time series data with non-matching intervals

Date/time	Temperature (°C)	Join	Date/Time	No. of cases
04/10/2019 6:00 AM	23		04/10/2019 6:00 AM	3
04/10/2019 7:00 AM	23.3		04/10/2019 7:30 AM	5
04/10/2019 8:00 AM	23.8		04/10/2019 9:00 AM	4
04/10/2019 9:00 AM	24.5		04/10/2019 9:30 AM	1
04/10/2019 10:00 AM	25		…	…
…				

Table 4 The resulting join of the two time series of Table 3

Date/time	Temperature (°C)	No. of cases (cumulative)
04/10/2019 6:00 AM	23	3
04/10/2019 7:00 AM	23.3	6
04/10/2019 8:00 AM	23.8	10
04/10/2019 9:00 AM	24.5	12
04/10/2019 10:00 AM	25	13
…	…	

paper, we have attempted to combine multiple time series obtained from different sources. The general machine learning techniques are applied to the resultant time series and compared it with the time series forecasting results.

The dataset needs to be partitioned if the number of attributes and size of data increases. For instance, partition can be done based on features or rows or it can be done with features and rows together. We chose the approach of partitioning based on rows, as features can be well fit into a memory and also it was suitable to find the correlations between them.

The data stored in the proposed solution is self-organizing in two ways [14]: (1) to accommodate changes in the cluster (for storage and execution) and (2) provide with replication and failure handling. These features are provided by the underlying layer of HDFS.

5 Results and Evaluation

The proposed architecture is tested with Dengue dataset from Mumbai region for the duration of three years from Jan 2012 to Dec 2015, see Table 5. This dataset is obtained from KEM Hospital, Mumbai. Other datasets of Climatic values such as temperature, humidity and precipitation given in Table 6 are taken as input [15].

Table 5 Sample dataset of dengue cases in Mumbai

Date of admission	No. of cases admitted
30/07/2014	02
08/08/2014	03
11/08/2014	05
12/08/2014	04
13/08/2014	01
16/08/2014	08
18/08/2014	06
19/08/2014	04
21/08/2014	02
23/08/2014	04
26/08/2014	07
...	...

Table 6 Climate dataset of Mumbai

Date	Temperature (°C)	Wind (km/h)	Humidity (%)	Pressure (mbar)	Visibility (km)
01/08/2014	27.70	17.55	0.90	1002.20	2.35
02/08/2014	27.55	18.45	0.91	1002.88	2.20
03/08/2014	27.38	18.82	0.92	1003.04	2.20
04/08/2014	27.30	18.84	0.92	1002.99	2.14
05/08/2014	27.35	18.78	0.92	1002.99	2.17
06/08/2014	27.34	19.05	0.92	1003.15	2.20
07/08/2014	27.45	19.06	0.91	1003.27	2.28
08/08/2014	27.49	18.74	0.90	1003.42	2.35
09/08/2014	27.45	18.07	0.90	1003.70	2.43
10/08/2014	27.53	17.67	0.90	1004.04	2.50
11/08/2014	27.60	17.58	0.89	1004.23	2.59
12/08/2014	27.61	17.49	0.89	1004.41	2.58
13/08/2014	27.68	17.50	0.89	1004.66	2.68
14/08/2014	27.72	17.48	0.89	1004.92	2.71
15/08/2014	27.74	17.54	0.88	1005.09	2.72
...	...	...	...	...	...

5.1 *Data and Preprocessing*

The time series with time attribute and at least one value attribute is given as an input to the system. Dengue time series consists of daily data of the number of cases

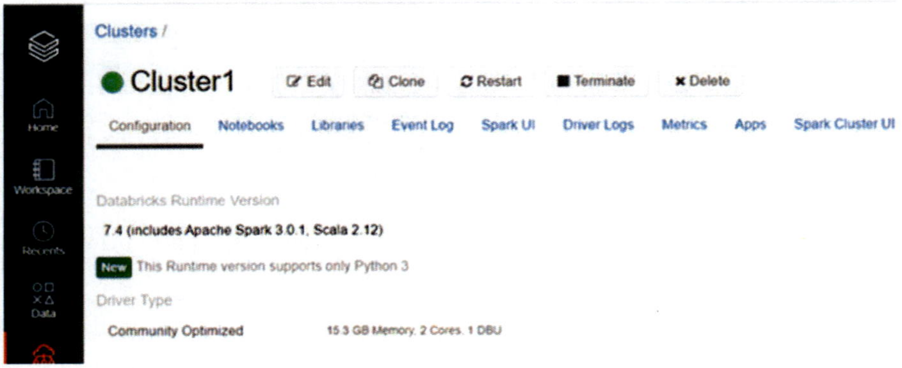

Fig. 2 Experimental setup of the cluster

identified with dengue. Also, a climate data that comprises temperature, humidity, etc. are considered on daily intervals.

There are few missing values in the cases time series thus making the intervals uneven. The datasets obtained was preprocessed to transform into a common date format. Also, the data obtained from hospital was with inconsistent date and time format. The data was preprocessed, and relevant attributes of date and incidence cases were stored.

5.2 Experimental Setup

The data executed on databricks platform [16] with Spark Cluster of 2 core, 1 DBU and 15 GB of memory. The screenshot of the platform is shown in Fig. 2. The dataset is uploaded on the platform and given as input to the program.

5.3 Results

The time series datasets of dengue incidence cases and two parameters of climatic data of temperature and Precipitation were used in the experiment.

These three time series data (incidence case, temperature, and precipitation) were uploaded on the platform. For our experiments with join operator, we have used the normal RDD structure for performing join on two uneven time series (Incidence case and temperature). Then the same join was performed by using the proposed Temporal RDD. As shown in Fig. 3, it is observed that the Temporal RDD took less amount of time to perform the operation as compared to RDD. The same experiment was repeated for joining three uneven time series and it can be verified from Fig. 4 that Temporal RDD took less amount of time for execution of three series data as well.

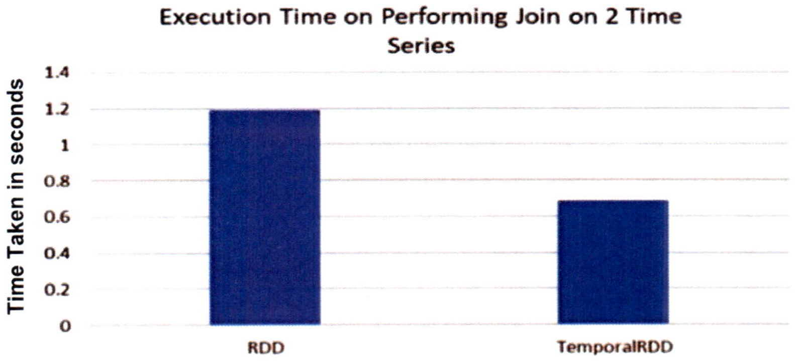

Fig. 3 Execution time required to join two time series dataset

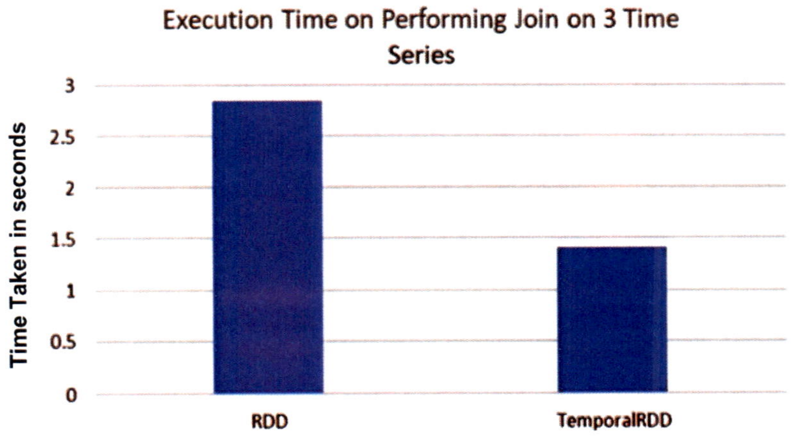

Fig. 4 Execution time required to join three time series dataset

6 Conclusion

In this paper, we have proposed a data representation and a Temporal RDD for execution of time series data in the distributed Spark environment. The proposed Temporal RDD performs join operations faster as compared to traditional RDD. Such representation can be extended to other types of data such as spatio-temporal data. Also, its capabilities can be extended by including other operations on to it like test of stationarity, differencing, and forecasting.

References

1. D. Mourtzis, E. Vlachou, N. Milas, Industrial big data as a result of IoT adoption in manufacturing. Procedia CIRP **55**, 290–295 (2016). ISSN 2212-8271
2. Spark.com: Apache Spark unified analytics engine for large scale data processing. https://spark.apache.org/docs/latest/. Accessed December 15, 2020
3. J. Dean, S. Ghemawat, Mapreduce: simplified data processing on large clusters. Commun. ACM **51**(1), 107–113 (2008)
4. S. Hagedorn, P. Gotze, K.-U. Sattler, The STARK framework for spatiotemporal data analytics on SPARK, *Datenbanksysteme für Business, Technologie und Web (BTW 2017)*, pp 123–142
5. S. Hagedorn, T. Rath, Efficient spatio-temporal event processing with STARK. In 20th International Conference on Extending Database Technology (EDBT), 21–24 March 2017, pp 570–573
6. M. Zaharia, M. Chowdhury, T. Das, A. Dave, J. Ma, M. McCauley, M.J. Franklin, S. Shenker, I. Stoica, Resilient distributed datasets: a fault-tolerant abstraction for in-memory cluster computing, in *Proceedings of the 9th USENIX conference on Networked Systems Design and Implementation*, (USENIX Association, 2012), p. 2
7. M. Tahmassebpour, A new method for time-series big data effective storage. IEEE Access **5**, 10694–10699 (2017)
8. Cloudera Blog on Spark Timeseries Library. https://blog.cloudera.com/blog/2015/12/spark-ts-a-new-library-for-analyzing-time-series-data-with-apache-spark/
9. Time series for Spark—A library for analyzes of large scale time series datasets. https://sryza.github.io/spark-timeseries/0.3.0/index.html
10. Huohua—Framework for Distributed Time Series. https://databricks.com/session/huohua-a-distributed-time-series-analysis-framework-for-spark
11. Flint—Library for time series. https://www.twosigma.com/insights/article/introducing-flint-a-time-series-library-for-apache-spark/.
12. Flint—Library for time series. https://ts-flint.readthedocs.io/en/latest/
13. K. Shvachko, H. Kuang, S. Radia, R. Chansler, The hadoop distributed file system, in 2010 IEEE 26th Symposium on Mass Storage Systems and Technologies (MSST). (IEEE, 2010), pp. 1–10
14. H. Tang, A. Gulbeden, J. Zhou, W. Strathearn, T. Yang, L. Chu, A self-organizing storage cluster for parallel data-intensive applications, in *SC '04: Proceedings of the 2004 ACM/IEEE Conference on Supercomputing*, Pittsburgh, PA, USA, 2004, pp. 52–52. https://doi.org/10.1109/SC.2004.9
15. Climate data retrieved from https://www.timeanddate.com/weather/india/mumbai
16. Databricks Platform for execution. https://databricks.com/

An Experimental Investigation of PCA-Based Intrusion Detection Approach Utilizing Machine Learning Algorithms

G. Ravi Kumar, K. Venkata Seshanna, S. Rahamat Basha, and G. Anjan Babu

Abstract Intrusion detection system (IDS) is a framework that screens and investigates information to distinguish any interruption in the computer system network. The enormous measure of information is traversing the computer networks organizes that lead to the weakness of information trustworthiness, secrecy and dependability. Along these lines, organize security is a consuming issue to keep the honesty of frameworks and information. To address the downsides of customary IDSs, machine learning-based models open up new chance to arrange strange traffic as oddity with a self-learning ability. The motivation behind this investigation is to distinguish significant decreased information highlights in building IDS that is computationally productive and viable. In this paper, we propose a system for features decrease utilizing principal component analysis (PCA) to eliminate immaterial features, choosing applicable features without influencing the data contained in the first information and afterward utilizing three machine learning calculations like decision tree, KNN and Naïve Bayes to arrange interruptions information. In journey to choose a decent learning model regarding exactness, accuracy, recall and F-score, this paper shows the exhibition correlation between decision tree, K-nearest neighbor (KNN) and Naïve Bayes order models. These models are prepared and tried on decreased subsets of highlights got from the first benchmark organize interruption discovery dataset, NSL-KDD utilizing PCA. Exact outcomes show that chose decreased credits give better execution to plan IDS that is productive and viable for arrange interruption recognition.

Keywords Machine learning · IDS · PCA · Decision tree · Naive Bayes · KNN and NSL-KDD

G. Ravi Kumar (✉) · K. Venkata Seshanna
Department of Computer Science, Rayalasema Unversity, Kurnool, India

S. Rahamat Basha
Department of CSE, RGMCET, Nandyal, Andhra Pradesh, India

G. Anjan Babu
Department of Computer Science, S. V. University, Tirupati, India

S. Shakya et al. (eds.), *Mobile Computing and Sustainable Informatics*, Lecture Notes on Data Engineering and Communications Technologies 68,
https://doi.org/10.1007/978-981-16-1866-6_18

1 Introduction

Interruptions are characterized as endeavors to bargain the secrecy, respectability or accessibility of computer network or system [1, 2]. They are brought about by aggressors getting to a framework from the Web, by approved user of the frameworks who endeavor to increase extra benefits for which they are not approved and by approved client who abuse the benefits given to them. This is commonly cultivated via naturally gathering data from an assortment of frameworks and system sources, and afterward examining the data for conceivable security issues [3].

Probably, the greatest test in interruption discovery is the broad measure of information gathered from the system. In this way, the current methodologies of interruption recognition have concentrated on the issues of feature choice or dimensionality decrease. Feature choice or decrease keeps the first features thusly and select subset of features that predicts the objective class variable with greatest arrangement precision [4, 5].

Interruption detection permits observing and examining client movement and framework, checking framework designs and weaknesses, assessing the framework respectability and information documents, factual examination of model action dependent on realized assault coordinating, investigation of inconsistencies and running the review framework. Accordingly, intrusion identification framework (IDS) is expected to distinguish vindictive or unapproved exercises on the organization and square the gatecrasher traffic association with keep the framework from additional harm. IDS initially dissected all the organization traffic and raised a caution to help the organization head if noxious endeavors are found. Interruption detection systems can utilize an alternate sort of strategies to recognize dubious exercises. It tends to be extensively separated into two kinds.

1.1 Kinds of IDS

i. **Signature-based interruption location**: These frameworks contrast the approaching traffic and a previous information base of referred to assault designs known as marks. Recognizing new assaults is troublesome. The merchants providing the frameworks effectively discharge new names.
ii. **Anomaly-based interruption location**: It utilizes insights to frame a benchmark use of the organizations at various time stretches. They were acquainted with distinguish obscure assaults. This framework utilizes AI to make a model reenacting customary movement and afterward contrasts new conduct and the current model.

The target of IDS is to identify any odd and unordinary action which breaks the verification and security strategy of computer network systems. In this work, we intend to sift through excess data and altogether lessen number of information features needed to identify assaults. This paper proposes principal segment investigation

(PCA) as a decrease method and back proliferation calculation as a learning device for the created framework, initially, we diminish the features and afterward apply the learning calculations.

Feature extraction incorporates features development, space dimensionality decrease, scanty portrayals and feature determination every one of these procedures are ordinarily utilized as pre-preparing to machine learning and insights undertakings of expectation, including design acknowledgment [6–8]. In our proposed calculation, PCA change utilized for dimensionality decrease which is generally utilized advance, particularly, when managing high-dimensional space of features. PCA-based methodologies improve framework exhibitions and a prepared fake neural system to distinguish any obscure assaults.

The target of our work is to assess the exhibition of administered learning models for organize interruption arrangement by considering fundamental measurements, for example, exactness, accuracy, review, and f-score. To assess the order precision, we make input include subsets of the NSL-KDD unique dataset with the assistance of PCA.

The destinations of this paper are:

- To study different features of info dataset.
- To apply the principal component analysis (PCA) for lessening the quantity of qualities.
- To group information utilizing three characterization models like decision tree, KNN and Naïve Bayes.

The remaining of this article is organized as. The related work is highlighted in Sect. 2. The methodology is clarified in Sect. 3. The experimental results and evaluation are analyzed in Sect. 4. The final research is concluded by Sect. 5.

2 Related Work

Ravi Kumar et al. [9] proposes the usage of principal component analysis (PCA) to diminish high-dimensional data and to upgrade the farsighted execution of the neural network machine learning model. The specific results show neural network request with PCA is a beneficial portrayal for broad datasets.

Benadd et al. [10] recommend another proposed PCA-fuzzy clustering-KNN technique that implies group of analysis of principal component and fuzzy clustering with K-nearest neighbor feature determination methods. They dissect NSL-KDD dataset utilizing PCA-fluffy clustering-KNN investigative and attempt to define the exhibition of occurrence utilizing AI calculations, and the calculation realizes what kind of assaults are found in which classes so as to improve the classification exactness and lessen high bogus alert rate and identifies the limit of recognition rate from dataset as appeared by the mathematical outcomes.

Aggarwal et al. [11] have depicted that the intrusion discovery framework manages huge measure of information which contains different superfluous and repetitive highlights bringing about expanded preparing time and low location rate. In this way, feature determination assumes a significant job in interruption discovery. They played out a near examination of various element choice strategies are introduced on KDDCUP'99 benchmark dataset and their presentation are assessed as far as identification rate; root mean square blunder and computational time.

Nskh et al. [12] has expressed that the presentation of various parts of support vector machine (SVM) are thought about in contrast to Knowledge Discovery in Databases Cup'99 (KDD) informational index and location precision, recognition time are analyzed. The location time is diminished by receiving principal component analysis (PCA) which shortens higher dimensional dataset to bring down dimensional dataset. The investigations which are led in this examination show that Gaussian radial basis function portion of SVM has higher recognition exactness.

3 Methodology

This section gives the concise thought of chosen administered models of decision tree, Naive Bayes and KNN.

3.1 Machine Learning Techniques

Machine learning (ML) is a part of computerized reasoning that secures information from preparing information dependent on well-established realities. ML is characterized as an examination that permits computers to learn information without being modified [13]. There are a few ML strategies embraced to anticipate the assaults in the test datasets which was utilized to prepare the framework. These calculations were utilized to order the assaults in other to discover an effective strategy in anticipating and arranging assaults. ML strategies are ordered into three general classes, for example, regulated learning and unaided learning [14]. Directed calculations learn for foreseeing the article class from pre-named (characterized) objects. Nonetheless, the unaided calculation finds the characteristic gathering of items given as unlabeled information. In this work, the premium is with the accompanying regulated learning calculations like decision tree, KNN and Naïve Bayes techniques are assessed.

3.1.1 Decision Tree

A decision tree is spoken to as a tree. It speaks to a lot of the decision and these choices are utilized to produce rules for the classification of information designs. The fundamental favorable circumstances of decision tree are that they are easy to comprehend and decipher. A hub of a decision tree identifies a trait by which the example is to be parceled [15]. Each hub has a few edges, which are marked by the plausible estimation of the quality in the parent hub. An edge connects either two hubs of a tree or a hub with a leaf. Leaf hubs are named with class marks for classification of the case.

3.1.2 Naive Bayes

The Naive Bayes classifier is a grouping technique dependent on the Bayes hypothesis. It significantly improves learning by expecting that features are free given class. In spite of the fact that autonomy is commonly a helpless presumption, by and by innocent Bayes regularly contends well with more refined classifier [8]. Naive Bayes classifier is known to be superior to some other characterization techniques. Since first, the fundamental quality of Naive Bayes is an extremely solid (Naive) supposition of autonomy from each condition or occasion. Second, its model is straightforward and simple to make. Third, the model can be executed for huge informational indexes.

Bayesian classifiers dole out the most probable class to a given model portrayed by its element vector. Learning such classifiers can be incredibly rearranged by expecting that highlights are autonomous given class, that is, $P(X|C) = \prod_{i=1}^{n} P(Xi|C)$, where $X = (X1, X2, \ldots, Xn)$ is a component vector and C is a class.

3.1.3 K-Nearest Neighbor (KNN)

The KNN is a straightforward yet compelling technique for arrangement. The KNN calculation is a technique for grouping objects dependent on nearest preparing models in the component space. KNN is a sort of example-based learning, or lethargic realizing where the capacity is just approximated locally and all calculation is conceded until order [2].

For an information record D to be arranged, its K closest neighbors are recovered, and this structure an area of D. Larger part casting a ballot among the information records in the area is generally used to choose the characterization for D with or without thought of separation-based weighting. In any case, to apply KNN, we have to pick a fitting an incentive for K, and the accomplishment of characterization is a lot of subject to this worth. The significant disadvantages as for KNN are (1) its low effectiveness—being a lethargic learning strategy denies it in numerous applications, for example, dynamic Web digging for a huge vault, and (2) its reliance on the choice of a great worth for K.

3.2 *Principal Component Analysis (PCA)*

PCA is a dimensionality decrease strategy used to change high-dimensional datasets into a dataset with less factors, where the arrangement of coming about factors clarifies the greatest difference inside the dataset [7, 16]. PCA is utilized preceding solo and administered machine learning steps to diminish the quantity of features utilized in the examination, in this way, decreasing the probability of blunder. It goes about as an incredible model to examine the information.

The essential segments are a straight line, and the primary head segment holds the most difference in the information. Each ensuing head part is symmetrical to the last and has a lesser fluctuation. PCA is dominatingly utilized as a dimensionality decrease method in spaces like facial acknowledgment, computer vision and picture pressure [17, 18]. The fundamental direct method for dimensionality decrease, head part examination, plays out a straight planning of the information to a lower-dimensional space so that the difference of the information in the low-dimensional portrayal is expanded [19]. It goes about as an incredible model to investigate the information. PCA will take all the first preparing set factors and disintegrate them in a way to make another arrangement of factors with high clarified difference.

4 Experimental Evaluation

The goal of this segment is to assess our proposed calculation regarding precision, number of chosen highlights and learning exactness on chose highlights. In this analysis, the PCA is utilized to decrease unimportant highlights. The proposed PCA-based choice tree has been tried NSL-KDD informational collection, which is a changed variant of KDD'99 informational collection [20]. The explanation behind utilizing NSL-KDD dataset for our analyses is that the KDD'99 informational collection has countless repetitive records in the preparation and testing informational index. For paired arrangement, the NSL-KDD groups the system traffic into two classes, to be specific, ordinary and abnormality.

The trials were performed on full preparing informational index having 125,973 records and test informational collection having 22,544 records. The preparation dataset is comprised of 21 unique assaults out of the 37 presents in the test dataset. The assault types are assembled into four classifications: DoS, Probe, U2R and R2L. Table 1 shows the quantity of individual records in four kinds of assaults for both preparing and testing for NSL-KDD dataset.

Table 1 Total instances of NSL-KDD data

Training data (1,25,973)		Testing data (22,544)	
Type of attack	Total no. of instances	Type of attack	Total no. of instances
Normal	67,343	Normal	9711
DOS	45,927	DOS	7456
Probe	11,656	Probe	2421
R2L	52	R2L	200
U2R	995	U2R	2756

4.1 Result and Discussion

The dataset is huge and high-dimensional in nature. Henceforth, both datasets go through dimensionality decrease utilizing PCA. Figure 1 shows the difference assessment of the 12 head segments acquired for both datasets as they contain similar 41 highlights. The essential segments having a combined variety of over 90% are held and the staying ones are disposed of based on total variety which is over 90% while the rest of the parts are disposed of as a result of the staying 10% variety which is immaterial. The parts with higher change will be held in the boundary determination results after dimensionality decrease utilizing PCA.

The information in NSL-KDD dataset is either named as ordinary or as one of the 24 various types of assault. From 41 credit, we have separated to 12 component vectors by utilizing PCA strategy to get an ideal choice from complete dataset for preparing just as for testing tests. Figures 2 and 3 show the test precision that accomplished by utilizing the three calculations for the full measurement information, and furthermore, after the component decrease with PCA method. The fundamental 12

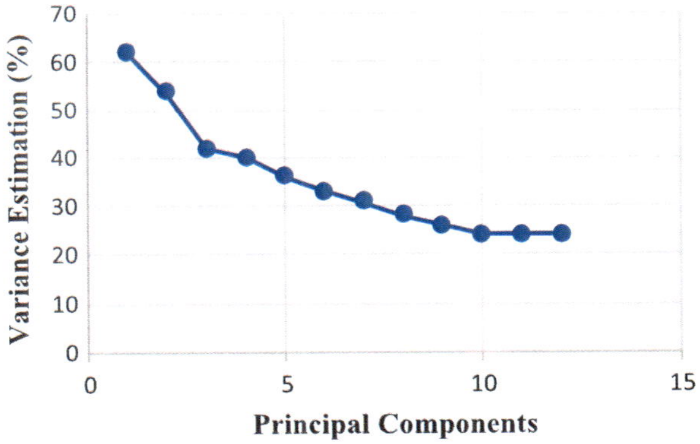

Fig. 1 Variance of principal components

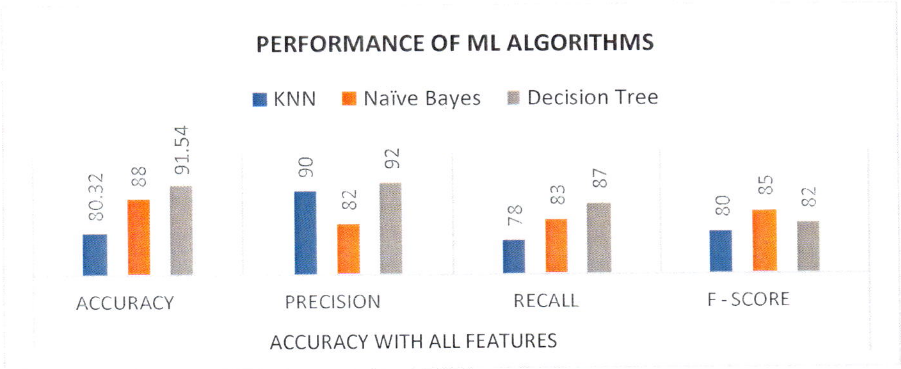

Fig. 2 Performance of ML algorithms without PCA

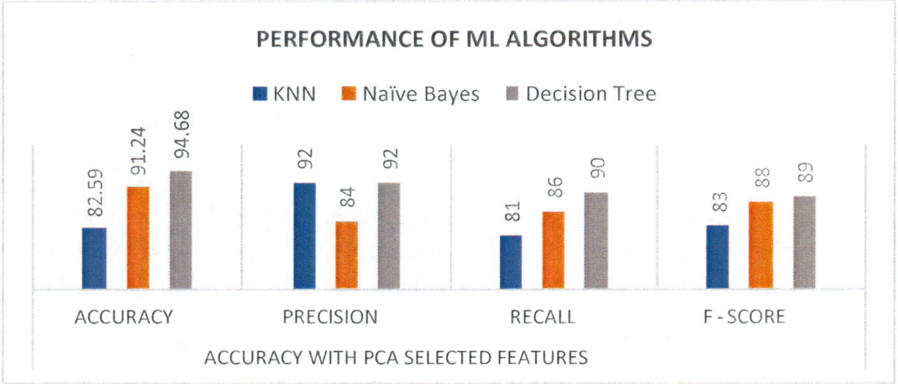

Fig. 3 Performance ML algorithms with PCA

highlights eliminated by PCA add to altogether less change, and thus the staying 29 characteristics chose. Figure 2 shows the test exactness on class normal assault that contrasted and 41 highlights, and Fig. 3 shows with the diminished arrangement of highlights by utilizing PCA procedure.

The presentation of model decision tree without PCA dependent on exactness, exactness, review and f-score estimations of choice tree are 91.54, 92, 87 and 82% individually, while the exhibition of choice tree with PCA dependent on precision, distinction and accuracy esteems are 94.68, 92, 90 and 89%. Here, the choice tree with PCA calculation shows the most elevated precision contrasted and choice tree without PCA.

We watch the presentation of three ML calculations as appeared in Fig. 2 without PCA dependent on precision of choice tree classifier calculation gives noteworthy improvement in the exactness when contrasted with a KNN and Naive Bayes classifier. For decision tree characterization, 11.22% exactness was improved over KNN, while 3.54% precision was enhanced Naive Bayes. So, the decision tree execution measurements are more than KNN and innocent Bayes calculations.

We watch the exhibition of three ML calculations as appeared in Fig. 3 with PCA dependent on precision of decision tree classifier calculation gives noteworthy improvement in the exactness when contrasted with a KNN and Naive Bayes classifier. For choice tree characterization, 12.09% exactness was improved over KNN, while 3.44% precision was enhanced innocent Bayes. So, the choice tree execution measurements are more than KNN and innocent Bayes calculations. From Figures 2 to 3, it is seen that the presentation of these models relies upon the feature determination. In this way, ideal component choice is a significant boundary to build up an effective IDS.

5 Conclusion

In our proposed work, the PCA is to decrease the dimensionality of the information while holding, however, much as could be expected of the variety present in the first dataset and prepared fake neural system to recognize any sort of new assaults. Tests and examination are done on NSL-KDD dataset. Our test results indicated that the proposed choice tree model with PCA gives better grouping exactness accomplished in identifying new assaults when contrasted with KNN and innocent Bayes models. Observational examinations demonstrate that the highlight decrease method is equipped for lessening the size of the dataset.

References

1. H.-J. Liao, C.-H.R. Lin, Y.-C. Lin, K.-Y. Tung, Intrusion detection system: a comprehensive review. J. Netw. Comput. Appl. **36**(1), 16–24 (2013)
2. L. Li, Y. Yu, S. Bai, Y. Hou, X. Chen, An effective two-step intrusion detection approach based on binary classification and k-NN. IEEE Access **6**, 12060–12073 (2018)
3. Y. Li et al., An efficient intrusion detection system based on support vector machines and gradually feature removal method. Expert Syst. Appl. **39**, 424–430 (2012). https://doi.org/10.1016/j.eswa.2011.07.032
4. E. Baraneetharan, Role of machine learning algorithms intrusion detection in WSNs: a survey. J. Inform. Technol. **2**(03), 161–173 (2020)
5. F. Kuang, W. Xu, S. Zhang, A novel hybrid KPCA and SVM with GA model for intrusion detection. Appl. Soft Comput. **18**, 178184 (2014)
6. S. Rahamat Basha et al., Impact of feature selection techniques in text classification: an experimental study. J. Mechan. Continua Math. Sci. **3**, 39–51 (2019). ISSN: 0973-8975. https://doi.org/10.26782/jmcms.spl.3/2019.09.00004
7. S. Rahamat Basha, J. Keziya Rani, A comparative approach of dimensionality reduction techniques in text classification. Eng. Technol. Appl. Sci. Res. **9**(6), 4974–4979 (2019). ISSN-1792-8036
8. Z. Li, P. Batta, L. Trajkovic, Comparison of machine learning algorithms for detection of network intrusions, in Proceedings of the IEEE International Conference on Systetm, Man and Cybernetics, Miyazaki, Japan (2018), pp. 4248–4253
9. K. Nagamani, et al., A framework of dimensionality reduction utilizing PCA for neural network prediction. Lecture Notes on Data Engineering and Communications Technologies. vol 37, pp 173–180 (2020) Springer Nature Singapore Pte Ltd.. ISBN 978-981-15-0977-3
10. H. Benadd, K. Ibrahimi, A. Benslimane, Improving the ıntrusion detection system for NSL-KDD dataset based on PCA-fuzzy clustering-KNN. IEEE (2018) 978-1-5386-7330-0/18. https://www.researchgate.net/publication/330786768
11. M. Aggarwal, Amrita, Performance analysis of different feature selection methods in ıntrusion detection. Int. J. Sci. Technol. Res. **2**(6), 225–231 (2013). ISSN 2277-8616
12. P. Nskh, M.N. Varma, R.R. Naik, Principle component analysis based intrusion detection system using support vector machine. 2016 IEEE International Conference on Recent Trends in Electronics, Information & Communication Technology (RTEICT) (2016), pp. 1344–1350. https://doi.org/10.1109/RTEICT.2016.7808050
13. G.Ravi Kumar, et al., A summarization on text mining techniques for ınformation extracting from applications and ıssues. J. Mechan. Contin. Math. Sci. **5**, 324–332 (2020). shttps://doi.org/10.26782/jmcms.spl.5/2020.01.00026

14. G. Ravi Kumar, V.S. Kongara, G.A. Ramachandra, An efficient ensemble based classification techniques for medical diagnosis. Int. J. Latest Technol. Eng. Manage. Appl. Sci. **II**(VIII), 5–9 (2013). ISSN-2278-2540
15. T. Ahmed, B. Oreshkin, M. Coates, Machine learning approaches to network anomaly detection, in *Proceedings of USENIX Workshop Tackling Computer Systems Problems with Machine Learning Techniques*, Cambridge, MA, USA, Apr. 2007, pp 1–6
16. J. Yan, B. Zhang, N. Liu, S. Yan, Z. Chen, Effective and efficient dimensionality reduction for large scale and streaming data processing. IEEE Trans. Knowl. Data Eng. **18**(3), 320–333 (2006)
17. S.R. Basha et al, A novel summarization-based approach for feature reduction, enhancing text classification accuracy. Eng. Technol. Appl. Sci. Res. **9**(6), 5001–5005 (2019). ISSN-1792-8036
18. S. Bhupal Rao, et al., A comparative approach of text mining: classification, clustering and extraction techniques. J. Mechan. Continua Math. Sci. **5**, 120–131 (2020). https://doi.org/10.26782/jmcms.spl.5/2020.01.00010
19. X. Lei, A novel feature extraction method assembled with PCA and ICA for network intrusion detection. Comput. Sci. Technol. Appl. IFCSTA **3**, 31–34 (2003)
20. Nsl-kdd data set for network-based intrusion detection systems. Available on: https://nsl.cs.unb.ca/KDD/NSL-KDD.html (March 2009)

OpenFlow-Based Dynamic Traffic Distribution in Software-Defined Networks

Duryodhan Chaulagain, Kumar Pudashine, Rajendra Paudyal, Sagar Mishra, and Subarna Shakya

Abstract Traditional load balancer has no flexibility to manage growing network traffic in real time and is expensive too. Software-defined networking (SDN) is a promising solution for handling today's growing network architecture. To address issues such as network congestion and overloading, service providers use multiple replicas in the server cluster to provide the same services where network virtualization and effective load balancing is very important. This work proposes the implementation of real-time traffic management strategy that distributes the network requests among multiple servers based on a fuzzy logic. The fuzzy membership characteristic that affects the performance parameters of server load has been analyzed, and load state of virtual servers in real-time is evaluated through fuzzy logic. Considering major parameters of Web server (CPU, RAM and bandwidth utilization) as an input of fuzzy system, the clients requests are forwarded to the server having minimum load at real time. The proposed load balance algorithm written in Python is simulated on Mininet, and performance of each Web servers is compared with existing round robbin (RR) and least connections (LC) load balancing schemes. The results demonstrate that the proposed scheme has improved response time and higher throughput as compared to the other load balance solutions.

Keywords Software-defined networking · OpenFlow · Load balancer · Server cluster

1 Introduction

The expansion growth of cloud ecosystem has influenced service providers to develop applications on the virtualized platforms and hosts all services on various server virtual instances. Due to the rapid growth of network traffic, some network problems

D. Chaulagain (✉) · K. Pudashine · S. Mishra
Nepal College of Information Technology, Pokhara University, Pokhara, Nepal

R. Paudyal · S. Shakya
Pulchowk Campus, Institute of Engineering, Tribhuvan University, Kirtipur, Nepal
e-mail: drss@ioe.edu.np

S. Shakya et al. (eds.), *Mobile Computing and Sustainable Informatics*, Lecture Notes on Data Engineering and Communications Technologies 68,
https://doi.org/10.1007/978-981-16-1866-6_19

such as server's overload and network congestion are involved. Thus, load balancer (LB) emerged as a technique used for distributing the incoming traffic among multiple Web servers to achieve minimum response time and maximum utilization of the servers. The control plane of the SDN network is a logical entity, which computes and takes all routing decisions of OpenFlow packets that are generated in the data forwarding plane. In this study, we have proposed a dynamic load distribution algorithm, where input parameters (CPU, memory and bandwidth utilization) of the Web servers have been considered for analysis of the performance of server in real time. The proposed dynamic traffic management algorithm follows the layered architecture of SDN consisting control plane and data forwarding plane. To practically evaluate and compare the performance of the proposed algorithm with round robin and least connections algorithm, we use Mininet with Docker containers to setup a virtual environment for simulation. Experimental results show proposed algorithm has improved performance than round robin and least connections in a heterogeneous environment with a large number of concurrent users. This paper is structured as follows. Section 2 covers the SDN-related load balancing architectures that are relevant to the proposed work. Section 3 explains the main framework development and implementation of the proposed algorithm. Section 4 shows the results obtained from different experiment scenarios. Section 5 aims to draw the final remarks and conclusions of the proposed work and describes our future recommendations.

2 Background and Related Studies

2.1 Software-Defined Network (SDN) and Openflow

SDN architecture consists of three major components: SDN controller, a network brain, SDN switches, routers, etc. In SDN architecture, the control plane and data plane are completely isolated. In addition, customized applications can be developed at the application layer using a northbound interface. These applications can be used to instruct the switches using OpenFlow protocol via a secure channel [1]. In SDN network, each switch has a flow meter; a collection of process data streams for all actions, such as looking up and forward. The main flow table contains headers, counters and actions, three fields in which the actions field is represented, forwarding rules and flow meters, which are updated intermittently [2].

2.2 Related Works

Most of the popular and widely used algorithms are round robin algorithm, weighted round robin algorithm, random allocation and so on. The round robin (RR) load balancing is a popular and simpler load scheduling algorithm that distributes the request from the clients one by one in cyclic order [3]. Most of the researches predict that RR algorithm performs efficiently where servers configuration are homogeneous. Similarly, in weighted round robin (WRR) algorithm [3], load balancer distributes clients requests based on the priority given to weights of the server cluster. The server having higher weight is assigned with more Web requests than Web servers with less weightage. This algorithm performs better than round robbin for heterogeneous server clusters. Sharma et al. [4] proposed a random load balancing strategy where resource allocation among server clusters is on a random basis. In dynamic load balancing, the Web requests are distributed on the basis of the network real-time status. In distributed load management, all servers within a cluster share status information with each other. Liu et al. [5] implemented least loading (LL) load balancing strategy that distributes requests to back-end servers by monitoring the current usage of servers CPU, memory and disk I/O in the server cluster. The server load is calculated using linear function by assigning the administrative setting values as the weight for all attributes as per their influences. Saifullah et al. [6] proposed a server load balancing approach in SDN networks that uses current server health status monitoring. The CPU and memory load factors are taken into consideration as server's health parameters. This scheduling algorithm provides an improvement on the overall throughput of Web servers compared to existing static algorithms. Chen et al. [7] proposed a security model for detecting malicious behavior for social multimedia networks to improve the efficiency of the system. Butt et al. [8] presented a fuzzy logic control system to enhance the processor resource scheduling mechanism in a distributed environment. Mason et al. [9] presented evolutionary neural networks (NN) model where the neural networks are used to predict the usage of CPU of the servers in advance. Datrois et al. [10] presented a machine learning approach for prediction of resource availability at the server node level. The prediction is based on the quantile regression method to estimate unused resources. Anand [11] proposed a secure and sustainable SDN model using replicated clustered of the controller for solving a single point of failure in the control plane.

3 System Design and Implementation

This research work is to manage traffic efficiently among server clusters which use the current utilization of each server such as CPU usage, bandwidth usage and memory usage as an input parameter to find the best candidate server with the least loading. This research purposes a fuzzy-based algorithm to calculate the server load status in real time. With a scenario with non-uniform load, there may be a noticeable difference in response time between large and small workloads of a server. In such cases, migration of load from heavily loaded server to the less loaded server can improve response time [12]. The main factors affecting the cut-off thresholds include:

- Request completion time of server.
- Homogeneous or heterogeneous type of server.
- The existing load of the server.

Based on the predefined cut-off threshold, the overloaded servers do not participate for further load handling and only the servers within the normal limits are used for serving the client requests. With continuous monitoring, the overloaded servers will participate in serving requests as soon as they have a low load.

3.1 System Architecture

The proposed dynamic traffic management algorithm follows the layered architecture of SDN consisting of control plane and data forwarding plane and has been presented in Fig. 1 consisting of different Python modules. Whenever any user sends a request for access to a particular service, the DNS server resolves the load balancer IP address (i.e., virtual IP address).

Upon receiving requests from the client, the load balancer then sends an acknowledgment of the redirection page with the IP address of the least-loaded server from the server cluster to the requesting client.

3.2 Load Balancing Algorithm

For calculating the best candidate server, three parameters such as CPU usage, RAM usage and bandwidth usage are considered. As soon as any request arrives from the client to load balancer, the background process running inside the load balancer will allocate the appropriate least-loaded server and send flow rules to the OpenFlow

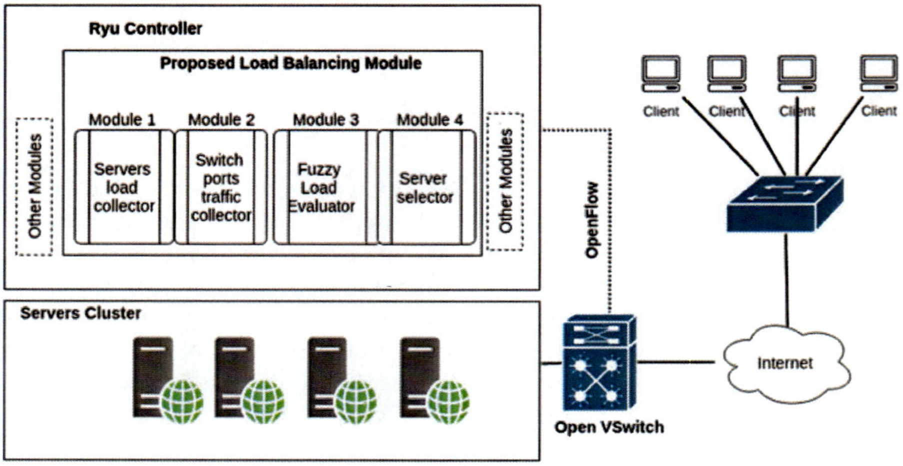

Fig. 1 Proposed architecture of dynamic traffic management

switches. Each request generated simultaneously from the client is put in a queue and executed on first come first served (FCFS) basis. The main load computation algorithm is based on the fuzzy decision-making system. The fuzzy logic system takes crisp input and produces the crisp output with the help of a knowledge base. The empirical threshold cut-off value [12] for the server is considered 70% for this thesis. If any server load exceeds the threshold, then the server does not participate in serving the requests. Only the servers that are below the threshold will participate and the least-loaded server is chosen among all. The load monitoring process is a continuing process and hence the utilization of all servers within the clusters will be within the normal range. The detailed working of the proposed algorithm consists of different phases as described below.

Servers Data Collection Module The daemon application residing inside the servers will periodically update the load status of each server to the central controller from open flow switches. The update period is set to 5s to avoid the controller from overloading. As shown in Fig. 1, the controller has a "servers load collector" module, which collects the real-time CPU and memory usages of individual Web server and stores data in JSON format.

Switch Port Traffic Collector The OpenFlow switches in the SDN network have built-in counters to record statistics of every flow of packets that passes via switch ports. The received counters collected at two-time intervals t_1 time and t_2 are Bt_1 and Bt_2. Bandwidth utilization (BU) is calculated as:

$$\text{BU} = \frac{(Bt_2 - Bt_1)}{P} * [8 \text{ bits}] \tag{1}$$

Load Evaluation Phase The SDN controller sets a thread for the collection of port statistics periodically. The current load status of the servers is calculated using the input parameters (CPU, MEM and bandwidth usage) with help of fuzzy logic theory. The northbound server selection application written in Python computes the load of servers using Mamdani [13] fuzzy inference system model (FIS). The fuzzy inference system is divided into three parts for analysis: fuzzifier, fuzzy inference engine and defuzzifier (Fig. 2).

Fuzzifier Let X be a universe of discourse. A fuzzy set A in X is denoted with the help of membership function$\mu_A(x)$ that associates with each point x, a real number in the interval [0,1], representing the degree of membership of X in A.

$$\text{A} =(x, \mu_A(x)); x \in X)$$

where,

$$\mu_A(x) : X \implies [0, 1]$$

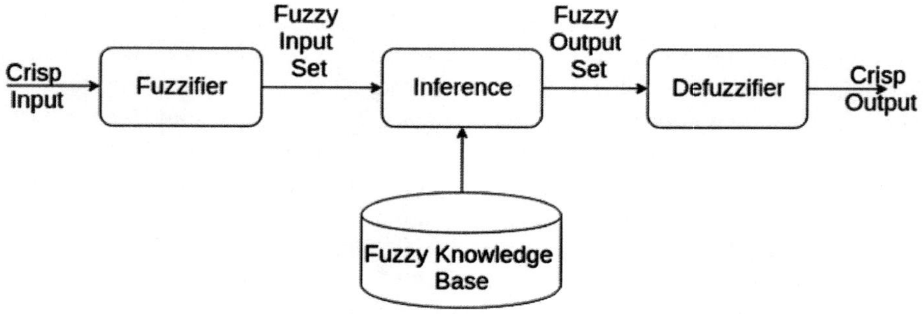

Fig. 2 Proposed fuzzy inference system

The value of $\mu_A(x)$ closer to unity represents higher grade of membership of x in A. If $\mu_A(x) = 1$ then x fully belongs to A, and if $\mu_A(x) = 0$, then x fully not belong to A.

In a fuzzy logic-based system, any parameter can take a whole range of values within an interval and also belong to multiple classes in the interval with different probabilities. Typically, sigmoid, Gaussian and Pi functions are used to represent a fuzzy sets. However, these functions increase the time of computation. Therefore, linear fit functions are used in general. Input and output variables are defined with the help of triangular and trapezoidal membership function as shown in Fig. 3.

CPU, Bandwidth and Memory Utilization

The current servers CPU, bandwidth and memory utilization are used as domain. The fuzzy membership for all parameters are defined as "high," "medium" and "low" and are computed as follows:

$$\mu_l(X) = \begin{cases} 1 & \text{if } X \in [0, 25], \\ 1.5 - 2X & \text{if } X \in [25, 75], \\ 0 & \text{else} \end{cases} \tag{2}$$

Fig. 3 Membership function for input variables

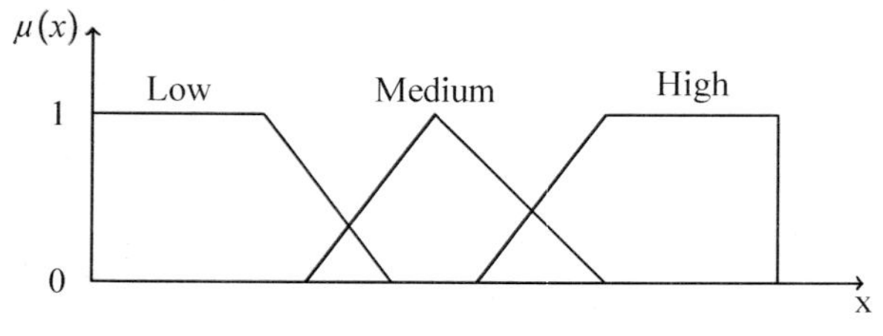

$$\mu_m(X) = \begin{cases} 0 & \text{if } X \in [0, 25], \\ 4X - 1 & \text{if } X \in [25, 50], \\ 3 - 4X & \text{if } X \in [50, 75], \\ 0 & \text{else} \end{cases} \tag{3}$$

$$\mu_h(X) = \begin{cases} 0 & \text{if } X \in [0, 50], \\ 2X - 0.5 & \text{if } X \in [50, 75], \\ 1 & \text{else} \end{cases} \tag{4}$$

Server Load Status

The current servers status is used as a domain for output variable. The fuzzy memberships are defined as parts of fuzzy subset of current server load status indicating membership as "extreme," "strong," "normal" and "light" and are computed as follows (Fig. 4):

$$\mu_l(L) = \begin{cases} 1 & \text{if } L \in [0, 10], \\ 1.5 - 5L & \text{if } L \in [10, 30], \\ 0 & \text{else} \end{cases} \tag{5}$$

$$\mu_n(L) = \begin{cases} 0 & \text{if } L \in [0, 10], \\ 5L - 0.5 & \text{if } L \in [10, 30], \\ 1 & \text{if } L \in [30, 40], \\ 3 - 5L & \text{if } L \in [40, 60], \\ 0 & \text{else} \end{cases} \tag{6}$$

$$\mu_s(L) = \begin{cases} 0 & \text{if } L \in [0, 40], \\ 5L - 2 & \text{if } L \in [40, 60], \\ 1 & \text{if } L \in [60, 70], \\ 4.5 - 5L & \text{if } L \in [70, 90], \\ 0 & \text{else} \end{cases} \tag{7}$$

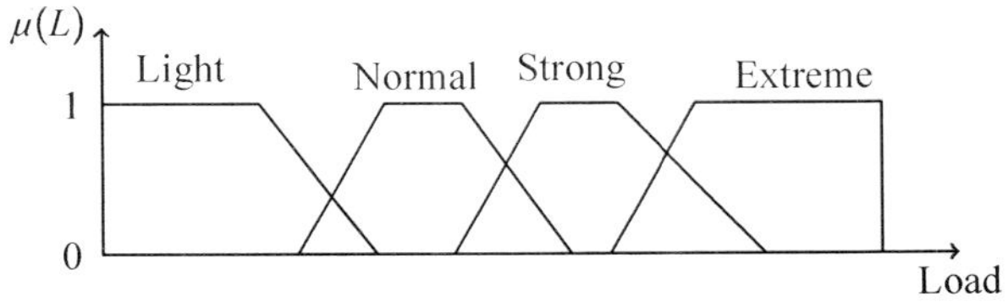

Fig. 4 Membership function for output variable

$$\mu_e(L) = \begin{cases} 0 & \text{if } L \in [0, 70], \\ 5L - 3.5 & \text{if } L \in [70, 90], \\ 1 & \text{else} \end{cases} \quad (8)$$

Fuzzy Inference Engine For the implementation of the above membership functions, we create the fuzzy sets based on heuristics by dividing the domain of discourse into trapezoidal and triangular fuzzy sets with 10–20% overlap typically. In this research, for easy deployment and understating, we have considered combinations of two input variables t_1 and t_2 to be considered as memory, bandwidth or CPU usage in a cyclic order. These two input parameters along with the server load status are mapped and illustrated in fuzzy associative memory with appropriate linguistic notation as shown in Fig. 5. Here, we have taken the Mamdani method which captures expert's knowledge and is widely used in developing decision support systems. For three crisp input CPU, memory and bandwidth, there are 27 possible rules defined.

For the proposed algorithm, let t1 and t2 be the inputs and L be the output control variable; then, the fuzzy rules are generally designed as follows:
If (antecedents), then (conclusion)
The fuzzy inference rules are written in an equivalent conditional proposition as If-THEN clauses.
If (t1,t2) is $X \times Y$, then L is Z,
where

$$[X \times Y](p, q) = \min[X(p), Y(q)] \quad (9)$$

for all $p \in [0, a]$ and $q \in [0, b]$
The fuzzy rule base consists of n fuzzy inference values, then
Rule 1: If (t1,t2) is X_1 x Y_1, then L is Z_1,
Rule 2: If (t1,t2) is X_2 x Y_2, then L is Z_2,
...
Rule n: If (t1,t2) is X_n x Y_n, then L is Z_n,
The symbols X_i,Y_i and Z_i ($i = 1, 2, ...n$) denote fuzzy sets that represent the lin-

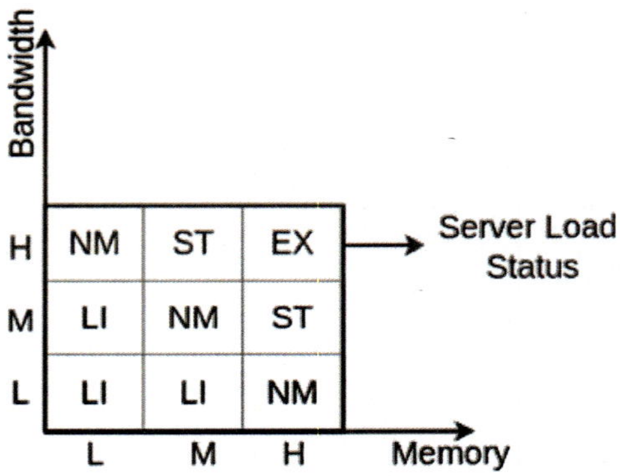

Fig. 5 Fuzzy associative memory rule

guistic states of variables t1, t2 and L. The rules are explained in terms of relation R_i and are disjunctive in nature. The pseudocode is mentioned in Algorithm 1.

Algorithm 1 Fuzzify Inference Engine Algorithm

1: **Input**
 $k = 3$ Number of fuzzy inputs
 $N = 3$ Number of membership function per input
 $R = N^k$ Number of fuzzy rules
 $R = N^k$ Number divisions for centroid calculation
 $\mu[R][1...k]$ Matrix of membership values for each MF
 $\mu_o[R][G]$ Matrix of output membership function values.
2: **Variables**
 $\mu_R[R]$ Array of membership values for each rule
 $\mu_A[G]$ Array of activation values aggregated over all rules
3: **for** x:1 step 1 until R **do**
 $\mu_R[x]$:= min{$\mu[x][1]$, $\mu[x][2]$, .., $\mu[x][k]$}
4: **end for**
5: **for** x:1 step 1 until G **do**
 $\mu_A[x]$:=max{min{$\mu_R[1]$, $\mu_o[1][x]$},..,min{$\mu_R[R]$, $\mu_o[R][x]$}}
6: **end for**

Defuzzifier The defuzzification is a final process of fuzzy system to produce a quantifiable crisp value from the fuzzy input set. As single quantitative value is required, the crisp output of the system is defuzzified using weighted average method defined as follows

$$\mathrm{X}^* = \frac{\sum_{i=1}^{n} \mu(x_i).x_i}{\sum_{i=1}^{n} \mu(x_i)} \tag{10}$$

where $\mu(x_i)$ is the weighted strength and x_i is the center point of the ith output membership function and n is the total number of output membership functions. The above equation results a crisp output value regarding the load of the server.

Decision Phase This Python module will now fetch the information of server load in JSON format and check whether the current load for each server is under a threshold (δ) or not and discard the server from the member list if a load is greater than 70% and sort the normal servers with "index 0" being a least loaded server. The pseudo-code of the algorithm is mentioned below.

Load balancer knows the least-loaded server prior to the requests from clients, as these monitoring and selection processes are already running inside the data center. As soon as the load balancer VIP gets requests, then it updates flow rules on OpenFlow switches and clients requests are assigned to the server dynamically.

Algorithm 2 Server Selection Module

for Every-time System Startup **do**
Fetch Server Load Data in JSON Format
if load of $S_i > \delta$ **then**
Remove server from active member list
else
Add and sort servers with least load at beginning
end if
Fetch Sid with index [0]
end for

4 Performance Evaluation and Results

4.1 Performance Parameters

Throughput: Throughput refers to how much bandwidth gets used during performance testing. It is the most common indicator for checking the Website's capacity to handle concurrent requests.

Response time: The amount of time between a specific request and a corresponding response is the response time. This parameter should be kept as small as possible. But, response times can vary drastically for different actions under different conditions.

Error Rates: When the load to the system is high, then this gives the measure of percentage of the problem requests relative to all requests. Common error HTTP status codes are 4xx and 5xx. So, it is the measures of performance failure in the application at a particular point of time.

Memory utilization: It is the portion of memory used by any application or services while processing a request. Usually, it is calculated in percentage, which is ratio of the resident set size to the physical memory.

CPU utilization: It is the amount of CPU time used by Web application while processing a request. It is also used to estimate system performance.

4.2 Experimental Environment

Suitable network topology for testing load balancing is simulated in Mininet [14] using Python API. It consists of two OpenFlow switches and three docker hosts [15] connected on edge switches which are configured as Apache Web servers; clients are connected with core switch and a remote controller on port: 6633 as shown in Fig. 6

As Mininet in-built hosts do not provide complete isolation of servers, we deployed Mininet environment with docker containers to ensure that the applications running on the servers are completely isolated from each other [16].

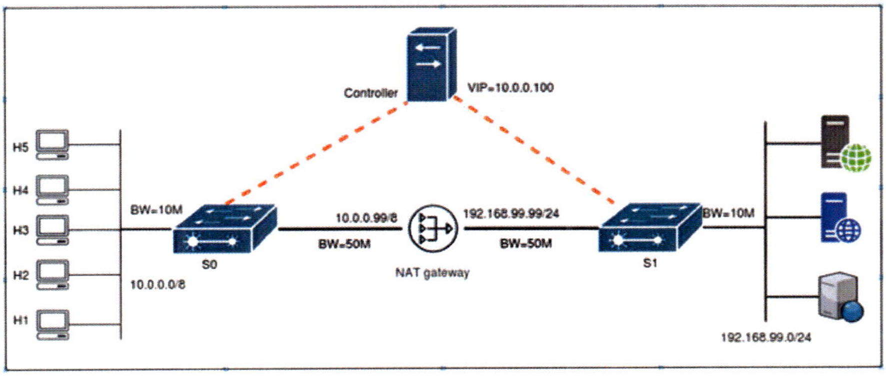

Fig. 6 Topology used for simulation

4.3 *Experimental Design*

The Python script is written on each backend server for sending the load status to the central controller periodically. Siege benchmarking tool [17] is used to measure the performance of RR, LC and the proposed dynamic traffic management algorithms. During simulation, a large number of requests through concurrent connections are sent to the server clusters using virtual IP. The virtual machines have been restarted for every test carried to avoid the influence of caching on performance. For the validation purpose, the simulation was carried out three times with concurrency of users as 2000, 4000 and 8000 for 30 min time interval.

4.4 *Experimental Results*

If the servers deployed within the clusters have the same specifications, the load balancing algorithms round robbin and least connections generally do not require extra calculations for server load and thus are expected to obtain better performance, whereas real-time network scenarios consist of heterogeneous servers; the controller can know the real-time server status within an entire pool of servers and have a high probability to distribute requests efficiently. Therefore, the proposed dynamic traffic management algorithm is expected to have improved performance than existing RR and LC. The test result shows that as the load on the Web server increases, there is a noticeable performance indication change for different load balancing mechanisms, whereas there is no significant difference between these load balancing methods for a lower number of Web requests. The line graphs in Figs. 7 and 8 compare the three load balancing algorithms in terms of their throughput and average response time for the concurrent users starting from 2000 to 20000. The average throughput of the proposed system decreases as the number of requests increases, but with lower

Fig. 7 Comparison of throughput tests results

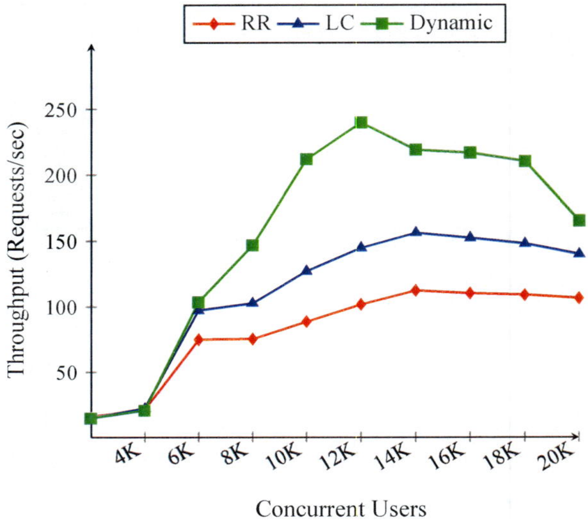

Fig. 8 Comparison of response time tests results

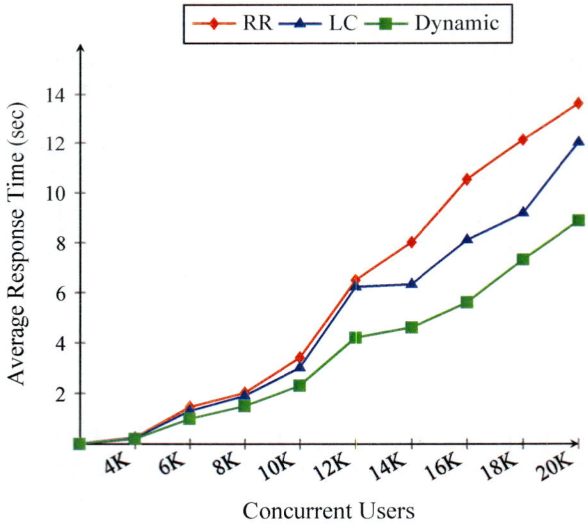

rates than RR and LC algorithms. Since Web servers are installed with minimal configuration, the throughput graph remains almost flat after reaching its maximum value.

Similarly, average response time increases as the number of requests increases but slower than RR and LC algorithms. In other words, the proposed algorithm has better results for response time as this parameter should be kept as small as possible. From Fig. 9, it is observed that CPU utilization is much closer for the proposed

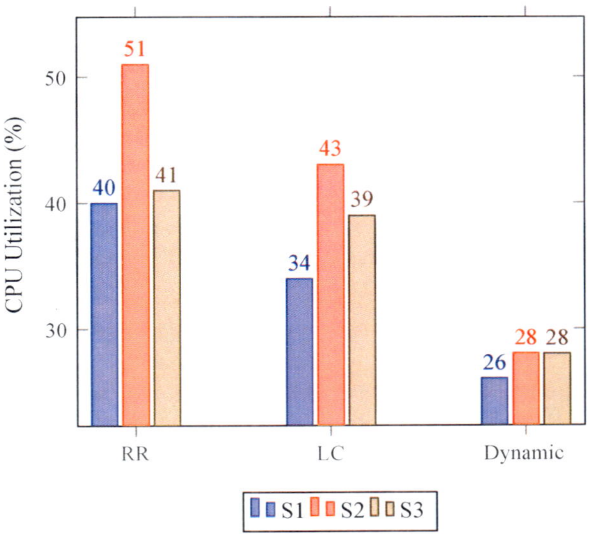

Fig. 9 Comparing overall server's CPU utilization

algorithm, whereas there is a significant gap for RR and LC algorithms. It means that the proposed dynamic traffic management algorithm can allocate user requests better and make rational utilization of CPU for each server.

5 Conclusions and Future Works

An efficient dynamic traffic management algorithm based on the fuzzy inference system has been proposed to distribute Web traffic load to overcome congestion of the Web servers considering CPU, RAM and bandwidth resources. Further, considering a heterogeneous environment of servers, a comparative analysis of the proposed load balancing technique has been performed with the existing algorithms (RR and LC). The results show that the proposed load balancing algorithm provides improved results for a large number of users simultaneously accessing the Web pages than the existing algorithms. In the future, the proposed work can be rerun in a real hardware environment and could simulate with a large number of Web servers. Another possible feature that may be added is to use multiple controllers in master–slave configurations with the proposed mechanism to avoid a single point of the failure problem.

References

1. Z. Bozakov, V. Sander, *Openflow: A Perspective for Building Versatile Networks, in Network-Embedded Management and Applications* (Springer, Berlin, 2013), pp. 217–245

2. G. Li, T. Gao, Z. Zhang, Y. Chen, Fuzzy logic load-balancing strategy based on software-defined networking. International Wireless Internet Conference. Springer, Berlin (2017), pp. 471–482
3. S. Kaur, K. Kumar, J. Singh, N.S. Ghumman, Round-robin based load balancing in software defined networking. 2nd International Conference on Computing for Sustainable Global Development (INDIACom), IEEE (2015), pp. 2136–2139
4. S. Sharma, S. Singh, M. Sharma, Performance analysis of load balancing algorithms. World Acad. Sci. Eng. Technol. **38**(3), 269–272 (2008)
5. H.-Y. Liu, C.-Y. Chiang, H.-S. Cheng, M.-L. Chiang, Openflow-based server cluster with dynamic load balancing. 19th IEEE/ACIS International Conference on Software Engineering, Artificial Intelligence, Networking and Parallel/Distributed Computing (SNPD), IEEE (2018), pp. 99–104
6. M.A. Saifullah, M.M. Mohamed, Open flow-based server load balancing using improved server health reports. 2nd International Conference on Advances in Electrical, Electronics, Information, Communication and Bio Informatics (AEEICB). IEEE (2016), pp. 649–651
7. J.I.Z. Chen, S. Smys, Social multimedia security and suspicious activity detection in SDN using hybrid deep learning technique. J. Inf. Technol. **2**, 108–115 (2020)
8. M.A. Butt, M. Akram, A novel fuzzy decision-making system for cpu scheduling algorithm. Neural Comput. Appl. **27**(7), 1927–1939 (2016)
9. K. Mason, M. Duggan, E. Barrett, J. Duggan, E. Howley, Predicting host cpu utilization in the cloud using evolutionary neural networks. Fut. Gener. Comp. Syst. **86**, 162–173 (2018)
10. J.-E. Dartois, A. Knefati, J. Boukhobza, O. Barais, Using quantile regression for reclaiming unused cloud resources while achieving SLA. International Conference on Cloud Computing Technology and Science (CloudCom). IEEE (2018), pp. 89–98
11. J.V. Anand, Design and development of secure and sustainable software defined networks. J. Ubiquit. Comput. Commun. Technol. (UCCT) **2**, 110–120 (2020)
12. J. Balasubramanian, D.C. Schmidt, L. Dowdy, O. Othman, Evaluating the performance of middle ware load balancing strategies. International Enterprise Distributed Object Computing Conference. EDOC IEEE (2004), pp. 135–46
13. M.S. Khan, Fuzzy time control modeling of discrete event systems. ICIAR-51. WCECS (2008), pp. 683–688
14. B. Lantz, N. Handigol, B. Heller, V. Jeyakumar, Introduction to mininet, in *Mininet Project* (2014). Available: https://github.com/mininet/mininet/wiki/Introduction-to-Mininet. Accessed on: 12 Dec 2019
15. M. Peuster, H. Karl, S. van Rossem, Medicine: Rapid prototyping of production-ready network services in multi-pop environments. IEEE Conference on Network Function Virtualization and Software Defined Networks (NFV-SDN) (2016). pp. 148–153
16. C. Calcaterra, A. Carmenini, A. Marotta, D. Cassioli, Hadoop performance evaluation in software defined data center networks. International Conference on Computing, Networking and Communications (ICNC). IEEE (2019), pp. 56-61
17. Siege benchmarking tool. Available: https://www.tecmint.com/load-testingweb-servers-with-siege-benchmarking-tool/

A Comparative Study of Classification Algorithms Over Images Using Machine Learning and TensorFlow

Subhadra Kompella, B. Likith Vishal, and G. Sivalaya

Abstract In today's world of technology, the fields of image processing are the trending research field of image classification aimed at image analysis. Image classification focuses on the labeling of images corresponding to the respective class label. Many researchers have applied several classification methods for image classification. The fundamental aim of classification is to achieve maximum precision, concentrating on the study of few classification algorithms and the comparison of classification accuracies. A comparative analysis of machine learning algorithms such as artificial neural networks and convolutionary neural networks applied to image data downloaded from an open-source data repository called Kaggle was carried out in this article. The implementation of the algorithms in this paper has been carried out using *TensorFlow*. It is evident from the experimental results that the classification is achieved better through convolutional neural networks.

Keywords Image classification · Artificial neural networks · Convolutional neural networks · TensorFlow

1 Introduction

Many algorithms have been developed to work on image data for classifying the images. Such classification algorithms include support vector machines, logistic regression, and Naïve Bayes classification to mention a few. In today's advanced research, machine learning algorithms are proven to be effective in image processing. As stated in the article [1], the implementation of neural networks is a milestone in the field of machine learning algorithms. This article [2] presents the classification of images using machine learning algorithms such as neural networks and shows efficiency through experimental results. The proposed work aimed at implementing neural networks through TensorFlow. As TensorFlow is open-source and provides a

S. Kompella · B. Likith Vishal (✉) · G. Sivalaya (✉)
Department of CSE, GITAM Institute of Technology, Visakhapatnam, India

S. Kompella
e-mail: skompell@gitam.edu

S. Shakya et al. (eds.), *Mobile Computing and Sustainable Informatics*, Lecture Notes on Data Engineering and Communications Technologies 68,
https://doi.org/10.1007/978-981-16-1866-6_20

large range of APIs and GPUs and since an efficient GPU gives faster computation as mentioned in article [3], it incorporates the classification for this purpose. This article implements the neural networks in two phases, one phase for training the data and the other is for testing the data. In the first phase of the model, the network is trained with the existing data. And in the second phase, the model is tested by feeding the network with unused or untrained data to evaluate the network's efficiency. Compared with the implementation of the artificial neural network over the same data collection, the performance of the proposed work was measured. Artificial neural networks (ANN) are the famous machine learning algorithms as mentioned in paper [4, 5] that aim at simulating human intelligence for a computing machine. Basically, ANNs are trained with the dataset in the form of an input which thereby the model or the network that has been trained is tested with some untrained data. The accuracy of the network is then measured for finding the accuracy of the model.

Furthermore, authors in articles [6–8], have demonstrated that convolutional neural networks (CNN) are the best-suited technique for images and pixels or computer vision from paper [9], and speech recognition is mentioned in paper [10, 11]. Tweaking between layers discovered that changing certain parameters results in high accuracy. Initially, building a machine learning model that functions precisely given any condition requires successful completion of two phases: *preparing* and *testing*. During preparation, the model is fed with existing information and trained accordingly. Testing, on the other hand, requires the use of new or unused data to verify the accuracy of the qualified model (not a part of the training data). In general, proficient models employ 80% of collected information in the preparation phase and the residual 20% for the testing phase. In all models, the percentage of data that is fed to training and testing network is basically consistent to provide precise results and to provide an enhanced learning rate.

1.1 Convolutional Neural Networks

In this paper, convolutional neural networks are utilized to construct a framework for image processing. The CNN comprises neurons and ordinarily has many convolutional layers and max-pooling layers, few dense layers for data handling, and an output layer presenting the results of classification that closes the network configuration. CNN can have a varying number of hidden layers to handle complex computation depending on the data to be handled. To perform image processing, the absolute first step is to gather information. This article adopts information from FASHION_MNST that is fed to the neural network. The dataset is recovered from Kaggle, a repository of datasets that are promptly accessible as Python datasets. The framework of the model in this paper is implemented in Python language with the assistance of the TensorFlow framework facilitating an advanced level application program interface (API). Gigantic multidimensional information present in the dataset and the enormous size of exhibit-related activities is productively taken care

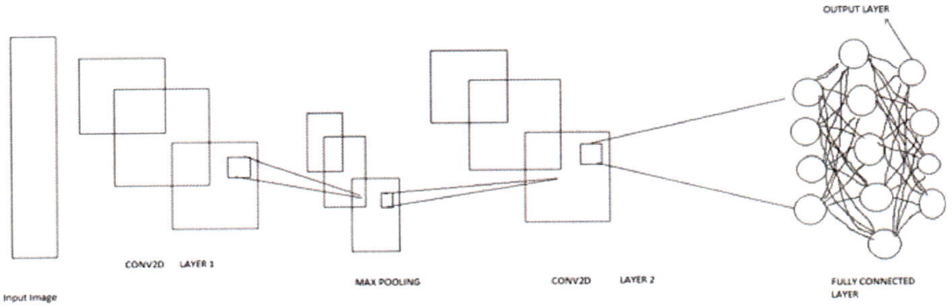

Fig. 1 Convolutional neural network model of the proposed work

of through Numpy. Authors in [7] also have worked on various machine learning algorithms for detecting diseased plant leaf by collecting images of plant leaves.

The architecture of a CNN model implemented in this work has been presented in Fig. 1 and embeds one input layer, one dense layer, and one output layer. Adding a convolutional layer is discretionary. However, it is highly recommended to add a few Conv2D layers.

Initially, the data is flattened in which the shape of the input image is transformed into a vector, and activation function ReLU is used between the layers which will consider the computation value and return the weight of the neuron to the next layer if the value is positive else it will return zero. Finally, a softmax activation function is used in which the probability of each class is obtained. In max-pooling, the maximum value of pixel will be chosen and sent to the next following Conv2D layer. This is repeated till all the pixels are covered within the image. Conv2D basically acts as a filter that a convolutional layer can have. It is an integer value and can vary from each layer. It allows convolutional layers to learn from those filters.

The architecture typically consists of three convolutional layers and a fully connected dense layer to perform image classification. The next section introduces the framework of the neural network used in this article which has been followed by the discussion on implementation and conclusion.

2 Methodology

As suggested in article [12], neural network techniques generally incorporate three stages: preprocessing, dimensionality reduction, and classification. The implementation of this framework starts with processing the data. The data used in this research work is 28×28 pixel images similar to any MNIST dataset referred to in the article [13], indicating that the dataset stores a list of 28 rows, each row further containing 28 values represented as: [[1, 2, 3, 4, 5, 6… x28], [1, 2, 3, 4, 5, 6… x28],… [x28]]. However, supplying images in an array-like format are not the ideal way of data representation in the case of neurons. To overcome this problem, data flattening is

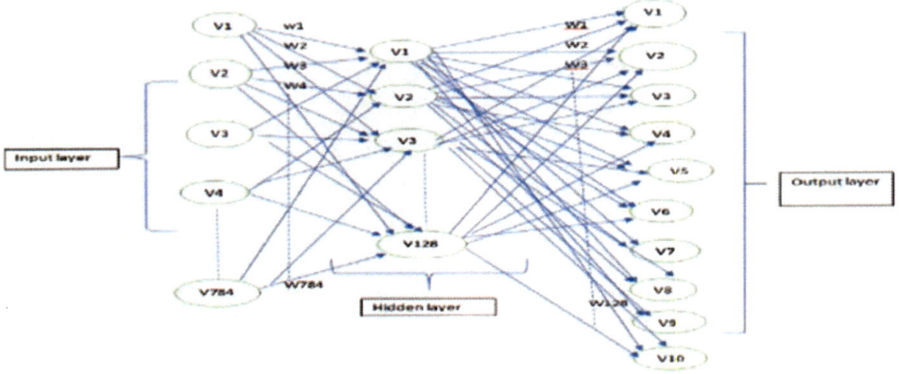

Fig. 2 Neural network model for the proposed framework

a recommended technique, as stated in the article [14]. By flattening the data, the multidimensional columns will be converted into a single vector turning the 28 × 28 pixel images into a vector of 784 rows as shown in Fig. 2 of the flattened vector. Thereafter, each pixel will act as an input neuron in the neural network model. The hidden layers are implemented to recognize the data patterns with the help of the *activation function* that is in turn used to predict the nearest values of image pixels to the expected values of images by adjusting the weights of each neuron. This article [14] explains in detail how flattening and usage of a fully connected neural network gives better performance by adjusting the weights between neurons such that the neural network model prediction will be nearer to the expected value, thus, making it highly accurate and reliable.

The next section presents the experimentation process of the framework and gives details about the results achieved after implementing various classification algorithms over the image dataset downloaded from the famous data repository—*Kaggle*.

3 Experimental Results

This section discusses the implementation of the framework through experimental results. The layers of the network in this work have been organized with the assistance of Sequential () method. This article consecutively organizes the layers to fabricate the model. As the first step of implementation, images will be given as input to the convolutional layers and later be passed to the deeply associated network to shape the output.

In the current work, an inbuilt dataset from Keras is imported which contains all style formats of fashionable costumes like sneakers, dresses, pants, shirts, coats, boots, shoes, satchels, etc. The dataset contains around 60,000 pictures of different style stock. The entire instructive record is isolated into an 80% training set and

20% test set of the dataset. The dataset for training and testing data is sufficient for preprocessing in the context of partitioning. Here, one-hot encoding is performed on the testing data and reshaped the pictures in preparing to procure a vector. At that point, standardized the picture pixels to get the estimation of pixels in the scope of 0–1. At this stage, the data that has been preprocessed is now ready for engineering the convolutional neural network presented in the framework. The input image will be passed to the first convolutional layer and concentrates the pixel esteems and with a portion size of (3,3). The main Conv2D layer will have 32 convolutional layers. Subsequently, the partial intermediate results of Conv2D layers are passed to the max-pooling layer where the most extreme estimation of the pixel will be selected and transferred after the Conv2D layer. This is reshaped till all the pixels are covered inside the picture. In each convolutional layer, an activation function is utilized to be added with is called ReLU.

After passing the image through convolutional layers, the next process is to straighten the data which frames a vector. The input shape of our picture is (28,28). So in the wake of straightening the data, there will be 784 neurons in the vector. Since the data has been leveled, it is then passed to hidden layers. In these hidden layers, the leveled data undergoes the application of an activation function such as ReLU as mentioned above. Now, the processed information will be passed to the output layer which produces the classified images. To predict the accuracy, this framework also uses an activation function called softmax. This framework presents the greatest incentive among the probabilities of each class and returns it as the output class. Also, this model has been implemented with 13 epochs and validated the loss and accuracy using test data which can be clearly observed in Fig. 3.

Fig. 3 Number of epochs trained and test accuracy and loss are shown

3.1 Model Development

The framework has been implemented in Python and the following presents a part of code developed in building the neural network for classifying images. The later part of the sample code gives the details about the packages included in developing the code.

```
IMG_ROWS = 28
IMG_COLS = 28
NUM_CLASSES = 10
TEST_SIZE = 0.2
RANDOM_STATE = 2018
#Model
NO_EPOCHS = 50
BATCH_SIZE = 128

IS_LOCAL = False

model = Sequential()
# Add convolution 2D
model.add(Conv2D(32, kernel_size=(3, 3),
                 activation='relu',
                 kernel_initializer='he_normal',
                 input_shape=(IMG_ROWS, IMG_COLS, 1)))
model.add(MaxPooling2D((2, 2)))
model.add(Conv2D(64,
                 kernel_size=(3, 3),
                 activation='relu'))
model.add(MaxPooling2D(pool_size=(2, 2)))

model.add(Conv2D(128, (3, 3), activation='relu'))
model.add(Flatten())
model.add(Dense(128, activation='relu'))
model.add(Dense(NUM_CLASSES, activation='softmax'))


model.compile(loss=keras.losses.categorical_crossentropy,
              optimizer='adam',
              metrics=['accuracy'])
```

3.2 Packages Required

```
import numpy as np
import pandas as pd
from sklearn.model_selection import train_test_split
from sklearn.metrics import classification_report
from tensorflow.python import keras
from tensorflow.python.keras.models import Sequential
from tensorflow.python.keras.layers import Dense, Flatten, Conv2D, Dropout, MaxPooling2D
from IPython.display import SVG
from keras.utils.vis_utils import model_to_dot
from keras.utils import plot_model
import seaborn as sns
import matplotlib.pyplot as plt
%matplotlib inline
import plotly.graph_objs as go
import plotly.figure_factory as ff
from plotly import tools
from plotly.offline import download_plotlyjs, init_notebook_mode, plot, iplot
```

The framework achieved an accuracy of 91.51% from the above figure, which indicates that when compared to the accuracy, and the accuracy of the model created is increased. The visual representation of the results of the work presented in this model is shown in Fig. 4. In the work of PNV, Syamala Rao et al. [9], the advantages of data analysis and maintenance were well justified.

The framework has been contrasted with the implementation of neural artificial networks with the same efficiency analysis data collection. Table 1 clearly shows the accuracy obtained by CNN is more effective than that of ANN.

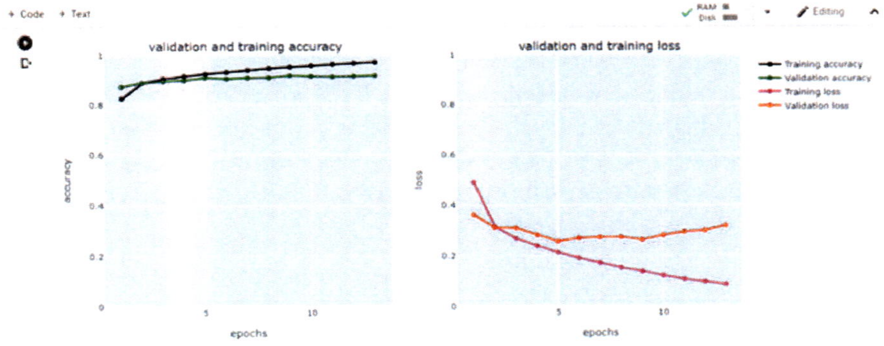

Fig. 4 Comparing the accuracy of training, testing, and the loss

Table 1 Comparison of ANN and CNN accuracies

	Test_accuracy	Test_loss	Train_accuracy	Train_loss
ANN	87.66	34.5	88.94	29.95
CNN	91.51	31.6	96.95	8.25

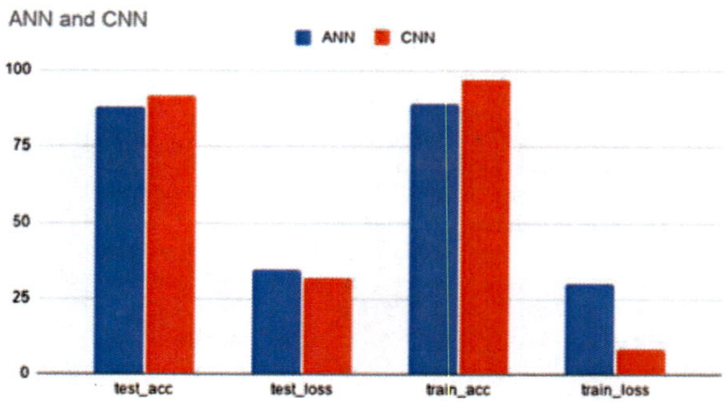

Fig. 5 Comparative results between ANN and CNN

The same has been plotted over a bar graph for visual understanding. Hence, it has been proven from the proposed work that the classification of images can be performed well efficiently using convolutional neural networks (Fig. 5).

The next sections conclude the framework with a slight focus on future work and presenting references.

4 Conclusion

The framework presented discusses the concepts of machine learning algorithms and their implementation. This article also introduces the importance of using an activation function during the processing of neural networks. A comparative study has been done over two most common machine learning algorithms used for classification such as artificial neural networks and convolutional neural networks which are applied over image dataset obtained from a famous repository such as Kaggle. From the experimental results, it is very obvious that CNNs performance has proven to be reliable by achieving an accuracy of 91.51%. By implementing CNN to define the postures of the videos, the structure of this proposed work can be further expanded and can also be checked for classifying video files.

References

1. S. Indolia, A.K. Goswami, S.P. Mishra, P. Asopa, Conceptual understanding of convolutional neural network—a deep learning approach. Procedia Comput. Sci. **132**, 679–688 (2018). ISSN 1877-0509. https://doi.org/10.1016/j.procs.2018.05.069.
2. B. Likith Vishal et al., Image classification using neural networks and tensor-flow. Test Eng. Manage. **83**, 20087–20091 (2020). ISSN: 0193-4120
3. A. Krizhevsky, Imagenet classification with deep convolutional neural networks. Adv. Neural Inform. Process. Syst. **25**(2) (2012). NIPS
4. S. Smys, DDoS attack detection in telecommunication network using machine learning. J. Ubiquitous Comput. Commun. Technol. (UCCT) **1**(1), 33–44 (2019)

5. S. Shakya, Machine learning based nonlinearity determination for optical fiber communication-review. J. Ubiquitous Comput. Commun. Technol. (UCCT) **1**(2), 121–127 (2019)
6. M.A. Abu, A study on image classification based on Deep Learning and Tensorflow. Int. J. Eng. Res. Technol. **12**, 2019. ISSN 0974-3154
7. J. Yim, J. Ju, H. Jung, J. Kim, Image classification using convolutional neural networks with multi-stage feature. In: Kim JH, Yang W, Jo J, Sincak P, Myung H (eds) *Robot intelligence technology and applications 3. Advances in intelligent systems and computing* (Springer, Cham, 2015) vol 345
8. K. Chauhan, Image classification with deep learning and comparison between different convolutional neural network structures using Tensorflow and Keras. Int. J. Adv. Eng. Res. **5**(02), 533–538 (2018)
9. R. Yamashita, M. Nishio, R.K.G. Do et al., Convolutional neural networks: an overview and application in radiology. Insights Imaging **9**, 611–629 (2018). https://doi.org/10.1007/s13244-018-0639-9
10. R. Swastika, et al, On comparison of deep learning architectures for distant speech recognition. IEEE, **17577316**, 1–2 (2017)
11. M. Rastegar et al., XNOR-Net: ImageNet classification using binary convolutional neural networks. European Conference on Computer Vision (2016)
12. H. Ibrahim et al., MRI brain image classification using neural networks. IEEE **13850628**, 26–28 (2013)
13. M. Courbariaux et al., Training deep neural networks with low precision multiplications. Accepted as a workshop contribution at ICLR 201, ICLR, December 2015
14. S. Kompella et al, An efficient detection system of plant leaf disease to provide better remedy. Int. J. Innov. Technol. Explor. Eng. (IJITEE), Google Scholar, APR-2020

Intelligent Routing to Enhance Energy Consumption in Wireless Sensor Network: A Survey

Yasameen Sajid Razooqi and Muntasir Al-Asfoor

Abstract Nowadays, the network and the Internet applications have gained a substantial importance in the digital world, due to the great impact which it provides for health and community services. Among the most important services that have been provided are smart devices and vital factor measurement devices for patients, whether in a hospital or outside the hospital. Furthermore, sensors collect medical data or measurements of temperature and humidity, in various critical environments. The proper types of network that may be used in such difficult environment are wireless sensor networks that used to sense and process data. Additionally, the wireless sensor networks have been used in the environment of Internet of Things and smart cities in the general services and health fields. All these reasons have made researchers focus on wireless sensor networks and addressing the challenges that face them. The most important challenge facing this type of network is energy consumption and increase battery life. This paper discusses the methodologies used in energy conservation in wireless sensor networks, such as data reduction technology, shortest path selection and artificial intelligence algorithms used in smart routing and energy saving. Besides, we have introduced comparisons between the standard algorithms which are suggested by the researchers, to make a clear picture of the energy consumption problem and propose some effective solutions in wireless sensor networks field.

Keywords Wireless sensor network (WSN) · AI · Routing · IoT · Energy consumption

Y. S. Razooqi (✉) · M. Al-Asfoor
Computer Science Department, College of Computer Science and IT, University of Al-Qadisiyah, Al-Qadisiyah, Iraq
e-mail: yasameen.sajid@qu.edu.iq

M. Al-Asfoor
e-mail: muntasir.al-asfoor@qu.edu.iq

S. Shakya et al. (eds.), *Mobile Computing and Sustainable Informatics*, Lecture Notes on Data Engineering and Communications Technologies 68,
https://doi.org/10.1007/978-981-16-1866-6_21

1 Introduction

In recent years have shown a significant interest in WSNs, in academia and industry due to their small size, low price, and its ability of self-configuring and organizing [1]. Those characteristics of sensors in addition to its main function of sensing physical actions made it possible to deploy a massive number of sensors in locations difficult to be managed manually [2]. The wireless sensor network is a type of ad hoc networks, in it, the sensed data transferred to the sink or base station in multi-hop routing method through several sensors [3]. Ad hoc network has no specific infrastructure nor centralization as in traditional network, and also it is subject to topology change in the event of node move, node death or new node joins [4]. WSNs applications ranging on wide and varied fields such as monitoring environment phenomena, target tracking and intrusion observation in the military, underwater searches, forests, health care, industrial machines, monitoring big structures like ships, aircraft, nuclear reactor, etc., and other applications [1]. According to data transfer requirement, applications have been classified into: (i) event-driven in which data transmission start after a certain event, (ii) time-driven sensed data transmitted to the destination in a specific time and (iii) query-driven data transmitted after some request reached to the sensor [5]. The most important investment of WSNs is the integration with IoT, this integration allows WSN to connect to the Internet that facilitates the role of WSN in many applications. The most known applications used for this integration are smart city and e-healthcare, what happens in it is accessing sensors' sensed data by the user via mobile or PCs connected to the Internet [6]. Because of the importance of critical data sent in these applications, harsh monitored area and sensor operation depending on battery, it became important to discover ways for extending WSN life span and developing methods for energy consumption management [7]. In this paper, specifics of WSNs and sensors will discuss in Sect. 2, different methods of energy maintaining in Sect. 3, discussion of the limitations and advantages of the methods in Sect. 4, and conclusion in Sect. 5.

2 Wireless Sensor Networks Structure

Wireless sensor networks "WSNs" consisting of a large number of small size devices with constrained capabilities named sensors deployed in some area for monitoring and measuring different physical actions, connected by wireless connections for working with each other as a network [8]. The physical actions vary according to the required application, actions like pressure, light, temperature, humidity, and others utilized in applications such as health, environment observation, industry, and military [9]. Thousands to billions of sensors disseminate in a large and far geographic environment or complicated systems that cannot be reached by a human, wherefore sensors must be small, inexpensive, and self-management [10]. Besides, sensors depend on the battery as a power supply which cannot be replaced nor recharged for

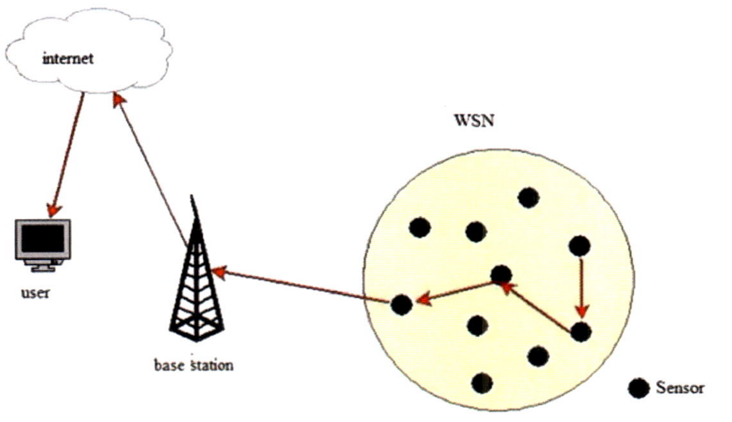

Fig. 1 The basic components of WSN

the same fore mentioned reasons. All these conditions led to limiting in processing, communication, storage, sensing and all other abilities [8]. The main task of the sensor is to sensing or measuring surrounding activities, then transform them into signals to transfer those signals in multi-hop communication to a base station called a sink [10]. One or more sinks connected to the WSN for collecting sensed data from sensors for more processing, analyzing or sending to the application, end-user or cloud through the Internet [11]. The sink or base station has much more capabilities than sensors in terms of processing, communication, storage and power [12], Fig. 1 shows the WSN infrastructure and its connection with an overlay network.

To accomplish that scenario a sensor must at least consist of three main components:

1. Sensing unit: may consist of one or more sensors for acquiring data, sensor type is application dependent [13], and may consist of analogue-to-digital converters (ADC) [10].
2. Processing unit: consist of controller and memory [10].
3. Communication unit or transceiver for sending and receive sensed data between nodes [14].

As mentioned above the battery is the source of power to the sensor which have a finite age before stop working that subsequently affecting sensor lifetime [8]. In addition to the main components, an ideal sensor has other components such as a location finding system (i.e., GPS) and mobilizer for mobility management (MM) in case of the mobile sensor, and both of them are expensive and consume sensor great energy [14]. Solar cells rarely added as a source of energy in sensor [8], Fig. 2 shows the ideal structure of a sensor.

There are two modes for sensor distribution in the monitoring area: (i) pre-planned mode is the easier situation in it the monitoring area can be reached and manage manually which provide good coverage with less number of sensors. (ii) Ad hoc mode is very significant in the harsh or far environment or systems that cannot be

Fig. 2 Sensor's structure

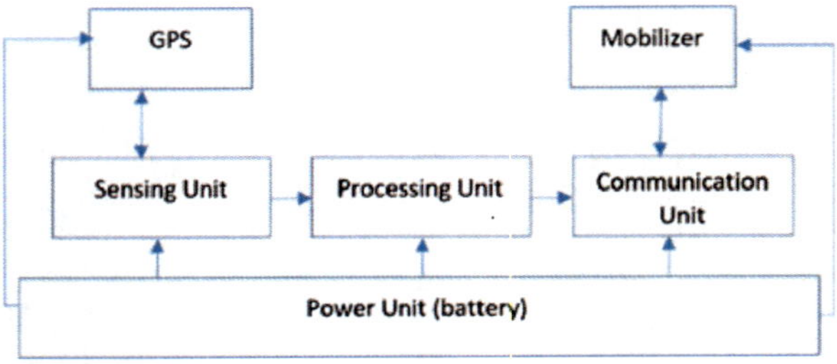

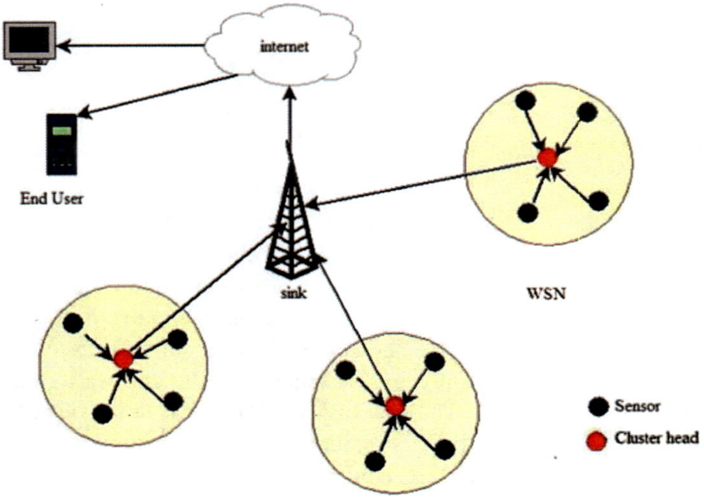

Fig. 3 Clustering approach in WSN

reached Which requires more sensors to provide the required coverage [3]. The latter case is the most common application of WSNs with existing a non-rechargeable or replaceable battery, and thus, the energy loss in some node or node death leads to all the network damage [15]. Node death and sensor mobility in case of the mobile sensor lead to a change in network topology as well as new node joining or node duty cycle [8]. All mentioned above are the reasons for researchers' orientation toward energy efficiency, prolong lifetime and energy balancing of WSNs to overcome power limitation and difficulty accessing the network [16].

3 The Approaches to Energy Consumption Managment

The sensor's energy restriction is the major problem in WSN because it is responsible for sensor working time, which in turn affecting the network performance and lifetime [17]. Communication consumes much more energy than processing or sensing, the energy needed for transferring one bit probably enough for processing thousands of instructions [14]. Therefore, most of the methods proposed by researchers aim to perform additional computations causing to reduce the needed communication to decrease the overall wasted energy [8]. While in another side some researchers proposed solutions to treat coverage holes caused by node death using mobile sensors [18], both sides targeting energy consumption management in WSNs and prolong the network lifetime as possible. The methods and techniques generally classified into sub-approach as follows:

3.1 Intelligent Routing Protocols

Designing and developing the routing protocols in WSN is a decisive and serious process for several reasons, including energy constraints in sensor, sensor mobility and multi-hops routing due to the large distance between source and destination. Besides, the absence of centralized communication and using peer-to-peer communication between WSN nodes increases the importance of routing protocol to prevent congestion when all nodes try to communicate with each other to deliver their messages to the destination [19]. Recently, smart routing protocol development has begun taking advantage of artificial intelligent algorithms such as ant colony optimization (ACO), neural networks (NNs), fuzzy logic (FL), and genetic algorithm (GA) in finding best path. These algorithms improve network performance by providing adaptability to suit the change of the WSN topology, energy problem and environment complexity and changing through intelligent behavior [20]. In general, routing protocols can be classified according to the WSN architecture into:

3.1.1 Hierarchical Protocols

In this type, network divided into sets called clusters, each of which contains many sensors. The sensors in one cluster sense data and send them to a special node in the cluster named Cluster Head (CH). CH in turn collects data from all cluster's ordinary sensors and send them to the sink in one hop or multi-hop communication through other CHs, Fig. 3 illustrates clustering approach in WSN. Different techniques are used for cluster forming, electing cluster head and perhaps changing CHs periodically [21]. Clustering benefits are energy consumption balancing in network and prevent coverage hole occurred due to nodes death near sink. The reason for that is as follows, WSN disseminates in large scale with a very huge number of sensors and various distances from the sink, therefore most sensors cannot connect to sink in one-hop. Instead, far nodes transfer data across other nodes closest to the sink in multi-hop scenario. That leads to exhaust energy in the nodes close to the sink causing node death and then network damage [19].

In [22]; the authors introduced two methods to optimize LEACH routing protocol for enhancing the energy consumption of the nodes in WSN. The first proposed novel method enhanced the chosen method of the cluster heads (CH) which selects the appropriate CH for every cluster at each round. While the second one enhanced TDMA schedule to modify the mechanism of transmission making nodes to send nearly the same bulk of data. While Govindaraj and Deepa [23] used Capsule neural network (CNN) for developing a novel learning model to perform energy optimization of WSN. The model consists of a wireless sensor layer with many sensor nodes, servers as cloud layer and sink nodes as an interconnecting layer. In Capsule Learning detecting the nodes that are appropriate to select them as cluster heads inside the cluster by identity records and also to select the forward nodes that are outside the clusters using shortest path selection thereby reducing the energy consumption. In [19] the researchers introduced a novel routing technique for clustering and routing management. They used gravitational force in addition to fuzzy rules for constructing clusters perfectly and accurately to increase the lifetime of the network. Gravitational clustering method and a gravitational routing algorithm have been proposed for finding an optimal solution for clustering and routing. Furthermore, they used a deductive inference system based on fuzzy logic for choosing suitable nodes as cluster head CH. Whereas, Muhammad Kamran Khan and his research group [24] suggested Energy-Efficient Multistage Routing Protocol EE-MRP based on clustering algorithms. In it, they divided the network area into various stages and disseminated cluster heads equally. The cluster head selection mechanism was enhanced through averting needless re-clustering by assuming a threshold in CHs selection and examination. To diminish the distance between CHs and BS, they inserted the forwarder node idea, that has extra power and its battery can be changed or recharged. All those concepts boost the network throughput and prolong the network lifetime. Greatest connectivity of nodes is the principle of Connectivity Cost Energy (CCE) Virtual Back Bone Protocol [25]. To perform that, firstly Information Table was created by Message Flooding Technique, the sink does that periodically to all nodes. Then Message Success Rate (MSR) was calculated, and if the sensor node needs to send data to the sink selects a parent node that has the highest MSR in the former hop. Secondly, Virtual backbone paths (VBPs) created by the sink through select nodes from all hops that have the highest connectivity than its neighbors, these nodes called concrete nodes. After that cluster head was chosen from the concrete nodes depending on connectivity. CH send join message to all sensors, the sensor node that receives it calculates the communication cost before joining that cluster. A threshold was put to the residual energy of the CH; CH would be changed by another concrete node if its energy was less than the threshold. Energy-efficient scalable clustering protocol (EESCP) takes into account distances between clusters and inside the cluster to produce equiponderant clusters [26]. In this protocol, a novel Particle swarm optimization technology based on Dragonfly's algorithm (DA–PSO) was utilized to choose cluster heads. In it also, a new energy-efficient fitness function was used to choose an optimal collection of CHs. EESCP works in three stages: 1- data collection in which the base station gathers details of residual energy and ID from sensor nodes. 2- cluster building to choose an optimal collection of CHs using DA–PSO by the base station, distances inside cluster reduced to lowering the communication of sending sensed data, however, distances between clusters increased to the good distribution of the CHs across the network to considerate scalability. Then the ID and role (member node or cluster head) sent to the sensor nodes, and those, in turn, choose the closest CH, CHs run TDMA scheduler to prohibit collisions inside clusters. 3- data transmission: sensors send sensed data to CHs according to TDMA scheduler, while CHs in their role transmit data to the base station. The researchers in [27] produce clustering protocol works in rounds and each round with phases aiming to achieve energy efficiency. Cluster head election is the first phase, in this phase two evolutionary algorithms was used, the first is harmony search to select a primary group of interim CHs and the second was a firefly algorithm to select and announce the final group of CHs from the candidates in the first algorithm for the present round. Cluster formation is the second phase in which normal nodes relate to CHs according to the closest distance and higher residual energy. Where the sensed data sent to the sink in the last phase after collecting data by CHs from normal nodes in the cluster using TDMA scheduler. In firefly algorithm, they employed energy prediction, node density, and cluster compactness as parameters of a fitness function to guarantee to choose the final group of CHs those have high residual energy and expend low energy within clustering process. However, other researchers using the concept of clustering without having cluster head such as balanced energy adaptive routing (BEAR) [28].

3.1.2 Flat Protocols

Within these protocols, all sensors in the network send sensed data to the sink without using the cluster head in the middle for aggregating data. Sensors deliver data to the sink in either one-hop in the small networks or multi-hop connection in the wide networks. The challenge in the second case is the choice of best next relay node in the path from source to destination that consumes least possible energy or shortest path taking into account latency and other Quality of Service (QoS) aspects [10]. The researchers in [29] produced a new routing protocol called Intelligent Opportunistic Protocol (IOP) make use of artificial intelligent for choosing the probable relay node by utilizing naïve Baye's classifier to attain both energy efficiency and reliability which is the main objective within WSN. Packet Reception Ratio, Residual Energy of nodes, and Distance were the features used for getting the probability of a node to be a forwarder for next-hop. They show as a result that IOP optimizes the network lifetime, constancy, and throughput of the WSN. While Jianpeng Zhang [30] decrease energy consumption and delay, balancing energy consumption and find global optimum in a neighbor choosing strategy of WSN through proposes a routing algorithm by means of optimizing the intelligent chaotic ant colony algorithm and using strategies such as neighbor chose and global optimum solution with other characteristics and conditions. In the same concern for choosing the next proper relay node, multi-objective integer problem (MOIP) was presented in [31] to solve the tradeoff between energy consumption and reliability optimization in WSNs routing. The concept of Pareto efficient solutions was used in multi-objective optimization, which introduces a series of solutions to meet halfway between the two objectives rather than the conventional thought of optimal solutions. They proposed an evolutionary algorithm based on the Non-dominated Sorting Genetic Algorithm II (NSGA-II) using the shortest paths and breadth-first search (BFS) methods to produce Pareto set in a small computational time. The researchers used real case studies to show NSGA-II's results in medium and wide size scenarios, and they prove the compromise between energy efficiency and delivery ratio. Furthermore, the performance of NSGA-II has not affected by the scalability.

3.2 Duty Cycle

In WSN, it is possible to sensors switch from active to a sleep state and vice versa to save power because in the sleeping state the sensor stops sensing and communication. Therefore, active mode consumes the sensor twice more times energy than sleeping, for that sensor must turn off as it can. Duty cycle is the percentage of the time the node is active for the length of the entire period [32]. Optimization duty cycle and size of forwarding node set (ODC-SFNS) schema was suggested in [33] to minimize delay and energy consumption. The authors supposed that the duty cycle (DC) value and the size of the forwarding node set (SFNS) effect on latency and energy-consuming. With small DC, waiting for the forwarding node to wake up is the source of delay, but fewer collisions thus low energy consumption. In the case of large DC, more than one candidate node is active, which causes collisions leading to delay. Concerning the SFNS, the great SFNS and low duty cycle, increase the number of active forwarding nodes when data is sent that causes less latency. In contrast, great SFNS and high DC mean that many active forwarding nodes at the same time, resulting in a collision, energy consumption and delay. To overcome this tradeoff, tow strategies were adopted (1) they decreased the SFNS in dense and large DC networks, which leads to lowering one-hop delay and the number of hops necessary for the data to reach the sink, thus lowering the end-to-end delay. (2) In parse and high DC networks, they increased the DC in the far region from the sink to reduce delay, prolong network life, and enhance energy efficiency. Finally, they combined strategy 1 and 2 to perform a comprehensive strategy.

3.3 Data Manipulation

The main purpose of data manipulation in WSNs is to reduce the data volume that is sent over the network in order to reduce the power spent on communications because a large amount of energy drains in communications compared to other sensor functions [14]. From the most used methods in this regard:

3.3.1 Data Reduction

Instead of sending all the sensed data from sensors to the sink, data reduction techniques are used to remove redundant or unnecessary data before sending. This process accomplished inside sensor nodes without neglecting data accuracy issue. Data Transmission (DaT) protocol introduced in [34] for energy management in WSN. DaT was executed locally at each sensor node periodically with two phases in every period. In the first phase, the sensed data were classified into classes using the modified K-nearest neighbor method, then the similar classes were combined into one class. While the second phase was a transmission of the data from the first phase to the sink node after they choose the best readings from the combined class rather than sending all the sensed data. The proposed protocol reduces the size of sent data which leads to decrease the energy consumption and thus prolongs the network lifetime. The same group of researchers developed their aforementioned work [35] by producing a two-layer data reduction protocol for achieving energy efficiency. In the first layer, DaT protocol, which was proposed in [34], was used in the sensor node level. Whereas in the second layer (ETDTR) protocol was implemented at the cluster head Which into it further reducing the redundant data sent from various nodes and merging them into a smaller size instead of sending all data forwarded from all nodes.

3.3.2 Data Compression

Data compression algorithms are used in sensors to minimize the amount of data before sending, that will subsequently minimize the communication cost as it is the biggest energy consumer in WSNs. But it is important to choose a proper compression algorithm which its execution does not consume larger energy than energy conserved in communication that leading to the same or more amount of the overall energy [36]. Adaptive Lossless Data compression (ALDC) algorithm was applied in [37] to compress sensed data in sensor before transmission using varied code choices. Two codes were developed based on Huffman coding algorithm called 2-Huffman Table ALEC and 3-Huffman Table ALEC. The proposed method was evaluated on different types of data sets considering real-time demands.

3.3.3 Data Fusion

Sensor readings may be affected by environmental factors like temperature, pressure, radiation or other noise, causing incorrect data. Sensor damage also gives inaccurate readings, furthermore, sensing range overlaps leads to duplicate data. Data Fusion is utilized in WSNs to remove those incorrect or duplicated data which burdens the network with needless transmission and thus an unnecessary waste of energy [38]. An improvement for LEACH clustering algorithm was performed in [39] by employing distance ratio and weighted energy, that prevents choosing the nodes with long-distance and low energy as cluster heads. Also, they used the data fusion rate to get better performance of data transmission, this was done by cluster heads, and CHs fused data before sending to the base station.

3.4 *Based on Mobility*

In general mobility in WSNs can be achieved either by mobilizer unit, but it consumes high energy to perform sensor moving, or by a mobile entity as a vehicle, people or animals. According to application requirements or environment of the monitoring area, sink and sensors can be static or mobile. Mobility is one of the most important methods to prolong the network lifetime by reducing and balancing energy consumption and repair coverage holes [14].

3.4.1 Sink Mobility

Several advantages gained from sink mobility in WSN. The first is to prevent holes' formation in the network due to node death near the static sink because those nodes work in transfer other nodes data, not just their own sensed data which drains their energy faster. The second is to reduce energy consumption because the sink movement reduces the transmission distance from sensors to sink. Moreover, it enhances network performance in terms of latency, throughput and connectivity [16]. The researchers in [40] discussed a solution for collecting big data from large-scale wireless sensor network (LS-WSNs) using Mobile Data Collectors (MDCs) rather than static sinks for achieving energy efficiency because their power is renewable. They asserted that MDCs can collect data in spatially detached geographies, and reduce energy consumption in the node after divided the network into groups and clusters. For mobile data collection in WSNs, two models have proposed in this paper: data mule (MULE) for the multi-hop approach, and sensor network with mobile access point (SENMA) for the single-hop approach. These analytical approaches were employed to calculate the energy consumption in the node and to decide the optimal number of clusters for lowering energy consumption. They used a stop-and-collect protocol in which data collection takes place when the mobile collector is laid immobile at the centroids of the clusters. The proposed model showed that MULE is preferable when the number of clusters is less, and SENMA is more efficient than the MULE when the number of clusters is larger. In [41], the authors performed an enhancement to Zone-based Energy-Aware Data collection (ZEAL) routing protocol which makes efficient energy consumption and data delivery. ZEAL protocol distinguishes network into mobile-sink node, sub-sink node and member node. The first gathers data from sub-sink nodes by changing its location between them. The second transmit packets received from the sensor nodes to the mobile-sink node. The third, sense data and transfers to the sub-sink node. E-ZEAL works in three stages: (1) preprocessing to find the optimal path for the mobile sink using the K-means clustering algorithm that minimizes energy consumption, (2) setup to enhance the choosing of sub-sink nodes that improve the data delivery, and (3) data collection. End-to-end delay, lifetime, remaining energy and throughput were the performance metrics used in comparing E-ZEAL with ZEAL, and E-ZEAL showed better results in all of them. Due to traffic loads in the region near to sink, sensors in this region lose their energy fast, that will eventually cause network splits and sink separation. To treat this issue, a novel routing algorithm was presented in [42], using a mobile sink that changing its location in the network, which leads to an adjustment in network energy consumption and solve sink separation by changing of its neighbors. The suggested contribution was the determining, updating and knowing the latest location of the sink by sensors to send sensed data to it, telling all nodes about that variable location consumes a lot of energy and increases latency. The proposed protocol adopted the idea of Virtual Grid-Based Infrastructure (VGB), in it, choosing some nodes called intersection nodes to save the latest location of the sink while the rest nodes could know the last location of the sink by sending a message to closest nodes in the virtual infrastructure. In addition, periodically replacing the intersection nodes if their energy diminished. Energy-efficient geographic routing algorithm was used, in which each node must know the sink location to send the sensed data to it.

3.4.2 Node Mobility

Data mule and message ferrying schemes are the most used in the mobile nodes approach. in both schemes, there existed static sensors for sensing data and mobile nodes for collecting data from those sensors and send them to the destination. These methods useful in a large-scale network with sparse nodes to minimize transmission range between sensors, hence lowering communication energy [14].

Another using of mobile nodes is filling coverage holes by the movement of the near sensors to extend the network lifetime. Yijie Zhang and Mandan Liu [18] proposed a model of node positioning to resolve the coverage hole occurred by node death during the WSN working. The general idea based on a regional optimization concept, first they assessed if the node position requires to be re-optimized when some node dies, after that extract a suitable optimization region and nodes to be moved. Surrounding, redundant and mixed strategies were used to for optimization region Determining. The dynamic algorithm for regional 1 Algorithm (MRDA). They suggested two objectives for the problem of node positioning optimization problem: coverage and energy consumption, with performance parameters: overall Score, the aggregate moving distance of nodes, and the rate of increasing coverage distance. They concluded that ADENSGA (a regional optimization dynamic algorithm) based MRDA (MR-ADENSGA) highly decrease moving distance for node movement which consumes great energy, improves the performance of searching and convergence velocity in WSNs with different node density. In addition, MR-ADENSGA achieved a result better in large-scale WSNs. While in [43], Neighbor Intervention by Farthest Point (NIFP) algorithm consumed extra energy in moving to cover the hole occurred by node death because of energy-draining or physical jam.

But this novel algorithm increases coverage with trying to minimize node moving range as it could. After calculating the size of the hole, it chooses one of the one-hop nodes in the direct neighborhood of the dead node to handle the hole area in simple computations considering the multitude of neighbors, the overlapping area with neighbors, residual energy and moving distance. NIFP compared with MDL and FSHC to show that tradeoff between energy and coverage, and high performance of NIFP in coverage with accepted energy consumption.

4 Disscusions

All aforementioned approaches of energy conservation in WSNs were used on the ground and proved their ability to lower energy expending and network lifetime expansion. Nevertheless, each one has drawbacks or limitations for using in some application. Nodes clustering method in WSNs has several benefits due to CHs using, that facilitate network operation, especially in wide and scalable ones as its managing cluster local routing rather than traditional routing over the whole network. In addition, data collection in CHs enables them to perform data manipulation inside the CH such as data reduction. Those factors help to save network energy as much as possible. The main limitation of this mechanism is an extra workload that leads to fast death, that because of CH work as data collector of the cluster normal nodes beside his function. Cluster formation and the most proper CH electing among cluster members are considered as challenges in hierarchal protocols, at the same time they must avoid burdening the network with great extra computation. Flat routing protocols utilize multi-hop communication to transfer packets from sensors to the sink over intermediate sensors. This scenario exhausting energy of the sink closest sensors, leaving coverage holes or causes sink separation, especially in large networks. Though these protocols suitable for medium and small networks, and those have high sensors density near the sink. The challenge in this kind of protocols is a next relay node selection that leads to the destination with fewer hops, minimum energy and delay. Also, the frequent chose to some bath more than others must be avoided to prevent unbalance energy consumption. The duty cycle method provides a direct factor in an energy saving of WSNs because of the inactive state for the sensor that reserves communication and sensing energy. The common use of this model is for dense networks in which sensors deployed randomly by helicopters in very large numbers due to anticipating some sensor damage. In this case, it's useful to turn off some nodes that have a cross sensing range alternately. Besides, it's possible to use this type for adapting network topology in some applications and changing paths as needed. The duty cycle amount must be well adjusted to avoid collision and delay and hence extra energy-consuming. If the duty cycle was very small, it will increase the probability of sleepy receivers that make the sender wait to sensors' waking up causing a delay in packets delivery that in turn causing energy loss. While the high duty cycle increases the probability of the active sensors at the same time, which leads to conflict or collision hence increase the delay and energy loss. The data reduction, compression and fusion have the same goal of minimizing the amount of data that sent over the network to save energy caused by needless data transmission. Also, minimizing the data amount can reduce delays and the size of the memory needed to store data in the sink. These techniques are undesirable in some applications that required high data accuracy as they may be causes reducing data precision or losses few data. From another side, processing of these techniques should not take a lot of time or energy that is equivalent to what already saved through communications lowering. Mobile sensor or sink is a very suitable solution for energy loss in sparse or low-density networks. In this network, mobile sinks moving across the network to collect data from static sensors or mobile sensors changing their location for collecting and sensing data. In both cases, the routing distance and the number of hops will be reduced in packets sending leads to energy saving. Changing the sink location will prevent nodes death in the nearest region to the sink because of the workload resulting from multi-hop data transmission. Nodes death causes holes, can be fixed by changing mobile sensor location. All that prolongs the network lifetime due to energy consumption balancing and provide good coverage. The drawback of this way is making the sensor waiting for the mobile collector to reach their communication range before sending data that will cause a delay. Another drawback is when using mobilizer unit to perform the movement, this expensive unit consumes the sensor great energy unless reducing the moving distance as possible or uses another way such as animals or vehicles. Finally, some environmental obstacles may obstruct node or sink movement which makes this approach inappropriate for some applications as adaptive the shape of the topology of overlay network [44]. Table 1 shows a summary of the papers that have been discussed in this survey with their distinguishing characteristics.

5 Conclusions

In this review paper, we have introduced the main approaches of energy consumption management in wireless sensor network, and some methods used to avoid energy-wasting. Special attention was devoted to the latest methods proposed by researchers to solve energy issues in such kind of network. The survey classified those methods into smart routing, duty cycle, data manipulation and mobile sensors or sink to minimize the communications because it's wasting energy more than computation. This literature figures out the limitations and advantages of using each method and compare the latest suggested solutions of the WSN energy conservation. The choosing of the best method for lowering energy consumption is application dependent. The purpose of the network, the location of the monitoring area, the network size, environmental factors, network density, and sensor and sink kind (static or mobile) are the most the factors that must be noticed in choosing the energy preservation approach.

Table 1 Summary of the papers that have been discussed in this survey

Authors	Algorithm	Protocol	Objectives	Monitoring area	Number & kind of nodes	Kind and number of sink	The initial energy of each node	Energy conservation approach
Elshrkawey et al. [22]	TDMA schedules	LEACH	Energy efficiency, network lifetime	100 m * 100 m	100 static	One static sink	5 J	Hierarchical protocols
Govindaraj and Deepa [23]	CNN	–	Energy efficiency, network lifetime	1000 m * 1000 m	200 static	One static sink	2 J	Hierarchical protocols
Selvi et al. [19]	Gravitational clustering method, fuzzy rules	Gravitational routing	Energy efficiency, network lifetime, delay	200 m * 200 m	20–1000 static	One static sink	0.5 J	Hierarchical protocols
Kamran Khan et al. [24]	–	EE-MRP	Throughput, network lifetime, energy efficiency	150 m * 150 m	100 static	One static sink	0.5 J	Hierarchical protocols
Julie et al. [25]	–	CCE Virtual Backbone cluster-based routing protocol	Reliability, delay, energy consumption	500 m * 500 m	50 to 200 static	One static sink	1 J	Hierarchical protocols
Singh et al. [26]	DA–PSO	EESCP	Energy efficiency, scalability	100 * 100 m^2 200 * 200 m^2 300 * 300 m^2	100 static 200 300	One static sink	2 J	Hierarchical protocols
Bongale et al. [27]	Firefly and harmony search algorithms	–	Energy efficiency, coverage	250 * 250 m^2	100 static 200 300 400	One static sink	2 J	Hierarchical protocols

(continued)

Table 1 (continued)

Authors	Algorithm	Protocol	Objectives	Monitoring area	Number & kind of nodes	Kind and number of sink	The initial energy of each node	Energy conservation approach
Javaid et al. [28]	Intra _SEB, inter _SEB	BEAR	Energy efficiency, network lifetime	Radius: 0.2–1 km 1–5 km	80 static 160	One static sink	0.5 J 10 J	Hierarchical protocols
Bangotra et al. [29]	Naïve Bayes conditional probability	IOR	Reliability, energy efficiency	500 m * 500 m	100 static	One static sink	0.5 J	Flat protocol
Zhang [30]	Intelligent chaotic ant colony algorithm	–	Energy efficiency, network lifetime, delay	–	0–60 static	One static sink	–	Flat protocol
Jeske et al. [31]	NSGA-II BFS shortest paths	CTP	Energy efficiency, reliability	10, 000 m^2	100 static	One static sink	100 J	Flat protocol
Wang et al. [33]	ODC-SFNS	–	Delay, energy consumption, network lifetime	Radius: 500 m	Varied density, static	One static sink	–	Duty cycle
Alhussaini et al. [34]	K-nearest neighbor method	DaT	Energy efficiency, network lifetime, data accuracy	–	54 static	One static sink	–	Data reduction
Idrees et al. [35]	K-nearest neighbor method	DaT, ETDTR	Energy efficiency, network lifetime, data accuracy	–	47 static	One static sink	–	Hierarchical protocols and data reduction
Kolo et al. [37]	ALDC, Huffman coding	–	Compression, energy efficiency	–	–	One static sink	–	Data compression

(continued)

Table 1 (continued)

Authors	Algorithm	Protocol	Objectives	Monitoring area	Number & kind of nodes	Kind and number of sink	The initial energy of each node	Energy conservation approach
Wu et al. [39]	Data fusion rate	LEACH	Energy efficiency, network lifetime	100 m * 100 m	100 static	One static sink	0.5 J	Hierarchical protocols and data fusion
Ang et al. [40]	MULE, SENMA	Termite-hill, MCBR_Flooding	Energy consumption, network lifetime	10 * 10 km^2	10,000 static	One mobile sink	–	Sink mobility
Allam et al. [41]	K-means clustering algorithm	ZEAL	Data delivery, energy consumption, delay	400 m * 200 m	120 static	One mobile sink	3000 J	Sink mobility
Zhang and Liu [18]	MR-ADENSGA	LEACH	Coverage, energy efficiency	100 m * 100 m 140 m * 140 m 170 m * 170 m 200 m * 200 m	50 100 150 200, mobile	One static sink	0.5 J	Sensor mobility
Khalifa et al. [43]	NIFP	–	Coverage, energy efficiency	300 m * 300 m	90, mobile	One static sink	1–100 J	Sensor mobility

References

1. H. Kdouh, G. Zaharia, C. Brousseau, H. Farhat, G. Grunfelder, G. El Zein, Application of wireless sensor network for the monitoring systems of vessels, wireless sensor networks, in *Technology and Applications*, ed. by M. Matin (IntechOpen, July 18th 2012)
2. B. Rashid, M.H. Rehmani, Applications of wireless sensor networks for urban areas: a survey. J. Netw. Comput. Appl. **60**, 192–219 (2015)
3. Z. Al Aghbari, A.M. Khedr, W. Osamy, et al., Routing in wireless sensor networks using optimization techniques: a survey. Wirel. Pers. Commun. **111**, 2407–2434 (2020)
4. O. Younis, Distributed clustering in ad-hoc sensor networks: a hybrid, energy-efficient approach. 3rd Annual Conference on IEEE ((2004)), pp. 1–12.
5. R.E. Mohamed, A.I. Saleh, M. Abdelrazzak et al., Survey on wireless sensor network applications and energy efficient routing protocols. Wirel. Pers. Commun. **101**, 1019–1055 (2018)
6. K. Bajaj, B. Sharma, R. Singh, Integration of WSN with IoT applications: a vision, architecture, and future challenges, *EAI/Springer Innovations in Communication and Computing* (2020)
7. N. Zaman, L.T. Jung, M.M. Yasin, Enhancing energy efficiency of wireless sensor network through the design of energy efficient routing protocol. J. Sens. 1–16. Article ID 9278701
8. V. Sharma, A. Pughat, Chapter 1: Introduction to energy-efficient wireless sensor networks, in *Energy-Efficient Wireless Sensor Networks*, 1st edn. (Taylor & Francis, CRC Press, August 1, 2017)
9. P. Rawat, K.D. Singh, H. Chaouchi et al., Wireless sensor networks: a survey on recent developments and potential synergies. J. Supercomput. **68**, 1–48 (2014)
10. P. Kuila, P.K. Jana, *Clustering and Routing Algorithms for Wireless Sensor Networks Energy Efficient Approaches* (CRC Press, Taylor & Francis Group, LLC, 2018)
11. T.K. Jain, D.S. Saini, S.V. Bhooshan, Lifetime optimization of a multiple sink wireless sensor network through energy balancing. J. Sens. **2015**, 1–6 (2015)
12. M. Elhoseny, A.E. Hassanien, Dynamic wireless sensor networks. Stud. Syst. Decis. Control 165 (2019)
13. B. Krishnamachari, *Networking Wireless Sensors* (Cambridge University Press, Cambridge, 2005)
14. G. Anastasi, M. Conti, M. Di Francesco, A. Passarella, Energy conservation in wireless sensor networks: a survey. Ad Hoc Netw. (2009)
15. A. Javadpour, N. Adelpour, G. Wang, T. Peng, Combing fuzzy clustering and PSO algorithms to optimize energy consumption in WSN networks, in *2018 IEEE SmartWorld, Ubiquitous Intelligence & Computing, Advanced & Trusted Computing, Scalable Computing & Communications, Cloud & Big Data Computing, Internet of People and Smart City Innovations*
16. J. Wang, Y. Gao, X. Yin, F. Li, H.-J. Kim, An enhanced PEGASIS algorithm with mobile sink support for wireless sensor networks. Hindawi Wirel. Commun. Mob. Comput. (2018)
17. Xenakis, F. Foukalas, G. Stamoulis, Crosslayer energy-aware topology control through simulated annealing for WSNs. Comput. Electr. Eng. **56**, 576–590 (2016)
18. Y. Zhang, M. Liu, Regional optimization dynamic algorithm for node placement in wireless sensor networks. Sensors (Basel) **20**(15), 4216 (2020)
19. M. Selvi, S.V.N. Santhosh Kumar, S. Ganapathy, et al., An energy efficient clustered gravitational and fuzzy based routing algorithm in WSNs. Wirel. Pers. Commun. **116**, 61–90 (2020)
20. W. Guo, W. Zhang, A survey on intelligent routing protocols in wireless sensor networks. J. Netw. Comput. Appl. (2013)
21. F. Kiani, E. Amiri, M. Zamani, T. Khodadadi, A.A. Manaf, Efficient intelligent energy routing protocol in wireless sensor networks. J. Distrib. Sens. Netw. **11** (2014) Article ID 618072
22. M. Elshrkawey, S.M. Elsherif, M. Elsayed Wahed, An enhancement approach for reducing the energy consumption in wireless sensor networks. J. King Saud Univ. Comput. Inform. Sci. **30**, 259–267 (2017)

23. S. Govindaraj, S.N. Deepa, Network energy optimization of IOTs in wireless sensor networks using capsule neural network learning model. Wirel. Pers. Commun. **115**, 2415–2436 (2020)
24. M.K. Khan, M. Shiraz, K.Z. Ghafoor, S. Khan, A.S. Sadiq, G. Ahmed, EE-MRP: energy-efficient multistage routing protocol for wireless sensor networks. Hindawi Wirel. Commun. Mob. Comput. (2018)
25. E.G. Julie, S. Tamilselvi, Y.H. Robinson, Performance analysis of energy efficient virtual back bone path based cluster routing protocol for WSN. Wirel. Pers. Commun. **91**, 1171–1189 (2016)
26. H. Singh, D. Singh, An energy efficient scalable clustering protocol for dynamic wireless sensor networks. Wirel. Pers. Commun. **109**, 2637–2662 (2019)
27. A.M. Bongale, C.R. Nirmala, A.M. Bongale, Hybrid cluster head election for WSN based on firefly and harmony search algorithms. Wirel. Pers. Commun. **106**, 275–306 (2019)
28. N. Javaid, S. Cheema, M. Akbar, N. Alrajeh, M.S. Alabed, N. Guizani, Balanced energy consumption based adaptive routing for IoT enabling underwater WSNs. IEEE Access **5**, 10040–10051 (2017)
29. D.K. Bangotra, Y. Singh, A. Selwal, N. Kumar, P.K. Singh, W.-C. Hong, An intelligent opportunistic routing algorithm for wireless sensor networks and its application towards e-healthcare. Sensors **20**, 3887 (2020)
30. J. Zhang, Real-time detection of energy consumption of IoT network nodes based on artificial intelligence. Comput. Commun. **153**, 188–195 (2020)
31. M. Jeske, V. Rosset, M.C.V. Nascimento, Determining the trade-offs between data delivery and energy consumption in large-scale WSNs by multi-objective evolutionary optimization. Comput. Netw. **179**, 107347 (9 October 2020)
32. Y. Zhang, W.W. Li, Energy consumption analysis of a duty cycle wireless sensor network model. IEEE Access **7**, 33405–33413 (2019)
33. F. Wang et al., To reduce delay, energy consumption and collision through optimization duty-cycle and size of forwarding node set in WSNs. IEEE Access **7**, 55983–56015 (2019)
34. R. Alhussaini, A.K. Idrees, M.A. Salman, Data transmission protocol for reducing the energy consumption in wireless sensor networks, *New Trends in Information and Communications Technology Applications. NTICT 2018. Communications in Computer and Information Science*, vol. 938, ed. by S. Al-mamory, J. Alwan, A. Hussein (Springer, Cham, 2018)
35. A.K. Idrees, R. Alhussaini, M.A. Salman, Energy-efficient two-layer data transmission reduction protocol in periodic sensor networks of IoTs. Pers. Ubiquit. Comput. (2020)
36. F. Marcelloni, M. Vecchio, A simple algorithm for data compression in wireless sensor networks. IEEE Commun. Lett. **12**(6), 411–413 (2008)
37. J. Gana Kolo, S. Anandan Shanmugam, D.W. Gin Lim, L.-M. Ang, K.P. Seng, An adaptive lossless data compression scheme for wireless sensor networks. J. Sens. **2012**
38. D. Izadi, J.H. Abawajy, S. Ghanavati, T. Herawan, A data fusion method in wireless sensor networks. Sensors (2015)
39. W. Wu, N. Xiong, C. Wu, Improved clustering algorithm based on energy consumption in wireless sensor networks. IET Netw. (2017)
40. K.L. Ang, J.K.P. Seng, A.M. Zungeru, Optimizing energy consumption for big data collection in large-scale wireless sensor networks with mobile collectors. IEEE Syst. J. **12**(1), 616–626 (2018)
41. A.H. Allam, M. Taha, H.H. Zayed, Enhanced zone -based energy aware data collection protocol for WSNS (E-ZEAL). J. King Saud Univ. Comput. Inform. Sci. (2019)
42. R. Yarinezhad, A. Sarabi, Reducing delay and energy consumption in wireless sensor networks by making virtual grid infrastructure and using mobile sink. Int. J. Electron. Commun. (AEÜ) (2018)
43. B. Khalifa, Z. Al Aghbari, A.M. Khedr, J.H. Abawajy, Coverage hole repair in WSNs using cascaded neighbor intervention. IEEE Sens. J. **17**(21), 7209–7216 (2017)
44. M. Al-Asfoor, M.H. Abed, The effect of the topology adaptation on search performance in overlay network (2020s). arXiv preprint arXiv:2012.13146

Deep Residual Learning for Facial Emotion Recognition

Sagar Mishra, Basanta Joshi, Rajendra Paudyal, Duryodhan Chaulagain, and Subarna Shakya

Abstract Facial emotion recognition has been an active research topic with its extensive applications in the field of health care, videoconferencing, security and authentication and many more. Human facial emotion recognition (anger, happiness, sadness, etc.) is still a challenging task, and many approaches have been developed. This thesis work proposed a deep residual learning network to recognize human facial emotions which defines approach to train very deep network. ResNet50 with 50 layers is the base model for this thesis work, and the performance is compared with convolutional neural network (CNN) in terms of accuracy, training time and training loss. Two publicly available datasets CK+ and FER were chosen, and the performance of the network was compared using these datasets. The result of the proposed model shows that the deep residual learning network (ResNet50) performs better than CNN for facial emotion recognition in both FER and CK+ datasets. The overall accuracy of the network was 89.8% in comparison with CNN 62% in CK+ dataset. Similarly for FER dataset, the accuracy was 81.9% compared to CNN 66%.

Keywords Facial emotion recognition · Deep residual learning

S. Mishra (✉) · D. Chaulagain
Nepal College of Information Technology, Balkumari, Nepal
URL: http://ncit.edu.np/

D. Chaulagain
URL: http://ncit.edu.np/

B. Joshi · R. Paudyal · S. Shakya
Institute of Engineering (IOE), Tribhuvan University, Pulchowk, Nepal
e-mail: basanta@ioe.edu.np
URL: http://ioe.edu.np/

S. Shakya
e-mail: drss@ioe.edu.np
URL: http://ioe.edu.np/

S. Shakya et al. (eds.), *Mobile Computing and Sustainable Informatics*, Lecture Notes on Data Engineering and Communications Technologies 68,
https://doi.org/10.1007/978-981-16-1866-6_22

1 Introduction

1.1 Background

Facial emotion recognition is a fundamental and important problem in computer vision, which has been widely studied over the past few decades. Facial emotion recognition is one of the important key steps toward many subsequent face-related applications, such as face authentication system, face clustering and videoconferencing. Human facial emotions are nonverbal way of expressing human psychology in the present state. Emotions like anger, surprise, fear, sad and happy are nonverbal ways in which human interacts with the society. With this usefulness of human emotions, it has been a active research topic to correctly depict the emotions of human being.

2 Related Works

Multiple approach has been proposed in the past for facial expression recognition. He [1] proposed a new approach of residual learning where he proposed a identity shortcut connection which leads to training of very deep network. The paper showed the approach to train 152 layers of residual network which was 8 times deeper than VGG nets and showed significant improvements in accuracy of the model. Karen et al. [2] explore and explain the effect in accuracy of convolution network when depth is increased. They proposed CNN with maximum depth of 19 layers for large-scale image classification which achieved higher accuracy compared to the previous generation of models. With increase in depth, the classification error decreased significantly but the error rate got saturated with the depth of 19 layers. With the success of deep learning and convolution neural network [3–5], several works were proposed for facial expression recognition. Among these works, some of the promising works are of Kuang et al. [3] who proposed a CNN-based approach for facial expression recognition which uses different structured subnets. The whole network was formed by combining these subnets together. FER dataset was used in this proposed model and whole model scored 65.03%. The proposed model did not explore different vision tasks. Mayya et al. [6] proposed using DCNN features for facial expression recognition. They extracted features for ImageNet and obtained 9216-dimensional vector for validation with SVM. Their experiment used CK+ and JAFFEE datasets to obtain accuracy of 96.02 and 98.12%, respectively. Their approach was difficult and time consuming to train. Behzad et al. [4] showed that the deep neural networks (DNNs) performed better in various visual recognition tasks including facial expression recognition (FER). Facial landmarks were extracted and used as input to the network for generating facial expression, as facial region does not contribute better for FER. Four well-known databases were used to evaluate the method: CK+, MMI, FERA and DISFA. Facial landmarks were multiplied with the input tensor in the residual module, and better performance was achieved than traditional methods. Ruiz-Garcia et al. [7] suggested the approach for emotion recognition with combination of deep convolutional neural networks (CNNs) and support vector machine (SVM) as emotion classifier. The proposed model was efficient in recognition of human emotions with good accuracy which learned a downsampled feature vector that retained spatial information using filter in convolution layers. In comparison with other very deep complex approaches, this hybrid model produced accuracy of 96.26% with KDEF dataset. In this paper based on the work of [1], we have proposed an approach for facial emotion recognition using deep residual network using 50 layers. The model was trained using ADAM optimizer [8], and accuracy was compared with two datasets FER and CK+.

3 Proposed Methodology

Many research and studies have shown that network depth plays vital role for providing better result in the field of recognition and classification. With the increase of depth in a network, there arises problem of vanishing/exploding gradients which hampers convergence. Adding more layers to a network leads to higher training error. Plain network with 34 layers has higher training error and less performance than 19-layer plain network. Kaming He proposed residual learning approach [1] which solves the degradation problem and allows training of deep network with better performance. The major idea of residual learning is to let the network to fit residual mapping rather than to learn the underlying mapping. The approach is to add a shortcut or skip connection for information to flow from upper layers directly to the deeper layers. Normal convolution network path is bypassed to add the shortcut to the next layer as shown in Fig. 1. Instead of $H(x)$, initial mapping, the approach lets network to learn $F(x) = H(x) - x$.

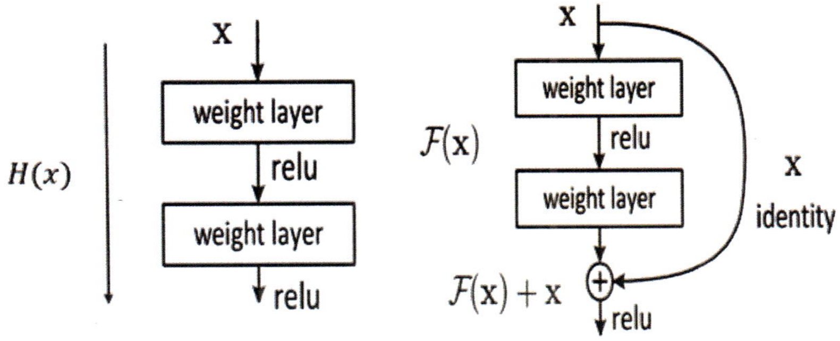

Fig. 1 Residual block

Figure 1 shows a comparison of normal layer and residual block, where a shortcut connection is introduced with identity connection to the deep layer. Formulation for this process can be defined as:

$$F(x) = W2\sigma(W1x) + x \tag{1}$$

where $W1$, $W1$ are the weights for the convolutional layers (plain network) and are the activation functions, in this case a ReLU function. $F + x$ is obtained by a shortcut connection and element-wise addition. The result is added to the activation function σ.

The resulting formulation for a residual block is:

$$y(x) = \sigma(W2\sigma(W1x) + x) \tag{2}$$

3.1 Proposed System Model

System model implemented in this paper is shown in Fig. 2. Set of images in CK+ and Kaggle FER datasets were used as input to the proposed model. These images of the standard datasets were preprocessed to fed in the training model. Preprocessing task involved cropping, normalization and resizing. Datasets were split into two sets, respectively, training and testing sets. Eighty percentage of images were split to training set, and 20% of data were split to testing set. Softmax was used as a

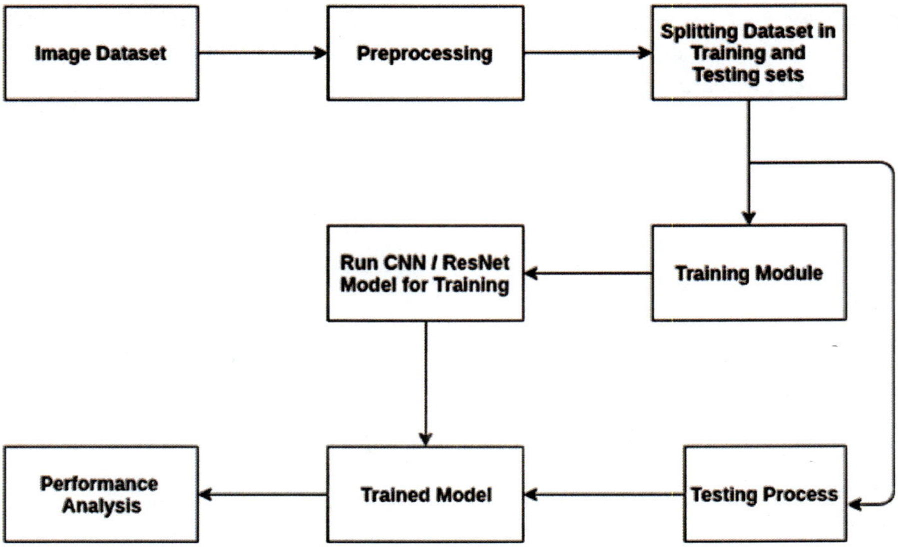

Fig. 2 System model

classifier which is implemented just before the output layer to classify the input to fall in one of the classes. It assigned decimal probabilities to each class that must sum up to 1.0 at the end. Batch size, no. of epoch, learning rate and dropout [9] for the model were varied for tuning the hyperparameters. The final output was image with emotion detected in the image.

ResNet module was designed, and in this process, the module was trained with the training set of data with a different batch size and epoch. This process gave trained ResNet module as output which was used in the further process for testing. In the testing process, test set was passed in the trained model and accuracy of model was evaluated with fivefold cross-validation method.

Figure 3 shows 50-layer residual network. Shortcut connection [1] was inserted, and 50 layers of network were realized in the paper. The 50-layer architecture is given in Table 1. The model was designed with stacking of convolution block and identity block in five stages. 256, 512, 1024 and 2048 filters were used in five stages with different shapes as shown in Table 1. 2D average pooling was used after the five stages with window of shape (2,2). Finally, fully connected (dense) layer was used with softmax activation to reduce the layer input to seven classes of emotions.

3.2 *Datasets*

The datasets that were used in this research are CK+ and FER.
CK+: CK+ dataset was released to help in the work of facial expression recognition. CK+ is a dataset which has labeled images for seven human emotions.

1. Anger
2. Disgust
3. Happy
4. Neutral
5. Surprise
6. Sadness
7. Fear.

It is a publicly available dataset of human facial expression recognition which has dataset of 7000 images of size $640 * 490$ for seven above-mentioned categories. The original dataset was divided into training and testing class (90 training – 10 testing). Figure 4 shows images of CK+ datasets for different facial expressions.

FER Dataset: FER dataset is a Kaggle dataset that consists of 48×48 pixel grayscale images of faces. Dataset is categorized based on the emotion of human faces in seven categories.

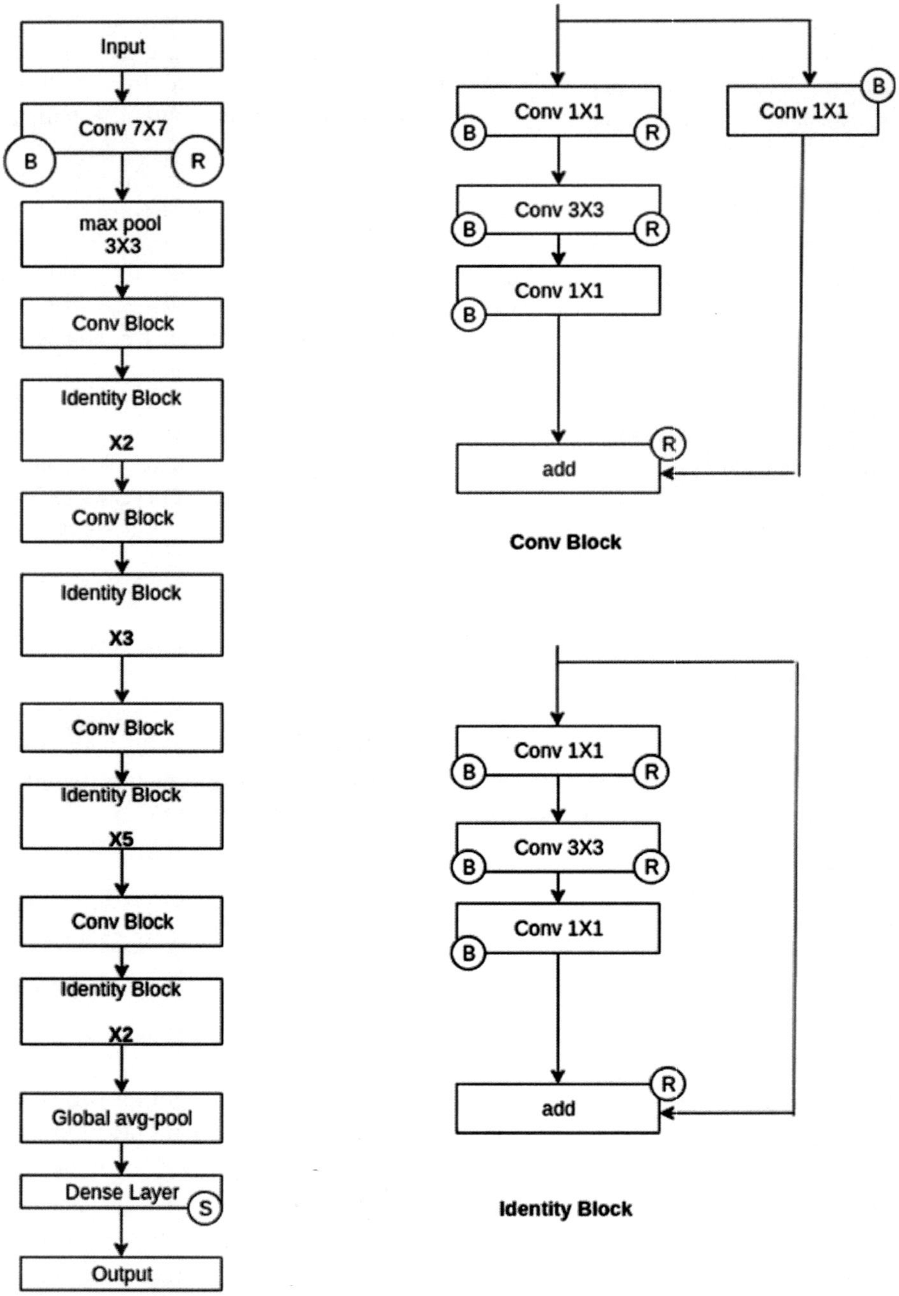

Fig. 3 Network architecture for facial emotion recognition

Table 1 Architecture of ResNet for facial emotion recognition

Layer name	Output size	50-layer ResNet
conv1	24 × 24	7 × 7.64, stride 2
conv2	11 × 11	$\begin{bmatrix} 1 \times 1, 64 \\ 3 \times 3, 64 \\ 1 \times 1, 256 \end{bmatrix} \times 3$
conv3	6 × 6	$\begin{bmatrix} 1 \times 1, 128 \\ 3 \times 3, 128 \\ 1 \times 1, 512 \end{bmatrix} \times 4$
conv4	3 × 3	$\begin{bmatrix} 1 \times 1, 256 \\ 3 \times 3, 256 \\ 1 \times 1, 1024 \end{bmatrix} \times 6$
conv5	2 × 2	$\begin{bmatrix} 1 \times 1, 512 \\ 3 \times 3, 512 \\ 1 \times 1, 2048 \end{bmatrix} \times 3$
	1 × 1	Average pool, FC-softmax

Fig. 4 CK+ dataset

1. Anger—4953 images
2. Disgust—547 images
3. Happy—8989 images
4. Neutral—6077 images
5. Surprise—4002 images
6. Sadness—6298 images
7. Fear—5121 images.

Dataset was split into training set consisting of 28,709 images and testing set consisting of 3589 images. Figure 5 shows images of CK+ datasets for different facial expressions.

Fig. 5 FER dataset

4 Result and Discussion

The training and validation was conducted in Google Colab, and the results for CK+ and FER dataset were obtained which are discussed in this section. Hyperparameters (learning rate, epoch, batch size and dropout) were varied in the work. Learning rate was varied from 0.1 to 0.0001 and best result was obtained in 0.001, whereas the network was trained for 100 epochs and batch size of 64 was used for this work. Figure 6 shows the accuracy comparison of two approaches CNN and ResNet with learning rate 0.001. For CNN model, accuracy of CK+ dataset was 62%, whereas for FER dataset 66% accuracy was obtained. Similarly for proposed model, the accuracy was 89.8 and 81.9%, respectively.

5 Confusion Matrix of ResNet Model

Confusion matrix was used to analyze the performance of the ResNet model that described the performance of the model on set of test data for which the true value

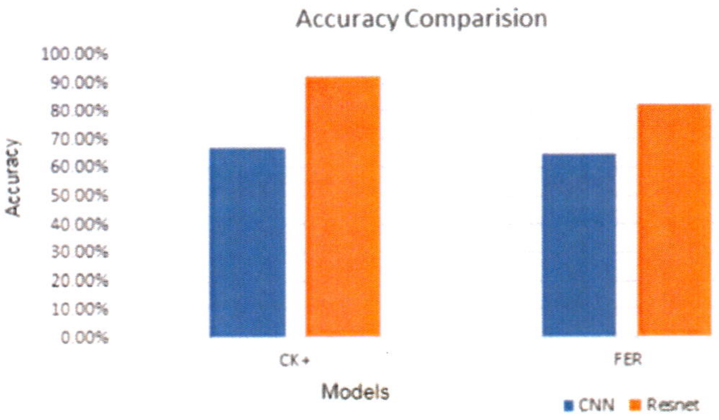

Fig. 6 Accuracy comparison

was known.

$$\text{Accuracy} = (\text{TP} + \text{TN})/(\text{TP} + \text{TN} + \text{FP} + \text{FN}) \quad (3)$$

Figures 7 and 8 show the result of the confusion matrix, which were evaluated with the detection of true positive (TP), true negative (TN), false positive (FP) and false negative (FN) values.

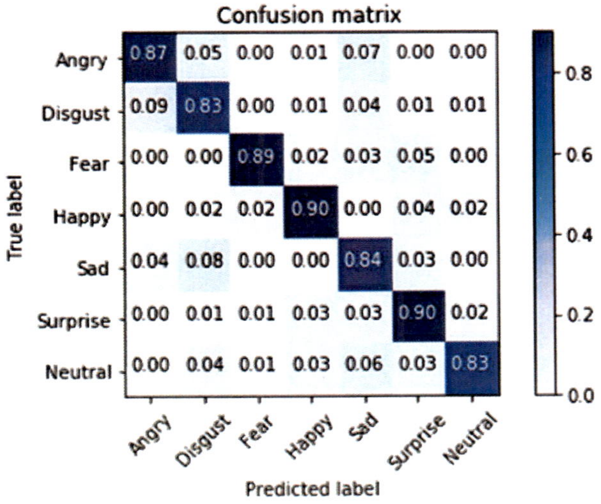

Fig. 7 Confusion matrix for CK+ dataset

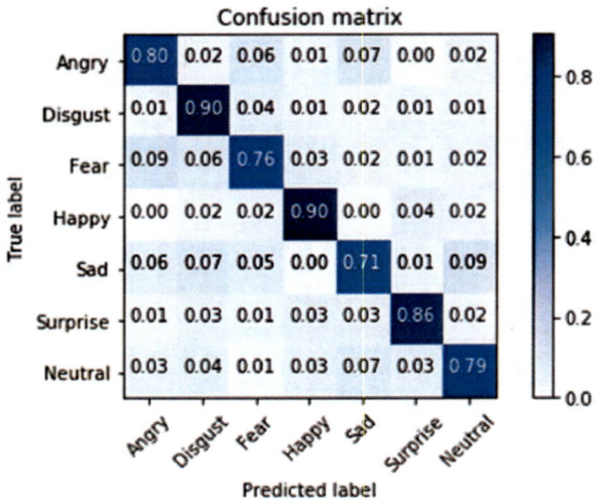

Fig. 8 Confusion matrix for FER dataset

6 ROC Curves

The ROC curve for the thesis work was done, and the result is highlighted in Fig. 9. The curve plots two parameters true-positive rate and false-positive rate where the area under ROC curve (AUC) was calculated for each emotion. These areas were used to identify and calculate the accuracy of proposed model. AUC ranged from 0 to 1, and different classes showed different values giving the accuracy at which the labels were correctly identified by the proposed model.

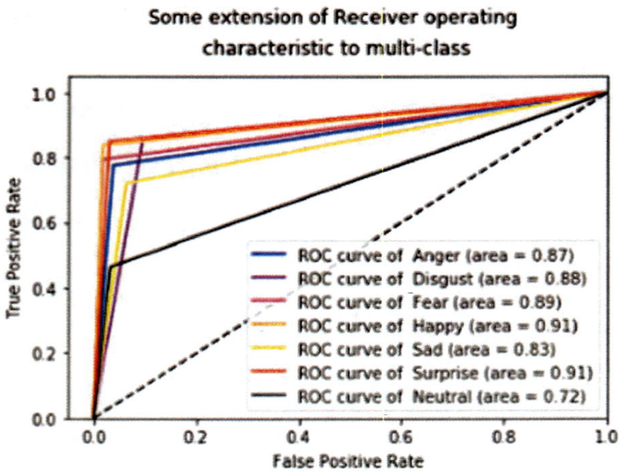

Fig. 9 ROC curve

7 Fivefold Cross-validation of ResNet50 Model for FER Dataset

Fivefold cross validation was performed on the ResNet50 model using FER dataset, and the result is shown below. SGD and ADAM optimizer algorithms were used to compare the optimizer performance as well. The training was done for 100 epochs using learning rate 0.001. Here, the dataset (FER) was split into fivefold 7177 images per fold. In the first iteration, first fold (fold 0) consisted of 7177 images as testing set and remaining 28,710 images were used for training purpose. Similarly in each iteration, each fold was used as testing set and remaining were used as training sets. This process is repeated until each fold of the fivefold has been used as testing set. Table 2 shows the accuracy in every fold for the model. As the fold increases, the accuracy was obtained to be increasing.

Average accuracy of the ResNet50 model was found to be **81.94** from fivefold cross-validation process.

8 Comparison with Standard Approaches

ResNet50 results were compared with best result from CNN models [2, 10]. Table 3 shows that the proposed ResNet50 model outperforms the results of CNN model-based approaches for both CK+ and FER datasets.

9 Comparison of Depth with Standard Approaches

ResNet50 results are compared with other approaches in depth. ResNet50 consists of 50 layers and 23 million parameters [11] (Table 4).

Table 2 Fivefold cross-validation result for FER dataset using ADAM optimizer

Learning rate	Fold	Accuracy (%)
0.001	Fold 0	80.51
0.001	Fold 1	81.7
0.001	Fold 2	81.91
0.001	Fold 3	82.71
0.001	Fold 4	82.88

Table 3 Comparison of ResNet with other standard approaches

Researcher	Dataset	Method	Accuracy (%)	Accuracy of proposed model (%)
Liu et al. (2017) [2]	FER	CNN	65.01	81.9
Pramerdorfer et al. (2016) [10]	FER	CNN	75.02	81.9
Liu et al. (2017) [2]	CK+	CNN	65.01	89.8
Pramerdorfer et al. (2016) [10]	CK+	CNN	75.02	89.8

Table 4 Comparison of depth and accuracy with standard approaches

Approach	Depth	Parameters (m)	Accuracy (%)	Our result (%)
VGG [12]	10	1.8	72.7	81.9
Inception [12]	16	1.2	71.06	81.9

10 Conclusion

The proposed ResNet model was developed using deep residual learning algorithm using Python programming language and TensorFlow for machine learning. The model was trained using CK+ and Kaggle FER dataset for different epochs using different learning rates. The performance of the model was evaluated using the two datasets, and accuracy was validated using confusion matrix and fivefold validation techniques.

The training accuracy for ResNet50 model was 88.3% for CK+ and 81% for Kaggle FER dataset. Similarly, these accuracies were 67% and 64%, respectively, for CK+ and FER dataset using CNN. With this comparison, it was seen that the accuracy and performance of ResNet50 model are far better than CNN when the depth of the network was increased for learning. With the increase of the depth and using identity shortcut connections, better feature learning can be achieved ultimately delivering a high performing network.

Hence, the research concludes that ResNet network has better facial expression recognition capacity than CNN. More than 50 layers can be used for better performance with high GPU training environment. Other datasets with large image size can be used with the very deep residual network. 101- and 152-layer residual model can be realized and tested with high performing GPU environment. Real-time datasets can be used for real-time facial emotion recognition [13, 14] . And facial emotion can be used to recognize emotions in videos, webcams and CCTV footage [11, 15].

References

1. K. He, X. Zhang, S. Ren, J. Sun, *Deep Residual Learning for Image Recognition*. ICCV (2015)
2. K. Simonyan, A. Zisserman. *Very Deep Convolutional Networks for Large-scale Image Recognition*. ICLR (2015)
3. K. Liu, M. Zhang, Z. Pan, *Facial Expression Recognition with CNN Ensemble*. IEEE (2016)
4. J.B. Hasani, M.H. Mahoor, Facial expression recognition using enhanced deep 3D convolutional neural networks. Conference on Computer Vision and Pattern Recognition Workshops (CVPRW). IEEE (2018)
5. A. Ucar, Deep convolutional neural networks for facial expression recognition. Innovations in Intelligent Systems and Applications (INISTA), IEEE International Conference (2017)
6. V. Mayya, R.M. Pai, M.M.M. Pai, Automatic facial expressionrecognition using dcnn. Proc. Comput. Sci. **93**, 453–461 (2016)
7. A. Ruiz-Garcia, M. Elshaw, A. Altahhan, V. Palade, *A Hybrid Deep Learning Neural Approach for Emotion Recognition from Facial Expressions for Socially Assistive Robots, Neural Computing and Application* (Springer, Berlin, 2018)
8. D.P. Kingma, J.L. Ba, *ADAM: A Method for Stochastic Optimization*. ICLR (2015)
9. N. Srivastava, G.E. Hinton, A. Krizhevsky, I. Sutskever, R. Salakhutdinov, Dropout: a simple way to prevent neural networks from overfitting. J. Mach. Learn. Res. (2014)
10. C. Pramerdorfer, M. Kampel, *Facial Expression Recognition using Convolutional Neural Networks: State of the Art*. ICLR (2016)
11. D. Meng, X. Peng, K. Wang, Y. Qiao, Frame attention networks for facial expression recognition in videos. IEEE 2019 International Conference on Image Processing (2019)
12. A. Vyas, H.K. Prajapati, V.K. Dalbhi, *Survey on Face Expression Recognition using CNN*. IEEE (2019)
13. S. Manoharan, Geospatial and social media analytics for emotion analysis of theme park visitors using text mining and GIS. J. Inf. Technol. **2**(02), 100–107 (2020)
14. J.I. Jeena, Capsule network based biometric recognition system. J. Artif. Intell. **1**(2), 83–94 (2019)
15. H. Li, H. Xu, *Deep Reinforcement Learning for Robust Emotional Classification in Facial Expression Recognition, in Knowledge Based System* (Elsevier, Amsterdam, 2020)

A Review on Various Routing Protocol Designing Features for Flying Ad Hoc Networks

J. Vijitha Ananthi and P. Subha Hency Jose

Abstract Flying Ad Hoc Network is derived from the mobile ad hoc network and it consists of unmanned aerial vehicles (UAV) for high-speed communication. Flying Ad Hoc Network has high mobility, and the users can communicate without the help of human intervention. Due to high mobility in FANET, network performance is to be concentrated more. Designing a routing protocol is the most important metric for continuous monitoring, analyzing the network performance, and improve efficient network communication. Design of routing protocol should fulfill the important criteria like neighbor nodes selection, shortest path, traffic control, high scalability, high reliability, high data delivery, low drop rate, lesser delay, and high throughput. Due to the high speed in FANET, routing protocols are to be focused on the improvement of network performance and quality of services. This research work studies the detailed review of suitable routing protocols for Flying Ad hoc Networks and discussed the possible outcomes of the different routing strategies such as source-initiated data-centric, table-driven, hybrid, multipath, location-aware, multicast, geographical multicast, power-aware, and energy-aware. This research study suggested the important metrics be concentrated on designing the efficient routing protocol for Flying Ad Hoc Networks and on improving the quality of services.

Keywords Routing protocols · Flying Ad Hoc Network · Quality of services · Network performance · Unmanned aerial vehicles

---

J. Vijitha Ananthi (✉) · P. Subha Hency Jose
Department of Biomedical Engineering, Karunya Institute of Technology and Sciences, Coimbatore, India
e-mail: vijithaananthi@karunya.edu.in; vijithaananthij@gmail.com

P. Subha Hency Jose
e-mail: hency20002000@karunya.edu

S. Shakya et al. (eds.), *Mobile Computing and Sustainable Informatics*, Lecture Notes on Data Engineering and Communications Technologies 68,
https://doi.org/10.1007/978-981-16-1866-6_23

1 Introduction

Flying Ad Hoc Network is the network derived from mobile ad hoc networks, and it consists of distributed infrastructure. Unmanned aerial vehicles are transported by the Flying Ad Hoc Network (FANET) for communication purposes to increase mobility and safety. Unmanned Aerial Vehicles (UAVs) are not human-dependent and can carry out data transmission without human assistance. The challenges involved in the Flying Ad Hoc Network are involved as

1. Node mobility
2. Surroundings constraints
3. Synchronization
4. Bandwidth utilization
5. Low latency

In the Flying Ad Hoc Network, decentralized architecture [1] is suitable for unmanned aerial vehicle communication, and centralized architecture is not suitable but produces more power constraints. Each transmission between unmanned aerial vehicles in a centralized architecture is dependent on the centralized unmanned aerial vehicle, and this leads to high power and energy constraints. The distributed architecture for efficient data communication between unmanned aerial vehicles is preferred by most of the Flying Ad Hoc Networks to avoid this problem. The Flying Ad Hoc Networks prefer a mesh network for multihop transmission [2]. The use of multihop transmission expands the network's size, increasing the likelihood of collisions and high traffic conditions. To do such incredible things that living things can do, Flying Ad Hoc Networks [3] have been added. Unmanned aerial vehicles will do data transmission without human support. This network will help the fast communication between the unmanned aerial vehicles and acts like human mobility. Due to high mobility changes [4] in Flying Ad Hoc Networks, there is a possibility of abrupt change in the topology and should be maintained by using a routing protocol. The most important area for improving the quality of services, distribution rate, and throughput is the design of the routing protocol for Flying Ad Hoc Networks [5].

Figure 1 shows that the Flying Ad Hoc Network architecture consists of unmanned aerial vehicles and base stations. Unmanned aerial vehicles are connected via a wireless link to the base station. Based on the routing protocol, communication will be initiated.

Kumar and Sharma [6] designed an ad hoc on-demand multipath routing protocol for efficient data communication between nodes. Multipath ad hoc on-demand routing protocol finds a different path and that leads to improving the network efficiency. Zemrane et al. [7] used the optimized link-state routing (OLSR) algorithm and temporally ordered routing algorithm (TORA). OLSR helps to make the efficient path between source and destination, and TORA maintains the effective routing table and lists out the possible routes between source and destination. Amponis et al. [8] explained the process of inter-UAV routing schemes, and it helps to reduce the channel bit error rate and signal to noise ratio. Sharma et al. [9] evaluate the five different routing protocols such as

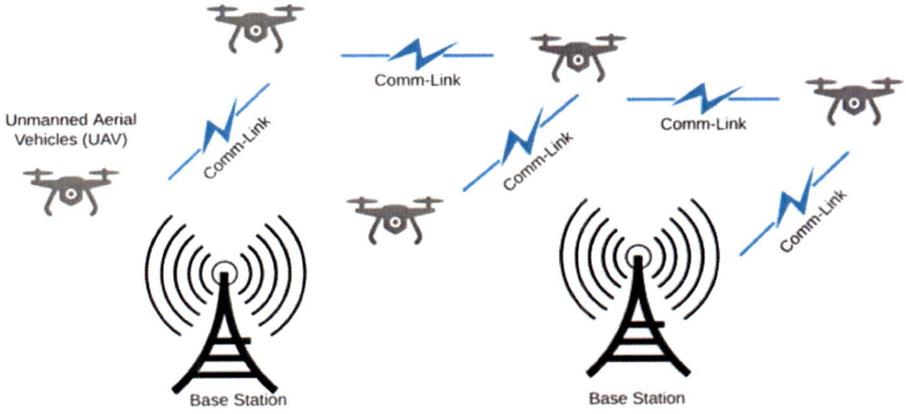

Fig. 1 Flying Ad Hoc Network architecture

1. AOMDV—Ad Hoc On-Demand Multipath Distance Vector
   It helps to avoid frequent route discovery and provides multipath for effective communication.
2. DSDV—Destination Sequence Distance Vector
   It helps to create and update a routing table frequently for better communication and provides lesser delay.
3. AODV—Ad Hoc On-Demand Distance vector
   It will create the routing table on-demand basis for efficient communication between source and destination.
4. DSR—Dynamic Source Routing
   It has a similar process to AODV, and it will be identified by using flow-id with a hop-by-hop basis.
5. OLSR—Optimized Link-State Routing
   It helps to find the open shortest path between the users. This will help to reduce the delay and communication overhead.

When compared to these five routing protocols, the AODV performs and adapts well for wireless ad hoc networks. Haglan et al. [10] discussed the merits and demerits of AOMDV and AODV applied in high-speed networks.

2 Routing Protocols in FANET

Al-Emadi and Al-Mohannadi [11] implemented a new routing protocol for Flying Ad Hoc Networks and that routing protocols are an advanced version of proactive, reactive, and hybrid routing protocols used in mobile ad hoc networks. Temporarily order routing algorithm (TORA) is used in the Flying Ad Hoc Network to minimize the network overhead, and this routing protocol is a combination of both reactive and proactive routing protocol. AlKhatieb et al. [12] introduced the ad hoc routing protocol for Flying Ad Hoc Network, and it compares the performance of network throughput, end-to-end delay, and data-dropped ratio with open link-state routing protocol for topology control, ad hoc on-demand distance vector for low-power communication, dynamic state routing for the identification of intermediate node, geographical routing protocol for the identification of position, and temporarily ordered routing algorithm for route creation, maintenance, and erasure. Among these five classical routing models, OLSR performs well in high mobility conditions, and finalized OLSR is suitable for FANET. Lakew et al. [13] discussed the different routing protocols used in the Flying Ad Hoc Network under different mobility conditions like paparazzi mobility model, random waypoint, random direction, Gauss Markova, spring mobility, distributed pheromone repel, smooth turn, and semi-random circular movement. Sang et al. [14] explained the process of different routing mechanisms utilized in FANET and helped to improve the quality factors of the network. This proposed scheme helps to identify the routing directions and avoids the routing issues in Flying Ad Hoc Networks. Liu et al. [15] proposed a new multi-objective routing protocol that depends on the Q-learning parameters. This Q-learning parameter identifies the neighbor relationship between the nodes and routing decisions based on the nearby neighbor nodes available in the networks.

Khan et al. [3] proposed smart Internet of things (IoT)-based nature-inspired routing protocol for Flying Ad Hoc Networks to avoid the diversity issues in the spatial and temporal models. This new efficient routing scheme used the ant colony optimization algorithm for legacy routing selections and optimal routing paths in the Flying Ad Hoc Networks. Kumar et al. [16] efficiently utilized the AODV and position-based protocols for finding the optimal routing path between the unmanned aerial vehicle (UAV) nodes in Flying Ad Hoc Networks. This efficient routing protocol improves the packet delivery ratio and controls the end-to-end delay. Kaur and Verma [17] designed the new routing protocol and that should satisfy the traffic control mechanism, continuous monitoring, and load balancing conditions. This new routing protocol is applied in different mobility scenarios, and it obtains the optimal routing path for Flying Ad Hoc Networks. Raj [18] introduced the hybrid routing protocol that improves the link discovery process by using the optimal link-state and dynamic state routing protocol, and the nature bee colony algorithm improves the efficient routing path in Flying Ad Hoc Networks. Surzhik et al. [19] designed an adaptive routing protocol by using continuous piecewise linear functions (CPLF) with two different conditions which are fixed and non-fixed step conditions. This adaptive routing protocol increases the accuracy level of the trajectory path of unmanned aerial vehicles in Flying Ad Hoc Networks.

Ibrahim and Shanmugaraja [20] suggested the open link-state routing protocol is the better choice for Flying Ad Hoc Networks to avoid an end-to-end delay, control overhead, routing overhead, network throughput, traffic control, and retransmission process. In result sections, the Flying Ad Hoc Network analyzes the four different scenarios with different data rates and obtains efficient results by using an open link-state routing protocol. Hassan et al. [21] introduced the Fisheye State Routing protocol for Flying Ad Hoc Networks, and it mainly focuses on the improvement of channel utilization. This FSR protocol is suitable for different mobility models and provides low latency. Kaur et al. [22] introduced the hybrid mobility models such as random waypoint and Gaussian Markov model to apply in Flying Ad Hoc Networks with different routing protocols such as AODV, DSR, and DSDV. This proposed model focused on the performance of QoS improvement in Flying Ad Hoc Networks. Hua and Xing [23] designed a new routing algorithm to improve the routing quality in Flying Ad Hoc Networks. Usman et al. [24] proposed the lifetime improvement through suitable next-hop nodes (LISN) routing protocol process applied in Flying Ad Hoc Networks for topology maintenance.

Table 1 shows that the different routing protocols and their outcome applied in Flying Ad Hoc Networks. Bhardwaj and Kaur [25] introduced the optimized route discovery and node registration approach in Flying Ad Hoc Networks and used ad hoc on-demand distance vector routing protocol to improve the throughput and reduce the jitter and delay. Kumar et al. [26] suggested the 3D cone-shaped location-aided routing algorithm which helps to identify the location of drones by using a position-based routing protocol. This scheme mainly focuses on reducing the routing overhead

Table 1 Different routing protocols and their outcome

Author[s]	Techniques	Routing protocol	Outcome
Bhardwaj and Kaur [25]	Optimized route discovery and node registration	AODV	High throughput, lesser jitter
Kumar et al. [26]	3D cone-shaped location-aided routing algorithm	Position-based routing protocol	Lesser routing overhead
Kumar and Bansal [27]	Topology-based routing protocol	AODV, DSR, DSDV	Lesser drop rate
Bhardwaj and Kaur [28]	Chaotic artificial algae algorithm	Efficient routing protocol	Improved QoS and QoE metrics
Tropea et al. [4]	UAV/drone behavior model	FANET simulator	High dynamic connectivity
Kakamoukas et al. [29]	iFORP-3DD	One-to-one direct communication	Determines the operation time
Park and Lee [30]	Closed-loop FANET	Destination path flow model	Avoid link breakage
Li et al. [31]	FANET in agriculture	Stop carry and forward technique and multipath technique	Improves QoS
Watanabe et al. [32]	Lightweight security	Reactive routing protocol	Resist the attacks
Sadiq and Salawudeen [33]	Intermittently connected FANETs	Mobility-assisted adaptive routing	Reduce the routing overhead

in Flying Ad Hoc Networks. Kumar and Bansal [27] explained the topology-based routing protocol, and it depends on the AODV, DSR, and DSDV. This technique mainly concentrates on the reduction of drop rate, and it maintains the topology even in high mobility conditions. Bhardwaj and Kaur [28] explained the process of the chaotic artificial algae algorithm, and it implements the efficient routing protocol for improvement of QoS and QoE metrics. Tropea et al. [4] discussed the unmanned aerial vehicles/drone behavior model (UAV) for the FANET simulator to improve the dynamic connectivity between the drones. Kakamoukas et al. [29] implemented the iFORP-3DD model for one-to-one direct communication, and it determines the operation time. Park and Lee [30] proposed the closed-loop FANET to avoid the link breakage between the drones with the help of the destination path flow model. Li et al. [31] introduced the FANET in agriculture applications and used many routing protocols such as the storage carry and forward technique and multipath technique for the improvement of QoS in Flying Ad Hoc Networks. Watanabe et al. [32] discussed the lightweight security model with the reactive routing protocol to avoid the attacks and identify the intruders in Flying Ad Hoc Networks. Sadiq and Salawudeen [33] explained the process of intermittently connected FANETs for high mobile nodes with mobility-assisted adaptive routing, and it helps to reduce the routing overhead.

3 Routing Protocols Classification

The routing protocols are classified based on function and performance. There are four different routing protocols such as proactive, reactive, hybrid, and hierarchical which are shown in Fig. 2.

1. Proactive routing protocol:
   This approach is called the table-driven routing protocol, and it will create a routing table for data transmission. This routing table contains all required information such as node id, distance, vector, and sequence number to select the best path for data communication.
   Example: DSDV.
2. Reactive routing protocol:
   This approach is called an on-demand routing protocol, and it will create a routing table whenever required. It will maintain the route discovery and route maintenance phase. The route discovery process helps to discover the possible routes between the nodes. Route maintenance helps to maintain the route while data transmission. This helps to avoid link failure.
   Example: AODV.
3. Hybrid routing protocol:
   This approach is a combination of a proactive and reactive routing protocol. It will create a routing table based on-demand basis. The best example of a hybrid routing protocol is a zone-based routing protocol. If the data transmission between two different zones, the reactive routing protocol will be implemented.

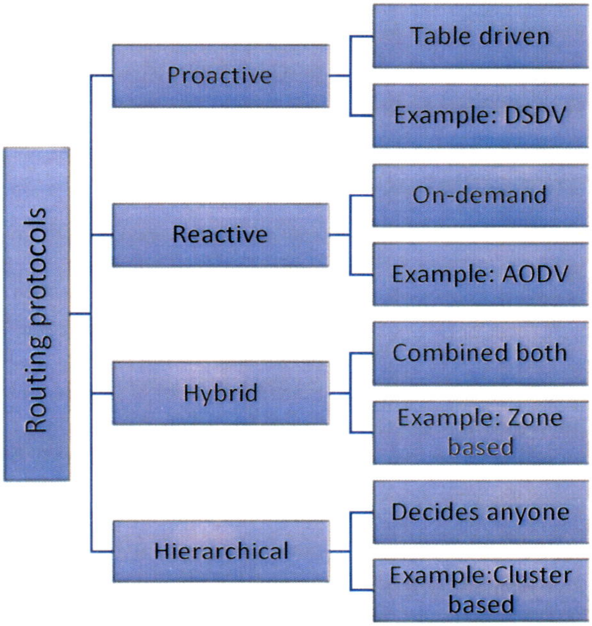

Fig. 2 Routing protocol classification

If the data transmission within the zone, a proactive routing protocol will be implemented.
Example: Zone-based routing protocol.

4. Hierarchical routing protocol
This approach will decide either proactive or reactive routing protocol based on the requirement of the network. The proactive routing protocol helps to create the routing table for the unstructured network based on the requisition of nearby nodes. The reactive protocol helps to provide accurate attribution to maintain the routing paths between the nodes.
Example: Cluster-based routing protocol.

4 Discussions and Possible Outcomes

Boukerche et al. [34] discussed all the possible routing protocols for multipath communication and explained the steps included in the routing discovery phase. The parameters included in the multiple routes are route metrics, link availability, route expiration time, error count, and hop count. The general routing table contains the required information to transfer the data from source to destination. Saputro et al. [35] elaborated on the designing of routing protocols and explained the process of routing information. This scheme introduced the route discovery process, route maintenance phase, and attack detection that is suitable for all type of network and especially for secure networks. This type of routing protocol helps to detect the attackers and suggests an alternate path for secure communication. Jain and Sahu [36] explain the topology and position-based routing protocol, and it mainly focuses on the source-initiated, table-driven, and location-aware routing strategies. The common goal of this scheme is to reduce the control overhead, drop rate, and improve the throughput and delivery rate. Ibrahim et al. [37] proposed the energy-efficient routing protocols for wireless sensor networks with a clustering approach. Energy-efficient routing protocol applied in cluster head and the cluster node details will be stored in the energy-efficient sensor-based routing table. This routing table contains the power and energy level of all cluster nodes and that helps to reduce the energy consumption of the cluster nodes. Senouci et al. [38] proposed the extended hybrid energy-efficient distributed routing protocol for wireless sensor networks. This routing protocol concentrates on the energy level of sensor nodes and helps to overcome the energy consumption level of sensor networks.

In the Flying Ad Hoc Network, all the unmanned aerial vehicles had high network mobility for high-speed communications. The designing of routing protocols is concentrated on the improvement of network performance by analyzing the network parameters such as the end-to-end delay, throughput, delivery ratio, drop rate, control overhead, normalized routing overhead, and congestion control. Figure 3 shows that the delivery rate and drop rate of different routing protocols such as ad hoc on-demand distance vector [25] achieves 90% delivery rate and 10% of drop rate, ad hoc on-demand multihop distance vector [10] achieves 80% of delivery rate and 20% of drop rate, dynamic source routing protocol [22] achieves 72% of delivery rate and 28% of drop rate, destination sequence distance vector routing protocol achieves 75% delivery rate and 25% of drop rate, and optimized link-state routing protocol achieves 85% of delivery rate and 15% of drop rate. This routing protocol examines under the Flying Ad Hoc Network simulation environment. Compared to other routing protocols, the AODV routing protocol is the one best suitable for the improvement of packet delivery ratio and reduces the drop rate.

Fig. 3 Delivery and drop rate comparison

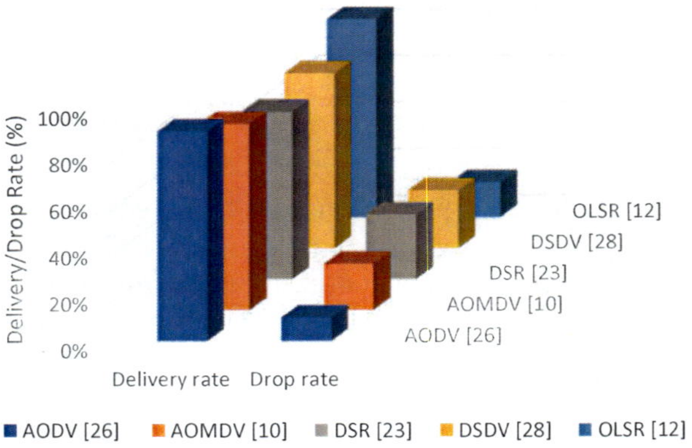

While designing the routing protocol, the following criteria need to be concentrated for the efficient Flying Ad Hoc Network.

1. Topology aware
2. Location-aware
3. Energy level
4. Link failure detection
5. Attacker detection
6. Shortest path
7. Power level
8. Possible multipaths
9. Regular updates
10. Clustering info
11. Fault detection

The topology-aware scheme needs to add the structure type of network. Location-aware needs to add the location of users with (x, y) coordinates. This helps to estimate the distance between the users. The energy level of each node should be added to the routing table and based on the energy level of each node can reduce the energy consumption. Link failure detection is to be added based on the drop rate level. Attacker detection is to be added based on the malicious function between the users. The shortest path helps to reduce the end-to-end delay. Power level identifies the residual energy level of each sensor node. Possible multiple paths lead to identifying the possible route between the users for the fastest and secure communication. Regular updates are required due to the high mobility in the network. If any clustering network is formed, cluster information should be added to the routing table. Clustering head and their respective clustered members are to be updated. Fault detection is to list the link failure, error rate, and transmission data rate to be addressed. If the above points are considered for the designing of routing protocol, it achieves effective and efficient network performance.

5 Conclusion

Based on the survey, this review work suggests a routing protocol design with strong requirements for network performance improvement and concerning the QoS metrics. In the existing approaches, many routing protocols are designed such as ad hoc on-demand distance vector routing protocol, temporally ordered routing algorithm, zone-based routing protocol, destination sequence distance vector routing protocol, hierarchical-based routing protocol, optimized link-state routing protocol, topology-based routing protocol, dynamic source routing protocol, congestion-based routing protocol, and ad hoc on-demand multihop distance vector routing protocol. Each has its own merits and demerits while applied in different types of ad hoc networks. In Flying Ad Hoc Networks, security, energy level, and efficient communication are to be maintained for high network performance even in high mobility conditions. For that case, all the metrics include identification of topology, location, energy, power, fault, attackers, multipath, shortest path, clustering info, and regular updates need to be addressed. In a future implementation, Flying Ad Hoc Networks can follow the important metrics based on the survey for the designing of routing protocol helps to improve the network performance and satisfy the QoS requirements in all aspects.

References

1. N.I. Mowla, N.H. Tran, I. Doh, K. Chae, AFRL: adaptive federated reinforcement learning for intelligent jamming defense in FANET. J. Commun. Netw. **22**(3), 244–258 (2020)
2. A.G. Orozco-Lugo, D.C. McLernon, M. Lara, S.A.R. Zaidi, B.J. González, O. Illescas, C.I. Pérez-Macías et al., Monitoring of water quality in a shrimp farm using a FANET. Internet Things 100170 (2020)
3. I.U. Khan, I.M. Qureshi, M.A. Aziz, T.A. Cheema, S.B.H. Shah, Smart IoT control-based nature inspired energy efficient routing protocol for flying adhoc network (FANET). IEEE Access **8**, 56371–56378 (2020)
4. M. Tropea, P. Fazio, F. De Rango, N. Cordeschi, A new FANET simulator for managing drone networks and providing dynamic connectivity. Electronics **9**(4), 543 (2020)
5. S. Radley, C.J. Sybi, K. Premkumar, Multi information amount movement aware-routing in FANET: flying ad-hoc networks. Mob. Netw. Appl. **25**(2), 596–608 (2020)
6. S. Kumar, D. Sharma, Evaluation of AOMDV routing protocol for optimum transmitted power in a designed ad-hoc wireless sensor network, in *Innovations in Computational Intelligence and Computer Vision* (Springer, Singapore, 2021), pp. 100–108
7. H. Zemrane, Y. Baddi, A. Hasbi, VOIP in MANETs based on the routing protocols OLSR and TORA, in *Advances on Smart and Soft Computing* (Springer, Singapore, 2021), pp. 443–453
8. G. Amponis, T. Lagkas, P. Sarigiannidis, V. Vitsas, P. Fouliras, Inter-UAV routing scheme testbeds. Drones **5**(1), 2 (2021)
9. S. Sharma, A.N. Mahajan, R.C. Poonia, A comparative study of modified DSRs routing protocols for MANET. Int. J. Innov. Sci. Eng. Technol. **7**(2) (2020)
10. H.M. Haglan, S.A. Mostafa, N.Z.M. Safar, A. Mustapha, M.Z. Saringatb, H. Alhakami, W. Alhakami, Analyzing the impact of the number of nodes on the performance of the routing protocols in MANET environment. Bull. Electr. Eng. Inf. **10**(1), 434–440 (2020)
11. S. Al-Emadi, A. Al-Mohannadi, Towards enhancement of network communication architectures and routing protocols for FANETs: a survey, in *2020 3rd International Conference on Advanced Communication Technologies and Networking (CommNet)* (IEEE, 2020), pp. 1–10
12. A. AlKhatieb, E. Felemban, A. Naseer,Performance evaluation of ad-hoc routing protocols in (FANETs), in *2020 IEEE Wireless Communications and Networking Conference Workshops (WCNCW)* (IEEE, 2020), pp. 1–6
13. D.S. Lakew, U. Sa'ad, N.-N. Dao, W. Na, S. Cho, Routing in flying adhoc networks: a comprehensive survey. IEEE Commun. Surv. Tutorials **22**(2), 1071–1120 (2020)
14. Q. Sang, Wu. Honghai, L. Xing, P. Xie, Review and comparison of emerging routing protocols in flying adhoc networks. Symmetry **12**(6), 971 (2020)
15. J. Liu, Q. Wang, C.T. He, K. Jaffrès-Runser, Y. Xu, Z. Li, Y.J. Xu, QMR: Q-learning based multi-objective optimization routing protocol for flying adhoc networks. Comput. Commun. **150**, 304–316 (2020)
16. S. Kumar, A. Bansal, R.S. Raw, Analysis of effective routing protocols for flying ad-hoc networks. Int. J. Smart Veh. Smart Transp. (IJSVST) **3**(2), 1–18 (2020)
17. M. Kaur, S. Verma, Flying ad-hoc network (FANET): challenges and routing protocols. J. Comput. Theor. Nanosci. **17**(6), 2575–2581 (2020)

18. J.S. Raj, A novel hybrid secure routing for flying ad-hoc networks. J. Trends Comput. Sci. Smart Technol. (TCSST) **2**(03), 155–164 (2020)
19. D.I. Surzhik, G.S. Vasilyev, O.R. Kuzichkin, Development of UAV trajectory approximation techniques for adaptive routing in FANET networks, in *2020 7th International Conference on Control, Decision and Information Technologies (CoDIT)*, vol. 1 (IEEE, 2020), pp. 1226–1230
20. M.M.S. Ibrahim, P. Shanmugaraja, Optimized link state routing protocol performance in flying ad-hoc networks for various data rates of unmanned aerial network. Mater Today Proc. (2020)
21. M.A. Hassan, S.I. Ullah, A. Salam, A.W. Ullah, M. Imad, F. Ullah, Energy efficient hierarchical based fish eye state routing protocol for flying ad-hoc networks
22. P. Kaur, A. Singh, S.S. Gill, RGIM: an integrated approach to improve QoS in AODV, DSR and DSDV routing protocols for FANETS using the chain mobility model. Comput. J. (2020)
23. Y.A.N.G. Hua, W.E.I. Xing, An application of CHNN for FANETs routing optimization. J. Web Eng. 830–844 (2020)
24. Q. Usman, O. Chughtai, N. Nawaz, Z. Kaleem, K.A. Khaliq, L.D. Nguyen, Lifetime improvement through suitable next hop nodes using forwarding angle in FANET, in *2020 4th International Conference on Recent Advances in Signal Processing, Telecommunications & Computing (SigTelCom)* (IEEE, 2020), pp. 50–55
25. V. Bhardwaj, N. Kaur, Optimized route discovery and node registration for FANET, in *Evolving Technologies for Computing, Communication and Smart World* (Springer, Singapore), pp. 223–237
26. S. Kumar, R.S. Raw, A. Bansal, Minimize the routing overhead through 3D cone shaped location-aided routing protocol for FANETs. Int. J. Inf. Technol. 1–7 (2020)
27. S. Kumar, A. Bansal, Performance investigation of topology-based routing protocols in flying ad-hoc networks using NS-2, in *IoT and Cloud Computing Advancements in Vehicular Ad-Hoc Networks* (IGI Global, 2020), pp. 243–267
28. V. Bhardwaj, N. Kaur, An efficient routing protocol for FANET based on hybrid optimization algorithm, in *2020 International Conference on Intelligent Engineering and Management (ICIEM)* (IEEE, 2020), pp. 252–255
29. G.A. Kakamoukas, P.G. Sarigiannidis, A.A. Economides, FANETs in agriculture-A routing protocol survey. Internet Things 100183 (2020)
30. Y.-G. Park, S.-J. Lee, PUF-based secure FANET routing protocol for multi-drone. J. Korea Soc. Comput. Inf. **25**(9), 81–90 (2020)
31. X. Li, F. Deng, J. Yan, Mobility-assisted adaptive routing for intermittently connected FANETs. IOP Conf. Ser. Mater. Sci. Eng. **715**(1), 012028 (2020)
32. N. Watanabe, O. Oyakhiire, K. Gyoda, Proposition to improve the performance of routing protocols iFORP-3DD for drone adhoc network, in *2020 35th International Technical Conference on Circuits/Systems, Computers and Communications (ITC-CSCC)* (IEEE, 2020), pp. 149–154
33. B.O. Sadiq, A.T. Salawudeen, FANET optimization: a destination path flow model. Int. J. Electr. Comput. Eng. (IJECE) **10**(4), 4381–4389 (2020)
34. A. Boukerche, B. Turgut, N. Aydin, M.Z. Ahmad, L. Bölöni, D. Turgut, Routing protocols in adhoc networks: a survey. Comput. Netw. **55**(13), 3032–3080 (2011)
35. N. Saputro, K. Akkaya, S. Uludag, A survey of routing protocols for smart grid communications. Comput. Netw. **56**(11), 2742–2771 (2012)
36. S. Jain, S. Sahu, Topology vs. position based routing protocols in mobile adhoc networks: a survey. Int. J. Eng. Res. Technol. (IJERT) **1**(3), 2278–3181 (2012)
37. A. Ibrahim, M.K. Sis, S. Cakir, Integrated comparison of energy efficient routing protocols in wireless sensor network: a survey, in *2011 IEEE Symposium on Business, Engineering and Industrial Applications (ISBEIA)* (IEEE, 2011), pp. 237–242
38. M.R. Senouci, A. Mellouk, H. Senouci, A. Aissani, Performance evaluation of network lifetime spatial-temporal distribution for WSN routing protocols. J. Netw. Comput. Appl. **35**(4), 1317–1328 (2012)

Parkinson's Disease Data Analysis and Prediction Using Ensemble Machine Learning Techniques

Rubash Mali, Sushila Sipai, Drish Mali, and Subarna Shakya

Abstract Parkinson's disease is an incurable neurodegenerative disorder which causes problems with communication, thinking, behavioral problems, and other difficulties. Detecting the disease at an early stage can help uplift the living quality of the patient. This paper presents the findings of data analysis performed on patients' speech signals measurement to understand the insights and uses ensemble models (GBDT, random forest, voting classifier, and stacking classifier) to predict whether a patient is suffering from Parkinson or not. Voting and stacking models use KNN, logistic regression, SVM, random forest, and GBDT as base learners. It can be noted that fundamental frequencies (minimum and average) have higher values for positive cases. Four metrics, F_1 score, accuracy, log loss, and ROC score, were used to report the performance of the model. The best model was SVM which out performed ensemble models in three out of the four performance metrics (F_1 score, accuracy, log loss). However, the stacking and voting ensemble models had better ROC scores than that of the base learners.

Keywords Parkinson's disease · Ensemble model · Stacking model · Log loss · ROC score

R. Mali (✉) · S. Sipai
Kantipur Engineering College, Lalitpur, Nepal
e-mail: rubashmali@kec.edu.np

D. Mali
College of Science and Engineering, The University of Edinburgh, Edinburgh, Scotland
e-mail: D.Mali@sms.ed.ac.uk

S. Shakya
Institute of Engineering, Tribhuvan University, Kirtipur, Nepal
e-mail: drss@ioe.edu.np

S. Shakya et al. (eds.), *Mobile Computing and Sustainable Informatics*, Lecture Notes on Data Engineering and Communications Technologies 68,
https://doi.org/10.1007/978-981-16-1866-6_24

1 Introduction

Neurodegeneration is the progressive loss of structure or function of neurons that includes diseases such as amyotrophic lateral sclerosis. Some common neurodegenerative diseases are Parkinson's disease, Alzheimer's disease, fatal familial insomnia, etc. Parkinson's disease (PD), or simply Parkinson's is a long-term degenerative disorder of the central nervous system that mainly affects the motor system. Parkinson's is the most common neurodegenerative disorder second to Alzheimer's [1]. The most obvious early symptoms are tremor, rigidity, slowness of movement, and difficulty in walking, but cognitive and behavioral problems might also occur. Dementia is commonly seen in the advanced stages of the disease. The prevalence of Parkinson ranges from 41 people per 100,000 in their 40s to 1900 people per 100,000 among those who are 80 and older [2]. There is no cure for PD; however, early diagnosis aims to improve the overall health of the patient. Initial treatment is typically done with the medication levodopa (L-DOPA), for further advanced cases dopamine agonists are used. More than 10 million people worldwide are living with PD. The combined direct and indirect cost of Parkinson's including treatment, social security payments, and lost income is estimated to be nearly $52 billion per year in the USA alone [3].

Medical institutions all around the world have collected various data related to numerous health-related diseases. This data can be leveraged using various traditional and emerging machine learning techniques to find hidden patterns and gain crucial insights. Normally, humans cannot apprehend these complex datasets which raises the need for machine learning algorithms to accurately predict the presence of different medical conditions.

2 Related Study of Disease Prediction Using Ensemble Techniques

Ensemble learning is the process of using a collection of machine learning models with the aim of using strengths of various models to improve the predictive performance [4]. The main objective of ensemble models is to tackle the bias variance trade-off problem by using different machine learning models, as every model has its own drawbacks. However, ensemble learning has the main two drawbacks: lack of interpretability and computational cost. Ensemble models can be broadly classified into four categories: bagging, boosting, stacking, and cascading.

Voting classifiers simply use the predictions from two or more models and using statistical techniques like mean , median or vote counting both regression and classification problems can be tackled. For classification, the voting classifier may use hard and soft voting strategies. Hard voting only considers the vote number while soft voting considers probabilities as well. The core idea boosting is to improve the performance of a weak learner by iteratively invoking a weak learner. It uses the mechanism of resampling the sample so that the most useful sample is used for each consecutive iteration [5].

Stacking is an ensemble technique that uses fine tuned classifiers to generate a high-level final model. Stacking is a two-level prediction technique where initially in level 0 the base learners predict according to the original data and the meta classifier performs level 1 (final) prediction using the prediction of base learners [6].

Esfahani and Ghazanfari [7] proposed a hybrid algorithm system consisting of decision tree, neural network, rough set, naive Bayes, and SVM algorithm for cardiovascular disease detection. They used UCL laboratory-based data with 74 features and took a subset of 14 features selected using Pearson's correlation coefficient. The hybrid algorithm showed better F-measure (86.8%) as compared to other individual classifiers. Tenfold cross-validation was used to evaluate this proposed method.

Mandal and Sairam [8] proposed multiple machine learning models such as multinomial logistic regression, artificial neural networks, rotation forest ensemble with support vector machines and principal component analysis and other boosting methods for prediction of Parkinson's disease. Accuracy, sensitivity, and specificity were considered as the performance metrics. Feature selection and ranking were done using a new ensemble method that used the Bayesian network. It was optimized using Haar wavelets as a projection filter and Tabu search algorithm as classifier. Multinomial logistic regression showed highest accuracy at 100% with specificity and sensitivity at 99.6% and 98.3%, respectively. All these experiments were validated using t-test at 95% confidence.

Wroge et al. [9] extracted two feature sets from mPower voice dataset, namely the Geneva Minimalistic Acoustic Parameter Set (GeMaps) and AVEC. Various supervised learning techniques were applied on these feature set of which CNN had the highest overall accuracy at 86% with AUC score of 0.823 and 0.915 for GeMaps and AVEC dataset, respectively.

Mathur et al. [10] applied "cfsSubsetEval" attribute evaluator and "BestFirstSearch" method to reduce the number of features from 24 to 11 using WEKA tool. On this new dataset SMO, KNN, random forest, AdaBoost.M1, bagging, MLP, and DT algorithms were applied. It was found that a combination of ANN and KNN algorithms gave the best accuracy (91.78%).

Sriram et al. [11] proposed a technique using voice dataset for diagnosis of PD. This dataset consists of voices from 31 people of which 23 had PD. This dataset had 26 attributes and 5875 instances. Weka V3.4.10 and Orange V2.0b software were used for carrying out the experiment. Best accuracy was achieved using random forest algorithm (90.2%).

3 Methodology

3.1 Data Collection and Pre-processing

The dataset which is used for the analysis and prediction was created by Max Little of the University of Oxford together with the National Centre for Voice and

Speech, where the speech signals were recorded. It consists of a range of biomedical voice measurements with 24 features which are name of patients, three types of fundamental frequencies (high, average, and low), several measures of variation in fundamental frequency (jitter and its type), several measures of variation in amplitude (shimmer and its type), two measures of ratio of noise to tonal components within the voice (NHR and HNR), status (0 for healthy and 1 for Parkinson positive case), two nonlinear dynamical complexity measures (RPDE, D2), DFA a signal fractal scaling exponent and three nonlinear measures of fundamental frequency variation (spread1, spread2, PPE) [12]. The dataset has about 75% of cases suffering from Parkinson's disease and 25% of cases which are healthy. As for data preprocessing, data standardization was performed.

3.2 *Data Analysis*

Before building a predictive model, the data is analyzed which helps to understand the data trend and insights on the data, which is presented in summary in this paper.

3.2.1 Performing Analysis on Fundamental Frequencies

The fundamental frequencies recorded are categorized into three types: high, low, and average. In this analysis, all the three categories are considered and for each of them, a distribution plot and a decile table are constructed which can be seen in Figs. 1, 2 and 3, and in Tables 1, 2 and 3. It is clear that the minimum and average vocal fundamental frequency tend to be higher in case of Parkinson positive cases than in the negative ones, but the same argument is not clear for maximum vocal fundamental frequency.

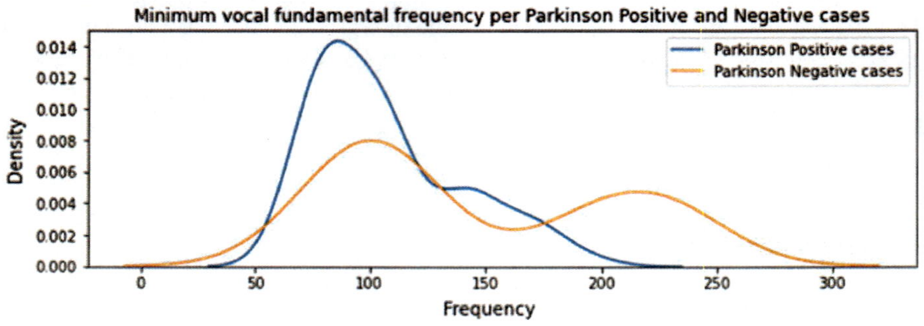

Fig. 1 Distribution plot for minimum frequency

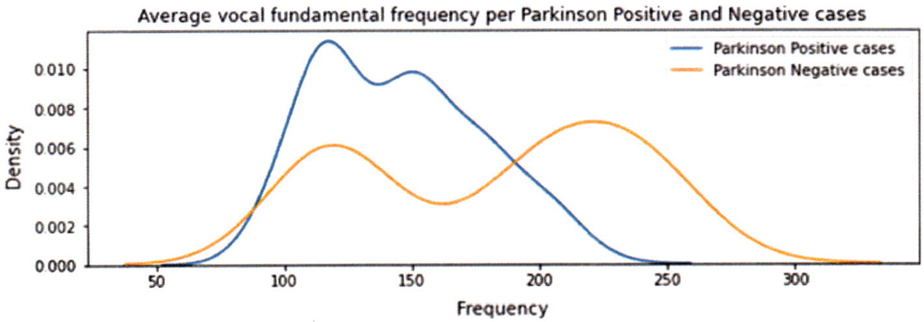

Fig. 2 Distribution plot for average frequency

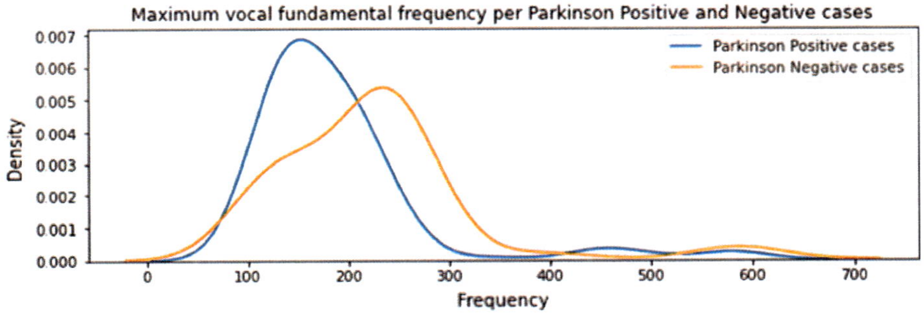

Fig. 3 Distribution plot for maximum frequency

Table 1 Decile values for minimum frequency

Percentile	Parkinson positive cases	Parkinson negative cases
0	65.476	74.287
10	75.347	86.062
20	79.092	95.322
30	84.05	99.945
40	90.607	108.23
50	99.77	113.938
60	106.917	134.227
70	112.496	195.448
80	140.636	220.607
90	157.833	230.034
100	199.02	239.171

Table 2 Decile values for average frequency

Percentile	Parkinson positive cases	Parkinson negative cases
0	88.33	110.379
10	109.439	116.014
20	114.942	116.91
30	120.061	124.635
40	128.138	174.588
50	145.174	198.996
60	151.75	204.376
70	159.806	223.252
80	176.715	236.816
90	190.299	243.028
100	223.361	260.105

Table 3 Decile values for maximum frequency

Percentile	Parkinson positive cases	Parkinson negative cases
0	102.145	113.597
10	125.269	127.547
20	131.076	134.566
30	139.697	180.252
40	157.317	211.588
50	163.335	231.162
60	189.882	239.634
70	200.923	247.107
80	216.712	254.227
90	233.252	263.371
100	588.518	592.03

3.2.2 Performing Analysis on Shimmer and Jitter

Jitter and shimmer are acoustic characteristics of voice signals, and they are quantified as the cycle-to-cycle variations of fundamental frequency and waveform amplitude respectively [13]. The dataset consists of four variations of jitter measurement (MDVP: Jitter (%), MDVP: Jitter (Abs), MDVP: RAP, MDVP: PPQ, Jitter: DDP) and as for shimmer, the variation measurements are MDVP: Shimmer, MDVP: Shimmer (dB), Shimmer: APQ3, Shimmer: APQ5, MDVP: APQ, Shimmer: DDA. For simplicity purpose, the study of MDVP: Shimmer (dB) and MDVP: Jitter (Abs). This study has considered the distribution plot for both positive and negative cases

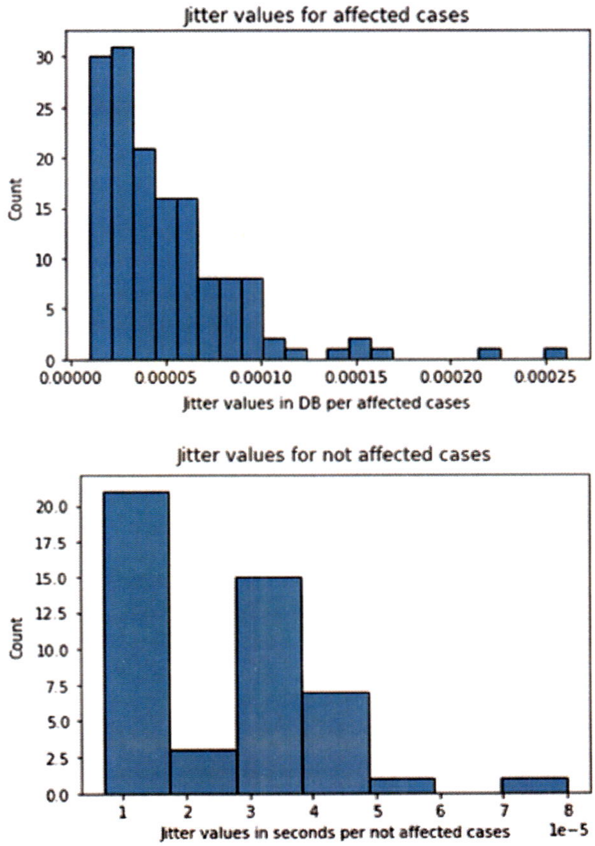

Fig. 4 Histogram of jitter values for affected and unaffected cases

of Parkinson's disease for both MDVP: Shimmer (dB) and MDVP: Jitter (Abs). In case of jitter and shimmer, patients suffering from Parkinson's disease tend to follow log distribution (Figs. 4 and 5).

3.2.3 Visualization of the Dataset Using PCA

Principal component analysis (PCA) is a technique to project data in a higher dimensional into a lower-dimensional space by maximizing the variance of each dimension [14]. The data consist of 24 columns where the first attribute is the name and id of the patient and the other is the status of the patient. The name and status columns are removed, and then the data is standardized as to perform PCA with the remaining 22 components. Finally, a graph of cumulative variance is plotted which shows variance with respect to number of components (Fig. 6).

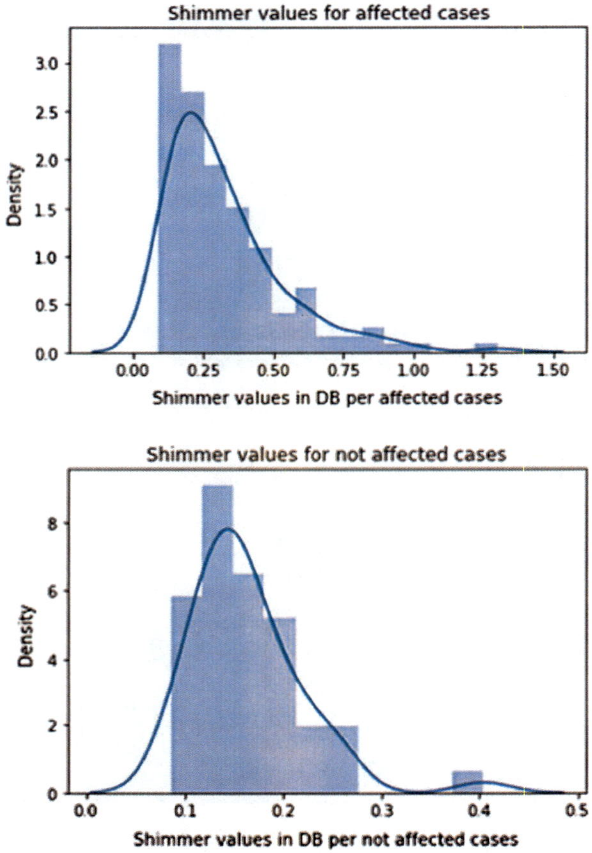

Fig. 5 Distribution plot of shimmer values for affected and unaffected cases

3.3 Proposed Architecture

This paper divides into ensemble models using bagging, boosting, stacking methods. A random forest model is used as bagging, and gradient boosting (GBDT) is used as boosting method and as for stacking, the combination of five different base models (KNN, SVM, random forest, GBDT, and logistic regression) and SVM was used as a meta classifier. Furthermore, a voting classifier was also created and the above-mentioned five different models were used. The core idea of this experiment is to utilize different models having their own weakness and strengths and combine them to study how each model performs on various performance metrics (accuracy, F_1 score, AUC score, and log loss) individually and when ensembled. Furthermore, the experiment tries to look into the issue that is ensemble necessary or just simple models are sufficient enough for Parkinson's disease detection.

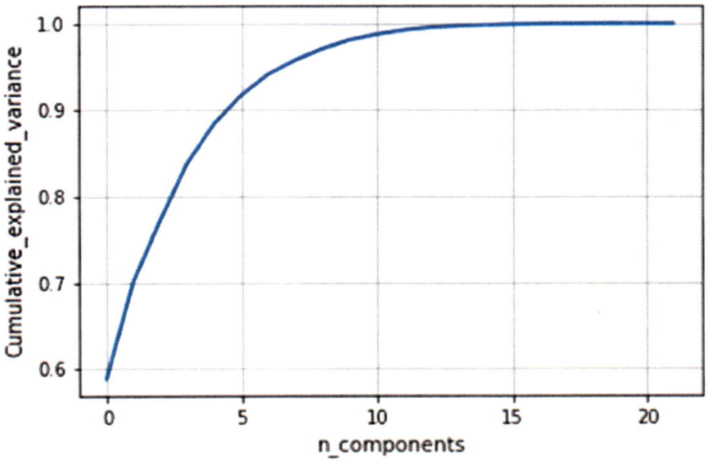

Fig. 6 Components versus cumulative variance explanation graph

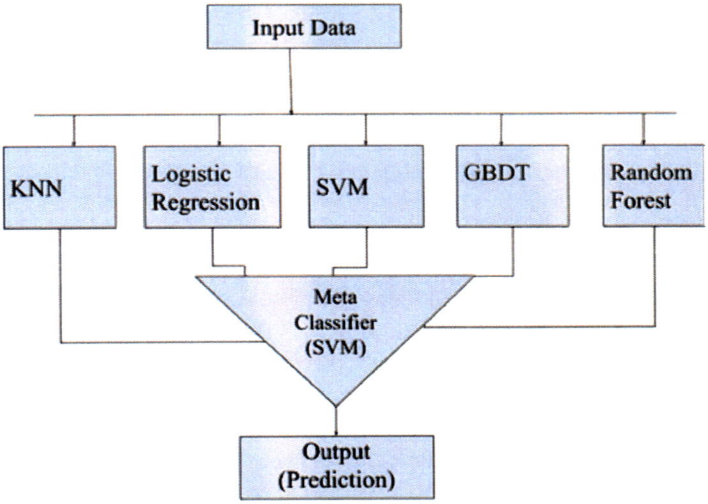

Fig. 7 Proposed architecture of stacking classifier

The training set was divided into three sets so that threefold validation could be done for hyperparameter tuning of the base models. The parameters (k-neighbors) for KNN, C (Regularization parameter) and gamma for SVM, C (Regularization parameter) with L2 regularizer for logistic regression and max depth for both random forest and GBDT was tuned accordingly. The base model was stacked using SVM as meta classifier to create a stacking classifier, and voting classifier was also created as shown in Figs. 7 and 8, respectively.

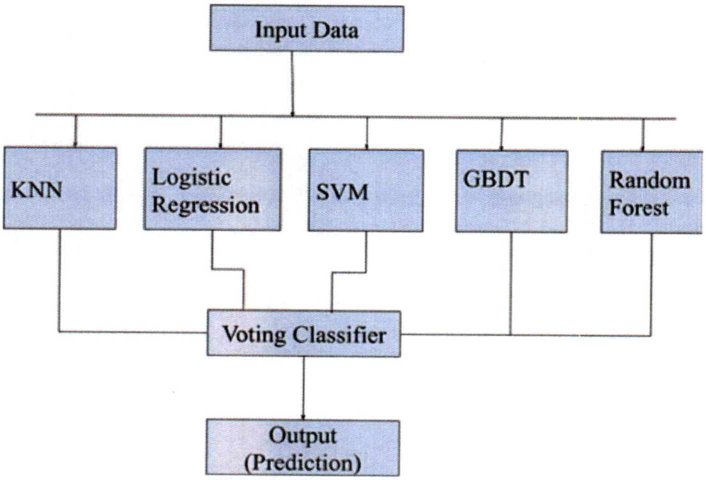

Fig. 8 Proposed architecture of voting classifier

4 Experimental Results

After all the models were trained, the accuracy, F_1 Score, log loss, and AUC score were noted. F_1 score metric is significant as it is the harmonic mean of recall and precision and thus takes both of them into consideration [15]. While AUC ROC score considers both true positive rate and false negative rate, finally, log loss is chosen as a metric as the problem is two-class classification. Table 4 shows all the models and the values of parameters for each model. After analyzing the base model, it was clear that SVC model was the best model in terms of all the parameters with an accuracy of 91.5%, F_1 score of 94.5%, log loss of 2.927, and AUC score of 93.48% followed by random forest with three parameters: accuracy, F_1 and log loss but AUC ROC score slightly less than that of KNN model while the logistic regression model under performed in all the mentioned parameters. In case of the four ensemble models, the voting classifier seems to be the best classifier in terms of the mentioned parameters. Also, random forest and voting classifier seem to have very similar performance. It can also be noted that both the stacking classifier and voting classifier have better AUC ROC scores than the base learners (Fig. 9). The ROC curves for all the ensemble models are shown in Figs. 10, 11, and 12.

5 Conclusion

From the above study, it can be observed that the minimum and average vocal fundamental frequency tend to be higher for Parkinson positive cases than in negative ones. Secondly, in terms of jitter and shimmer, using only MDVP: Shimmer (dB)

Table 4 Performance of various machine learning models

Model	Accuracy (%)	F_1 score (%)	ROC score (%)	Log loss
Logistic regression	81.35	87.05	84.39	6.439
KNN	88.13	92.13	93.18	4.097
SVM	91.52	94.50	93.48	2.927
Random forest	89.83	93.33	91.96	3.512
GBDT	88.12	88.13	90.30	4.097
Voting classifier	89.37	93.32	96.66	3.512
Stacking classifier	88.13	92.13	95.45	4.097

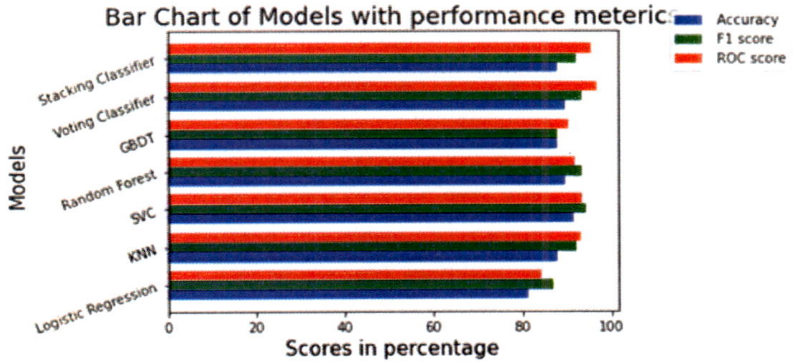

Fig. 9 Bar chart showing performance of various machine learning models

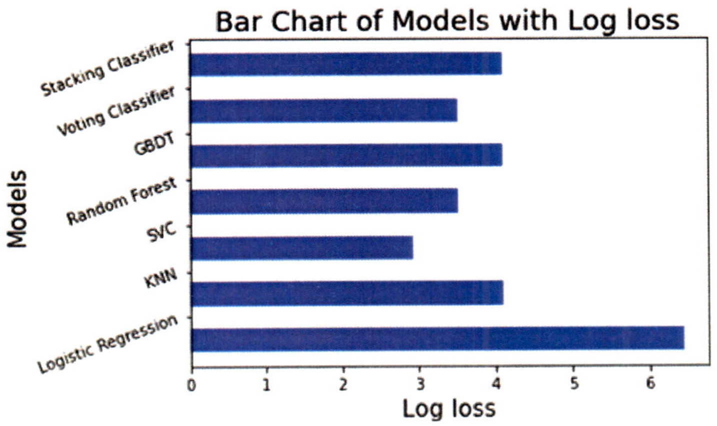

Fig. 10 Log loss of various machine learning models

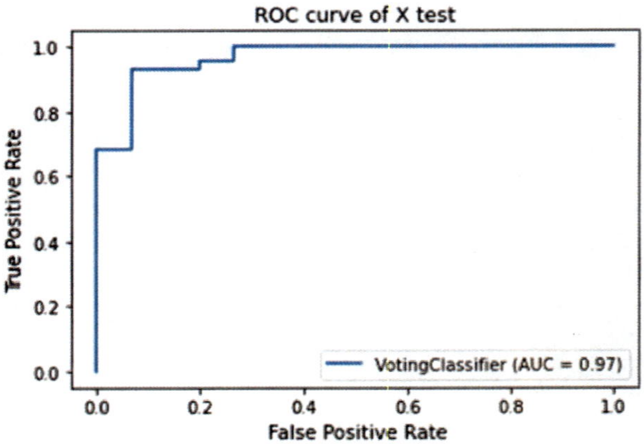

Fig. 11 ROC curve of stacking classifier

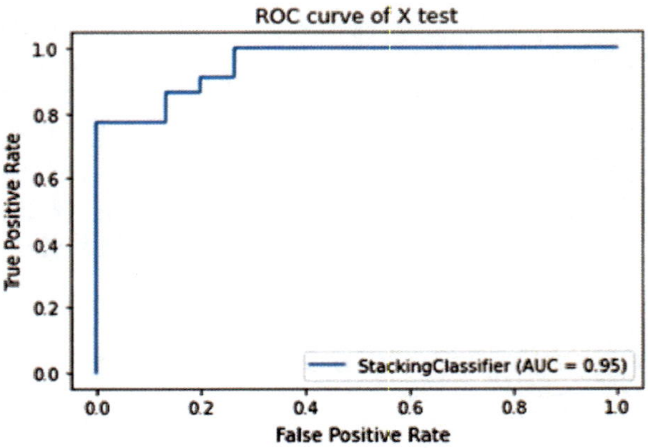

Fig. 12 ROC curve of voting classifier

and MVP: Jitter Features (Abs) do not provide much insight on Parkinson's cases. Therefore, some variations in both measures in the dataset may be helpful for further analysis. Finally, using PCA we can see that all 22 features can be reduced to a smaller number to reduce the dimension of the data. Even after the reduction to five features, it can be seen that more than 90% of the variance is explained.

As for the predictive models, it is clear that SVC is the best base model and it outperforms even the ensemble models in terms of accuracy, F_1 score, and log loss. However, the ensemble models have a better AUC ROC score than any of the base learners. Hence, it can be noted that even after using various ensemble models like stacking model, bagging model, boosting model, or voting model, in some cases, the base learners can perform better than that of the final ensemble model, which can be seen in this experiment. So it can be concluded that using ensemble does not necessarily increase performance in all aspects; however, for this experiment ensemble actually provides decent performance and the AUC score has increased significantly. Although only classical machine learning methods are used in this study, this approach can be extended by using deep neural networks and RNN. Adopting such measures can increase the computational performance of the base model significantly which may increase the predictive power of the ensembles as well.

References

1. A. Tsanas et al., Accurate telemonitoring of Parkinson's disease progression by noninvasive speech tests. IEEE Trans. Biomed. Eng. **57**(4), 884–893 (2010)
2. E. Naqvi, Parkinson's disease statistics. Parkinson's News Today (2021). https://parkinsonsnewstoday.com/parkinsons-disease-statistics/
3. Statistics. Parkinson's Foundation (2020). https://www.parkinson.org/Understanding/Parkinsons/Statistics
4. G. Kyriakides, K.G. Margaritis, *Hands on Ensemble Learning with Python* (Packt Publishing Limited, 2019)
5. L. Rokach, *Pattern Classification Using Ensemble Methods* (World Scientific, 2010)
6. S.-A.N. Alexandropoulos et al., Stacking strong ensembles of classifiers, in *Artificial Intelligence Applications and Innovations* (Springer International Publishing, Cham, 2019), pp. 545–556. (IFIP Advances in Information and Communication Technology)
7. H.A. Esfahani, M. Ghazanfari, Cardiovascular disease detection using a new ensemble classifier, in *2017 IEEE 4th International Conference on Knowledge-Based Engineering and Innovation (KBEI)*, Tehran (2017), pp. 1011–1014. https://doi.org/10.1109/KBEI.2017.8324946
8. I. Mandal, N. Sairam, New machine-learning algorithms for prediction of Parkinson's disease. Int. J. Syst. Sci. **45**(3), 647–666 (2012)
9. T.J. Wroge et al., Parkinson's disease diagnosis using machine learning and voice, in *2018 IEEE Signal Processing in Medicine and Biology Symposium (SPMB)* (2018), pp. 1–7
10. R. Mathur et al., Parkinson disease prediction using machine learning algorithm, in *Emerging Trends in Expert Applications and Security* (Springer Singapore, Singapore, 2018), pp. 357–363. (Advances in Intelligent Systems and Computing)
11. T.V.S. Sriram et al., Diagnosis of Parkinson disease using machine learning and data mining systems from voice dataset, in *Proceedings of the 3rd International Conference on Frontiers of Intelligent Computing: Theory and Applications (FICTA)* (Springer International Publishing, Cham, 2014), pp. 151–157. (Advances in Intelligent Systems and Computing)
12. M.A. Little et al., Exploiting nonlinear recurrence and fractal scaling properties for voice disorder detection. Biomed. Eng. Online **6**(1), 23 (2007)
13. M. Farrús, Jitter and shimmer measurements for speaker recognition (2007)
14. E. lpaydin, *Introduction to Machine Learning* (MIT Press, 2004)
15. Y. Sasaki, The truth of the F-measure. Teach. Tutor Mater. (2007)

Facial Expression Recognition System

Harshit Kumar, Ayush Elhance, Vansh Nagpal, N. Partheeban, K. M. Baalamurugan, and Srinivasan Sriramulu

Abstract Facial gestures help to understand what is really going on in a visual way in someone's mind. A different meaning is expressed by different expressions. Facial expression recognition system plays an important role in the field of machine learning and deep learning; it helps us to understand a deeper meaning of human–machine interaction. Recognition of human expression with high accuracy is still a difficult task for computers. Facial expression recognition is categorized into four stages they are preprocessing, registration of face, feature extraction of face, and classification of expressions. It is used in several applications such as understanding mental disorders, analyzing what is going on in a human's mind, lie detection, etc. Here, a convolutional neural network is utilized which is very effective for image processing. The data used is supervised data. The FER2013 dataset is used for analysis and popular deep learning frameworks such as Keras and OpenCV for the detection of facial expressions in an accurate manner.

Keywords Deep learning · Machine learning · Image classification · OpenCV

1 Introduction

One of the foremost common kinds of non-verbal communication from which someone understands through facial expressions in one's mood/attitude [1]. Facial expression recognition is one of the most interesting and challenging fields of machine learning and computer vision. Facial expressions can be categorized as happy, angry, sad, fear, and surprise [2]. In the field of computer vision, facial expression systems are not only limited to understanding psychological analysis,

H. Kumar · A. Elhance · V. Nagpal
CSE, Galgotias University, Greater Noida, UP, India

N. Partheeban (✉) · K. M. Baalamurugan · S. Sriramulu
School of Computing Science and Engineering, Galgotias University, Greater Noida, UP, India
e-mail: n.partheeban@galgotiasuniversity.edu.in

S. Sriramulu
e-mail: s.srinivasan@galgotiasuniversity.edu.in

S. Shakya et al. (eds.), *Mobile Computing and Sustainable Informatics*, Lecture Notes on Data Engineering and Communications Technologies 68,
https://doi.org/10.1007/978-981-16-1866-6_25

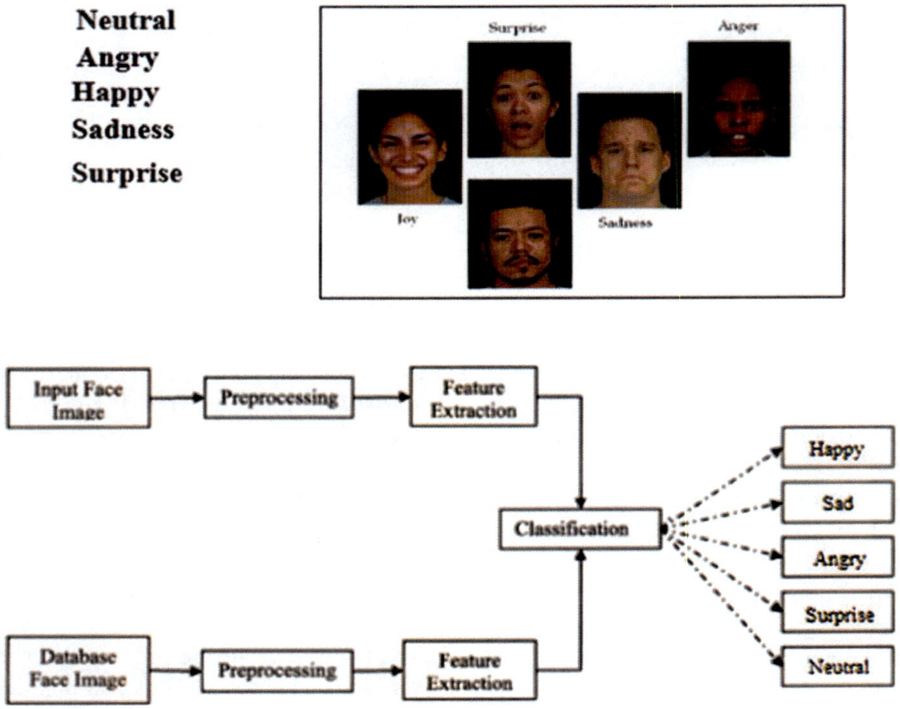

Fig. 1 Facial expression recognition

but also, they can be used in other fields such as a lie detector, movie recommendation according to someone mood, fatigue detection of drivers, music recommendation as per mood, etc. [3, 4]. Figure 1 depicts the facial expression recognition.

1.1 Motivation

The main motivation behind choosing this proposed work is that many MNCs investing millions and billions to get the feedback of their product for more sales and profit but somewhat they are not getting the favorable result as expected. A facial expression recognition system is a design that aims to enhance the product and service performance by extracting the mood of their customers after getting or using the product.

1.2 Examples

1. Disney is used to find the feedback of its customers.

2. Kellogg's used the attractive software to find the audience mood for its cereal advertisement.

3. Unilever used HireVue's for recruitment purpose to get better employees for its company.

2 Literature Review

As per various literature surveys from different sources, it is found that for implementing this proposed work includes four basic steps that are required to be performed.

1. Preprocessing
2. Registration of face
3. Feature extraction of face
4. Classification of emotion

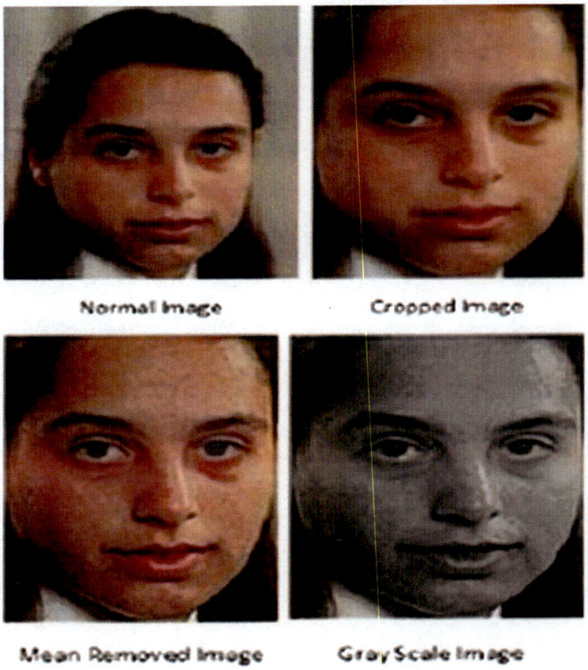

Fig. 2 Facial expression of different images

2.1 Preprocessing

For deepest level of abstraction of an image, preprocessing is utilized.

The preprocessing steps are

1. Reducing the noise or error
2. Converting the image or photo into grayscale
3. Transformation of pixel brightness
4. Transformation of given output (Fig. 2).

2.2 Facial Expression

The face is detected in this step, and the calculation is performed based on the value of the function. The number of pixels below the gray rectangle can be calculated by subtracting the sum of the pixels occupied by the white rectangle. The position between two is contained in three rectangle features.

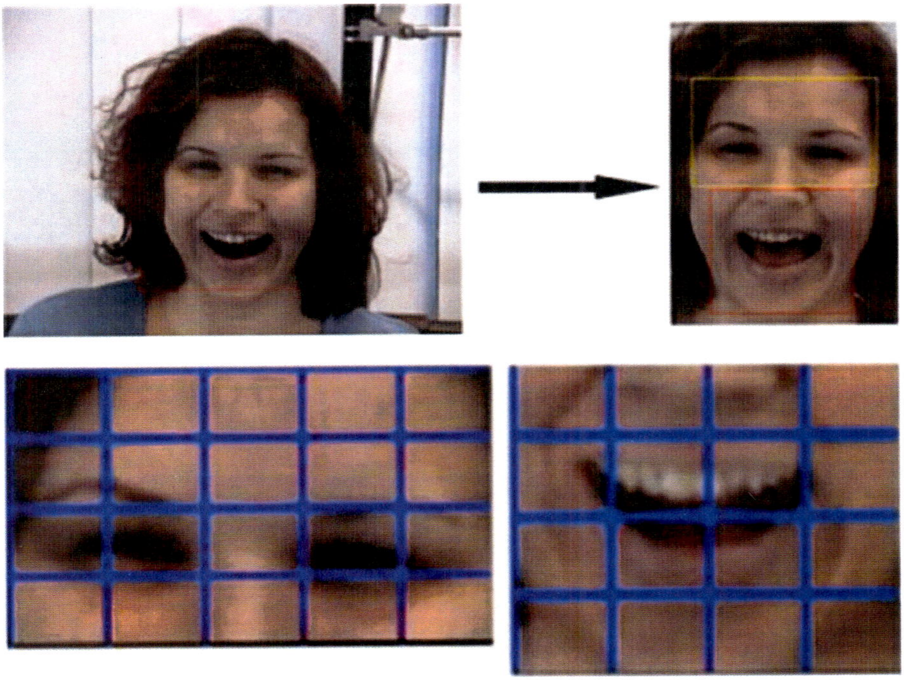

Fig. 3 Emotional classification

2.3 Feature Extraction

When the face is detected from the image, various operations for the expression are utilized. There are two types of features such as appearance feature and geometric feature. In the geometric feature, the face is presented in terms of face pattern. The appearance feature measures the appearance changes based upon texture analysis (Fig. 3).

2.4 Emotion Classification

In this step, the expression of the given photo is recognized out of these five facial expressions (Fig. 4).

Fig. 4 Five different facial expressions

3 Formulation of Problem

People's expressions are grouped into five emotions such as anger, sad, happy, surprise, and neutral (Fig. 5).

Fig. 5 Steps in facial expression processing

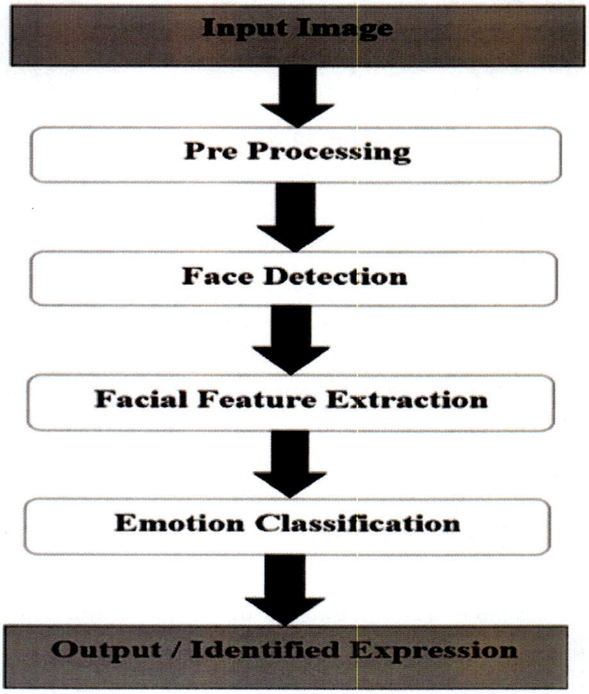

4 Software Requirement Analysis

4.1 *TensorFlow*

It is an open-source library to perform the numerical calculations and machine learning algorithms. It combines a wide range of machine learning and neural networking models algorithms to make them useful in the form of standard simulations. It uses Python to provide a pre-programmed API for building applications in the framework, while making those applications with high C++ performance.

4.2 *Keras*

Keras is the API of TensorFlow 2.0: an accessible, highly productive interface for solving machine learning problems. It also provides critical releases and blocks for building learning and solutions with high velocity.

4.3 *OpenCV*

Open computer vision (OpenCV) provides infrastructure or platform for computer vision applications which are very useful in machine learning (ML) applications.

5 Feasibility Analysis

5.1 *Performance of Model*

It uses various algorithms related to efficiency in order to increase the performance of the model.

5.2 *Technical Considerations*

This work is performed by a large dataset.

5.3 Cost Feasibility

The model is not too costly and takes information from free-to-access government sources.

5.4 Resource Feasibility

The model is primarily based on large datasets. So, the productive outcomes would be maximized by providing massive resources.

6 Goal

The main objective is to obtain the image from FER and then find the emotion of the person present in the image. That emotion can be anger, happy, neutral, sad, and surprise. There can be more than one result, but it is required to identify the most favorable outcome.

7 Analysis

The steps while developing this proposed work are as follows:

1. Analyze the stated problem
2. Gather the required data of model
3. Check the feasibility of the model
4. Formation of layout
5. Going through the research paper and journals
6. Choose the method to develop algo
7. Analyze benefits and limitations
8. Development of project (start)
9. Install various software: Python, Visual Studio, Keras, etc.
10. Analyze the algo by guide (Figs. 6 and 7).

Fig. 6 Architectural diagram

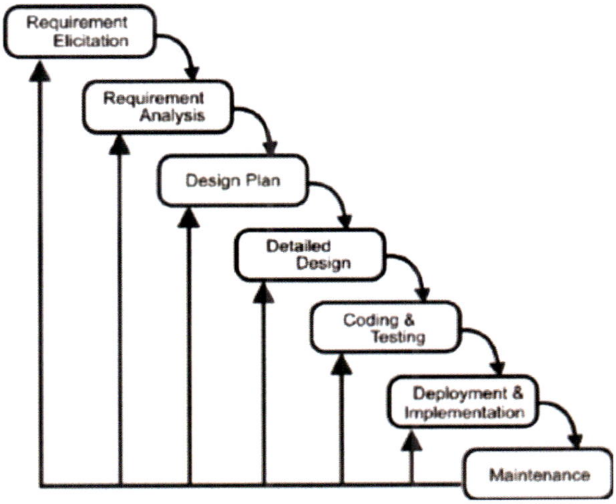

8 Training, Implementation, and Testing

8.1 The Database

The dataset used for training the model is taken from Kaggle. It consists the photos of 48 × 48 pixel greyscale. The main task is to categorize the images based upon the five categories (0 = Angry, 1 = Happy, 2 = Sad, 3 = Surprise, 4 = Neutral).

The training set consists of thousands of examples.

Emotion labels in the dataset:

0: 4539 images—***Angry***
1: 9898 images—***Happy***
2: 6770 images—***Sad***
3: 4020 images—***Surprise***
4: 6189 images—***Neutral*** (Fig. 8)

8.2 Training

ANN and CNN use parameters that are available to obtain the desired output. Backpropagation is the most common algorithm used by ANN and CNN. It attempts to measure the feature gradient that can be used to decide how these parameters can be used to minimize errors that affect the efficiency of the model. The most common problem is over fitting (Fig. 9).

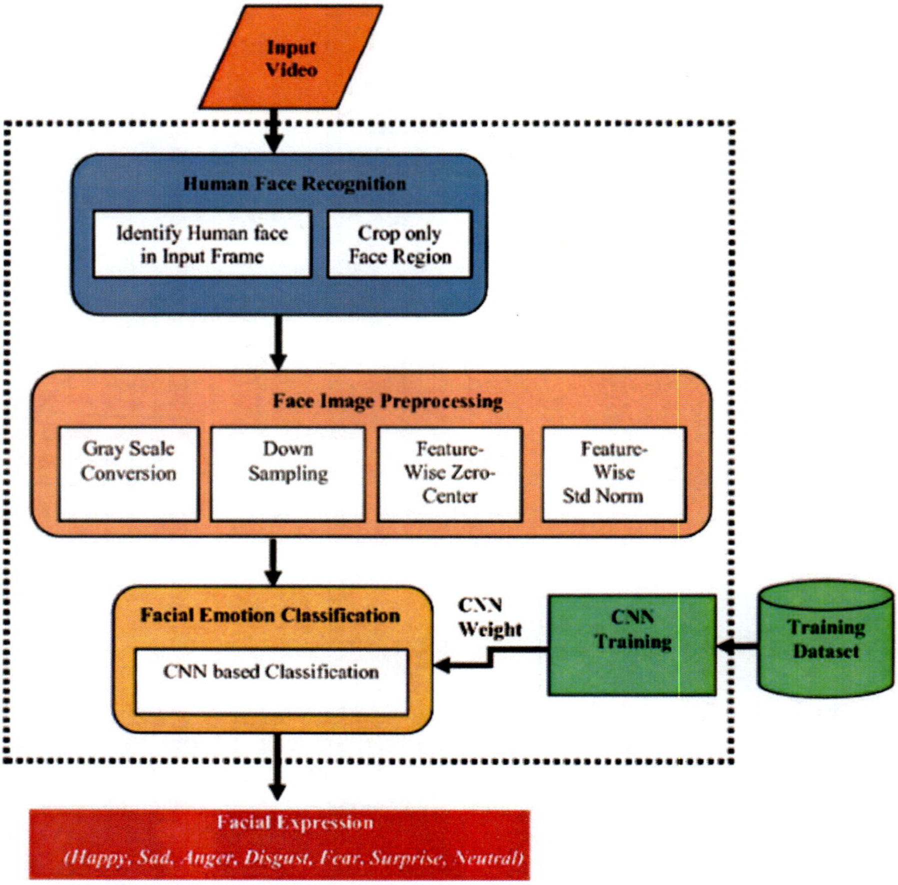

Fig. 7 Flowchart for facial recognition

Fig. 8 Different emotional expressions

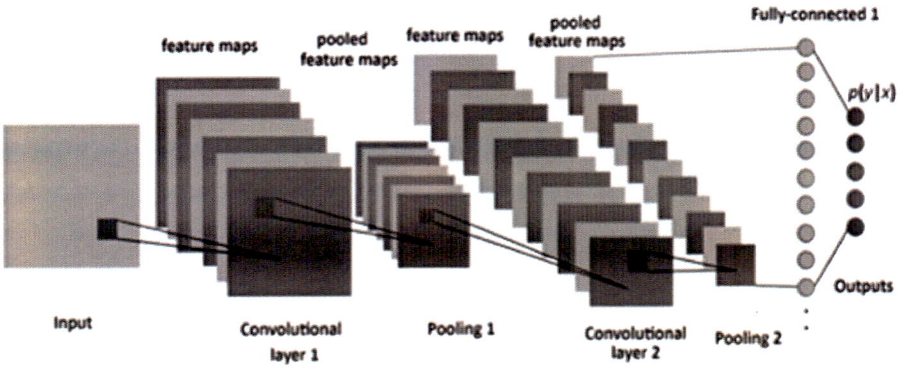

Fig. 9 Layers of convolutional neural network (CNN)

8.3 Output

See Fig. 10.

9 Limitations and Future Scope

1. In the future, it is estimated that this proposed work will increase from 21.6 billion dollars in 2019 to 56 billion dollars by 2024 in the market at a CAGR of 21% during this period. In support of this proposed work for the betterment of society, technical advancements in machine learning and artificial intelligence are promoted.
2. With an accuracy of about 86%, it produces real-time results, and further studies will achieve more than 95%, as an expression is the mixture of the cognitive state of another, a personality that is reflected.

10 Conclusion

In the proposed work, the implementations will also be supported by developments in the field of facial and emotion recognition systems. The potential growth in this field is supposed to make user experiences even better for all users. This will encourage the consumer to use this system effectively. The benefit of using this system will be provided to both the client and the company.

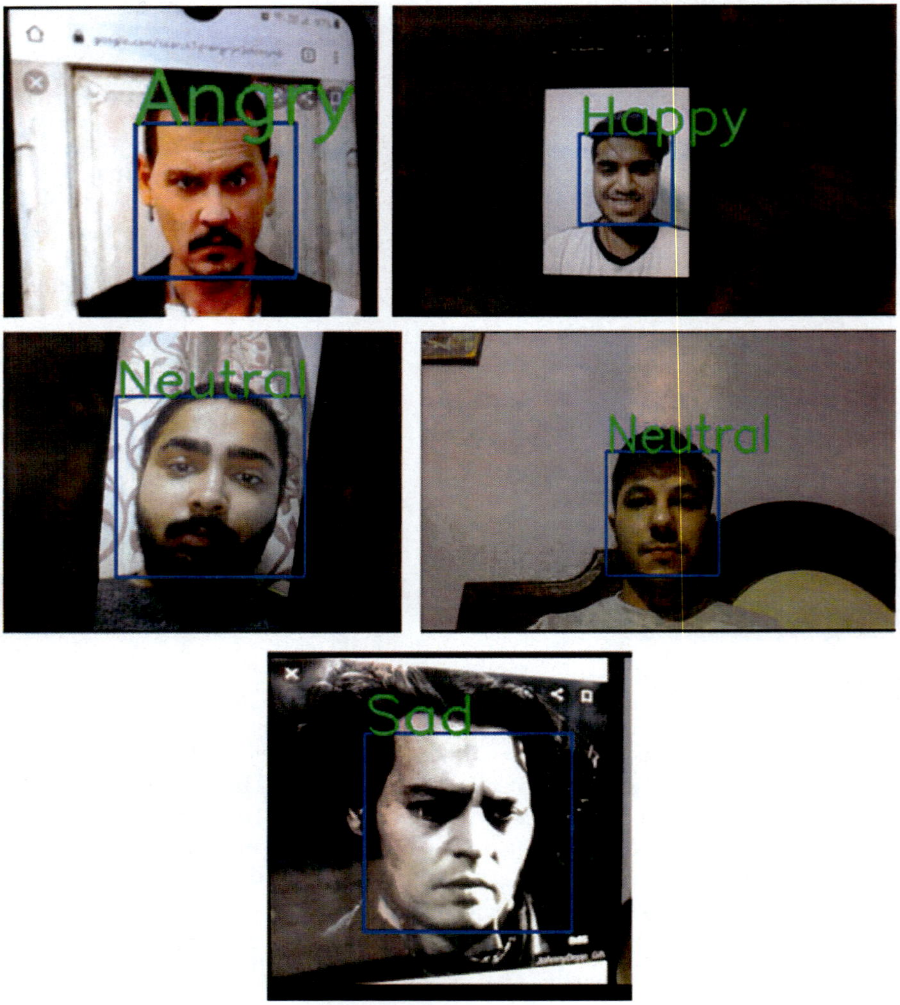

Fig. 10 Different outputs of the proposed work

References

1. https://www.coursera.org/projects/facial-expression-recognition-keras
2. https://medium.com/themlblog/how-to-do-facial-emotion-recognition-using-a-cnn-b7bbae79cd8f
3. https://iopscience.iop.org/article/10.1088/1742-6596/1193/1/012004/pdf
4. https://www.kaggle.com/ashishpatel26/tutorial-facial-expression-classification-keras

Design of IP Core for ZigBee Transmitter and ECG Signal Analysis

K. Sarvesh, S. Hema Chitra, and S. Mohandass

Abstract In the modern day, the footprint of science and technology on human life has been Herculean, and the positive impact brought out to the human society has been unparalleled. In spite of these advancements, cardiac diseases are still a major cause of concern. Hospitals are in need of a constant supervision of the patients with such heart problems. This requires a system which should monitor different patients simultaneously and detect abnormalities. One such solution to the problem is designing a communication system to monitor the ECG signal of several patients simultaneously, from a remote location and detect any abnormality. The objective of this paper is to create an intellectual property (IP) core of ZigBee (IEEE 802.15.4) Transmitter and analysis of ECG signals to detect abnormalities. The ECG signals of the patients are transmitted through the ZigBee Transmitter. The ZigBee Receiver receives the ECG signals from transmitter, and heartbeat analysis is done at the receiver side. Verilog modeling of ZigBee Transmitter is done, IPs of ZigBee Transmitter modules have been created and integrated together to form a ZigBee Transmitter IP. For simulation, ECG signals are taken from MIT/BIH database and are analyzed using Pan–Tompkins algorithm by which abnormalities are detected. ZigBee Transmitter IP core creation is helpful in developing low power, low cost, and reduced time-to-market designs for future applications.

Keywords Intellectual property (IP) core · Heartbeat analysis · ZigBee Transmitter · Pan–Tompkins algorithm

K. Sarvesh · S. Hema Chitra · S. Mohandass (✉)
Department of Electronics and Communication Engineering, PSG College of Technology, Peelamedu, Coimbatore, India
e-mail: smd.ece@psgtech.ac.in

S. Hema Chitra
e-mail: shc.ece@psgtech.ac.in

S. Shakya et al. (eds.), *Mobile Computing and Sustainable Informatics*, Lecture Notes on Data Engineering and Communications Technologies 68,
https://doi.org/10.1007/978-981-16-1866-6_26

1 Introduction

The world is not what it was a few years back. It has evolved drastically, thanks to the technological developments. These advances are apparent in the medical field. The technology involved in the medical field has changed for the better in the last few decades. Access to medical care has become a reality for almost everyone. Many maladies that were once classified incurable are now treatable. This has increased the average life expectancy of the general population. Still the Cardiovascular diseases are one of the major causes of death globally. It is more predominant in the developing and under developed countries. These heart-related diseases cannot be removed completely, but they can be treated with the proper techniques. The existing systems which utilize this method generally use wired medium of communication to transfer the data recorded to a central location to process [1]. But this method is not effective, as the wired mode of communication is slow and are most likely to be affected by powerline noise which can disrupt the important ECG signals, thus rendering it useless. So, wireless communication is preferred over wired communication.

Wi-Fi, ZigBee, and Bluetooth are different protocols for the wireless communication. These protocols help the devices to connect and communicate with each other. Even though all these are used for the data transmission purpose, they are different in their specifications. They are used in different applications based on the requirements. These protocols can be differentiated based on their data rate, range, power consumption, etc. For example, Bluetooth technology has serious disadvantages, as it is limited by its range. So, it cannot cover a large distance, making the central location very close to the router and thereby limiting the number of patients monitored [2]. A possible solution to overcome this problem is by adapting a wireless technology that has better noise immunity, data transfer speed, and a better range. This can be achieved by using ZigBee technology [3]. ZigBee communication protocol is simpler as well as economical than the other wireless technologies. It operates by creating a personal area network (PAN). A ZigBee pair can be configured to act as a transmitter and receiver, and data can be transferred through them [4].

ZigBee also offers the added advantage of noise immunity [5]. This module can be connected to a SPP communication of a personal computer (PC). Thus, data can be transmitted serially. At the receiver end, the ZigBee module is connected to a central system, where the processing is done. During the transfer of the signal, the data can be distorted by noise and other interferences [6]. This might affect the output of the system. Hence, de-noising is carried out to minimize the impact of noise. There are various methods to do this. A popular method is to pass the signal through a combination of low-pass filter and high-pass filter [7]. ZigBee finds its application in home automation and medical data transfer [8, 9]. It is used in low data rate applications which require low-power requirements. IP of the ZigBee Transmitter is created because IPs have the advantages such as re-usability, portability, and reducing design time. By having an IP which has been verified and its results being reliable, it drastically reduces the time-to-market and also provides a platform for the future works that can be done using these predesigned IP cores.

The electrocardiogram (ECG) is the compilation of the electrical activity of the heart. It is measured by the electrodes placed on the skin which measure the tiny changes in the potential caused by the depolarizations of the heart muscles. The rapid depolarization of the left and right ventricles results in the formation of the QRS wave also known as the QRS complex. This part has the highest amplitude in a single heart beat which can be explained by the larger muscle mass of the ventricles in comparison with the atria. The repolarization of the ventricles is the reason behind the formation of the T-wave. The amplitude of the T-wave is generally greater than the P-wave but is definitely shorter than the R-peak. The very low amplitude but normally absent U-wave comes after the T-wave. The hypothesis behind the formation of this wave is the repolarization of the intraventricular septum. The features of the ECG signals are analyzed for detecting abnormalities in the functioning of heart [10].

Among the many algorithms for QRS detection, the ones devised by Pan and Tompkins [11], Fraden and Neuman [12], Gritzali [13], and Okada [14] are most popular. On many conditions, it was inferred that the algorithm devised by the Pan and Tompkins is the most efficient one barring a few. The most appropriate algorithm can be selected by taking into account the context of parameter determination. One major advantage of the Pan–Tompkins algorithm over the others is due to its low complexity with respect to computations and reliability in ambulatory conditions. Quantitative investigation of QRS detection and analysis is discussed in [15].

2 ZigBee Standard and Heartbeat Analysis

ZigBee is a global wireless standard that is used to connect various devices. The name ZigBee is derived from the motion of the honey bees. It operates by creating a wireless personal area network (WPAN). It falls under IEEE 802.15.4 specification. It is created with the intention of achieving low-power consumption than its counterparts. ZigBee technology is economic and simpler. ZigBee is widely used in the industrial, scientific, and the medical fields because of its low-power requirement and its extended battery life. ZigBee supports mesh networking, thereby connecting a lot of devices together and supports the working of a system that needs a complicated circuitry. The range of the module is limited to 10–100 m Line Of Sight (LOS). The ZigBee standard is controlled by the ZigBee alliance, a group of companies that owns the trademark of the standard. It is available for free in the market for non-commercial applications. Microchip and Texas Instrument are one of the prominent companies that produce commercial SOCs that use ZigBee. Typical application of the protocol includes home applications and control, wireless sensor network, industrial control, and medical data collection.

In a ZigBee-based ECG monitoring system, the ECG samples are sent to the ZigBee Transmitter as inputs, and the data is transmitted wirelessly to the ZigBee Receiver in appropriate format. In the receiver side, ECG samples are analyzed for abnormalities as shown in Fig. 1. The ZigBee Transmitter is modeled in this work.

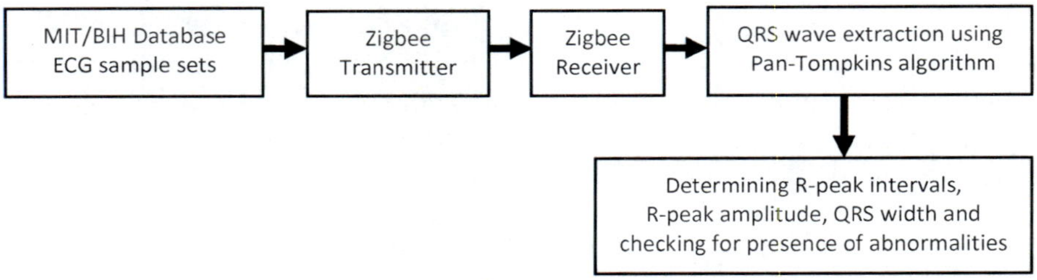

Fig. 1 Process flow of the proposed ECG monitoring system

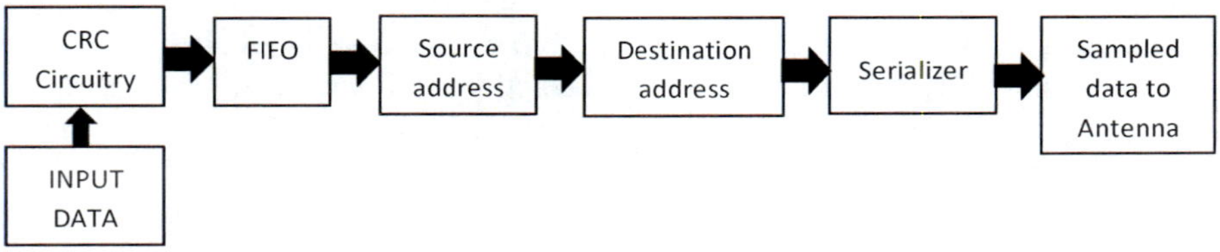

Fig. 2 ZigBee Transmitter

2.1 *ZigBee Transmitter*

The ZigBee Transmitter has the modules such as CRC circuitry, FIFO, source address, destination address, and Serializer. Figure 2 shows the individual modules of a ZigBee Transmitter. The design flow is as follows: (1) The input ECG signals are fed into the CRC circuit. (2) CRC checker works like an LFSR. If signals or data are repeated, then this module indicates the repetition. (3) FIFO makes sure that the data sent are sent out orderly, First In First Out. (4) Source address shows from which address the data sample comes. (5) Destination address shows, where the data will be stored. (6) Serializer makes sure that the data from the destination address are sent in the correct order. (7) Sampled data sent to the antenna which is nothing but ZigBee Transmitter.

2.2 *Pan–Tompkins Algorithm*

The Pan–Tompkins algorithm was developed by Jiapu Pan and Willis J. Tompkins. This Pan–Tompkins algorithm consists of several steps to remove artifacts [11] and to filter the signal to the requirements, in this case the QRS complex. This algorithm is suitable for detecting abnormalities in the ECG signals and has to be incorporated in the ZigBee Receiver. The algorithm is diagrammatically represented in Fig. 3. The series of steps involved are as follows: (i) DC offset cancelation (ii) low-pass filter (LPF) (iii) high-pass filter (HPF) (iv) derivative (v) squaring and (vi) moving window integrator. In this paper, this algorithm is modeled and simulated in MATLAB 2015,

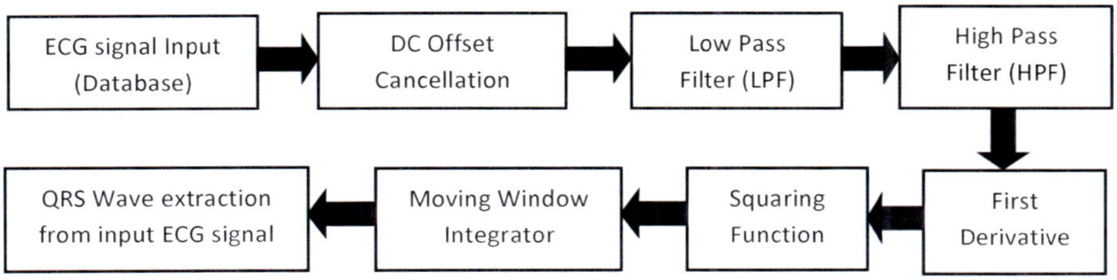

Fig. 3 Pan–Tompkins algorithm

and for simulation purpose, the ECG signals are taken from MIT/BIH database. The ECG samples available are for 10 s of duration.

Baseline wandering is the effect where the base axis of any signal viewed on a screen of a CRO appears to move up or down rather than be straight. This results in the entire signal shifting from its normal base. This kind of an effect is very prominent while extracting ECG measurement. Figure 4a shows the input ECG signal which has time as the *X*-axis and amplitude peak as the *Y*-axis. Figure 4b shows ECG signal after DC offset cancelation. The next step in the Pan–Tompkins algorithm for QRS detection is the application of low-pass filter (LPF) of cut-off frequency 20 Hz. This is done by generating the desired filter coefficients for a cut-off frequency of 20 Hz and then convolving this filter with the signal obtained after DC drift cancelation. The output of LPF is shown in Fig. 4c. The application of a high-pass filter (HPF) to the signal obtained after the LPF constitutes the next step in QRS detection in Pan–Tompkins algorithm. This high-pass filter has a cut-off frequency of 5 Hz. This high-pass filter cascaded with the low-pass filter in the previous step together make the band-pass filter. This cascade, thus, has a pass band of 5–20 Hz. This band-pass filter reduces the effects of muscle noise, 50 Hz interference, T-wave interference, and baseline wander. The ECG signal after HPF is shown in Fig. 4d. The differentiation of the signal obtained after the high-pass filter constitutes the next step. This step takes the slope of the signal as its value and, thus, results in the partial elimination of the P-wave and the T-wave, as they have lesser amplitudes with lesser slopes. In other words, the higher frequency portions of the signal are taken, and the lower frequency parts of the signal are almost eliminated. These higher frequency portions are represented in Fig. 4e.

The samples are then squared after differentiation. This effectively removes all the low amplitude parts of the ECG signal, thus signaling the end of the road for the P-wave and the T-wave in this processing. The function does this by amplifying all the greater slopes and by making all the lesser slope portions of the signal negligible. Thus, effectively what remains of the original signal is only the QRS portion and maybe a few other high-frequency portions of the signal. By the application of the squaring function, all the data points are also made positive. The output after squaring is shown in Fig. 4f. The output of the squaring function produces multiple peaks for a single QRS complex. This is due to the fact that Q and S portions which may have negative values are squared thus making it positive, thus resulting in multiple

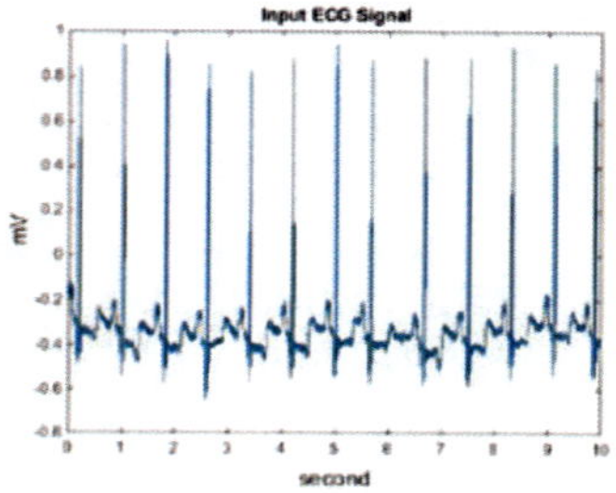

(a) Input ECG Signal

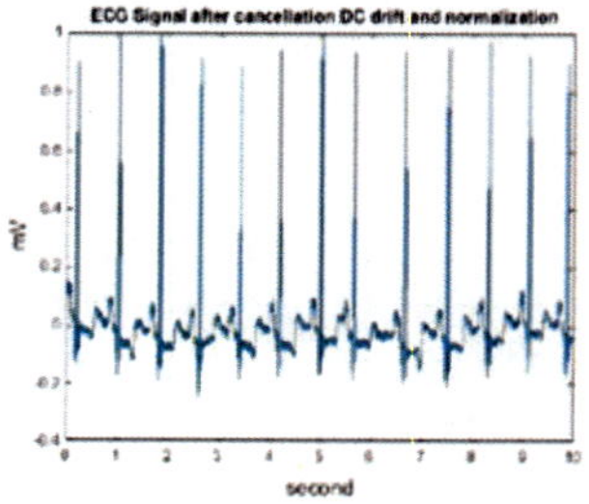

(b) ECG Signal after cancellation of DC drift

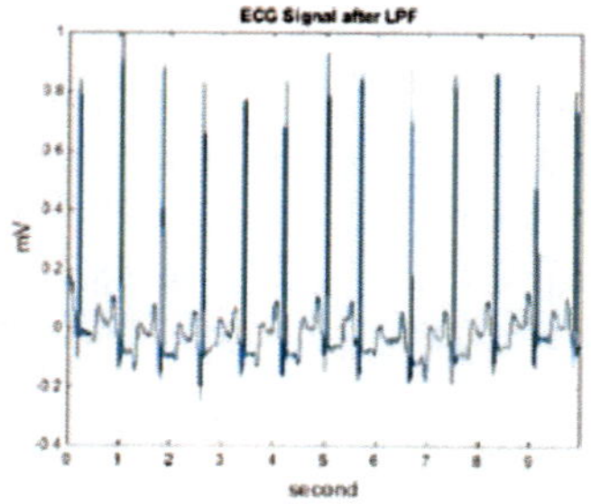

(c) ECG Signal after LPF

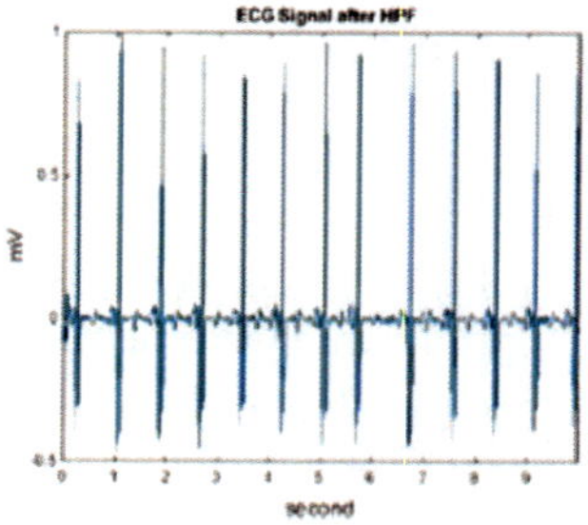

(d) ECG Signal after HPF

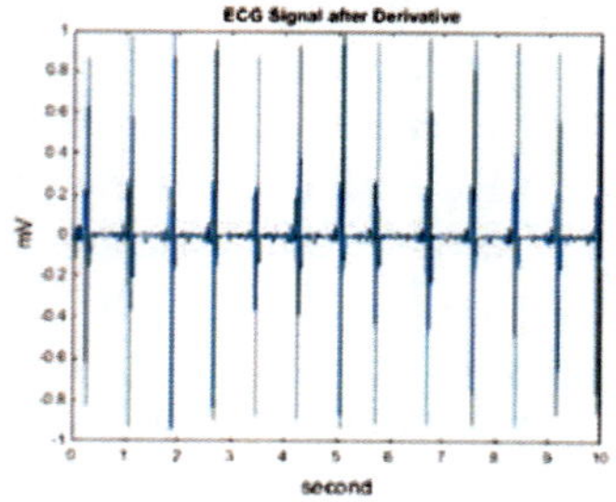

(e) ECG Signal after Derivation

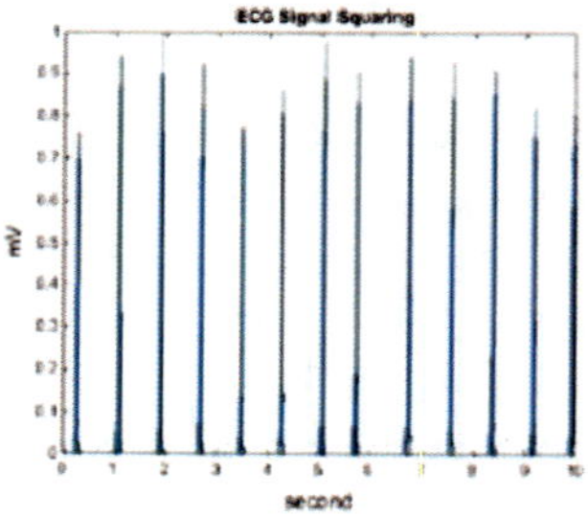

(f) ECG Signal after squaring

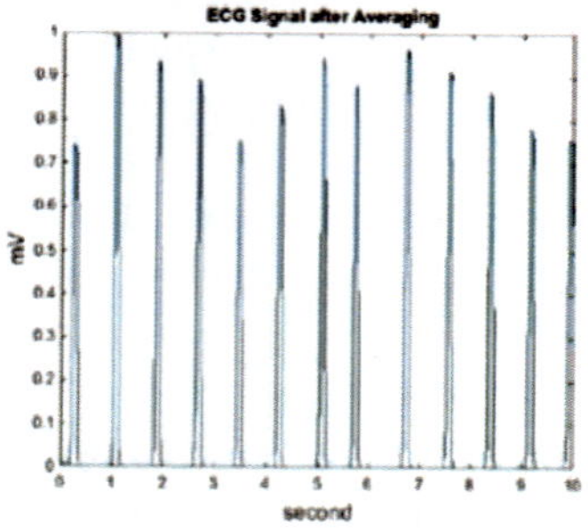

(g) ECG Signal after averaging

Fig. 4 ECG signals at each step of Pan–Tompkins algorithm peaks. The filter coefficients for the moving window integrator are convolved with the signal to obtain the final signal that contains only the QRS complex. The output is represented in Fig. 4g.

2.3 *Abnormality Detection Results*

The three parameters taken into account are the RR interval, R-peak amplitude, and QRS width. The abnormality detection is performed using MATLAB 2015. The **ideal QRS** width should be 0.12–0.20 s duration, **RR-interval** should be 0.6–1.2 s, and **R-peak** should be around 2.5–3.0 mV. Variation in these parameters indicates abnormalities. Table 1 indicates the number of times the abnormalities are detected within the 10 s duration for each different sample. The abnormality is based on different parameters such as QRS width, RR-interval, and R-peak. The ideal abnormality percentage is 10; but in this paper, the percentages 8 and 12 have also been taken into consideration to increase the probability of the abnormalities detected. For instance, sample number 100 is taken, and abnormalities have been detected. As shown in the table, sample 100 appears to be normal according to the parameter QRS width, but abnormalities have been detected based on the other two parameters. Thus, taking into consideration, multiple parameters improve the probability of abnormality detection.

Table 1 Number of abnormalities detected with different margins

Sample	QRS width			RR-interval			R-peak		
	8%	10%	12%	8%	10%	12%	8%	10%	12%
100	0	0	0	7	5	5	2	2	2
102	10	10	9	8	8	6	0	0	0
112	1	1	0	2	2	2	0	0	0
114	0	0	0	4	4	4	0	0	0
121	0	0	0	5	4	4	0	0	0
141	2	2	2	1	1	1	2	2	2
424	2	1	1	0	0	0	2	1	1
1628	0	0	0	1	1	1	1	1	1
1825	0	0	0	3	3	3	2	2	3
2060	0	0	1	3	3	3	2	2	2

3 ZigBee Transmitter and Its IP Creation

The individual modules of the ZigBee Transmitter have been modeled in Verilog and implemented in Xilinx Vivado2017. The IPs of the individual modules are created in Vivado, and then, they are integrated to form the ZigBee Transmitter IP.

3.1 Cyclic Redundancy Checker

The simulation waveforms of the CRC module are shown in Fig. 5a, and the schematic of the IP core created is shown in Fig. 5b. If there is a redundancy in din (data in), there will be a 1 in output register q[15:0], since there is no redundancy a **0000** is obtained.

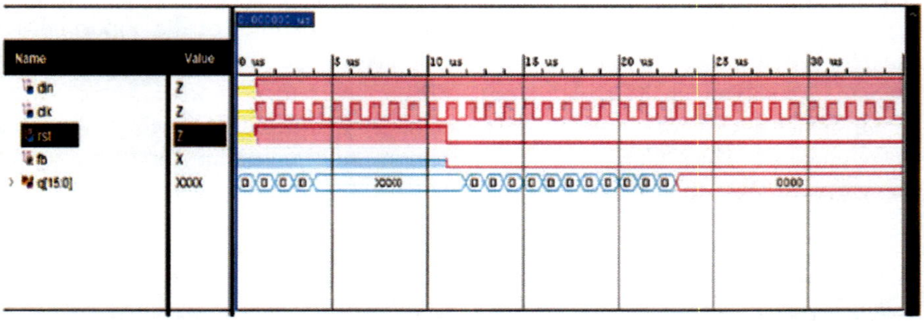

(a)

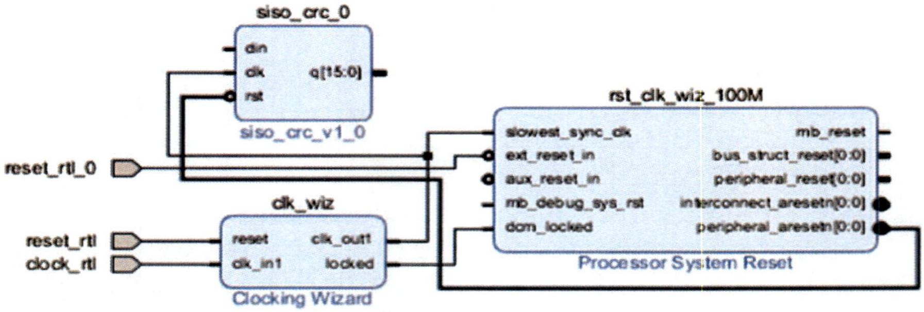

(b)

Fig. 5 CRC results: **a** simulation waveforms **b** IP core

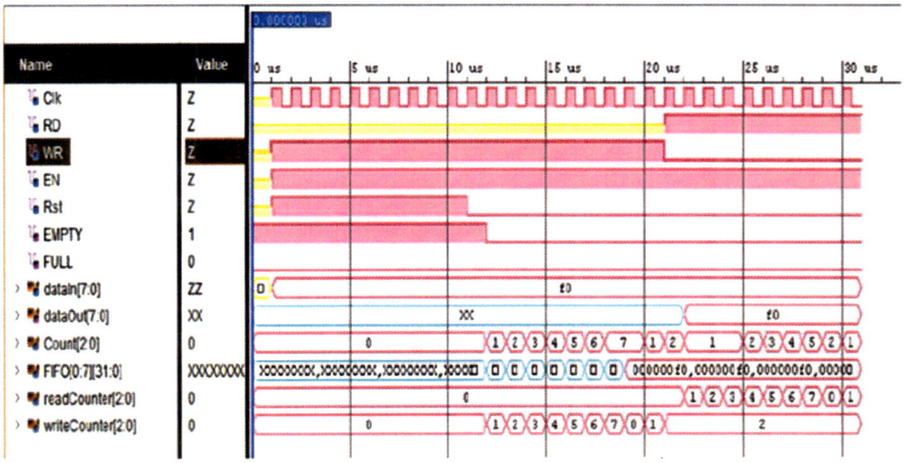

(a)

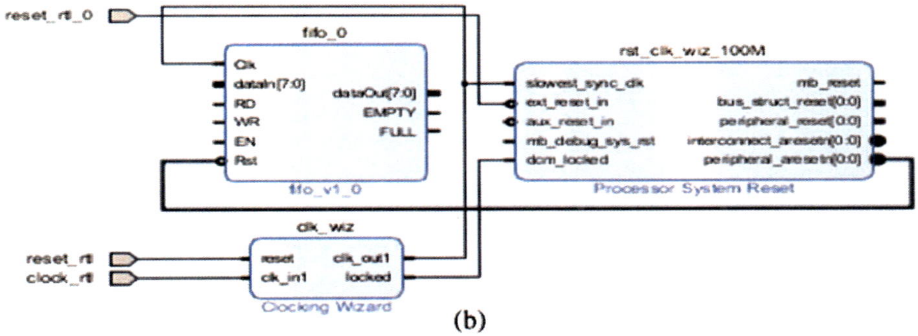

(b)

Fig. 6 FIFO results: **a** simulation waveforms **b** IP core

3.2 *FIFO*

The simulation waveforms and IP of the First In First Out is shown in Fig. 6a, b. The written data $\mathbf{F0}_{\mathbf{H}}$ is read and output signal Dataout reads **FO** as well. Thus, the transmitted data is received on the output. The IP core of FIFO is generated as shown in Fig. 6b.

3.3 *Source Address*

The address from which the data is transmitted is the source address. The simulation waveforms and IP of the source address module are shown in Fig. 7a, b. In Fig. 7a,

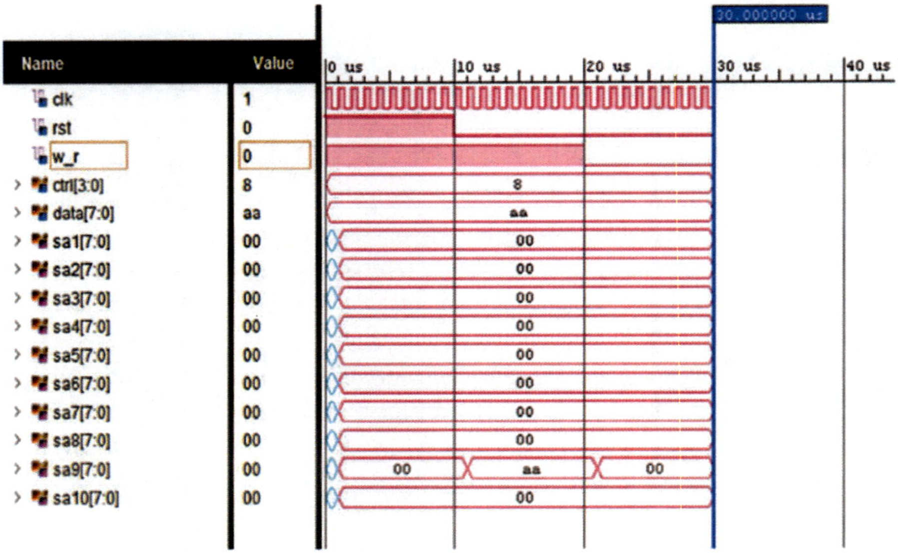

(a)

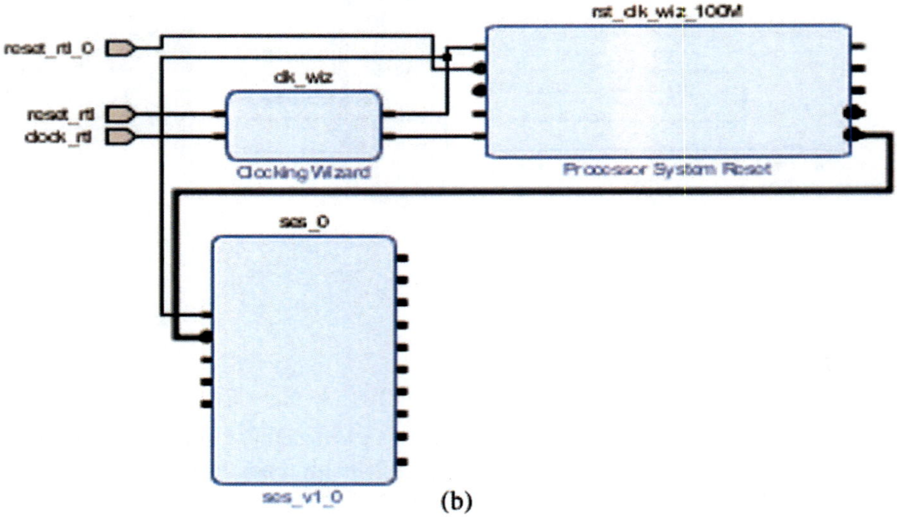

(b)

Fig. 7 Source address results: **a** simulation waveforms **b** IP core

the source_address signal line **sa9** is chosen as the source address because value 8 is given in the Cntrl (Control) signal. The data **AA** is transmitted to the respective address successfully. The IP core of source address module is generated as shown in Fig. 7b.

3.4 Destination Address

The address where the data is stored is the destination address. The loaded data is received as output in the respective destination_address signal line **da10** since we have chosen nine in the Cntrl (Control). The data **AA** is transmitted to the respective address successfully as shown in Fig. 8a. The IP core of destination address module is generated as shown in Fig. 8b.

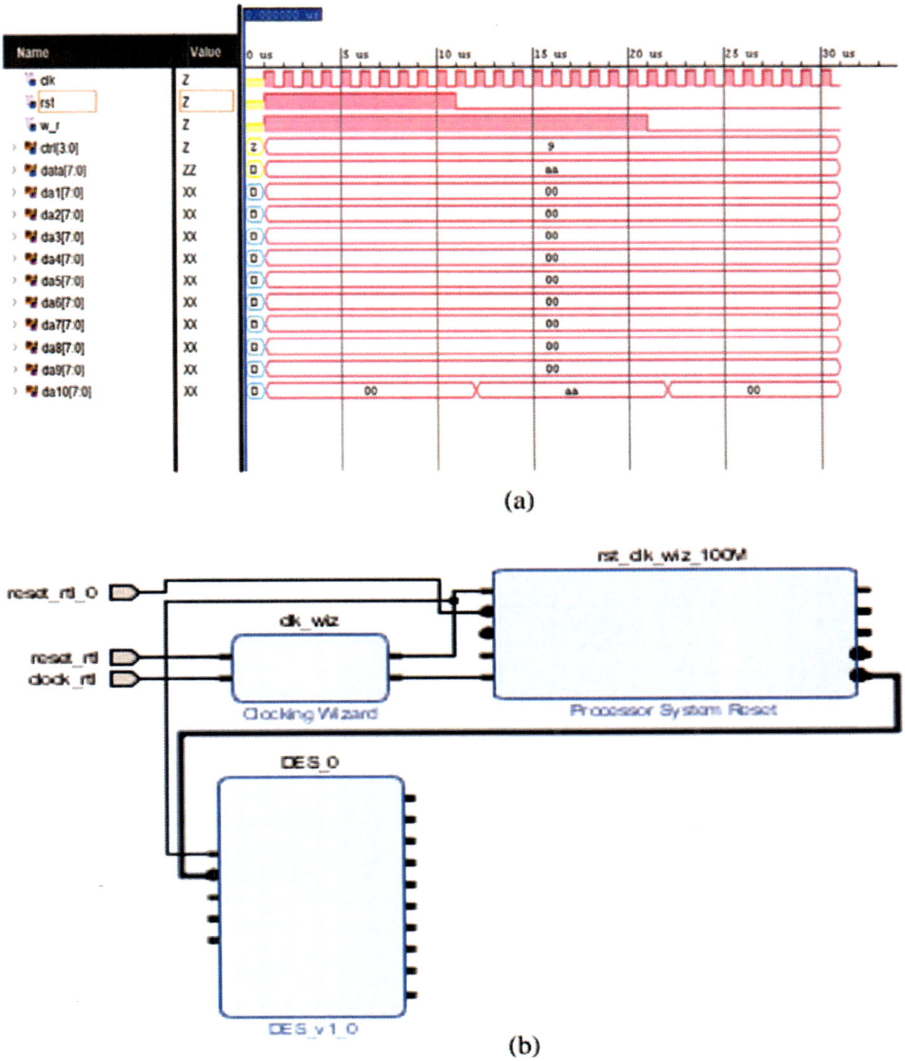

(a)

(b)

Fig. 8 Destination address results: **a** simulation waveforms **b** IP schematic

3.5 Serializer

The serializer makes sure that the data is transmitted serially and in the correct order. The simulation waveforms and IP of the serializer module are shown in Fig. 9a, b. If there is any change in the serial transmission, a high signal in Treg (indicates if signals are not in order) is obtained. Since, here there is no change in order of data, value **CC** is transmitted to the output. The IP core of serializer module is generated as shown in Fig. 9b.

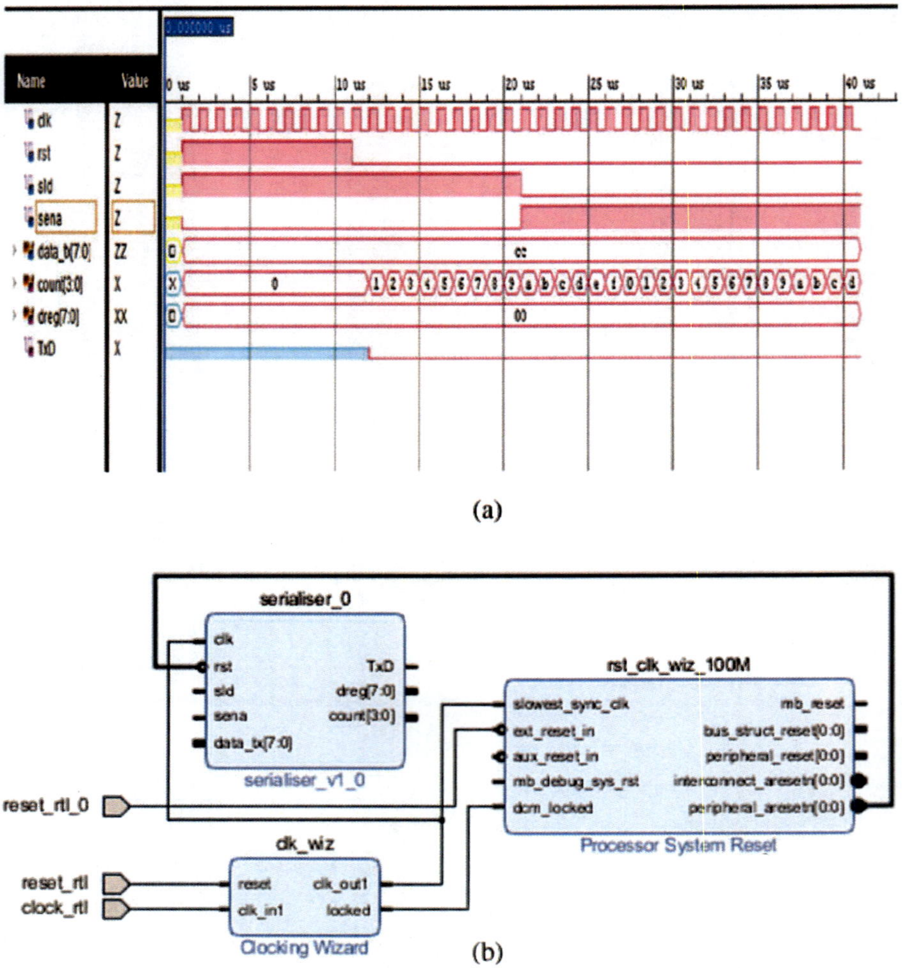

Fig. 9 Serializer results: **a** simulation waveforms **b** IP schematic

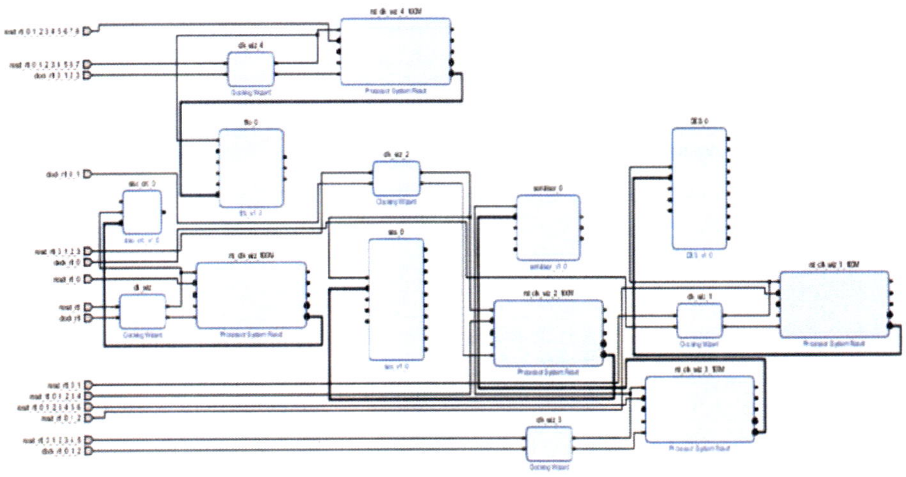

Fig. 10 ZigBee Transmitter IP schematic after integration

3.6 ZigBee Transmitter IP

The individual IPs are instantiated, and the IPs are integrated by the "Run Block Automation" option in Vivado 2017. Once these blocks are automated, the ZigBee Transmitter IP gets created. The ZigBee Transmitter IP has been successfully created as shown in Fig. 10. This IP core functionality has been verified. Now by just instantiating the IP in any FPGA platform, the IP acts as a ZigBee Transmitter. In future, this ZigBee Transmitter IP can be used for other data transmitting applications, and hence, this would reduce the time-to-market. It is portable in multiple FPGA platforms also.

4 Conclusion

The work outlined in this paper involves creation of IP core for a ZigBee Transmitter. Firstly, the individual modules of the ZigBee Transmitter have been modeled using Verilog, and IP cores of individual modules have been created. All the individual modules have been integrated together to form a ZigBee Transmitter IP core. The implementation is done using Xilinx Vivado 2017.1. The ZigBee Transmitter IP can be ported to any FPGA platform and can be instantiated for any other application. Thus, the IP is reusable. Though the IP creation is done in Xilinx Vivado 2017, it can be ported to any other FPGA platform which support IPs. Thus, the IP creation enables low cost, low-power, and reduced design time for portable applications. Heartbeat analysis of ECG signals taken from MIT/BIH database has been implemented using MATLAB. During simulation, ECG parameters such as RR intervals, R-peak amplitude, and QRS width were computed and were compared with a threshold value and abnormalities if any are detected. The usage of three parameters simultaneously increases the probability of detection of abnormalities. The IP core of ZigBee Receiver will be generated in future. Telemedicine and telerobotics will play a vital role in future medical applications [16]. So, the proposed work can be extended to support telemedicine and telerobotics.

References

1. A. Secerbegovic, A. Mujcic, N. Suljanovic, M. Nurkic, J. Tasic, The research mHealth platform for ECG monitoring, in *International Conference on Telecommunications—ConTEL*, Austria (2011), pp. 103–108
2. J.S. Choi, M. Zhou, Performance analysis of Zigbee-based bodysensor networks, in *IEEE International Conference on Systems, Man and Cybernetics (SMC)*, Istanbul, Turkey (2010), pp. 2427–2433
3. E. Jovanov, A. O'Donnel, A. Morgan, B. Priddy, R. Hormigo, Prolonged telemetric monitoring of heart rate variability using wireless intelligent sensors and a mobile gateway, in *The Second Joint 24th Annual Conference and the Annual Fall Meeting of the Biomedical Engineering Society*, Houston, USA (2002), pp. 1875–1876
4. Y.R.K. Paramahamsa, S.U. Basha, S. Mallika, S. Naveed, N. Ambika, Implementation of Zigbee transmitter using Verilog. Int. J. Res. Appl. Sci. Eng. Technol. (IJRASET) **5**(3), 1010–1017 (2017)
5. S.K. Chen, T. Kao, C.T. Chan, C.N. Huang, C.Y. Chiang, C.Y. Lai, T.H. Tung, P.C. Wang, A reliable transmission protocol for ZigBee-based wireless patient monitoring. IEEE Trans. Inf. Technol. Biomed. **16**(1), 6–16 (2012)
6. K. Cai, X. Liang, A Zigbee based mesh network for ECG monitoring system, in *International Conference on Bioinformatics and Biomedical Engineering (iCBBE)*, China (2010)
7. J. Jung, J. Lee, Device access control and reliable data transmission in Zigbee based health monitoring system, in *IEEE 10th International Conference on Advanced Communication Technology* (Electron. & Telecommun. Res. Inst., Seoul, 2008), pp. 795–97
8. U. Varshney, Improving wireless health monitoring using incentive-based router cooperation. IEEE Comput. **41**, 56–62 (2008)
9. Y. Gu, A. Lo, I.G. Niemegeers, A survey of indoor positioning systems for wireless personal networks. IEEE Commun. Surv. Tutorials **11**(1), 13–32 (2009)
10. A. Szczepan'ski, K. Saeed, A. Ferscha, A new method for ECG signal feature extraction, in *International Conference on Computer Vision and Graphics*, Warsaw, Poland (2010), pp. 334–341
11. J. Pan, W.J. Tompkins, A real-time QRS detection algorithm. IEEE Trans. Biomed. Eng. **BME-32**(3) (1985)
12. J. Fraden, M. Neuman, QRS wave detection. Med. Biol. Eng. Comput. **18**, 125–132 (1980)
13. F. Gritzali, Towards a generalized scheme for QRS detection in ECG waveforms. Signal Process. **15**, 183–192 (1988)
14. M. Okada, A digital filter for the QRS complex detection. IEEE Trans. Biomed. Eng. **BME-26**, 700–703 (1979)
15. P. François, A.I. Hernández, G. Carrault, Evaluation of real-time QRS detection algorithms in variable contexts. Med. Biol. Eng. Comput. **43**(3), 379–385 (2005)
16. S. Manoharan, N. Ponraj, Precision improvement and delay reduction in surgical telerobotics. J. Artif. Intell. **1**(1), 28–36 (2019)

Modeling and Simulation of 1 × 4 Linear Phased Array Antenna Operating at 2.45 GHz in ISM Band Applications

Barbadekar Aparna and Patıl Pradeep

Abstract The article introduces modeling and simulations of phased array using analog phase shifter IC 2484 using s_p^2 touchstone files and its equivalent transmission line model. Four-element array with equal amplitude and spacing is designed using a single element and simulated with CST software. The feed network is calibrated when used in phased arrays. This is because the behavior of individual IC 2484 is greatly influenced due to the surrounding components in the array. The phased array is steered for an excitation angle of 90° and 0°. The performance parameters of both simulated methods are compared with the parameters calculated using a standard set of equations. Performance parameters of simulated phase array are measured for 60° and 90° phase excitation angle and radiation patterns in polar and in Cartesian plot are plotted. A good performance antenna with gain 11 dB and maximum directivity 12.8 is designed. Side lobe level (SLL) is less than theoretically calculated value of −13.5 dB.

Keywords Printed dipole antenna · Array antenna · s_p^2 touchstone files · Transmission line · Beam steering · Gain · Directivity

1 Introduction

Phased array is used to reinforce the radiation pattern in a desired direction using phase shifters which change the phase of RF input signals at the output port. With the advancement of the technology, mechanical phase shifters were slowly substituted by PIN diodes, field effect transistors (FET) and ferrite-type components that solved all the problems facing in mechanical phase shifters and improved phase shift accuracy. PIN diodes are extensively utilized in fabrication for high performance RF switches. The bias current of PIN diodes switches from forward to reverse mode and

B. Aparna (✉)
Department of Electronics and Telecommunication, AISSMS IOIT, Pune, India

Department of Electronics and Telecommunication, VIIT, Pune, India

P. Pradeep
Department of Electronics and Telecommunication, JSPM/TSSM'S COE, Pune, India

S. Shakya et al. (eds.), *Mobile Computing and Sustainable Informatics*, Lecture Notes on Data Engineering and Communications Technologies 68,
https://doi.org/10.1007/978-981-16-1866-6_27

generates phase shifter. It needs DC bias circuitry uncoupled from RF signal propagation which leads to complicated structure. FET is another electronic switch which provided large improvement over PIN diodes performance [1]. It is fast switching, low DC power consumption and simple DC bias circuit and decoupled from the RF path at lower frequencies. A great amount of loss in the front end is introduced which is around 1.6 dB at 12–18 GHz [2, 3]. Barium strontium titanate (BST) undergoes reverse polarization when an external electric field is applied. It results in a nonlinear polarization state and is used in BST-based tunable capacitors. Further, BST shows ferroelectric properties and widely used in ferroelectric phase shifters. The advantages of BST phase shifters are low loss, high power-handling capacity, but they have greater loss compared to PIN diodes or FET switches. The cost of fabricating a good quality BST pallet is high, requiring DC high bias voltage and high cost and hence preventing their use in military applications. Due to technological developments, there is a remarkable change in phase shifter design process. Monolithic microwave integrated circuit (MMIC) [4–6] phase shifters are reported recently. Due to less sensitivity to power supply, temperature variations and high insertion loss, their use is limited. Radio frequency microelectro mechanical systems (RF MEMS) [7–11] developed a series of phase shifters such as switched line, loaded line, reflection type and distributed line phase shifters. Switch line phase shifters are simple and easy to design. It switches to unequal transmission line sections and provides the phase shift desired. But these phase shifters need large substrates and introduce high insertion loss. In loaded line phase shifters, the transmission line is loaded with two different reactive impedance networks [12] either inductive or capacitive in nature. These phase shifters are simple, small size with low loss but easy modeling performance. However, these phase shifters are only useful for small phase delays only. Reflective-type phase shifters are used in both variants, i.e., digital and analog. However, analog format is rarely used due to its design complications.

The main drawback of a designed and fabricated phased array system is the complications involved in the design process. Further, the precise performance measurement of the antenna element is extremely problematic because of the indifferent performance due to surrounding effects of elements in the array [13]. The number of modifications and measurements is required during the design of an element. It is extremely costly and time consuming when it is required to design an array of hundred or more of elements [14]. From the above discussions, it is true that all the phase shifters need simulations as an indispensable part of the design process. Giorgetti et al. [15] designed a phased array with six circular antenna elements using HFSS software. In [16], the author presented an antenna design with 12 passive elements using electronically controlled phase shifters. Fingerprinting algorithm is used with directional main beam to improve the accuracy. Kamarudin et al. [17] designed reconfigurable four-element antennas using PIN diodes switches, but this method has to be improved to correct possible error due to simultaneous arrival of the signal. Bui et al. [18] use of 4×4 Butler matrix in CST is used to design a switched beams array. The author has used CST simulation software which presented few accurately simulated elements to present a large (in fact, infinity) number of elements in a regular array. Here, analog phase shifter IC 2484 is simulated by two methods:

transmission line method and use of s_p^2 files of IC 2484. This paper describes new array design functionality in CST simulation software which makes the design of phased arrays easy. The various parameters such as half power beamwidth (HPBW), voltage standing wave ratio (VSWR), gain, return loss, radiation pattern and main lobe direction are measured and compared with fabricated phased array antenna.

2 Design Methodology

The system has been designed to observe the antenna performance when used in a 4 × 1 array and perform beam steering. Various elements are designed, simulated and evaluated before use in the phase array system.

2.1 Design of Printed Dipole Antenna

The printed dipole antenna via balun is designed with input impedance of 50 Ω. FR4 substrate with relative dielectric permittivity (ε_r) of 4.4, low loss (tan δ) and substrate height (h) = 1.6 mm are used for operating frequency 2.45 GHz. The following set of Eqs. (1–5) is used to design the antenna dipole elements as shown in Fig. 1.

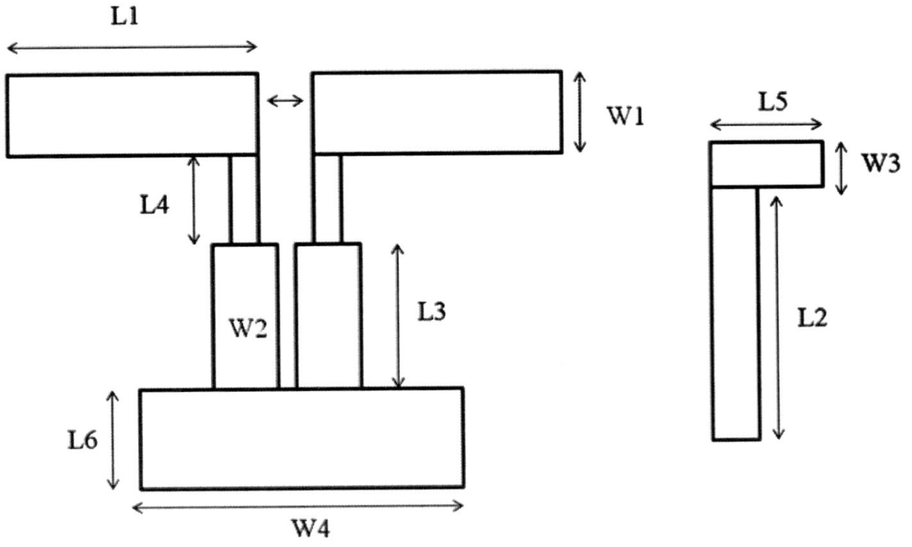

Fig. 1 Configuration of printed dipole antenna via L-shaped balun

The balun is designed using Marchand third-order method using three stubs which is a microstrip to coplanar strip (CPS) line transformation. The balun is an imbalance to balance transforming network with six degrees of freedom corresponding to wavelength βl, where $\beta = 2\Pi/\lambda_g$, and l is the physical length of the stub (50 Ω corresponds to 3 mm strip). The lower limit for the microstrip line is decided by size limits. The ground level is approximately three times as compared to the width of microstrip.

$$\varepsilon_{\text{reff}} = \frac{\varepsilon_r + 1}{2} + \frac{\varepsilon_r - 1}{2}\left[\left(1 + \frac{12h}{W}\right)^{-\frac{1}{2}}\right] \tag{1}$$

$$\Delta L = 0.412 \times h\left\{\frac{\varepsilon_r + 0.3}{\varepsilon_r - 0.258}\right\}\left\{\frac{\frac{W}{h} + 0.264}{\frac{W}{h} + 0.813}\right\} \tag{2}$$

$$W = \frac{c}{2f_r\sqrt{\frac{\varepsilon_r + 1}{2}}} \tag{3}$$

$$L = \left[\frac{c}{2f_r\sqrt{\varepsilon_{\text{reff}}}}\right] - 2\Delta L \tag{4}$$

$$W_b = \frac{\lambda}{4} \tag{5}$$

From the above calculations, we found the values of the parameters required for antenna design and are presented in Fig. 2, and the parameters shown are $L_1 = 20$ mm, $L_2 = 29$ mm, $L_3 = 11$ mm, $L_4 = 8$ mm, $L_5 = 6$ mm, $L_6 = 20$ mm, $W_1 = 6$ mm, $W_2 = 5$ mm, $W_3 = 3$ mm, $W_4 = 70$ mm, gap $g = 3$ mm, via radius = 0.33 mm.

The S_{11} and VSWR of dipole antenna versus frequency are shown in Figs. 3 and 4. The maximum −10 dB return loss bandwidth is 46.40% at operating frequency of 2.2253 GHz. When it operates at 2.815 GHz, −10 dB return loss bandwidth is 36.68%. It shows that the microstrip dipole antenna does not resonate at 2.45 GHz as desired.

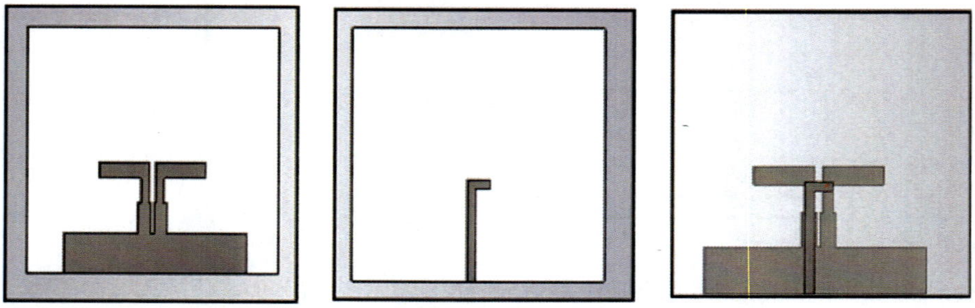

Fig. 2 Simulated dipole antenna

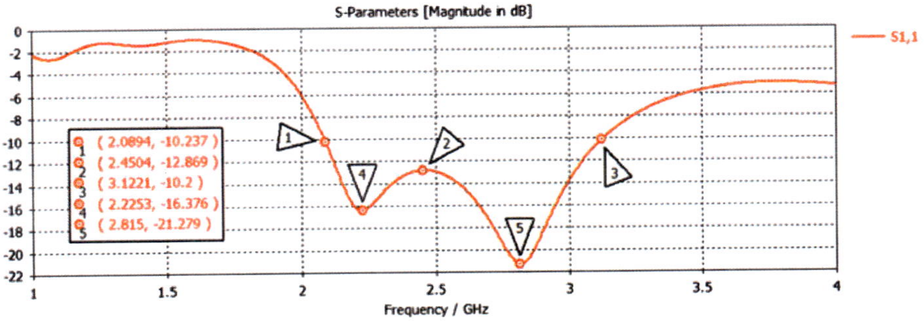

Fig. 3 Return loss (S_{11}) of printed dipole antenna

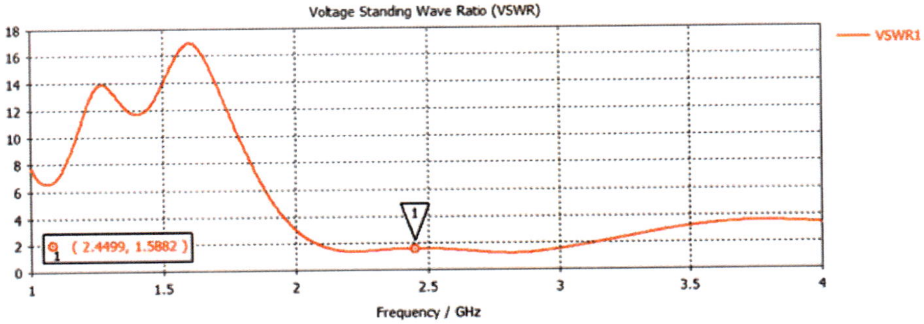

Fig. 4 VSWR of printed dipole antenna

The reflection coefficient calculated from Fig. 4 for VSWR 1.582 is 0.227 which is a desirable parameter for dipole antennas.

The parameters obtained from radiation pattern shown in Fig. 5a–c are far field gain Abs = 5.54 dB, half power beamwidth (HPBW) = 75.2°, directivity = 3.52 and

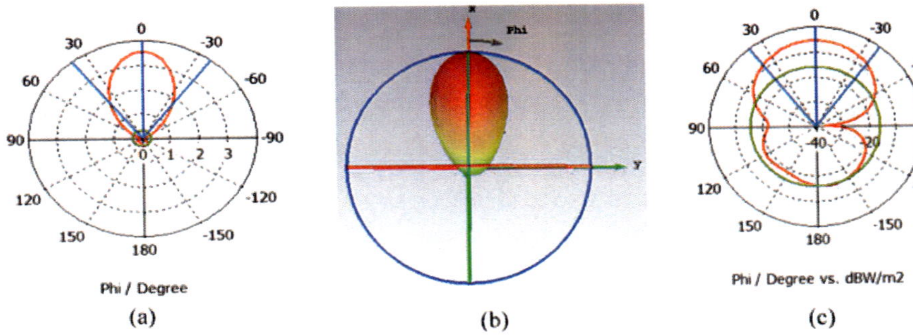

Fig. 5 **a** 2D radiation pattern directivity, **b** 3D radiation pattern, **c** 2D field pattern

side lobe level (SLL) $= -10.1$ dB. Shift in main lobe direction is $-1°$. To improve these parameters, the single dipole antenna should be used in array form.

2.2 Beam Steering Mechanism

The working principle of steering of four-element arrays is as shown in Fig. 6. If the phase delay of $\phi, 2\phi, 3\phi, 4\phi$ is introduced into the first, second, third element, respectively, then beam will shift by an angle θ toward right side. The phase delay of $(N-1)\phi$ is introduced for the N element array. The path difference of d will be introduced due to the phase difference ϕ.

The ratio of path d to wavelength is equal to the phase difference (excitation angle) π to 2π, as 2π period is related with wavelength

$$\frac{2\pi}{\phi} = \frac{\lambda}{d}, \text{ where } \phi = \frac{2\pi d \sin\theta}{\lambda} \tag{6}$$

Thus, by changing the excitation angle ϕ, the steering angle θ can be changed and hence the steering of the beam. The radiation parameters [19] can be calculated from the following set of Eqs. (6–10).

Magnitude of SLL for any array of an isotropic point sources is

$$\text{AF} = \left|\frac{1}{n \times \left(\frac{(2k+1)\pi}{2n}\right)}\right| = \frac{2}{(2k+1)\pi} = 0.212 \quad \text{For } K = 1 \text{(First SLL)}$$

$$\text{SLL in dB} = 20\text{Log}0.212 = -13.5\,\text{dB} \tag{7}$$

Total bandwidth of the main lobe between the first nulls for a long broad side array. i.e.,

$$\text{FNBW} = \gamma_{01} = \frac{2\lambda}{nd}$$

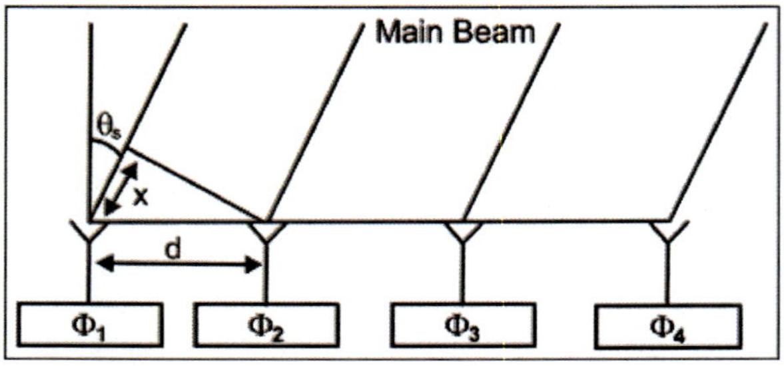

Fig. 6 Beam steering mechanism

$$\mathrm{HPBW} = \frac{\mathrm{FNBW}}{2.25} \tag{8}$$

Scan angle, when the field pattern is maximum in any arbitrary direction

$$\begin{aligned} 0 &= d_r \cos\phi_1 + \delta \\ \delta &= -\pi \cos\phi_1 \end{aligned} \tag{9}$$

where $d_r = 2\pi d/\lambda$ and $d = \lambda/2$.

The gain of an antenna

$$G = kD \tag{10}$$

where $k =$ efficiency factor ($0 \le k \le 1$) dimensionless.

2.3 *Feed Network*

The feed network is a backbone of any phased array system. Typically, it consists of a power divider and phase shifter so as to excite each element of the phased array antenna. The power divider separates out the input signal into four parts with equal phase and amplitude at the output ports. The phase shifter provides the appropriate phase shift to antenna elements.

2.3.1 Wilkinson Power Divider (WPD)

The format of the WPD, as presented in Fig. 7a–b, composed of two $\frac{\lambda}{4}$ transmission lines. The branch impedances $\sqrt{2Z_0}$ and $2Z_0$ are used in the circuit design. The formulae (11–15) used in calculation of WPD parameters. Figure 8 shows two-way and three-stage structure of WPD along with simulated design as shown in Fig. 9.

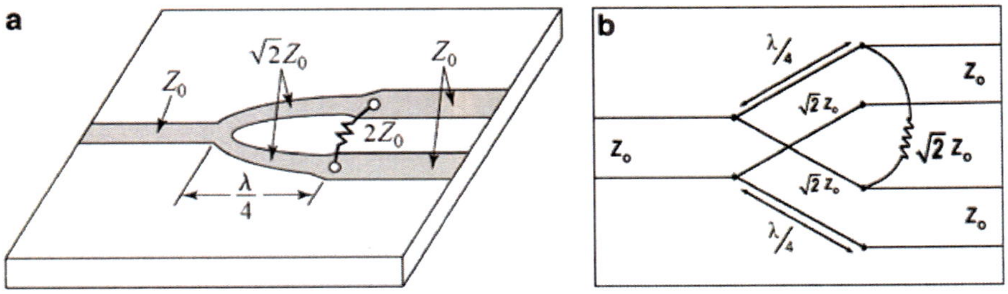

Fig. 7 **a** Layout of WPD, **b** equivalent transmission line circuit

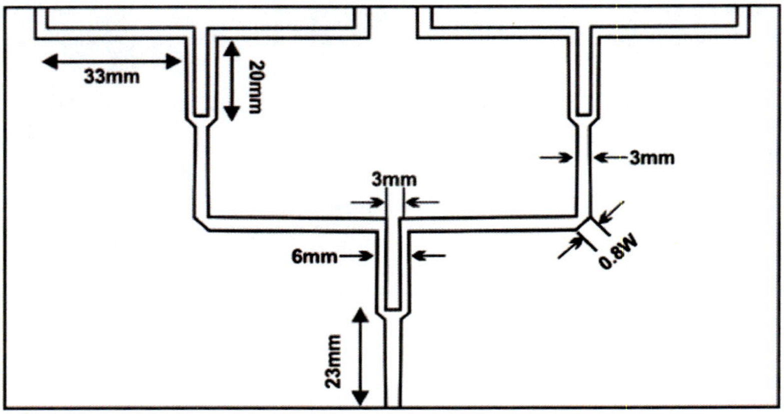

Fig. 8 Configuration of Wilkinson power divider

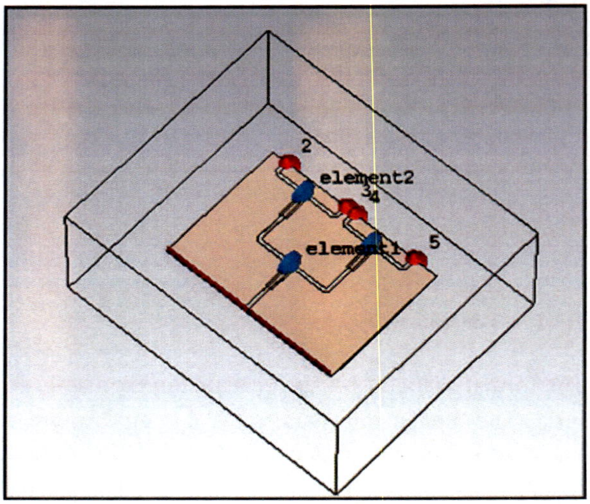

Fig. 9 Simulated design of Wilkinson power divider

Wavelength

$$\lambda_d = \frac{C}{f\sqrt{\varepsilon_r}} \tag{11}$$

Effective dielectric constant $(1 < \varepsilon < \varepsilon_r)$

$$\varepsilon_e = \frac{\varepsilon_r + 1}{2} + \frac{\varepsilon_r - 1}{2}\frac{1}{\sqrt{1 + 12d/W}}, \quad W/d \geq 1$$

$$\varepsilon_e = \frac{\varepsilon_r + 1}{2} + \frac{\varepsilon_r - 1}{2}\frac{1}{\sqrt{1 + 12d/W}} + 0.04\left(1 + \frac{W}{h}\right)^2, \quad W/d \geq 1 \tag{12}$$

W/d Ratio

$$\frac{W}{d}=\left[\frac{2}{\pi}\left[B-1-\ln(2B-1)+\frac{\varepsilon_r-1}{2\varepsilon_r}\ln(B-1)+0.39-\frac{0.6T}{\varepsilon_r}\right]\right],\quad \frac{w}{d}>2 \tag{13}$$

With

$$A=\frac{Z_0}{60}\sqrt{\frac{\varepsilon_r+1}{2}}+\frac{\varepsilon_r-1}{\varepsilon_r+1}\left[0.23+\frac{0.11}{\varepsilon_r}\right] \tag{14}$$

$$B=\frac{377\pi}{2z_0\sqrt{\varepsilon_r}} \tag{15}$$

Figure 8 shows the simulated WPD with a two way and three stage structure connected with 50 Ω transmission lines which arbitrarily adjust the distance between two output ports in order to meet the needs of the dimensions of the phase shifters and element spacing between antennas in the phase array system.

When signal power is reflected due to the discontinuities in a transmission line, the power loss is known as return loss or reflection loss measured in dB. The average return loss from the simulated graphs shown in Fig. 10 is −19.069 dB with operating frequency of 2.45 GHz.

If a device is inserted in a transmission line, the loss of signal is termed as insertion loss. The ideal value of insertion loss is 0 dB, but for all practical purposes, the maximum −3 dB insertion loss admissible is $10\log N=-6$ dB. From the simulation graphs as shown in Fig. 11, the average insertion loss is −7.829 dB at frequency of 2.45 GHz.

Isolation is the insertion loss in the open path between any two output ports. The ideal value of isolation should be infinity. However, the isolation loss reduces as the number of output ports of the power divider for the highest theoretical value permitted

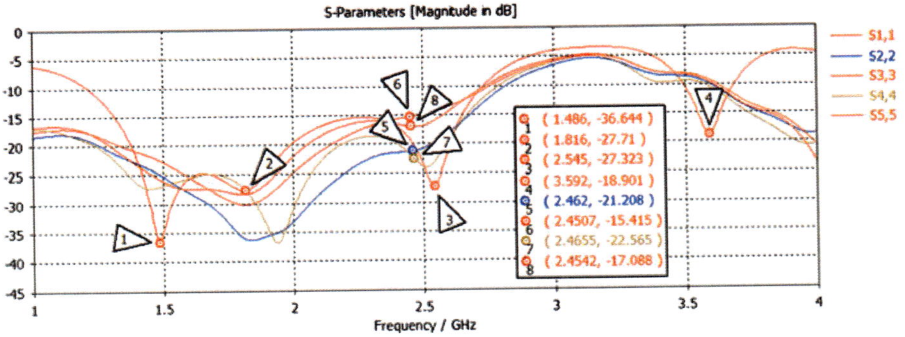

Fig. 10 Return loss (S_{11}) of simulated WPD

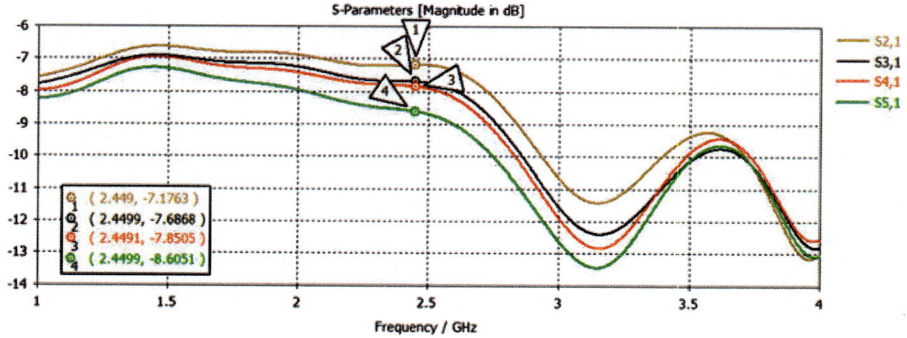

Fig. 11 Insertion loss of WPD

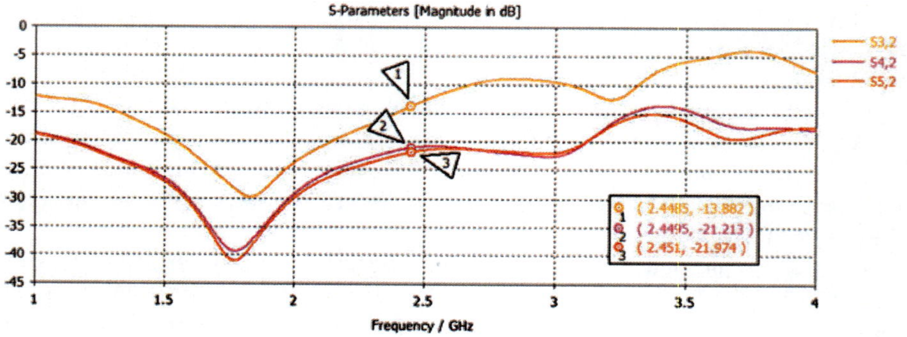

Fig. 12 Isolation loss of WPD

is less than −20 dB. The isolation obtained from the simulated graph presented in Fig. 12 is −19.023 dB which matches the theoretical value.

Thus, the WPD satisfies all the characteristics, namely simple configuration and the matching of impedance and isolation at output ports.

2.4 Simulated Phased Array Antenna Using Transmission Lines

Phase shift is a function of either dielectric constant, permeability, frequency or length. The theoretical phase shift can be determined by Eq. (16)

$$\theta = \sqrt{\epsilon_{\text{reff}}}\, K \Delta L \tag{16}$$

where $K = \frac{2\pi f}{c}$, ΔL = change in transmission line length.

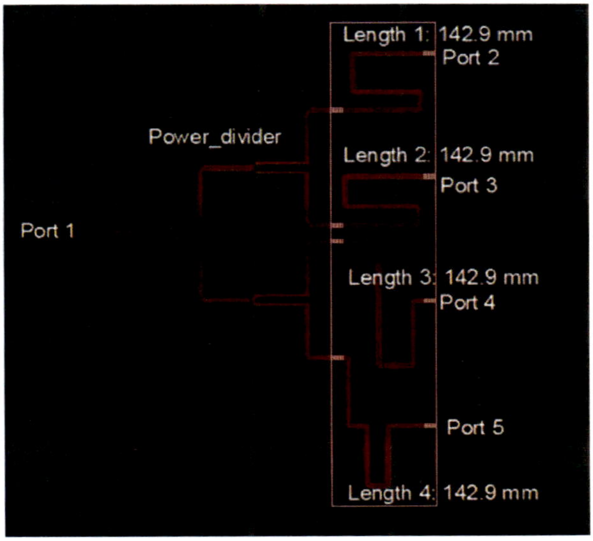

Fig. 13 Layout of transmission lines for scan angle 90°

CST microwave simulation software is used to simulate the feed network including WPD and phase shifter. The simulation is carried out on FR4 substrate with thickness 1.6 mm, dielectric constant ($\varepsilon_r = 4.44$) and loss tangent as 0.01 mm with operating frequency of 2.45 GHz. The calculated value of electrical guided wavelength (λ_g) is 66.8 mm for 360°. Hence, the electrical length of transmission line required for 90° phase shift is 16.7 mm. Thus, with this arrangement as shown in Fig. 12, each antenna element is fed with a phase angle of 90° for excitation angle is 0°. This condition is termed as default position of the phase array. The layout of transmission lines for scan angle of 90° and 60° is shown in Figs. 13 and 14. The 2D and 3D views of simulated phased array antennas are shown in Figs. 15 and 16.

The resulting radiation patterns and Cartesian plots for 60° and 90° scan angle are as shown in Fig. 17a–d.

It proves that the steering of the beam is possible simply by changing the length of the transmission line.

2.5 *Simulated Phased Array Antenna Using S_p^2 File of Phase Shifter IC 2484*

A touchstone file (ASCII proprietary file) format is used for detailing the *n*-port network specifications for frequency-domain linear circuit simulator. The touchstone files incorporate *S*-parameters specified by frequency-dependent linear network specification for 1–10 port components. These files are used in the system design and simulation using the CST software to observe the antenna performance when used in a 4 × 1 array. This system ideally needs a power divider, a phase shifter and an

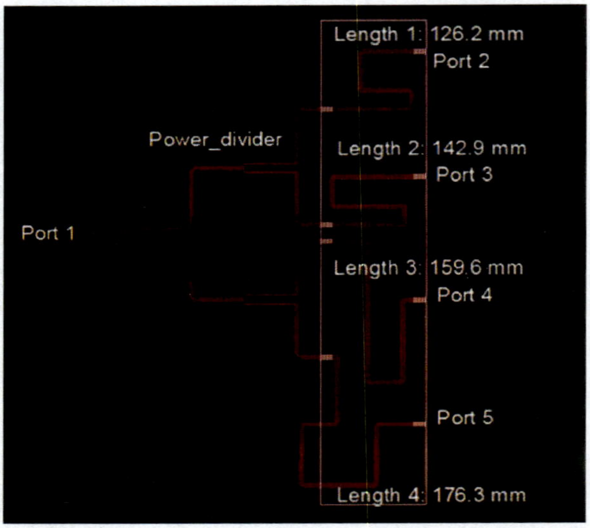

Fig. 14 Layout of transmission lines for scan angle 60°

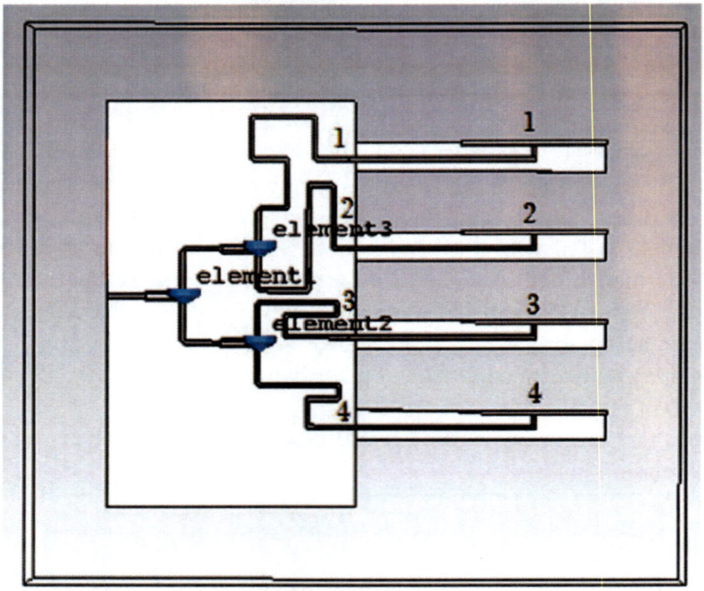

Fig. 15 Simulated phased array antenna using transmission lines

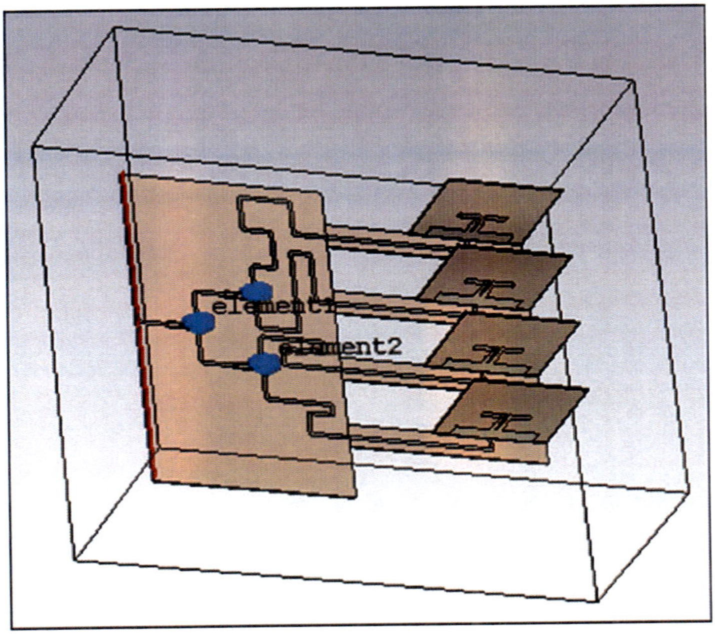

Fig. 16 3D view simulated phased array antenna using transmission lines

array antenna to perform this operation. In this simulation system, each component is simulated individually, and parameters like radiation pattern gain, directivity and half power bandwidth (HPBW) are measured. A schematic view of the simulated phased array antenna using an S_p^2 touchstone file of IC 2484 is as shown in Fig. 18a, and a 3D aspect of the simulated phased array antenna is as shown in Fig. 18b.

The resulting radiation patterns and Cartesian plots for 60° and 90° scan angle are as shown in Fig. 19a–d.

The scanning of beam is implemented by providing each antenna element with progressive phase shift generated in CST simulation software, and radiation patterns for excitation angle of 90° and 0° are as shown in Figs. 17a–d and 18a–d. When the phase difference between two conjugative elements is zero, the radiation pattern is directed toward the broadband side at 90°. This is termed as the default position of the phase array. The drift in DOA observed is of ± 1° as shown in Table 1 when compared with numerical calculations. When the beam is scanned through 60°, the drift for transmission line method is 1°, and for s_p^2 touchstone files, it is + 5°. Further, it is observed that the beam width is widened from 25.45° to 28.8° max when scanned through 60°, and result is 19.50% decrease in directivity when compared with reference to default position of phase array. The gain is also reduced considerably due to different losses involved in the antenna. The prominent loss is due to substrate material and mismatch in feeding networks. The side lobe levels (SLL) usually represent unwanted reduction in undesired directions. The maximum

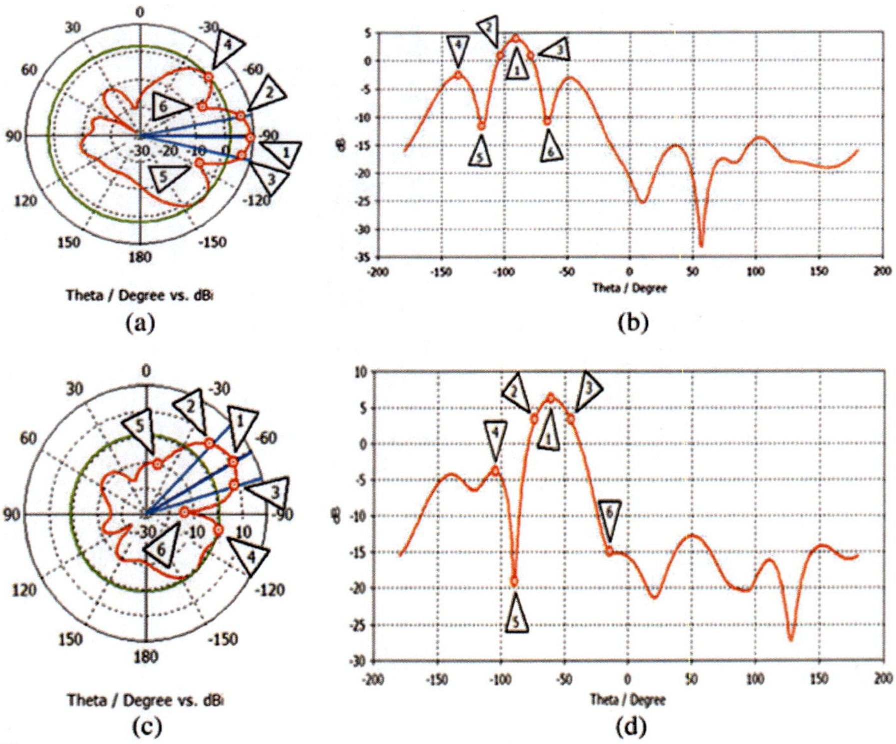

Fig. 17 **a** Simulated radiation pattern at 90°, **b** Cartesian plot at 90°, **c** simulated radiation pattern at 60°, **d** Cartesian plot at 60°

SLL allowed is −13.6 dB, and from Tables 1 and 2, it is observed that the SLL in case of both the methods is permissible.

3 Conclusion

A linear phased array with multiple antenna elements has a lot of potential in communication and radar application. Initially, the modeling of phased array models of antenna with two types of simulation methods using transmission line model and S_p^2 touchstone file of IC 2484 for steering the antenna beam is presented in this paper. Performance analysis based on key elements such as HPBW, FNBW, directivity, gain and SLL is carried out. The drift in DOA observed is ± 1° when compared with numerical calculations. The directivity is decreased by 19.50% when the beam is scanned from default position to 60°. The maximum SLL as compared to main lobe level is −13.6 dB lower. The simulation is limited to one axis and can be expanded

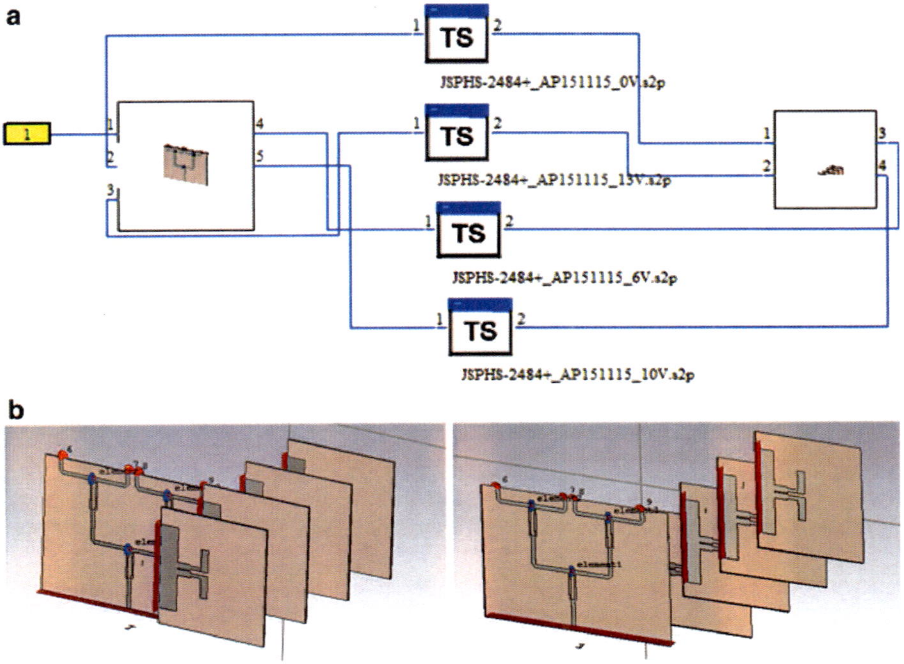

Fig. 18 **a** Schematic view of simulated phased array antenna with S_p^2 file of IC 2484. **b** 3D view of simulated phased array antenna

across the other axis (elevation plane) and carries out the beam steering along both planes at a given angle.

The future work of antenna design would involve replacing the current analog components and algorithms-based systems with a fully IC-based equivalence. This will reduce cost and size of the phased array and enhance the practical use in the consumer market. With the combined use of analog and digital circuits, it is possible to consider a novel, holistic approach to communication and radar systems operating at mm wave frequencies.

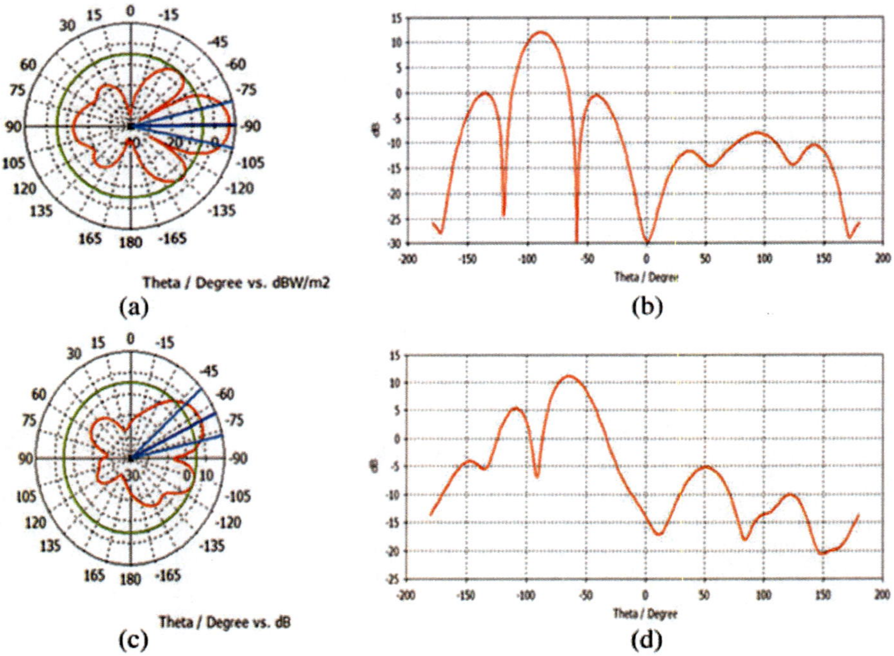

Fig. 19 **a** Simulated radiation pattern at 90°, **b** Cartesian plot at 90°, **c** simulated radiation pattern at 60°, **d** Cartesian plot at 60°

Table 1 Comparison between simulated antenna arrays when estimated DOA is at 90°

Performance parameter	Simulated phased array using		Theoretical analysis
	s_p^2 File of IC 2484	Theoretical analysis	
DoA in degrees	89	91	90
HPBW in degrees	26.4	24.5	25.46
FNBW in degrees	60	52	57.3
Gain in dB	6.99	8.6	14.20
Directivity in dBi	15.9	7.25	15.75
Side lobe level (SLL) in dB	−24.1	−11.5	−13.6

Table 2 Comparison between simulated antenna arrays when estimated DOA is at 60°

Performance parameter	Simulated phased array using		Theoretical analysis
	s_p^2 File of IC 2484	Transmission line	
DoA in degrees	65	61	60
HPBW in degrees	28.8	28.4	25.46
FNBW in degrees	76	76.45	57.3
Gain in dB	11	11	14.20
Directivity in dBi	12.8	12.5	15.75
Side lobe level (SLL) in dB	−15.7	−17.1	−13.6

References

1. V.K. Varadan, K.J. Vinoy, K.A. Jose, *RF MEMS and Their Applications* (Wiley, London, 2003).
2. J. Wallace, H. Redd, R. Furlow, Low cost MMIC DBS chip-sets for phased array application, in *IEEE MTT-S International Microwave Symposium Digest*, Anaheim, CA (1999), pp. 677–680
3. C.F. Campbell, S.A. Brown, A compact 5-bit phase-shifter MMIC for K-band satellite communication systems. IEEE Trans. Microwave Theory Tech. **48**(12), 2652–2656 (2000)
4. I.J. Bahl, D. Conway, L- and S-band compact octave bandwidth 4-bit MMIC phase shifters. IEEE Trans. Microwave Theory Technol. **56**(2), 293–299 (2008)
5. J.-H. Tsai, C.-K. Liu, J.-Y. Lin, A 12 GHz 6-bit switch-type phase shifter MMIC, in *Proceedings of 44th European Microwave Conference* (2014), pp. 1916–1919
6. Q. Zheng, Z. Wang, K. Wang, G. Wang, H. Xu, L. Wang, W. Chen, M. Zhou, Z. Huang, F. Yu, Design and performance of a wideband Ka-band 5-b MMIC phase shifter. IEEE Microwave Wirel. Compon. Lett. **27**(5), 482–484 (2017)
7. S. Dey, S.K. Koul, A.K. Poddar, U.L. Rohde, Reliable and compact 3- and 4-bit phase shifters using MEMS SP4T and SP8T switches. J. Microelectromech. Syst. **27**(1), 113–124 (2018)
8. I. Kalyoncu, E. Ozeren, A. Burak, O. Ceylan, Y. Gurbuz, A phase calibration method for vector-sum phase shifters using a self-generated LUT. IEEE Trans. Circ. Syst. I: Regul. Pap. **66**(4), 1632–1642 (2019)
9. G.L. Tan, R.E. Mihailovich, J.B. Hacker, J.F. DeNatale, G.M. Rebeiz, A very-low-loss 2-bit X-band RF MEMS phase shifter, in *IEEE MTT-S International Microwave Symposium Digest* (2002), pp. 333–335
10. A.S. Abdellatif, A. Abdel Aziz, N. Ranjkesh, A. Taeb, S. Gigoyan, R.R. Mansour, S. Safavi-Naeini, Wide-band phase shifter for mm-wave phased array applications, in *IEEE Global Symposium on Millimeter-Waves (GSMM), Digest* (2015), p. 3
11. T.W. Yoo, J.H. Song, M.S. Park, 360° reflection-type analogue phase shifter implemented with a single 90 branch-line coupler. IET Digit. Libr. **33**(3), 224–226 (1997)
12. D. Uttamchandani, *Handbook of MEMS for Wireless and Mobile Applications* (Woodhead, Cambridge, 2013), pp. 136–175
13. P.W. Hannan, M.A. Balfour, Simulation of a phased array antenna in waveguide. IEEE Trans. Antennas Propag. (1964)
14. P.W. Hannan, P.J. Meier, M.A. Balfour, Simulation of phased array antenna impedance in waveguide. IEEE Trans. Antennas Propag. (Commun.) **AP-11**, 715–716 (1963)
15. G. Giorgetti, A. Cidronali, S.K. Gupta, G. Manes, Single-anchor indoor localization using a switched-beam antenna. IEEE Commun. Lett. **13**(1), 58–60 (2009)

16. M. Rzymowski, P. Woznica, L. Kulas, Single-anchor indoor localization using ESPAR antenna. IEEE Antennas Wirel. Propag. Lett. **15**, 1183–1186 (2016)
17. M.R. Kamarudin, Y.I. Nechayev, P.S. Hall, Onbody diversity and angle-of-arrival measurement using a pattern switching antenna. IEEE Trans. Antennas Propag. **57**(4), 964–971 (2009)
18. T.D. Bui, B.H. Nguyen, Q.C. Nguyen, M.T. Le, Design of beam steering antenna for localization applications, in *2016 International Symposium on Antennas and Propagation (ISAP)* (2016), pp. 956–957
19. J.D. Kraus, R.J. Marhefka, A.S. Khan, *Antenna and Wave Propagation*

Execution Improvement of Intrusion Detection System Through Dimensionality Reduction for UNSW-NB15 Information

P. G. V. Suresh Kumar and Shaheda Akthar

Abstract Computer security is a significant issue in the present networking conditions. These days, with the improvement of Web advances administrations on the planet, the gatecrashers have been expanded promptly. Consequently, the need for an intrusion detection system (IDS) framework in the security of the organization field keeps interlopers from approaching the data. IDS is a gadget or programming application that screens the organization or framework for malevolent action or strategy infringement and sends caution to framework and managers at a legitimate time. IDS screens both inbound and outbound traffics and is used to recognize potential interruptions. In this article, IDSs are assembled utilizing machine learning (ML) methods. IDSs dependent on ML techniques are viable and exact in identifying network assaults. In any case, the presentation of these frameworks diminishes for high-dimensional information spaces. It is necessary to perform a fitting component extraction strategy that can prune a portion of the highlights that does not have an incredible effect in the characterization cycle. Feature choice is expected to choose the ideal subset of highlights that delivers to the whole dataset to expand the precision and the characterization execution of IDS. In this work, the UNSW-NB15 interruption discovery dataset is examined to prepare and test the models. Additionally, a covering-based element is applied to decrease the strategy by utilizing the SVM-RFE calculation for IDS to choose the best subset of highlights. At that point, the two ML approaches are executed to utilize the diminished component space such as decision tree and support vector machine (SVM). The outcomes exhibited that the support vector machine RFE-based component determination strategy takes into consideration techniques; for example, the decision tree is build to test the exactness from 98.89 to 99.99% for the twofold order plot. The investigation results affirmed the adequacy of the proposed feature determination strategy in improving organization IDS.

P. G. V. Suresh Kumar (✉)
Department of CSE, Acharya Nagarjuna University, Guntur, Andhra Pradesh, India

S. Akthar
Department of Computer Science, Government College for Women, Guntur, Andhra Pradesh, India

S. Shakya et al. (eds.), *Mobile Computing and Sustainable Informatics*, Lecture Notes on Data Engineering and Communications Technologies 68,
https://doi.org/10.1007/978-981-16-1866-6_28

Keywords IDS · SVM-RFE · Decision tree · ML · UNSW-NB

1 Introduction

Admittance to the Internet has become a major part of the daily life of individuals with the advancement of computer network innovation. Also, the number of people interfacing with the Web is progressively developing, and this makes the network security a difficult issue. Computer networks are viably utilized for business information handling, training and learning, cooperation, boundless, and information securing. The computer network convention stack that is being used today was created with a rationale to make it straightforward and easy to use. Computer network assaults are a collection of underhanded exercises to harm, deny, decline, or obliterate data and administration inside the computer networks. The universe of figuring is currently faced with the prospect of unintended vacations through numerous assaults and security breaches.

This led to the improvement of a powerful correspondence convention stack. The adaptability of the convention has made it powerless against the assaults dispatched by the interlopers. This makes the prerequisite for the computer network organizations to be consistently observed and ensured. Hence, the need for IDS in the security of the organization field keeps interlopers from approaching the data. IDS is a gadget or programming application that screens the organization or framework for malignant movement or strategy infringement and sends alarms to framework chairmen at the appropriate time [1]. IDS screens both inbound and outbound traffics and is used to recognize potential interruptions. Today, IDSs are utilized to shield computer frameworks from the danger of dangers. Subsequently, the need for IDS in the security of the organization field keeps interlopers from approaching the data [2].

The fundamental task of IDS's is the organizations can be powerless against be assaulted by both interior and outer gatecrashers. An IDS is the major parts of computer security to identify these noxious dangers with the point of shielding frameworks from basic damages and gathering weaknesses [1, 2].

The main issue of network IDS is distinguished known as the administrations or organization assaults, which is called abuse discovery, by utilizing the design of coordinating methodologies. However, an inconsistency location framework recognizes assaults by making profiles of ordinary organizations or framework practices and later distinguishes the assaults if the practices are fundamentally veered off from the typical framework or organization profiles.

Large IDSs are partitioned into signature-based and peculiarity-based recognitions. In mark-based discovery, bundles are filtered to search for a collection of recently distinguished assaults. In any case, in oddity-based location, the interruption discovery frameworks misuse standards of conduct. Indeed, a profile of ordinary conduct is made, and any deviation from this conduct is viewed as a peculiarity.

Beginning interruption discovery frameworks broadly utilized mark-based interruption location strategies. Notwithstanding, they had a high bogus caution rate. Subsequently, developed strategies depend on conduct demonstrating and utilize information mining strategies, factual examinations, and man-made brainpower procedures to distinguish inconsistencies [3]. Machine learning (ML)-based IDSs have arisen as the main frameworks in the interruption location research area. ML gives frameworks the capacity of learning and improving by utilizing the past information. In essence, ML based PC programs don't should be unequivocally designed [4].

Machine learning, a part of man-made brainpower, is a logical control about the plan and advancement of calculations that permit computers to advance practices dependent on observed information, for example, from sensor information or information basis. A significant focal point of AI research is to naturally figure out how to perceive complex examples and settle on smart choices dependent on information. ML has a wide scope of utilizations, including Web indexes, clinical analysis, text and penmanship acknowledgment, picture screening, load anticipating, promoting, deal conclusion, etc. AI strategies can be utilized to discover and bring data by the methods for models which cannot be distinguished effectively by human perception. These components are classifiers which arrange the organization information approaching into the framework to choose whether the movement is an assault or some typical action.

This exploration at the features recognized for UNSW-NB15 dataset distinguishes the huge features and lessens the quantity of features in the dataset. Hence, a subset of huge features in identifying interruption can be proposed by utilizing AI methods. These features at that point can be utilized in the plan of IDS, pursuing robotizing oddity recognition with less overhead [5, 6].

2 Feature Reduction

Feature choice is a cycle that generally utilizes machine learning (ML) and data mining (DM) applications. It means to lessen the impact of the scourge of dimensionality by eliminating excess and unimportant features to improve the indicator execution [1, 7]. The forecast precision relies upon the choice of the subset of features that productively conveys the information. In this manner, there is a need to remove superfluous and repetitive features that contain lower segregation data about the classes. Some informational indexes have high component measurements and even measurements that surpass the number of tests in the informational collection [8, 9]. An unnecessary number of features imply the disadvantageous in the demonstrating cycle. The primary explanation is that there is a great deal of excess data and commotion in high-dimensional information. Therefore, how to remove the useful data from high-dimensional information plays an important role in the subsequent demonstration [3]. Feature selection calculation can viably remove the excess information and commotion information in order to select the most applicable element factors, so it can successfully lessen the components of data [10, 11].

The element determination strategy is categorized into three sorts, namely covering, channel, and mixture techniques [12, 13]. A preprocessing step is performed in the channel technique to choose a subset of features based on the selected models. In this preprocessing step, highlights are chosen without suggesting about their presentation of the classifier. In comparison with the covering technique, channel strategy is therefore considered less cumbersome since the covering strategy tests the part choice strategy based on the classifier results [14, 15]. Even though the covering technique performs well in contrast with channel strategy regarding classifier exactness, when the classifier is changed, the outcomes acquired become not appropriate in a similar circumstance.

2.1 *Support Vector Machine–Recursive Element Disposal (SVM-RFE)*

SVM-RFE is an effective element choice strategy dependent on a retrogressive component erasure technique that has been utilized in numerous applications for high-dimensional information. It positions the features as indicated by the recursive component erasure arrangement dependent on SVM. SVM-RFE [16] is a component determination calculation dependent on SVM. While the SVM learning model is assembled, loads of the features are likewise figured. The fundamental cycle of SVM-RFE is to recursively take out features of low loads, utilizing SVM to decide include loads. Beginning from the full arrangement of features, at every emphasis, the calculation prepares a direct SVM classifier dependent on the excess arrangement of features, positions features as per the outright estimations of feature loads in the ideal hyperplane, and disposes of at least one feature with the most minimal loads. This RFE cycle stops until all features have been taken out or an ideal number of features are reached.

The primary motivation behind SVM-RFE is to register the positioning loads for all highlights and sort the highlights as indicated by weight vectors as the arrangement premise. SVM-RFE is an emphasis cycle of the retrogressive evacuation of features. It includes a set of determination have appeared as follows.

1. Use the UNSW-NB15 dataset to prepare the classifier.
2. Compute the positioning loads for all the features.
3. Delete the component with the littlest weight.

3 Methodology

In this suggested method, the reduction strategy using SVM-RFE is driven as an underlying step toward reducing the number of properties without lacking the primary purpose and target data from the very first data. The subsequent stage is building up an indicator with an improved precision to group information collection utilizing decision tree and SVM calculations. There are various stages in the proposed design for an efficient IDS. Another model is proposed for effective element determination and interruption discovery. This methodology is of the accompanying strides as follows:

Stage 1: Read the UNSW-NB15 dataset.
Stage 2: Preprocess the dataset.
Stage 3: Select the critical features utilizing SVM-RFE calculation.
Stage 4: Perform classification utilizing decision tree and SVM calculation on the dataset to choose the best features.
Stage 5: Evaluate execution of the classifier.

3.1 Decision Tree

Decision trees are a basic recursive structure for communicating a consecutive characterization measure in which a case, depicted by a bunch of properties, is allocated to one of the disjoint arrangements of classes [17]. A decision tree is a tree structure that characterizes an info test into one of its potential classes. Decision trees are utilized to separate information by settling on choice guidelines from the enormous measure of accessible data. A decision tree classifier has a straightforward structure that can minimize the productive group's new information.

Decision trees comprise hubs and leaves. Every hub in the tree includes testing a specific trait, and each leaf of the tree signifies a class. Normally, the test contrasts trait esteem and a steady. Leaf hubs give an order that applies to all examples that arrive at the leaf or a bunch of arrangements or a likelihood conveyance over every conceivable characterization. To arrange an obscure occurrence, it is directed down the tree as indicated by the estimations of the characteristics tried in progressive hubs, and when a leaf is reached, the occasion is characterized by the class allocated to the leaf.

3.2 Support Vector Machine

SVM is another sort of AI strategy dependent on the factual learning hypothesis. On account of conventional advancement and higher precision, SVM has become the examination focal point of the AI people group. SVMs are a set of related administered learning techniques utilized for order and relapse. A few late examinations have detailed that the SVM by and large is equipped for conveying better grouping precision than the other information characterization calculations. SVM is build a measurable learning hypothesis by Mukherjee and Sharma [18] proposed another learning strategy, which is based on a predetermined number of tests in the data contained in the current preparing text to get the best characterization results.

An extraordinary property of SVM will limit the experimental arrangement mistake and amplify the mathematical edge. SVM is also called maximum margin classifiers. SVM depends on the structural danger minimization. SVM maps input vector to a higher-dimensional space where a maximal isolating hyperplane is developed. Two equal hyperplanes are built on each side of the hyperplane that differs the information. The isolating hyperplane is the hyperplane that augments the distance between the two equal hyperplanes. An assumption is made that the greater the edge or distance between these equal hyperplanes the stronger the classifier's speculation blunder [18].

4 Experimental Results

The UNSW-NB15 dataset was utilized to assess the exhibition of the proposed technique. The usage of the proposed technique was composed predominantly of Python 3.6 programming language and a machine learning library known as scikit-realize which has a few learning calculations. The climate on which the application is fabricated is called Jupyter, which is a workbench given by an Anaconda stage. The ML models are assembled, prepared, assessed, and tried on the scikit-learn ML Python system. Scikit-learn is a flexible open-source stage that is built on top of matplotlib, NumPy, and SciPy Python libraries. Besides, classification, regression, and clustering undertakings would all be able to be led utilizing scikit-learn. Concerning the equipment utilized, a PC having an arbitrary access memory of 4 GB RAM, stockpiling size 1 TB with processor type Intel (R) Core (TM) i3 CPU at 2.53 GHz speed, was utilized.

4.1 UNSW-NB15 Dataset

For the test measures, the UNSW-NB15 is used to identify the assaults dataset [19]. The UNSW-NB15 informational index was made at the Cyber Range Lab of the Australian Center for Cyber Security (ACCS) utilizing the AXIA Perfect Storm device to make a half and half of the present-day ordinary and unusual organization traffic [20, 21] distributed in 2015. This dataset is a crossover dataset that comprises a genuine current typical organization activity and engineered adjusted assault. This informational index included 49 features and nine assault families: Fuzzers, Analysis, Backdoors, DoS, Exploits, Generic, Reconnaissance, Shellcode, and Worms. These assaults are dispatched against various workers. The creators got tcpdump hints of the organization traffic toward the start of 2015 for a complete time of 31 h. They made a dataset from these network logs, which comprised 49 highlights for each organization stream.

The UNSW-NB15 dataset deteriorated into two parcels, the preparation and the testing sets, including 175,340 and 82,000 records individually. Features of the

UNSW-NB15 informational index are ordered into six gatherings, in particular, basic features, flow features, time features, content features, additional generated features, and labeled features. Two features are utilized as a mark: attack_cat which demonstrates the class of the assault and the typical state and name which takes 0 for ordinary and 1 for assault. In the preparation information (175,340), there are 56,000 typical classes and 119,341 assaults, while in testing information (82,000), there are 37,000 ordinary classes and 45,332 assaults.

4.2 *Preprocessing*

Additional information is eliminated in both the preparation and testing sets. The UNSW-NB15 dataset accompanies 45 credits and 2 classes ascribes. The majority of the machine learning calculations cannot deal with entire features except if they are changed over to mathematical qualities. The absolute features can be ostensible or ordinal. In this dataset, the qualities of proto, administration, and state and attack_cat are ostensible features. Along these lines, the dataset is encoded from unmitigated to mathematical. The entire features across the datasets are one-hot encoded. Besides, the class property attack_cat reveals the classifications of assaults, and ordinary traffic is taken out before feature determination. Expulsion of the features is named "ID." Property ID is only the quantity of lines in the dataset, so this quality is pointless here.

An arbitrarily selection is performed for a preparation set (70%) and a testing set (30%) of three informational indexes. To approve the forecast aftereffects of the proposed technique, *k*-crease hybrid approval is utilized. Usually, the *k*-crease hybrid approval is used to minimize the blunder that occurred in the correlation of the rightness of different forecast models due to arbitrary checking. The entire arrangement of information is haphazardly separated into k folds with a similar number of cases in each crease. The preparation and testing are performed for *k* occasions, and onefold is chosen for additional testing, while the rest are chosen for additional preparation. The current investigation partitioned the information into tenfolds where onefold was for trying and ninefolds were for preparing for the ten-crease hybrid approval.

4.3 *Feature Selection*

In this analysis, the SVM-RFE is used to include reduction calculation for determining the significant features. There are 45 features which are accessible in the UNSW-NB15 dataset. Subsequent to applying the calculation, merely 22 features were chosen related to the order.

dload, sload, sttl, swin, sloss, dloss, ct_srv_dst, ct_dst_sport_ltm, ct_src_dport_ltm, ct_state_ttl,
ct_dst_src_ltm, is_ftp_login, is_sm_ips_ports, ct_ftp_cmd, ct_flw_http_mthd, response_body_len, dtcpb, tcprtt, dwin, trans_depth, sinpkt and dinpkt.

4.4 Performance Metrics

This article breaks down the impacts of feature determination for decision tree and SVM characterization for the UNSW-NB15 dataset. Accuracy, precision, recall, and F-measure are utilized to assess and resemble the model [17]. There exist a number of measurements to assess ML-based IDS frameworks:

1. **Accuracy**: Accuracy is the measure of effectively arranged examples of the absolute occurrences, characterized as the proportion number of most immeasurable expectations to the complete number of forecasts [17]. It is appropriate to use on a dataset with symmetric objective classes and equivalent class significance.

$$\text{Accuracy} = \frac{\text{TP} + \text{TN}}{\text{TP} + \text{TN} + \text{FP} + \text{FN}} \tag{1}$$

2. **Precision**: Precision articulates how accurate the assumptions of the classifier are because it specifies the measure of genuine positives that were expected by the classifier from all the names relegated to the examples. The level of optimistic expectations that are right is precision.

$$\text{Precision} = \frac{\text{TP}}{\text{TP} + \text{FP}} \tag{2}$$

3. **Recall**: The recall is the extent of positive examples that are accurately anticipated positive. It shows the measure of really anticipated positive classes out of the measure of the entire classes.

$$\text{Recall} = \frac{\text{TP}}{\text{TP} + \text{FN}} \tag{3}$$

$$F\text{-Measure} = \frac{\text{TP} + \text{TN}}{\text{TP} + \text{TN} + \text{FP} + \text{FN}} \tag{4}$$

where

1. True positive (TP) = the number of positive examples accurately anticipated.
2. False negative (FN) = the number of positive examples wrongly anticipated.

3. False positive (FP) = the number of negative examples wrongly anticipated as certain.
4. True negative (TN) = the number of negative examples effectively anticipated.

4.5 *Results and Discussion*

An arbitrarily selection is made for a preparation set (70%) and a testing set (30%) of two UNSW-NS15 informational collections. The experimentation was ordered in two fundamental cycles. In the main stage, select tree and SVM learning calculations is prepared on the first arrangement of highlights was utilized in the analysis. In the second stage, an SVM-RFE is utilized to calculate the satisfactory number of features to recognize the feature chosen. The outcomes obtained from the decision tree and SVM without feature determination and with feature determination are depicted in Tables 1 and 2 with their comparing characterization esteems.

Table 1 Confusion matrix of UNSW-NB15 training (52,602)

Decision tree all features			Decision tree selected features		
Actual attack	Predicted		Actual attack	Predicted	
	Attack	Normal		Attack	Normal
Attack	35,099	710	Attack	35,619	190
Normal	623	16,170	Normal	264	16,529
SVM All features			SVM Selected features		
Actual attack	Predicted		Actual attack	Predicted	
	Attack	Normal		Attack	Normal
Attack	34,890	919	Attack	35,128	681
Normal	2240	14,553	Normal	1573	15,220

Table 2 Confusion matrix of UNSW-NB15 testing (24,700)

Decision tree all features			Decision tree selected features		
Actual attack	Predicted		Actual attack	Predicted	
	Attack	Normal		Attack	Normal
Attack	13,481	127	Attack	13,608	0
Normal	146	10,946	Normal	2	11,090
SVM all features			SVM selected features		
Actual attack	Predicted		Actual attack	Predicted	
	Attack	Normal		Attack	Normal
Attack	13,034	574	Attack	35,128	681
Normal	592	10,500	Normal	1573	15,220

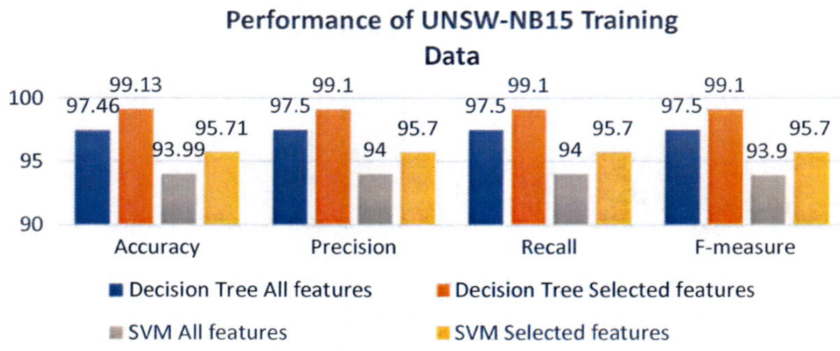

Fig. 1 Performance of UNSW-NB15 training data

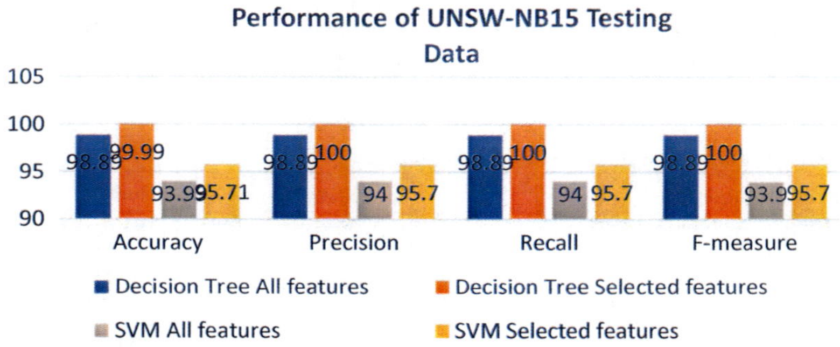

Fig. 2 Performance of UNSW-NB15 testing data

The properties to quantify the demonstration of the techniques (e.g., exactness, accuracy, review, and *F*-measure) are obtained from the disarray lattices and are detailed in Figs. 1 and 2.

The training execution of the data is calculated in Fig. 1, and the results obtained are 97.46 and 99.13% of the accuracy rates over the test set regarding the selection tree order using the entire and decreased features (22) individually.

Figure 2 determines that the training execution of the expected data and the results obtained are 98.89 and 99.99% of the accuracy rates over the test set using the entire and decreased features (22) individually for the option tree grouping. SVM's outcomes achieved a test accuracy of 93.99 and 95.71% using the complete and decreased features separately (22). Therefore, the performance of classifiers with complete features and reduced characteristics has been enhanced in both preparing and testing information collections.

5 Conclusion

This analysis examined the use of the SVM-RFE calculation to include the option of two choice tree ML methods and SVM to execute correct IDSs. To test the demonstration of these techniques, the UNSW-NB15 dataset was used. The paired arrangement was thought of in this work. In addition, the quality preference technique based on SVM-RFE was used over the UNSW-NB15, and thus, 22 ideal features were selected. Specifically, gathered a depiction of the results of the exhibition obtained by the two classifiers with complete features and compared them with those acquired without providing the concept of our proposed method. Initially, over the full element space of the UNSW-NB15 dataset, the investigations were performed using the suggested ML approaches. From that point on, the analyses used the reduced component vector created by the extraction calculation of SVM-RFE features proposed in this work. The test results showed that the use of a diminished component vector has its advantages in terms of expanding the accuracy of recognition on test data. The aim is to use multiclassification in actual effort to identify different assaults.

References

1. L. Dhanabal, S. Shantharajah, A study on NSL-KDD dataset for intrusion detection system based on classification algorithms. Int. J. Adv. Res. Comput. Commun. Eng. **4**(6), 446–452 (2015)
2. H. Nkiama, S.Z.M. Said, M. Saidu, A subset feature elimination mechanism for intrusion detection system. Int. J. Adv. Comput. Sci. Appl. **7**(4), 148–157 (2016)
3. N. Noureldien, I. Yousif, Accuracy of machine learning algorithms in detecting DoS attacks types. Sci. Technol. **6**(4), 89–92 (2016)
4. N. Moustafa, J. Slay, The evaluation of network anomaly detection systems: statistical analysis of the unsw-nb15 data set and the comparison with the kdd99 data set. Inf. Secur. J. **25**(1–3), 18–31 (2016)
5. Y. Li et al., An efficient intrusion detection system based on support vector machines and gradually feature removal method. Expert Syst. Appl. **39**, 424–430 (2012). https://doi.org/10.1016/j.eswa.2011.07.032
6. Y. Pang, Y. Yuan, X. Li, Effective feature extraction in high dimensional space. IEEE Trans. Syst. (2008)
7. V. Boln-Canedo, N. Snchez-Maroo, A. Alonso-Betanzos, *Feature Selection for High-Dimensional Data* (Springer, Berlin, 2016).
8. UNSW-NB15 Dataset; UNSW Canberra at the Australian Defence Force Academy, Canberra, Australia (2015). Available from: https://www.unsw.adfa.edu.au/australian-centre-for-cybersecurity/cybersecurity/ADFA-NB15-Datasets/
9. V. Vapnik, *The Nature of Statistical Learning Theory* (Springer, New York, 1995).
10. S. Lakhina, S. Joseph, B. Verma, Feature reduction using principal component analysis for effective anomaly-based intrusion detection on NSL-KDD. Int. J. Eng. Sci. Technol. **2**(6), 1790–1799 (2010)
11. M. Aggarwal, Amrita, Performance analysis of different feature selection methods in intrusion detection. Int. J. Sci. Technol. Res. **2**(6), 225–231 (2013). ISSN 2277-8616
12. I. Guyon, J. Weston, S. Barnhill, V. Vapnik, Gene selection for cancer classification using support vector machines. Mach. Learn. **46**, 389–422 (2002)

13. H.-J. Liao, C.-H.R. Lin, Y.-C. Lin, K.-Y. Tung, Intrusion detection system: a comprehensive review. J. Netw. Comput. Appl. **36**(1), 16–24 (2013)
14. J. Han, M. Kamber, *Data Mining Concepts and Techniques*. The Morgan Kaufmann Series in Data Management Systems, 2nd edn. (Morgan Kaufmann, San Mateo, 2006)
15. J. Yan, B. Zhang, N. Liu, S. Yan, Z. Chen, Effective and efficient dimensionality reduction for large scale and streaming data processing. IEEETrans. Knowl. Data Eng. **18**(3), 320–333 (2006)
16. G. Ravi Kumar, V.S. Kongara, G.A. Ramachandra, An efficient ensemble based classification techniques for medical diagnosis. Int. J. Latest Technol. Eng. Manage. Appl. Sci. **II**(VIII), 5–9 (2013). ISSN-2278-2540
17. I.S. Thaseen, C.A. Kumar, Intrusion detection model using fusion of chi-square feature selection and multi class SVM. J. King Saud Univ. Comput. Inf. Sci. (2016)
18. S. Mukherjee, N. Sharma, Intrusion detection using Naive Bayes classifier with feature reduction. Procedia Technol. **4**, 119–128 (2012)
19. R. Bace, P. Mell, NIST Special Publication on Intrusion Detection Systems (2001)
20. M.A. Ambusaidi, X. He, P. Nanda, Z. Tan, Building an intrusion detection system using a filter-based feature selection algorithm. IEEE Trans. Comput. **65**(10), 2986–2998 (2016)
21. N. Moustafa, J. Slay, Unsw-nb15: a comprehensive data set for network intrusion detection systems (unsw-nb15 network data set), in *Military Communications and Information Systems Conference (MilCIS)* (IEEE, 2015), pp. 1–6

Weight Optimization in Artificial Neural Network Training by Improved Monarch Butterfly Algorithm

Nebojsa Bacanin, Timea Bezdan, Miodrag Zivkovic, and Amit Chhabra

Abstract Artificial neural networks, especially deep neural networks, are the promising and current research domain as they showed great potential in classification and regression tasks. The process of training artificial neural network (weight optimization), as an NP-hard challenge, is typically performed by back-propagation algorithms such as stochastic gradient descent. However, these types of algorithms are susceptible to trapping the local optimum. Recent studies show that, the metaheuristics-based approaches like swarm intelligence can be efficiently utilized in training the artificial neural network. This paper presents an improved version of swarm intelligence and monarch butterfly optimization algorithm for training the feed-forward artificial neural network. Since the basic monarch butterfly optimization suffers from some deficiencies, improved implementation, that enhances exploration ability and intensification–diversification balance, is devised. Proposed method is validated against 8 well-known classification datasets and compared to similar approaches that were tested within the same environment and simulation setup. Obtained results indicate that, the method proposed in this work outperforms other state-of-the-art algorithms that are shown in the recent outstanding computer science literature.

Keywords Optimization · Neural network · Metaheuristics · Monarch butterfly optimization · Classification.

N. Bacanin (✉) · T. Bezdan · M. Zivkovic
Singidunum University, Danijelova 32, Belgrade 11000, Serbia
e-mail: nbacanin@singidunum.ac.rs

T. Bezdan
e-mail: tbezdan@singidunum.ac.rs

M. Zivkovic
e-mail: mzivkovic@singidunum.ac.rs

A. Chhabra
Guru Nanak Dev University, Amritsar, India
e-mail: amit.cse@gndu.ac.in

S. Shakya et al. (eds.), *Mobile Computing and Sustainable Informatics*, Lecture Notes on Data Engineering and Communications Technologies 68,
https://doi.org/10.1007/978-981-16-1866-6_29

1 Introduction

Artificial neural networks (ANNs) are proven to be very successful in solving various very complex issues in the engineering and computer science domain. These networks can successfully deal with different issues such as image recognition, object detection, clustering, classification tasks, and time-series forecast. For instance, in the domain of medicine, ANNs successfully utilized in helping of classifying medical data and disease detection [6, 9]. The main benefit of neural networks can be credited to the ability to process huge amount of data during the training process and consequently reduces the time required for diagnostics drastically [11].

Neural networks, in general, mimic the behavior of the human brain. Similar to the neurons and connections between them in the brain, an artificial neural network (ANN) simulate this structure. ANN basically consists of layers, and each layer has a specific number of nodes, known as hidden units (or neurons), which are sending signals to other nodes in the network, similar to the neurotransmitters in the organic brain. Other artificial neurons can receive these signals and perform processing. In other words, ANN takes a set of inputs, processes them, and produces the set of outputs as a result. Signals in ANN are denoted by real numbers, while the yield of each individual neuron in the network is calculated according to a defined nonlinear function.

Neurons and their connections in ANN are being adjusted in the process of the network learning for each particular problem, as there is no general solution that fits all possible ANN applications. During the learning process, ANN will start producing results which will be close to the result of the actual output of the network. After certain number of adjustment, the training process is finished when predefined exit criteria has been met. Therefore, one of the most difficult tasks in ANN training is to finding the right parameter values, in other words, to optimize the parameter values which will give the best result. This is a very time-consuming process, which in general consists of large amount of trials and errors. This kind of optimization problem belongs to the class of NP-hard.

The traditional deterministic approaches, such as back-propagation [18], are widely utilized optimization algorithms in ANNs, which has a downside of vanishing gradient and slow convergence. On the other hand, stochastic approaches, such as metaheuristic algorithms, are derivative-free algorithms and they are widely used in complex optimization problems. They will result in an optimal solution (although not guaranteed to be the fittest solution) in an acceptable amount of time. To avoid the issues of deterministic algorithms, researchers have been implemented different metaheuristic-based algorithms [12]. Nature-inspired metaheuristics have proven to be a very powerful tool for solving complex optimization problems, as they can handle large multimodal and noncontinuous search space. Those features make them appropriate for the efficient process of the ANN training. Even though numerous algorithms are proposed for the neural network learning process in the literature, there is still a need for the development of a more efficient algorithm that will overcome the issues of vanishing gradient, slow, early convergence.

Based on the stated above, in this work, we propose an improvement of monarch butterfly optimization (MBO) [28] swarm intelligence-based algorithm applied for neural network training. The MBO was selected because it is one of the novel algorithms, which have already proven to be a very powerful optimizer. Studies show that some algorithms have better exploration, while others are stronger in exploitation process; thus, improvement and hybridization can reduce existing weaknesses of the algorithm and make it more robust for a specific real-world optimization problem.

The rest of the paper is organized as follows: Sect. 2 describes the application of nature-inspired metaheuristics in the ANN optimization, and the proposed method is presented in Sect. 3, Sect. 4 presents the results of the experiment, and Sect. 5 summarizes this research and provides future work.

2 Background and Related Work

Nature-inspired stochastic methods are utilized in training neural networks. These algorithms are divided into two major classes, swarm intelligence metaheuristics (SI) and evolutionary-based algorithms (EA).

The SI metaheuristics are motivated by collective behavior such as swarms of insects, birds, fish, and other animals. When these individuals form large groups, they tend to show highly intelligent behavior when they search for food, for instance. The scientists have made appropriate mathematical models of this kind of behavior and created a vast number of SI algorithms. One of the most notable representatives of SI algorithms is artificial bee colony (ABC) [4, 14], bat algorithm (BA) [26, 30], and firefly algorithm (FA) [13, 29]. ABC algorithm is one of the first SI algorithms, and it was inspired by the behavior of the bee swarm when foraging food sources. It is a population-based algorithm, where food source location denotes a possible solution for the given optimization problem, while the quality of each solution depends of the quantity of nectar of the observed source. On the other hand, BA is based on the echolocation used by the small bats during hunting and finding the prey. Finally, FA models the flashing behavior of swarms of fireflies, where each firefly has its brightness, and less bright fireflies tend to fly towards the brighter ones. Novel SI algorithms such as moth search algorithm (MSA) [19, 27] and monarch butterfly optimization (MBO) [20, 28] showed excellent performances and great potential in solving problems from various domains, both in original and hybridized implementations.

Some of the most recent SI algorithms implementations include enhanced dragonfly algorithm [34], gray wolf optimizer (GWO) [2, 33], flower pollination algorithm (FPA) [5], whale optimization algorithm (WOA) [3], and tree growth algorithm (TGA) [23]. SI algorithms have been successfully applied for global optimization problem [24], and also in various practical real-life engineering problems, such as task scheduling in the cloud computing [7], localization in wireless sensor networks [3], wireless sensor networks energy efficiency, and lifetime optimization [31, 33].

SI metaheuristics have also proven to bee very successful in ANNs optimization. Research published in [32] proposed a hybrid machine learning and SI method for the prediction of new cases of COVID-19, by applying the enhanced beetle antennae search (BAS) algorithm to adaptive network-based fuzzy inference system (ANFIS), with promising results. ANFIS is a special type of the ANN which is based on the fuzzy inference systems; therefore, it represents a hybrid between ANN and fuzzy logic. Improved salp swarm algorithm for the purpose of the neural network training was proposed in [15]. Tree-seed algorithm was utilized in feed-forward multilayer perceptron (MLP) ANNs, as described in [8]. Monarch butterfly optimization (MBO) was used for tuning the hyperparameters of the convolutional neural networks [1]. Finally, in [6] authors proposed a modified firefly algorithm (FA) to design a convolutional neural network (CNN) that will help in the classification process of the glioma brain tumors from MRI.

3 Improved MBO

The monarch butterfly optimization algorithm (MBO) is developed in 2015 by Wang and the method showed great performance on tested 38 standard benchmark functions [28]. The algorithm is inspired by the process of monarch butterfly species migration. In the paper that introduced MBO [28], it was compared to five state-of-the-art algorithms and established as robust NP-hard challenges optimization approach. It is considered to be one of the most promising metaheuristics, with large number of applications in different domains.

The algorithm was inspired by the behavior of the swarms of Northern American monarch butterflies, who migrate every year from the north to the south, on the route that spans for thousands of kilometers. The MBO metaheuristics models monarch butterflies migration and adjusting behaviors by determining search direction of each population member. As in the case of every other nature-inspired approach, in modeling real-world systems, MBO uses the following approximation rules [28]:

1. monarch butterflies populations reside in two locations, denoted as Land1 and Land2;
2. by applying migration operator, offsprings are generated in both locations;
3. greedy selection between parent and offspring is applied; i.e., if offspring has higher fitness than the parent, parent is replaced, otherwise, parent remains in the population;

In two following subsections, migration and adjusting operators are briefly described.

Migration Operator. Initially generated population N of solutions is first divided into two subpopulations, subpopulation1 and subpopulation2, denoted as Sp_1 and Sp_2, respectively. These subpopulations model lands where monarch butterflies are located. In the original MBO, the migration is simplified in a way that the butterflies stay in Land1 for five months, and in Land2 for seven months during the year.

The Sp_1 and Sp_2 are generated by using the following expressions [28]:

$$Sp_1 = \text{ceil}(p \cdot N) \tag{1}$$

$$Sp_2 = N - Sp_1, \tag{2}$$

where parameter p denotes migration ratio of individual in Sp_1 and ceil represents ceiling function.

Migration process for individual i in each iteration is expressed in the following way [28]:

$$x_{i,j}^{t+1} = \begin{cases} x_{r_1,j}^t & \text{if } r \leq p, \\ x_{r_2,j}^t & \text{otherwise} \end{cases} \tag{3}$$

where $x_{i,j}^{t+1}$ denotes j component of ith solution (individual), and $x_{r_1,j}^t$ and $x_{r_2,j}^t$ mark positions of individuals r_1 and r_2, respectively, at iteration t.

As can be seen from Eq. (3), additional parameter r is used for determining if the jth components of the new individual i will be selection from Sp_1 or Sp_2. The additional parameter r is obtained as:

$$r = \text{rand} \cdot \text{peri}, \tag{4}$$

where rand is pseudo-random number between 0 and 1. Suggested value of peri that generated best results is 1.2 [28].

Adjusting Operator. The butterfly adjusting operator is used to perform search process either towards the current fittest individual, or towards the randomly chosen individual from population.

Adjusting operator functions in the following way: if rand $\leq p$, position of individual i will be updated towards the current best solution from the entire population [28]:

$$x_{i,j}^{t+1} = x_{\text{best},j}^{t} \tag{5}$$

where the j-th parameter of the current fittest individual at iteration t is marked by $x_{\text{best},j}^{t}$, and new position of individual i is indicated as $x_{i,j}^{t+1}$.

In the opposite case, if rand $> p$, then the location of individual i will be updated by moving towards the random solution [28]:

$$x_{i,j}^{t+1} = x_{r_3,j}^{t} \tag{6}$$

where $x_{r_3,j}^{t}$ denotes the j-th component of a randomly chosen individual from the Sp_2.

If a random number in range [0,1] is larger from the value of BAR, the position of solution i is updated according to the following expression [28]:

$$x_{i,j}^{t+1} = x_{i,j}^{t+1} + \alpha \times (\mathrm{d}x_j - 0.5), \tag{7}$$

where $\mathrm{d}x_j$ denotes the step size of the i-th individual. $\mathrm{d}x$ is calculated by Lévy flight function as:

$$\mathrm{d}x = Levy(x_i^t) \tag{8}$$

The scaling factor α is calculated as follows:

$$\alpha = S_{\max}/t^2 \tag{9}$$

where $S_{\max}$ denotes the upper bound of step size.

The value of α is responsible for making proper balance between exploration and exploitation.

3.1 *Observed Drawbacks of Original MBO*

Based on extensive practical experiments with classical unconstrained and constrained benchmark instances, as well as on the previously conducted research with the MBO algorithm [21, 22, 25], two most crucial deficiencies of the basic MBO were observed: inappropriate intensification and diversification trade-off and the absence of strong exploration mechanism.

Diversification in the original MBO is conducted in the direction of existing individuals from the population and new, previously undiscovered regions of the search space are not explored. Consequently, in some runs, if existing individuals are located far from the global optimum region, algorithm misses the right part of the search space and worse average values are obtained. If the search process gets "lucky" and random solutions are generated in the promising domain of the search space, then the algorithm manages to converge to optimum or near optimum solution.

Moreover, the lack of exploration power leads to the inadequate exploitation–exploration balance which is biased in favor to exploitation. Utilization of parameter α is used to adjust this balance, which is not enough, and some other mechanisms are required.

3.2 *Proposed Improved Method*

Quasi-reflection-based learning (QRL), as an effective mechanism that can improve both exploration, exploitation–exploration balance and convergence speed, has been recently proposed in the modern literature [10, 16, 17].

The quasi-reflected component j of solution x, x_j^{qr}, is generated in the following way [16]:

$$x_j^{qr} = \text{rnd}\left(\frac{lb_j + ub_j}{2}, x_j\right) \tag{10}$$

where the mean of j-th component's lower bound (lb_j) and upper bound (ub_j) is expressed as $\frac{lb_j + ub_j}{2}$, and $\text{rnd}\left(\frac{lb_j + ub_j}{2}, x_j\right)$ generates random number from uniform distribution in range $\left[\frac{lb_j + ub_j}{2}, x_j\right]$.

At the end of each iteration, for each individual i from the entire population P, that consists of N individuals, quasi-reflected solution x_i^{qr} is created. Ultimately, quasi-reflected population P^{qr} is generated. Finally, both population P and P^{qr} are merged and solutions from such merged population are sorted according to fitness in descending order and best N individuals are select to be propagated to the next iteration.

Inspired by the incorporated enhancements, proposed approach is named MBO with quasi-reflection learning (MBOQRL). Pseudo-code of proposed method is shown Algorithm 1.

Algorithm 1 MBOQRL pseudo-code

```
Initialize random initial population P of N solutions.
Set the initial values for the parameters: p, peri, BAR, and S_max
Evaluate fitness of all N solutions
Initiate t with 1 and determine MaxIter in one run
while t < MaxIter do
  Sort entire population P based on individuals' fitness
  Split population P into two subpopulations (Sp_1 and Sp_2)
  for all i = 1 to Sp_1 do
    for all j = 1 to D do
      if r ≤ p then
        Select individual from Sp_1 and generated new j-th component by applying Eq. (3)
      else
        Select solution from Sp_2 and generated new j-th component by applying Eq. (3)
      end if
    end for
  end for
  for all i = 1 to Sp_2 do
    for all j = 1 to D do
      if r ≤ p then
        Update the j-th component in the newly generated individual by using Eq. (5)
      else
        Select solution from Sp_2 and generate the j-th component of the new individual by using Eq. (6)
        if r > BAR then
          Update the position by using Eq. (7)
        end if
      end if
    end for
  end for
  Create new population by merging the two subpopulations
  Evaluate the new individuals' fitness value
  for i = 1 to N do
    for j = 1 to D do
      if x_{i,j} < (lb_j+ub_j)/2 then
        x^{qr}_{i,j} = x_{i,j} + ((lb_j+ub_j)/2 - x_{i,j}) · rand
      else
        x^{qr}_{i,j} = (lb_j+ub_j)/2 + (x_{i,j} - (lb_j+ub_j)/2) · rand
      end if
    end for
    Add solution x^{qr}_{i,j} to P^{qr}
  end for
  Merge P and P^{qr} (P ∪ P^{qr})
  Select N best solutions as new population P
  Increase the iteration counter by 1
end while
Output the fittest individual and perform visualization
```

4 Simulations and Discussion

This section presents the experimental and the datasets utilized in the simulations. Next, the results of the simulations are showed and the comparative analysis is provided, followed by the discussion.

The experiment is performed on an ANN with one hidden layer. The proposed MBOQRL is evaluated on eight benchmark datasets. Table 1 summarizes each dataset.

Table 1 Dataset details

Dataset	No. of classes	No. of features	No. of instances
Balloon	2	4	36
Cancer	2	9	803
Diabetes	2	8	918
Heart	2	13	330
Ionosphere	2	10	421
Iris	2	4	300
Liver	2	6	415
Wine	3	13	229

The values of the weights and biases are encoded into a one-dimensional vector which is optimized by MBOQRL, and the values are bounded between -1 and 1. Initially, the solutions are generated randomly and the neural network is evaluated by the value of fitness.

The output of one unit in the neural network is calculated as follows:

$$Z_j = a\left(\sum_{i=1}^{n} W_{ij}x_i + b_j\right) \tag{11}$$

where Z_j denotes the output value, the weigh is denoted by W, which is multiplied by the input x, and b is the bias. The activation function is denoted by $a()$. As the activation function, the well-known sigmoid function is utilized, and it is calculated as:

$$a_j(z) = \frac{1}{1 + e^{-z}} \tag{12}$$

The length of the solution vector n_w is calculated according to the following formula:

$$n_w = (n_x \times n_h + n_h) + (n_h \times n_o + n_o) \tag{13}$$

where n_x denotes the size of the input feature. n_h and n_o denote the hidden unit number and output layer size, respectively.

In ANN, the most important indicator of the accuracy of the model is the error measure RMSE (root-mean-square error), which represents the difference between the derived and real values, and helps in choosing the most suitable ANN configuration for the given problem. Therefore, the fitness function is the loss function, the RMSE function, which can be obtained by the following formula:

$$\text{RMSE} = \sqrt{\frac{1}{n}\sum_{i=1}^{n}(y_i - \hat{y}_i)^2} \tag{14}$$

Table 2 MBOQRL control parameters

Parameter	Notation	Value
Population of the solutions	N	50
Subpopulation 1	Sp_1	21
Subpopulation 2	Sp_2	29
Ratio of migration	p	5/12
Period of migration	peri	1.2
Max step size	S_{max}	1.0
Butterfly adjusting rate	BAR	5/12

Table 3 Simulation results and comparison

Dataset	DE	GA	GWO	SSA	STS-SSA	MBO	MBOQRL
Balloon	**1.000**	**1.000**	**1.000**	**1.000**	**1.000**	**1.000**	**1.000**
Cancer	0.984	0.984	0.984	0.984	0.984	0.984	**0.989**
Diabetes	0.743	0.736	0.736	0.758	0.763	0.765	**0.782**
Heart	0.778	0.797	0.797	0.834	**0.852**	0.835	0.848
Iris	0.767	0.867	0.967	0.967	0.967	0.966	**0.977**
Ionosphere	0.472	0.472	0.686	0.698	0.715	0.721	**0.738**
Liver	0.589	0.574	0.633	0.633	0.652	0.662	**0.693**
Wine	0.726	0.726	0.765	0.770	0.782	0.782	**0.783**

where the predicted and the actual class are denoted by $\hat{y}_i$ and y_i.

The size of the population is set to 50, and the number of maximum iteration is 250, the experiment is repeated in 30 runs. The summary of the MBOQRL control parameters is presented in Table 2.

The experiments are conducted on the original MBO and on the proposed improved version of MBO. The simulation results are compared to other metaheuristic methods, where the simulations are performed under similar conditions and on the same datasets. The results of DE, GA, GWO, SSA, and STS-SSA are taken from [15]. We included MBO in the comparative analysis in order to measure the accuracy between the original algorithm and the improved version. The accuracy result on each dataset and comparison with other methods are presented in Table 3 and visually shown in 1. It can be clearly seen from the given results that the improved MBO overcomes the deficiencies of the original approach and achieves superior results for the majority of the datasets included in the research.

The experimental results proves the efficiency of the improved MBOQRL method, which outperforms other metaheuristic-based optimization techniques in terms of classification accuracy. All methods achieved 100% classification accuracy on the Balloon dataset. On the cancer dataset, MBOQRL has the best results, while other methods achieved the same accuracy of 98.4%. In the case of the diabetes dataset, MBOQRL has 78.2% accuracy, and the second-best resulted method is MBO. STS-

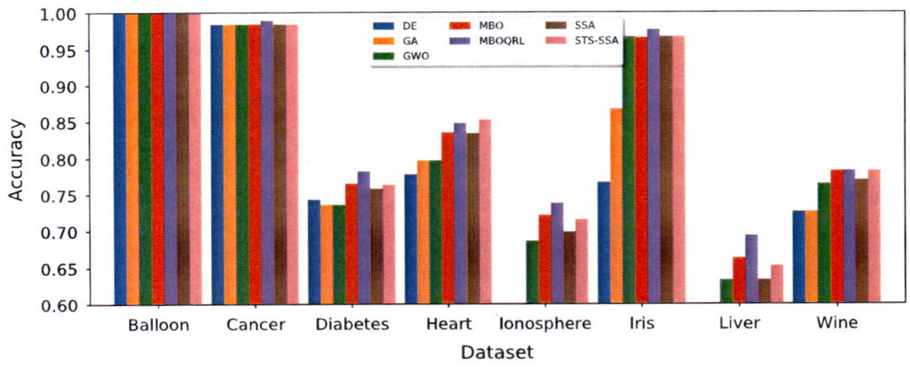

Fig. 1 Accuracy comparison

SSA resulted in the highest accuracy on the Heart dataset, where the accuracy is 85.2% and MBOQRL is the second best with 84.8%. On the Iris dataset, MBOQRL achieved the highest accuracy of 97.7%, while GA, GWO, SSA, and STS-SSA have the same accuracy of 96.7%. MBOQRL has the best result on the Ionosphere, Liver, and Wine dataset as well. For the Liver dataset, in particular, the proposed algorithm obtains the accuracy 3% better than the second best algorithm.

5 Conclusion

This study proposes an improved MBO method, named as MBOQRL for training neural network. The proposed method overcomes the issue of the gradient-based neural network training method, which can make the training difficult to the vanishing gradient problem and slow convergence. We utilize derivative-free approach for finding the optimal value of weights and biases in the neural network learning process. We improved the MBO algorithm by incorporating QRL mechanism, which makes better diversification and intensification balance in the algorithm. The proposed improved MBO was developed with a clear goal to overcome the deficiencies of the original method, which can be clearly seen from the simulation results.

The proposed method is evaluated on eight benchmark datasets, Ballon, Cancer, Diabetes, Heat, Ionosphere, Iris, Liver, and wine. The experimental results are compared to other metaheuristic-based results, to DE, GA, GWO, SSA, and STS-SSA. The proposed MBOQRL approach outperforms other current approaches in terms of accuracy rate.

In future work, we plan to extend the experiment to more datasets and other metaheuristic-based optimization algorithms.

Acknowledgements The paper is supported by the Ministry of Education, Science and Technological Development of Republic of Serbia, Grant No. III-44006.

References

1. N. Bacanin, T. Bezdan, E. Tuba, I. Strumberger, M. Tuba, Monarch butterfly optimization based convolutional neural network design. Mathematics **8**(6), 936 (2020)
2. N. Bacanin, T. Bezdan, E. Tuba, I. Strumberger, M. Tuba, M. Zivkovic, Task scheduling in cloud computing environment by grey wolf optimizer, in *2019 27th Telecommunications Forum (TELFOR)* (IEEE, 2019), pp. 1–4
3. N. Bacanin, E. Tuba, M. Zivkovic, I. Strumberger, M. Tuba, Whale optimization algorithm with exploratory move for wireless sensor networks localization, in *International Conference on Hybrid Intelligent Systems* (Springer, 2019). pp. 328–338
4. N. Bacanin, M. Tuba, Artificial bee colony (ABC) algorithm for constrained optimization improved with genetic operators. Stud. Inform. Control **21**(2), 137–146 (2012)
5. T. Bezdan, M. Zivkovic, M. Antonijevic, T. Zivkovic, N. Bacanin, Enhanced flower pollination algorithm for task scheduling in cloud computing environment, in *Machine Learning for Predictive Analysis* (Springer, 2020), pp. 163–171
6. T. Bezdan, M. Zivkovic, E. Tuba, I. Strumberger, N. Bacanin, M. Tuba, Glioma brain tumor grade classification from mri using convolutional neural networks designed by modified fa, in *International Conference on Intelligent and Fuzzy Systems* (Springer, 2020). pp. 955–963
7. T. Bezdan, M. Zivkovic, E. Tuba, I. Strumberger, N. Bacanin, M. Tuba, Multi-objective task scheduling in cloud computing environment by hybridized bat algorithm, in *International Conference on Intelligent and Fuzzy Systems* (Springer, 2020). pp. 718–725
8. A.C. Cinar, Training feed-forward multi-layer perceptron artificial neural networks with a tree-seed algorithm. Arabian J. Sci. Eng. **45**(12), 10915–10938 (2020)
9. A. Esteva, B. Kuprel, R.A. Novoa, J. Ko, S.M. Swetter, H.M. Blau, S. Thrun, Dermatologist-level classification of skin cancer with deep neural networks. Nature **542**(7639), 115–118 (2017)
10. A.A. Ewees, M. Abd Elaziz, E.H. Houssein, Improved grasshopper optimization algorithm using opposition-based learning. Expert Syst. Appl. **112**, 156–172 (2018). https://doi.org/10.1016/j.eswa.2018.06.023, http://www.sciencedirect.com/science/article/pii/S0957417418303701
11. H. Faris, I. Aljarah, N. Al-Madi, S. Mirjalili, Optimizing the learning process of feedforward neural networks using lightning search algorithm. Int. J. Artif. Intell. Tools **25**(06), 1650033 (2016)
12. H. Hakli, M.S. Kiran, An improved artificial bee colony algorithm for balancing local and global search behaviors in continuous optimization. Int. J. Mach. Learn. Cybern. 1–26 (2020)
13. R.C. Hrosik, E. Tuba, E. Dolicanin, R. Jovanovic, M. Tuba, Brain image segmentation based on firefly algorithm combined with k-means clustering. Stud. Inform. Control **28**(2), 167–176 (2019). https://doi.org/10.24846/v28i2y201905
14. D. Karaboga, B. Basturk, On the performance of artificial bee colony (abc) algorithm. Appl. Soft Comput. **8**(1), 687–697 (2008)
15. N. Panda, S.K. Majhi, Improved salp swarm algorithm with space transformation search for training neural network. Arabian J. Sci. Eng. 1–19 (2019)
16. Q. Fan, Z. Chen, Z. Xia, A novel quasi-reflected harris hawks optimization algorithm for global optimization problems. Soft Comput. (2020). https://doi.org/10.1007/s00500-020-04834-7
17. S. Rahnamayan, H.R. Tizhoosh, M.M.A Salama, Quasi-oppositional differential evolution, in *2007 IEEE Congress on Evolutionary Computation*, pp. 2229–2236 (2007). https://doi.org/10.1109/CEC.2007.4424748
18. D.E. Rumelhart, G.E. Hinton, R.J. Williams, Learning representations by back-propagating errors. Nature **323**(6088), 533–536 (1986)
19. I. Strumberger, E. Tuba, N. Bacanin, M. Beko, M. Tuba, Hybridized moth search algorithm for constrained optimization problems, in *2018 International Young Engineers Forum (YEF-ECE)*, pp. 1–5 (May 2018). https://doi.org/10.1109/YEF-ECE.2018.8368930

20. I. Strumberger, E. Tuba, N. Bacanin, M. Beko, M. Tuba, Modified monarch butterfly optimization algorithm for rfid network planning, in *2018 6th International Conference on Multimedia Computing and Systems (ICMCS)*, pp. 1–6 (2018). https://doi.org/10.1109/ICMCS.2018.8525930
21. I. Strumberger, M. Sarac, D. Markovic, N. Bacanin, Hybridized monarch butterfly algorithm for global optimization problems (2018)
22. I. Strumberger, E. Tuba, N. Bacanin, M. Beko, M. Tuba, Modified and hybridized monarch butterfly algorithms for multi-objective optimization, in *International Conference on Hybrid Intelligent Systems* (Springer, 2018), pp. 449–458
23. I. Strumberger, E. Tuba, M. Zivkovic, N. Bacanin, M. Beko, M. Tuba, Dynamic search tree growth algorithm for global optimization, in *Doctoral Conference on Computing, Electrical and Industrial Systems* (Springer, 2019). pp. 143–153
24. I. Strumberger, E. Tuba, M. Zivkovic, N. Bacanin, M. Beko, M. Tuba, Dynamic search tree growth algorithm for global optimization, in *Technological Innovation for Industry and Service Systems*, ed. by L.M. Camarinha-Matos, R. Almeida, J. Oliveira (Springer International Publishing, Cham, 2019), pp. 143–153
25. I. Strumberger, M. Tuba, N. Bacanin, E. Tuba, Cloudlet scheduling by hybridized monarch butterfly optimization algorithm. J. Sensor and Actuator Networks **8**(3), 44 (2019). https://doi.org/10.3390/jsan8030044
26. M. Tuba, A. Alihodzic, N. Bacanin, in *Cuckoo Search and Bat Algorithm Applied to Training Feed-Forward Neural Networks* (Springer International Publishing, Cham, 2015), pp. 139–162. https://doi.org/10.1007/978-3-319-13826-8_8
27. G.G. Wang, Moth search algorithm: a bio-inspired metaheuristic algorithm for global optimization problems. Memetic Comput. (2016). https://doi.org/10.1007/s12293-016-0212-3
28. G.G. Wang, S. Deb, Z. Cui, Monarch butterfly optimization. Neural Comput. Appl. 1–20 (2015). https://doi.org/10.1007/s00521-015-1923-y
29. X.S. Yang, Firefly algorithms for multimodal optimization, in *Stochastic Algorithms: Foundations and Applications*. ed. by O. Watanabe, T. Zeugmann (Springer, Berlin Heidelberg, Berlin, Heidelberg, 2009), pp. 169–178
30. X.S. Yang, A. Hossein Gandomi, Bat algorithm: a novel approach for global engineering optimization. Eng. Comput. **29**(5), 464–483 (2012)
31. M. Zivkovic, N. Bacanin, E. Tuba, I. Strumberger, T. Bezdan, M. Tuba, Wireless sensor networks life time optimization based on the improved firefly algorithm, in *2020 International Wireless Communications and Mobile Computing (IWCMC)* (IEEE, 2020). pp. 1176–1181
32. M. Zivkovic, N. Bacanin, K. Venkatachalam, A. Nayyar, A. Djordjevic, I. Strumberger, F. Al-Turjman, Covid-19 cases prediction by using hybrid machine learning and beetle antennae search approach. Sustain. Cities Soc. **66**, 102669 (2021). https://doi.org/10.1016/j.scs.2020.102669, http://www.sciencedirect.com/science/article/pii/S2210670720308842
33. M. Zivkovic, N. Bacanin, T. Zivkovic, I. Strumberger, E. Tuba, M. Tuba, Enhanced grey wolf algorithm for energy efficient wireless sensor networks, in *2020 Zooming Innovation in Consumer Technologies Conference (ZINC)* (IEEE, 2020). pp. 87–92
34. M. Zivkovic, T. Zivkovic, K. Venkatachalam, N. Bacanin, Enhanced dragonfly algorithm adapted for wireless sensor network lifetime optimization, in *Data Intelligence and Cognitive Informatics* (Springer, 2021), pp. 803–817

Deep Learning Modeling Using Normal Mammograms for Predicting Breast Cancer

A. Sivasangari, D. Deepa, T. Anandhi, Suja Cherukullapurath Mana, R. Vignesh, and B. Keerthi Samhitha

Abstract Breast cancer is the most dangerous leading cause of cancer-related death among women in the world. Internet of things (IoT) connected device that can help detect breast cancer earlier. Mammograms are the main way to detect breast cancer. But it is difficult to detect the cancer cell in the breast tissue. Because it has more tissue and less fat, the digitized mammography images are used for analyzing the abnormal areas of density, mass, and calcification that indicate the presence of cancer. In cases where adequate labeled data is available, deep learning approaches show considerable progress, and many advanced deep learning approaches have been proposed that have shown efficiency in the last few years in various modalities of computer vision and medical imaging. The small size of the medical image dataset also lowers the efficiency and robustness of computer-aided detection and/or diagnosis (CAD) systems based on deep learning. This research aims to build and test a new hybrid deep learning-based CAD method to predict the probability of a breast lesion found on a mammogram being malignant in an attempt to overcome this technological challenge. In this method, a deep convolutional neural network (CNN) will first be pre-trained using the ImageNet dataset. This work aims to develop and test a new CAD method based on hybrid transfer learning for mammographic mass classification. By combining the initial mammogram images with their morphological and texture variations, it is possible to produce more pseudo-color images to be added with information. Additional filters are used to remove unrelated transferred CNN characteristics on the network.

Keywords Breast cancer · SVM · Classification · Accuracy

1 Introduction

According to a 2018 survey of breast cancer statistics, a total of 162,468 people have died due to the lack of development techniques, and also breast cancer survival is

A. Sivasangari (✉) · D. Deepa · T. Anandhi · S. C. Mana · R. Vignesh · B. Keerthi Samhitha
Department of IT, Sathyabama Institute of Science and Technology, Chennai, India

Department of CSE, Sathyabama Institute of Science and Technology, Chennai, India

S. Shakya et al. (eds.), *Mobile Computing and Sustainable Informatics*, Lecture Notes on Data Engineering and Communications Technologies 68,
https://doi.org/10.1007/978-981-16-1866-6_30

more difficult in the higher-order stage levels. Moreover, most of the women in India suffer from stage 3 and stage 4 breast cancer. It is because of a lack of awareness to detect and to screening and diagnosis rates. Similarly, breast cancer also occurs in both men and women but it is far more prevalent in women. Further, in women, it is a common type of cancer worldwide. According to the USA, breast cancer incidence in a woman is 1 in 8 (about 13%) which is women have a 3% chance of breast cancer which causes their death.

In the last decades, a lot of research projects have been aiming to develop computational systems in specialists in the task of detecting cancer and categorizing cancer are very complicated. In the current specialized period, mechanized indicative frameworks broadly use mammogram screening techniques to characterize the breast tumor. The computer-aided diagnosis (CAD) framework regularly depends on AI strategies to distinguish tumors in digitized mammogram pictures. Such methods need to work with discriminant and distinct highlights to order pictures into numerous classes. In the previous decade, various techniques have been proposed to arrange the mammogram pictures and to accomplish better exactness, proficiency, strength, and accuracy.

Researchers have developed efficiency diagnosis CAD systems that present a microscopic histopathological image used in philosophy to assist authorities in assigning mammogram images to differentiate masses. This method utilizes cell neural organizations to separate the premium regions, removes form highlights from those locations. At that point, through Ripley's K ability, Moran's, and Geary's lists, it removes shape highlights from those locations and depicts their surface. Finally, the support vector machine (SVM) directed learning approach is used to group the competitor regions as masses or non-masses. In this specific situation, any district mass is considered that relates to a neoplasm, regardless of whether it is dangerous or favorable. Moreover, in mammographic photographs, the capacity of cell neural organizations partially assumes districts and suggests an approach that integrates an innovative use of geostatistical capabilities as surface marks for mass discovery. The results obtained by this research will allow this approach to be applied to a clinical zone PC unit, providing assistance particularly in cases where perception is troublesome.

Numerous scientists have contemplated mammogram images for two-class (ordinary versus irregular) characterization and accomplished critical outcomes. Some of the researchers proposed a layout dependent on an acknowledgment calculation for bosom masses. Their information index depended on 1852 Digital Database using NN has accomplished best in class over the last couple of years brings about various fields of PC vision, such as object recognition and order. Profound learning models are likewise applied to different clinical imaging fields like tissue grouping in histopathology and histology images. In any case, a predetermined number of studies are accessible utilizing profound learning for mammogram picture grouping. CNNs were utilized to portion the bosom tissue of mammographic surface. Multistage highlights autoencoders were applied to ascertain bosom thickness score. CNNs were utilized to group the miniature calcifications yet the informational collection was exceptionally little. Some of the articles stated that unaided profound learning applied to bosom thickness division utilized adaptive deconvolutional networks (ADN) to describe the bosom. Breast cancer using IoT connected device and classifying mammograms and grouping strategy for bosom malignancy identification dependent on CNN is used. The initial calculation depends on 2D discrete wavelet change which was proposed while the other depends on curvelet change. A profound learning strategy can be utilized to detect bosom disease. The results show that taking in highlights from the informational collection before contributing the information to the CNN is more useful for disease identification. According to mammograms, it will classify into two categories: malignant and benign. Both are very contrast toward shapes. In benign, tumors shape around oval or round. But in malignant tumors halfway adjusts with the sporadic layout. Also, the malignant mass seems more than any tissue encompassing it.

This paper is organized in the following manner. Section 2 describes related kinds of literatures, and Sect. 3 proposed works contains the flow of the proposed idea. In Sect. 4, performance analyses with comparison graphs are explained, and finally, in Sect. 5, the conclusion of the research work is presented.

2 Related Work

Paul et al. [1] the author suggested level of comprehension of danger factors and screening strategies, particularly bosom self-assessment (BSE), given the non-practicality of analytic apparatuses, for example, mammography in the sacred city of Varanasi, Uttar Pradesh, India, for bosom malignant growth screening procedures. The investigation found that the view of danger components and BSE practice among ladies in Varanasi is incredibly poor contrasted with different urban areas and nations.

Sun et al. [2] For the most recent two years, examinations on pathogenesis, related changes, hazard elements, and avoidance measures for bosom malignancy have been completed. In the long fight against bosom disease, these outcomes mark a little stage. Bosom malignancy ought to be dodged. The two key strides for lessening bosom malignant growth are the end of danger factors and the usage of chemoprevention. There is far to go, however, to assemble broad comprehension about bosom malignancy.

Lee et al. [3] Significant contemplations and standards identifying with the treatment of YABC patients. Finally, more young women require more psychosocial assistance because they experience more enthusiastic agony and reduced personal satisfaction than more developed ladies when they are treated. More women regularly need reasonable direction and different intercessions regarding the support of ripeness and future origination. To choose whether they have acquired bosom disease, these ladies should be screened for BRCA1/2 changes.

Gupta et al. [4] Breast cancer detection and prognosis are two clinical applications that give scientists an incredible test. The whole strategy for bosom disease discovery and forecast has been altered by the utilization of fake learning and information mining methods. Breast cancer diagnosis determines sort of from threatening bosom protuberances and breast cancer prognosis predictions in women who have had their tumors removed that it is necessary to repeat breast cancer. A new examination is being completed utilizing information handling strategies to improve the determination and anticipation of bosom disease.

Kumar et al. [5] This postulation centers around the use of numerous characterization techniques for information mining to figure dangerous and favorable bosom malignancy. As a test dataset, the Breast Cancer Wisconsin information assortment from the UCI vault was utilized while the bunch thickness boundary was utilized as an evaluation class. In the cancer prediction method, sometime after of these twelve equations are taken into account.

Verma et al. [6] the author proposed two infection datasets from the UCI AI library for bosom malignant growth and diabetes. On the WEKA Explorer and WEKA Experimenter GUI, five characterization calculations were performed. A solid order technique utilized in this paper is the WEKA instrument. Credulous Bayes, SMO, REP Tree, J48, and MLP calculations are utilized on the WEKA GUI to recognize the dataset of bosom disease and diabetes. In the bosom malignant growth and diabetes dataset, the yield of these five calculations was assessed utilizing the preparation information test mode. In the wake of assessing all calculation results, it was discovered that guileless Bayes gives 72.70% exactness on the dataset of bosom malignancy and SMO gives 76.80% precision on the diabetes dataset. Shawarib et al. [7] as far as exactness in the Wisconsin Breast Cancer dataset, they propose an ANN model that beats all past administered learning approaches by hitting 99.57. As far as precision, test discoveries on Haberman's Breast Cancer Survival dataset show the matchless quality of the proposed approach by accomplishing 88.24%. The discoveries are the best-recorded outcomes got utilizing the JNN climate from the artificial neural network with no pre-handling of the dataset.

Alshammari et al. [8] The Waikato Ecosystem for Knowledge Research (WEKA) strategy is utilized to lead a sensible information mining test utilizing a scholarly trial bosom disease dataset. Thirteen examinations plot the strategy for arrangement utilized in this investigation. The outcomes reason that, among different classifiers, Lazy IBK classifier k-NN can accomplish 98% exactness. The primary preferences of the investigation were the smallness of 13 distinctive information mining models and 10 unique estimations of execution, and the plotting of characterization blunder figures. Shaper et al. [9]

Breast cancer is an applied bosom disease study roused by the creator's experience as a patient with bosom malignancy. It plans to show the present importance of philosophical deduction in medication and offers exhortation on the best way to manage the vulnerability of analysis, therapy, and endurance of bosom malignancy.

Camille et al. [10] The patient had a negative malignant growth family ancestry and for 2.5 years had been on testosterone treatment. In trans under androgen therapy basically examine the expected effect of chemical treatment on the advancement of bosom malignant growth. In addition, forecast of past studies on bosom malignant growth in the population of trans men post-mastectomy is presented. It is vital to report such cases and gather epidemiological information due to the shortage of this disease event, and thinking about the mental effect of such conclusion. Expanded consciousness of bosom disease in transsexual patients will empower to improve screening and give better rules to the executives.

Curigliano et al. [11] a redesign of emergency clinic organizations was presented. For patients with disease, more grounded individual security has been made accessible. Antitumor treatment was begun simply in the wake of assessing quiet execution status, comorbidities, infection science, and the conceivable impact of treatment on the result, prompting the outrageous pressing factor made by COVID-19. In the foundation of a multidisciplinary tumor board, remedial alternatives considered as dynamic adjusted danger and treatment benefits, dependent upon the situation, and also in the feeling of the individual pandemic level.

Zhou et al. [12] The proof remaining parts are negligible and conflicting, particularly in different menopausal conditions. Higher birth weight in premenopausal women is linked to an increased risk of BC, especially when the birth weight is above 3.5 kg.

DeSantis et al. [13] A biennial update by the American Cancer Society on insights on the female bosom disease in the USA, including rate, mortality, endurance, and screening information. By stretching out admittance to a great preventive, early conclusion, and recuperation projects to all ladies, decreases in bosom disease mortality might be expanded.

Fahim et al. [14] The significant part of HSP70 articulation and its capability is to be a demonstrative and prognostic marker in Egyptian bosom malignancy (BC) patients. Our outcomes show that breath-holding utilizing the RPM framework may bring about an extra decrease of the heart portion in patients treated with digressive fields accepting inclined left-bosom RT.U. Goyal et al. [15] Early-stage bosom disease results are improving. To diminish heart portions for left-sided bosom radiotherapy, the blend of inclined position is explored to profound motivation for breath-hold. It shows that breath-holding with the RPM framework may bring about an extra decrease of the heart portion in patients treated with unrelated fields getting inclined left-bosom RT. Moreover, Co et al. [16] breast malignancy has been proposed to be more pervasive in airline stewards than in the overall population. There is inadequate information to show an association between infinite illumination of airline stewards, circadian unsettling influence, and bosom disease. The occurrence and mortality of bosom malignancy in airline stewards were identical to those of the overall population.

Gu et al. [17] Pathologic total reduction after neoadjuvant chemotherapy has a role in managing the administration of bosom malignancy. The present meta-examination investigated the accuracy of differential enhanced attractive reverberation imaging (CE-MRI) and distribution weighted attractive reverberation imaging (DW-MRI) and comparing CE-MRI, ultrasonography, mammography, and positron outflow tomography/processed tomography (PET/CT) in identifying the reaction to neoadjuvant chemotherapy. Morikawa et al. [18] Patients and methods three hundred and eighteen straight bosom malignancy patients determined to have leptomeningeal metastases were accounted for at the Memorial Sloan Kettering Cancer Center between January 1998 and December 2013. To upgrade the treatment of these conditions, a nearer audit to decide causes, for example, the execution of new foundational medicines that could prompt longer-term endurance could give direction. Elements that corresponded with the life span that has set up might be seen as defining factors in the plan of possible randomized clinical preliminaries locally. Ritter et al. [19] The current analysis shows that for 30-year-old serum research, the most recent sequencing breakthrough of the following decade is genuinely steady and explicit. It is also possible to store serum tests for quite a while for longitudinal exploration, thereby removing the need for fast and non-stop extraction and sequencing of cfDNA. Tests ought to be joined and investigated without a moment's delay utilizing the most modern innovations accessible at the finish of the preliminary, where case conglomeration empowers clinical outcomes to be associated with a fitting limit. Munford et al. [20] Identify preoperative influences from invisible socioeconomics, preoperative workup, or biopsy findings that may be characteristic of postoperative edge status. Most of the dissected variables were comparative, but varieties were noted in high-grade tumors, in situ status, and progesterone receptor-negative (PR-) tumors. Patients with in situ (stage 0) status tumors were 1.986 occasions more probable than patients with obtrusive tumors to have positive edges ($P = 0.030$). Similarly, Samhitha et al. [21] proposed utilizing convolutional neural organization to classify tumors and distinguish them from tomography channels.

3 Proposed Work

A few class-deep training models is to organize the issue accomplice's in earlier ordinary mammography images to predict the outcome of bosom disease versus bosom emission as no proof of disease status. This shows a brief denomination of the probability/danger of producing bosom malignancy (specifically the gap between the season of obtaining typical images and also hour dependent on result). In particular, the framework was modified in a variety of plans with a GoogLeNet structure-based CNN model and with an effective strength evaluation (LDA) investigation (implied as GoogLeNet-LDA). In addition, the proposed framework was updated. Figure 1 represents the step-by-step diagram of the prospective demonstrating plans.

The suggested plans (Fig. 1) for the short-term bosom malignancy threat forecast for deep learning-based are displayed. The first half is an end-to-end expectation framework using GoogLeNet calibration and is modified in the GoogLeNet-LDA framework as a disconnected deep view highlight separator (base one).

The end-to-end framework assumes that initial configuration of the GoogLeNet, whereas double-way softmax model is utilized to provide standardized likelihood to double order between bosom disease and bosom cancer-free result. The end-to-end profound learning expectation model does not separate imaging highlights for disconnected examination. Conversely, the GoogLeNet-LDA model concentrates profound imaging highlights disconnected utilizing the calibrate GoogLeNet framework. The GoogLeNet calibration in this design has been modified as a part separator and deep highlights are removed from the layer before the entirely attached layer. In this framework, the investigation based on the end-to-end framework is analyzed

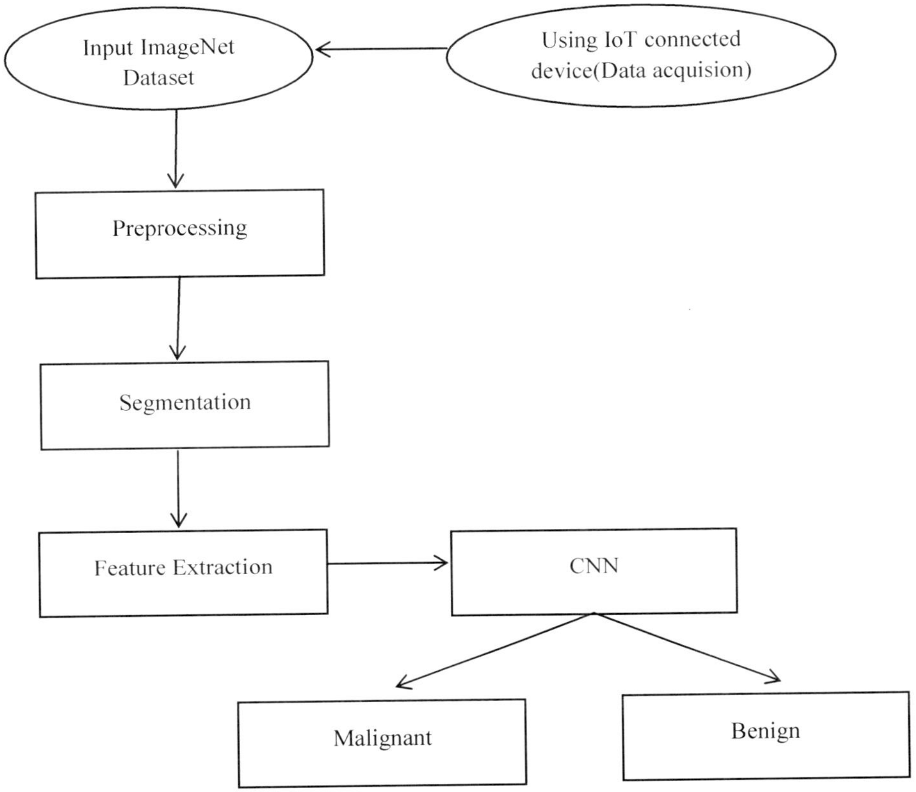

Fig. 1 Proposed flow diagram

along with the GoogLeNet-LDA framework. This connection initially attempts to collect bits of information among the end-to-end demonstrating part based on the disconnected deep training highlight separation system on the impacts of bosom disease risk forecast.

GoogLeNet framework was implemented by transferring training of the pre-trained framework from huge image datasets (especially from ImageNet, 29 consisting of 1,000,000 additionally named normal images) from the basic segment based on the plans and then fine-tuned to use our own personal mammogram imaging information. From the end-to-end forecast framework, the calibrated GoogLeNet is legitimately loaded in. In the conspiracy of GoogLeNet-LDA, all preparation tests to the fine-tuned GoogLeNet framework are to extract 1024 deep figuring features based on the layer just before the completely related layer, trailed over the most un-outright shrinkage. Determination administrator includes selection and after the preparation of the LDA classifier in the GoogLeNet-LDA system test phase, the equivalent GoogLeNet calibrate framework is used to delete the 1024 deep image highlights and then recognized a similar subset as recently selected in the preparation stage. The highlights of the subset were then taken care of in the LDA model for testing.

In the proposed framework, the GoogLeNet is selected to be used overwhelmingly because it has been shown feasible in various applications yet moreover has a large size of limits (particularly, ~ 5 million) interestingly with another notable CNN models (e.g., ~ 60 million limits as AlexNet31) to conceivably diminish chances of over fitting on our dataset. Our model was actualized utilizing MATLAB (R2018b) contrasting over the super PC framework as well as accompanying specifications32: 9 HPE Apollo 6500 workers, severally with 8 NVIDIA Tesla V100 Graphics Processing Units (GPUs) and 16 GB GPU memory, associated over the NVLink 2.0. The equal GPU programming component is used to fasten process step preparing. A stochastic slope drop with force (SGDM) streamlining agent was utilized to discover ideal framework boundaries. In the proposed structure started with a preparation speed of 0.01 and dropped the preparation rate as factor 0.2 every 10 ages. Our gathering size was 30. For preprocessing, histogram evening out was rushed to change as power contrast over all photographs, and each picture was downsampled utilizing standard bicubic consideration to 224 × 224 pixels from the fundamental goal to oblige the necessities of the pre-arranged GoogLeNet model.

Convolutionary neural networks are known for our particular measuring role as functioning mystery pieces and as a motion to other than appreciate the colossal being evident. The possible imaging features/locale that are usually relevant to the need for the short-term hazard are attempted to be formulated. Complementary and alternative medicine (CAM) framework, where test mammograms were forward in-wrinkled via the fine-tuned GoogLeNet model and the started incorporate finders were broadened again into the critical picture space to imagine the most notable imaging features/zones. The most basic convolutional layer is used in the fine-tuned GoogLeNet (i.e., inception_5b-pool_proj) for CAM consolidates control description.

4 Performance Evaluation

4.1 Analysis of Model Performance

In this section, the proposed model is compared and analyzed for the performance model. The proposed model is compared with the other existing models, and the results are obtained. In the proposed methodology, deep learning mammograms CAD model is applied to classify malignant and benign tumors. Compare to the traditional detection model, CNN models perform better because of their dense nature and ability to learn features. By selecting efficient activation functions and max-pooling functions, better performance is possible. The proposed model is evaluated in terms of precision, recall, accuracy, average accuracy, and $F1$-score. The model accuracy is tested at various tests and train data split rate. Accordingly, accuracy and dropout rate are evaluated as shown in the below figure.

The precision is calculated as the formula given below

$$\text{Precision} = \frac{\text{TP}}{\text{TP} + \text{FP}}$$

where TP is true positive results which are the positive truly predicted as positive values. And FP is false positive is the results which are the values predicted as true, but actually they are not true.

Recall is calculated as shown in the equation below. From that $F1$-score is calculated with precision and recall weighted average.

$$\text{Recall} = \frac{\text{TP}}{\text{TP} + \text{FN}}$$

The graphical representation of the accuracy data is given in Fig. 2

As shown in the graph (Fig. 2), the model has an accuracy of 91 at 30% split, 80 at 25% split, and 75 at 20% split. The model parameters are also tested at various dropout rates. The corresponding graph is given in Fig. 3 at dense layers with a dropout accuracy is 80%.

Similarly, the average recall and $f1$-score comparison is given in the graph Fig. 4.

From these parametric evaluations, it is clear that the proposed model can give promising results. It provides better performances when compared to similar existing models.

Similarly, the graph (Fig. 5) shows the cancer detection model accuracy plotting for both training and testing data. And also it displays the difference based on the different parameters.

The ROC graph (Fig. 6) is showing the comparison between the specificity and sensitivity model of the region shown. The proposed model displays optimal performance.

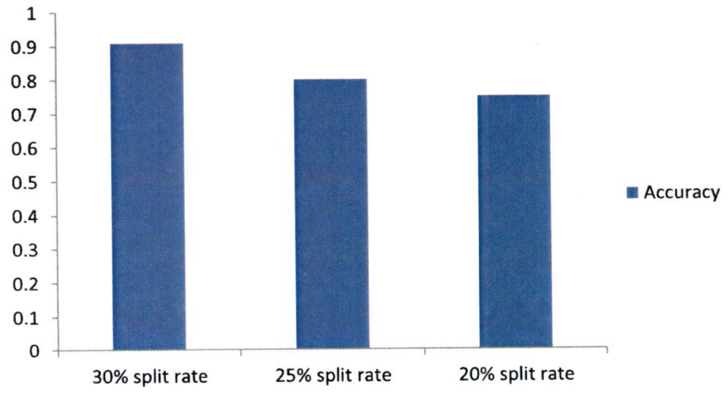

Fig. 2 Accuracy rates

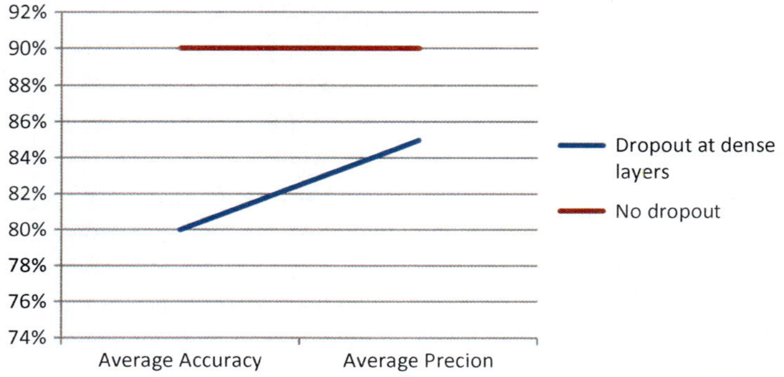

Fig. 3 Average accuracy and precision rates

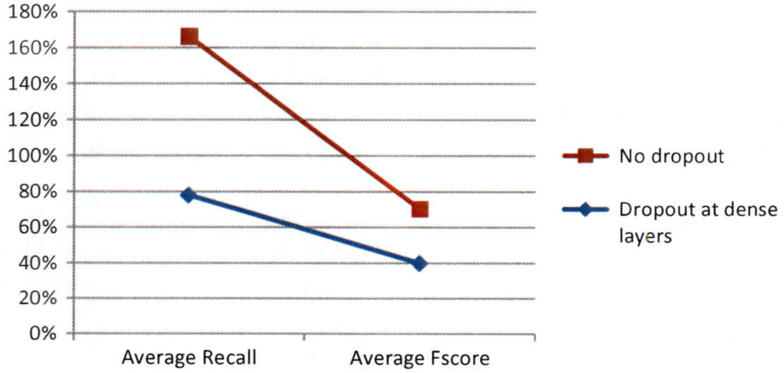

Fig. 4 Average recall and $f1$-score rates

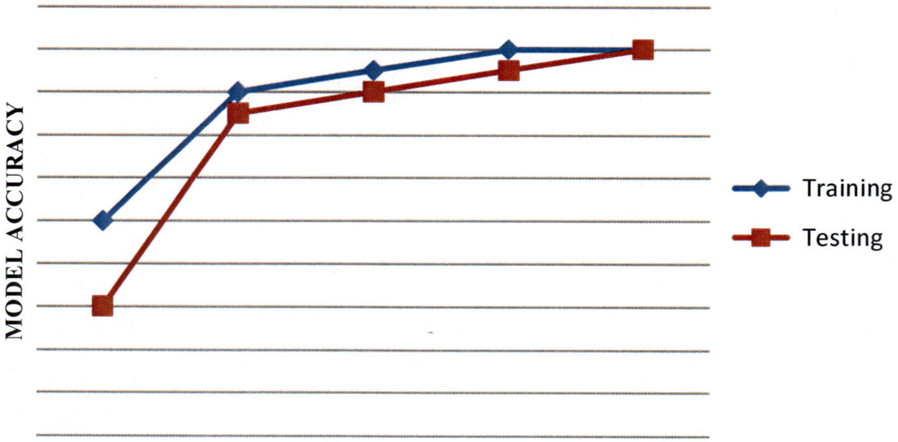

Fig. 5 Model accuracy plot for testing and training data

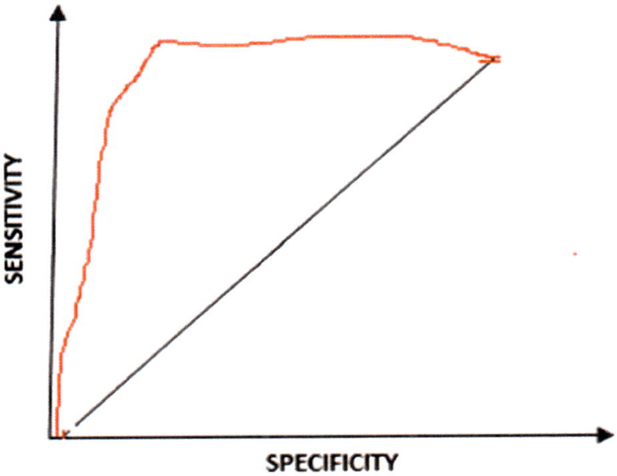

Fig. 6 ROC curve for the model

5 Conclusion

Breast cancer is the most important and crucial stage in the medical field. This paper discussed the ImageNet input dataset and categorizing malignant or benign cancer types. The proposed approach considers input from IoT connected devices to collect different data from various parameters and that will be preprocessed and further applied using the CNN algorithm. Later, classification based on the fully connected layers can classify whether that cancer belongs to the benign type of malignant variety tumor. The comparison results show that the improvement of the area development and CNN segmentation and feature extraction are segmented tumor divisions and the use of shape highlights of the divided tumors produces palatable frameworks to analyze benign and malignant tumors.

References

1. S. Paul, P.P. Solanki, U.P. Shahi, S. Srikrishna, Epidemiological study on breast cancer associated risk factors and screening practices among women in the holy city of Varanasi, Uttar Pradesh, India. Asian Pac. J. Cancer Prev. **16**(18), 8163–8171 (2016)
2. Y.S. Sun, Z. Zhao, Z.N. Yang, F. Xu, H.J. Lu, Z.Y. Zhu, W. Shi, J. Jiang, P.P. Yao, H.P. Zhu, Risk factors and preventions of breast cancer. Int. J. Biol. Sci. **13**(11), 1387 (2017)
3. H.B. Lee, W. Han, Unique features of young age breast cancer and its management. J. Breast Cancer **17**(4), 301–307 (2014)
4. S. Gupta, D. Kumar, A. Sharma, Data mining classification techniques applied for breast cancer diagnosis and prognosis. Indian J. Comput. Sci. Eng. (IJCSE) **2**(2), 188–195 (2011)
5. V. Kumar, B.K. Mishra, M. Mazzara, D.N. Thanh, A. Verma, Prediction of malignant and benign breast cancer: a data mining approach in healthcare applications, in *Advances in Data Science and Management* (Springer, Singapore, 2020), pp. 435–442

6. D. Verma, N. Mishra, Analysis and prediction of breast cancer and diabetes disease datasets using data mining classification techniques, in *2017 International Conference on Intelligent Sustainable Systems (ICISS)* (IEEE, 2017 Dec), pp. 533–538
7. M.Z.A. Shawarib, A.E.A. Latif, B.E.E.D. Al-Zatmah, S.S. Abu-Naser, *Breast Cancer Diagnosis and Survival Prediction Using JNN* (2020)
8. M. Alshammari, M. Mezher, *A Comparative Analysis of Data Mining Techniques on Breast Cancer Diagnosis Data using WEKA Toolbox* (2020)
9. J.A. Marcum, *Mary Ann G. Cutter: Thinking Through Breast Cancer: A Philosophical Exploration of Diagnosis, Treatment, and Survival* (Oxford University Press, 2018), xxiv+ 223 pp, $34.95 (cloth). ISBN: 978-0-19-063703-3
10. K. Camille, S. Clara, Z. Khalil, L. Aikaterini, B. Olivier, *Invasive Breast Cancer in a Female-to-Male Transgender After Bilateral Mastectomy: Case-Report and Literature Review* (Clinical Breast Cancer, 2020)
11. G. Curigliano, How to guarantee the best of care to patients with cancer during the COVID-19 epidemic: the Italian experience. Oncologist **25**(6), 463 (2020)
12. W. Zhou, X. Chen, H. Huang, S. Liu, A. Xie, L. Lan, Birth weight and ıncident of breast cancer: a dose-response meta-analysis of prospective studies. Clin. Breast Cancer **20**(5), e555–e568
13. C.E. DeSantis, J. Ma, M.M. Gaudet, L.A. Newman, K.D. Miller, A. Goding Sauer, A. Jemal, R.L. Siegel, Breast cancer statistics. CA A Cancer J. Clin. **69**(6), 438–451 (2019)
14. H.H. Fahim, G. Mohamed, G. Safwat, A. Abo-bakr, M.H. Ibraheem, S. Al-Mofty, M.M. Kamel, S. Abdel-Moneim, A.M. Gameel, HSP70 as a diagnostic and prognostic marker in Egyptian breast cancer women. Clin. Breast Cancer (2020)
15. U. Goyal, K. Saboda, D. Roe, V. Gonzalez, Prone positioning with deep ınspiration breath hold for left breast radiotherapy. Clin. Breast Cancer (2020)
16. M. Co, A. Kwong, Breast cancer rate and mortality in female flight attendants—a systematic review and pooled analysis. Clin. Breast Cancer (2020)
17. Y.L. Gu, S.M. Pan, J. Ren, Z.X. Yang, G.Q. Jiang, Role of magnetic resonance imaging in detection of pathologic complete remission in breast cancer patients treated with neoadjuvant chemotherapy: a meta-analysis. Clin. Breast Cancer **17**(4), 245–255 (2017)
18. A. Morikawa, L. Jordan, R. Rozner, S. Patil, A. Boire, E. Pentsova, A.D. Seidman, Characteristics and outcomes of patients with breast cancer with leptomeningeal metastasis. Clin. Breast Cancer **17**(1), 23–28 (2017)
19. M. Ritter, V. Paradiso, P. Widmer, A. Garofoli, L. Quagliata, S. Eppenberger-Castori, S.D. Soysal, S. Muenst, C.K. Ng, S. Piscuoglio, W. Weber, Identification of somatic mutations in thirty-year-old serum cell-free DNA from patients with breast cancer: a feasibility study. Clin. Breast Cancer (2020)
20. C. Munford, E. Clayton, D. Ji, J. Eskew, W. Burak, Preoperative predictive factors of positive margin status after breast conserving therapy (BCT) with ıntraoperative radiation therapy (IORT): a retrospective study of 400 patients. Clin. Breast Cancer (2020)
21. B.K. Samhitha, S.C. Mana, J. Jose, R. Vignesh, D. Deepa, Prediction of lung cancer using convolutional neural network (CNN). Int. J. **9**(3) (2020)

A Citation Recommendation System Using Deep Reinforcement Learning

Akhil M. Nair, Nibir Kumar Paul, and Jossy P. George

Abstract Recommender systems have seen tremendous growth in the last few years due to the emergence of web services like YouTube, Netflix, and Amazon, etc. An excessive amount of data is being utilized to give proper recommendations to the users. The number of research articles getting published every day is increasing exponentially and thus an efficient model is required to provide accurate and relevant recommendations to the research scholars. The proposed Deep Reinforcement Recommender for Citations (DRRC) model uses reinforcement learning to train the available citation network to achieve the most relevant recommendations. The proposed DRRC model outperforms the state-of-the-art models.

Keywords Reinforcement learning · Recommender system · Deep learning model

1 Introduction

Recommender systems are algorithms that help in suggesting relevant items to the users, such as, movies to watch, products to buy, etc. In the last few decades, companies like YouTube, Netflix, and Amazon have risen to power and the need for a better recommender system has also increased exponentially to provide the best recommendations according to the user preferences. It predicts the user preferences based on the previous data. Recommender systems have been proven to be efficient in decision making, improving quality, and also in reducing the transaction costs in an online shopping environment [1]. They find a wide range of applications in various fields including economic, education, and scientific research.

In the last few years, the number of research papers published has increased exponentially. Due to this huge bulk of data, the search engines used earlier for recommending citation are not able to give accurate enough recommendations any more [2]. Usually, citation recommendation falls under three categories—Collaborative Filtering, Content-Based, and Hybrid Recommender System. The collaborative

A. M. Nair (✉) · N. K. Paul · J. P. George
CHRIST (Deemed to be University), Lavasa, India
e-mail: akhil.nair@christuniversity.in

S. Shakya et al. (eds.), *Mobile Computing and Sustainable Informatics*, Lecture Notes on Data Engineering and Communications Technologies 68,
https://doi.org/10.1007/978-981-16-1866-6_31

filtering-based recommender systems predominantly process the similarity analysis a paper-citation matrix for recommendations whereas content-based systems consider the content and meta-data of the paper like title, keywords [3].

Due to the flexibility of citation analysis, a lot of graph-based and network-based citation recommendations have been done currently but very few have models that have used deep learning methods to increase the accuracy of the model. Thus, the main objective of this study is to build an efficient citation recommender system with the help of Deep Reinforcement Learning to improve the accuracy of recommending citations for a given paper. The scope of this project currently focus on citation recommendations since the datasets contain a graph-based citation network, if in other domains the dataset can be transformed to a similar network format then it will also have great application in those.

The major outcome that we intend to achieve is that our model will give better recommendations for scientific articles than most of the state-of-the-art recommender systems in existence.

2 Literature Review

Steinert and Ulrich Hoppe [1] discussed three scientific similarity measures that are used for scientific paper recommendation. These measures are namely, bibliographic coupling, co-citation coupling, and cosine similarity. All three measures are based on the connections of papers in citation networks. These citation networks are adjacency matrices. Similarity values are calculated on either the rows or columns of these matrices. The evaluation and analysis is performed on a detailed mathematical as well as an empirical level. The mathematical evaluation indicates that few steps are similar when compared with the structure, if the nodes are systematically arranged according to their similarity values obtained.

Yang et al., [2] proposed a graph-based citation recommendation model that uses a link prediction method to generate links between citations. The model used various network representation methods such as Deep Walk, Node2vec, and LINE for extracting structure features in the network. First, a bibliographical heterogeneous network is created and then the Network Representation Based Edge Prediction (NREP) model incorporates node content data, observed structure, and hidden edge. There are three separate functions being created which analyze this information and then they are optimized to learn the embeddings of the bibliographical network. When applied to a dataset, the experimental output shows that the NREP based approach outperforms the other Network based approaches in terms of Recall, MAP, and NDCG.

Waheed et al., [3] proposed a hybrid approach that uses centrality measures and author rankings for research paper recommendation. As search engines nowadays are unable to give the best results for recommended articles due to the huge amount of data, they have taken a graph-based filtering approach where they use citation network analysis along with the author's relationship network to give the best-recommended papers. When a user puts a query for a research paper the model identifies the papers which cite and are cited by the user's paper. Then it proceeds to calculate the candidate score of each paper and selects relevant papers based on the candidate score. The average rank of each paper is considered and the authors from the top papers are listed. The centrality measures (Betweenness, Closeness, Degree, and Eigen Vector) of each paper are calculated and converted into ranks. An author's collaboration network is generated where the top similar authors are chosen based on their collaboration score and finally, papers published by the top authors are shown to the user.

Cai et al., [4] proposed a three-layered graph-based filtering model where the recommendations are based on the paper, author, and venue. Usually, citation recommendations use paper and author papers and authors as a base for filtering out research papers. By adding a third layer of venues to it the clarity of the model increases. At first similar vertices are clustered together and then a deep learning walk is passed on these clusters according to the user's requirements for accurate paper recommendations. Evaluation matrices such as Recall@*N*, MAP, and MPR suggest that the model can give proper recommendations. Experiments carried out on the three different datasets determines the efficiency of the proposed methodology in contrast with the other citation recommendation approaches.

Chen et al., [5] proposed a graph-based filtering citation recommendation by adding weights to a heterogeneous information network. While creating the weighted HIN two types of vertices (paper and author) and four types of links (semantic linking, citing, writing and co-author) are used. Then by using the network representation learning method, the model can generate the future exemplification of each vertex thus able to calculate the similarity between the nodes. Thus, the model can generate citation recommendations. When experimented on two datasets, it is seen that the model has better recommendations than baseline approaches.

Guo et al., [6] the authors instead of just using citation information also used paper content thus creating a collaborative-based filtering method for citation recommendation. The Bag-of-words model generates a paper vector according to the paper content by utilizing the similarities between the papers with similar content and then on a citation information level, it integrates the assumed reference relationship along with the candidate papers citation with the help of a Deep Walk approach. Finally, while examining on a live dataset, it is understood that the performance of Citation Recommendation method with Content-Aware bibliographic network representation (CRCA) when compared with base algorithms like Metric, Pw, APW, LINE, Node2vec performs better with more recommendations than others.

Asim and Khusro [7] proposed a model to create a content-based recommender system that uses a specific amount of selected features according to the researchers' preferences. By doing this the authors hope to address the previous issues regarding content-based recommender systems which faced problems like the limited content analysis. The model tries to match a user profile with a CSP profile and wherever the similarity is highest, the model recommends those CFPs to the user. The performance of the proposed system is assessed with respect to precision, recall, and f-measure. The estimation results illustrate a minor degradation with the increased number of users.

Liu et al., [8] proposed a collaborative filtering model with a hypothesis that if two citing papers are significantly occurring together with same citing paper(s) then those two should be a lot similar. Thus, an association rule mining technique is deployed to determine the co-occurred citing papers where each citing paper is represented by some other citing paper and forming a link in the process. Then the similarities between the pairs are computed by the pair-wise representation of the papers. Finally, when the citation of citing papers is similar to a target paper, those papers are recommended to the user. When an experimental study is done on two datasets, it is seen that the CCF method outperforms the baseline approaches in terms of MAP and MPR.

Mohammadi et al., [9] proposed a graph-based model that will give citation recommendations on query related articles. The model approaches to build a co-citation network using various statistical significance tests. The similarity between the papers is calculated by using an NLP based approach where the content of the first paper is linked to other papers with similar subject similarity. Once the network is created random walks are done to give paper recommendations. The dedicated network has extremely higher prediction capability of choosing related items. The proposed framework is general and applies to different recommender systems to personalize recommendations according to the subject of interest.

Tantanasiriwong and Haruechaiyasak [10] proposed a collaborative based approach for citation recommendation based on the selected co-citations. The model tries to find patents with similar topics as most of the time similar topics have a similar set of publications. The similarity score between the two documents are calculated using various similarity measures and the one with the highest similarity score is used for citation recommendation. When compared with other baseline approaches using the evaluation matrix, it was seen that the proposed model had a better recommendation.

A deep learning-based study was done by Mohamed Hassan [11] that used the Recurrent Neural Network (RNN) for modeling a recommendation system for research articles. The explicit and implicit feedback of the users collected is used for recommendation purposes and the model uses the semantic representation of the paper's title and abstract for matching the interest of users from the feedback. Word2vec transforms the text into vector format and later parallelization is done which results in a lesser training time compared to other deep learning models. The Long Short Term Memory (LSTM) model is also used for representing an article based on the semantic representation of the words. The evaluation of the model could be done only after getting exact results after the implementation on which the work in progress.

Suglia et al., [12] introduced deep learning content-based recommendations using Recurrent Neural Network (RNN) where the authors used Long Short Term Memory (LSTM) to learn word embeddings of items and user preferences. A logistic regression layer followed by the LSTM units calculates the relevancy of each item to the given user's preference and top *N* recommendations are returned. The proposed methodology named Ask Me Any Rating (AMAR) was experimented on Movielens and DBbook datasets and was found that the method is outperforming other state of art methods. The paper also introduced AMAR Extended Architecture where along with user profile and item representation, item genres were also considered for recommendation. On calculating the $F1@10$ score, AMAR secured a value of 0.66 which is the highest compared to other baseline models.

Bulut et al., [13] conducted a study on recommending papers based on user profile and bibliographic approach. The dataset created from the metadata is used for creating a user profile after undergoing certain preprocessing. Candidate recommendation articles were clustered by the K-Means clustering method based on their citations. K-NN algorithm is used to calculate the closeness of the cluster to the user's profile. It was observed that the accuracy of the proposed method reduces with the increase in the number of Top N predictions, but the Recall value and F-1 value increases. From the experiments conducted, it is concluded that it is more accurate to recommend Top 10 articles for which the F-1 metric is higher according to the paper's work.

Yanga et al., [14] discussed a network correlation-based query-oriented citation recommendation approach. By constructing a semantic and citation network, correlations are built to these networks and top features are selected to calculate the semantic similarity for recommending citations. Original network semantic structure is maintained and the performance of the model is affected by the number of features. The model acquires an accuracy of nearly 95% as the number of features is increased.

Henderson et al., [15] discuss the importance of reinforcement learning in recent years, most recently in video games. The author examines the outcomes of certain hyperparameters on the performance of algorithms if it is not tuned properly and differences due to codebases when most of the other factors are held constant. It was seen that ReLu or Leaky ReLU activations achieve the best accuracy across environments and models.

François-Lavet et al., [16] in their book deeply talks about Deep Reinforcement Learning. Reinforcement Learning differs from Machine Learning and Deep Learning including the benchmark, perspective, and gradient methods. Finally, the study states how Deep Reinforcement Learning can be the future of recommendation systems due to their ability to learn by themselves from their past mistakes thus enhancing the recommendations. All these techniques could be done in a semantic manner using the very basic binary encoding to embed the content to the network to include more information and weight to the network [17].

3 Proposed Framework

The proposed model can be categorized into two phases—(a) to generate the threshold for the Reinforcement Learning algorithm and (b) to apply reinforcement learning with the obtained threshold on the citation network.

To get the threshold for the Reinforcement Learning algorithm, the model generates the respective adjacency matrix from the citation network. It then calculates the similarity values of the matrix by the method of cosine similarity. Once the similarity matrix is retrieved, the mean of that matrix is taken as a threshold for the Reinforcement Learning algorithm.

The reinforcement learning algorithm consists of three parts—state, action, and reward. The state of the model is to calculate the similarity value. The action is to compare the values of similarity and the threshold. If the similarity value is greater than the threshold, then the model is awarded a positive reward 1, else a negative reward 0. Based on these rewards, a new adjacency matrix is be formed which - has an improved number of connections than the previous one. Then the similarity value is again calculated for the newly obtained matrix. This procedure repeats for 1 lakh iterations randomly selecting two nodes every time which were not connected previously. The mean of the similarity matrix is calculated at each iteration of the model which shows the learning output of the model. Finally, the model is evaluated by calculating its Recall@N value where $N = [20, 40, 60, 80, 100]$.

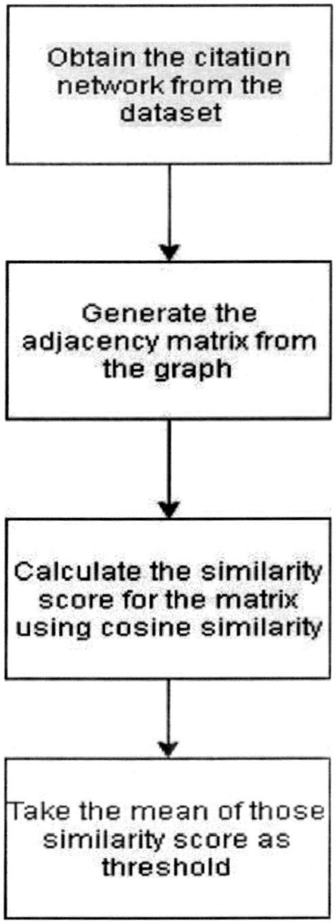

Fig. 1 Calculation of threshold for reinforcement learning model

In Fig. 1, the model calculates the threshold value for the reinforcement learning parameter. The model first obtains the adjacency matrix from the citation network and then generates its similarity matrix by calculating the cosine similarity between each point. The mean of this similarity is then taken as the threshold.

In Fig. 2, a new adjacency matrix is generated on top of the previously generated matrix. The new adjacency matrix will be based on the fact that if the cosine similarity value of any two points is greater than the threshold then a new connection would be made between those two points. ACL Anthology Network (AAN) dataset is a collection of ACL (The Association of Computer Linguistics) papers. It contains several networks including paper citation, author citation, and author collaboration. The latest version of the dataset has a collection of papers that were released in 2016.

The dataset that would be used in the proposed model is the AAN 2014 released dataset which is a 387 MB size file. The file consists of various text documents such as author-citation-network which shows the relations between which authors have cited which ones, paper-ids which contains all the papers along with their titles, and so on. The text document named paper-citation-network is used for this proposed model where it contains the paper ids of all 23,766 papers along with the papers it cited giving in a total of 1,24,857 paper citations. Each paper is considered as a node and each citation between two papers is considered as an edge.

The algorithm in Table 1 provides a step by step guide of how the proposed model works after the dataset has been formatted accordingly. It starts with selecting the threshold for the Reinforcement Learning algorithm followed by the state, action, and rewards system required for the algorithm. The state is the calculation of the cosine similarity between 2 nodes, the action is checking if the obtained cosine similarity is greater than the threshold and finally giving it a positive reward if it is greater than the threshold else assigning a negative reward. Finally, the mean of the cosine similarity matrix at each iteration is taken for evaluation to the plot and see the learning rate or growth of our model, and the Recall@N value is calculated for the newly obtained similarity matrix.

4 Experimental Setup

Since the dataset is a graph network and a Reinforcement Learning algorithm is being used, there is no separate train or test dataset for this model. The model learns based on the reward it gained in its previous iteration. The threshold of gaining this reward is the mean of the cosine similarity values from the adjacency matrix of the graph. If the calculated cosine similarity between two nodes is greater than the mean the model gets a positive reward else it gets a negative reward.

The DRRC model will be scored and validated by calculating the mean similarity value of the adjacency matrix at each iteration and finally plotting it. This will show us how good of a similarity the model is generating after every iteration along with the rate at which the model is learning.

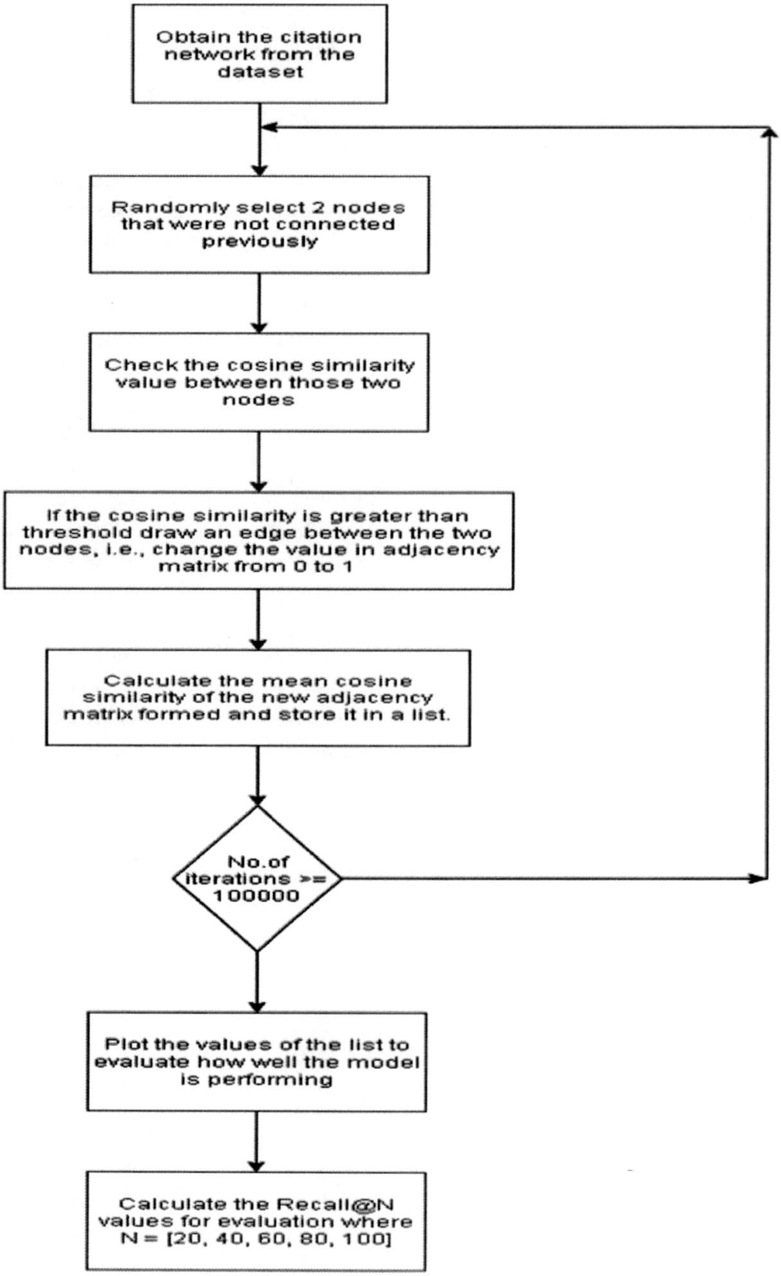

Fig. 2 Reinforcement learning framework

Table 1 Proposed model framework

Step 1	Create the citation network graph from the formatted dataset
Step 2	Create the adjacency matrix of the graph obtained
Step 3	Obtain the similarity matrix which consists of the similarity value between two nodes. The similarity values are calculated with the help of cosine similarity from the adjacency matrix
Step 4	Take the mean of the similarity values as the threshold
Step 5	Randomly select 2 unconnected nodes
Step 6	If the similarity value in Step 5 is above the obtained threshold, change the corresponding value in the adjacency matrix at Step 2–1 else to 0
Step 7	Recalculate the mean of cosine similarity of the new adjacency matrix
Step 8	Keep repeating Step 5–7 for 1 lakh iterations
Step 9	Plot all the newly obtained similarity values after every iteration of the model for evaluation
Step 10	Check Recall@N values for the newly obtained matrix

In Fig. 3, it can be seen that originally before 20,000 iterations the model was fumbling to learn which nodes to connect but after that, there is a steep increase in the reward stating that it had learned from its mistake and now is on the correct path of recommendation.

There are various performance metrics available for the evaluation of a recommender system such as [2].

(a) Recall@N calculates the mean of the percentage of papers from the test set that appears among the top N ranked list, for some given N.

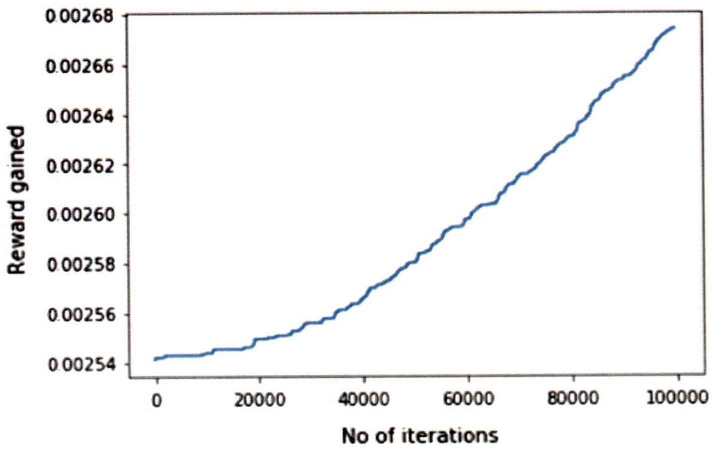

Fig. 3 Score of the DRRC model

Table 2 Comparison with other state-of-the-art models

	Recall@20	Recall@40	Recall@60	Recall@80	Recall@100
DRRC	0.552	0.591	0.619	0.641	0.655
QR	0.242	0.331	0.401	0.425	0.479
CQR	0.239	0.328	0.396	0.421	0.475
QR-PV	0.219	0.285	0.348	0.367	0.411
QR-PA	0.215	0.281	0.343	0.362	0.402
QR-P	0.206	0.273	0.339	0.351	0.396
CQR-P	0.206	0.269	0.332	0.349	0.395
Cosine	0.196	0.253	0.319	0.337	0.382
CF	0.196	0.250	0.318	0.335	0.379
Okapi	0.193	0.250	0.315	0.334	0.377
K-L	0.191	0.249	0.312	0.331	0.376

(b) Mean Average Precision (MAP) is the average precision scores for each manuscript in a set of manuscripts.
(c) Normalized Discounted Cumulative Gain (NDCG) measures the accuracy of rankings when there are several levels of relevance judgment.

The performance metric chosen for the DRRC model is Recall@N with N values as 20, 40, 60, 80, and 100.

Table 2 shows the metric values of the proposed DRRC model compared with the other state-of-the-art models. It can be easily seen that our DRRC model outperforms all the state-of-the-art models for Recall@N values.

5 Conclusion

As information overload is one of the major problems in the current era, it is a tedious task to obtain relevant information from the available data. Recommender systems provide filtered information from the available data based on user preferences. The citation recommender system aims at providing relevant articles to a research scholar. DRRC (Deep Reinforcement Recommender for Citations) framework uses the paper*paper citation network to provide recommendations to a given user thereby increasing the efficiency of the existing systems. The similarity between the papers is calculated with the help of cosine similarity. The framework uses a complete Deep Learning-based Reinforcement Learning to generate recommendations from the citation network. The threshold for Reinforcement Learning is calculated from the mean of the similarity scores. Recall@N values where N = [20, 40, 60, 80, 100], shows that DRRC not only improves the citation accuracy but also gives more recommendations than other models compared in the study.

References

1. L. Steinert, H.U. Hoppe, A comparative analysis of network-based similarity measures for scientific paper recommendations, in *Third European Network Intelligence Conference*, Wroclaw, 2016, pp. 17–24
2. L. Yang, Z. Zhang, X. Cai, L. Guo, Citation recommendation as edge prediction in heterogeneous bibliographic network: a network representation approach. IEEE Access **7**, 23232–23239 (2019)
3. W. Waheed, M. Imran, B. Raza, A.K. Malik, H.A. Khattak, A hybrid approach toward research paper recommendation using centrality measures and author ranking. IEEE Access **7**, 33145–33158 (2019)
4. X. Cai, J. Han, W. Li, R. Zhang, S. Pan, L. Yang, A three-layered mutually reinforced model for personalized citation recommendation. IEEE Trans. Neural Netw. Learn. Syst. **29**(12), 6026–6037 (2018)
5. J. Chen, Y. Liu, S. Zhao, Y. Zhang, Citation recommendation based on weighted heterogeneous information network containing semantic linking, in *IEEE International Conference on Multimedia and Expo*, Shanghai, China, 2019, pp. 31–36
6. L. Guo, X. Cai, H. Qin, Y. Guo, F. Li, G. Tian, Citation recommendation with a content-sensitive DeepWalk based approach, in *International Conference on Data Mining Workshops*, Beijing, China, 2019, pp. 538–543
7. M. Asim, S. Khusro, Content-based call for papers recommendation to researchers, in *12th International Conference on Open Source Systems and Technologies*, Lahore, Pakistan, 2018, pp. 42–47
8. H. Liu, X. Kong, X. Bai, W. Wang, T.M. Bekele, F. Xia, Context-based collaborative filtering for citation recommendation. IEEE Access **3**, 1695–1703 (2015)
9. S. Mohammadi, S. Kylasa, G. Kollias, A. Grama, Context-specific recommendation system for predicting similar Pubmed articles, in *IEEE 16th International Conference on Data Mining Workshops*, Barcelona, 2016, pp. 1007–1014
10. S. Tantanasiriwong, C. Haruechaiyasak, Cross-domain citation recommendation based on Co-citation selection, in *11th International Conference on Electrical Engineering/Electronics, Computer, Telecommunications and Information Technology*, Nakhon Ratchasima, 2014, pp. 1–4
11. A.M. Hassan, Hebatallah, Personalized research paper recommendation using deep learning, in *Proceedings of the 25th Conference on User Modeling, Adaptation and Personalization*, Jul 2017, pp. 327–330
12. A. Suglia, C. Greco, C. Musto, M. Gemmis, P. Lops, G. Semeraro, A deep architecture for content-based recommendations exploiting recurrent neural networks, in *Proceedings of the 25th Conference on User Modeling, Adaptation and Personalization*, Jul 2017, pp. 202–211
13. B. Bulut, B. Kaya, M. Kaya, A paper recommendation system based on user interest and citations, in *1st International Informatics and Software Engineering Conference*, Turkey, 2019, pp. 1–5
14. L. Yanga, Y. Zhengb, X. Caia, S. Panc, T. Daid, Query-oriented citation recommendation based on network correlation. J. Intell. Fuzzy Syst. **35**, 4621–4628 (2018)
15. P. Henderson, R. Islam, P. Bachman, J. Pineau, D. Precup, D. Meger, Deep reinforcement learning that matters, in *The Thirty-Second AAAI Conference on Artificial Intelligence (AAAI-18)* (2018)
16. V. François-Lavet, P. Henderson, R. Islam, M.G. Bellemare, J. Pineau, An introduction to deep reinforcement learning. Found. Trends Mach. Learn. **11**(3–4), 219–354
17. A.M Nair, J. George, S. Gaekwad, Similarity analysis for citation recommendation system using binary encoded data, in *2020 International Conference on Electrical, Communication, and Computer Engineering (ICECCE)*, Turkey, 2021

Influence of Schema Design in NoSQL Document Stores

Monika Shah, Amit Kothari, and Samir Patel

Abstract Schema-free (flexible schema) is an attractive feature in the NoSQL document store, where users can access the database without referring to its schema. It reduces the stress of developers to design database schema and applications. Data modeling is defined as arranging application data using acceptable data structuring patterns and is the primary component of the schema design in the document store. In addition, studies have continued examining NoSQL's energy consumption pattern. But, our recent survey does not observe any energy-efficienct analysis for NoSQL schema. This work therefore designs alternative schemes using a particular combination of document store structuring patterns and analyzes storage criteria for everything. In NoSQL, the use of indexing and application-level joints is widely adopted. This article also compares performance and energy requirements for query processing with/without indexing, and application-level join operations and also helps to design query optimization.

Keywords Schema · NoSQL · Document store · MongoDB · Energy consumption · Performance · Proportionality · Query optimization

1 Introduction

NoSQL was designed to discover an open-source, schema-less, distributed, and highly horizontal scalable database over commodity hardware. Besides, NoSQL systems have other attractive features like the simple interface, high availability, better performance through eventual consistency, and reducing the need for "join"

M. Shah (✉)
Department of Computer Science and Engineering, Nirma University, Ahmedabad, India

A. Kothari
Gujarat Technological University, Ahmedabad, India

S. Patel
Department of Computer Science and Engineering, Pandit Deendayal Petroleum University, Gandhinagar, India
e-mail: samir.patel@sot.pdpu.ac.in

S. Shakya et al. (eds.), *Mobile Computing and Sustainable Informatics*, Lecture Notes on Data Engineering and Communications Technologies 68,
https://doi.org/10.1007/978-981-16-1866-6_32

operations. This makes NoSQL a prime component of the majority of today's applications. One of the most attractive features of NoSQL is its schema-less or flexible schema design. To be specific, users have the flexibility to add data without referring to data types and schema of existing data records belong to the same class.

There are many research and developer articles [4, 9, 20, 21, 27, 30, 40, 45] guiding schema design of document store to achieve better application performance. The flexible schema supports on one side, and articles on schema design for improved result on the other side. This elevates many questions for NoSQL developers. Does schema matter in NoSQL? What are the costs of flexible schema support in NoSQL? Is there anything else than application performance affected by choosing any random schema? Is there any schema uniformly cost-effective for all applications or all queries of an application? This article helps NoSQL developers to get answers to these questions.

Sustainable development through energy conservancy is the most popular research topic these days. With continuously increasing computation demand, energy efficiency has drawn more attention in the computing world. [43] The NoSQL data store installations require massive computing infrastructure, which consumes a significant amount of energy. Li et al. [29] also shows that despite the NoSQL Database is widely used as a solution for "green cloud computing," NoSQL databases still lack mature solutions to evaluate and reduce Energy Consumption (EC). Some articles involve attempts to evaluate energy consumption at different NoSQL levels, such as database generation, query optimization, consistency management, and load balancing. The authors [12, 23] suggest that data modeling influences query performance, consistency, maintainability, and a few other aspects. Despite schema design being a prime component affecting other features, there is no comprehensive study to analyze the influence of schema design on energy consumption. Hence, Energy Consumption (EC) and performance costs are analyzed for alternate schema designs in MongoDB. Modeling relationships using different structuring patterns, identifying index attributes, selecting sharding key are the major components of schema design. Here, relationship models and indexing are analyzed.

In addition to schema-less, No Join has created more interest in application developers to use NoSQL. This elevates some queries to NoSQL users. For example, Does the NoSQL document store remove the need of join operation for all alternate schema design? If not, what type of schema need to join two collections? What are alternates to implement join operations in the document store? What is the influence of each alternative method of joining for each schema design? This article has included an experiment to find answers to these questions also. The contribution toward this article can be summarized as follows:

(a) Comprehensive literature survey on schema design in NoSQL document store
(b) Analyze the influence of different data structures and index(es) on storage space, query performance, and energy consumption
(c) Analyze the influence of using database-level join and application-level join with and without indexing.

The rest of the article is organized as follows. Section 2 describes the alternate approach of modeling data relationships as part of schema design in a document store. Section 3 discusses the literature review on related work. Section 4 lists data models that need to use join operations, and an alternate way to implement join operations in MongoDB. Section 5 describes the experiment setup. Result of the experiment is presented and analyzed along with interpreted research opportunities in Sect. 6. Finally, the work concluded with future ideas in Sect. 7.

2 Schema Design in Document Store

Today, the document store is popular in handling large and complex data in Big Data, Web applications, and IoT-based application [46]. Document store is widely flexible among NoSQL database systems to manage the schema [2, 17, 45]. But this flexibility may lead to improper design in aspect of the application's query workload execution [45, 48]. Modeling data relationships is the prime component of schema design in document store. It is a process to organize data as per the data relationships, where it uses different data structuring patterns. Mostly, developers model data in document stores with a mindset of relational database design. Unfortunately, this approach produces an incorrect schema. [19, 20, 48] describe relationship cardinalities for the document store.

The word "Many" is another breakdown of the words Few (up to a few hundred), Many (up to a few thousand), and Squillion (more than thousands with no limit).

The key categories for model data relationships are embedding, referencing, bucketing, and de-normalization, which are shown in Fig. 1. De-normalization merges all relationships and organizes all data into a single document. It is simplest in modeling as well as query processing approach. Every facility comes at a cost. De-normalization has a cost of larger storage size and redundancy. Embedding is the process to create sub-documents (nested documents) for the associated child/parent data entity. De-normalization and embedding are widely used to avoid "Join" operation. The dark side of embedding is increasing the storage space of a parent document, and expensive query processing on sub-documents. Every document has a limit in size. Document store uses multiple memory chunks to store such documents. Query processing will take more time to collect multiple memory chunks for a single document. The system facilitates handling large documents by modeling such a relationship using a bucketing approach, where sub-documents are split into smaller manageable sizes either considering the count of documents, aggregation to units like hours, days, and years. Few efforts are guiding for schema design considering given entity-relationships and query patterns [1, 20, 21, 39, 42, 48]. Most of them are aiming for better performance. The author [20] presented a long list of guidelines for the selection of appropriate structuring pattern considering experts' review. To design proper NoSQL document schema using this long list of guidelines, both the expertise and experience required. Table 1 summarizes alternate structuring patterns

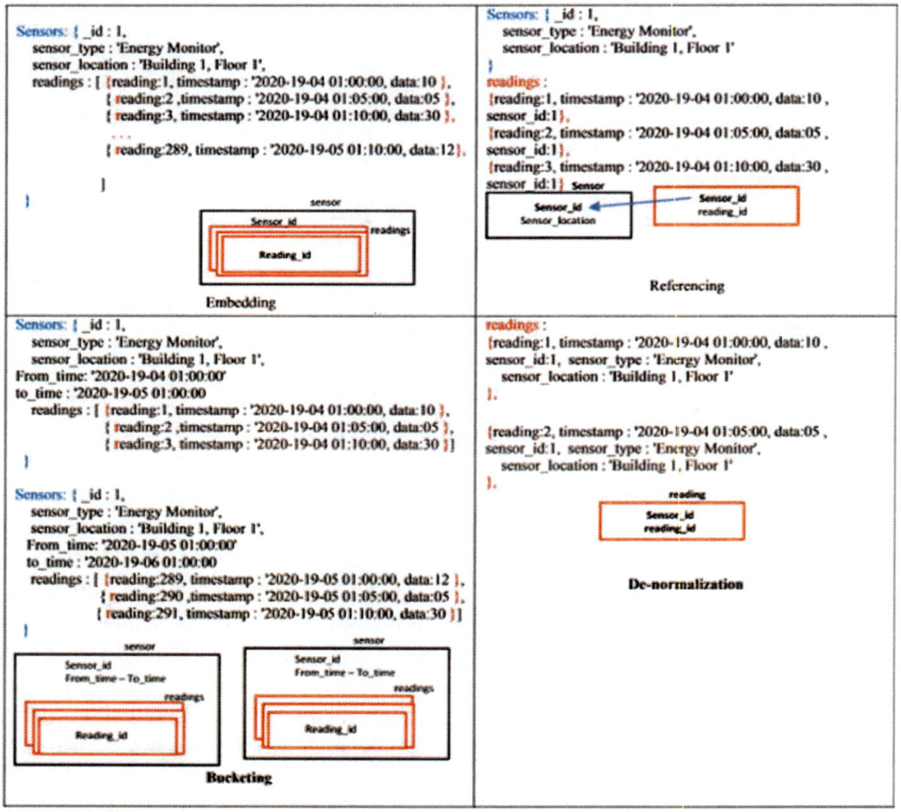

Fig. 1 Primitive data structures for data modeling in document store

Table 1 Relationship structuring style in document store

Structuring pattern	Description
Child embedding (CE)	Document for parent entity embeds its child entities
Parent embedding (PE)	Document for child entity embeds its parent entities
Two-way embedding (TE)	Both CE and PE applied
Child referencing (CR)	Document for parent entity store id of documents for child entities
Parent referencing (PR)	Document for child entity store id of documents for parent entities
Two-way referencing (TR)	Both CR and PR applied
Relationship (R)	A new document stores id of each related entity that participated in the relationship
Bucketing (BUK)	Splitting subdocuments into smaller manageable sizes either by count, aggregation unit like an hour, a day, a year, etc
De-normalize (DN)	De-normalize data into a single collection

available to represent various data-relationship scenarios, which is summarized from [19, 20, 42, 48].

3 Literature Review

Recently, all application developers are witnessing an increasing trend of NoSQL databases because of its schema-less or flexible schema feature. Despite NoSQL claimed to be schema-less, the application presents data in some structure, which is the prime modeling component. Atzeni [3] argues that a modeling-based approach can provide the base of the physical design of NoSQL. Atzeni et al. [4] Despite the flexibility to organize data in any of the available structuring patterns, they are not equivalent in supporting performance, scalability, and consistency. At this time, NoSQL data modeling is based on best practices and guidelines [25], which are related to specific systems [8, 16, 26]. Conventional schema design at NoSQL is typically a very difficult and costly process [40], but schema optimization is highly applicable for NoSQL databases [34].

With relational databases, NoSQL differs a lot, including cardinality, information organizations, distributed architecture, and the use of "join" operations. Still, in the document store, the behavior of schema design is found to be very similar to a relational database. For example, [15] adopts the traditional approach of transforming the E-R model to the relational model. The author [14] presents an anti-normalization-based schema design. References [24, 27] propose an idea of de-normalizing relational database schema using primary key and foreign key. With time, NoSQL specific schema designs are also discussed. The author [9] presented an UMLto[No]SQL approach mapping conceptual schema to data storage solutions. The author [21] has extracted guidelines made available by various NoSQL document store development team and provided extensive guidelines. The article intends to direct the improvement of application functionality, the minimization of errors, the improvement of security, and the fulfillment of desired system requirements such as consistency and availability. The author [3] proposes an approach from the conceptual modeling phase to aggregate design. In addition to application performance, [4, 20, 21] have also considered consistency, and availability. There are guidelines [21, 22, 30] suggest users to understand data operations and query access patterns and follow guidelines accordingly for good schema design. The author [45] has proposed a novel approach to design schema using formal concept analysis.

Despite such good collections of articles guiding for good schema design, there are several issues: (i) It is difficult to find a single article covering guidelines for all relationship structuring pattern. (ii) It is difficult to find a single article considering various parameters like data relationships, query access patterns, system requirements, application requirements, and to name a few. (iii) The most critical issue is expertise, and good experience is required to understand this extensive guideline and follow it for the application. Hence, [4–6, 37] presented the need for systematic data modeling approaches or automatic data modeling tools to design good schema in

NoSQL for better performance of the applications. To form a good schema design or automate the process of schema designing, it is required to identify application factors that can be affected by the choice of the data model. The author [13] presented metrics to reveal structure complexity in the document store. The author [18] presents various criteria to evaluate document-oriented NoSQL data models.

Related efforts have been observed in articles. Al Shekh Yassin [1] the performance, and that too for only three data models. The author [41] has investigated the influence of different aggregate data models on the performance of analytical queries. The author [12] has tried to present the influence of different schema on database size and query performance. Energy-efficient system design is an essential requirement today. Researchers started analyzing energy consumption patterns at various functional modules of MongoDB like query optimization [32, 33], query processing [10, 11], sharding [7], MapReduce process scheduling [44], and load balancing [31, 44]. But, there is not work analyzing the influence of schema design on energy consumption. This may because of some belief that energy is spent in proportion to execution time [11]. The author has performed a study of energy performance proportionality for selected query operations.

4 Join Operations in Document Store

NoSQL document stores suggest designing schema such that the use of join operations can be avoided as far as possible. This indirectly suggests using denormalization and embedding structuring patterns only to represent data relationships. Even after ignoring the use of referencing structuring patterns, some query patterns need to join operations. For example, execution of a query "identify the employee in each company drawing maximum salary" over a single collection employee (id, name, salary, company). The implementation logic requires joining operations to connect these two phases: (i) find the maximum salary of employees in each company (ii) identify employees matching this highest salary and company resulted from phase 1. Observing the need of join operations, every NoSQL document supports a different mechanism to implement join operations. For example, MongoDB supports $lookup in aggregate operation, MapReduce in CouchDB, JOIN clause by Couchbase, and DocumentDB. Despite facilitating the functionality of joins operations, it is suggested to avoid this database-level join and use application-level join. This article has also incorporated experiments to demonstrate the influence of application-level join and database-level join on performance and energy consumption.

5 Experiment Setup

The focus of the work to make developers and researchers understand the influence of alternate schema designs on various parameters including database size, storage space, application performance, and energy consumption. Looking toward the popularity of MongoDB in the world of the NoSQL document store, experiments are conducted on MongoDB 4.4. It is deployed on a cluster of two servers and one application node. Each of these two servers has MongoDB shards and a config daemon. MongoS is installed on the first server. Each of these servers has Intel Xeon(R) CPU E5-2630 v3@2.40 GHz × 16, Memory:16G (15.5MiB), Disk partition: 294.2 GB, and Ubuntu 18.04.5 LTS, 64 bit. Application node has 64 bit Ubuntu 18.04 bionic operating system and Intel Core i3-3110M CPU@2.4 GHz × 4 Ivybridge system. PowerAPI is a well-known toolkit for building software-defined power meters. From the literature survey, it is observed that the power consumption of monitor and disk are mostly stable during execution. Docker-based PowerAPI is configured in the system to monitor the energy consumption of the CPU every second (Watt per second). Energy consumption for a query is computed as a product of Pxt, where *P* is the average energy reading at every second during the execution period and t is the average execution time for each query.

5.1 Database

This work uses a sample database covering company, department, and employee information, which is generated using the python faker library. It covers 1:F cardinality between Company and Department, 1:M cardinality between Department and Employee, and 1:S cardinality between Company and Employee. Considering all structuring patterns and their combination would make the search space very large. Hence, this article chooses structuring patterns recommended in various research articles. These 10 schemas and their relation with the structuring pattern is described in Table 2

For example, [1] concludes to use a single-level embedding structure, where another relationship is preferred to be structured using referencing. But, there is no guidance for selecting embedding or referencing to model different data relationships. Besides, it also not guiding for selection of parent embedding or child embedding, parent referencing, or child referencing. To experience this recommendation, schema S2, S7, S8, and S9 are designed with a different combination of referencing and embedding relationship. References [14, 20, 21, 42, 48] prefer to use embedding for better performance, specifically CE has been given higher preference. Schema S3, S5, and S6 have used child embedding and parent embedding structures to represent 1:F and 1:M relationships. De-normalization is a widely used structuring pattern in NoSQL, special when read to update ratio is high. S10 schema shows de-normalized structuring only. Schema S1, S2, S7, S8, and S9 use different

Table 2 Structuring patterns used in test schema

Schema	Collections				Structuring pattern	
	CMP	DEPT	EMP	CDES	Cmp-Dept (1:F)	Dept-Emp (1:M)
S1	Y	Y	Y		PR	PR
S2	Y	Y			PR	CE
S3	Y				CE	CE
S4	Y	Y	Y	Y	R	R
S5			Y		PE	PE
S6	Y	Y	Y		PE + I	PE + I
S7	Y		Y		CE	CR
S8	Y		Y		CE	TR
S9	Y		Y		CE	PR
S10			Y		DN	DN

Table 3 Test query set

No.	Query pattern	Query
1	Exact matching records	Employees getting a salary equal to h*salary*i
2	Range query	Employees getting a salary higher than h*salary*i
3	Aggregate-based matching	Employees getting the highest salary.
4	Group on join key, aggregate	Employees getting the highest salary per company id
5	Group on join non-key, aggregate	Employees getting the highest salary per company name
6	Aggregate operation	The highest salary offered
7	Join operation	Information about the companies and their departments

referencing styles like CR, PR, and TR. Schema S4 is much analogous to relational. Therefore, S4 is used only in the first experiment that executes on a non-sharded MongoDB server. Query workload describes the relationship between an application and the database schema. Test queries with different query patterns are chosen here to find answers to these questions. Table 3 lists these queries. These queries are implemented using Python 3 and Pymongo 3.8.

6 Results and Analysis

This experiment observes storage size, execution time, and energy consumption of application queries defined in Sect. 5 for all schema structures described in Table 2.

6.1 *Influence of Schema Design on Database Size*

Every schema may have one more collections from a set based on schema designs (EMP, DEPT, CMP, CDES). For each schema without and with additional index(es), respectively, Fig. 2a, b indicates individual collection sizes. Additional index(es) means the index created for each attribute reference and attribute query. The set and database size of the different schemas is shown in Fig. 2. Parent embedding is adopted by Schema S5 and S6 and creates the largest database size. In database size, Schema S10 is the next tallest. Size of S2 and S3 is found optimal. It helps us to conclude that child embedding has minimum storage requirement. Sharding is much famous to handle increasing database size. But, in Fig. 2c shows sharding increase storage almost 2 times for clustering of 2 nodes.

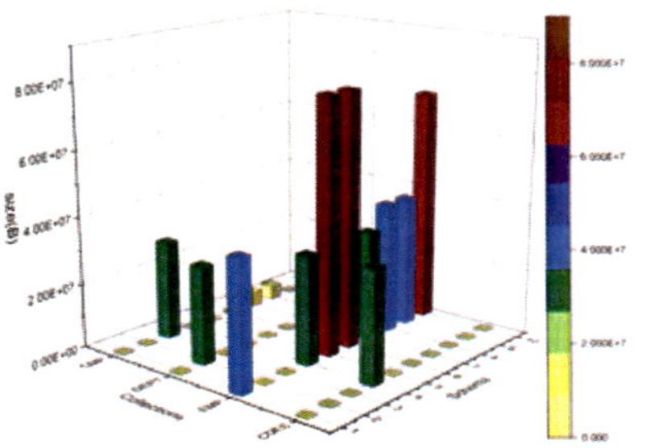

(a) Size of collections without indexing

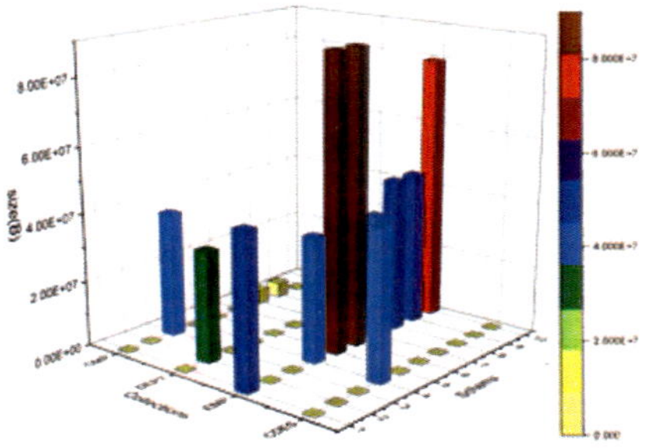

(b) Size of collections with indexing

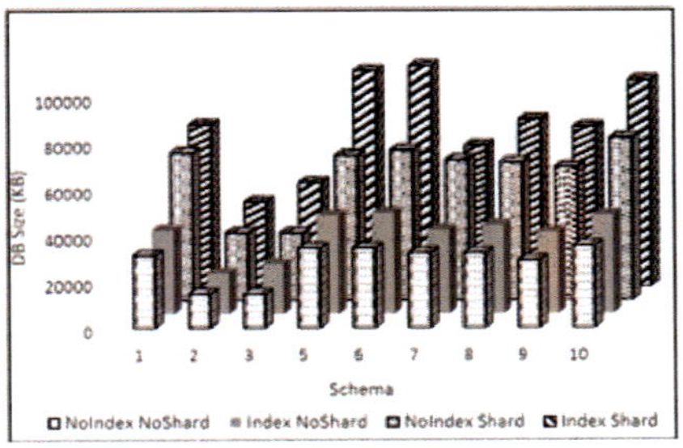

(c) Database Size before/after indexing and sharding

Fig. 2 Impact of indexing and sharding on database size in schema 1–10

6.2 Influence of Structuring Pattern on Performance and Energy Consumption

All queries listed in Table 3 were executed using single MongoDB statement by performing database-level join and unwind operations for accessing attributes from multiple collections and accessing nested elements respectively. Figure 3 shows execution time, and energy consumption by these queries on different schema described in Table 2. Figures 3a and 4a show energy consumption by different schema on the unshaded database without and with index(es), respectively. The main focus of developers is to identify the schema consuming best performance and optimum energy. Therefore, schema consuming near to minimum energy is magnified in the graph as shown in Figs. 3c and 4c. Similar to performance, no schema designs found neither uniformly best nor uniformly worst for the entire application query workload. Schema S7 is consuming minimum energy for query 6 without additional indexing, but it is not giving the best performance. Similarly, Schema S4 gives Query 3 the best efficiency, but consumes more energy. These examples show that energy consumption is not proportional to query performance. It requires detailed analysis to identify scenarios when energy optimization is a result of performance optimization, and when not.

6.3 Comparison of Database-Level Join and Application-Level Join

This section compares the influence of database-level join and application-level join on execution time and energy consumption. The requirement of join operations depends on the query and data model. For example, query-7 requests to find company details along with its associated department details. Database-level join operations are implemented using the "$lookup" function in the "aggregate" operation to identify such associations. Database-level join for query 7 on schema S2 is implemented using the following statement:

```
db.cmp.aggregate([{"$lookup":{"from":"dept","localField":"id",
"foreignField":"cmp id","as":"depts"}}]
```

Python code to implement application-level Join is described as:

```
result = db.cmp.find() for rec in resultC:
resultD = db.dept.find({"cmp id":rec["id"]})
```

Figure 5 shows the impact of applying to index, database-level join, and application-level join for query 5. The result proves that the use of application-level join is beneficial in improving query performance as well as reducing energy consumption in most cases. There are some exception cases like application-level join consume more time by schema 6 to execute query 5. Indexing is well-known to reduce execution time. But, Fig. 5 shows exceptions highlighted with red color label.

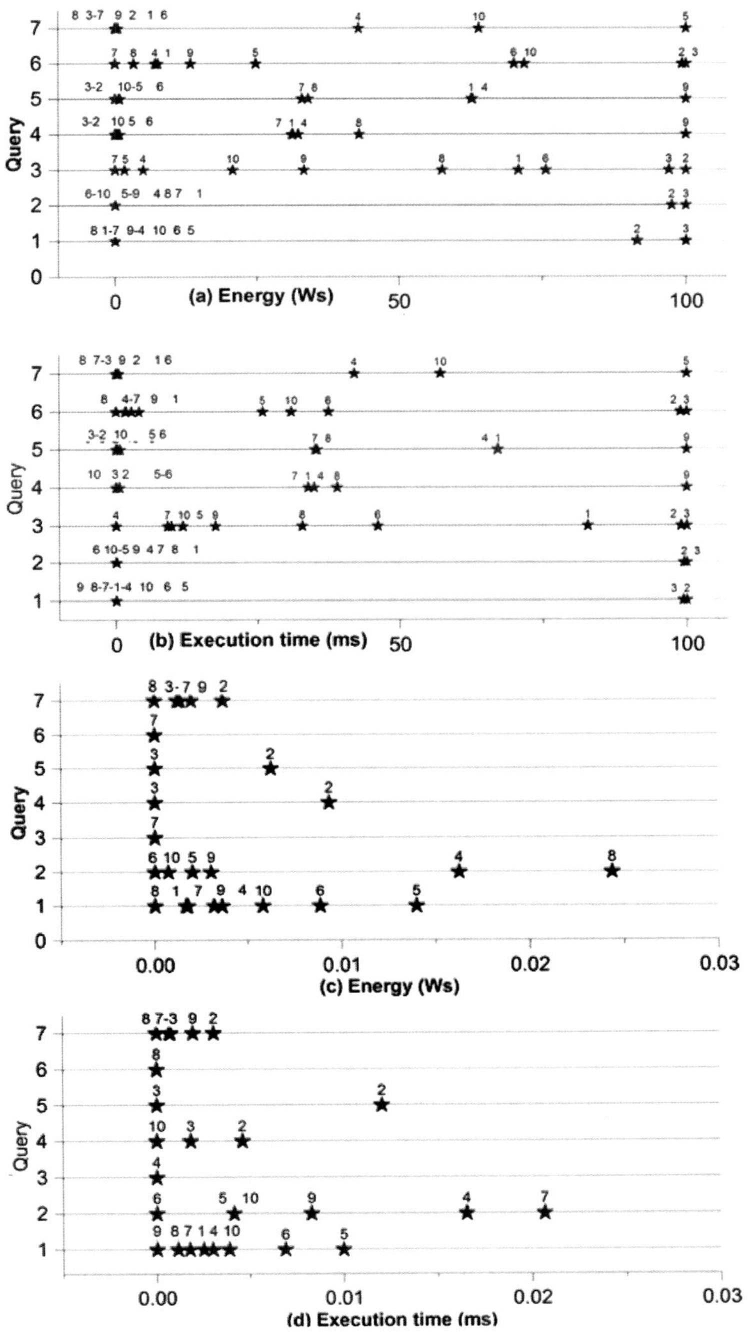

Fig. 3 Execution time and energy consumption for queries without index(es)

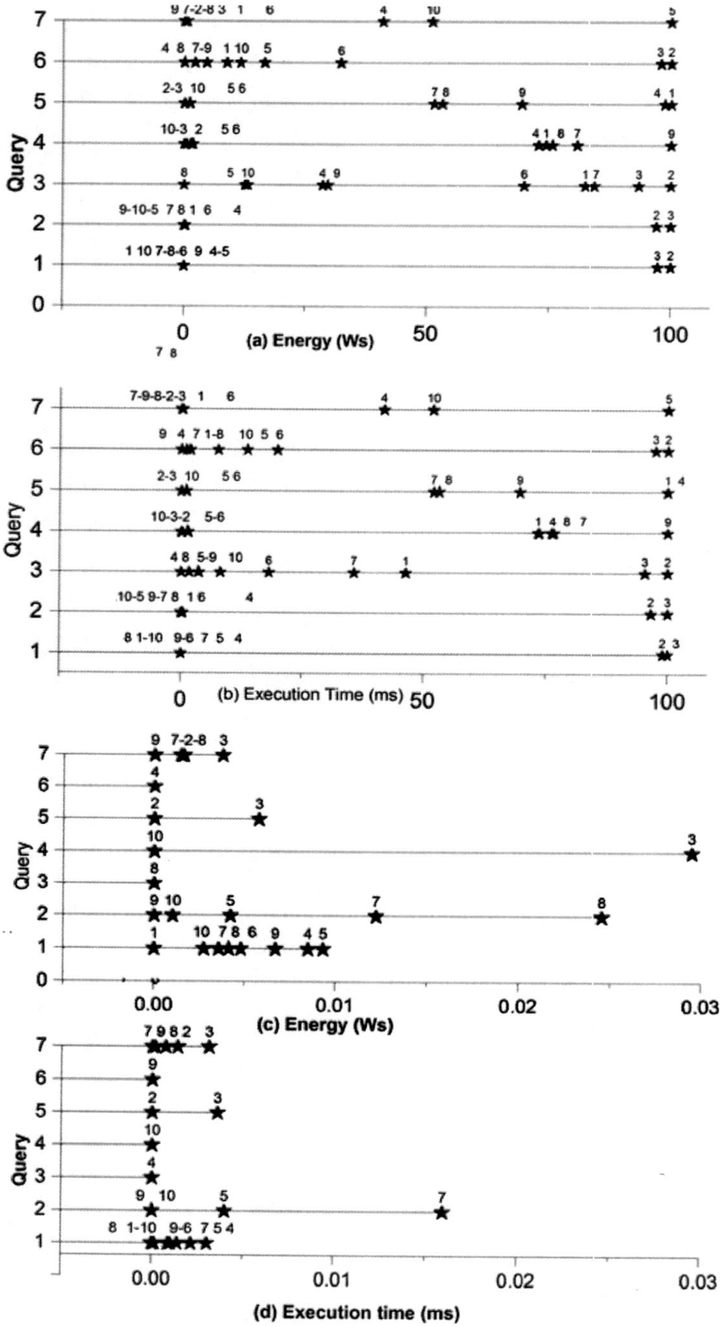

Fig. 4 Execution time and energy consumption for queries with index(es)

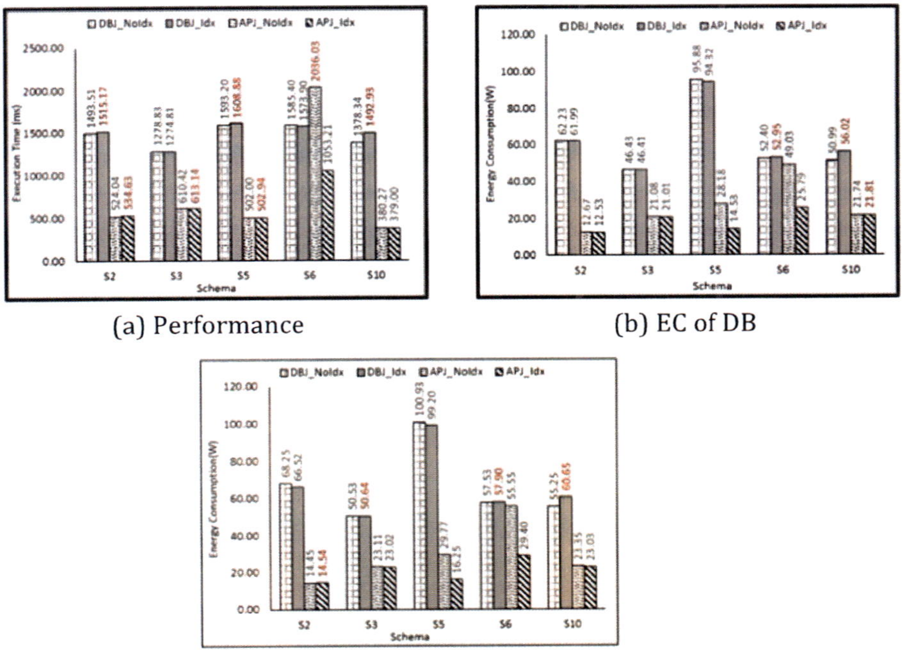

(a) Performance (b) EC of DB

(c) EC of (DB + App)

Fig. 5 Impact of join type, indexing, and sharding for query 5

6.4 Query Optimization and Schema Design

Execution time is considered as an objective for query optimization in most cases. These days, reducing energy consumption is another important objective, which is still in research. Application-level joins distribute significant computing to application nodes, which consumes noticeable energy consumption on application nodes.

The results shown in Fig. 5 show that optimization of energy consumption of database servers or/and application servers is not always returned as a result of performance optimization. Therefore, on document store, query optimization is implemented using these three goals, which determine the use of indexing and type of join (application-level join and database-level join). Result of using cost factors (execution time, energy consumption of database servers, and energy consumption of database server + application node) for each query on different schema is presented in Fig. 6a, b respectively. For each schema, query plan (indexing or not, application-level join or database-level join) returning minimum execution time is selected for each query. Figure 6a shows the sum of execution time and energy consumption of each query. If query plans are selected based on execution time, S2 can save energy consumption by (82.33 – 66.35 = 15.98 W) at cost of delay query workload

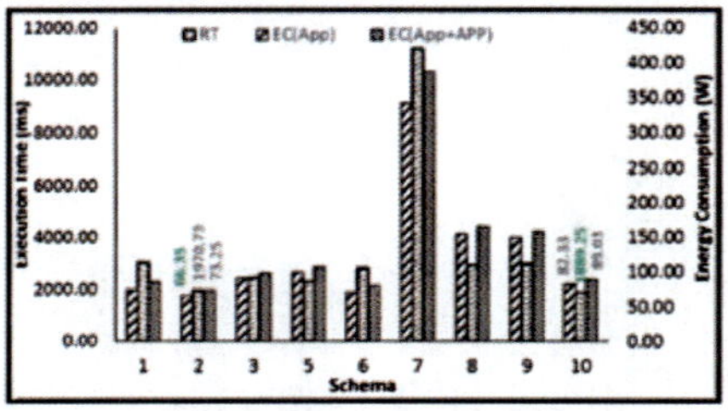

(a) Performance based

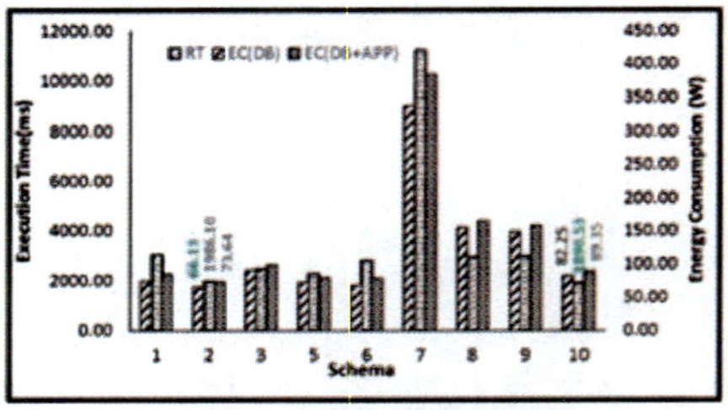

(b) EC (DB) based

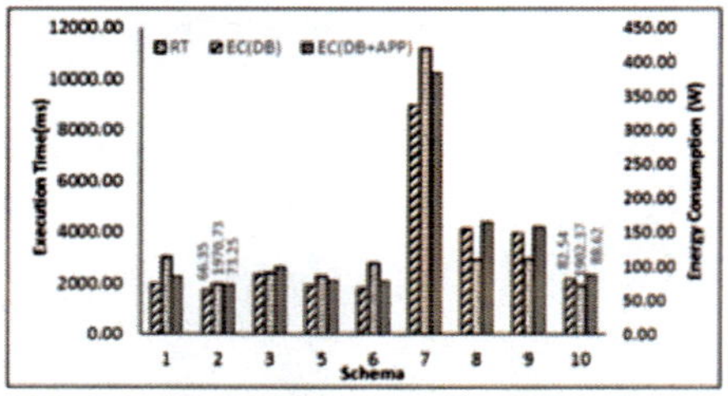

(c) EC (DB + App) based

Fig. 6 Comparison of query optimization factors

by ((1970.73 – 1889.25)/1000 = 0.081 s) only. Detailed analysis of S2 and S10 is described in Fig. 7a, b, which shows the impact of S2 and S10 on energy consumption and performance for different types of queries. De-normalizing structure pattern is found optimal at performance and energy both for queries on individual attributes. On another side, child embedding (CE) is more energy-efficient for join and aggregate operations. While the result of queries, Figs. 4 and 5 show that energy optimizations are not always a side effect of performance optimization. CE consumes less energy while taking more execution time.

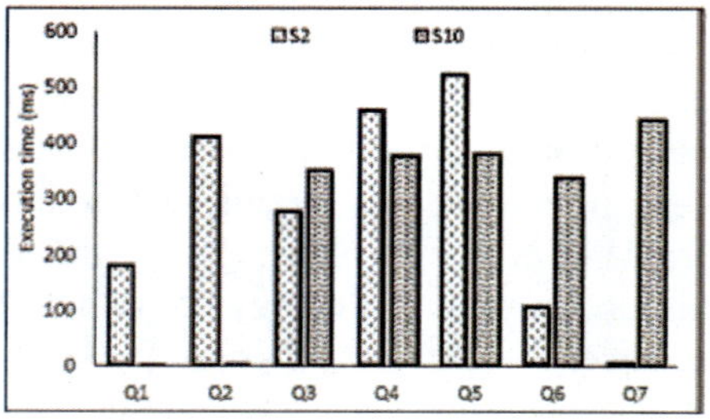

(a) Performance

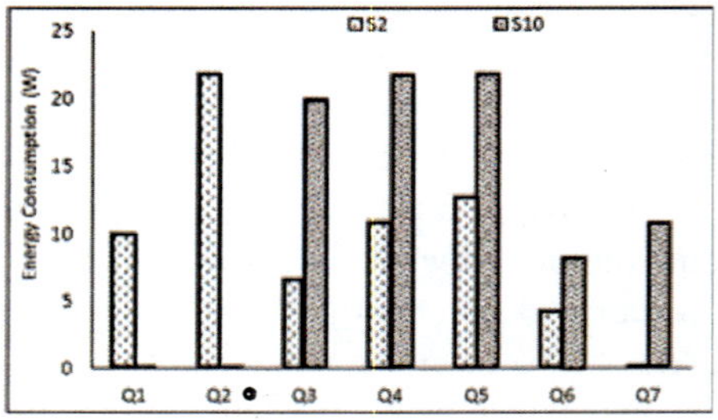

(b) EC (DB) based

Fig. 7 Performance and energy consumption of optimized plan for S2 and S10

6.5 *Interpreted Research Scope*

Experiment results and literature review show that expertise is required to structure (organize) data in document store for various aspects like storage space, query performance, energy consumption, consistency, availability, and load balancing. Nowadays, many application developers use cloud-based NoSQL services to get better scalability in their applications. Based on storage space requirements, multiple inserts, updating, removing, and reading requests, etc., cloud database services are paid. The above empirical analysis shows that the size of the storage and number of query transactions depend on query patterns, schema, and application-level join use. This shows us direction to many other research opportunities like guidelines for selecting structuring patterns for given size constraint, structuring pattern, performance constraint, and energy budget. Tools or framework can be presented to select proper schema design based on given application workload, system requirement, and data relationships. Application workload needs to be specified in detail like query pattern, the significance of query (e.g., query 1 needs better performance), the priority of parameters like query performance, energy consumption, storage size. For multi-objective, possibilities of trade-off and way of trade-off are required to identify.

7 Conclusion

This article presents an analysis of schema design in the NoSQL document store. To evaluate the database size, efficiency, and energy consumption of queries, it considered the structuring pattern and indexing portion of schema design. This article reveals that the choice of structuring pattern is to use index matters. This empirical analysis reveals several important points: (i) Embedding and de-normalizing on many sides has more storage requirement. (ii) There is no schema design neither uniformly best nor worst for all queries. (iii) Energy consumption is not proportional to query performance for all queries on any schema design, (iv) Application-level join optimizes performance as well as energy consumption of join operations. But, the use of indexing is not always efficient in the application-level join. (v) Query workload and optimization cost factors help to select energy-efficient schema.

References

1. F.J. Al Shekh Yassin, *Migrating From SQL to NoSQL Database: Practices and Analysis*. Accounting Dissertations (2017)
2. T.A. Alhaj, M.M. Taha, F.M. Alim, Synchronization wireless algorithm based on message digest (SWAMD) for mobile device database, in *International Conference on Computing, Electrical and Electronic Engineering (ICCEEE)*, 2013, pp. 259–262
3. P. Atzeni, Data modelling in the NoSQL world: a contradiction? in *Proceedings of the 17th International Conference on Computer Systems and Technologies 2016* (ACM, 2016), pp. 1–4

4. P. Atzeni, F. Bugiotti, L. Cabibbo, R. Torlone, *Data modeling in the NoSQLworld. Computer Standards and Interfaces* (2016)
5. P. Atzeni, C.S Jensen, G. Orsi, S. Ram, L. Tanca, R. Riccardo, The relational model is dead, SQL is dead, and I don't feel so good myself. ACM SIGMOD Rec. (2013)
6. A. Badia, D. Lemire, A call to arms. ACM SIGMOD Rec. (2011)
7. B. Bani, F. Khomh, F., Y.G. Guéhéneuc, A study of the energy consumption of databases and cloud patterns. Service-Oriented Comput. **9936**, 606–614 (2016)
8. A. Chebotko, A. Kashlev, S. Lu, A big data modeling methodology for Apache Cassandra, in *2015 IEEE International Congress on Big Data*, 2015, pp. 238–245
9. G. Daniel, A. Gómez, J. Cabot, UMLto[No]SQL: mapping conceptual schemas to heterogeneous datastores, in *2019 13th International Conference on Research Challenges in Information Science (RCIS)*, 2019, pp. 1–13
10. D. Duarte, O. Belo, Evaluating query energy consumption in document stores, in *Emerging Technologies for Developing Countries* (Springer International Publishing, 2018), pp. 79–88
11. C. Gomes, E. Tavares, M.N.D.O. Junior, Energy consumption evaluation of nosql dbmss, in *Anais do XV Workshop em Desempenho de Sistemas Computacionais e de Comunicação, SBC*, 2020, pp. 71–81
12. P. Gómez, R. Casallas, C. Roncancio, Data schema does matter, even in NoSQL systems!, in *2016 IEEE Tenth International Conference on Research Challenges in Information Science (RCIS)*, 2016, pp. 1–6
13. P. Gomez, C. Roncancio, R. Casallas, Towards quality analysis for document oriented bases, in *International Conference on Conceptual Modeling* (Springer, 2018), pp. 200–216
14. Y. Gu, X. Wang, S. Shen, J. Wang, J.U. Kim, Analysis of data storage mechanism in NoSQL database MongoDB, in *2015 IEEE International Conference on Consumer Electronics—Taiwan* (IEEE, 2015), pp. 70–71
15. S. Hamouda, Z. Zainol, document-oriented data schema for relational database migration to NoSQL, in *2017 International Conference on Big Data Innovations and Applications (Innovate-Data)*, 2017, pp. 43–50
16. M. Hamrah, *Data Modeling at Scale: MongoDb + Mongoid, Callbacks, and Denormalizing Data for Efficiency*. Adventures in HttpContext
17. J. Han, E. Haihong, G. Le, J. Du, Survey on NoSQL database, in *6th International Conference on Pervasive Computing and Applications*, 2011, pp. 363–366
18. H. Hashem, D. Ranc, Evaluating NoSQL document oriented data model, in *2016 IEEE 4th International Conference on Future Internet of Things and Cloud Workshops (FiCloudW)*, 2016, pp. 51–56
19. A.A. Imam, S. Basri, R. Ahmad, N. Aziz, M.T. Gonzalez-Aparicio, New cardinality notations and styles for modeling NoSQL document-store databases, in *TENCON 2017—2017 IEEE Region 10 Conference*, 2017, pp. 2765–2770
20. A.A. Imam, S. Basri, R. Ahmad, J. Watada, M.T. González Aparicio, M.A. Almomani, Data modeling guidelines for NoSQL document-store databases. Int. J. Adv. Comput. Sci. Appl. **9**(10), 544–555 (2018)
21. A.A. Imam, S. Basri, R. Ahmad, J. Watada, M.T. Gonzalez-Aparicio, Automatic schema suggestion model for NoSQL document-stores databases. J. Big Data **5**(1), 1–17 (2018)
22. T. Jia, X. Zhao, Z. Wang, D. Gong, G. Ding, Model transformation and data migration from relational database to MongoDB, in *2016 IEEE International Congress on Big Data (BigData Congress)*, 2016, pp. 60–67
23. A. Kanade, A. Gopal, S. Kanade, A study of normalization and embedding in MongoDB, in *2014 IEEE International Advance Computing Conference (IACC)*, 2014, pp. 416–421
24. G. Karnitis, G. Arnicans, Migration of relational database to document oriented database: structure denormalization and data transformation, in *2015 7th International Conference on Computational Intelligence, Communication Systems and Networks*, 2015, pp. 113–118
25. Katsov, *NoSQL Data Modeling Techniques* (2012), https://highlyscalable.wordpress.com/2012/03/01/nosql-data-modelingtechniques/
26. A. Khurana, *Introduction to HBase Schema Design*. White Paper, Cloudera (2012)

27. C. Lee, Y. Zheng, SQL-to-NoSQL schema denormalization and migration: a study on content management systems, in *2015 IEEE International Conference on Systems, Man, and Cybernetics*, 2015, pp. 2022–2026
28. W. Lee, B.W. On, I. Lee, J. Choi, A big data management system for energy consumption prediction models, in *Ninth International Conference on Digital Information Management (ICDIM 2014)* (IEEE, 2014), pp. 156–161
29. T. Li, G. Yu, X. Liu, J. Song, Analyzing the waiting energy consumption of NoSQL databases, in *2014 IEEE 12th International Conference on Dependable, Autonomic and Secure Computing* (IEEE, 2014), pp. 277–282
30. X. Li, Z. Ma, H. Chen, QODM: a query-oriented data modeling approach for NoSQL databases, in *2014 IEEE Workshop on Advanced Research and Technology in Industry Applications (WARTIA)*, 2014, pp. 338–345
31. S. Lin, H. Chen, F. Hu, A workload-driven approach to dynamic data balancing in MongoDB, in *2015 IEEE International Conference on Smart City/SocialCom/SustainCom (SmartCity)*, 2015, pp. 786–791
32. D. Mahajan, C. Blakeney, Z. Zong, Improving the energy efficiency of relational and NoSQL databases via query optimizations. Sustain. Comput. Inform. Syst. (2019)
33. D. Mahajan, Z. Zong, Energy efficiency analysis of query optimizations on MongoDB and Cassandra, in *2017 Eighth International Green and Sustainable Computing Conference (IGSC)* (IEEE, 2017), pp. 1–6
34. M.J. Mior, Automated schema design for NoSQL databases, in *Proceedings of the 2014 SIGMOD PhD Symposium* (ACM, 2014), pp. 41–45
35. M.J. Mior, K. Salem, A. Aboulnaga, R. Liu, NoSE: schema design for NoSQL applications, in *2016 IEEE 32nd International Conference on Data Engineering (ICDE)*, 2016, pp. 181–192
36. M.J. Mior, K. Salem, A. Aboulnaga, R. Liu, NoSE: schema design for NoSQL applications. IEEE Trans. Knowl. Data Eng. **29**(10), 2275–2289 (2017)
37. Mohan, History repeats itself, in *Proceedings of the 16th International Conference on Extending Database Technology* (2013), https://doi.org/10.1145/2452376.2452378. Accessed 2020-05-24
38. S. Mohan, Teaching NoSQL databases to undergraduate students: a novel approach, in *Proceedings of the 49th ACM Technical Symposium on Computer Science Education* (ACM, 2018), pp. 314–319
39. D. Pasqualin, G. Souza, E.L. Buratti, E.C. de Almeida, M.D. Del Fabro, D. Weingaertner, A case study of the aggregation query model in read-mostly NoSQL document stores, in *Proceedings of the 20th International Database Engineering & Applications Symposium* (ACM, 2016), pp. 224–229
40. A. Rasin, S. Zdonik, An automatic physical design tool for clustered column stores, in *Proceedings of the 16th International Conference on Extending Database Technology* (Association for Computing Machinery, 2013), pp. 203–214
41. R.A.S.N. Soransso, M.C. Cavalcanti, Data modeling for analytical queries on document-oriented DBMS, in *Proceedings of the 33rd Annual ACM Symposium on Applied Computing* (ACM, 2018), pp. 541–548
42. L. Stanescu, M. Brezovan, D.D. Burdescu, Automatic mapping of MySQL databases to NoSQL MongoDB, in *2016 Federated Conference on Computer Science and Information Systems (FedCSIS)*, 2016, pp. 837–840
43. B. Subramaniam, W.C. Feng, *On the Energy Proportionality of Distributed NoSQL Data Stores. High Performance Computing Systems. Performance Modeling, Benchmarking, and Simulation*, vol. 8966 (Springer International Publishing, 2015), pp. 264–274
44. P. Sugandhi, H. Karnewar, C. Joseph, P.J. Chaudhari, Improving energy efficiency of MapReduce framework using dynamic scheduling of work. Int. J. Recent Innov. Trends Comput. Commun. **4**(5), 373–377 (2016)
45. V. Varga, K.T. Jánosi, B. Kálmán, Conceptual design of document NoSQL database with formal concept analysis. Acta Polytech. Hungarica **13**(2), 229–248 (2016)

46. J.S. van der Veen, B. van der Waaij, R.J. Meijer, Sensor data storage performance: SQL or NoSQL, physical or virtual, in *2012 IEEE Fifth International Conference on Cloud Computing*, 2012, pp. 431–438
47. H. Vera, W. Boaventura, M. Holanda, V. Guimaraes, F. Hondo, Data modeling for NoSQL document-oriented databases. CEUR Workshop Proc. **1478**, 129–135 (2015)
48. William, Rules of Thumb for MongoDB Schema Design. Keon's Blog (2016), https://keon.io/mongodb-schema-design/. Accessed 2019-10-14

A Survey on Wireless Sensor Networks and Instrumentation Techniques for Smart Agriculture

R. Madhumathi, T. Arumuganathan, and R. Shruthi

Abstract Agriculture is the primary source for the development of our country's economy. It involves production, processing, marketing, and distribution of agricultural produce. Smart agriculture is one of the modern farming practices that increase the agricultural production in a most efficient manner. Smart agriculture involves wireless sensor networks (WSN) for various operations. Wide variety of sensors are used in WSN that assists farmers to know the statistical details of their field which helps them to take accurate decisions and provides instantaneous feedback for various agricultural parameters. Numerous researchers and manufacturers aim to develop real-time sensors based on different approaches, namely electromagnetic, electric, optical, mechanistic, acoustic, or electrochemical, which calculates the physical and chemical properties of soil in agricultural fields. The main objective of this study is to review the application of wireless sensors used in agriculture, the principle and various instrumentation techniques in measuring different parameters in agriculture.

Keywords Smart agriculture · Sensors · Wireless sensor networks · Instrumentation techniques · Soil properties

1 Introduction

Agriculture is the premier source of production which is the basis for livelihood. The primary objective in agriculture is to increase the crop productivity with improved product quality. The soil parameters like soil moisture, pH, temperature, humidity, nutrient levels, electrical conductivity, salinity level, crop and weather conditions are

R. Madhumathi (✉) · R. Shruthi
Department of Computer Science and Engineering, Sri Ramakrishna Engineering College, Coimbatore, India
e-mail: madhumathi.r@srec.ac.in

R. Shruthi
e-mail: shruthi.1801217@srec.ac.in

T. Arumuganathan
ICAR-Sugarcane Breeding Institute, Coimbatore, India

S. Shakya et al. (eds.), *Mobile Computing and Sustainable Informatics*, Lecture Notes on Data Engineering and Communications Technologies 68,
https://doi.org/10.1007/978-981-16-1866-6_33

to be monitored continuously to ensure proper growth of crops. Smart agriculture is a technique that uses computer technology to ascertain optimal growth of crops. It encompasses a set of leading-edge advancements which combines sensors, information systems, and information management to enhance the crop cultivation. Smart agricultural systems improve productivity, reduce waste, and enable management of greater number of resources.

With the current advancements, various on-the-go sensors are available to measure the soil, crop, and weather attributes of the field [1]. The development of sensors increases the effectiveness of precision agriculture. These sensors provide data which assists farmers to monitor and optimize the cultivation of crops by adapting to the environmental changes. Many support systems are developed with sensor networks in the area of application under Internet of things (IoT) [2]. For agriculturists, the innovation in technology is a boon in enhancing the productivity with minimum efforts and expenditure. The recent ideas for precise measurement of soil or plant characteristics using sensors and various instrumentation methods in smart agriculture are reviewed and its applications are discussed.

2 Sensors for Precision Agriculture

WSN consists of embedded devices such as sensors and gateways which are connected to empower several facilities for measurement at low power and minimized effort. The sensing methods in smart agriculture measures the in-situ soil and plant characteristics [3]. These include

- Electromagnetic
- Optical
- Mechanical
- Acoustic and Airflow
- Electrochemical

In addition to this, there are many individual sensors which directly measures the parameters like soil moisture, soil pH, oxides and nitrates, soil electrical conductivity, soil compaction, leaf turgidity, and weather factors. Figure 1 describes the network connectivity of sensors with several hardware components.

Various networking and technological solutions like wireless sensor networks, wireless multimedia networks, and mobile application-based systems are being developed which promotes precision agriculture. The applications of IoT using microcontrollers [4] for sensor-based monitoring have led to the development of smart agriculture [5]. Emerging technologies like cloud computing, data analytics, and artificial intelligence (AI) also promote precision agriculture. An e-learning system based on cloud model to educate the farmers about modern farming methods was developed which empowers farmers to obtain real-time information through different devices like laptops and smart phones [6]. Table 1 represents the sensing methods targeted with soil characteristics.

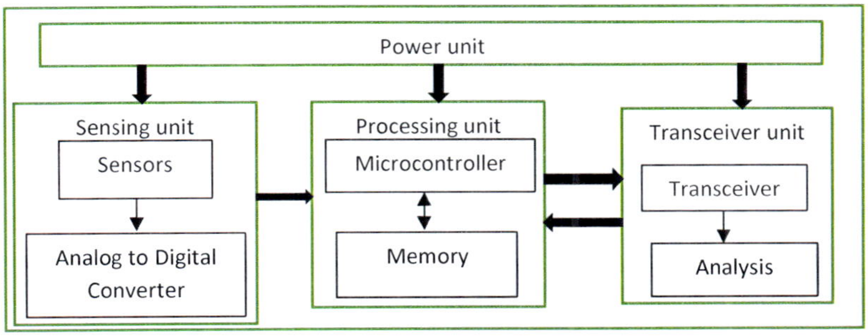

Fig. 1 Sensor node hardware components

2.1 Electrical and Electromagnetic Sensors

In order to obtain the soil physical and chemical properties such as soil electrical conductivity, cation exchange capacity (CEC), soil roughness and smoothness, total carbon content, soil drainage conditions and salinity levels, electromagnetic sensors are used. To determine the potential of soil particles and the electrical conductivity, electrical circuits are used in these sensors. Numerous approaches using contact and non-contact methods for using electromagnetic sensors have been attempted. Based on the parameters measured, electromagnetic sensors commercially available are classified as soil electrical conductivity determination sensors and electromagnetic response mapping sensors.

Soil Electrical Conductivity Sensor: The soil electrical conductivity (EC) is used to map the soil variables and helps in evaluating the concentration of the soil solute. Various soil properties are analyzed by measuring the electrical conductivity of the soil. EC is weighed in milli-Siemens/meter. Numerous factors affect the EC of the soil. They are the moisture level, temperature, porosity, cation exchange, and the soil salinity. These factors are to be considered when measuring the EC value. Figure 2 represents the principle of electrical conductivity in a coil inducing magnetic flux where an electromagnetic force (EMF) is generated.

2.2 Optical Sensors

In order to obtain the absorption, reflectance, and the transmitting characteristics of the soil, optical sensors are used. These sensors operate at various wavelengths ranging from monochromatic to multispectral and hyperspectral. The light reflectance principle helps in measuring the nutrients, pH, moisture, and cation exchange capacity (CEC), organic matter, and total carbon content present in the soil. Leaf area index (LAI), crop nitrogen status, leaf chlorophyll, and yield are measured by the leaf sensor which senses the green light wavelengths. The development of these optical systems has robust calibration and validation methodologies

Table 1 Soil characteristics focused with several on-the-go sensing approaches

S. No.	Sensing methods	Soil texture (clay, silt, and sand content)	Soil organic matter or total carbon content	Soil electrical conductivity	Soil compaction or bulk density	Soil pH	Soil moisture content	Depth variability (depth of topsoil or hard pan detection)	Potassium content	Total Nitrogen
1	Electrical and electromagnetic	✔	✔	✔			✔	✔		✔
2	Optical	✔	✔			✔	✔			✔
3	Mechanical				✔			✔		
4	Acoustic and pneumatic	✔			✔			✔		
5	Electrochemical			✔		✔	✔		✔	✔

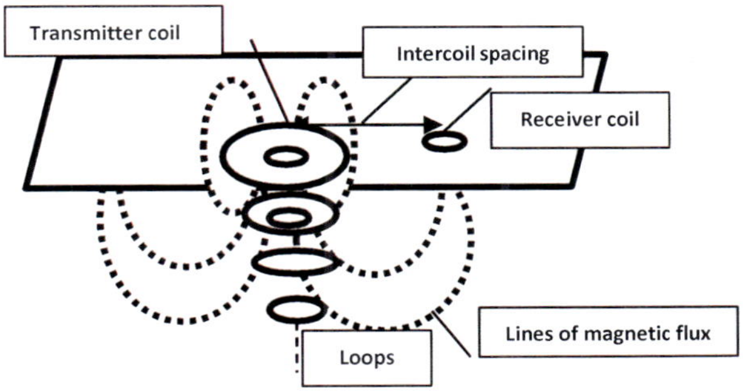

Fig. 2 Principle of electrical conductivity sensor

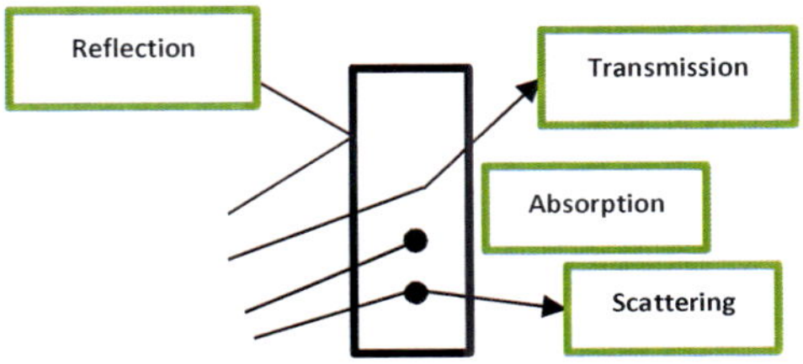

Fig. 3 Principle of optical sensor

to ensure accuracy in the result obtained. Figure 3 represents the principle of optical sensors that include absorption, reflection, transmission, and scattering.

2.3 Mechanical Sensors for Crop

Mechanical sensors estimate soil mechanical resistance and operate based on soil compactness for the growth of crop's environment. The mechanism involves recording the potential determined by the strain gauges or by penetration of the soil. The ratio of pressure required to diffuse into the soil medium is the mechanical resistance of the soil.

Soil Compaction Sensor: The major factor affecting the root growth is the soil compaction. When soil particles are held together compaction occurs, thereby reducing the pore space. This sensor is equipped with a rod, and a cone is attached to it. When this sensor is inserted in the ground, the values are measured. A digital hand-pushed penetrometer was designed [7] to measure the soil compaction and is attached to a tractor horizontally, and the penetration resistance values were obtained. A 2D/3D compaction maps for different depth layers of soil were designed, and the same principle of penetrometer was performed [8]. The surface nuclear gauge (SNG) is used for detecting the compaction in soils where the organic matter is less than 5%. Using SNG, the soil texture, moisture, and compaction level is obtained. Penetrometers are very useful for preliminary evaluation of soil compaction but their readings are less reliable in compacted soils [9]. Another alternate methodology to cone penetrometer involves generation of maps of soil mechanical resistance using three instrumented blades, GPS, and data logger which is connected to a tractor [10].

Sensors to Identify Water Potential in the Crop: This sensor measures the changes in the turgor pressure of the leaf. When the leaf is placed in between the two magnets, it measures the difference between the turgor of the plant and magnetic pressure. When the plant turgor is low, the magnetic pressure increases, thereby confirming that the plant needs to be watered.

Sensors for Measuring Biomass: This sensor consists of a pendulum which moves over the crop and detects differences in the biomass of the crop. Therefore, by measuring the biomass of the crop, this sensor facilitates the application of nitrogen on certain areas, apart from ensuring protection of the plant.

2.4 Acoustic and Air Flow Sensors

In order to determine the soil texture (silt, sand, clay), soil depth variability, soil depth variability, soil compaction, crop herbicide application, fruit harvesting, grain and forage harvesting, weeding, pest detection, acoustic sensors are utilized. The texture of soil is measured by the change in the noise levels when the sensor comes into contact with the soil. The sensor emits the signal with certain frequency (50–100 kHz), and several echoes occur. By comparing the echo signal with the features of the foliage structure, various data like size, shape, positioning, and orientations of leaves are obtained. Wireless acoustic sensors for pest detection are located in the field which picks up the sound waves of insects and prevents the pests from damaging the crops. Figure 4 represents the sound waves produced and the distance covered by the wave signal of the acoustic sensor.

Pneumatic Sensor: In order to measure the soil air/water permeability, pneumatic sensors are used. The permeability is measured by the pressure which is necessary to thrust a specific quantity of air/water into the soil.

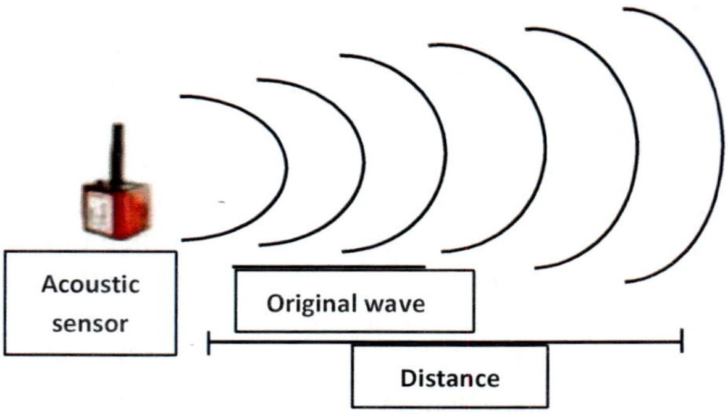

Fig. 4 Principle of acoustic sensor

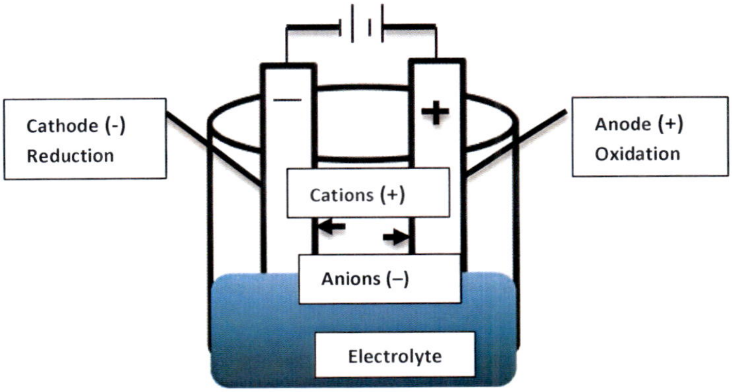

Fig. 5 Principle of electrochemical sensor

2.5 Electrochemical Sensors

In order to determine the concentration of soil nutrients, pH and moisture, electrochemical sensors are used. These sensors consist of an ion-selective electrode (ISE) or an ion-selective field effect transistor (ISFET) for measuring the voltage which is then mapped with its range to determine the values. The amount of moisture, nitrates, phosphates, potassium, hydrogen, and pH are detected with electrochemical systems. Figure 5 represents an electrochemical chamber consisting of an anode, cathode, electrolyte, and a DC source.

2.6 Soil Moisture Sensor

The soil moisture is measured by determining the soil properties such as dielectric constant or the electrical resistance. This supports farmers to maintain the irrigation system efficiently. Increase in yield and the quality of the crop can be enhanced by efficient management of water content. There are two probes in the soil moisture sensor which allows the current to pass through the soil. The electrodes are inserted into the soil and output indicating the resistance of the soil is obtained which is then mapped to get the moisture percentage in the soil. When there is high soil moisture, the soil conducts more electricity and offers negligible resistance. When there is less moisture, the soil conducts less electricity and the resistance offered is high. Listed below are the types of moisture sensors available.

Tensiometer: An instrument which determines the matric water potential (soil moisture tension) of soil is tensiometer. This instrument is shoved into the soil, and the partial vacuum is pulled out using a hand pump. The vacuum increases when water is removed from the soil and decreases when water is included in the soil. The gauge determines the matric water potential. Tensiometer manages the irrigation system through irrigation scheduling and measures the depth of the roots. Farmers can determine the wetness in the soil from the readings obtained. The tensiometer has a tube which is filled with deaerated water and is immersed into the soil. Vacuum is created inside the tensiometer due to the movement of the water, and the matric water potential is displayed on the dial gauge. This reading gives the measure of force needed to extract water from the soil.

Watermark Sensor: This is a permanent sensor, which is kept in the soil for continuous monitoring of soil water tension. A pair of very high resistant electrodes are immersed within a granular matrix, and current is applied to it. When the tension changes with respect to the variation in moisture level, the resistance changes accordingly. Two electrodes are kept in a gypsum block, and the resistance between them is noted. This gypsum block supports in measuring the soil moisture content. The sensor is then immersed in the soil. If the soil is dry, there would be less water in the gypsum block resulting in high resistance between the electrodes. Alternatively, if the soil is wet, there would be more water and the resistance will be less. The soil moisture content is directly proportional to the electrode resistance.

An electrical resistance is developed which is mapped and converted to obtain the moisture level in the soil. A limited linear relationship is maintained. The observed resistance is normalized to soil temperature in terms of °C by Eq. (1).

$$R_{21} = \frac{R_s}{1 - (0.018 \times \mathrm{d}T)} \tag{1}$$

where

R_{21} denotes the resistance in Ohm (Ω) at 21 °C,
R_s denotes the measured resistance in Ω,
$\mathrm{d}T$ denotes the soil temperature in °C,

The moisture content in the soil is calculated using Eq. (2).

$$\text{SoilMoisture} = (0.07407 \times R_{21}) - 0.03704 \tag{2}$$

Soil Moisture Capacitive Sensor: Soil moisture can be measured using the dielectric permittivity of the soil, and the change in the capacitance value indicates the moistness of the soil. The sensor must be calibrated for different types of soil, as the measurement is soil dependent. When the sensor is immersed in the soil, the water content is determined. A probe is immersed into the soil at a depth of 1–3 m, and the dielectric permittivity is calculated. Since water has higher dielectric constant than the other soil components, changes in the soil moisture content will show a change in the soil capacitance. The capacitance is determined using Eq. (3).

$$C = (\varepsilon_r \varepsilon_0 A/d) \tag{3}$$

where

C denotes the capacitance in Farad (F),
A denotes the overlapping area of the plates in m^2,
ε_r denotes the relative permittivity,
ε_0 denotes the permittivity of free space,
d denotes the distance between the positive and negative plate in cm.

The voltage developed and the charge on the plates can be obtained by Eq. (4).

$$C = q/V \tag{4}$$

where

q denotes the charge in Coulomb (C),
V denotes the potential difference in Volts (V).

2.7 Soil pH Sensor

The soil pH is measured using different techniques that use either electrodes or colorimetry or a spectroscopy. The potential of hydrogen present in the soil determines whether the soil is acidic or alkaline or neutral. The optimal pH value for a plant growth ranges between 5.5 and 7.5. The availability of microorganisms and other essential nutrients are determined by their pH values. The pH is mostly alkaline or neutral in dry weather, whereas during warm conditions, the acidification of the soil happens. A pH sensor is made up of two electrodes and a battery. A potential is developed which is proportional to the hydrogen ion concentration in the solution. The pH value is calculated using Eq. (5).

$$\text{pH} = -\log(a_{H+}) \tag{5}$$

The potentiometric pH value is obtained by Eq. (6).

$$\text{pH}(Y) = \text{pH}(\text{Std}) + \frac{E(\text{Std}) - E(Y)}{R \times \frac{T}{F} \log 10} \tag{6}$$

where

R denotes the gas constant,
T denotes the temperature in K,
F denotes the Faraday constant,
pH(Std) denotes the pH of standard solution,
pH(Y) denotes the pH of unknown solution,
$E(\text{Std})$ and $E(Y)$ denote the potential of electrodes in mV in the pH cell.

In addition to this, spectroscopic methods are also designed to determine soil pH.

2.8 Soil Temperature Sensor

In order to measure the soil temperature, the soil temperature sensor is used. By measuring the temperature over a period of time, we can determine the crop type. Temperature sensors have different designs which use thermocouple and thermistors. The sensors transmit electric signals, which can be converted into the different units of measurement like Kelvin, Celsius and Fahrenheit. A temperature sensor contains a *p–n* junction diode. The Eq. (7) gives the diode's forward current.

$$I_{\mathrm{f}} = \exp \frac{qV_{\mathrm{f}}}{nkT} \tag{7}$$

where

I_{f} denotes Forward current in mA,
k denotes Boltzmann's constant,
V_{f} denotes Forward voltage in V,
T denotes the absolute temperature in K,
q denotes the magnitude of charge in C,
n denotes the values ranging from 1 to 2.

It is observed that the output voltage varies linearly with temperature. On rearranging Eq. 7, we obtain the absolute temperature using Eq. (8).

$$T = \frac{q \times V_{\mathrm{f}}}{nk \times \log I_{\mathrm{f}}} \tag{8}$$

A multimodal sensor which simultaneously measures temperature, pH, and electrical conductivity was developed using CMOS technology. The temperature is measured based on the output voltage obtained [11].

Measuring Temperature using Thermocouple: The temperature can also be determined using a thermoelectric device called a thermocouple. Temperatures ranging between 200 and 2000 °C can be measured using appropriate thermocouple. Thermocouple consists of two metal wires connected to both ends, and a potential difference is formed between the two junctions which is directly related to the difference in temperature. The resolution of this sensor ranges from 9 to 12 bits, and the analog output value is converted into digital value.

2.9 Sensors for Measuring Soil and Plant Characteristics

The different sensors for determining soil and leaf/plant characteristics are listed in Tables 2 and 3.

Table 2 Sensors for determining soil characteristics

S. No.	Sensors	Temperature	Moisture	Dielectric permittivity	Rain/water flow	Conductivity	Salinity
1	Hydra probe II soil sensor	✔	✔	✔	✔	✔	✔
2	CS 650 soil sensor	✔	✔	✔		✔	
3	Hydra GO	✔	✔	✔		✔	
4	ECH_20 EC-5	✔	✔	✔		✔	
5	EC 250 sensor	✔	✔		✔	✔	✔
6	107 L temperature sensor	✔					
7	Tipping bucket rain gauge				✔		

Table 3 Sensors for determining leaf/plant characteristics

.	Sensors	Photosynthesis	Hydrogen	CO_2	Temperature	Moisture
1	237 leaf wetness sensor				✔	✔
2	Davis leaf wetness sensor—6420				✔	✔
3	Leaf wetness sensor LM-100				✔	✔
4	YSI 6025 chlorophyll sensor	✔				
5	PM-11 phytomonitor	✔			✔	✔
6	TP2 portable photosynthesis	✔		✔	✔	✔
7	PTM-48 photosynthesis monitor	✔		✔	✔	✔
8	CL-340 handheld photosynthesis	✔	✔	✔	✔	

3 Instrumentation Techniques in Smart Agriculture

3.1 Electrical and Electromagnetic Techniques

The spatial variability of soil distinguished by the electromagnetic induction (EMI) surveys is used to assess field-scale soil salinity, soil moisture, soil texture, EC, compaction, and pH. Geospatial information is easily derived through these surveys, and this information improves the overall supervision of the agricultural fields [12]. For differentiating soil types, a study was proposed to estimate that whether EC alone has greater accuracy in determining the soil type or EC along with the yield data of crop has high accuracy. The study proves that EC along with the yield data of crop has high accuracy which scales over 90% [13]. EMI technique is proved to be more advantageous for soil EC sensing as it is a non-contact approach and evaluates soil salinity directly by using depth-weighted apparent EC measurements. Another approach to distinguish spatial variability of soil which assesses EC, salinity, coarseness, moisture content, and bulk density is field-scale apparent soil EC directed soil sampling [14]. The soil properties are monitored dynamically to perform soil surveys, selection of crop, site-specific management, and assessment of soil quality.

3.2 Optical Techniques for Precision Agriculture

An optical sensor based on photometric principle using IoT devices to detect soil nutrients was developed which measured six soil nutrients: ammonia nitrogen, nitrate nitrogen, available phosphorous, iron, manganese, and calcium [15]. An atomic absorption spectroscopy technique involving laser-induced breakdown spectroscopy (LIBS) was implemented, where a spectrum is formed due to the excitation of ions when the laser beam is made to fall on the sample. The LIBS technique combined with multivariate methods like Gaussian process regression (GPR), partial least squares regression (PLSR), and least absolute shrinkage and selection operator regression (LASSO) is an alternative for simultaneous detection of pH, CEC, and nutrients [16]. A soil reflectance spectrum was acquired by the spectroradiometer which measures the soil carbon (C) and nitrogen (N) [17]. The model obtained from the spectrum was trained to calculate clay, sand and other soil organic matter (SOM) in the field using multivariate statistical analysis. A system for prediction of nitrogen using the tractor mounted-N sensor based on SPAD algorithm was developed [18]. The amount of light reflectance is obtained using spectrometers and is calibrated. For rapid detection of NPK, the single pulse and the double pulse laser-induced spectroscopy methods are compared which ensures to be a reliable method. Collinear double pulse laser method showed high accuracy than that of single pulse method [19]. Also, the reflectance spectrum obtained is manipulated to determine the vegetation features of plants and detects pests.

3.3 *Acoustic and Pneumatic Techniques*

A novel sensor was designed for the measurement of soil texture and degree of compactness [20]. These findings concluded that with the passive acoustic emission, the soil particle size can be distributed. A model was employed in porous media and the speed-moisture curve which produced a suitable sound frequency was derived, and this method proved to be inexpensive [21]. An on-the-go multiprobe acoustic compaction system was developed, and the soil compaction was estimated in three layers based on the power spectral density (PSD) of the sound which can be easily mounted on tillage tines [22]. A triaxial permeameter was used to determine the co-efficient of permeability (CP) of deformable low permeability soils [23]. The CP was indirectly measured by soil–water characteristic curves (SWCC) using Van Genuchten model, Fredlund and Xing model. The CP at different depths using falling head permeability test was calculated, and this proved to be an efficient method to calculate CP.

3.4 *Electrochemical Techniques*

A direct soil measurement (DSM) using ion-selective electrodes is implemented for the soil pH, available potassium, nitrogen, and sodium contents using the root mean squared error (RMSE) [24]. The influence of different electrolytes on the suspension effect of pH in calcareous soils was studied. There was a variation in soil pH due to the release of ions [25]. Soil pH is directly measured by a sensor with an ion-selective antimony electrode which had higher accuracy. There was a high linearity obtained between the sensor readings and the standard laboratory measurements [26]. Voltammetry sensor based on Nafion modified screen-printed carbon electrode was designed to determine the soil potassium. The analytes are detected based on the voltammetry principle which is an electrochemical technique [27].

4 Conclusion

Agriculture requires technical solutions to increase crop production, and due to the advancements in technology, the usage of sensors is applied in the farm to monitor the agricultural parameters effectively. There exists a wide range of sensors due to which selection has been a difficult task because of various characteristics. Hence, it is preferable to consider the on-the-go mapping of electrical, electrochemical, and electromagnetic sensors to discover the differences in the field. Since development of many on-the-go sensors are in process, only electrochemical, electrical, electromagnetic sensors, and optical techniques are highly used in precision agriculture along with various real-time and map-based techniques are applied in smaller fields.

Some of the commercially available sensors cannot be directly used to measure physical/chemical properties such as moisture, salinity, organic matter, texture. In contrast to this, various electrical and electromagnetic sensors provide useful data about the soil differences, and they are mapped to deduce the physical variables. Studies investigating these sensors are ongoing, and more data will be available soon for precision agriculture.

Acknowledgements The authors sincerely thank the Science and Engineering Research Board (SERB) of DST for the financial support and the Director, ICAR-SBI, Coimbatore, and the Management and Principal of Sri Ramakrishna Engineering College, Coimbatore, for extending the required facilities.

References

1. A.Z. Aqueel-ur-Rehman, N.I. Abbasi, Z.A. Shaikh, A review of wireless sensors and networks' applications in agriculture. Comput. Stand. Interf. **36**, 263–270 (2014)
2. R.K. Kodali, N. Rawat, L. Boppana, *WSN sensors for precision agriculture*, in *IEEE Region 10 Symposium*, 2014, pp. 651–656
3. V.I. Adamchuk, J.W. Hummel, M.T. Morgan, S.K. Upadhyaya, On-the-go soil sensors for precision agriculture. Comput. Electron. Agric. **44**, 71–91 (2004)
4. S. Veena, M. Rajesh, K. Mahesh, S. Salmon, Survey on smart agriculture using IoT. Int. J. Innov. Res. Eng. Manage. **5**, 63–66 (2018)
5. M.S. Mekala, V. Perumal, A survey: smart agriculture IoT with cloud computing, in *International Conference on Microelectronic Devices, Circuits and Systems* (IEEE, Vellore, India, 2017), pp. 1–7
6. R. Madhumathi, Elucidating farmers towards smart agricultural farm building through cloud model, in *10th International Conference on Computing Communication and Networking Technologies* (IEEE, Kanpur, 2019), pp. 1–4
7. A. Rezaee, Design, construction and evaluation of a digital hand-pushed penetrometer. Int. J. Adv. Smart Sensor Netw. Syst. (IJASSN) **7**, 1–10 (2017)
8. Y. Tekin, B. Kul, R. Okursoy, Sensing and 3D mapping of soil compaction. Sensors **8**, 3447–3459 (2008)
9. T.B. Randrup, J.M. Lichter, Measuring soil compaction on construction sites: a review of surface nuclear gauges and penetrometers. J. Arboricult. **27**, 109–117 (2001)
10. V.I. Adamchuk, P. José, Molin: instrumented shanks for soil mechanical resistance measurements. Engenharia Agrícola **26**, 161–169 (2006)
11. M. Futagawa, H. Takao, M. Ishida, K. Sawada, Fabrication of a multi-modal sensor with pH, EC and temperature sensing areas for agriculture application, in *IEEE Sensors* (New Zealand, 2009), pp. 2013–2016
12. S.M. Lesch, D.L. Corwin, D.A. Robinson, Apparent soil electrical conductivity mapping as an agricultural management tool in arid zone soils. Comput. Electron. Agricult. **46**, 351–378 (2005)
13. C.M. Anderson-Cook, M.M. Alley, J.K.F. Roygard, R. Khosla, R.B. Noble, J.A. Doolittle, Differentiating soil types using electromagnetic conductivity and crop yield maps. Soil Sci. Soc. Am. J. **66**, 1562–1570 (2002)
14. D.L. Corwin, E. Scudiero, Field-scale apparent soil electrical conductivity, in *Methods of Soil Analysis*, vol. 1, ed. by S. Logsdon (Soil Science Society of America, Madison, WI, 2016), pp. 1–29

15. M. Yokota, T. Okada, I. Yamaguchi, An optical sensor for analysis of soil nutrients by using LED light sources. Measur. Sci. Technol. **18**, 2197–2201 (2007)
16. A. Erler, D. Riebe, T. Beitz, H.-G. Lohmannsroben, R. Gebbers, Soil nutrient detection for precision agriculture using handheld laser-induced breakdown spectroscopy (LIBS) and multivariate regression methods (PLSR, Lasso and GPR). Sensors **20**, 1–17 (2020)
17. A.-M. Fortuna, P.J. Patrick, A.M. Nelson, J.L. Steiner, Prediction of soil carbon fractions using a field spectroradiometer equipped with an illuminating contact probe. Soil Syst. **3** (2019)
18. M. Singh, R. Kumar, A. Sharma, B. Singh, S.K. Thind, *Calibration and algorithm development for estimation of nitrogen in wheat crop using tractor mounted N-sensor*. Sci. World J. **2015** (2015)
19. Y. He, X. Liu, Y. Lv, F. Liu, J. Peng, T. Shen, Y. Zhao, Y. Tang, S. Luo, Quantitative analysis of nutrient elements in soil using single and double pulse laser-induced breakdown spectroscopy. Sensors **18** (2018)
20. M. Naderi-Boldaji, M.Z. Tekeste, RA. Nordstorm, D.J. Bernard, S.J. Birrell, A mechanical-dielectric-high frequency acoustic sensor fusion for soil physical characterization. Comput. Electron. Agric. **156**, 10–23 (2019)
21. F. Adamo, G. Andria, F. Attivissimo, N. Giaquinto, An acoustic method for soil moisture measurement. IEEE Trans. Instrument. Measur. **53**, 891–898 (2004)
22. A. Moallemi-Oreh, S. Minaei, A. Sharifi-Malvajerdi, A.M. Borghaee, *Multiprobe soil compaction sensor: an acoustical approach*. J. Food Agric. Environ **8**, 747–750 (2010)
23. S. Huang, D.G. Fredlund, S.L. Barbour, Measurement of the coefficient of permeability for a deformable unsaturated soil using a triaxial permeameter. Canadian Geotechn. J. **35**, 436–432 (1998)
24. V.I. Adamchuk, E.D. Lund, B. Sethuramasamyraja, B., M.T. Morgan, A. Dobermann, D.B. Marx, Direct measurement of soil chemical properties on-the-go using ion-selective electrodes. Comput. Electron. Agricult. **48**, 272–294 (2005)
25. A. Al-Busaidi, P. Cookson, T. Yamamoto, Methods of pH determination in calcareous soils: use of electrolytes and suspension effect. Aust. J. Soil Res. **43**, 541–545 (2005)
26. M. Schirrmann, R. Gebbers, E. Kramer, J. Seidel, Soil pH mapping with an on-the-go sensor. Sensors **11**, 573–598 (2011)
27. S. Bhandari, U. Singh, S. Kumbhat, Nafion-modified carbon-based sensor for soil potassium detection. Electroanalysis **31**, 1–8 (2019)

Effective Spam Bot Detection Using Glow Worm-Based Generalized Regression Neural Network

A. Praveena and S. Smys

Abstract In today's world, social network plays a major role for communicating all over the world. The available popular social networking platforms are Twitter and Facebook. In both the platforms, the users can communicate with any user irrespective of the location and time. But this platform also has a disadvantage which is the spreading of false information and also false communication between the users. The false information spread is due to the fake users and bot users. To overcome this drawback in the social network, several machine learning algorithms were proposed for identifying the malicious users. Hence, in this, the spam users are detected in the Twitter social network platform. To perform the classification of users as original and fake users, the user and social attribute features were extracted. Then, the features are reduced using the optimization approach called glow worm optimization. Then, the reduced features are trained using the generalized regression neural network. The reduction of features is based on minimizing the error rate of the classifier. This approach helps to improve the classification as compared to the deep Q-learning algorithm. The whole process is performed on the social honeypot data set, and it is implemented using MATLAB R2018a version. The proposed approach outperforms the conventional algorithm in terms of accuracy, sensitivity, specificity and *F*-measure.

Keywords Social network · Twitter · User features · Social features · Reduction · Glow worm · Generalized regression neural network

1 Introduction

In the last two decades, the evolution of the social online networking sites is increasing day to day. The online platforms provide various performances like chatting, video

A. Praveena (✉)
Computer Science and Engineering, Jansons Institute of Technology, Coimbatore, India
e-mail: praveena.s@jit.ac.in

S. Smys
Computer Science and Engineering, RVS Technical Campus, Coimbatore, India

S. Shakya et al. (eds.), *Mobile Computing and Sustainable Informatics*, Lecture Notes on Data Engineering and Communications Technologies 68,
https://doi.org/10.1007/978-981-16-1866-6_34

call, marketing, online survey, online donations and awareness and so on. All these activities can be performed at any time and at any place without any direct contact with the users.

Despite of various advantages, the remote access itself results in spam messages and fake news propagation in fast manner. A single user can create numerous accounts in the online network without any ideal verification of the user. This results in fake accounts and spam bots. The fake user and spam bots result in spreading of rumour and anonymous messages to the user.

Several research works were carried out to determine the nature of the user based on their activities and friend's circles. Those works were discussed below in this section and in the literature survey section. Spam bots are determined based on their activities like tweets, friends, followers and followers' and tweet structure. Based on this, several methods were introduced to detect the spam bots. A survey on different techniques in detecting the malicious users in social network is discussed in [1]. A hybrid approach called content and graph-based features for detecting the malicious users in Twitter [2]. Here, the graph based is used to depict the social network of the user and content based to depict the tweets nature.

A complete survey on the bot detection and its shortcomings are discussed in [3]. It discussed about the bot detection in three groups. One is machine learning, second is based on crowd sourcing and the third one is based on the structure-based. In each group, different techniques discussed and its problem and suggestions to overcome those problems are discussed. Generally, the bots are detected, but [4] defined the nature of spammer in the future using DNA sequence of the existing spammer with the help of genetic algorithm. It helps to predict the attacks for the spammers. A trust model is designed based on the features extract from the original users. Based on this, the spam users are separated from the original users [5]. Here, the URL feature alone is used for the classification of the users using Bayesian learning and Dempster–Shafter theory. Topic modelling using Latent Dirchlet allocation method is used for social bot detection [6]. Here, the LDA is applied on the features extracted from the Twitter users using natural language processing is performed. It performed well for the smaller data sets.

Machine learning techniques were used for the malicious user detection in Twitter based on the friends and follower count [7]. Ersahin et al. [8] performed the fake account detection in Twitter [8]. Here, the data set is created from the crawlers. Then, the data set is processed using the entropy minimization discretization for the feature extraction process. Then, the extracted features are trained and tested using the Naïve Bayes classifier. This approach is able to classify the accounts effectively as compared to without discretization process.

Elazab [9] performed the classification of original and fake users in the Twitter social online platform [9]. Here, the features from different research works have been collected for classification process. From that, totally 22 attributes were totally collected. These attributes were processed in the first stage, and it is weighted for the second stage of processing. Then, based on the weights, the topmost seven features were used for the classification. For the classification process, the different classifiers like Naïve Bayes, support vector machine and so on. Helmester and Paulheim [10] performed the classification of tweets as original and fake [10]. Here, the large source of data set is collected from the web sources. Then, the features are tweet, users, social and text features. Then, these features are trained with the different classification algorithms. From the evaluation of the approach, tweet and source features play a major role in the classification of the tweets as original and fake with accuracy of 93.6%.

Jia et al. [11] performed the fake account detection in Twitter platform [11]. Here, the data set comprises of both the original and malicious users. Then, the random samples from both the data are used for the classification. To perform the classification process, the Sybil walk, Sybil Rank, Sybil walk-variant and CIA approach is used. From the evaluation of results, Sybil walk able to classify even with the weaker data set. Swe and Myo [12] performed the classification as Twitter accounts as legitimate and illegitimate users using blacklist approach [11]. Here, the traditional spam worm-based classification is not performed. Instead of that, the keyword extraction and the features from the tweets were used for classification. Then, the extracted features were trained and tested more than 10 k-based tweet accounts and social honeypot data set using the metalearner classification.

Gheewala and Patel [13] survey about the various machine learning approaches for the classification of fake account users [13]. Here, the Twitter accounts are classified using different types of features namely graph-based features, user-based features and social-based features. Then, these features are trained using different types of classifiers namely SVM, regression network, and other machine learning approaches were studied.

Gurjala et al. [14] proposed a new feature for identifying the fake users in the Twitter account [14]. Here, the profile characteristics of the account holders are collected like URL, profile creation time, friends and followers details and so on. Based on this, a ratio between the friends to followers is calculated. For the fake users, this ratio keeps on increasing and it is used for the classification process. Mateen et al. [15] proposed the fake user detection using hybrid feature set [15]. The feature set used for the classification is both the content-based features and graph-based features. This proposed features help to improve the classification ratio using decorate and J48 classifiers.

Alarifi et al. [16] performed the fake account detection with the help of characteristics obtained from the fake accounts [17]. Generally, the original user's data are used for the classification process. But, here, the fake user's data were used for the classification process. It helps to observe the fake user features effectively as compared to the original user features. Gracia et al. [18] performed the identification of defaming users in the Twitter network [19]. To perform this identification process, the defaming comments made by the user for the real users is studied and trained using the network. Based on this, the cyberbullying users are determined and they can be blocked.

From the above studies, it is observed that either the tweets or the user or the graph features were used for the classification. But, only few works were processed all the features and it requires from five to ten features for the classification process. The classification process is performed through the basic classifier or by the deep learning process. Those types of networks result in the high computational time for the spam bot detection.

Hence, in this paper, this problem is overcome by minimizing the number of features using optimization algorithm called glow worm optimization. This results in improved performance of the classifier and reduced computational time.

The remaining sections of the paper are organized as follows. The various approaches for performing spam bot detection are discussed in Sect. 2. The conventional approach for spam bot detection in Twitter platform is discussed in Sect. 3. Sections 4 and 5 explained about the proposed approach and its implementation process. Section 6 and 7 summarized about the proposed approach and future work.

2 Literature Survey

In this section, the techniques recently used for the spam bot detection in social network are discussed, and it is shown in the table format in Table 1.

3 Existing Method

The Twitter users are classified as social bots and influencer bots using the deep Q-learning algorithm [31]. Here, the users are classified based on the following attributes namely social attributes, tweet attributes and user attributes. The tweet attributes are based on the nature of the tweets by the user. The user attributes describe about the usage of Twitter and the friends circle of the user. The social attributes indicate the properties of the users in the form of graph. The social-based attributes are varying every time and hence it is normalized using the z-score. Then, these attributes are used for classifying the users as bots and active users using deep Q-learning algorithm. This approach is tested on the three types of data sets namely fake project data set, social honeypot data set and user popularity band data set. The performance is evaluated using the following metrics namely accuracy, precision, recall, true positive rate, false positive rate and f-measure.

The Q-learning method is able to classify the users better in all types of data sets by maintains the accuracy above 90. But, it has the following drawbacks:

- The number of features for classifying the users is high.
- The computational time is high.
- The accuracy is also less even with large number of features.

Table 1 Survey of spam bot detection techniques in social network

Author	Technique	Findings	Shortcomings
Mazza et al. [20]	Long short-term memory for retweet buster	Retweets is the only feature used for the malicious users	Chances for considering original users as malicious is high
Pakaya et al. [21]	XGBoost based on tf-idf component of Twitter	Word2vec is used for the feature extraction	The feature extraction solely depends on the word2vec results
Yue et al. [17]	Autoencoder	Autoencoder itself performs both classification and feature extraction	Time for training is high
Adewole et al. [19]	Analysis of spam messages using different classifier	Evolutionary search algorithm is used Logiboost is best	Processed only on short messages
Patil et al. [22]	Naïve Bayes classifier for tweets classification	Tweets are classified as original and fake with the help of retweets, location and authenticity nature	Able to perform well for smaller data sets
Zervoplus et al. [23]	Analysis of classifier in Hongkong protest tweets	Natural language processing is used for the feature extraction. Word entropy is the best feature	Translation of words faces a drawback in tweet classification
Karakaşlı et al. [24]	Twitter user classification	Features are extracted from the user. K-nearest neighbour chose eight features between them. Support vector machine is used for the classification	KNN not able to predict the exact feature due to the dynamic nature of Twitter users
Sahoo and Gupta [25]	Petrinet is used for the classification	Twenty-one features are extracted and genetic algorithm is used for the reduction process	Data augmentation is used in the petrinet which requires larger training period
Adewole et al. [26]	Analysis of classifier and K-means clustering for Twitter user classification	K-means cluster is to group the users. SVM, RF and MLP are used for the classification. Random forest is the best	URL and Hash tag is only considered for the classification
Hamdi et al. [27]	Spam user detection using graph-based approach	Follower and followees determined using node2vec for classification	Limited number of features results in poor estimation for larger data sets

(continued)

Table 1 (continued)

Author	Technique	Findings	Shortcomings
Sungheetha and Sharma [28]	Transcapsule network for sentiment classification	N-gram is used for the feature extraction in the feature capsule layer Pretrained word is used as look-up table for classification	Look-up table should be properly updated for better analysis in long-term manner
Vijayakumar and Vinothkanna [29]	Capsule network for font classification	Basic pre-processing steps are performed on the word to split into letter and trained using capsule network along with labels. Simultaneously, the features are trained with other classifiers	Training data is lesser and requires more pre-processing steps to improve the performance
Zhao et al. [30]	Graph-based attention network for spam bot detection in social network	Directed graph is used for feature extraction. User feature and the relation between the users is used	Hidden features of the user has to be analysed

3.1 Novelty in the Proposed Method

In this, the shortcomings of the existing method are overcome by the proposed method, and its reason to choose the method is shown in Table 2.

Hence, to overcome these drawbacks, an optimization-based approach is proposed for classifying the Twitter users in this paper. The proposed method uses the glow worm optimization for selecting the dominant features. Then, the dominant features are trained and classified using the generalized regression neural network.

4 Proposed Method

In this work, the spam accounts are detected with the help of optimization process and machine learning approach. The optimization process used in the proposed method is the glow worm optimization. It is used to select the optimal features to reduce the computational time as well as to increase the accuracy of the classifier. For the classification process, the generalized regression neural network is used. The flow chart of the proposed method is shown in Fig. 1.

Table 2 Novelty of the proposed method

Existing method (deep Q-learning)	Proposed method (GWO-GRNN)	Reasons
Three attributes for the features extraction • Social • User • tweet	Only two attributes which comprise all information regarding the user • User • tweets	User attributes hold the information about the account holder and its circle Tweet attributes denotes the participation level
Normalization is performed for the social attributes	No normalization is performed	User attribute itself resembles its social behaviour. So this process is omitted
Utilized all features for training and classification	Feature reduction is performed using glow worm optimization	Glow worm is used because it updates its neighbour with highest luciferin level Due to this, the optimal feature from glow worm is efficient as compared to the existing technique
Deep Q-learning process is used	Generalized regression neural network is used	GRNN able to perform both classification as well as able to estimate the approximate future tweets of the account holder if required
Training time is high	Training time is less	Minimum features and no normalization process is performed

4.1 *Input Data*

The proposed method is tested on the publicly available data set called social honeypot data set from. https://infolab.tamu.edu/data/. The data set comprises of information of both the original users of 19,276 and their corresponding tweets are 3,259,693 and spam users of 22,223 and their tweets are 2,353,473.

4.2 *Pre-processing*

The legitimate and non-legitimate users were sorted and their information were collected from the raw data set and the missing details are removed. The complete data set of both the users are sorted and then it is used for the feature extraction process.

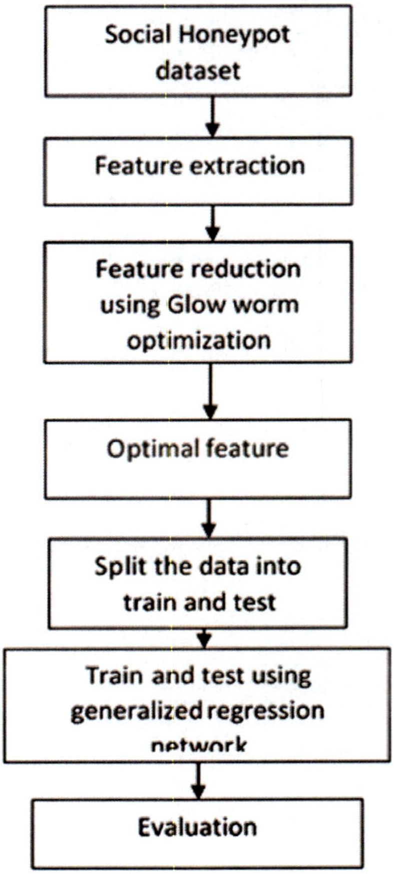

Fig. 1 GWO-GRNN flow chart for spam bot detection

4.3 Feature Extraction

The pre-processed data is subjected to feature extraction process based on the user profile criteria and tweets information. Table 3 is the features extracted from the data set.

Table 3 shows that totally, twelve features are extracted from both the legitimate and non-legitimate users. Training of twelve features for the smaller data set is easier process. But, for the larger network like Twitter application results in longer time period for the spam bot detection. To overcome this problem, the optimal features which are able to classify the users as spam or original user is determined through the optimization process.

Table 3 Feature extraction from the tweets

Attributes	Feature	Uses	Range
Based on the user profile	Number of friends	It indicates the number of users linked to the single tweet account holder	Gradual increasing in their connection to the other account holders
	Number of following users	It indicates the number of users followed by the account holder	Gradual increasing
	Number of followers	It indicates the number of users following the account holder	Lesser followers percentage as compared to following indicates the spam user
	Profile creation time	Time taken by the account holder to create a profile	Lesser time period for non-legitimate users
	Follower to following ratio	Ratio between the followers to following ratio	It should be greater than 0.5
	Active time	The account holder online availability	Smaller time period
Based on the tweets	Number of tweets	Total number of tweets responded by the account holder	Optimal number of tweets
	Number of retweets	Retweet refers to share the other account holder tweets. Total number of retweets in their tweets by account holder Non-legitimate users mostly use more retweets	Retweets should be minimum
	Number of URL used in the tweets	Legitimate users only limited URLs in the tweet as compared to non-legitimate users	The redirection of the URL in the tweet should be less
	Number of hashtags in the tweets	Non-legitimate users use more hashtags which are irrelevant to them	Legitimate users use limited hash tags in the tweets
	URL ratio	URL ratio is the ratio of URL to the tweets	It should be less for the legitimate users
	Hashtag ratio	Hashtag ratio is the ratio of number of hashtags to the tweets	It should be less for the legitimate users

4.4 Glow Worm-Based Feature Reduction

The role of optimization algorithm is to reduce the trial and error method for finding the optimal solution for a problem. It is performed based on processing the objective or fitness function of the problem. For each application, the objective function and parameters used for processing will vary. Here, the objective function of the glow worm optimization is to reduce the misclassification rate of the classifier.

The misclassification ratio is calculated based on the following Eq. 1.

$$\mathrm{MC} = 1 - \frac{\text{correctly Identified classes}}{\text{Actual classes}} \tag{1}$$

The glow worm is placed in certain position, and it move search space for reducing the above objective function. The solution of the glow worm indicates the minimum misclassification that can be obtained from the classifier. The corresponding position is the optimal features that can be used for the classifier.

4.4.1 Glow Worm Parameters

In this step, the parameters to perform the optimization process are initialized. These parameters are responsible for finding the optimal feature for training the classifier. The lower boundary of the glow worm optimization is one. The upper boundary of the glow worm optimization is twelve. The optimal feature is selected as five. These five features will be used for the classification process. The solution will be found out in fifty iterations.

The initialization of the glow worm is given in Eq. 2.

$$G = \left\{G_1, G_2, \ldots, G_g\right\};\ 1 < g < N_{\mathrm{GLW}} \tag{2}$$

4.4.2 Objective Function Evaluation

Initially, the positions of the glow worms are placed in the random position. The solution of the glow worms are calculated based on Eq. 1. The solutions of the glow worms keep on updating till the iterations stops.

4.4.3 Glow Worm Attraction Phase

The glow worms update its position based on the attraction process of the glow worms. Because, the glow worms are attracted in such a way to increase its brightness. The attraction is based on the level of luciferin in the glow worms. The luciferin level is important for increasing the brightness of the glow worms. Hence, the luciferin level of the glow worm also keeps on changing as the process continues. The luciferin update is based on Eq. 3. The update is based on the optimal solution, decay and enhancement level of the luciferin level.

$$LL_{t+1}^{g} = (1-\rho)LL_{t}^{g} + \gamma MC\left(G_{t+1}^{a}\right) \tag{3}$$

4.4.4 Glow Worm Position Update

The glow worm has to update its position for each iteration and each search agents. Because, then only the optimal feature for the spam bot detection in Twitter can be determined. The glow worm update its position based on the two factors namely luciferin level and its neighbour range.

The selection of glow worm for the attraction is based on the probability Eq. 4.

$$\mathrm{PB}_t = \frac{LL_t^z - LL_t^g}{\sum_{d \in NE_t^g} LL_t^d - LL_t^g} \tag{4}$$

The position of the neighbour glow worm based on the following Eq. 5.

$$\mathrm{NB}_t^g = \left\{p;\, D_t^{gp} < R_t^g;\, LL_t^g < LL_t^p\right\} \forall p \in \mathrm{NE}_t^g \tag{5}$$

The D is the Euclidean distance between the two glow worms. The R_t^g is the space to select the neighbour glow worms.

Then, the overall position update equation is given in Eq. 6.

$$\mathrm{Pos}_{t+1}^g = \mathrm{Pos}_t^g + \delta\left(\frac{\mathrm{Pos}_t^p - \mathrm{Pos}_t^g}{\left\|\mathrm{Pos}_t^p - \mathrm{Pos}_t^g\right\|}\right) \tag{6}$$

4.4.5 Neighbourhood Range Update

The position update of the glow worms is based on the neighbour glow worms. Hence, the search space for the neighbourhood range has also to be updated.

Based on that, the following Eq. 7 shows the neighbourhood range update equation.

$$R_{t+1}^g = \min\left\{R_s^1, \max[0, R_t^g + \beta(y - \left|\mathrm{NB}_t^g\right|\right\} \tag{7}$$

4.4.6 Elite Solutions

The glow worms update its solution for each search agents till the iterations complete. The solution for the glow worm in each iteration is called as the local elite or local best solution. Then the local elites are sorted and the topmost solution is the global best or global elite solution.

4.4.7 Termination

The glow worm stops its process till the total iterations are complete. The solution and position indicate the optimal features for the generalized regression neural network.

4.5 Spam Bot Detection Using Generalized Regression Neural Network

In this, the GRNN is used for spam bot detection based on the optimal features from the glow worm optimization. In this, the network introduced by Specht [32] is used for the classification process. Here, the network operation is similar to the radial basis function and kernel for the network is Gaussian function.

Unlike other neural network, this network can be used for all operations like regression, prediction and also to control an operation. Due to the regression nature and parallel operations, it can process large size of data effectively. The layers in the generalized regression neural network is shown in Fig. 2.

The layers in the generalized regression neural network are:

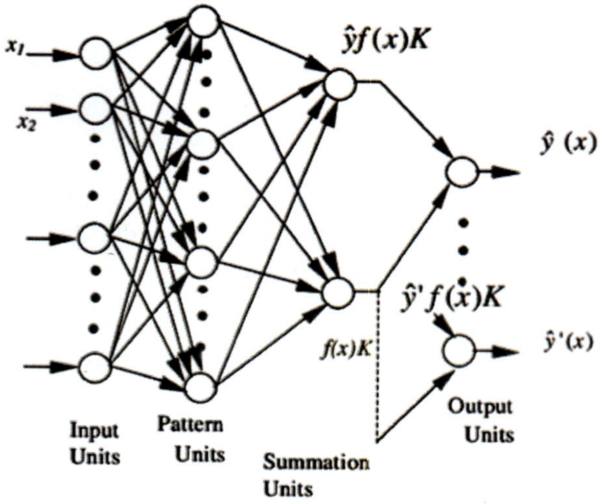

Fig. 2 Regression neural network

- Input layer
- Pattern layer
- Summation
- Output.

4.5.1 Input layer

In this, the input layer is the five optimal features from the glow worm optimization of the data set. It is given in the following equation

$$I = \{F1, F2, F3, F4 \text{ and } F5\} \tag{8}$$

4.5.2 Pattern Layer

In this, the input layer and the activation functions are combined to form a relationship between the output layer and input parameter.

$$\text{PL} = I. * \text{Gaussian function} \tag{9}$$

4.5.3 Summation

'In this layer, the pre-processed feature vectors are processed based on the weights and it is combined to make a connection between the input and outputs.

4.5.4 Output Layer

The final layer consists of label of the network. Here, the labels are mentioned as 0 for spam user and 1 for the original user.

4.6 Steps in GWO-GRNN for Spam Bot Detection

Step 1 The social honeypot data set is the input.
Step 2 The data is pre-processed to perform feature extraction.
Step 3 The features mentioned in 4.3 are extracted.
Step 4 The parameters for the glow worm optimization are initialized
Step 5 The solution for the initial glow worms are updated based on Eq. 1.
Step 6 The glow worms attract other glow worm based on Eq. 3.

Step 7	For each iteration, the glow worm updates its neighbour based on Eq. 5
Step 8	The glow worm update its position based on Eq. 6.
Step 9	The search space for the glow worm is based on the Eq. 7.
Step 10	The local elite solution for each iteration is calculated.
Step 11	is maximum iteration reached
Step 12	The optimal feature is calculated.
Step 13	else
Step 14	The process repeats from step 7.
Step 15	The optimal feature for the neural network is calculated.
Step 16	The feature set is divided into training and testing using hold out approach.
Step 17	The network is trained with the optimal features from the glow worm.
Step 18	The trained network is tested with optimal features from the testing data set.
Step 19	The nature of the users are determined.
Step 20	The process stops.

5 Experimental Set-Up and Discussion

In this, the proposed glow worm-based generalized regression neural network for detecting the spam bots is implemented using the Matrix Laboratory software R2018 a version. The whole process is simulated under Windows 10 environment.

The proposed method is tested on the publicly available data set called social honeypot data set. The data set is pre-processed and the features are classified based on the user profile attributes and tweet-based attributes. Then, the dominant feature is selected using the glow worm optimization. Then, these feature are trained and tested using the generalized regression neural network.

The above parameters are calculated based on the four values.

- True positive
- True Negative
- False positive
- False negative.

True Positive indicates identified the users correctly as original Twitter users.

True negative indicates identified the users correctly as Twitter spam bots.

False positive indicates identified the spam bots as original Twitter users.

False negative indicates identified the original users as spam bots.

Based on the above parameters, the proposed method is evaluated based on the following metrics:

- Accuracy
- Sensitivity
- Specificity
- Recall
- F-measure.

5.1 *Accuracy*

Accuracy evaluates the proposed method for the classifications. It determines the identification of users as spam bots or original users. It is calculated based on the following formula.

$$\text{Accuracy} = \frac{\text{TP} + \text{TN}}{\text{TP} + \text{FP} + \text{TN} + \text{FN}} \tag{10}$$

5.2 *Sensitivity*

Sensitivity measures the classification of users correctly identified as original users. It is calculated using the following formula.

$$\text{Sensitivity} = \frac{\text{TP}}{\text{TP} + \text{FN}} \tag{11}$$

5.3 *Specificity*

Specificity measures the classification of users correctly identified as spam bots. It is calculated using the following formula.

$$\text{Specificity} = \frac{\text{TN}}{\text{TN} + \text{FP}} \tag{12}$$

5.4 *Recall*

Recall is similar to the measure of sensitivity. It is calculated using the following formula.

$$\text{Recall} = \frac{\text{TP}}{\text{TP} + \text{FN}} \tag{13}$$

Table 4 Comparison of DQL versus GWO-GRNN for spam bot detection

Parameter	Deep Q-learning	GWO-GRNN
Accuracy	93.5	95
Sensitivity	88	90
Specificity	25	90
Precision	90	92
F-measure	89	92

5.5 F-measure

F-measure is a parameter which is superior to the accuracy. The F-measure value can be high only if the precision and recall values are high. It is calculated using the following formula.

$$F\text{ - meausure} = 2 * \frac{\text{precision} * \text{recall}}{\text{precision} + \text{recall}} \tag{14}$$

Based on the above evaluation metrics, the proposed glow worm-based generalized regression neural network for spam bot detection is compared with the deep Q-learning method. Spam separating is the most popular utilization of the NB classifier. It is an overall technique for separating unapproved messages for example spam from the legitimate ones, regularly alluded to as ham. Most mail customers execute Bayesian spam sifting nowadays. Though clients can introduce e-mail-sifting programming, worker side e-mail channels using Bayesian spam separating strategies are dug in inside programming that makes e-mail offices to perform viably. Those values are tabulated in Table 4 and Fig. 3.

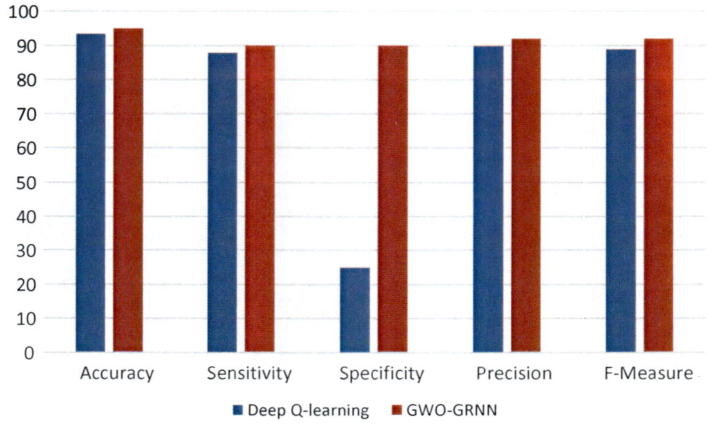

Fig. 3 Comparison of DQL versus GWO-GRNN for spam bot detection

Table 5 Computational time comparison

Method	Computational time (s)
Deep Q-learning	1707
GWO_GRNN	1000

From the results, it is observed that the proposed method able to outperform the existing approach in all parameters. Especially, the accuracy and F-measure of the proposed method are high which show that the proposed glow worm-based generalized regression neural network able to identify the spam bots in Twitter data set.

The proposed method not only improves its performance in the evaluation and it also utilizes the less amount of time for processing. The computational time comparison is shown in Table 5.

From the above table, it shows that the proposed approach requires lesser amount of time for identifying the nature of the users. It becomes possible due to the lesser amount of feature from the glow worm optimization. Due to this, the training and testing time of the network is reduced for the proposed GWO-GRNN-based spam bot detection.

Hence, the proposed GWO-GRNN is best approach in terms of processing as well as performance as compared to the existing approach deep Q-learning method.

6 Conclusions

Social networking plays a major role in spreading the information as well as to improve the communication all over the world in the last two decades. The fast communication itself is a disadvantage in spreading the rumours as well as the anonymous messages. To limit this problem, the fake users are identified using the deep Q-learning approach. But, the computational time is high due to the more number of features. This problem is overcome by reducing the features using the proposed glow worm optimization process. Then, the reduced features are classified using the generalized regression neural network with higher evaluation metrics as accuracy and F-measure in compared to the deep Q-learning process. It is due to the training of optimal features by minimizing the error rate of the classifier. Hence, the proposed glow worm-based GRNN is able to identify the nature of the Twitter users effectively.

7 Future Work

In the future, the proposed method is like to expand by implementing the text feature and also to reduce the computational time further.

References

1. E. Alothali, N. Zaki, E.A. Mohamed, H. Alashwal, Detecting social bots on twitter: a literature review, in *2018 International Conference on Innovations in Information Technology (IIT)* (IEEE, 2018 Nov), pp. 175–180
2. M. Mateen, M.A. Iqbal, M. Aleem, M.A. Islam, A hybrid approach for spam detection for Twitter, in *2017 14th International Bhurban Conference on Applied Sciences and Technology (IBCAST)* (IEEE, 2017 Jan), pp. 466–471
3. A. Karataş, S. Şahin, A review on social bot detection techniques and research directions, in *Proceedings of International Security and Cryptology Conference Turkey*, 2017 Oct, pp. 156–161
4. S. Cresci, M. Petrocchi, A. Spognardi, S. Tognazzi, From reaction to proaction: unexplored ways to the detection of evolving spambots, in *Companion Proceedings of the Web Conference 2018*, 2018 Apr, pp. 1469–1470
5. G. Lingam, R.R. Rout, D.V.L.N. Somayajulu, Detection of social botnet using a trust model based on spam content in Twitter network, in *2018 IEEE 13th International Conference on Industrial and Information Systems (ICIIS)* (IEEE, 2018 Dec), pp. 280–285
6. P. Andriotis, A. Takasu, Emotional bots: content-based spammer detection on social media, in *2018 IEEE International Workshop on Information Forensics and Security (WIFS)* (IEEE, 2018 Dec), pp. 1–8
7. N. Singh, T. Sharma, A. Thakral, T. Choudhury, Detection of fake profile in online social networks using machine learning, in *2018 International Conference on Advances in Computing and Communication Engineering (ICACCE)* (IEEE, 2018 June), pp. 231–234
8. B. Erşahin, Ö Aktaş, D. Kilinç, C Akyol, Twitter fake account detection, in *2017 International Conference on Computer Science and Engineering (UBMK)* (IEEE, 2017 Oct), pp. 388–392
9. A. ElAzab, Fake accounts detection in twitter based on minimum weighted feature, in *World* (2016)
10. S. Helmstetter, H. Paulheim, Weakly supervised learning for fake news detection on Twitter, in *2018 IEEE/ACM International Conference on Advances in Social Networks Analysis and Mining (ASONAM)* (IEEE, 2018 Aug), pp. 274–277
11. J. Jia, B. Wang, N.Z. Gong, Random walk based fake account detection in online social networks, in *2017 47th Annual IEEE/IFIP International Conference on Dependable Systems and Networks (DSN)* (IEEE, 2017 June), pp. 273–284
12. M.M. Swe, N.N. Myo, Fake accounts detection on twitter using blacklist, in *2018 IEEE/ACIS 17th International Conference on Computer and Information Science (ICIS)* (IEEE, 2018 June), pp. 562–566
13. S. Gheewala, R. Patel, Machine learning based Twitter Spam account detection: a review, in *2018 Second International Conference on Computing Methodologies and Communication (ICCMC)* (IEEE, 2018 Feb), pp. 79–84
14. S. Gurajala, J.S. White, B. Hudson, B.R. Voter, J.N. Matthews, Profile characteristics of fake Twitter accounts. Big Data Soc. **3**(2), 2053951716674236 (2016)
15. M. Mateen, M.A. Iqbal, M. Aleem, M.A. Islam, A hybrid approach for spam detection for Twitter, in *2017 14th International Bhurban Conference on Applied Sciences and Technology (IBCAST)* (IEEE, 2017 Jan), pp. 466–471
16. A. Alarifi, M. Alsaleh, A. Al-Salman, Twitter turing test: Identifying social machines. Inf. Sci. **372**, 332–346 (2016)
17. H. Yue, L. Zhou, K. Xue, H. Li, MADAFE: Malicious account detection on twitter with automated feature extraction, in *2019 11th International Conference on Wireless Communications and Signal Processing (WCSP)* (IEEE, 2019 Oct), pp. 1–6
18. P. Galán-García, J.G.D.L. Puerta, C.L. Gómez, I. Santos, P.G. Bringas, Supervised machine learning for the detection of troll profiles in twitter social network: application to a real case of cyberbullying. Logic J. IGPL **24**(1), 42–53 (2016)
19. K.S. Adewole, N.B. Anuar, A. Kamsin, A.K. Sangaiah, SMSAD: a framework for spam message and spam account detection. Multimedia Tools Appl. **78**(4), 3925–3960 (2019)

20. M. Mazza, S. Cresci, M. Avvenuti, W. Quattrociocchi, M. Tesconi, Rtbust: Exploiting temporal patterns for botnet detection on twitter, in *Proceedings of the 10th ACM Conference on Web Science*, 2019 June, pp. 183–192
21. F.N. Pakaya, M.O. Ibrohim, I. Budi, Malicious account detection on twitter based on Tweet account features using machine learning, in *2019 Fourth International Conference on Informatics and Computing (ICIC)* (IEEE, 2019 Oct), pp. 1–5
22. R.A. Patil, K. Gawande, S.N. Dhage, Identification of Rumors on Twitter, in *Recent Trends in Communication and Intelligent Systems* (Springer, Singapore, 2020)
23. A. Zervopoulos, A.G. Alvanou, K. Bezas, A. Papamichail, M. Maragoudakis, K. Kermanidis, Hong Kong protests: using natural language processing for fake news detection on Twitter, in *IFIP International Conference on Artificial Intelligence Applications and Innovations* (Springer, Cham, 2020 June)
24. M.S. Karakaşlı, M.A. Aydin, S. Yarkan, A. Boyaci, Dynamic feature selection for spam detection in Twitter, in *International Telecommunications Conference* (Springer, Singapore, 2019), pp. 239–250
25. S.R. Sahoo, B.B. Gupta, Hybrid approach for detection of malicious profiles in twitter. Comput. Electr. Eng. **76**, 65–81 (2019)
26. K.S. Adewole, T. Han, W. Wu, H. Song, A.K. Sangaiah, Twitter spam account detection based on clustering and classification methods. J. Supercomput. **76**(7), 4802–4837 (2020)
27. T. Hamdi, H. Slimi, I. Bounhas, Y. Slimani, A hybrid approach for fake news detection in Twitter based on user features and graph embedding, in *International Conference on Distributed Computing and Internet Technology* (Springer, Cham, 2020 Jan), pp. 266–280
28. A. Sungheetha, R. Sharma, TransCapsule model for sentiment classification. J. Artif. Intell. **2**(03), 163–169 (2020); C. Zhao, Y. Xin, X. Li, H. Zhu, Y. Yang, Y. Chen, An attention-based graph neural network for spam bot detection in social networks. Appl. Sci. **10**(22), 8160 (2020)
29. T. Vijayakumar, R. Vinothkanna, Capsule network on font style classification. J. Artif. Intell. **2**(02), 64–76 (2020)
30. C. Zhao, Y. Xin, X. Li, H. Zhu, Y. Yang, Y. Chen, An attention-based graph neural network for spam bot detection in social networks. Appl. Sci. **10**(22), 8160 (2020)
31. G. Lingam, R.R. Rout, D.V. Somayajulu, Adaptive deep Q-learning model for detecting social bots and influential users in online social networks. Appl. Intell. **49**(11), 3947–3964 (2019)
32. D.F. Specht, A general regression neural network. IEEE Trans. Neural Netw. **2**(6), 568–576 (1991)

Modelling of Flood Prediction by Optimizing Multimodal Data Using Regression Network

C. Rajeshkannan and S. V. Kogilavani

Abstract Natural disasters are an unpredictable one, but the damages caused by the disaster is severe. It causes hazards to both humans and their properties. Among many hazards like earthquake, eruption and flood, the flood prediction is a quite predictable one. But, it requires proper learning for predicting the floods. Because the flood occurs due to the overflow of water from dams and rivers which is caused by heavy rainfall. Due to this, proper learning based on the weather conditions and previous flood data, the possibility of flood range and area can be detected. Such a flood prediction is performed on the real-time data set collected from the Columbia province. It uses data mining approach on the collected information for forecasting the flood level. But, it is not only sufficient for real time. Hence, the image-based flood prediction is proposed to analyse the flood level in the particular region using a satellite image of the area. Yet, both the techniques offer good prediction in the individual unit it suffers from predicting the nearby flood areas. Hence, in this, a combined approach of image processing and data mining is proposed to forecast the flood level. For data mining, the regression learning approach is used to forecast future flood levels. It is combined with the corresponding image processing of the particular area to accurately the flood level based on extracting the depth of water in the area. Finally, the predicted level is broadcasted to the people through social media for immediate action to save their lives. The proposed regression-based data mining and image analysis help to forecast the level of flood in the area accurately along with broadcasting saves precious lives of people.

Keywords Natural disasters · Flood · Regression learning · Image processing · Social media network · Broadcast

C. Rajeshkannan (✉)
Suguna College of Engineering, Coimbatore, Tamil Nadu, India

S. V. Kogilavani
Kongu Engineering College, Erode, Tamil Nadu, India

S. Shakya et al. (eds.), *Mobile Computing and Sustainable Informatics*, Lecture Notes on Data Engineering and Communications Technologies 68,
https://doi.org/10.1007/978-981-16-1866-6_35

1 Aim

Hydraulic drainage model is used for predicting the outcome of the water from the three water source areas in Melbourne [1]. The estimation from this model is fed to the prediction model to estimate the level of water in the area. It is used to estimate the level of flood along with the damage caused by the flood.

The estimation of flood from the single model will result in wrong estimation due to its limited number of information for processing. This drawback is overcome by using the Bayesian average model for estimating the flood from 81 flood model [2]. Here, the averaging process is applied on the data collected from the Missouri and Arkansas. It is able to improve the estimation as compared to the single model results.

A study on the resolution of the image in the process of finding the flood is analysed in [3]. Here, the analysis is carried out by varying the resolution size in three categories < 50, 50 and 100. In these categories, the estimation is poor in the resolution < 50. When the resolution is 100, the estimation of the flood is good due to the clear regions. Hence, the resolution also plays a role in flood estimation.

A highly supervised learning approach called deep learning is used for the estimation of the depth of flood [4]. In this, the depth of the water is determined by the trained convolutional neural network. The network is trained with the help of data collected for 10 years of flood information from LISFLOOD-FP. Then, the trained network is tested on one part of the collected information. It is able to predict the depth accurately due to the more feature extraction by different layers.

A learning and clustering-based approach is proposed for predicting the floods in the urban areas [5]. In this, two-stage of process is performed for predicting the floods. In the first stage, the regressive learning and neural network are applied for learning the relationship between the rainfall and the water overflow from the river. Then, this learned relationship and the corresponding flood levels were grouped using the self-organizing map feature for the final flood prediction. Therefore, this two-stage process is able to predict the level of flood in the US environmental data set as compared to the single hydrodynamic model.

A survey about different techniques for predicting the flood is discussed in [6]. Here, the study is carried out by using the different techniques namely mathematical model, hydrodynamic model and concept-based models. In each model, there are different types of methods used for predicting the level of flood. It helps to learn the procedure for predicting the flood in each model and how it can be improved further for improving the prediction.

The role of spatial resolution is analysed by varying the pixel resolution to coarse resolution in the flood prediction process [7]. Here, this analysis is carried out on the hydrodynamic model called LISFLOOD-FP. Before, applying the model, the images obtained from the digital elevation module is subjected to the spatial changes and then it is applied to the hydrodynamic model. From the analysis, it is observed that the inflow graph and the spatial properties are important based on the area of flood prediction.

The prediction of flood is performed through the hydrodynamic models [8]. This hydrodynamic model can be of two types namely one-dimensional and two-dimensional models. The hydrodynamic models which come under these two categories were subjected to estimate the flood and also to determine the role of properties of surface is analysed. From the analysis, the one-dimensional requires uniform surface for the better analysis. Overall, the uniform surface requires less computation and better flood prediction.

The flood prediction in urban area is important due to the abundant population and using the river area for living and commercial purposes [9]. Due to this, it gets highly affected during the flood time. It is due to the poor drainage system in connection to the river in underground system in Korea. To analyse and estimate the flood, the SWMM model was applied on the drainage facility system. From the model, the relation between rainfall and overflow can be analysed but the drainage system cannot be improved. Hence, a separate pond and extension of sewer conduit is required.

Due to the development of technology, the tools were developed to estimate the flood in a region instead of analysing the hydrodynamic model [10]. Such a comparison is made between tools like AutoRoute and handy with hydrodynamic model of two dimensions namely HEC-RAS. The analysis is carried out on Black Warrior River and Cedar River. From the analysis, it is observed that, with proper resolution, the tool also produces better estimation in comparison with the hydrodynamic models.

From the above survey, it is observed that the hydrodynamic models were used for the flood estimation alone. But, the estimation alone will not save the lives, it requires proper and fast communication to reach the people.

Hence, in this paper, the social media platform is used for reaching the people to update the flood information in areas. The estimation is also improved by using the both learning and water depth of image to improve the flood estimation.

The paper is formatted in the following order. The next section describes about the other approaches in the flood estimation. Sections 3 and 4 describe the existing methods, working of proposed methods, demerits and merits, respectively. Sections 5 and 6 conclude the paper based on the discussion of results using proposed method compared with the existing technique.

2 Literature Survey

Rosser et al. [11] utilized the Bayesian model approach for the flood estimation [11]. Here, the model is analysed on the combination of three sources of information for estimating the flood in a region. The first source is satellite images for estimating the water depth. The second source is the tweets about the mapped flood location areas. The third is the information collected from the government departments regarding rainfall and overflow information. Then, the Bayesian model is applied on the three sources of information which are collected from the 2014 flood data set and to estimate the flood in the Oxford city area.

Gobeyn et al. [12] analysed the time role in the estimation of flood using SAR images [12]. Here, the SAR images of flood area are gathered in two forms: one is during the normal weather conditions and the other is during the flood period. These two types of images were used for estimating the flood in a region. This analysis is compared with the post-flood image and peak flood image. Among all the types of flood images, the pre-flood image is best in estimating the flood as compared to the other images. Hence, a proper time sequence data is required for estimating the flood.

Buahin et al. [13] utilized the learning of flood based on the inflow for the estimation of the flood in a region [13]. In this, a library is formed based on the study of the flood level concerning the inflow of the river. This library is called rating curve library. Based on this information, the estimation of the flood is done. It is a learning-based estimation. The rating curve library should be updated regularly for the proper results.

Mai and De Smedt [14] used both the hydrological and hydraulic models for estimating the flood in a river in Vietnam [14]. Here, the hydrological model is applied for the prediction of flow rate of the water in the upper basin of the river. It is performed by using the Wetspa model. Then, the flow in the down basin of the river is determined by using the HEC-RAS model. This combined approach is able to predict the flood level effectively as compared to the single approach. Because, it is able to determine the level of flood in both upstream and downstream directions.

French et al. [15] utilized the machine learning concept for predicting the tidal waves height at the port [15]. It is performed on the atmospheric data and tidal data collected from the port for the prediction process. Here, the artificial neural network is applied for the estimation. In this, the neural network can estimate the tidal level before eight hours and it helps to save the lives of the people.

Mcgrath et al. [16] used the basic physics model for the estimation of flood [16]. Generally, the hydrodynamic model is used for the estimation of the flood. But, it is not available for all the regions in Canada. So, a simple model is required for the flood estimation. Hence, the three types of digital terrain module are used for this to estimate the flood content. Among these three modules, the HAND module produced the better result as compared to the inclined mode and planar mode.

Lee et al. [17] proposed two metrics for evaluating the estimation of flood results [17]. Because, the flood estimation is performed through various approaches. But, the results from those approaches cannot be verified. To verify and determine the exact results, two metrics have been introduced namely receiver operating characteristics and Lee sallee shape index. These two metrics are evaluated in the estimation of flood by XP-SWMM model. From the evaluation, the method can estimate the flood accurately of about 85%. Hence, this two metrics can be used for the evaluation for flood estimation.

Mojaddadi et al. [18] utilized the support vector machine algorithm for estimating the flood in the river of Malaysian country [18]. In this, the flood occurring parameters were estimated from the DEM of the image and collected from the previous data. Based on this, the probability of the flood areas is estimated. These results are combined with the estimation of flood from the two-dimensional hydrodynamic model. By combining these two results, the final flood estimation is determined. This approach is able to estimate the flood regions effectively as compared to the single hydrodynamic model-based approach.

Chang et al. [19] utilized the machine learning techniques for estimating the flood before 12 h in the Kemmam downstream area [19]. In this, it utilized two algorithms namely self-organizing map and recurrent auto-regressive networks. It used the 4 * 4 SOM for first stage of estimation, and it acts as input to the recurrent network. The recurrent network can estimate flood prior from 3 to 12 h before occurrence. It reduces the need of two-dimensional hydrodynamic model for the inundation process.

Ngo et al. [20] proposed a swarm and artificial intelligence-based flood estimation in the Vietnam region [20]. Here, the information collected from the GIS and the information collected from the satellite images of the area before and after the flood events were processed. Then, both the information are fed into the firefly algorithm to determine the optimal parameter for the estimation process. Then, the flood estimation is performed with the help of hybrid backpropagation and artificial neural network. This approach requires limited information and reduced complexity for the flood inundation as compared to the two-dimensional hydrodynamic models. Arul raj and Rajaram suggested a new digital filter design using booth multipliers and it consumes less power for filter the signals.

3 Background

In the existing method, the Monte Carlo process was used for estimating the flood level in the area by processing hundred years of data. Then, the structural support vector machine techniques were used to predict the level of the water in the corresponding geographic image of the flood area. Then, the final flood level is predicted and it is telecasted to the social media for conveying the flood possible regions. This approach can predict and telecast the flood. But it faces the following drawbacks:

- It uses Monte Carlo method which requires high processing time for processing large number of information.
- The structural support vector machine is used for depth which uses only two classes for predicting the level.
- In practical, the level of water will be a different one.

To reduce the above shortcomings and improve the forecasting, the proposed method utilized the regression learning for the flood-level detection on the gauge data and the top hat filtering technique for determine the level of water in the image. The brief explanation of the proposed technique is given in further sections.

4 Experimental

In this, the estimation of flood is performed by combining the three modes of results. Here, the estimation of flood is performed by four data set. The first data set is based on the satellite images. The second data set is based on the gauge data. The third data set is based on the tweet points regarding the flood. The fourth data set is based on the record about the 100 years of flood. It is performed based on the information collected from the US Washington river data set. The flow diagram for the proposed method is shown in Fig. 1.

4.1 Global Mode

In this mode, the satellite image and gauge values are processed using the regression neural. Here, the image processing technique is applied on the satellite images of the river area. Then, the gauge value is processed using the digital elevation mode of the images.

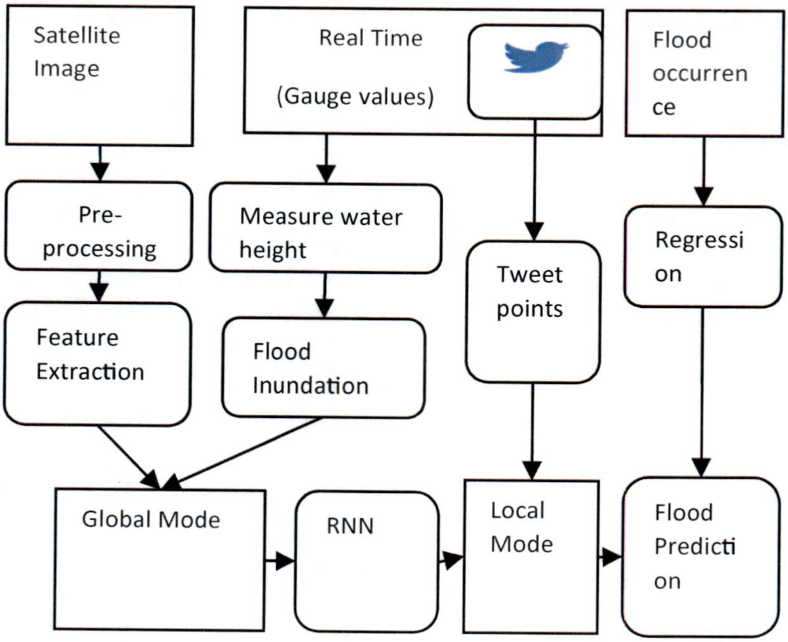

Fig. 1 Regression-based flood estimation

4.2 *Processing with Satellite Image*

In this, the satellite image of the flood region is taken as the input for the flood estimation. Here, the flood region is estimated by grouping the pixels in the image.

Here, the flood region is determined by finding the difference between the pre- and post-flood images of the same region. The aerial view of the satellite images is as follows.

a. **Input Images**

In this, the aerial view of the satellite image for both pre- and post-flood is taken as the input image [21].

b. **Noise removal process**

The image is filtered using the median filter to remove the noises present in the image and improve the quality of the image for further processing. Figure 2 shows the working of a median filter for the mask size of three.

The left side of Fig. 2 is the pixel values of the image and the right side shows the pixels in the mask size of 3 and the corresponding median value. This type of filtering is performed for both the images and starts from topmost pixel in the left side to the bottom down pixel in the right side. All these processes are performed on the grayscale and resized image.

c. **Flood region extraction**

In this, the flooded region is extracted by subtracting the filtered images. It is given in Eq. 1.

$$\text{Flood region} = AF_F - BF_F \tag{1}$$

$$AF_F = \text{after flood filtered image} \tag{2}$$

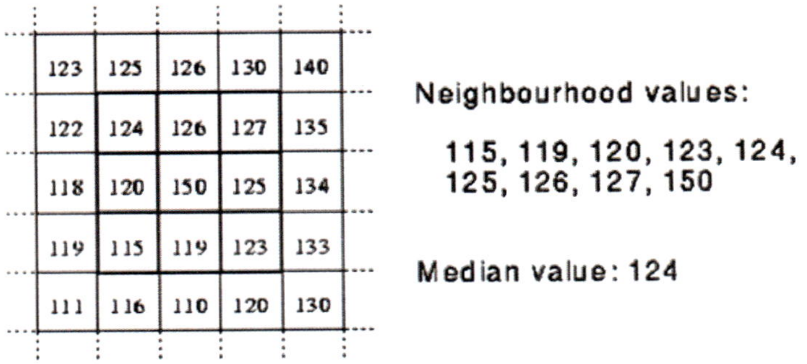

Fig. 2 Median filter

$$BF_F = \text{Before flood filtered image} \tag{3}$$

d. **Morphological process-based feature extraction**

Here, the flood region extracted from the previous section is opened by using the morphological operations of imopen on the image. This operation is performed to identify the flood region and to measure the properties of the flood region like perpendicular, slope, max distance, area and skew.

Based on this, the flood region is estimated. This information is combined with the gauge reading-based flood estimation to estimate the overall flood estimation.

In real time, this process is performed by subtracting the current view of the image to the image captured in the normal condition. Based on the area of the flood, the severity of the flood can be estimated.

4.2.1 Processing with Gauge Readings

In this, the water level of the river basin is determined to estimate the flood level. Generally, the stream flow rate is in the normal condition and the height of the basin will be within its maximum range.

But, during the rainfall season, there will be changes in the flow rate and height of the basin level. Based on this, the approximate level of height of the basin and the flow rate is estimated using the regression neural network. By this estimation of levels, the chances of flood can be estimated.

The chances of flood using regression neural network based on gauge readings is processed along with the real-time estimation of flood. Overall, the estimation of flood based on the river basin and its surrounding satellite images is performed in the global mode.

4.3 *Local Mode*

In this mode, the information from the social media platform is used for the flood estimation and giving alerts. Here, the number of tweets made by the users during the flood period is taken as the input. In this, the number of tweets made during the hurricane Florence period is collected from the openly available data set.

The hurricane Florence occurs between the time of August 31st and 18th September. Here, the data is collected from 5th September to 3rd October [22]. The tweets are the combination of various information but its major contribution is based on the hurricane Florence in different hashtags. The tweets collected in all the stages of flood like initial, peak and recover stage. The hurricane Florence causes a greater impact and changes in the North and South Carolina. Due to this, this data set is used for analysis.

4.3.1 Regression Neural Network

This network is used to estimate the severity of the flood based on the number of tweets in time series format. The process diagram for the regression-based neural network is shown in Fig. 3. Here, the regression-type neural network *k* is used to improve the learning in the time series format.

The RNN estimates its prediction by processing the input through two layers:

a. **First layer**

The first layer of the network is the input layer. The input layer consists of the both predictor and responses. The predictors are the number of tweets and date. The response is the flood severity. Equations 4 and 5 show the input for the neural network.

$$\text{Predictors } (P) = [\text{date no of tweets}] \tag{4}$$

$$\text{responses} = \text{flood severity} \tag{5}$$

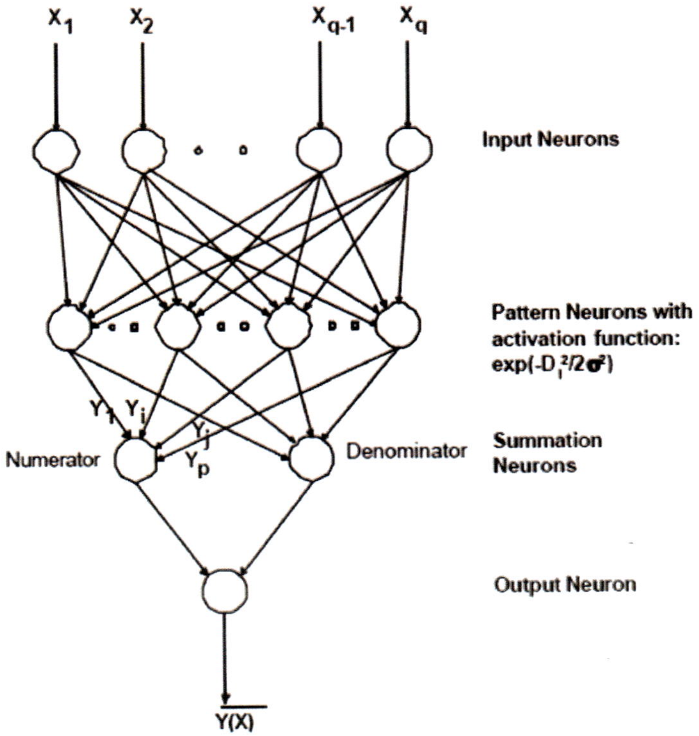

Fig. 3 RNN process diagram

b. **Second layer**

This layer is to define the relation between the training and input data by calculating the distance based on Euclidean formula, and it is given in Eq. 6. In this layer, the Euclidean distance between the input sample and training sample will be calculated using Eq. 6.

$$\text{Relation between Input and training (RE)} = (P - P_i)^T (P - P_i) \tag{6}$$

c. **Third layer**

It is used to define the relation between the predictors and responses through activation function. To define, it uses two neurons. The first neuron is calculated by combining the response and the activation function. The second neuron is about the activation function property. Both the workings of neurons are given in Eqs. 7 and 8.

$$\text{FN} = \sum_{i=1}^{T} R_i \text{e}^{-\left(\frac{\text{RE}}{2\sigma^2}\right)} \tag{7}$$

$$\text{SN} = \sum_{i=1}^{T} \text{e}^{-\left(\frac{\text{RE}}{2\sigma^2}\right)} \tag{8}$$

$$\text{Act\_function} = \text{e}^{-\left(\frac{\text{RE}}{2\sigma^2}\right)} \tag{9}$$

d. **Fourth layer**

It is the fourth and final layer to estimate the flood severity based on number of tweets by processing through two layers. It is the ratio of first and second neuron from the third layer results. It is given in Eq. 10

$$\text{Estimated responses} = \frac{\text{FN}}{\text{SN}} \tag{10}$$

Here, regression neural network is trained to classify the number of tweets as initial, peak and recover stage of flooding to alert the users.

4.4 Forecasting the Flood Based on Time Series Information

This section is to estimate the possible flood level by learning the hundred years of flood information. The learning is based on the linear regression model by fitting the years and the corresponding flood level. The working of linear regression is given below.

4.4.1 Linear Regression

The linear model is fitted between the predictors and responses through two intercept values *a* and *b*. The variables can be fitted into it only if both the variables have relation between them by having the correlation property as 1. Once, it correlates, then the two variables can be fitted into linear model through the following Eq. 11.

$$R = a + xP \tag{11}$$

This regression approach is applied on the gauge readings and the flood level of hundred year's data.

4.5 Steps for the Flood Prediction Using Regression Neural Network

Step 1 Select the region for flood estimation.
Step 2 Collect satellite image of the current aerial view of the region.
Step 3 Calculate the flood region alone by subtracting the current view from the standard image.
Step 4 Perform morphological operations to estimate the level of flood.
Step 5 Collect the gauge readings for the nearby river basins in the region.
Step 6 Apply linear regression to estimate the level of water in the river basin area.
Step 7 Estimate the flood by gauge readings.
Step 8 The results from step 4 and 7 are the first stage of flood estimation in the global mode of the region.
Step 9 The tweets based on the flood hashtags in the region are collected during different phases of the flood.
Step 10 The level of flood based on tweets is determined using regression neural network.
Step 11 The overall flood in the region for hundred years is collected.
Step 12 The linear regression model applied to estimate the approximate level of flood in the future year is estimated.
Step 13 The final probability of the flood in the region is estimated by averaging the results obtained from steps 4, 7, 10 and 12.
Step 14 Thus, the overall flood is estimated through the regression approach.

5 Results and Discussion

The whole process is analysed with the help of matrix simulation-based software called MATLAB in R2018a version. In this section, each output for the individual flood estimation is shown and discussed below.

5.1 Satellite Image-Based Outputs

First, the output for the estimation of flood based on the satellite images is shown in Figs. 4, 5, 6 and 7.

Figure 4a shows the region before the flood and 4b is the corresponding filtered grayscale image.

Similarly, the region after flood and corresponding filtered grayscale image is shown in Fig. 5.

Then, the difference between the before and after flood image is shown in Fig. 6.

Then, the difference image is subjected to the morphological process to estimate the area of the flood regions and its complete process is shown in Fig. 7.

Figure 7 shows the process in the flood region estimation in which the image processing method is displayed. In this, the last two figures show a straight line which shows the linear relation between the differences of the two images.

The properties of the flood-region is shown in Table 1.

The SSVM method is applied on the after flood image directly whereas the proposed morphological-based method is applied on the difference between the before and after flood images. Due to this, the values are reduced and it shows the exact flooded region alone as compared to the two other techniques values in Table 1.

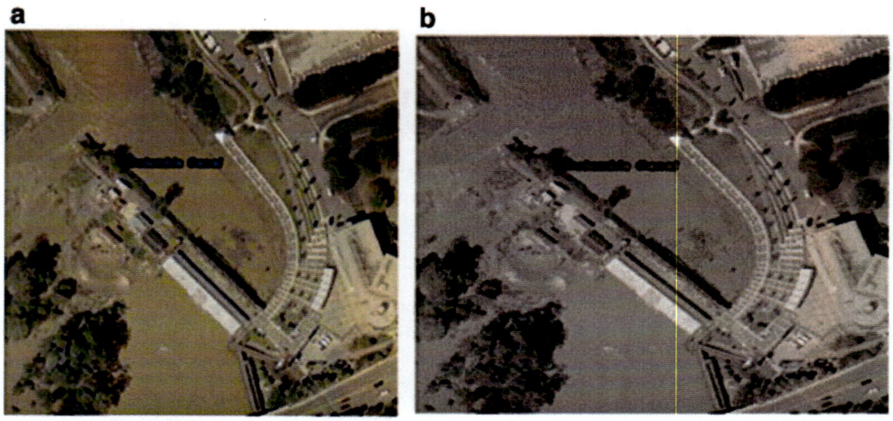

Fig. 4 **a** Before flood, **b** filtered before flood image

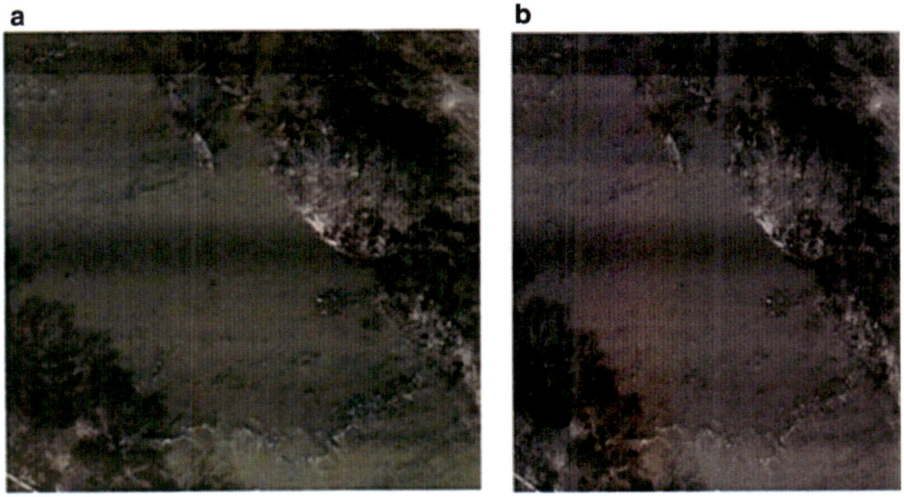

Fig. 5 **a** After flood, **b** filtered after flood image

Fig. 6 Difference image

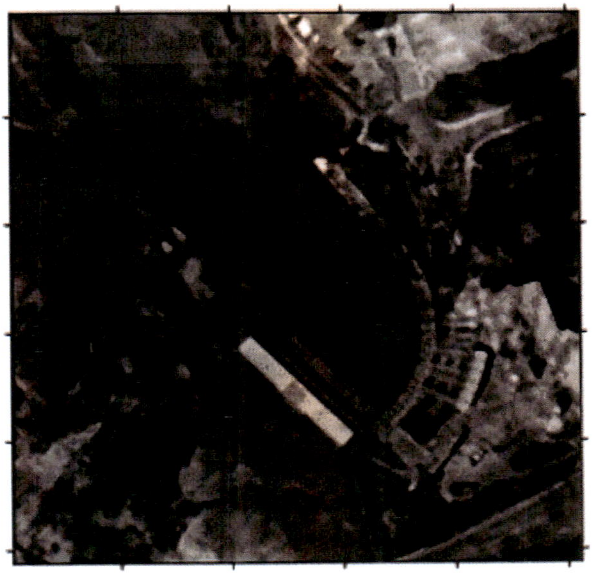

5.2 *Tweets-Based Estimation of Flood*

Third, the regression neural network-based estimation based on the Twitter data is shown in the below figures.

Here, thirty three days of tweet information is collected and it is divided into two sets as training and testing using hold out method. Then, the regression neural

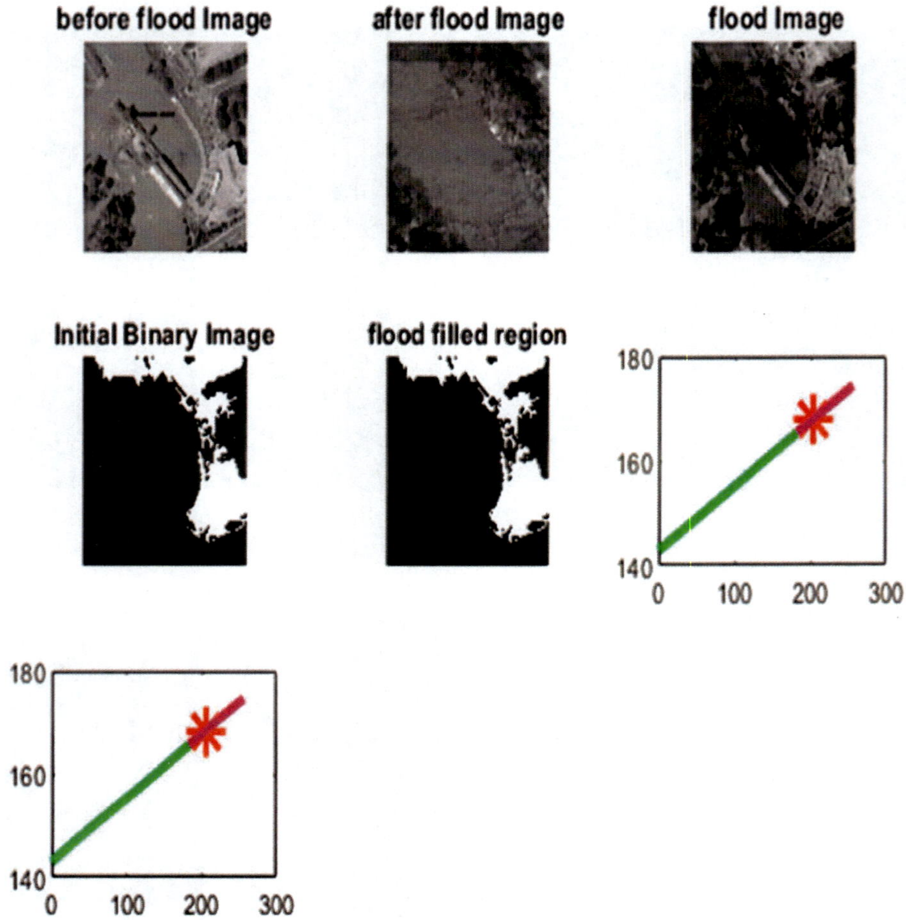

Fig. 7 Flood region estimation by image processing method

Table 1 Properties of flood region

Metric	Existing work	SSVM	Morphological-based
Long slope	0.75449	0.9331	− 8.133
Perpendicular slope	− 1.32539	− 1.0649	0.1230
Max distance	2020.61	270.25	122
Perpendicular distance	1520.00	138.22	73.81
Overall perpendicular width	182.71	96.34	53.87
Area covered	369,186	26,037	6622
Skew value	− 0.51472	− 0.5112	− 0.4999

network is used for training and estimate the level of flood. The corresponding outputs are shown from Figs. 8, 9, 10 and 11.

The input for the tweet-based estimation is shown in Fig. 8.

Command Window

names are saved in the VariableDescripti

date	noOfTweets	Level
'05-09-2018'	'13519'	'1'
'2018-09-06'	'16789'	'1'
'2018-09-07'	'18149'	'1'
'2018-09-08'	'31606'	'1'
'2018-09-09'	'60596'	'1'
'2018-09-10'	'259221'	'1'
'2018-09-11'	'430562'	'2'
'2018-09-12'	'747091'	'2'
'2018-09-13'	'680341'	'2'
'2018-09-14'	'786233'	'2'
'2018-09-15'	'530983'	'2'
'2018-09-16'	'314687'	'2'

Fig. 8 Tweets about flood

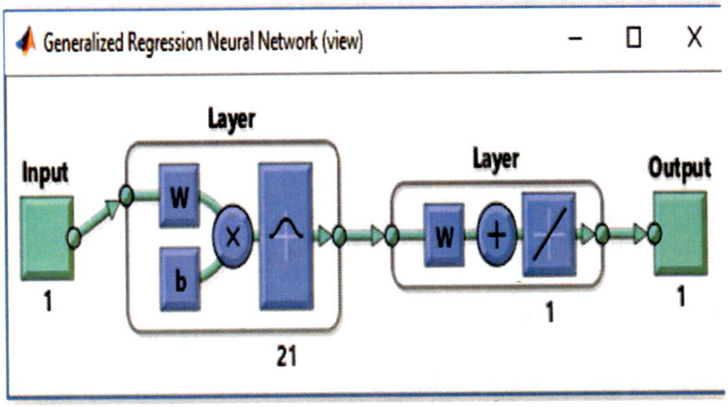

Fig. 9 Tweet-based RNN

```
        NumberOfObservations: 21
              ControlClasses: [2×1 double]
               TargetClasses: 1
           ValidationCounter: 1
          SampleDistribution: [21×1 double]
           ErrorDistribution: [21×1 double]
   SampleDistributionByClass: [3×1 double]
    ErrorDistributionByClass: [3×1 double]
              CountingMatrix: [4×3 double]
                 CorrectRate: 1
                   ErrorRate: 0
             LastCorrectRate: 1
               LastErrorRate: 0
            InconclusiveRate: 0
              ClassifiedRate: 1
                 Sensitivity: 1
                 Specificity: 1
```

Fig. 10 Verification of flood estimation using tweets

It shows that, it has three columns. The first two columns will be act as the input and the third column will be the output to estimate the level of flood in the region. The training of the above information with regression network is shown in Fig. 9.

Figure 9 shows the training set consists of twenty data and the test set consist of 12 and the network diagram for training is displayed. The corresponding results based on the flood estimation using tweet numbers is shown in Fig. 10.

The regression approach able to achieve accuracy above 100% for the data trained in the network and above 75% for the new data. The estimation for the new data is shown in Fig. 11.

The flood estimation based on tweets using regression learning can predict maximum 50% of the flood by providing larger data set for training.

5.3 *Linear Regression for Overall Flood Prediction*

In this, the linear regression model is applied on the hundred year collection of flood information. The corresponding results are shown in the below figures.

The input for the regression learning is shown in Fig. 12.

Figure 12 shows two columns of information. The first column is about the year and the second column is about the corresponding flood range in the year. In this,

```
NumberOfObservations: 8
       ControlClasses: [2×1 double]
        TargetClasses: 1
    ValidationCounter: 1
   SampleDistribution: [8×1 double]
    ErrorDistribution: [8×1 double]
SampleDistributionByClass: [3×1 double]
 ErrorDistributionByClass: [3×1 double]
       CountingMatrix: [4×3 double]
          CorrectRate: 0.7500
            ErrorRate: 0.2500
      LastCorrectRate: 0.7500
        LastErrorRate: 0.2500
     InconclusiveRate: 0
       ClassifiedRate: 1
          Sensitivity: 0
          Specificity: 0.8571
```

Fig. 11 Evaluation of flood estimation on test data

the linear relationship is achieved between the year and flood and it is fitted in the linear regression model and is shown in Fig. 13.

The model can define the linear relationship between the year and flood range by reducing the root mean square error value to 0.394 and it is shown in Fig. 13.

5.4 *Flood Estimation Based on Gauge Reading*

In this, the gauge reading for the river basin in the region is taken as the input. It is shown in Fig. 14.

Then, it is trained with the linear regression model to estimate the level of the water based on the flow of the water. The training of the linear regression model in the river basin is shown in Fig. 15.

From Fig. 15, it is observed that the root mean square value is 0.00178. The lesser number of error degree and root minimum square shows that the linear model is perfectly fit for the flow rate and height of the water level.

Based on the above training, the estimation of the gauge reading is performed and the result is estimated in terms of the difference between the original and estimated level of the water and it is shown in Fig. 16.

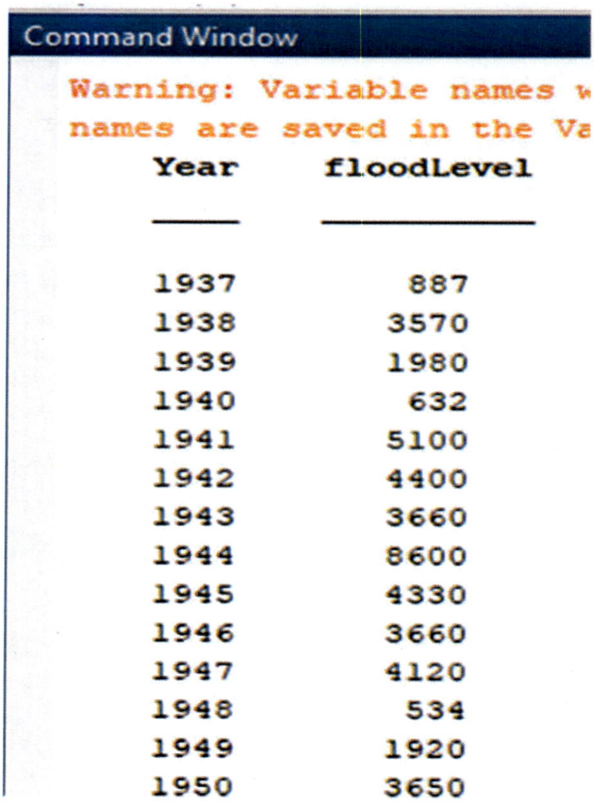

Fig. 12 Flood information for yearwise

Figure 16 shows that the linear model is well fitted between the flow and stage of the river basin. It is able to predict the level of river basin accurately with minimum overall error value of − 9.080 * 10 ^ − 5. This level estimation helps to predict the overflow rate of the basin during the rainfall period.

5.5 Overall Flood Estimation

In this, the flood is estimated from four modes of information. The mode of information is the image, gauge reading, tweets and previous years flood information. All these information are collected for one region. Because the estimation based on the single mode of information is not effective due to climate change and other atmospheric disturbances. Hence, in this, four modes of information are processed.

The flood estimation is done in each mode to predict the flood level. The final estimation is given if the flood estimation is positive in all four modes. Because it indicates the higher probability for occurring flood in that region. In this, the high

Fig. 13 Linear regression results

```
mdl =

Linear regression model:
    y ~ 1 + x1

Estimated Coefficients:
                    Estimate       SE        tStat       pValue
                   __________    ______    _______    __________

    (Intercept)    -2.8207e+05    48244    -5.8468      2.841e-07
    x1                  146.56   24.379     6.0117     1.5402e-07

Number of observations: 57, Error degrees of freedom: 55
Root Mean Squared Error: 4.38e+03
R-squared: 0.397,  Adjusted R-Squared 0.386
F-statistic vs. constant model: 36.1, p-value = 1.54e-07
>>
```

alert will be given to the all positive results of four modes. The medium alert will be given to the three positive results and minimum alert for the two results. This type of approach is not only for predicting the flood but also able to categorize the level of flood in the region.

6 Conclusion

Natural hazard prediction is a difficult one as compared to the hazard caused by chemical or human-made, because the nature is disturbed by the human in the form of pollution and destroying the forest and increase of electronic gadgets. Due to this, the climate change occurs and estimation of natural hazard flood becomes a tedious one. Initially, the flood is estimated through the atmospheric conditions. But, it results negative in today scenario. Hence, a proper estimation of flood is important to avoid the serious damages. In this, the flood is estimated through the

Command Window

Date	time	flow	stage
'01-11-2016 00:00'	23:45:00	5.17	11.08
'01-11-2016 00:00'	23:30:00	5.14	11.06
'01-11-2016 00:00'	23:15:00	5.15	11.07
'01-11-2016 00:00'	23:00:00	5.15	11.07
'01-11-2016 00:00'	22:45:00	5.15	11.07
'01-11-2016 00:00'	22:30:00	5.17	11.08
'01-11-2016 00:00'	22:15:00	5.18	11.09
'01-11-2016 00:00'	22:00:00	5.17	11.08
'01-11-2016 00:00'	21:45:00	5.15	11.07
'01-11-2016 00:00'	21:30:00	5.14	11.06
'01-11-2016 00:00'	21:15:00	5.2	11.1

Fig. 14 Gauge readings of a river basin in SC

data mining and image processing technique. The whole process is performed on a single region called South Carolina. The flood is estimated through four modes by image, gauge readings, tweets and hundred years of flood information of the regions. The regression learning is applied on all the modes to improve the learning in the time series format. In each mode, the regression network outperform the existing technique structured support vector machine and Markov chain Monte Carlo method. Hence, the flood prediction and estimation can be improved by four modes of analysis using regressive learning. In future, these four modes of operation are reduced by minimal mode through optimization approaches.

```
mdl =


Linear regression model:
    y ~ 1 + x1

Estimated Coefficients:
                   Estimate       SE         tStat       pValue
                   ________    _________     ______    __________

    (Intercept)     7.6886     0.0055576     1383.4    1.1858e-322
    x1             0.65616     0.0011272     582.13    1.1404e-263


Number of observations: 159, Error degrees of freedom: 157
Root Mean Squared Error: 0.00184
R-squared: 1,  Adjusted R-Squared 1
F-statistic vs. constant model: 3.39e+05, p-value = 1.14e-263
```

Fig. 15 Training of gauge reading in linear regression model

```
overall_error in estimating the flood
   -3.7280e-04


>>
```

Fig. 16 Evaluation of the gauge reading using linear regression

References

1. B. Jamali, R. Löwe, P.M. Bach, C. Urich, K. Arnbjerg-Nielsen, A. Deletic, A rapid urban flood inundation and damage assessment model. J. Hydrol. **564**, 1085–1098 (2018)
2. Z. Liu, V. Merwade, Accounting for model structure, parameter and input forcing uncertainty in flood inundation modeling using Bayesian model averaging. J. Hydrol. **565**, 138–149 (2018)
3. J.T.S. Savage, P. Bates, J. Freer, J. Neal, G. Aronica, When does spatial resolution become spurious in probabilistic flood inundation predictions? Hydrol. Process. **30**(13), 2014–2032 (2016)
4. S. Kabir, S. Patidar, X. Xia, Q. Liang, J. Neal, G. Pender, A deep convolutional neural network model for rapid prediction of fluvial flood inundation. J. Hydrol. **590**, 125481 (2020)
5. H.I. Kim, H.J. Keum, K.Y. Han, Real-time urban inundation prediction combining hydraulic and probabilistic methods. Water **11**(2), 293 (2019)
6. J. Teng, A.J. Jakeman, J. Vaze, B.F. Croke, D. Dutta, S. Kim, Flood inundation modelling: a review of methods, recent advances and uncertainty analysis. Environ. Model. Softw. **90**, 201–216 (2017)
7. J. Thomas Steven Savage, F. Pianosi, P. Bates, J. Freer, T. Wagener, Quantifying the importance of spatial resolution and other factors through global sensitivity analysis of a flood inundation model. Water Resour. Res. **52**(11), 9146–9163 (2016)
8. Z. Liu, V. Merwade, K. Jafarzadegan, Investigating the role of model structure and surface roughness in generating flood inundation extents using one-and two-dimensional hydraulic models. J. Flood Risk Manage. **12**(1), e12347 (2019)
9. Y.H. Tak, J.D. Kim, Y.D. Kim, B. Kang, A study on urban inundation prediction using urban runoff model and flood inundation model. J. Korean Soc. Civ. Eng. **36**(3), 395–406 (2016)
10. S. Afshari, A.A. Tavakoly, M.A. Rajib, X. Zheng, M.L. Follum, E. Omranian, B.M. Fekete, Comparison of new generation low-complexity flood inundation mapping tools with a hydrodynamic model. J. Hydrol. **556**, 539–556 (2018)
11. J.F. Rosser, D.G. Leibovici, M.J. Jackson, Rapid flood inundation mapping using social media, remote sensing and topographic data. Nat. Hazards **87**(1), 103–120 (2017)
12. S. Gobeyn, A. Wesemael, Van J. Neal, H. Lievens, K. Eerdenbrugh, Van N. Vleeschouwer, H. De Vernieuwe, G.J.P. Schumann, G. Di Baldassarre, B. Baets, P.D. De Bates, Impact of the timing of a SAR image acquisition on the calibration of a flood inundation model. Adv. Water Resour. **100**, 126–138 (2017)
13. C.A. Buahin, N. Sangwan, C. Fagan, D.R. Maidment, J.S. Horsburgh, E.J. Nelson, V. Merwade, C. Rae, Probabilistic flood inundation forecasting using rating curve libraries. JAWRA J. Am. Water Resour. Assoc. **53**(2), 300–315 (2017)
14. D.T. Mai, F. De Smedt, A combined hydrological and hydraulic model for flood prediction in Vietnam applied to the Huong river basin as a test case study. Water **9**(11), 879 (2017)
15. J. French, R. Mawdsley, T. Fujiyama, K. Achuthan, Combining machine learning with computational hydrodynamics for prediction of tidal surge inundation at estuarine ports. Proc. IUTAM **25**, 28–35 (2017)
16. H. McGrath, J.F. Bourgon, J.S. Proulx-Bourque, M. Nastev, A.A. El Ezz, A comparison of simplified conceptual models for rapid web-based flood inundation mapping. Nat. Hazards **93**(2), 905–920 (2018)
17. J.Y. Lee, D. Kim, K.W. Park, T.W. Kim, Establishment of accuracy criteria of flood inundation map using quantitative evaluation indices. J. Korean Soc. Civ. Eng. **39**(3), 381–389 (2019)
18. H. Mojaddadi, B. Pradhan, H. Nampak, N. Ahmad, A.H.B. Ghazali, Ensemble machine-learning-based geospatial approach for flood risk assessment using multi-sensor remote-sensing data and GIS. Geomatics Nat. Hazards Risk **8**(2), 1080–1102 (2017)
19. L.C. Chang, M.Z.M. Amin, S.N. Yang, F.J. Chang, Building ANN-based regional multi-step-ahead flood inundation forecast models. Water **10**(9), 1283 (2018)

20. P.T.T. Ngo, N.D. Hoang, B. Pradhan, Q.K. Nguyen, X.T. Tran, Q.M. Nguyen, P. Samui, D. Tien Bui, A novel hybrid swarm optimized multilayer neural network for spatial prediction of flash floods in tropical areas using sentinel-1 SAR imagery and geospatial data. Sensors **18**(11), 3704 (2018)
21. Story Map Swipe and Spyglass, 2020. Retrieved 19 Nov 2020, from https://scemd.maps.arcgis.com/apps/StorytellingSwipe/index.html?appid=d073b65eeee14704ae1fdbf034dc0b58
22. M.E. Phillips, *Hurricane Florence Twitter Dataset*, 2018-09-05/2018-10-03; https://digital.library.unt.edu/ark:/67531/metadc1259406/. Accessed 19 Nov 2020. University of North Texas Libraries, UNT Digital Library, https://digital.library.unt.edu

Trust-Based and Optimized RPL Routing in Social Internet of Things Network

S. Selvaraj, R. Thangarajan, and M. Saravanan

Abstract Internet of things plays a major role in all the applications that are utilized for automating the environment. In this, the searching of devices in the larger network is a tedious process. Further, it became easier by defining the social relationship among the devices in the network. Social Internet of things is the combination of real-world network and multiple devices in the Internet of things. In social Internet of things, the user can access information globally at any place based on its ID. But, the accessing of this ID in the large database is difficult. Several works were carried out to locate the ID of the devices in the social Internet of things [SIoT]. Community and effective search-based algorithm is one of the methods for effective localization of devices in the social Internet of things as the conventional method for device localization. The Routing protocol for Low power and Lossy networks (RPL) able to increase lifetime of network with high packet loss. In existing, the localization and routing is performed separately. This problem is overcome with the proposed trust-based community nodes and optimized RPL in social Internet of things network. Here, the community is formed with the combination of high trust and low trust nodes of similar taste. These trusted node-based community formation helps to locate the trusty nodes in the transmission process. The dragon fly optimization is used to improve the RPL performance by selecting trusty nodes as next hop in the routing process. Due to the combination of trust community and optimized RPL, the lifetime and success rate of the data transmission is increased by 90% as compared to the existing preference-based device localization and traditional RPL protocol. The whole process is realized with the help of MATLAB R2018a in windows 10 environment. Therefore, the proposed trust community and optimized RPL is best

S. Selvaraj (✉)
Department of Computer Science and Engineering, Kongu Engineering College, Erode, Tamil Nadu, India
e-mail: selvaraj.cse@kongu.edu

R. Thangarajan
Department of Information Technology, Kongu Engineering College, Erode, Tamil Nadu, India
e-mail: thangs_68@kongu.ac.in

M. Saravanan
Department of Computer Science and Engineering, AVS College of Technology, Salem, Tamil Nadu, India

S. Shakya et al. (eds.), *Mobile Computing and Sustainable Informatics*, Lecture Notes on Data Engineering and Communications Technologies 68,
https://doi.org/10.1007/978-981-16-1866-6_36

for the trust-based data transmission in social Internet of things network by having high alive nodes of 90% and time delay of 30 s for hop localization.

Keywords Internet of things · Social · Community · Trust · RPL · Dragon fly

1 Introduction

Social Internet of things describes about connecting Internet of things devices in the real-time network for sharing the information mutually between them.

Geetha [1] first described about the Internet of things concept to provide a smart environment with less human interactions. Based on this, when this information is connected to the Internet for global sharing or usage, it becomes social Internet of things. It discussed about the connection between the Internet of things object and the real-time world through gateway. Then, describe about the application of social Internet of things in domains like smart home, smart localization of people.

Jadhav and Patil [2] described about the role of social connections between the devices in the home monitoring system. In this, the Facebook is used as the social medium to provide the controlling operations of the users. Here, only granted users in Facebook can have access to control the operations in the home management system. In this, the smoke and temperature sensor is continuously monitored to maintain the room temperature, and here, the controlling operations are performed through the Arduino controller through water sprinkler and switching off the light. This mechanism is carried out through the users connected in the Facebook with granted access for modifying the controlling operations.

Girau et al. [3] suggested a mechanism for determining the neighbouring nodes in the Internet of things. The neighbouring detection mechanism is concentrated to improve the relationship between the objects in the social network. To perform this, it uses three modes for defining the relationship. Here, the three modes are analysing the communication channels, considering the past network information and attributes of the objects.

Arjunasamy and Ramasamy [4] analysed the next node based on the four parameters. Here, the four parameters are Giant component, average degree and path length and clustering coefficient. Among these, the giant component produced the best results for selecting the best hop for passing the information in the Internet of things. Here, the next hop is based on the friends of the network which needs to be improved in future.

Qiu et al. [5] proposed a routing scheme for the devices in the Internet of things. The necessity of the new routing scheme is to improve the communication link between the devices and also to include the correct path for the routing. Here, it also able to perform and reduces the data loss in the transmission. All these merits were achieved with the help of residual energy-based routing. Here, the residual energy of all the devices were considered to determine whether it is capable of transmitting packets. Due to this, it able to provide a minimum loss data transmission.

Zhong et al. [6] proposed a routing scheme for the devices in the Internet of things to improve its interconnection between the devices. Here, the routing scheme is implemented on the opportunistic access scheme of the cognitive radio network. First, the channel for the users is assigned based on the game theoretical approach. Then, the routing is based on the residual energy and trust of the node based on the previous transmission. Finally, the data is transmitted using network coding technique. It is the first routing scheme for the SIoT network in cognitive radio mode.

Truong et al. [7] proposed a new model to evaluate a trust in the social network of Internet of things devices. Here, the trust is based on two factors. One is direct trust and the other is indirect trust. The direct trust is based on the two parameters Self-knowledge and experience with the node in the network. Then, indirect trust is based on the reputation factor based on the evaluation of nodes by other nodes. Based on these factors, the trusty network is formed which able to provide a reliable and secure communication in exchanging the information.

Gadallah et al. [8] proposed a relation-based network for sharing the information between the devices in the Internet of things network. Here, the nodes are related in terms of normal human relationships like family, friends, colleagues and competitor. All these relations were based on its operation and owner of the operations. Due to this, it able to identify the possible threats in it and able to route the information securely. It is formed using the access information form MAC layer and create a table for it in the routing management [9].

Ning et al. [10] proposed the social-based scheme for the devices in the Internet of things. A game theory is proposed for monitor the power management and the node selection in the Internet of things application for communication. It supports the communication between the devices and also it able to improve the connectivity and high link capacity for the IoT applications. Trust model is performed with cloud service using meta-heuristic optimization; here it used for work flow scheduling approach [11].

Paper organization is as follows: the next two sections describes about the existing works in SIoT routing, the proposed trust-based and optimal routing in SIoT is explained in Sect. 4 and Sect. 5 discuss about the results of the proposed and existing method. Finally paper is concluded based on the summary in Sect. 6 and future work in Sect. 7.

2 Related Works

Kowshalya and Valarmathi [12] proposed a method to improve the trust of Internet of things network to improve the social relationship status of the devices. Here, the trust is calculated by the authenticator node. The authenticator node is determined by having more number of connections with the other devices. Because, the higher degree centrality node only able to access more information of the nodes. The authenticator node evaluate the other node based on the screening message shared information between them. Sometimes, the higher degree centrality node may also be malicious and drain out its energy more as compared to the other.

Lakshmanaprabu et al. [13] proposed an optimization algorithm for determine the optimal attribute for the classification of SIoT data. Here, the data is initially processed with the Gabor filter to remove noises in it. Then, the filtered data is processed with the elephant herd optimization algorithm for determine the sensitive data for the classification. Then, the selected data is classified linear kernel support vector machine and evaluated in terms of accuracy. The approach able to classify the information effectively but the head selection is single oriented process which limits the selection of attributes.

Han et al. [14] proposed a combination of three routing scheme for data transmission in the SIoT network. Here, the three routing schemes are Greedy routing, directed routing and cyclic routing. Each routing is based on the conditions of the device. It also tested under three attacks to check the security of the information. The three attacks are eavesdropping, hop by hop and direction-based attacks. The combination three schemes able to overcome the attacks and find the possible routing.

Chen et al. [15] proposed a method based on trust and connectivity between the devices in Internet of things network. Here, first the channel state information and unknown interference from the cellular to the base station network is assumed. Then, the locations of the devices are realized through Poisson Point process method. Then, based on the known locations, the trust between the devices is realized and the communication is between the device is realized through connectivity probability. This approach is able to achieve high trust across the devices in a distributed manner.

Atzori et al. [16] proposed a new storage format for the locator based on the splitting of the address in the Internet of things. Here, the new storage format is used to define the social relationship between the devices. To describe these relationships, it uses four field records. The four fields are virtual entity, repository for all the devices, relationship manager and browser to search the relationship between the devices. Based on these, it able to locate the devices easily but it require more memory space to perform its actions.

Hasan and Al-turjman [17] proposed a heuristic-based approach for routing the data in the social IoT. Here, the devices are considered in mesh topology with the different heuristic approach of particle swarm optimization. All the optimization process works to minimize the energy and delay of the network and maximize the throughput. The analysis is carried out on the two tier heterogeneous network. Among many optimization algorithms, the canonical particle swarm optimization produced best results by reducing delay with high throughput. It require tedious analysis to produce the best solutions.

Roopa et al. [18] surveyed about the social Internet of things in a brief manner. First, it described about the basic concepts in the Internet of things and its implementation in the real-world scenario. Second, it defined about the relation between the devices and then described about the social Internet of things. Third, it describes about the challenges in the social Internet of things in terms of relationship manager, security and trust worthiness. Finally, it describe about the methods used in SIoT and it possible extension to increase the efficiency of SIoT.

Amin et al. [19] suggested a clustering approach for easy navigation of devices in the social Internet of things network. Generally, the devices in the larger network is not able to access easily. For an easy access, the relationship between the devices are used for the easy locating the devices. But, it able to perform well for the smaller networks, for higher networks, it require an algorithm for locating the services easily. To accommodate this process, here, the clustering of devices based on relationship is performed. Then, the devices are located based on its clustering property. This approach is able to reduce the search time for localization but it does not include the trust factor in it. It is tested on the social sites data like Facebook.

Imran et al. [20] surveyed about the social Internet of things in a brief manner. First, it described about the basic concepts in the social Internet of things and its implementation in the real-world scenario. It surveyed about different techniques used in SIoT for routing as well as to improve the navigation of devices. It gives a brief summary and its short comings of different techniques used in SIoT to improve the data transmission.

Amin et al. [21] performed an analysis based on the trust worthiness of the devices in the SIoT. Here, it performed a survey on the techniques in SIoT network which works based on the trust and friendly relationship between the devices. Based on the trust, it analysed in terms of Bayesian, weighted sum and belief theory. Then, for the searching of devices, it is based on the direct and indirect search, distributed, centralized and service-based search. The trust on a device can be formed using a single device or based on the opinion of multiple devices. Then, it listed about the challenges in forming the trust-based friendly network in SIoT.

Chuknhoo et al. [22] suggested a twin for the devices in IoT to reduce the workload of the devices in the edge IoT. Here, the edge devices in IoT is placed to improve the data transmission and speed up the data exchange between them. It used the mixed integer linear programming to solve the placement of the devices in IoT. But, it increases the cost for the devices.

Kurle and Radhika [23] proposed a clustering and scanning-based technique for efficient routing in the IoT. Here, the devices are grouped using K-means randomly in three groups as high, medium and low as trust values. Then, the trust values are incremented or decremented based on the delivery of packets. Then, the DBSCAN algorithm is applied to determine the high trust nodes for efficient data transmission based on fixed hops and condition for the cluster formation. This fixed condition reduces the selection of nodes in the network.

3 Existing Method

Social Internet of things is the combination of real-world network and multiple devices in the Internet of things. In social Internet of things, the user can access information globally at any place based on its ID. But, the accessing of this ID in the large database is difficult. Several works were carried out to locate the ID of the devices in the social Internet of things. Community and effective search-based algorithm is one of the methods for effective localization of devices in the social Internet of things [24]. Here, the devices are first grouped in terms of three categories namely, choices, location and social similarity. Then, the searching of devices is performed by conducting a searching mechanism within the community and between the communities. This approach is able to locate the device effectively using this two searching mechanisms with success rate of 92% for intracommunity and 85% for intercommunity search-based device localization. But, it has the following limitations in it.

- Routing of the data is not addressed.
- Trust value of the devices is not considered.
- Device localization only concentrated.

For the social Internet of things, the device localization is important, but it requires all the above parameter for proper defining the relation between the devices in the network. It is achieved with our Similarity and trust-based routing in social Internet of things network.

4 Proposed Method

In this, the trust-based community is formed and also proposed an optimized approach for the devices in the network to route the packets in the network. The main motto of the proposed method is to achieve the following objectives:

- Effective localization of trusted devices.
- Efficient and shortest path for routing the information.
- Stable the devices in the network.

Let us consider an Internet of things network with N number of devices (like wearables, smart phones and so on) and it has different servers for processing the information. In this, the social relationship between the devices is defined in terms of the following parameters:

- Operations
- Location
- Trust

The network of devices in IoT are connected in the graph format $G = (V, E)$. The term V indicates the vertices of the graph that is the number of devices in the network and E is edges that indicates the connection between the devices in the network. In this, the effective localization of a device and routing the information in the network is proposed.

(a) **Localization of device using trust model**

In this, the devices in the IoT are grouped based on the relation between each node in terms of operations and Location of the nodes. Then, the grouped nodes are sorted based on the trust value within the community to improve the hop selection in the routing process.

Consider, an IoT consist of N devices which are interconnected to each other. Among this N devices, the devices which performs same operations are grouped O_D and the devices which are located in the same geographical space is denoted as G_D.

i. **Operation similarity**

The similarity between the devices operations is calculated based on the cosine similarity value of the device outputs is given in Eq. 1.

$$\mathrm{OS}(N_i, N_j) = \frac{\overrightarrow{N_i} \cdot \overrightarrow{N_j}}{N_i \cdot N_j} \tag{1}$$

ii. **Location similarity**

In this, the movement of the IoT and location of the devices is used for grouping the devices based on the Eq. 2.

$$\mathrm{LS}(N_i, N_j) = \frac{3\sum_{i=1}^{N}\sum_{j=1}^{N} V(i, j)b_i * b_j}{4N\pi r^3\sqrt{\sum_{i=1}^{N} b_i^2} * \sqrt{\sum_{j=1}^{N} b_j^2}} \tag{2}$$

In Eq. 2, the term V is the common region between the devices and b_i and b_j are the weights of the individual device regions.

Based on the above two equations, the combined similarity for the devices is given in Eq. 3.

$$\mathrm{CS}(N_i, N_j) = \alpha\,\mathrm{OS}(N_i, N_j) + \beta\mathrm{LS}(N_i, N_j) \tag{3}$$

Here, the terms α and β are the weights of the operational and locational similarity. By adding both the similarity, it should be equal to 1. To allocate a node in a community, this sum component should be greater than 0.6.

Then, grouping of nodes in the two community based on the trust value is performed with the help of following algorithm 1 (Fig. 1).

The device localization is performed using zone routing protocol mode by defining the devices within the community as intra-zone and devices in the other community as inter-zone. Here, the optimized RPL is used and it is explained as follows.

(b) **Optimization-based routing in IoT network**

In this, the optimized improved zone routing protocol is used for the routing of information between the nodes in the IoT network. Here, the zone routing protocol is used because in this communities are formed in terms of intra- and inter-community nodes.

General zone routing protocol utilize both proactive and reactive protocol for communication between the communities. But, in this, the common protocol is used

Fig. 1 Algorithm for trust-based community

Input: IoT devices N and neighbouring elements Nn
Output: Community list
Step 1: Initialize the OD and GD.
Step 2: For each Nn
Step 2.1: Assign Initial trust value as 10.
Step 2.2: Transmit Hello message
Step 2.3: if transmit successfully
Step 2.4 Trust value increment by 1
Step 2.5: else
Step 2.6: trust value decrement by 1.
Step 2.7: compute similarity based on operations and Locations using equation 3.
Step 3: if (similarity threshold > 0.6)
Step 3.1: add device to community list
Step 3.2: initiate the message for joining
Step 3.3: if accepted
Step 3.4: device added to the community
Step 4: community list with trusted nodes.
Step 5: Stop

based on the optimization algorithm. Because, the device is localized with the help of trust-based community network. Due to this, it reduces the necessity of two protocols for routing process. Here, the satin bower-based routing protocol for Low power and Lossy networks.

In this, dragon fly optimization is to maximize the throughput of the RPL by maximize its objective function. The objective function of the RPL is given in equation as follows

$$\mathrm{DF}_{\mathrm{obj}} = \text{Energy efficiency of RPL} \tag{4}$$

$$\text{Energy Efficiency} = \frac{\text{Througput}}{\text{Power consumption}} \tag{5}$$

The next hops in the RPL is determined with the help of dragonfly optimization by improving its energy efficiency of the network. The following operations are performed in the dragon fly process to select the next hop.

i. **Distancing mode**

Generally, the aim of dragon flies is to reach the food (i.e. objective function). But, it should maintain a distance between each dragon flies to determine the optimal solutions. Consider two dragon flies X and Y in a search space area of N. Then, the distance between the dragon flies is as follows in 6

$$D_j = -\sum_{j=1}^{N} X - Y_j \tag{6}$$

ii. **Matching mode**

Even though the dragon flies are separated, it should match its velocity with other flies to reach the optimal solution effectively. It is given in Eq. 7

$$M_j = \frac{\sum_{j=1}^{N} \text{Velocity}_j}{N} \tag{7}$$

iii. **Attraction mode**

When the velocities are aligned, at some point, the dragon fly can be attracted to its neighbourhood point. It is given in the Eq. 8.

$$A_j = \frac{\sum_{j=1}^{N} Y_j}{N} - X \tag{8}$$

iv. **Optimal solution**

The major motto of dragon fly is to determine the next hop by maximize its objective function. It is given in Eq. 9.

$$\text{OS}_i = Y^{+} - X \tag{9}$$

v. **Enemy**

To prevent the dragon fly from false results, the following Eq. 10 is used to remove the poor nodes in the IoT for hop selection based on trust values.

$$E_i = Y^{-} - X \tag{10}$$

vi. **New position**

The optimal solution is obtained only after many iterations. Hence, the position of dragon flies is also need to be update for every iteration. Based on that, the dragon fly position update equation is as follows in 11

$$P(X)_{t+1} = P(X)_t + \Delta P(X)_{t+1} \tag{11}$$

The term $\Delta P(X)_{t+1}$ is the step vector is calculated in the following Eq. 12

$$\Delta P(X)_{t+1} = m\left(M_j\right) + a\left(A_j\right) + o\left(\text{OS}_j\right) + e(E_i) \tag{12}$$

The terms m, a, o and e are the weights of the modes in the dragon fly.

The position update is further improved with the levy flight concept as in Eq. 13.

$$P(X)_{t+1} = P(X)_t + \text{levy}\ ((X)) \times P(X)_t \tag{13}$$

The optimal hops for the RPL-based routing with trusted nodes in the social-based Internet of things network is determined with the help of proposed trust-based community and dragon fly-based network.

Algorithm for trust-based and optimized RPL

Inputs: IoT devices.

Outputs: community devices, Optimal routing.

Step 1: Number of IoT devices and its ID.

Step 2: Community formation based on O, M, and B.

Step 3: transmit hello message.

Step 4: if (transmission successful).

Step 5: trust value increased.

Step 6: node is added to the community.

Step 7: else.

Step 8: trust value decreased.

Step 9: node remain in the community.

Step 10: Community of the node is formed.

Step 11: Trust value of all nodes calculated.

Step 12: if packet transmission needs.

Step 13: device located using inter- and intra-search algorithm.

Step 14: Trusty nodes are used for transmission.

Step 15: Dragon fly-based RPL is applied.

Step 16: optimal route obtained.

Step 17: packet transmission performs.

Step 18: else.

Step 19: device only located using inter- and intra-search algorithm.

Step 20: Stop.

The working of the above algorithm is represented in pictorial representation in Fig. 2.

5 Implementation and Discussion

In this, the proposed trust-based localization and optimization-based routing of devices in the Internet of things is implemented with the help of Matrix laboratory software under windows 10 environment in R2018a version.

A sample diagram for connections between the devices in the network is shown in Fig. 3.

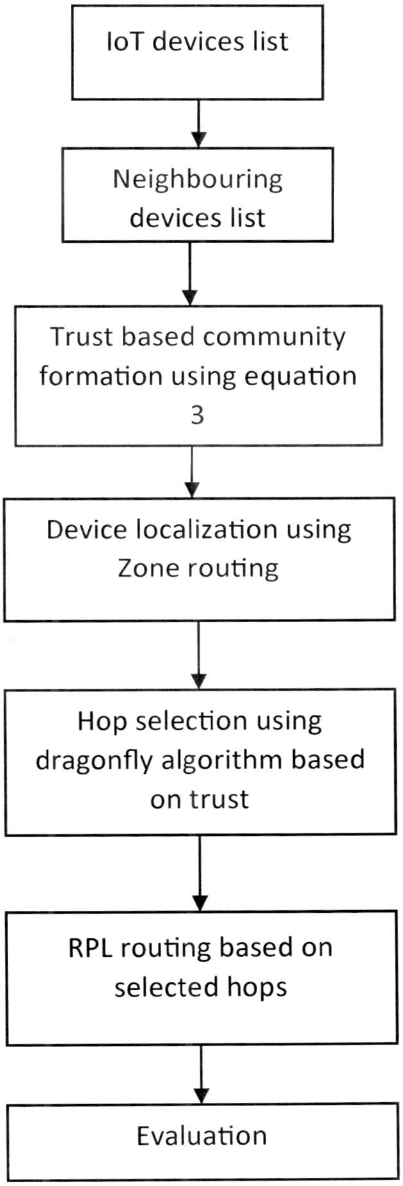

Fig. 2 Trust-based and optimized RPL routing in SIoT

Then, the above devices are grouped based on the trust score and similarity of the devices using our proposed method. The proposed method trust-based localization of devices is evaluated in terms of following metrics.

- Triangular participation ratio
- Modularity score

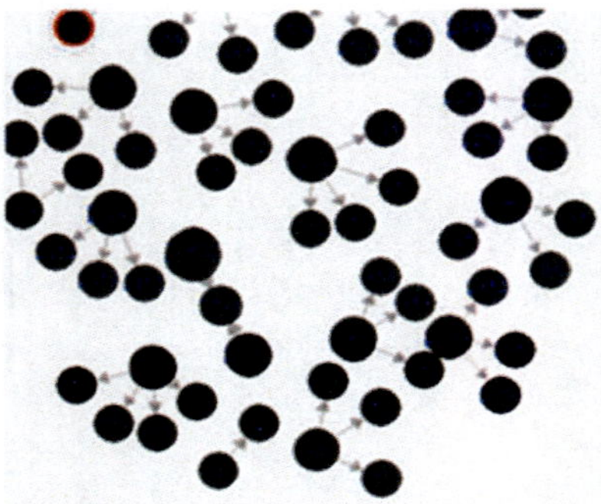

Fig. 3 Sample connection of devices in IoT

- Delay
- Success in device localization

(a) **Triangular participation ratio**

It is the number of triangles formed by the nodes in the community. For a stronger relationship between the devices in the network, the number of triangle formation in the network should be high. It is calculated using the following formula in 14.

$$\text{TPR} = \frac{\text{Number of triangles}}{\text{Number of nodes in the community}} \tag{14}$$

The evaluation of the community formation in terms of triangle formation in the community is compared with the existing methods like resource location based on preference and movement, inter- and intra-based search and the proposed trust-based community formation is shown in Fig. 4.

The TPR comparison for the community formation using three techniques is shown in Fig. 3. It depicts that the proposed trust-based community formation able to define the relationships between the devices in the IoT effectively as compared to existing techniques.

(b) **Modularity Score**

It is to evaluate the originality of the communities detected by the algorithm. Because, the proper community only can help to locate the device effectively.

The comparison of the modularity score of the trust-based community formation with the existing approaches is shown in Fig. 5.

The proposed trust-based community technique able to form a stronger community for 2500 devices with more than 150 communities by having high modularity score of 0.9 as compared to the existing techniques RLPM of 0.54 and I&I L of 0.63, respectively.

Fig. 4 TPR comparison

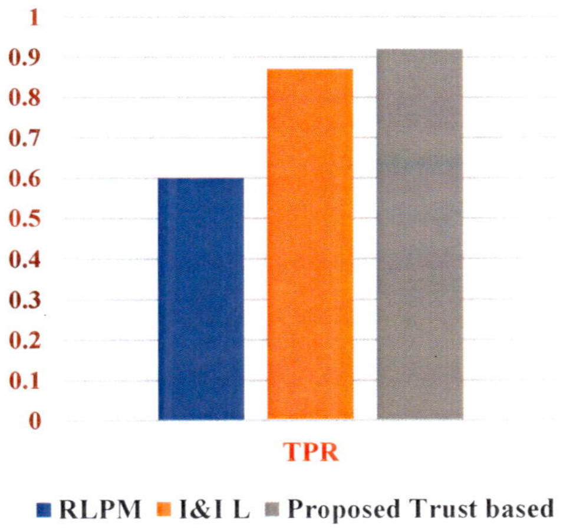

Fig. 5 Comparison result of modularity score

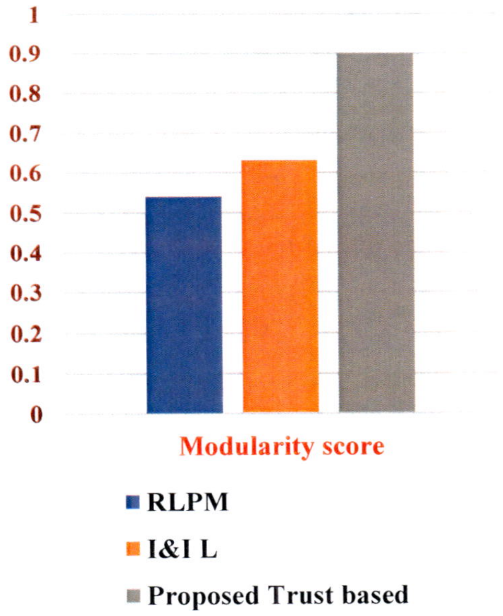

(c) **Success in device localization**

It indicates the number of devices successfully located by the algorithm. It is calculated using the following formula in 15.

$$S = \frac{\text{Device localized successfully}}{\text{total number of query devices}} \tag{15}$$

Based on the above formula, the comparison of device localization using trust-based method and other methods are compared in Fig. 6.

The proposed trust-based model able to locate the devices effectively in both intracommunity and intercommunity by having above 0.80 as its success rate in locating the devices. While, the existing I and I L method able to achieve only above 0.75 for device localization in both the communities.

(d) **Delay**

It is the time required for successful identification of devices queried by the user. For a better processing, the delay should be minimum. The comparison of delay for 1500 devices using proposed and existing techniques is shown in Fig. 7.

The proposed trust-based community able to detect more devices accurately in shorter time period of 29.7 s as compared to the existing techniques like RLPM and I&I L method requires above 30 s for 60–70% of successful device localization.

Therefore, based on the above four evaluations, it is observed that the proposed method is able to define the relationship between the devices effectively using trust-based approach as compared to the similarity-based techniques.

Then, the proposed method is evaluated for routing process in terms of energy consumption and network lifetime.

(e) **Energy Consumption**

The term energy consumption refers to the energy consumed by the IoT devices in the SIoT network to transfer an information from source to destination. The power consumption of the nodes should be minimum for prolong the network time period.

(f) **Network Lifetime**

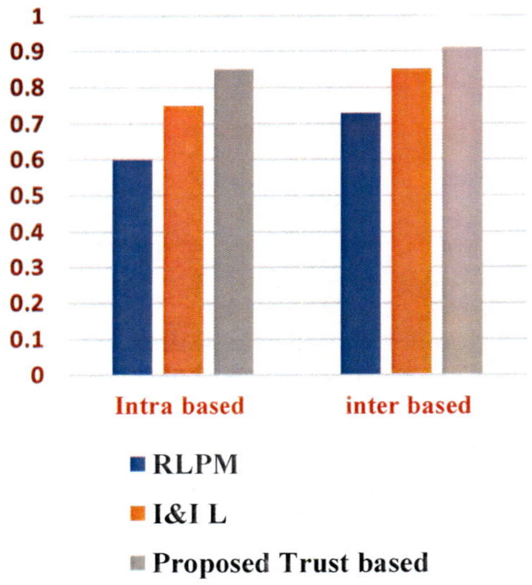

Fig. 6 Success rate comparison

Fig. 7 Comparison of delay for 1500 devices

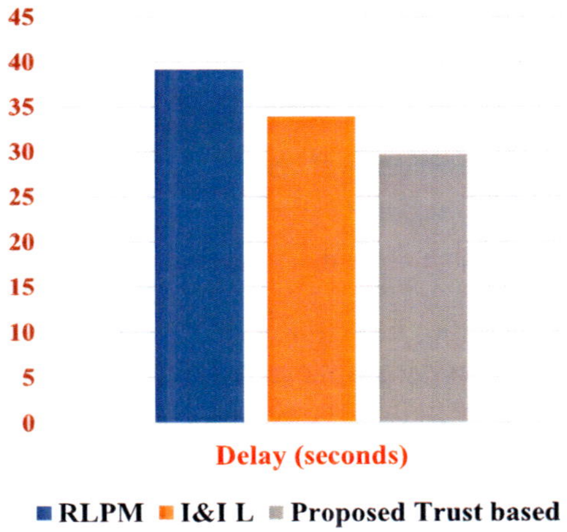

Table 1 Performance results of proposed approach

Method	Energy consumption (%)	Network lifetime (%)	Throughput (%)
Proposed trust-based service	15	92	91

The network lifetime indicates the higher number of stable devices in the social Internet of things network to provide an uninterrupted services. Here, the network lifetime is analysed in terms of alive nodes. Because the stable network can able to increase the lifetime of the network.

(g) **Throughput**

Throughput is the indication of successful transmission of packets from source to destination with minimal loss. The evaluation of the proposed routing method based on these two parameters is in Table 1.

Table 1 shows the energy consumed and the number of nodes alive after the one successful transmission using the proposed routing approach in Internet of things.

Based on both the community formation and the routing approach, the proposed trus-based optimal routing able to outperform the existing techniques and also able to provide services in longer time period by having more alive nodes.

6 Conclusion

Internet of things plays a major role in all applications for automating the environment. In this, the search of devices in the larger network is a tedious process. It became easier with defining the social relationship among the devices in the network. Many research works were carried out in defining the social relationship among the devices. Based on this, in this, the trust-based community formation defines the social relationship among the devices. Then, the optimization approach is applied on the network for efficient routing. The advantages of the proposed method over the existing techniques like RLPM and I & I L method are:

- It was able to define a stronger relationship among the nodes by having higher score in TPR, modularity score and success rate.
- The time taken for device localization is less than thirty seconds.
- The proposed method also able to increase the lifetime of the IoT network by optimal routing.

7 Future Work

In future, to propose a single approach to implement the trust and social behaviour-based routing in social Internet of things application.

References

1. S. Geetha, Social internet of things. World Sci. News **41**, 76 (2016)
2. B. Jadhav, S. C. Patil, Wireless home monitoring using social Internet of Things (SIoT), in In *2016 International Conference on Automatic Control and Dynamic Optimization Techniques (ICACDOT)*. IEEE, Sept 2016, pp. 925–929
3. R. Girau, S. Martis, L. Atzori, Neighbor discovery algorithms for friendship establishment in the social Internet of Things, in *2016 IEEE 3rd World Forum on Internet of Things (WF-IoT)*. IEEE, Dec 2016, pp. 165–170
4. A. Arjunasamy, T. Ramasamy, A proficient heuristic for selecting friends in social Internet of Things, in *2016 10th International Conference on Intelligent Systems and Control (ISCO)*. IEEE, Jan 2016, pp. 1–5
5. T. Qiu, Y. Lv, F. Xia, N. Chen, J. Wan, A. Tolba, ERGID: an efficient routing protocol for emergency response Internet of Things. J. Netw. Comput. Appl. **72**, 104–112 (2016)
6. X. Zhong, R. Lu, L. Li, S. Zhang, ETOR: energy and trust aware opportunistic routing in cognitive radio social internet of things, in *GLOBECOM 2017–2017 IEEE Global Communications Conference*. IEEE, Dec 2017, pp. 1–6
7. N.B. Truong, H. Lee, B. Askwith, G.M. Lee, Toward a trust evaluation mechanism in the social internet of things. Sensors **17**(6), 1346 (2017)
8. Y. Gadallah, M. el-Tager, Y. Ismail, A practical social relationship-based system for cooperative Internet of things operation, in *2017 IEEE International Black Sea Conference on Communications and Networking (BlackSeaCom)*. IEEE June 2017, pp. 1–5
9. D. Hussein, S.N. Han, G.M. Lee, N. Crespi, E. Bertin, Towards a dynamic discovery of smart services in the social internet of things. Comput. Electr. Eng. **58**, 429–443 (2017)
10. Z. Ning, X. Wang, X. Kong, W. Hou, A social-aware group formation framework for information diffusion in narrowband Internet of Things. IEEE Internet of Things J. **5**(3), 1527–1538 (2017)
11. G. Jeeva Rathanam, A. Rajaram, Trust based meta-heuristics workflow scheduling in cloud service environment. Circuits Syst. **7**, 520–531

12. A.M. Kowshalya, M.L. Valarmathi, Dynamic trust management for secure communications in social internet of things (SIoT). Sādhanā **43**(9), 1–8 (2018)
13. S.K. Lakshmanaprabu, K. Shankar, A. Khanna, D. Gupta, J.J. Rodrigues, P.R. Pinheiro, V.H.C. De Albuquerque, Effective features to classify big data using social internet of things. IEEE Access **6**, 24196–24204 (2018)
14. G. Han, L. Zhou, H. Wang, W. Zhang, S. Chan, A source location protection protocol based on dynamic routing in WSNs for the Social Internet of Things. Fut. Gener. Comput. Syst. **82**, 689–697 (2018)
15. G. Chen, J. Tang, J.P. Coon, Optimal routing for multihop social-based D2D communications in the Internet of Things. IEEE Internet of Things J. **5**(3), 1880–1889 (2018)
16. L. Atzori, C. Campolo, B. Da, R. Girau, A. Iera, G. Morabito, S. Quattropani, Enhancing identifier/locator splitting through social internet of things. IEEE Internet of Things J. **6**(2), 2974–2985 (2018)
17. M.Z. Hasan, F. Al-Turjman, SWARM-based data delivery in Social Internet of Things. Fut. Gener. Comput. Syst. **92**, 821–836 (2019)
18. M.S. Roopa, S. Pattar, R. Buyya, K.R. Venugopal, S.S. Iyengar, L.M. Patnaik, Social Internet of Things (SIoT): foundations, thrust areas, systematic review and future directions. Comput. Commun. **139**, 32–57 (2019)
19. F. Amin, R. Abbasi, A. Rehman, G.S. Choi, An advanced algorithm for higher network navigation in social Internet of Things using small-world networks. Sensors **19**(9), 2007 (2019)
20. M. Imran, S. Jabbar, N. Chilamkurti, J.J. Rodrigues, Enabling technologies for social internet of things. Fut. Gener. Comput. Syst. **92**(15), 715–717 (2019)
21. F. Amin, A. Ahmad, G. Sang Choi, Towards trust and friendliness approaches in the social Internet of Things. Appl. Sci. **9**(1), 166 (2019)
22. O. Chukhno, N. Chukhno, G. Araniti, C. Campolo, A. Iera, A. Molinaro, Optimal placement of social digital twins in edge IoT networks. Sensors **20**(21), 6181 (2020)
23. A. Kurle, K.R. Radhika, Machine learning based trust routing for clustered IoT devices. Mach. Learn. **4**(3) (2020)
24. A.M. Kowshalya, X.Z. Gao, M.L. Valarmathi, Efficient service search among Social Internet of Things through construction of communities. Cyber-Phys. Syst. **6**(1), 33–48 (2020)

Trivial Cryptographic Protocol for Resource-Constraint IoT Device Security Using OECC-KA

K. Raja Rajeshwari and M. Ramakrishnan

Abstract The IoT structures use data as a general rule; the data arrangement from contraptions can moreover be a goal of cyberattacks. It is a direct result of this that countermeasures dependent on encryption are as of now picking up in significance. Trivial (lightweight) cryptosystem is an encryption approach which includes slight impress with low-computational multifaceted nature. A few difficulties can block the fruitful execution of an IoT framework and its associated gadgets, including security, interoperability, power/handling capacities, adaptability, and accessibility. Many of these can be addressed with IoT gadget the board, for example, by accepting standard IoT conventions like lightweight M2M or utilising a comprehensive IoT gadget the executives stage offered by a reliable supplier as part of a SaaS model. In this paper, we propose a simple cryptography protocol for securing resource-constrained devices that uses Optimised ECC with Karatsuba algorithm (OECC-KA) for fast multiplication. Several efficient cryptographic algorithms have been widely used for resource constraint device; it may not be suitable with its limited computing power, bandwidth and CPU power. Elliptic curve cryptography is the best choice for limited power usage with high-level security. The traditional elliptic curve cryptography occupies 80% of execution time for scalar multiplication. The proposed optimized ECC with Karatsuba algorithm increases the execution time with fast *multiplication algorithm*.

Keywords IoT · Cryptography · ECC · Karatsuba algorithm · Lightweight · Scalar multiplication

1 Introduction

The Internet of Things (IoT) is driving the present computerized change. Depending on a blend of advances, conventions, and gadgets, for example, remote sensors and recently created wearable and embedded sensors, IoT is changing each part

K. Raja Rajeshwari (✉) · M. Ramakrishnan
Department of Computer Applications, School of Information Technology, Madurai Kamaraj University, Madurai, Tamil Nadu 625021, India

S. Shakya et al. (eds.), *Mobile Computing and Sustainable Informatics*, Lecture Notes on Data Engineering and Communications Technologies 68,
https://doi.org/10.1007/978-981-16-1866-6_37

of everyday life, particularly ongoing applications in computerized medical care. IoT joins different sorts of equipment, correspondence conventions, and administrations. This IoT variety can be seen as a twofold edged blade that gives solace to clients however can lead additionally to an enormous number of security dangers and assaults.

Gadget the board enables organisations to coordinate, arrange, screen, and remotely oversee brilliant resources at scale, providing essential highlights to keep up the health, availability, and security of IoT asset obliged gadgets all through their lifecycle. However, safeguarding IoT gadgets and the organisations with which they interact can be difficult due to the variety of gadgets and suppliers, as well as the difficulty of provisioning, ensuring security, and setting up dependable gadget worker correspondence for gadgets with limited assets. In an attempt to give a broad definition of the term, it may be stated that asset obliged gadgets are those that, by design, have limited handling and capacity abilities to provide the most information yield possible with a minimal force input while remaining savvy. Because of their successful application as keen hardware working in the field, in unforgiving conditions or/and difficult to-arrive at places, they ordinarily run on batteries to keep up the harmony between the viable range of their lifetime and the expected expenses of gadget substitution.

Cryptography is the utilization of codes and codes to ensure private correspondence and keep it hidden from everybody aside from the proposed beneficiaries. The most punctual known utilization of cryptography is from Egypt around 1900 BCE. Alongside, the advancement of cryptography has been an equal exertion in cryptanalysis, which is the study of "breaking" codes and codes. The Enigma was an electromechanical cryptographic machine utilized by Nazi Germany during World War II for secure correspondence. The Enigma code was ultimately "broken" with extraordinary trouble by an allied cryptanalysis group. Some accept that this abbreviated the battle by as much as two years.

Cryptography use was generally restricted to governments until 1977 when two occasions moved it into the public space—the formation of a U.S. government-endorsed figure, the Data Encryption Standard (DES), and the public presentation of RSA, which was the principal reasonable public-key cryptosystem.

The Data Encryption Standard (DES) is a symmetric-key calculation. DES has a 56-digit key size, which albeit secure when presented, no longer gives adequate security. In January 1999, it was exhibited that a 56-digit DES key could be broken utilizing a savage power assault in 22 h and 15 min. Breaking this key so immediately was finished utilizing an overall organization of almost 100,000 PCs that were trying keys at a pace of 245 billion keys for each second.

In a symmetric-key calculation, similar to DES, at least two gatherings have an indistinguishable (or almost indistinguishable) key. These methodology works however have the viable trouble that safely conveying the key from the main party to different gatherings can present a security hazard. Any individual who accesses the symmetric key can peruse any of the messages as well as alter and send on the messages without the beneficiaries realizing that the messages have been adjusted. Uneven cryptography or public-key cryptography successfully fixes these issues.

Public-key cryptography utilizes public keys and private keys that are numerically identified with one another. The public keys can be known by anybody and everybody, while the private keys are left well enough alone and are known to just their proprietors. Utilizing this framework, anybody can encode a message utilizing a public key and the scrambled message would then be able to be left on a public worker or sent over a public organization without security concerns. This message must be unscrambled by the planned beneficiary utilizing their private key. This framework depends on cryptographic calculations dependent on numerical issues that do not right now have effective arrangements, for example certain whole number factorizations, discrete logarithms, and circular bends. Public key/private key sets can be created generally effectively and can be utilized for encryption and decoding. The strength of the security lies in the way that it is amazingly hard to decide an appropriately produced private key from its public key. Likewise, private keys of adequate length cannot be gotten through savage power assaults utilizing the whole world's computational control throughout many years of passed time. NIST has decided prescribed key lengths to give compelling security during that time 2031 and past dependent on the extended pace of development in computational force.

The appropriate utilization of cryptography through cryptographic ICs is accordingly fundamental to making sure about all IoT gadgets, just as numerous other electronic items. In an earlier blog named, "Hearty IoT security costs short of what you think," I gave an illustration of how successful IoT security can be cultivated by a cryptographic IC that adds under $1 to an IoT gadget bill of material (BOM). Actualizing public-key cryptography for IoT gadgets requires huge ability, yet associations can go to IoT security accomplices who are fit for giving total, vigorous security arrangements, commonly very quickly, which is generally acted in corresponding with other improvement endeavors.

Clearly, any key can hypothetically be broken utilizing a savage power assault with adequate processing power. The down-to-earth approach of present-day cryptography is to utilize a key of adequate enough length that it cannot be broken without an exceptional measure of registering power that would be altogether in overabundance of the estimation of the substance that the cryptography ensures. Fortunately, the previously mentioned sub-$1 cryptographic IC uses 256-cycle elliptical curve cryptography (ECC) keys, every one of which is secure to the point that the computational capacity to break a solitary key would require registering assets equivalent in expense to 300 million times the whole world's yearly GDP (78 trillion USD) laboring for a whole year.

Current cryptographic ICs make IoT security so moderate that no IoT item creator has any substantial reason to keep on overlooking IoT security chances. Encryption is a compelling countermeasure, and the IoT is currently needed to apply encryption to sensor gadgets in conditions with different limitations that have not recently been dependent upon encryption. Trivial cryptography is an innovation investigated and created to react to this issue. In this paper, the creator will depict the security dangers of IoT and talk about the countermeasures that depend on encryption.

2 Security Threats for IoT, Countermeasures Based on Encryption

The best security-related peril of IoT structures begins in regular IT organisations regardless, by means of strategies designed for data arrangement since this current reality has turned into a target of cyberattacks, the solution behind smearing IoT to a plant is to continuously improve the effectiveness and suitability by social event data starting innumerable sensors presented. If device information distorted during this cycle, mixed up assessment outcome actuated and inaccurate switch on account of occasion in capacity of provoking huge mischief. Also, since assessment data and control orders are exclusive developments related with the mastery of creation and the chiefs, thwarting spillages is similarly critical from the viewpoint of earnestness. Whether or not there is no issue as of now, it is essential to consider the effect of perils that may get clear later on.

Spreading encoding to sensor devices implies the use of statistics affirmation order then reliability, an incredible counterpart in contrast to the fears (Fig. 1). However, for devices with limited resources, trivial cryptography measures limit the use of locked encoding.

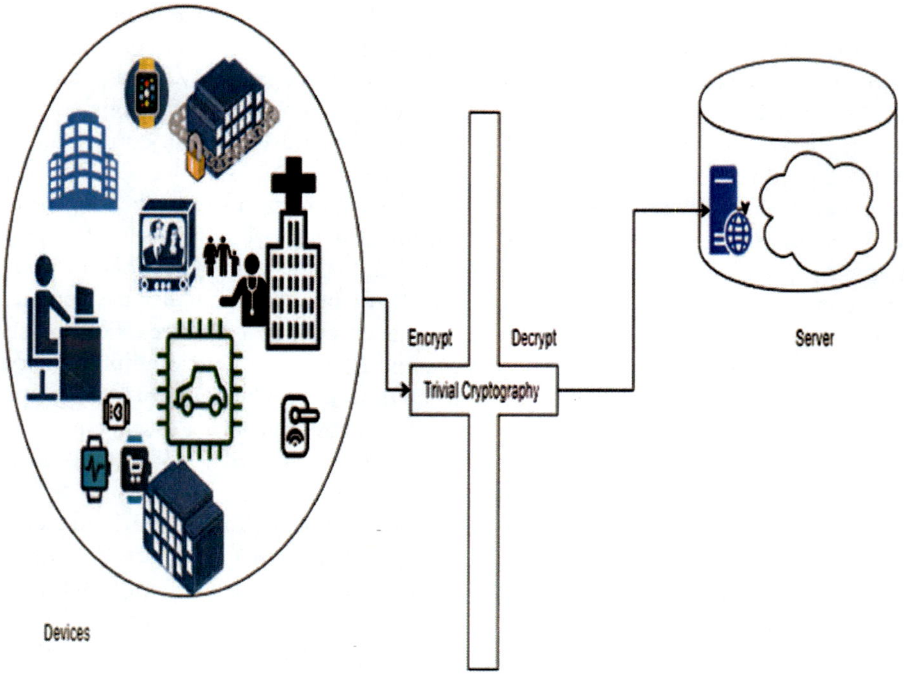

Fig. 1 Trivial cryptography

Requirements for Trivial Cryptography The accompanying variables on the usage are needed for trivial cryptography.

- Size
- Power
- Power utilization
- Communication

Trivial cryptography is the branch of cryptography that focuses on the development of calculations for use in devices that cannot provide the majority of the current codes but have adequate resources (memory, power, and size) for the activity. The utilization of trivial cryptography is the most current calculations intended for use as a component of program frameworks without equipment enhancement programming. This reality makes it difficult to utilize the vast majority of existing cryptographic calculations in gadgets with restricted preparing power, little volume, and low force utilization. Strategies for cryptographic security of information in frameworks with ease turned into the premise of trivial cryptography. Trivial cryptography gets important in the circumstance with "Web of Things," which is a remote self-designing organization between objects of various classes, that can incorporate apparatuses, vehicles, shrewd sensors, and RFID labels (RFID).

Regularly, designers of trivial calculations are compelled to pick between the two, in some cases commonly specific prerequisites for calculations: well-being, cost, and efficiency. By and by, it is anything but difficult to upgrade any two of the three plan objectives: security and cost, well-being and profitability, or the expense and execution; however, it is hard to enhance every one of the three plan objectives at the same time. In such a manner, there are numerous executions of lightweight cryptography calculations: both programming and equipment. They have extraordinary and now and again clashing qualities.

3 Proposed Algorithm

Elliptic curve cryptography (ECC) is a type of public-key cryptography that is based on the logarithmic structure of elliptic twists around restricted fields. ECC permits more unobtrusive keys looked inversely in relative to non-EC cryptography (considering plain Galois fields) equivalent safety.

EC has fitting key plan, progressed marks, pseudo-discretionary makers, and numerous responsibilities. An indirect method for obtaining basic simultaneity in conjunction with symmetric encryption. They are moreover used in a couple of entire number factorization figuring reliant on elliptic curves that have applications in cryptography, for instance, Lenstra elliptic-twist factorization.

The condition with the elliptic bend over the limited field speaks to as follows (Fig. 2).

$$Y^2 = x^3 + ax + b \tag{1}$$

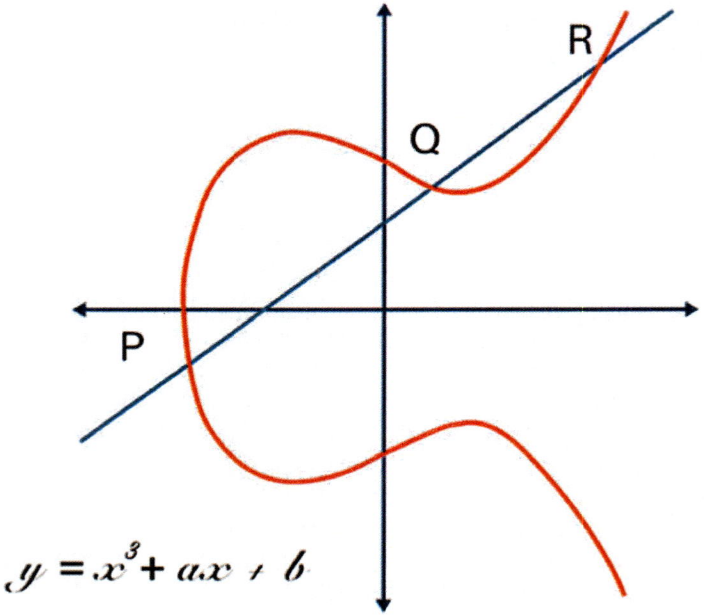

Fig. 2 Elliptic curve cryptograph

Karatsuba algorithm The Karatsuba computation is a brisk increment estimation. Karatsuba–Ofman's calculation (KOA) was the main number augmentation technique that broke the quadratic intricacy obstruction in positional number frameworks. Because of its effortlessness, its polynomial form is broadly embraced to plan VLSI equal multipliers in GF(2n)-based cryptosystems Two boundaries are regularly used to gauge the exhibition of a GF(2n) equal multiplier, namely the existence complexities. The space multifaceted nature is spoken to regarding the total number of two-input XOR AND doors utilized. The relating time unpredictability is given in terms of the most extreme defer looked by a sign because of these XOR AND entryways.

ECC optimization using the Karatsuba algorithm Elliptic scalar increase activity consistently accumulates an opinion lengthways an EC to itself. It is utilized in ECC as a method for delivering a single direction work. This activity is offered by the writing as a scalar duplication in the form of an EC. This activity is also known as EC point augmentation; however, it gives the unfavourable impression of being a duplication between two focuses.

$nP = P + P + P + \cdots + P$ for some scalar (integer) n and a point $P = (x, y)$ that lies on the curve, E.

Table 1 Comparison analysis of delay

Multipliers	Delay (ns)
ECC with conventional multiplier	1.860
ECC with Vedic multiplier	1.233
The proposed method (optimized ECC with Karatsuba algorithms)	1.121

Allow k to be a scalar that is increased with the guide P in order to obtain another point Q on the bend. For example, to discover $Q = kP$.

Step 1: M is the degree polynomial divided into two half X and Y.

Step 2: $X = XhRm/2 + Xl$ and $Y = YhRm/2 + Yl$.

Step 3: M is the padding that has been applied to make the X and Y the same size. Padding is represented by the letters Xh and Yh.

Step 4: Karatsuba has been applied for $(Xh \cdot Yh)(XI \cdot YI)$ $(Xh + Yh)(XI + YI)$.

4 Result and Analysis

In this part, we center around the FPGA usage of the proposed calculation. The proposed designs are coded in Verilog HDL and are orchestrated utilizing Xilinx ISE form 14.5 plan programming and are executed on Xilinx Spartan 3E FPGA (Table 1).

5 Discussion

In this work, the OECC-KA has been proposed to optimize the ECC for mathematical implementation. In general, the ECC provides high security with a lower-key size but the computational cost has been very high due to its the mathematical operations behind the ECC, i.e., scalar operation. The resource constraint device is playing a significant role in technological advancement. The storage capacity and memory space for the resource-constraint device come under minimal categories. This work has been designed to optimize the memory space for ECC operations with the Karatsuba algorithm.

The result of the work has been compared with the two implementations conventional ECC and Vedic ECC. The delay has been measured very less for the proposed implementation of 1.121 ns. In the future, the work has been extended for the real-time situation and compared the performance metrics of throughput, delay and communication, and security overheads.

6 Conclusion

Elliptic curve point increase is the major activity for elliptic bend cryptosystems. The scalar increase over the limited fieldfare re-enacted utilizing the recreation apparatus modalism. This paper gives productive arrangement which improves framework execution regarding speed, power, and area usage and lessens the multifaceted nature. This paper provides a productive arrangement that improves framework execution in terms of speed, power, and area usage while reducing the multifaceted nature.

File Security Using Hybrid Cryptography and Face Recognition

Shekhar Khadka, Niranjan Shah, Rabin Shrestha, Santosh Acharya, and Neha Karna

Abstract Currently, data security has been a major issue as data is usually transferred over the Internet between users. Some security flaw might leak sensitive data. For resolving this, we can use cryptography to secure data so that other unauthorized individuals can't access the data stored in the file. Use of one cryptography algorithm might not provide higher level security for sensitive data. In this paper, we have proposed a new security system which uses three separate cryptography algorithms to provide better security for the data and facial recognition to verify the user who is currently using the system to provide better security. The file to be secured is split into three parts and is encrypted using AES, Blowfish and Twofish, respectively. File and user information are added to the database. The user of the system is verified using facial recognition during the authorization process in decryption phase.

Keywords AES · Blowfish · Twofish · Hybrid cryptography · Facial recognition

1 Introduction

File Security is one of the most challenging jobs in today's world. Protecting data from unauthorized access and data corruption throughout its life cycle is mandatory. Everyone from the individual to organizations wants to maintain their data secure for privacy purpose. So set of standards and technologies should be maintained to protect data from intentional or accidental destruction, modification or disclosure. Finally, the major goal of the file security is to fulfill the confidentiality, integrity, and availability.

There are various methodologies followed for securing the files. Among them, cryptography is one of the most popular technique for securing communication and data in the presence of adversaries. There are two types of cryptography algorithms: symmetric and asymmetric key. Some of the symmetric key algorithms are AES,

---

S. Khadka (✉) · N. Shah · R. Shrestha · S. Acharya · N. Karna
Kathmandu, Nepal

N. Karna
e-mail: neha@acem.edu.np

S. Shakya et al. (eds.), *Mobile Computing and Sustainable Informatics*, Lecture Notes on Data Engineering and Communications Technologies 68,
https://doi.org/10.1007/978-981-16-1866-6_38

DES, 3DES, IDEA, Twofish and Blowfish. Some of the asymmetric key algorithms are RSA, ECC, RSA, Diffie-Hellman, etc. In symmetric key, receiver knows the encryption key and same key is used for encryption and decryption. But in asymmetric key, encryption key is known by everyone but receiver only knows the decryption key which make this algorithm more secure and difficult to crack. But this algorithm increase delay data encoding and decoding. As data is getting riskier day by day proper security measures must implement to protect the sensitive data of the organization. And here comes the concept of hybrid cryptography [1, 2]. It is just a combination of two or more than two cryptography algorithms which makes our data more secure and immune to vulnerable attacks.

In order to keep our file secure, we implemented this system. Our system will provide the two layers of security to our sensitive data, i.e., hybrid cryptography and face recognition. For hybrid cryptography [3, 4], we have used three symmetric key algorithms, i.e., AES, Blowfish [5] and Twofish [6]. And in face recognition [7], we have used real-time face recognition system.

At first for any file, we divide the file into three parts. These three split files are encrypted using three different algorithms. Since we are using single key, we have used Caesar's cipher to generate different keys as encrypting key for three algorithms. Finally, the files after encryption are placed inside database along with key which is hashed using SHA512.

During decryption process, at first user should verify his/her face for authentication and then only if he/she gets verified by face recognition system then he/she will get chance to proceed toward decryption step. For decryption, user needs a key which is compared with the hashed key stored in database. If key is correct, then he/she can decrypt the file or else system will deny the decryption.

2 System Description

In the proposed system, user1 and user2 are the users registered in the database containing file and user information. Any user registered in the database can encrypt a file, but only those users who are authenticated to decrypt the file can proceed forward to decrypt the file. A user such as user1 can encrypt a file, and during this process, he/she can specify usernames of the individuals who can decrypt the file, which will be stored in database and along with the hash of the key specified by user1. User1 can then provide the encrypted file and key to user2 through any medium they desire. In order to perform the decryption of file, user2 is first authenticated using face recognition to verify his/her identity, followed by checking in the database to see if he is authorized to decrypt the file. After the authentication phase is completed, the user2 can retrieve the decrypted file (Fig. 1).

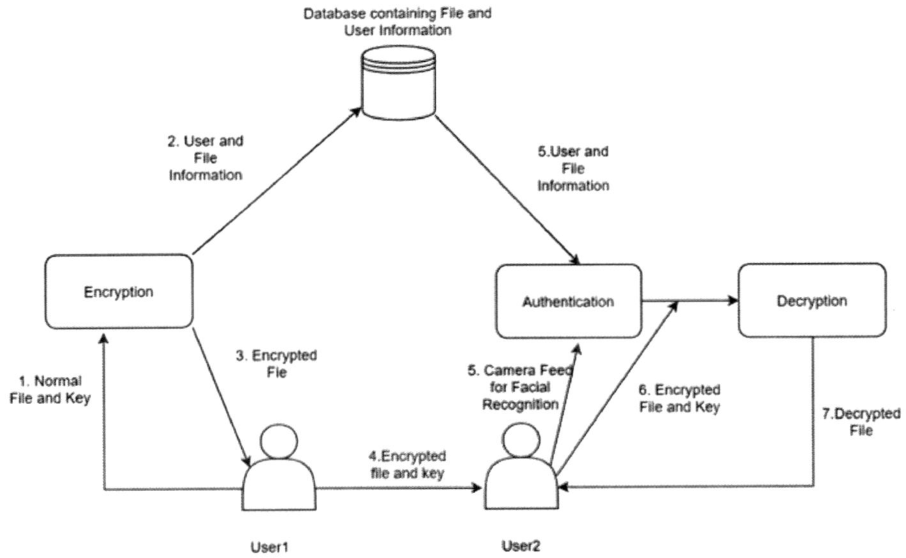

Fig. 1 System overview

2.1 *Encryption Phase*

In encryption phase, the user provides file to be encrypted and key to encrypt the file. The file is split into 3 parts, all of which will be encrypted by AES, Blowfish and Twofish. Three subkeys are generated by passing the original key to Caesar's cipher, which will be used as keys for AES, Blowfish and Twofish. Unique name of the encrypted file is generated and added to database along with the usernames of the user who can decrypt the file, and hash of the original key. The encrypted files are merged together in a zip file which is passed to user who are allowed to decrypt the file to get original data (Fig. 2).

2.2 *Decryption Phase*

In decryption phase, a user registered in the database first has to verify his identity by passing camera feed to the facial recognition system after which he can insert the encrypted file and key to the system. The key is then converted to hash using SHA512 which is then compared to the hash stored in the database, if the hash is same then the file unzipped into 3 files which are then decrypted using AES, Blowfish and Twofish, respectively. The keys for AES, Blowfish and Twofish are generated by passing the original key through Caesar's Cipher. After the three separate files have been decrypted, they are merged to give original file (Fig. 3).

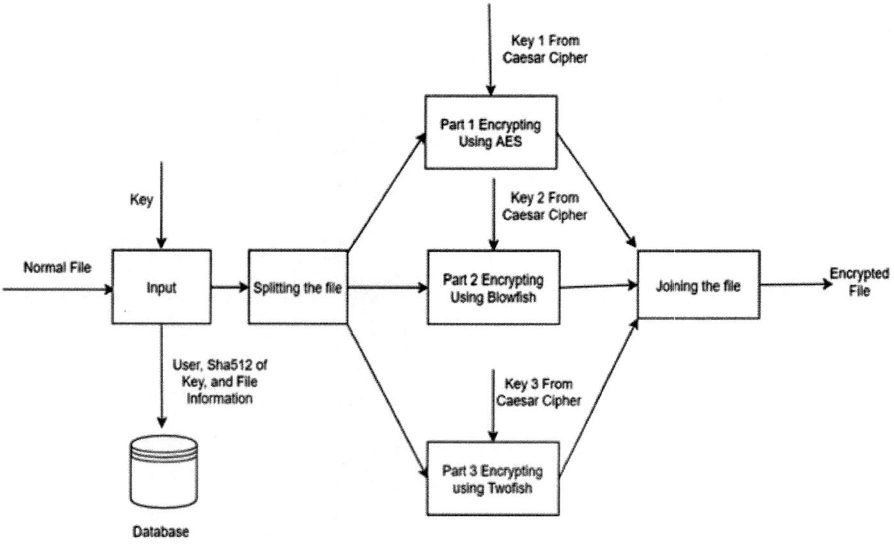

Fig. 2 Encryption process

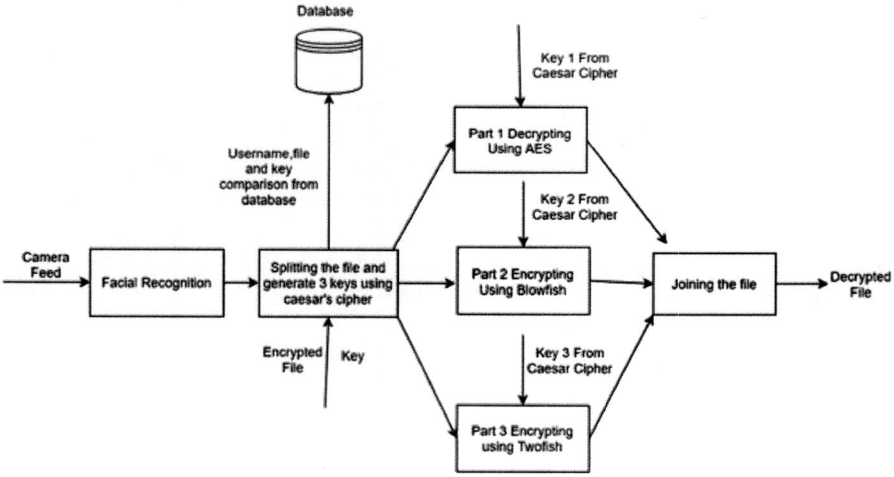

Fig. 3 Decryption process

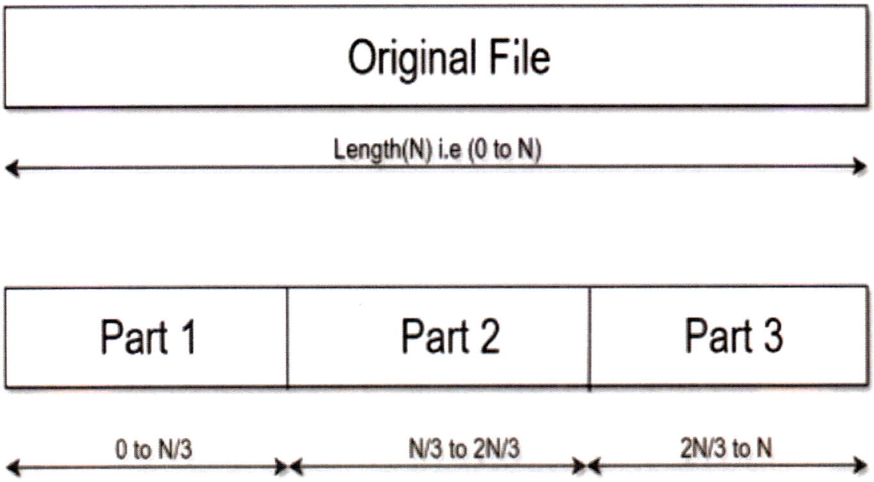

Fig. 4 Splitting the file

3 Methodology

3.1 Splitting the File

The encryption process starts after the original file has been split into three parts. To split any file we have to open the file in binary mode as it reads the content of the file without applying any kind of transformations such as new lines, or decoding/encoding. The length of the bytes after reading the file in binary mode is calculated which we consider as N. The first part is created using the data of the file from 0 to $N/3$ the second part of the file is created from $N/3$ to $2N/3$, and the last part of the file is created from $2N/3$ to N. And, thus the original file is split into smaller 3 part (Fig. 4).

3.2 Caesar Cipher

We use the Caesar cipher to generate the three different keys for encryption from a single key and use that three keys for three different algorithms (Fig. 5).

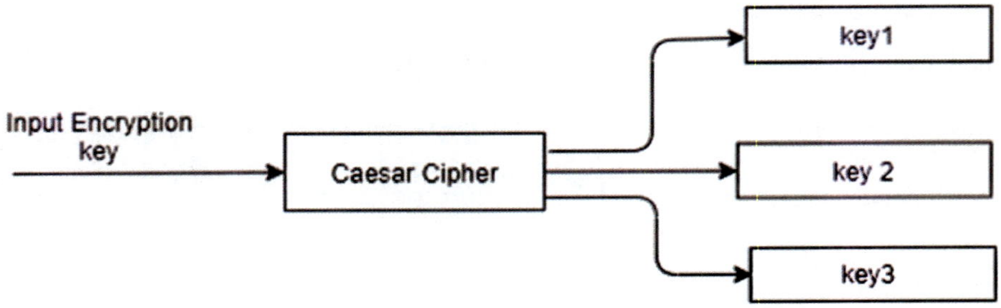

Fig. 5 Caesar cipher for generating three keys

3.3 Encryption Algorithm

3.3.1 AES

AES is a cryptographic algorithm, proposed by Rijndael, which works on block cipher technique, i.e., the size of the plaintext and ciphertext must be same. In AES, the length of block is 128 bits, and key lengths can be 128, 192 or 256. According to the length of key used, the AES can be of three types AES-128, AES-192 and AES-256, with the later on more secure than the former. Here, we have used key size of 256 bits with 14 round processing.

The four major steps in AES can be briefly described as below:

(i) Sub-bytes
The 16 input bytes in the matrix after initial round is substituted using the s-box

(ii) Shift Rows
The four rows of the state matrix are shifted to the left cyclically. The first row isn't shifted, the second row is shifted by one position, the third row is shifted by two positions and the fourth row is shifted by three positions.

(iii) Mix Column
The state matrix we get after shifting rows is then multiplied by a predefined matrix which gives us a new state matrix, on which further transformation occurs. This step is not performed during the final round.

(iv) Add Round Key
The elements of the state matrix are XORed with the different round key that are generated using the scheduling. After performing round key addition in the final round, we get ciphertext (Fig. 6).

3.3.2 Twofish

Twofish is a cryptographic algorithm which also works symmetric block cipher technique. The length of block in Twofish cryptography is 128 bit and the key size that

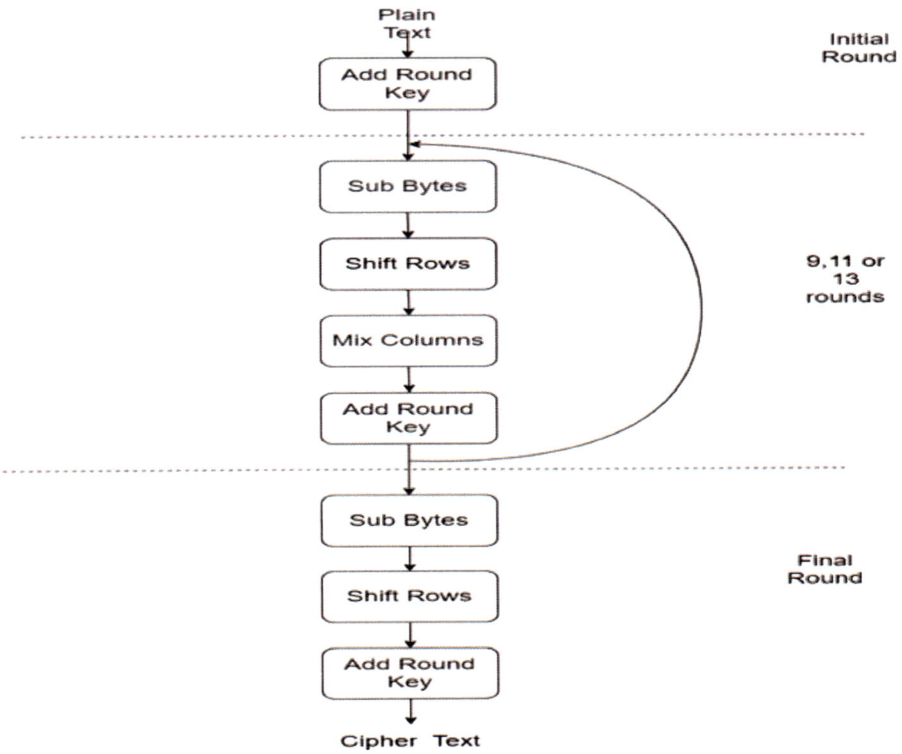

Fig. 6 AES blocks

can be used are 128, 192 or 256 bits. It was designed so as to be highly flexible and highly secure. It has complex key scheduling and has key-dependent s-boxes making the algorithm different from others. The key is divided into two halves, among which one half is used as actual encryption key, and the other is used to modify the encryption algorithm. The number of processing round in Twofish is found to be 16 rounds (Fig. 7).

The two fish working mechanisms are listed below.

- At first, the 128-bit text file is divided into two equal parts, i.e., $X1$ and $X2$.
- In each round, two 32-bit words serve as input into the F function. Such that each word is broken into four bytes.
- These four bytes are sent through four different key-dependent S-boxes
- And then four output bytes (the S-boxes have 8-bit input and output) are combined using a maximum distance separable (MDS) matrix and combined into a 32-bit word.
- The two 32-bit words are now combined using a pseudo-Hadamard transform (PHT), then added to two round subkeys, then XORed with the right half of the text.

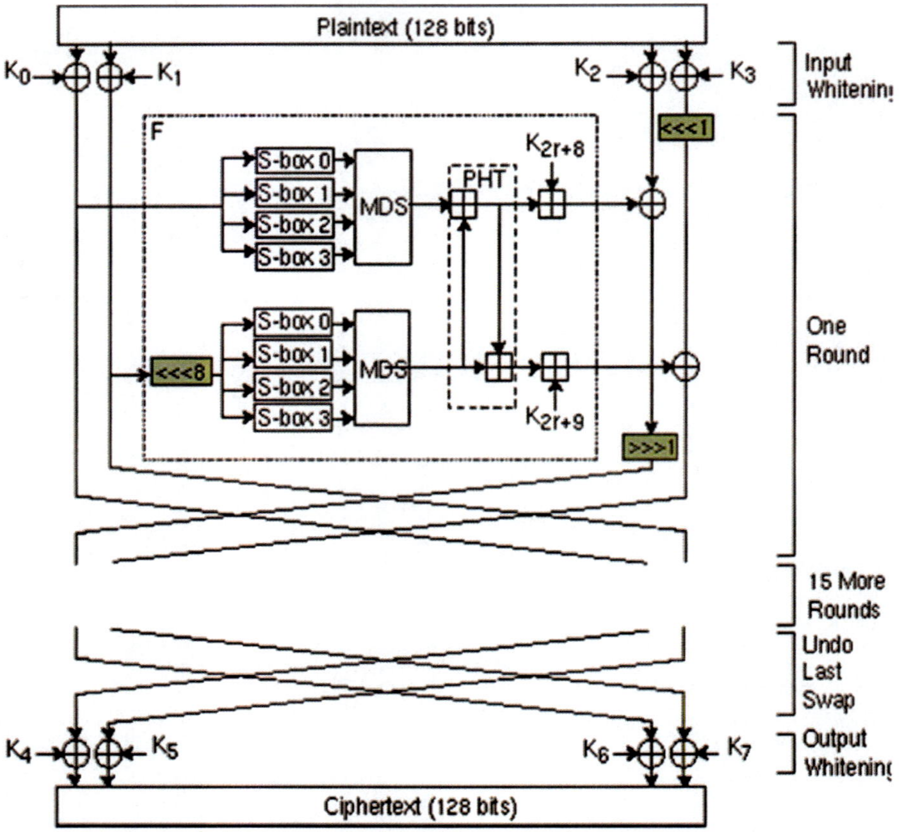

Fig. 7 Two fish algorithm structure

- Also, two 1-bit rotations are performed, one before and one after the XOR.
- Then pre-whitening and post-whitening is done, i.e., before the first and after the last round additional subkeys are XORed into the text block.

3.3.3 Blowfish

It is a symmetric block cipher algorithm for encryption and decryption. It is 64-bit block cipher. Blowfish is a variable-length key encryption algorithm whose key varies from 32 to 448 bits which makes it for securing data. It was designed in 1993 by Bruce Schneier as a fast, free alternative to existing encryption algorithms. It is a 16-round Feistel cipher which uses large key-dependent S-boxes, which is the main feature of this algorithm. S-boxes take 8-bit input and produce 32-bit output. The advantages of this algorithm are secure and easy to implement, but the demerit is that it requires more space for ciphertext because of difference in key size and block size.

In this algorithm, x is 64-bit input data, which is then divided into two equal parts namely $x1$ and $x2$. For the 15 rounds, we calculate $x1$ as the XOR of $x1$ and Pi (where i changes from 1 to 16), which are the subkeys needed for encryption), $x2$ is calculated as XOR of func($x1$) and $x2$, then $x1$ and $x2$ are swapped. In 16th round, we calculate $x1$ as XOR of $x2$ and $P18$, and $x2$ as XOR of $x2$ and $P17$. After this, $x1$ and $x2$ are combined together to get the ciphertext.

The steps in this algorithm are (Fig. 8):

- 18 subkeys are generated from $P1$ to $P18$ which are stored in a array of 32 bits entry.
- The input data x of 64 bits is divided into two equal parts namely $x1$ and $x2$.
- For the first 15 rounds, in each round we calculate $x1$ as $x1 = x1$ XOR Pi (where i changes from 1 to 16, which are the subkeys needed for encryption), $x2$ is calculated as $x2 =$ func($x1$) XOR $x2$, and $x1$ and $x2$ are swapped.

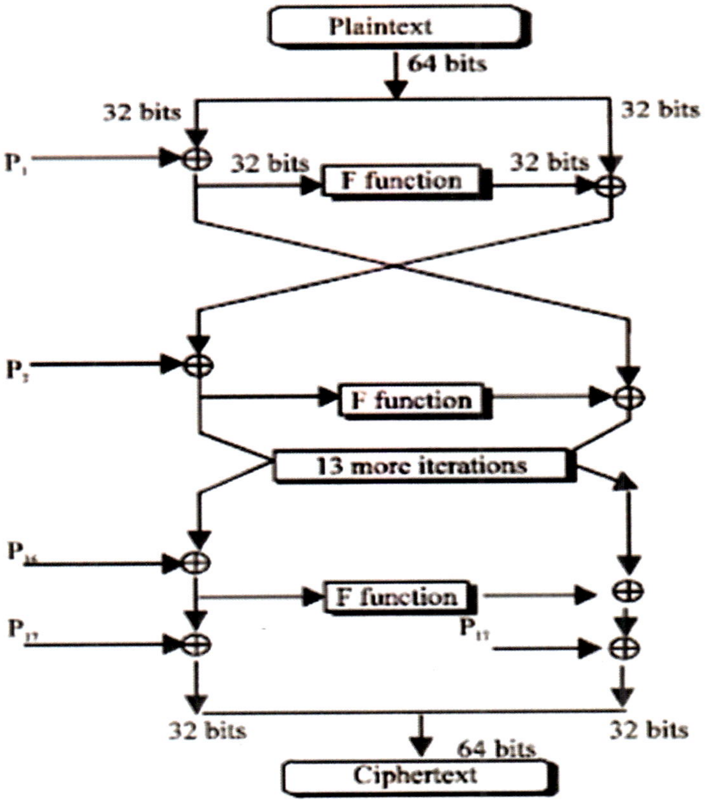

Fig. 8 Blowfish structure

- In 16th round, we calculate $x1$ as $x1 = x2$ XOR P18, and $x2$ as $x2 = x2$ XOR P17.
- After this $x1$ and $x2$ are combined together to get the ciphertext.
- The decryption process is similar to encryption process but the subkeys are used in reverse order.

3.4 Face Recognition

As we have double layer security for files, we have used one of the most advance biometric verification, i.e., face recognition, as the first layer security. Before decryption of file from database, at first the user should verify his face in a real-time (video). If user face is verified, then only he/she can proceed forward to the system.

For the face recognition, we have used one of the most popular neural networks called convolution neural network (CNN). Unlike other neural networks, it extracts the features from the image and uses that features to classify the distinct image. In our project, we had trained our datasets using CNN and use that trained model as transfer learning for real-time recognition.

There are five major steps,

(a) Data collection and labeling
(b) Preprocessing
(c) Training using CNN
d) Transfer learning
(e) SVM for classification
(f) Real-time recognitions

3.4.1 Data Collection and Labeling

Our dataset contains images of 11 person such that each person have around (150–180) images. Each person's image is placed inside separate directory, and the directory is labeled on the basis of their name (Fig. 9).

3.4.2 Preprocessing

Preprocessing of image is mandatory because it helps us to filter out unnecessary portion of the images. For this, we have to crop, align and rescale the facial portion of the images so that unnecessary portion of image and feed forward.

Training using CNN

Finally, we will train the preprocessed data using the CNN. We have used five convolution layers, and each of them followed by the batch normalization, ReLU, max pooling and dropout layer. And finally, 3 dense layer is used (Fig. 10).

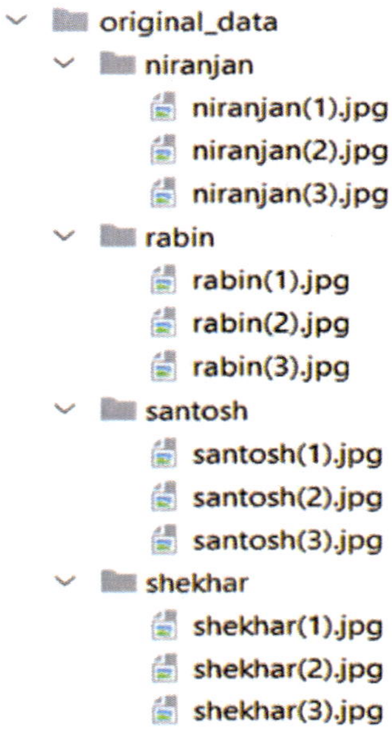

Fig. 9 Image labeling

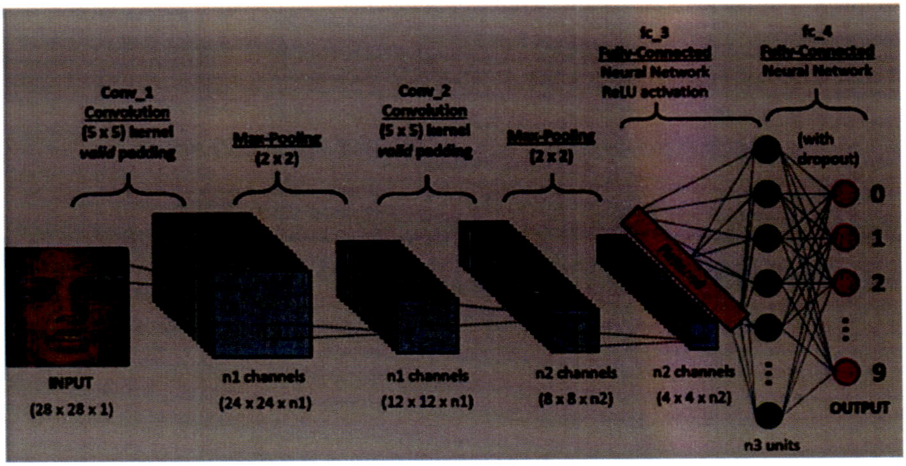

Fig. 10 Basic CNN architecture

3.4.3 Transfer Learning

So, we would use the per-trained model to generate the embedding vector of the images using transfer learning technique. Transfer learning is the concept of overcoming the isolated learning model by using knowledge gained for one problem to solve similar ones. It is one of the most popular in deep learning where per-trained models are used as a base on computer vision and natural language processing tasks. Transfer learning helps us to minimize the time and resources because it removes the burden of training the huge data of similar type.

3.4.4 SVM for Classification

Thus, the feature vectors of different images obtained from the transfer learning are classified using the SVM. SVM is a supervised machine learning model that uses classification algorithms for two-group classification problems. The main objective of the support vector machine algorithm is to find a hyperplane of the feature vector in an N-dimensional space that distinctly classifies the data points.

4 Result and Analysis

In the proposed system, we used three algorithms: AES, Blowfish and Twofish to perform hybrid cryptography on the file provided by the system. The given algorithms use three keys generated by Caesar's cipher using the key provided by the user. After the file is encrypted, the information (i.e., unique name of file generated, username of users who are allowed to access the file and hash of the key provided by user) are updated in the database. The recipient of the decrypted file is verified using facial recognition when he tries to decrypt the given decrypted file using the encryption, which then results in the generation of decrypted file containing the original file.

In the proposed system we used AES, Blowfish and Twofish algorithms to perform hybrid cryptography. Since we used multiple algorithms, the sensitive data is more secure and with addition of facial recognition for authorization of user, a second layer of security is added to the system. Implementation of the proposed system was done using Python programming language. For testing speed of the proposed system, we have compared the encryption and decryption speed of our system with existing system that uses Twofish algorithms, which is one of the algorithms used by us. Encryption and decryption were done on two files having size 239 and 467 MB.

As the proposed system uses multiprocessing to perform the hybrid cryptography on three parts of the file simultaneously, the speed of encryption as well as decryption is higher than compared to Twofish-based existing system which performs encryption and decryption on the whole file. Figure 11a, b shows the differences in the speed during the encryption and decryption of a file.

Similarly, while training the face recognition model, we have achieved the training accuracy of 97.31% and validation accuracy of 95.72% accuracy (Fig. 12).

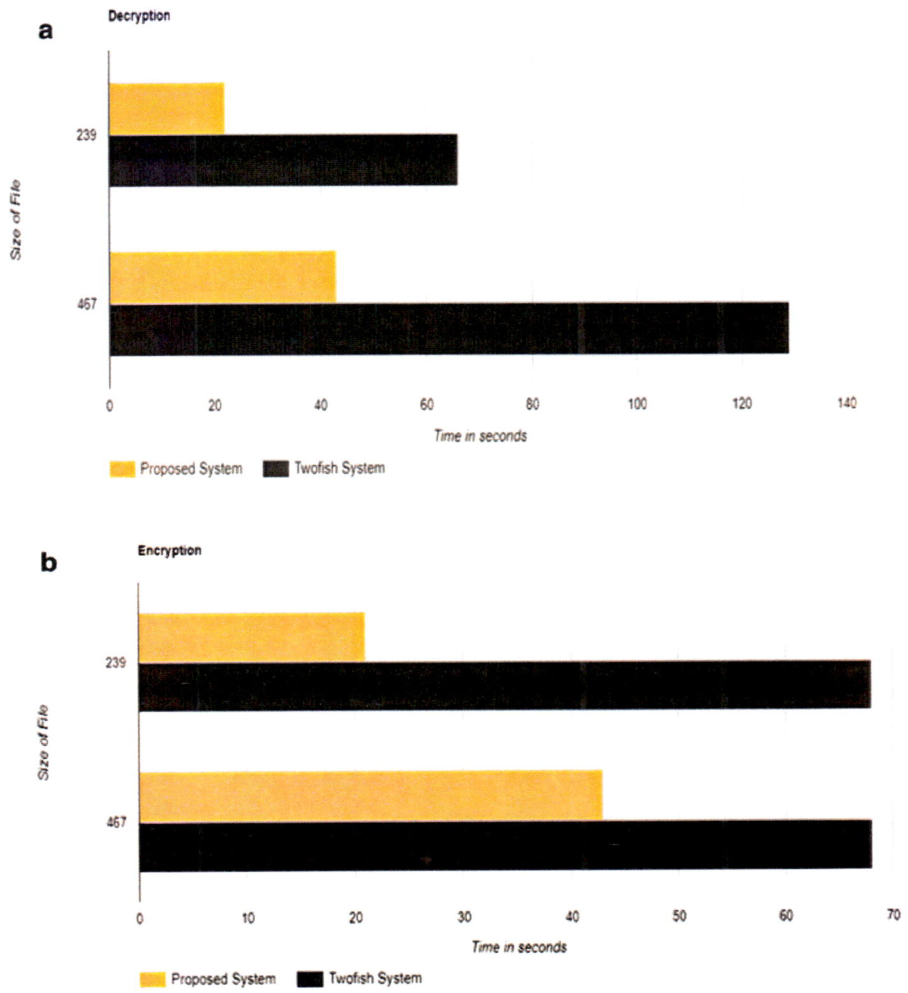

Fig. 11 **a** Comparison of encryption with between proposed system and Twofish-based system, **b** Comparison of decryption with between proposed system and Twofish-based system

5 Conclusion

Hybrid cryptography and Face recognition plays an important role in file security. This research work analyzed three different encryption algorithms, and it includes advanced encryption standard, Blowfish and Twofish. From the experimental result, we have identified the AES is better than other algorithms in the terms of speed and memory space. Also the addition of double layer security with face recognition can play a very crucial impact on file security. High protection, authentication and

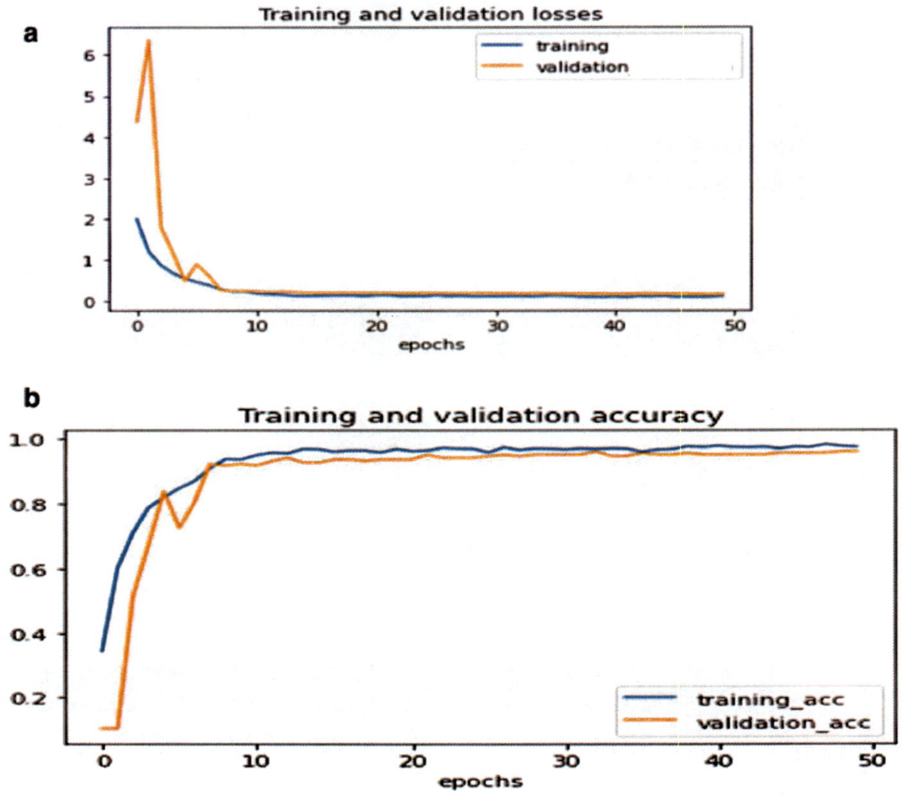

Fig. 12 **a** Training and validation losses, **b** training and validation accuracy

confidentiality criteria are achieved with the help of the proposed security mechanism data integrity.

References

1. M. Jain, A. Agrawal, implementation of hybrid cryptography algorithm. Int. J. Core Eng. Manag. **1**(3), 1–8 (2014)
2. P. Shaikh, V. Kaul, Enhanced security algorithm using hybrid encryption and ECC. IOSR J. Comput. Eng. (IOSRJCE) **16**(3), 80–85 (2014)
3. A.E. Takieldeen, Design and implementation of hybrid encryption algorithm. Int. J. Sci. Eng. Res. **4**(12), 669–673 (2013)
4. J.M. Najar, S.B. Dar, A new design of a hybrid encryption algorithm. Int. J. Eng. Comput. Sci. **3**(11), 9169–9171 (2014)
5. J. Thakur, N. Kumar, DES, AES and Blowfish: symmetric key cryptography algorithms simulation based performance analysis. Int. J. Emerg. Technol. Adv. Eng. **1**(2), 06–12 (2011)

6. H.K. Verma, R.K. Singh, Performance analysis of RC6, Twofish and Rijndael Block Cipher algorithms (2012)
7. F. Schroff, D. Kalenichenko, J. Philbin, FaceNet: a unified embedding for face recognition and clustering, 06–19 Oct 2015

BER Performance Comparison Between Different Combinations of STBC and DCSK of Independent Samples and Bits of Message Signal

Navya Holla K and K. L. Sudha

Abstract In recent times, Chaos communication plays a significant role in addressing the security challenges that are more common in the regular communication systems. By exploiting the chaotic waveform properties, variety of chaos-based communication systems are proposed. Considering the aforementioned facets of communication, this research objective at merging the space-time block code (STBC) and orthogonal differential chaos shift keying (ODCSK) methods to improve the security aspects of multiple-input multiple-output (MIMO) based wireless system. Chaos signal is created by using the method called logistic mapping to achieve better secrecy. STBC method is used during the modulation stage for arranging the chaos signal. This system with different types of STBCs is aimed at analysing the differential chaos shift keying (DCSK) and orthogonal DCSK (ODCSK) modulation in AWGN channel. MATLAB software is used to simulate the Bit Error Rate [BER]. Initial simulation proves that the combination of orthogonal DCSK and Alamouti STBC with MIMO system gives better results for AWGN channel when compared to other combinations.

Keywords Differential Chaos shift keying · Space-time block code · Multiple-input multiple-output

1 Introduction

Most of the modern wireless communication systems are restricted in performance due to the multipath propagation effects and inferences that are created inside the system. This may indirectly affect the security of the telecommunication applications. By representing the message signals with non-deterministic chaotic signals or

N. Holla K · K. L. Sudha (✉)
Department of ECE, Dayananda Sagar College of Engineering, Bengaluru, India
e-mail: drsudha-ece@dayanandasagar.edu

N. Holla K
e-mail: navya-ece@dayanandasagar.edu

S. Shakya et al. (eds.), *Mobile Computing and Sustainable Informatics*, Lecture Notes on Data Engineering and Communications Technologies 68,
https://doi.org/10.1007/978-981-16-1866-6_39

applying the technique called spread spectrum, the drawback of the aforementioned systems can be avoided [1].

Importance of employing robust modulation techniques in present wireless communication systems are exposed to severe adversarial factors such as fading, interference, impulsive and phase noise, which motivates the researchers in the direction of using chaotic carriers. Chaos systems are generated by using simple circuitry and are very sensitive to initial conditions. Hence, different chaos signals can be generated by using different initial value, which have the characteristics of low cross-correlation and impulse auto-correlation. Chaotic systems are represented by wideband power spectrum and they appear random in time.

The incoming symbols can be recovered at the receiver either with synchronization or without synchronization as in a conventional communication system. According to the recent studies, synchronization method is not suitable for most of the practical application systems as significant noise and filtering are additional at the channel. One more important point to be remembered is reference signal is necessary at the receiver side in order to retrieve the message signal. Therefore, non-coherent techniques, which do not need any synchronization, is considered for demodulation to get back the required transmitted information from the noisy received signal [2].

To allow the communication of several copies of a data through a number of antennas, STBC is used with MIMO system. Space-time block code gives temporal and spatial diversity during transmission. Due to this, the major channel problems like thermal noise and fading can be resolved. Before transmission, encoding of data sequence in block takes place and then divided into multiple antennas with proper spacing between them. Compared to single-input single-output (SISO), multiple-input multiple-output system gives higher data rate at the receiver section due to transmission of different set of data symbols [3, 4].

2 Literature Survey

The BER performance of a MIMO system using Chaos Shift Keying has been analysed in [5] using Almouti coding scheme. This system has been established to increase the security using reduced complexity and low cost. It is found that this performance is same as MIMO-BPSK system. The chaotic communication performance in a MIMO system is analysed in [6]. Space-time block code is used for a MIMO system (2 transmitters and 2 receivers) and Bit error rate (BER) is analysed in AWGN channel. Second-order Chebyshev Polynomial Function (CPF) generates chaotic signal.

Orthogonal multi-code (OMC) at the transmitter and equal gain combining (EGC) at the receiver are the techniques used in MIMO-DCSK system which is proposed in [7]. In the respective transmit antenna to spread the same bit of information, method called as orthogonal multi-code (OMC) is used and it is constructed by single chaotic sequence called orthogonal space-time block code (OSTBC). At the receiver section, the demodulation method called equal gain combining is used to detect the information bit. By combining M-ary orthogonal modulation and T-R technique, a new scheme called as multilevel modulation is proposed in [8]. Under different channel conditions, this scheme gives remarkable improvement in BER performance and also achieves spectral efficiency and more data rate. To achieve channel gain using multiple antennas, Space-time block code is used in communication system such as mobile system. By using STBC along with technique called continuous phase modulation (CPM), power efficiency and spectral efficiency can be achieved at the same time in [9] for 2×2 and 4×4 antennas. Alamouti STBC over MIMO-OFDM system is used in [10] to extend the orthogonal area time block code output and code rate output over 4×4 antennas. For different combination of transmit and receive antennas, BER values have been tabulated.

A new digital communication system called orthogonal chaotic vectors differential chaos shift keying (OCV DCSK) is presented in [11]. It is shown that by transferring M bits in non-coherent method, bit transformation rate and system can be improved. In order to produce orthogonal M chaotic signal, process called Gram Schmidt process is used. The performance of modulation scheme named orthogonal multilevel chaos shift keying along with MIMO has been evaluated under different channel coding and signal detection technique in [12]. Both Hilbert transform and Walsh Hadamard codes have been used resulting in a bit rate of 1 Gbps performance.

Alamouti STBC performance estimation with MIMO system for four transmit antenna and N number of receive antenna has been analysed in [13] to extend the orthogonal block code rate output and code rate. Importance of STBC and space-time code has been explained. Efficiency of the system is majorly affected by wireless channel impairments such as fading, inter-symbol interference and different types of noises. In order to overcome the above disadvantages, the author at [14] proposed MIMO technology which uses the concept of multipath propagation under four different equalization methods. In order to improve the factors like channel capacity, bit error rate and gain, a new technique named multiple access differential noise shift keying is proposed in [15]. This new system has been analysed under conventional detector and simulation results presented that the above-said factors have been improved compared to chaos shift keying techniques.

3 Proposed Method

Proposed methodology block diagram is shown in Fig. 1. At the transmitter side, randomly the message bits are generated. 2×2 MIMO system is considered for the experimentation purposes.

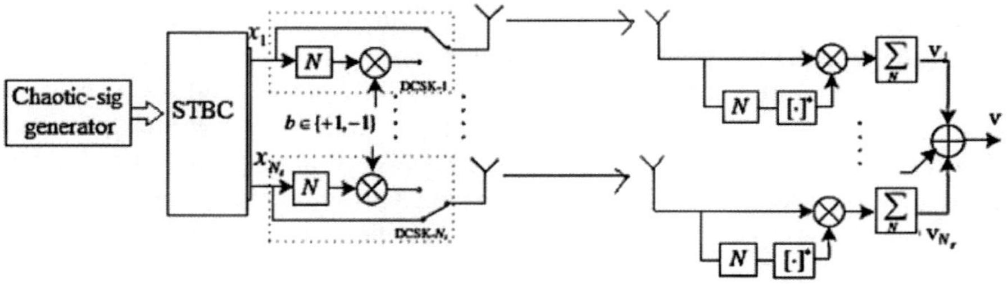

Fig. 1 General diagram for modulation scheme for MIMO using STBC and DCSK

3.1 *Chaos Signals*

Chaos signal is generated using logistic mapping method. This mapping uses differential equation as a logistic function which treats time as a continuous. To represent in discrete time step, nonlinear difference equation is used.

$$x_m + 1 = r x_m (1 - x_m)$$

where x_m is ratio of existing population to the maximum possible population and its value is between 0 and 1. To keep the x_m value bounded between 0 and 1, the parameter r should be lies between 0 and 4. When there is small variation in the value of r, there will be large variation in the chaos behaviour as shown in Fig. 2

3.2 *DCSK*

Period of the bit is distributed into two parts. Considering the first half bit duration of message bits, reference chaos signal is transferred which is generated from chaos generator. If the message bit is 1, then for the second slot, the same reference signal is transmitted. Suppose if the message bit is zero, either different chaos signal is transmitted or same chaos signal with different initial condition is transmitted. Let the chaos signal be represented as $x(t)$. Let t_b and t_l represent the bit duration and the time instant when the lth information bit is transmitted. If the message bit is 1, then the equation becomes,

$$x_{tx} = \begin{cases} x(t), & t_l \leq t < t_l + t_b \\ x(t - t_{b)}, & t_l + t_b \leq t < t_l + 2t_b \end{cases} \tag{1}$$

If the incoming binary information symbol is "0", then first the reference signal is transmitted, and then another chaotic signal $y(t)$ is transmitted

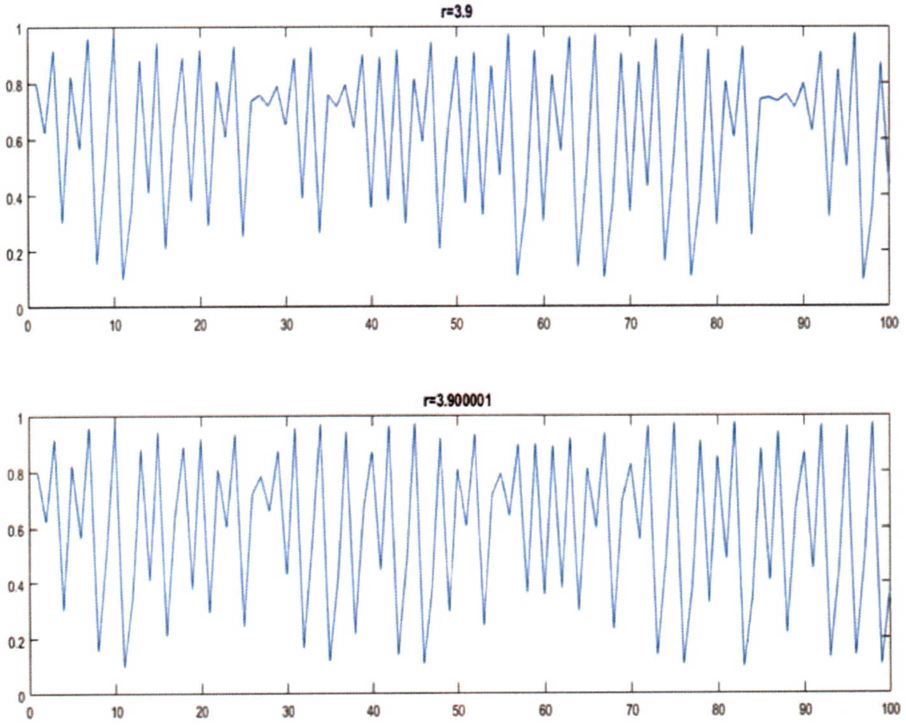

Fig. 2 Generation of chaos signal for $r = 3.9$ and $r = 3.900001$

$$x_{tx} = \begin{cases} x(t), & t_l \le t < t_l + t_b \\ y(t - t_b), & t_l + t_b \le t < t_l + 2t_b \end{cases} \quad (2)$$

3.3 ODCSK

The difference in ODCSK with respect to DCSK is instead of transmitting another chaos signal; the orthogonal version of original chaos signal is transmitted in the second slot of message bit 0.

$$x_{tx} = \begin{cases} x(t), & t_l \le t < t_l + t_b \\ -x(t - t_b), & t_l + t_b \le t < t_l + 2t_b \end{cases} \quad (3)$$

3.4 Space-Time Block Code (STBC)

To allow the communication of several copies of a data through a number of antennas, STBC is used with MIMO system. To retrieve maximum information from each of the antennas, space–time coding techniques combine all the copies of the received signal. STBC can be generated using either an independent sample or an independent bit approach.

First approach (independent bits): The message bits be represented as L and S be the chaotic sample values generated during the first half bit duration of the message bits. Let the set of message bits be represented for the message signal as $x = (x_1, x_2, x_3, \ldots x_L)$.

If the OSTBC scheme is used and 2 transmitting antennas are used as an example, the set of message bits used for modulation adopting appropriate DCSK scheme for transmitter1 is $x_1, x_2, x_3, x_4 \ldots x_{L-1}, x_L$.

At the same time, the sequence of message bits to be used for modulation by tranmitter2 is $-x_2, x_1, -x_4, x_3 \ldots - x_L, x_{L-1}$.

A similar approach can be adopted, if STBC is used in transmitters 1 and 2 respectively as shown below:

$$x_1, x_2, x_3, x_4 \ldots x_{L-1}, x_L \tag{4}$$

$$x_1, x_3, x_5, x_7 \ldots x_{L-1} \quad x_2, x_4, x_6, x_7 \ldots x_L$$

In the channel, the received signal from the transmitted part is added with AWGN. At the receiver, it is modelled as, first transmitting antenna signal is received directly by first receiving antenna and second transmitting antenna is received directly by second receiving antenna.

$$r = x + \text{AWGN} \tag{5}$$

The received signal at the receiver side can be evaluated by the formula,

$$v_j = \sum_{n=0}^{S-1} r_j(s)\big[[r_j(s+S)\big]^* \tag{6}$$

where $r_j(s)$ represents the reference signal and information bearing signal is represented as $r_j(s+S)$ by considering both have same delays and gains. The output signal is given as

$$v = \sum_{j=1}^{S_r} v_j \tag{7}$$

The above-received signal value v is decided by comparing it with threshold value in the experiment it is considered as 0. The bit 1 is considered if v is greater than 0 that is threshold value, otherwise it is considered as 0.

Using all above concepts, four combinations can be produced using STBC and DCSK. They are DCSK with STBC, DCSK with OSTBC, ODCSK with STBC and ODCSK with OSTBC.

Second approach (independent samples): If independent samples are taken during a bit interval for transmission, variations in the chaotic sample values for different antennas will not affect the bit errors in the receivers compared to the independent bit approach.

Orthogonal multi carriers (OMC) are generated from techniques called orthogonal STBC at the transmitter side to send the similar information bit. S chaotic sequence from jth modulator which is defined as $\boldsymbol{x_j} \equiv \{x_j(1), x_j(2), \ldots, x_j(S)\}$ is considered, where S is spreading factor. The resulting signal at the output of jth antenna is

$$t_i(s) = \begin{cases} \frac{1}{\sqrt{S_t}} x_j(s) & 0 < s \leq S \\ \frac{1}{\sqrt{S_t}} b x_j(s - S) & S < s \leq 2S \end{cases} \quad (8)$$

The S_t DCSK modulator uses orthogonal multi-spreading codes which are represented as $x = (x_1, x_2, x_3, \ldots x_L)$. Using the definition of OSTBC, its matrix is represented as

$$\begin{pmatrix} x_1 & x_2 \\ -x_2 & x_1 \end{pmatrix} \quad (9)$$

Orthogonal chaotic sequences can be written as

$$\begin{pmatrix} x_1 \\ x_2 \end{pmatrix} = \begin{pmatrix} x_1 & x_2 & x_3 & x_4 \ldots \\ -x_2 & x_1 & -x_4 & x_3 \ldots \end{pmatrix} \quad (10)$$

This concept is also called as ODCSK-OSTBC.

3.5 *Alamouti STBC*

Let two symbols x_1 and x_2 are considered at a time for the transmission in two successive time slots. Considering 1st time period, from the first antenna x_1 is transmitted and x_2 is transmitted from the antenna 2. $-x_2^*$ is transmitted in the 2nd time period, from the first antenna, while x_1^* is passed from the antenna 2 which is shown in Fig. 3.

Receiver operation is same as above method.

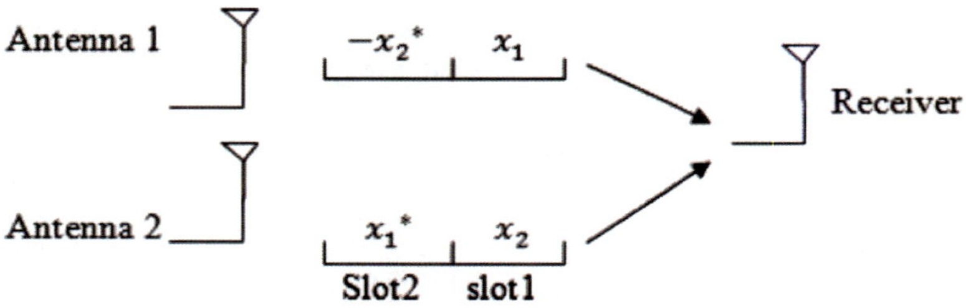

Fig. 3 The Alamouti scheme

4 Simulation Results

4.1 *First Approach (Independent Bits)*

DCSK with STBC, DCSK with OSTBC, ODCSK with STBC and ODCSK with OSTBC are four combinations of STBC and DCSK can be generated. By running all the above combination using MATLAB software, BER values for SNR 10 are obtained for message bits taken as 10,000 as shown in Fig. 4 (Tables 1 and 2).

Graph of SNR versus BER for all the combinations is shown in Fig. 5.

From the graph, it is observed that ODCSK with OSTBC of independent bits gives smaller BER (71) compared to remaining methods because of orthogonality to both modulation and STBC methods. If the modulation is orthogonal, it gives lesser

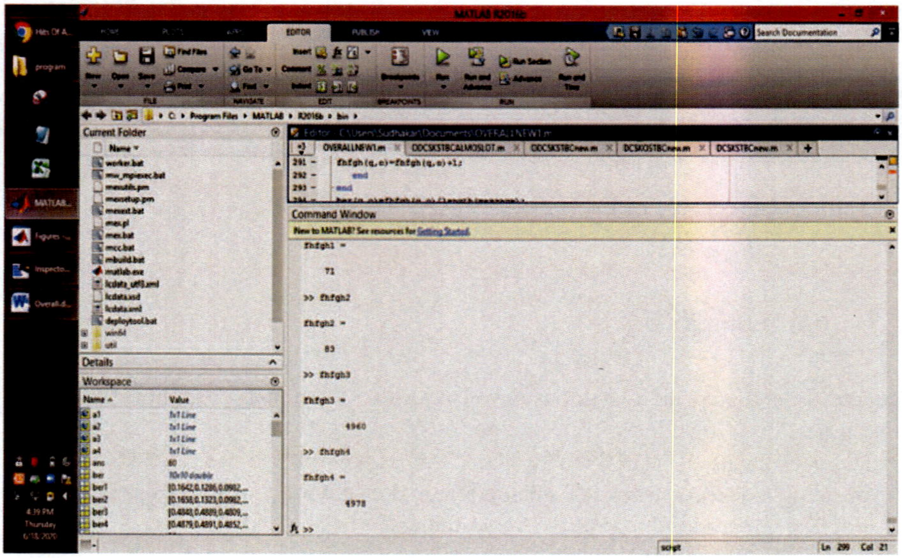

Fig. 4 BER values of independent bit

Table 1 Different BER values of independent bits

Combinations	Received errors out of 10,000 message bit for SNR 10	BER
DCSK with STBC	4978	0.4978
DCSK with OSTBC	4960	0.496
ODCSK with STBC	83	0.0083
ODCSK with OSTBC	71	0.0071

Table 2 Different BER values of independent bits

Combinations	Received errors out of 10,000 message bit for SNR 10	BER
DCSK with STBC	5033	0.503
DCSK with OSTBC	2533	0.253
ODCSK with STBC	324	0.0324
ODCSK with OSTBC	7	0.0007

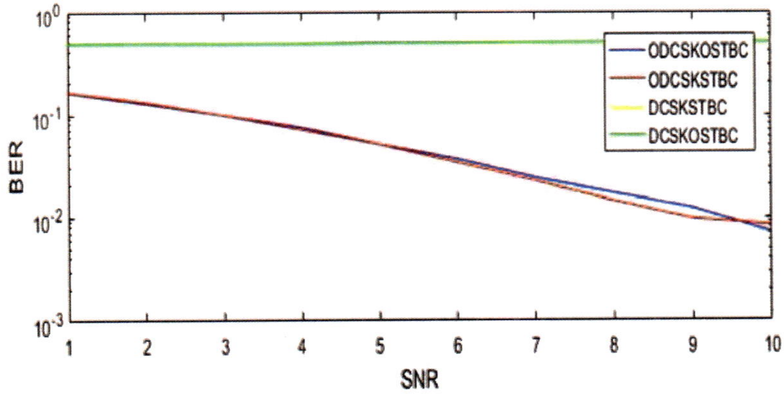

Fig. 5 SR versus BER of all combination of STBC and DCSK of independent bits

BER as seen in ODCSK-STBC method (83). Because orthogonality in modulation helps to extract original message bits in the receiver by fixing a constant threshold value. DCSK with OSTBC and DCSK with STBC combinations produce greater BER as modulation is applied only in DCSK. Even though orthogonality is applied at STBC side, it does not affect much on BER because orthogonality concepts alter only bit arrangements during transmission. Therefore, at the modulation side due to orthogonality gives good performance compared to remaining concepts.

4.2 Second Approach (Independent Sample)

Similar to first approach technique, in independent bit approach also four combinations of STBC and DCSK can be tabulated. The values of BER of all combination for message bits of 10,000 are shown in Fig. 6 and its graph versus SNR is shown in Fig. 7.

From the graph, observation can be made as ODCSK with OSTBC combination of independent samples gives smaller BER (7) compared to remaining methods because of orthogonality to both modulation and STBC methods. By comparing above two

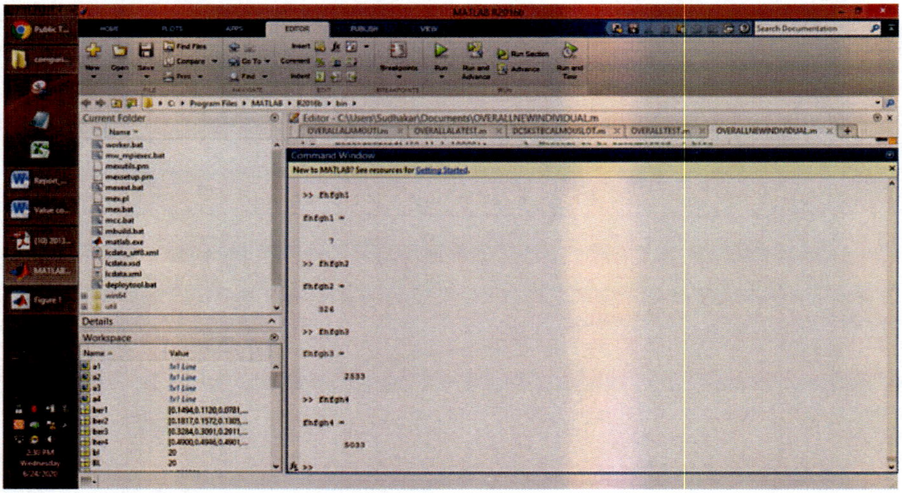

Fig. 6 BER values of independent samples

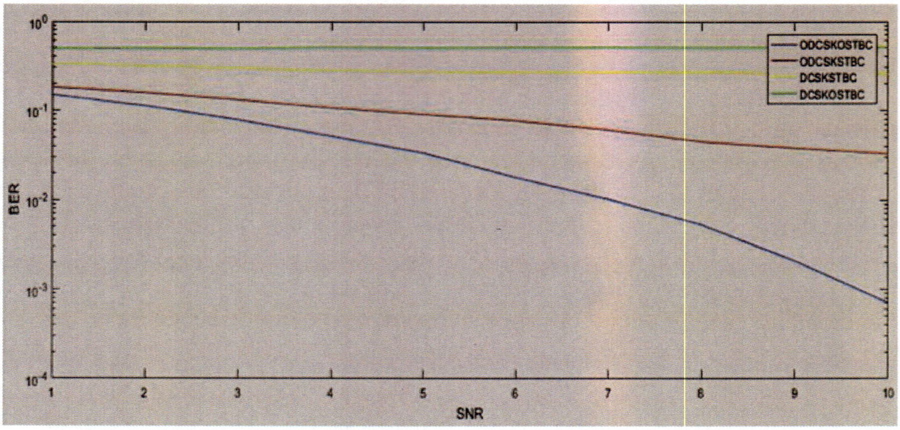

Fig. 7 SNR versus BER of all combinations of STBC and DCSK of independent samples

approaches, independent sample approach gives less BER compared to independent bit approach. But only for ODCSK-STBC case, it gives higher BER since random sequence of bits has been taken in case of STBC. Hence another method of STBC technique has been adopted.

In case of Alamouti STBC, applying Alamouti STBC to both DCSK and ODCSK techniques, its BER values for the same 10,000 message bits are shown in Fig. 8 and its graphs of BER versus SNR is given in Fig. 9 (Table 3).

The above all concepts have been performed by considering the direct propagation path between transmitters and receivers. Effect of fading will be considered in the future work.

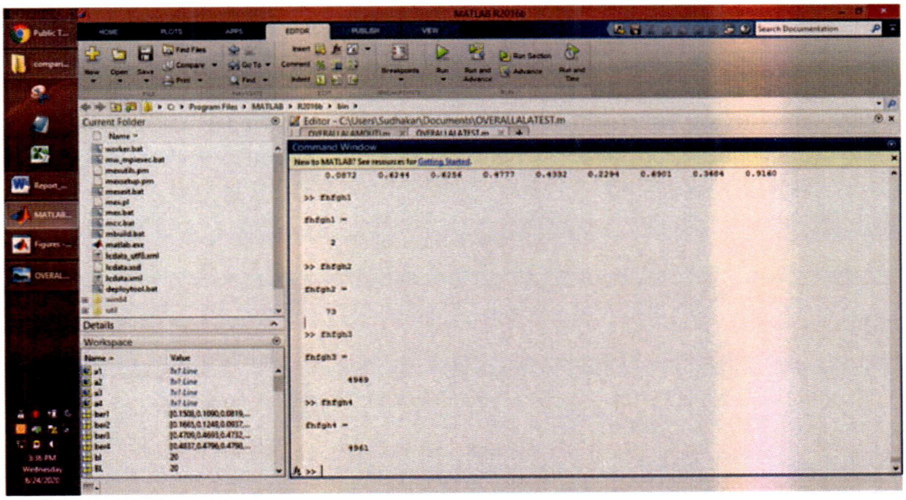

Fig. 8 BER values of DCSK and Alamouti scheme

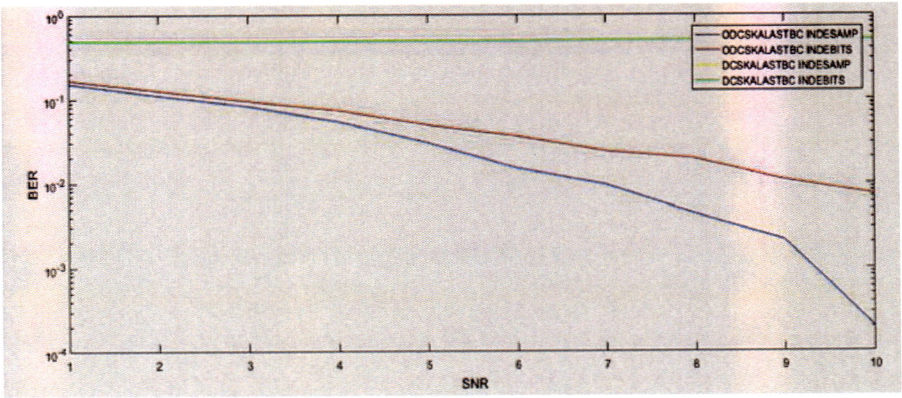

Fig. 9 SNR versus BER of combination of Alamouti STBC and DCSK

Table 3 Different BER values of DCSK and Alamouti scheme

	Combinations	Received errors out of 10,000 message bit for SNR 10	BER
Independents bits	DCSK with Alamouti STBC	4961	0.4961
	ODCSK with Alamouti STBC	73	0.007
Independents samples	DCSK with Alamouti STBC	4969	0.496
	ODCSK with Alamouti STBC	2	0.0002

5 Conclusion

By comparing different DCSK and STBC combinations for BER, it is found that ODCSK with Alamouti STBC combination gives improved performance (Lesser BER-2 errors for 10,000 message bits) compared to other combinations considered in this research. Also independent sample approach gives better performance compared to independent bits in ODCSK-Alamouti STBC combination. Further research can be carried out to analyse the performance considering different types of fading channels and different STBC type's techniques applied to ODCSK-MIMO systems.

References

1. G. Kaddoum, Wireless Chaos-based communication systems: a comprehensive survey. IEEE Access **4**, 2621–2648 (2016)
2. H. Yang, W.K.S. Tang, G. Chen, G.-P. Jiang, Multi-carrier chaos shift keying: system design and performance analysis. IEEE Trans. Circuits Syst. I (2017)
3. P. Pathak, P.R. Pandey, A novel Alamouti STBC technique for MIMO system using 16-QAM modulation and moving average filter. IJERA J. **4**(8), 49–55 (2014)
4. T. Hashem, M. Imdadul Islam.: Performance analysis of MIMO link under fading channel. IEEE International Conference on Computer and Information Technology, pp. 498–503 (2014)
5. S. Mrinalee, M.K. Mukul, Performance analysis of the MIMO system using the chaos shift keying, no. 1, pp. 1824–1828 (2016)
6. G. Kaddoum, M. Vu, F. Gagnon, Performance analysis of differential chaotic shift keying communications in MIMO systems. In: Proceedings of IEEE International Symposium Circuits System, no. 1, pp. 1580–1583 (2011). https://doi.org/10.1109/ISCAS.2011.5937879
7. S. Wang, S. Lu, E. Zhang, MIMO-DCSK communication scheme and its performance analysis over multipath fading channels. J. Syst. Eng. Electron. **24**(5), 729–733 (2013). https://doi.org/10.1109/JSEE.2013.00085
8. H. Yang, W.K.S. Tang, S. Member, G. Chen, System design and performance analysis of orthogonal multi-level differential chaos shift keying modulation scheme. IEEE Trans. Circuits Syst. I: Reg. Pap. **63**(1), 146–156 (2016)
9. K. Morioka, S. Yamazaki, D. Asano, Study on spectral efficiency for STBC-CPM with two and four transmit antennas, in *IEEE International Symposium on Signal Processing and Information Technology (ISSPIT)* (2017)
10. P.K. Bharti, P. Rawat, Alamouti-STBC based performance estimation of multi Tx and Rx antenna over MIMO OFDM, in IEEE proceedings of the 2nd international conference on trends in electronics and informatics (ICOEI). ISBN 97801-5386-3570-4 (2018)

11. F.S. Hasan, Design and analysis of an orthogonal chaotic vectors based differential chaos shift keying communication system. Al-Nahrain J. Eng. Sci. (NJES) **20**(4), 952–958 (2017)
12. M. Omor Faruk, S.E. Ullah, Performance evaluation of orthogonal multi-level chaos shift keying modulation scheme aided MIMO wireless communication system. Int. J. Netw. Commun. **8**(1), 10–17
13. P.K. Bharti, P. Rawat, Alamouti-STBC based performance estimation of multi Tx and Rx antenna over MIMO OFDM, IEEE international conference on trends in electronics and informatics. ISBN: 978-1-5386-3570-4 (2018)
14. A. Jain, P. Shukla, L. Tharani, Comparison of various equalization techniques for MIMO system under different fading channels, *IEEE International Conference on Communication and Electronics Systems*. ISBN:978-1-5090-5013-0 (2017)
15. V.P. Singh, A. Khare, A. Somkunwar, To improve the BER and channel capacity through the implementation of differentia noise shift keying using Matlab. Indian J. Sci. Technol. **10**(9) (2017)

A GUI to Analyze the Energy Consumption in Case of Static and Dynamic Nodes in WSN

Trupti Shripad Tagare, Rajashree Narendra, and T. C. Manjunath

Abstract Wireless sensor networks consist of spatially dispersed sensors for monitoring physical parameters in various applications and in organizing the collected data at centralized location. These days, energy is one of the main factors that need to be intelligently consumed in wireless sensor networks. Techniques which can perform energy efficient clustering are required to maintain reliable sensing. In this paper, we consider both static and dynamic nodes. For static nodes, direct communication and low energy adaptive clustering hierarchy algorithm calculations are examined and discussed. For dynamic nodes, the distance-based algorithm is presented which ensures that work is allocated evenly to the nodes. Further, the distance and energy-based algorithm ensures that the nodes drain out evenly. Simulations are carried out on MATLAB. An easy-to-understand graphical user interface allows us to handily analyze the energy proficiency for various amount of nodes and clusters. The results are shown for four nodes and one cluster.

Keywords Wireless sensor networks (WSNs) · Cluster head · Low energy adaptive clustering hierarchy (LEACH) · Graphical user interface (GUI) · Levy walk model · Threshold level · Sensor node · Base station

---

T. S. Tagare · T. C. Manjunath
Dayananda Sagar College of Engineering, Bengaluru, India
e-mail: truptitagare-ece@dayanandasagar.edu

T. C. Manjunath
e-mail: drtcmanjunath-ece@dayanandasagar.edu

R. Narendra (✉)
Dayananda Sagar University, Bengaluru, India
e-mail: rajashree-ece@dsu.edu.in

S. Shakya et al. (eds.), *Mobile Computing and Sustainable Informatics*, Lecture Notes on Data Engineering and Communications Technologies 68,
https://doi.org/10.1007/978-981-16-1866-6_40

1 Introduction

1.1 An Overview

The period from 1950 to 1960 witnessed in-depth research and speedy improvement in the field of the far-off sensing. The first international far-off sensing device appeared in 1990, and from then the research within the discipline gained momentum.

Currently, WSNs have acquired serious international interest, particularly in surveillance monitoring a bodily environment parameter like temperature, atmospheric strain in addition to pollutants and so on [1]. In WSN, nodes co-operate with each other, sense and even control the environmental physical parameters, additionally interaction among persons or computers with the encircling environment [2]. Many nodes are spatially dispersed in the area to be monitored. These nodes consist of sensing, processing devices, radio transceivers and a battery. All the operations in the node should be carried out with limited power consumption to enhance the lifetime of the nodes. So, energy efficient protocols are the need of the hour.

The detected information is imparted to the base station through the sensor hubs. Now, depending on distance between the two, varying amount of energy is required in the transmission process. Also, communication requires maximum amount of energy [3–5]. We, in our paper mainly focus on increasing the energy efficiency of WSN and the comparison between different techniques applied.

1.2 Literature Survey

- Paper [6] gave basic understanding of the wireless sensor networks (WSN). Basics of the WSN and its architecture, its definition, WSN classification, and more information about the physical layer of WSN.
- In [7], the author highlighted a method to find CH technique which is efficient in nature to saving energy and thus increasing the existence of the network. Here, nodes which are stationary in nature were considered and couple of methods were proposed to check as to which is better for continual use.
- While [8] explained a method called LEACH algorithm to find cluster head by random probability and how the energy is affected from it. Thus, building a basic understanding of how the energy parameter is dealt with.
- Paper [9] focused on taking into consideration the energy left in the node for increasing the life of the WSN and also explained Levy walk model.

1.3 Objective

A sensor node can be deployed in any harsh surrounding. So, it is impracticable if not impossible to change or recharge the batteries used in these nodes which power them. Therefore, to increase the lifetime of sensor nodes we should aim for efficient energy consumption.

1.4 Methodology

In case of LEACH protocol, the cluster head (CH) is selected at random. This selected node as CH maybe with low energy. So, in our paper, the CH choice depends absolutely on energy variable and mean distance of the nearby nodes. We focus on allocating the CH which will transmit the sensed data to a cellular base station by selecting an appropriate cluster head. Also, consider the relative distance of nodes in the network to make energy consumption more-fair. Finally, we aim at providing the results through MATLAB and Simulink.

1.5 Working Principle

The fact that the power source which is the battery is one time deployed in the nodes and if the energy drains out then the WSN is compromised. So mainly focusing on this issue we are trying to enhance current algorithms and give a comparison to the user through MATLAB GUI to the user as to which approach holds good for his/her application.

2 Efficient Energy Consumption Methods for Static Nodes

2.1 Direct Communication

Here, each node communicates the data with the base station straightforwardly. During communication, there is major energy consumption in WSN and consumption increases with distance. Let us consider the scenario when every node of the network directly establishes a communication with the base station as shown in Fig. 1.

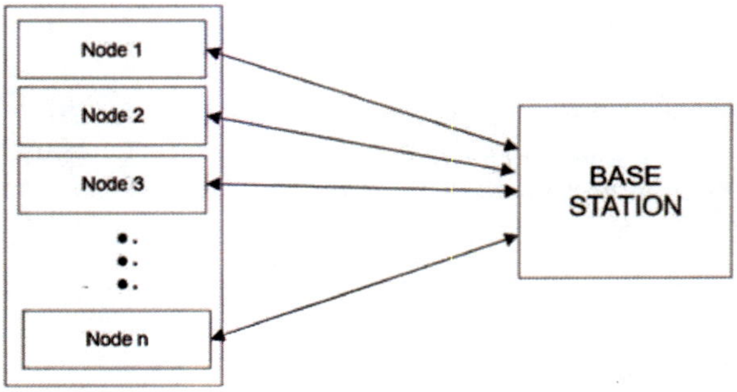

Fig. 1 Direct communication of data established between the nodes and base station

2.2 LEACH Algorithm

Low-energy adaptive clustering hierarchy (LEACH) is a TDMA-based MAC convention. It is co-ordinated with clustering and routing convention in WSNs. LEACH primary objective is to use minimum energy in the process of creating and maintaining the clusters. LEACH is a hierarchical protocol. Once a node is a CH for one round, it cannot be selected again for next *P* rounds. Here, *P* is the number of occasions a node becomes a CH in percentage units. Thus, each node has 1/*P* probability of being selected as the CH. As one round ends, each node that is not CH will search for the nearest CH and join its cluster. The CH creates a schedule for every node in that cluster group. Only during their time slot, the nodes keep their radio on. Hence, lot of energy is saved [10].

2.3 Energy Consumption Model

This paper utilizes a basic radio utilization model to check the energy utilization in the node. It estimates the energy consumed when node communicates information during each cycle. The equation for energy required during sending the information is given by

Energy consumed during Transmitting:

$$\mathrm{ET} = \mathrm{Ee} * k + \mathrm{Ea} * k \text{ in J units} \tag{1}$$

'*k*' is the total number of data bits.

Also, the energy consumed during receiving is given by

Energy consumed during Receiving:

$$\mathrm{ER} = \mathrm{Ee} * k \text{ in J units} \tag{2}$$

'Ea' is the energy scattered in transmit amplifier given in J/bit/m^2 units,

'Ee' is the circuit energy when sending or accepting *k* bit of data is J/bit units, '*d*' is the distance between a sensor node and its respective CH or between CH and base station.

Here, we focus on optimizing the consumption of energy only due to communication and not due to the sensing and processing operations. Also, we assume that the nodes are far from base station. We additionally accept that every sensor node has equivalent transmitter-receiver and data processing capacities, which sends the detected data each *S* seconds.

Distance between any two nodes say *A* and *B*, is represented by

$$dAB = \text{sqrt}\left((xA - xB)^2 + (yA - yB)^2\right) \quad (3)$$

The above equation estimates Euclidean distance.

Assumptions Made

- The nodes are far away from base central station and the distance is larger when compared to the distance between the nodes.
- No energy is consumed during sensing and processing operations.
- Every sensor node has equivalent transmitter–receiver and data processing capacities, which send the detected data each *S* seconds.
- Each sensor node initially has energy of 2 J (Battery)

2.4 Direct Communication Technique with Equations

Here, every node sends its detected information. Equation (4) below represents cost for direct communication [7]:

$$D = \left[\text{Ee} + \text{Ea} * (d)^2\right] * k \quad (4)$$

where d: represents the distance between sensor node and base station,
Ea: represents energy spent in transmit amplifier (J/bit/m^2),
Ee: represents the circuit energy when sending or accepting k bit of data (J/bit).
Figure 2 shows the equivalent model of direct communication using Simulink.

LEACH Communication

For CH selection, each node in the cluster will generate a number randomly in the fixed range between 0 and 1, then compare it with the set Threshold level (T). If the random number generated is less than the threshold value, then the node is made the CH for that specific round [11].

The threshold value T is calculated by the formula

$$T = \frac{p}{1 - p(r \bmod (1/p))} \quad (5)$$

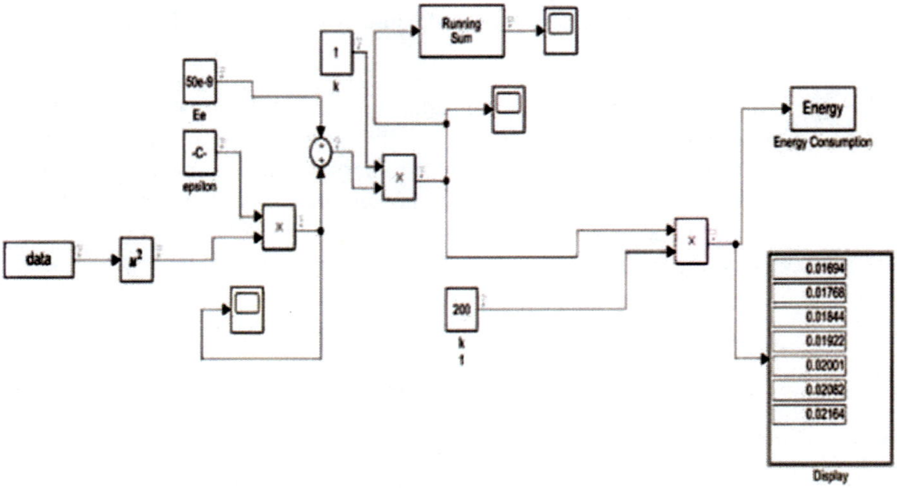

Fig. 2 Equivalent model of direct communication (Simulink)

Here, *p* is the probability of the node to be considered as CH.

r is no. of the election decision round.

The above procedure is repeated every election round. Once a CH is selected, its conveyed to the base station which in turn informs all other nodes about the CH. Data sensed by the nodes is sent to the nearest CH which in turn communicates the data to the base station. Hence, all the data bits will be transmitted to the base station by consuming lesser energy. Figure 3 depicts the flowchart explaining LEACH algorithm.

3 Efficient Energy Consumption Methods for Dynamic Nodes

3.1 Levy Walk Around Model

To incorporate the mobility of the nodes, levy walk around approach plays a vital role. Here using the animals and identifies the moves in random distance and direction [9]. Step size will be identified by using Levi walk function and based on the below equations.

$$\theta = \text{rand} * 2 * \text{pi}; \tag{6}$$

$$f = \text{rand}^{(-1/\alpha)}; \tag{7}$$

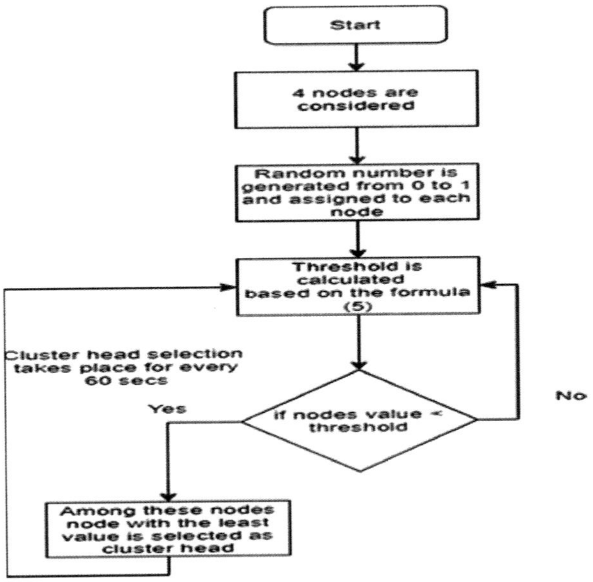

Fig. 3 Flowchart explaining LEACH algorithm

$$x(n) = x(n-1) + f * \cos(\theta); \tag{8}$$

$$y(n) = y(n-1) + f * \sin(\theta); \tag{9}$$

where α is stability exponent.

This distance of the node from the base station is read in every 60 s.

Assumptions made:

- Only four nodes are considered and they are mobile in nature
- The nodes move according to Levy walk pattern
- The nodes move at every S second
- Each node initially has energy of 10 J (Battery)
- Since the nodes will be mobile the energy consumption between the nodes changes every S second and hence it is not ignored.

3.2 *Distance-Based Cluster Head Election Algorithm*

In mobility nature, calculation of distance is most important for data transmission. Cluster head selection will be based on the distance vector of each node to the base station. Closer to the base station will be considered as cluster head and farer to the base station will not be considered as cluster head. The probability of the nodes

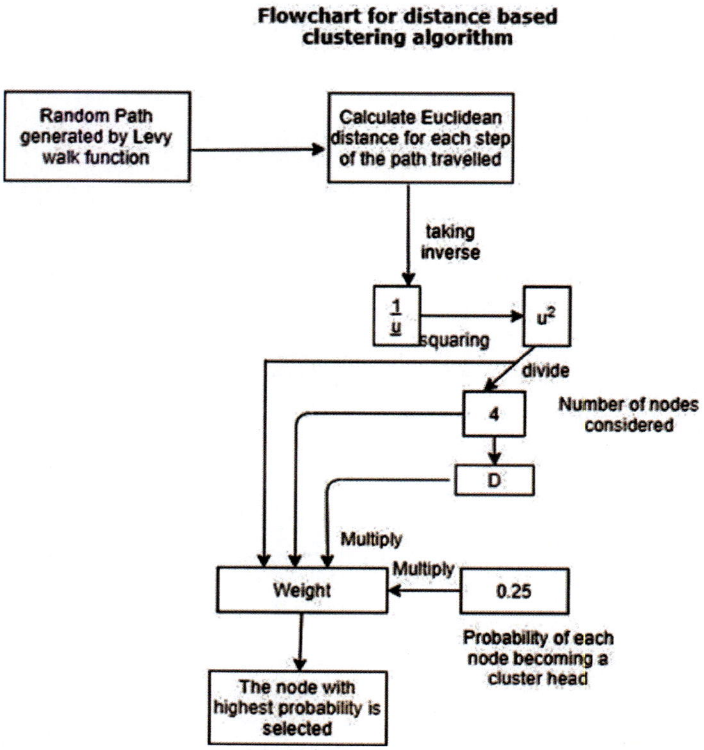

Fig. 4 Flowchart for distance-based cluster head algorithm

becoming Cluster Head is accumulated by a weighted value which is determined by distance between the nodes and the base station [12–14] (Fig. 4).

Here, we have assumed that all nodes start from (0, 0) and the base station is located at (500, 500).

3.3 *Distance and Energy-Based Cluster Head Election Algorithm*

Here, the main goal is to consider energy remaining in each node as a factor along with the distance between the node and the base station in calculating the probability of the node becoming a CH [13, 15, 16]. The probability of the nodes becoming a CH is influenced by the parameter termed as the DE factor. This DE factor considers both distance and energy remaining in the node and hence helps in efficient energy consumption to transmit data for a longer period of time. The mobile node with the highest DE factor value is considered as the CH during that round of election.

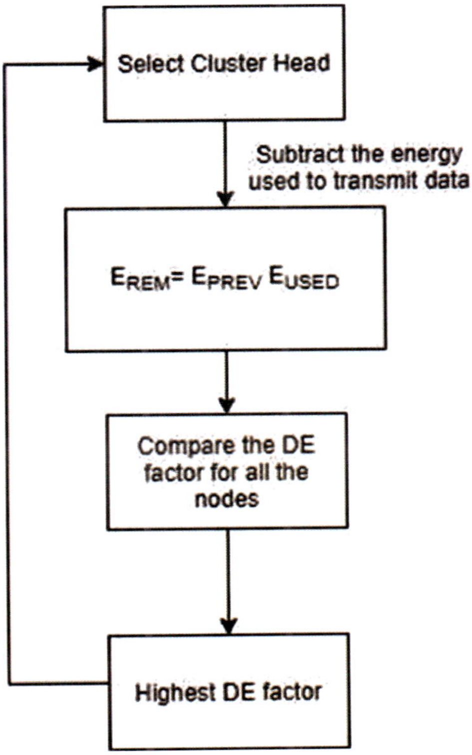

Fig. 5 Flowchart explaining distance and energy-based cluster head algorithm

DE factor is given by the formula:

$$\mathrm{DE} = \mathrm{Erem}/(d\mathrm{mean})^2 \tag{10}$$

where Erem is the energy remaining in the mobile node after previous CH election and dmean is the mean of the distances of all the mobile nodes to the base station (Fig. 5).

4 Results and Discussions

4.1 Graphical User Interface (GUI) Output

See Fig. 6.

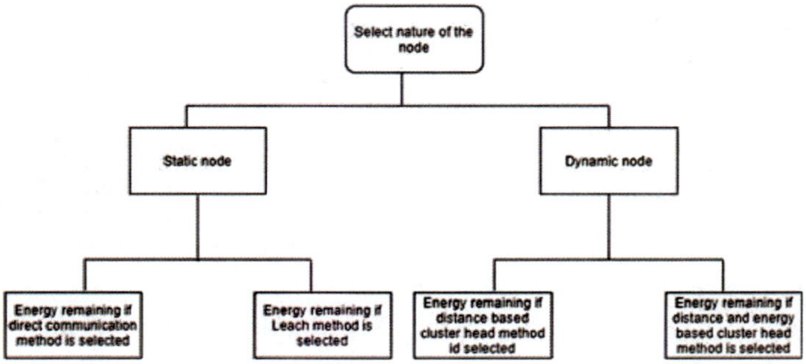

Fig. 6 Flowchart illustrating the design of GUI

Fig. 7 GUI output for static nodes

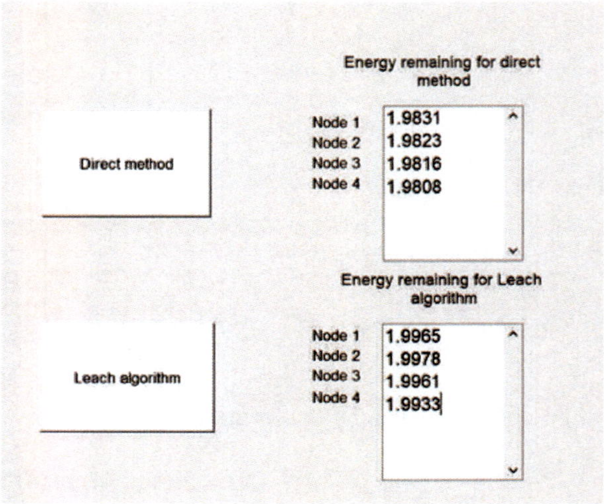

4.2 *For Nodes Which Are Static in Nature*

In Fig. 7, when the respective button 'Direct Method' or 'Leach algorithm' is pushed the energy remaining in each node after data transmission is displayed in the adjacent text field. From Fig. 7, it is seen that energy remaining in the nodes using LEACH algorithm is more than in case of direct method. So, it clearly depicts that for static nodes, LEACH algorithm provides longer lifetime for the nodes.

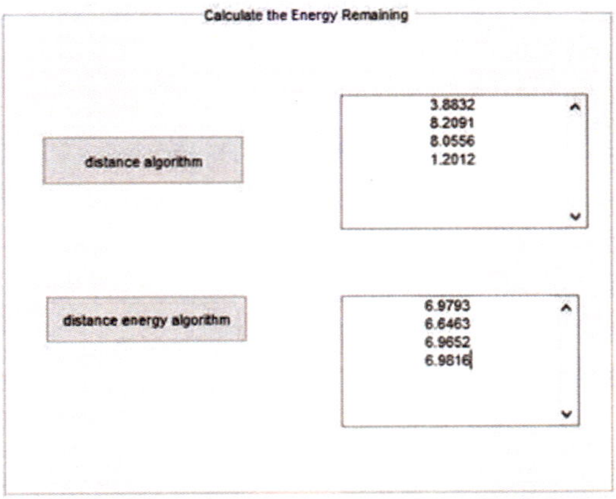

Fig. 8 GUI output for dynamic nodes

Table 1 Tabulated output of energy remaining in the node and the number of times a node is elected as cluster head

Node	Distance (in m)	Residual energy (in J)	Cluster head (no. of times)
1	920	1.9831	200
2	940	1.9823	200
3	960	1.9816	200
4	980	1.9808	200

4.3 For Nodes Which Are Dynamic in Nature

In Fig. 8, when 'distance algorithm' is pushed the energy remaining in each node after data transmission is displayed and when 'distance energy algorithm' is pushed the energy remaining in each node after data transmission is displayed in the adjacent text field. From Fig. 8, it is seen that energy remaining in the nodes using distance- and energy-based algorithm is more than in case of when only distance is considered. So, it clearly depicts that for dynamic nodes, distance and energy-based algorithm provides longer lifetime for the nodes.

4.4 Tabulated Results for Static Nodes

Direct communication

See Table 1.

Table 2 Tabulated output of energy remaining in the node and the number of times a node is elected as cluster head

Node	Distance (in m)	Residual energy (in J)	Cluster head (no. of times)
1	920	1.9964	41
2	940	1.9972	25
3	960	1.9961	42
4	980	1.993	69

LEACH Algorithm

See Table 2.

4.5 Tabulated Results for Dynamic Nodes

Distance-Based Algorithm

See Table 3.

Distance and Energy-Based Algorithm

See Table 4.

4.6 Observations Made

When nodes are static in nature

- In direct communication method, the energy in all the nodes drain out together but they last for a shorter period of time. From Table 1, it is seen that energy left in each node is around 1.98 J out of 2 J.
- In LEACH algorithm, the energy in the nodes is utilized efficiently and they last for a longer period of time. However, the chances that energy in all the nodes drain out together are quite thin. From Table 2, it is seen that energy left in each node is around 1.99 J out of 2 J.

  Therefore, LEACH algorithm is better for static nodes.

Table 3 Tabulated output of energy left remaining in the node and the number of times a node is elected as cluster head

Node	Residual energy (in J)	Cluster head (no. of times)
1	9.2104	16
2	3.1921	137
3	7.7171	46
4	9.9501	1

Table 4 Tabulated output of energy remaining in the node and the number of times a node is elected as cluster head

Node	Residual energy (in J)	Cluster head (no. of times)
1	7.4750	47
2	7.5230	47
3	7.2533	56
4	7.4716	50

When nodes are dynamic in nature

- In distance-based cluster head election algorithm, the results observed are very similar to LEACH that all the nodes utilize their energy efficiently but all drain unequally depending on the path taken by each node. From Table 3, it is seen that energy left in the node which is elected as CH 137 times has least energy remaining 3.19 J out of 10 J.
- In distance and energy-based cluster head algorithm, the energy remaining in the nodes is utilized efficiently and that all the nodes drain out almost together. From Table 4, it is seen that energy left in all the nodes has same amount of energy draining.

This is one of the main advantages of this method that it does both efficient energy utilization and energy of all nodes is drained uniformly.

5 Conclusion and Future Work

Energy is a major constraint for WSNs. To maintain reliable sensing, energy efficient network protocols are required. The results as obtained in Table 3, show that the distance-based algorithm allocates work to the nodes more evenly compared to LEACH and the node with maximum energy is elected as CH, and it drains out all its energy. Next, by considering the energy parameter along with the mean distance the CH loses its energy very slowly. Results show that all the mobile nodes spend almost equal amount of energy to transmit data to the base station. This method ensures that all the nodes almost drain together at once (as seen in Table 4) and they can all be replaced easily by the customer rather than making multiple rounds to replace a node at different times. This method is more efficient in conservation of energy in the nodes. In future work, event-driven cluster head election can be implemented for the nodes which is useful for emergency applications.

References

1. R.M. Zuhairy, M.G.H. Al Zamil, Energy-efficient load balancing in wireless sensor network: an application of multinomial regression analysis. Int. J. Distrib. Sensor Netw. **14**(3). https://doi.org/10.1177/1550147718764641. journals.sagepub.com/home/dsn
2. A. Broring, et al.: New generation sensor web enablement. Sensors **11**(3), 26532–2669. ISSN: 1424-8220 (2011). https://doi.org/10.3390/s110302652
3. M. El Fissaoui, A. Beni-Hssane, M. Saadi, Energy efficient and fault tolerant distributed algorithm for data aggregation in wireless sensor networks. J. Amb. Intell. Human. Comput. (2018). https://doi.org/10.1007/s12652-018-0704-8
4. R. Priyadarshi, P. Rawat, V. Nath, Energy dependent cluster formation in heterogeneous wireless sensor network. Microsyst. Technol. https://doi.org/10.1007/s00542-018-4116-7
5. Y.K. Yousif, R. Badlishah, N. Yaakob, A. Amir, An energy efficient and load balancing clustering scheme for wireless sensor network (WSN) based on distributed approach, *1st International Conference on Green and Sustainable Computing (ICoGeS) 2017*. IOP Publishing IOP Conference Series: Journal of Physics: Conference Series 1019, 012007 (2018). https://doi.org/10.1088/1742-6596/1019/1/012007
6. O.N. Al Khayat, *Wireless Sensor Networks Physical Layer Performance Investigation and Enhancement* (Ministry of Higher Education and Scientific Research, University of Technology, Cairo, Egypt)
7. D. Out, *Protocol for Energy Efficient Cluster Head Election for Collaborative Cluster Head Elections* (Dublin, 2018)
8. A. Krishna Kumar, V. Anuratha, An energy efficient cluster head selection of LEACH protocol for WSN (2010)
9. C. Fu, Z. Jiang, et al., An energy balanced algorithm of LEACH protocol in WSN. WEI IJCSI Int. J. Comput. Sci. **10**(1), No 1, 354–359 (2013)
10. M. Elshrkawey, S.M. Elsherif, M. Elsayed Wahed, An enhancement approach for reducing the energy consumption in wireless sensor networks. J. King Saud Univ. Comput. Inf. Sci. 4 April 2017 Networks, in *7th Annual International Conference on Mobile Computing and Networking* (2001)
11. A.H. Sodhro, L. Chen, A. Sekhari, Y. Ouzrout, W. Wu, Energy efficiency comparison between data rate control and transmission power control algorithms for wireless body sensor networks. Int. J. Distr. Sensor Netw. **14**(1) (2018). https://doi.org/10.1177/1550147717750030. sagepub.com/home/dsn
12. R. Banerjee, S. Chatterjee, S.D. Bit, Performance of a partial discrete wavelet transform based path merging compression technique for wireless multimedia sensor networks. Wireless Personal Commun. (2018) https://doi.org/10.1007/s11277-018-6008-7
13. Y. Liu, A. Liu, N. Zhang, X. Liu, M. Ma, Y. Hu, DDC: dynamic duty cycle for improving delay and energy efficiency in wireless sensor networks. J. Netw. Comput. Appl. https://doi.org/10.1016/j.jnca.2019.01.022
14. K.A. Darabkh, M.Z. El-Yabroudi, A.H. El-Mousa, BPA-CRP: a balanced power-aware clustering and routing protocol for wireless sensor networks. Ad Hoc Netw. (2018)
15. S. Jabbar, M. Ahmad, K.R. Malik, S. Khalid, J. Chaudhry, O. Aldabbas, Designing an energy-aware mechanism for lifetime improvement of wireless sensor networks: a comprehensive study. Mobile Netw. Appl. https://doi.org/10.1007/s11036-018-1021-3

16. T.A. Alghamdi, Secure and energy efficient path optimization technique in wireless sensor networks using DH method. IEEE Access.https://doi.org/10.1109/ACCESS.2018.2865909

Comparing Strategies for Post-Hoc Explanations in Machine Learning Models

Aabhas Vij and Preethi Nanjundan

Abstract Most of the machine learning models act as black boxes, and hence, the need for interpreting them is rising. There are multiple approaches to understand the outcomes of a model. But in order to be able to trust the interpretations, there is a need to have a closer look at these approaches. This project compared three such frameworks—ELI5, LIME and SHAP. ELI5 and LIME follow the same approach toward interpreting the outcomes of machine learning algorithms by building an explainable model in the vicinity of the datapoint that needs to be explained, whereas SHAP works with Shapley values, a game theory approach toward assigning feature attribution. LIME outputs an *R*-squared value along with its feature attribution reports which help in quantifying the trust one must have in those interpretations. The *R*-squared value for surrogate models within different machine learning models varies. SHAP trades-off accuracy with time (theoretically). Assigning SHAP values to features is a time and computationally consuming task, and hence, it might require sampling beforehand. SHAP triumphs over LIME with respect to optimization of different kinds of machine learning models as it has explainers for different types of machine learning models, and LIME has one generic explainer for all model types.

Keywords Interpretability · LIME · SHAP · Explainable AI · ELI5

1 Introduction

Artificial intelligence (AI) is playing a major role in automating day–to-day tasks in the real world. There is no denying that many business decisions have heavy dependency on artificial intelligence [1]. The dependency is justified by the accuracy of the models AI runs on. In the earlier era of machine learning, the models were simpler and easier to explain [2]. As this era advances, the models are getting more

A. Vij · P. Nanjundan (✉)
Christ (Deemed To Be University), Lavasa, India
e-mail: preethi.n@christuniversity.in

A. Vij
e-mail: aabhas.vij@science.christuniversity.in

S. Shakya et al. (eds.), *Mobile Computing and Sustainable Informatics*, Lecture Notes on Data Engineering and Communications Technologies 68,
https://doi.org/10.1007/978-981-16-1866-6_41

complex [3]. As the complexity increases, the models start acting as "black boxes," i.e., their outputs get tougher to explain [4–6]. This is where explainable artificial intelligence (XAI) comes into the picture. Explainable AI alludes to methods and techniques in the application of artificial intelligence (AI) such that the results of the solution can be understood by humans [7–9]. It appears complementary in relation to the idea of the "black box" in AI where even their modelers cannot clarify why the AI arrived at a particular decision [10]. XAI was implemented to arise the social right for explanation, and it does not have any legal rights for the explanation. XAI approach helps the customers to know quality of the products based on the user experience and sharing reviews [11]. The idea of interpreting machine learning models is relatively new, and as of today, there exists only a handful of frameworks working toward this purpose [12–15]. For this study, three such frameworks were chosen—LIME, SHAP and ELI5. LIME and ELI5 work using a similar approach of creating a simple regression-based surrogate model in the vicinity of the data point that needs explanation, whereas SHAP works by assigning Shapley values, a game theory approach, to all the features thus giving an output of feature attribution for a certain output with respect to a data point.

2 Proposed Framework

This paper proposes a study of three frameworks—LIME, ELI5 and SHAP. For this paper, four kinds of machine learning models were trained over a common dataset that deals with classifying whether or not a customer bought from a Portuguese bank marketing campaign.

The workflow in Fig. 1 briefly shows the work that needs to be done on each machine learning model. For each model built, each of the three interpretability frameworks are used to explain an output for a random data point. These reports are then observed and analyzed to understand how each of the frameworks respond to

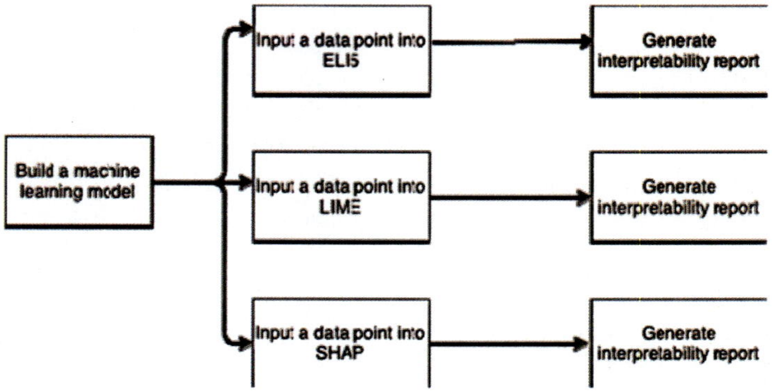

Fig. 1 Project workflow

different kinds of models. Four model types are chosen in order of their increasing complexities. Models chosen are as follows: logistic regression, decision tree, random forest and XGBoost. These models will be used for a binary class classification problem. Logistic regression makes use of a probabilistic threshold for classification. Random forest and decision tree models are tree-based models that make use of hierarchical decision making to arrive at a result, whereas XGBoost is an optimized machine learning library that implements machine learning algorithms under gradient boosting framework and provides a parallel tree boosting. After training and validating the mentioned models, an instance is chosen at random for explanation. For the output of that instance, interpretations are generated using ELI5, LIME and SHAP.

3 Experimental Setup

The machine learning models—logistic regression, decision tree, random forest and XGBoost were built over a dataset that classifies whether or not a customer bought from a Portuguese bank campaign. All of the models were explained using a random datapoint using all the three mentioned frameworks. The scoring metrics for the models are shown in Tables 1, 2, 3 and 4.

It is common knowledge that only a validated model will require explanations although no link between a model's validation score, and the surrogate model's explanations have been established. The same can be studied as the future work for this study.

Table 1 Classification report for logistic regression model

	Precision	Recall	$f1$-score	Support
0	0.95	0.86	0.90	10,965
1	0.36	0.65	0.46	1392

Table 2 Classification report for decision tree model

	Precision	Recall	$f1$-score	Support
0	0.95	0.89	0.92	10,965
1	0.41	0.62	0.49	1392

Table 3 Classification report for random forest model

	Precision	Recall	$f1$-score	Support
0	0.94	0.93	0.92	10,965
1	0.48	0.58	0.52	1392

Table 4 Classification report for XGBoost model

	Precision	Recall	$f1$-score	Support
0	0.91	0.99	0.95	10,965
1	0.71	0.21	0.32	1392

Fig. 2 Feature attribution reports by ELI5

After validating the models, a random data point is chosen whose output needs to be explained. That data point is explained using all the three chosen frameworks. A few highlights from the explanations generated using ELI5, LIME and SHAP will be attached in this paper. Feature attribution report in ELI5 shows all the positively and negatively impacting features for a particular output of the model.

Features colored in green impact the model's output toward a "1" prediction, whereas the ones in red impact the model's output toward a "0" prediction. The report on the right-hand side in Fig. 2 shows feature importance in decision trees model. ELI5 does not say in what direction a feature impacts the output of a model in case of decision trees.

In order to explain why the model classifies a particular observation as class 0 or 1, LimeTabularExplainer from the library lime was used (Python 3.7). It is the main explainer to use for tabular data. LIME also provides an explainer for text data, for images and for time series. LimeTabularExplainer shows attribution of top n features (where we can specify the value of n) and also shows probabilities of predicting either outcome (0 or 1) (Fig. 3).

LIME also generates the R-squared score of the surrogate model [1] that can be called using an appropriate method. It can help quantify the trust one must have in the explanations. High R-squared score implies a better fit surrogate linear regression model. The better the surrogate model fits, the better is the explanations generated through it. The following graph shows the R-squared score achieved with LIME for the four models.

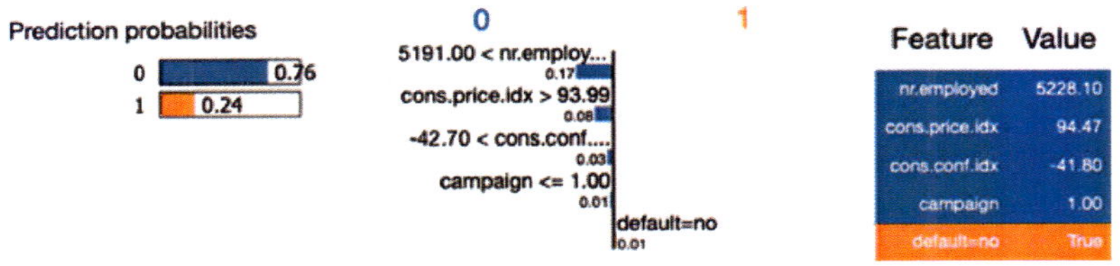

Fig. 3 Feature attribution report by LIME

Fig. 4 Achieved *R*-squared score for surrogate models

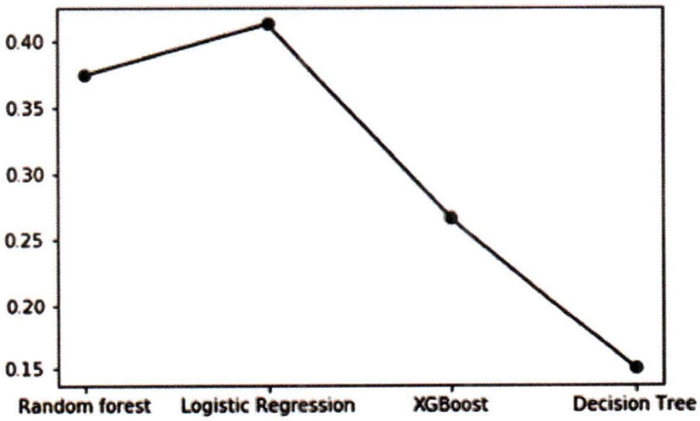

Highest *R*-squared score was achieved for the surrogate model in logistic regression model. The graph in Fig. 4 indicates that trust in LIME's explanations decreases as the complexity of the model increases. Further experimentation is required to comment further on the same and can be done as the future work for this project (Fig. 5).

SHAP framework works by assigning Shapley values [3] to each feature which is computationally expensive and is also time consuming. The above graph is a summary plot, available in SHAP library that shows feature importance which is evaluated by assigning Shapley values to the features. SHAP framework works by assigning Shapley values to each feature which is computationally expensive and is also time consuming. The above graph shows feature importance evaluated by assigning Shapley values to the columns. Theoretically, approach adopted by SHAP guarantees more accuracy than the one adopted by LIME and ELI5. SHAP provides three main explainers, namely TreeExplainer, DeepExplainer and KernelExplainer. First two specialize in generating Shapley values of tree-based models and neural network-based models, respectively. KernelExplainer is more of a generic explainer that can work for any kind of model (Fig. 6).

Features in red color influence the prediction positively, i.e., toward making the prediction as 1, whereas features in blue color affect the model's output in a negative manner, i.e., toward making the prediction as 0. SHAP force plot predicts a real-numbered value which is then transformed to a probability space hence giving out a value of 0 or 1 (if $p < 0.5$ or $p > 0.5$, respectively).

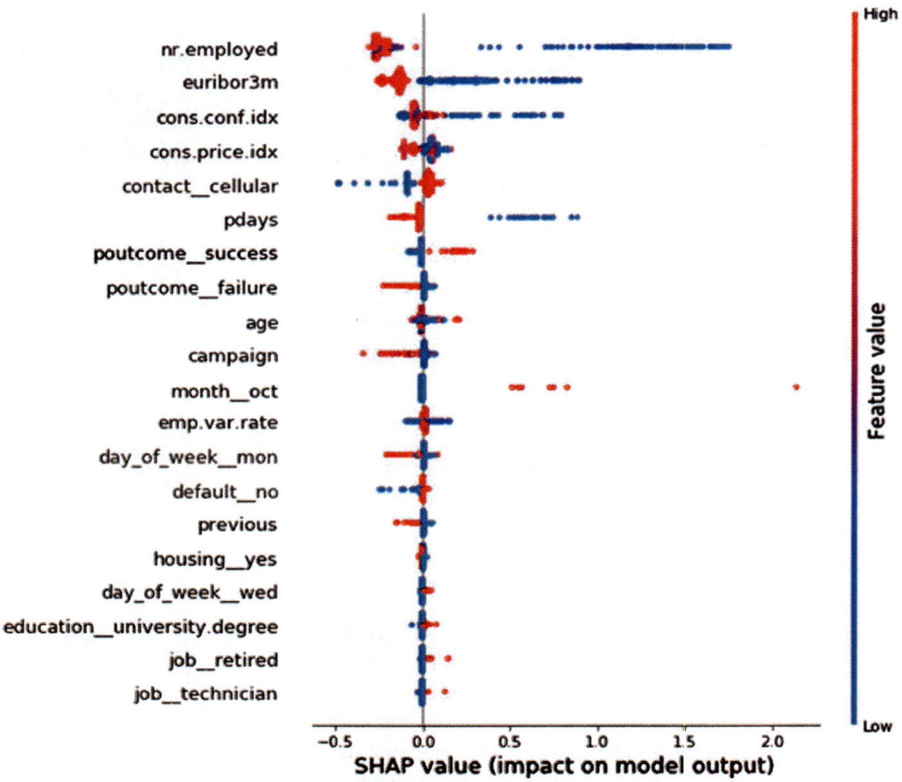

Fig. 5 SHAP summary plot for a random instance

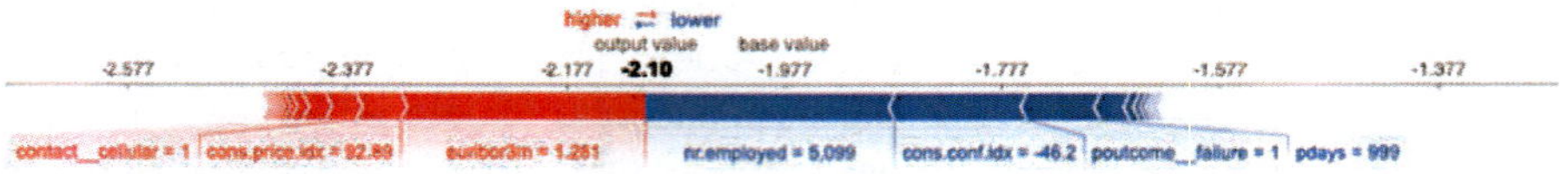

Fig. 6 SHAP force plot for a random instance

4 Conclusion

Importance of interpreting machine learning models increases by the day. There are only a handful of open-source frameworks that help in interpreting machine learning models. It is believed that interpretation of machine learning models will soon become a necessity for any AI project primarily for transparency-related reasons. The objective of this study was to identify the strengths and weaknesses of the three mentioned interpretability frameworks. ELI5 works with the same intuition as LIME. They build a simple surrogate model that tries to approximate the actual model around the vicinity of the data point that needs explanation. The *R*-squared scores generated by

LIME explainer help in quantifying the trust one that should have in the approximations generated by LIME. As observed, the least R-squared score was generated in the case of decision tree model followed by XGBoost model and then the random forest model. A relatively higher score for random forest model indicates a good approximation by LIME in this particular case but the approximation for XGBoost (a black box model), and decision trees is not as high as expected. On the other hand, SHAP makes use of Shapley values which can be computationally costly to generate for a big dataset. In this project, a sample of one thousand instances had to be taken at random since incorporating the whole training dataset consumed a lot of computing power and time. Despite the high computational cost, SHAP theoretically guarantees higher accuracy than LIME and ELI5, although this accuracy comes with a trade-off with time taken for execution. SHAP also comes with pre-optimized methods for explaining certain types of models, unlike LIME which only has one explainer for all kinds of models which adds to SHAP's advantages over LIME. SHAP has TreeExplainer for tree-based models, DeepExplainer for neural networks, KernelExplainer for generic explanations and a handful of more explainers for certain other kinds of models. The force plot and the summary plot features of SHAP library add to SHAP's advantages as they can help a human understand impact of all the features at once.

5 Future Work

As this project could not incorporate all kinds of machine learning models, other model types can be used as the future work for this project. Apart from machine learning models, neural networks can also be attempted to explain using the fore-mentioned frameworks. The problem type can also be changed from a classification type to a regression or clustering type that might enable a different point of view for these frameworks. Also, impact of a model's accuracy on the surrogate model's prediction can also be studied in the future. The interpretability frameworks are not just limited to the three previously mentioned algorithms. More interpretability algorithms like YellowBrick, Alibi and Lucid can be studied as the future work.

References

1. M.T. Ribeiro, S. Singh, S.C. Guestrin, Why should I trust you?: explaining the predictions of any classifier. arXiv:1602.04938
2. J. Zhang, Y. Wang, P. Molino, L. Li, D.S. Ebert, Manifold: a model-agnostic framework for interpretation and diagnosis of machine learning models. IEEE Trans. Visual. Comput. Graphics. https://doi.org/10.1109/TVCG.2018.2864499
3. S.M. Lundberg, S.I. Lee, A unified approach to interpreting model predictions, in *31st Conference on Neural Information Processing Systems (NIPS 2017)*, Long Beach, CA, USA

4. P. Schmidt, F. Biessmann, Quantifying interpretability and trust in machine learning systems. Amazon Res. arXiv:1901.08558
5. D.A. Melis, T.S. Jaakkola, On the robustness of interpretability methods. arXiv:1806.08049v1
6. A. White, A.D. Garcez, Measurable conterfactual local explanations for any classifier. arXiv: 1908.03020v2
7. I. Giurgiu, A. Schumann, Explainable failure predictions with rnn classifiersbased on time series data. arXiv 1901.08554
8. S. Shi, X. Zhang, W. Fan, A modified pertrubed sampling method for local interpretable model-agnostic explanation. arXiv:2002.07434v1
9. S. Shi, Y. Du, W. Fan, An extension of LIME with improvement of interpretability and fidelity. arXiv:2004.12277v1
10. A.K. Noor, *Potential of Cognitive Computing and Cognitive Systems* (De Gruyter, 2014)
11. L.H. Gilpin, D. Bau, B.Z. Yuan, A.Bajwa, M. Specter, L. Kagal, Explaining explanations: an overview of interpretability of machine learning, in *IEEE 5th International Conference on Data Science and Advanced Analytics (DSAA), Turin, Italy*, pp. 80–89 (2018). https://doi.org/10.1109/DSAA.2018.00018
12. C. Rudin, Stop explaining black box machine learning models for high stakes decisions and use interpretable models instead. Nat. Mach. Intell. **1**, 206–215 (2019)
13. D. Das, J. Ito, T. Kadowaki, K. Tsuda, An interpretable machine learning model for diagnosis of Alzheimer's disease. https://doi.org/10.7717/peerj.6543
14. R. Revetria, A. Catania, L. Cassettari, G. Guizzi, E. Romano, T. Murino, G. Improta, H. Fujita, Improving healthcare using cognitive computing based software: an application in emergency situation, in *Advanced Research in Applied Artificial Intelligence. IEA/AIE 2012*. Lecture Notes in Computer Science, vol. 7345 (Springer, Berlin)
15. D.V. Carvalho, E.M. Pereira, J.M. Cardoso, Machine learning interpretability: a survey on methods and metrics. Electronics **8**, 832 (2019). https://doi.org/10.3390/electronics8080832

Obstacle Avoidance and Ranging Using LIDAR

R. Kavitha and S. Nivetha

Abstract For many years, engineers are searching solutions for car accidents that are caused by human error due to drowsiness or by a person who comes out of nowhere in front of the vehicle. In such a situation, vehicle accident may occur. So, there is a need to develop a vehicle that saves human life from a dangerous accident. Autonomous vehicle have emerged with cameras and sensors to avoid the strategic accidents of human error. Autonomous vehicle systems get information about other vehicles, pedestrians, and other immediate surroundings that is necessary to detect the presence of pedestrians, vehicles and other related entity objects. In this paper, novel LIDAR technologies for automotive applications and the perception algorithms based on field of view (FOV) are used for obstacle avoidance, and ranging of object or vehicle nearby to avoid collision of data in crucial conditions is discussed. This paper is structured based on a vehicle detection and processing of 2D LIDAR sensor to avoid traffic accidents and save life of the people with the help of deep learning algorithm concept. The LIDAR serves as image sensor, and the output response is a steering response, based on the image decisions taken either stop or turn the vehicle depending on the object in front of the vehicle based on road boundary detection.

Keywords Autonomous vehicle · Deep learning neural network · 3D LIDAR · Road boundary detection

1 Introduction

Road traffic in India is increased day by day due to growth in number of vehicles. The majority of road accidents are due to road crashes and careless driving. The

R. Kavitha (✉)
Department of Electrical and Electronics Engineering, Kumaraguru College of Technology, Coimbatore, India
e-mail: kavitha.r.eee@kct.ac.in

S. Nivetha
Department of Electrical and Electronics Engineering, Kumaraguru College of Technology, Coimbatore, India
e-mail: nivetha.19mes@kct.ac.in

S. Shakya et al. (eds.), *Mobile Computing and Sustainable Informatics*, Lecture Notes on Data Engineering and Communications Technologies 68,
https://doi.org/10.1007/978-981-16-1866-6_42

major cause for this accidents is when the driver forgot to toggle or does not take the correct decision on moving the vehicle according to the statement by the NCRB, which function under the Union Home Ministry. The accidents occur due to the carelessness of the driver who may be in drowsiness or during night time where he cannot see clearly or may be due to fog or mist around by. To decrease and overcome the accident rate, many technocrats must pave the way to prevent accident rate due to the mistakes of driver [1–3]. One of the pathways was an autonomous vehicle termed as self-driving vehicles that navigate its path [4, 5] and are capable of sensing its environment and moving safely with a combination of the navigation system automatically [6, 7].

Nowadays, vehicle detection plays tedious job on determining oncoming vehicles and their motion of planning the tasks. So LIDAR technology is employed instead of sound and radio waves. Vehicle control unit, location system vehicle mapping, global path planning, radar perception, computer vision-based, and speed monitoring are some implementation done in autonomous vehicle. LIDAR is one such technology implemented for avoiding collision due to careless of the driver. LIDAR is a technology that senses precise distances using invisible laser beams. By measuring thousands of distances in a defined field of view for mapping and localization, 2D LIDAR data is pre-processed, by detecting the vehicle position, and noise points are deleted and filtered using the unique features of the boundary box algorithm for the measurement based on neural networks. To precisely identify the sizes of all the objects, an efficient calibration method is employed with multiple LIDAR'S using a simple rectangular board for the efficient calibration.

The obstacle avoidance module is done with data acquired process from a LIDAR sensor that provides 3-D geometry information of the object by extract feature of an image with depth information, long-range detection [8–11]. Image detection and segmentation are done by many neural networks [12–14]. Object tracking in the traditional approach [15–18] is by GPS, but GPS signal cannot detect the tress, road boundaries, and building in an urban area so LIDAR boundary box algorithm is used. The multi-hypothesis tracking system in LIDAR tracks and detects vehicles on frontal and birds eye projection trained on convolution neural network with boundary box algorithm.

2 Methodology

LIDAR laser pulses are forming point clouds and MATLAB software for digitalized representation of the world and algorithms are developed to detect and track objects with the mode to collision of vehicle.

The architecture is shown in Fig. 1 the block of autonomous vehicles which consist of a module for obstacle detection, obstacle avoidance, and driver module. The algorithms for object detection are: Image Recognization, Image Segmentation, Pattern Recognition Algorithms, Clustering to Identify Objects, Decision Matrix,

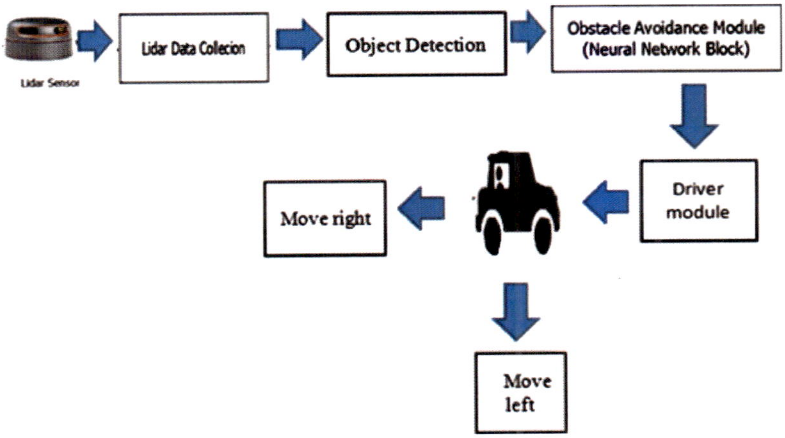

Fig. 1 Proposed block diagram

Regression Algorithm. The algorithm for obstacle avoidance is Convolutional Neural Network and for object mapping is SLAM-based algorithm.

2.1 *Algorithm for Object Detection*

- **Image Recognization**
  An image shown in Fig. 2 is an matrix, of square pixels arranged in columns and rows using imread() function, Velodyne file is import and image is segmented, based on down sampling, image is visualized, and data is stored in 2-D point cloud (LIDAR tool box) MATLAB.

Fig. 2 Image of typical traffic system

- **Segmentation of Image**
  Image segmentation is the method of splitting an image into multiple pixels that is determined to change or to simplify an image for easier verification. This technique is typically used to locate objects and boundaries in images and then process and assign label in an image that determine certain characteristic.
- **Algorithm for Pattern Recognition**
  Images from the sensor determines the pixel values which form a data and filters the images based on object category. The algorithms for pattern recognization rule out these unusual data points in a dataset which is an important step before classification of the objects with edges and fitting line segments aligned to edges and corner to form a new line segment. The support vector machines are with histograms of oriented gradients.
- **Clustering to Identify Objects**
  Clustering is an significant operation for tracking and object detection. HDL-64E LIDAR generates approximately 120,000 point-cloud data per frame, which depicts the spatial distribution of the obstacle.
- **Decision-Making Algorithm**
  Decision algorithm is used for systematically analyzing, identifying, and rating the performance of information in dataset. These algorithms used for decision making on the circumstance of object detection, and a car needs to take a right/left turn or it needs to stop it which depends on the algorithms that is classified, recognized and predicted based on the movement of objects trained independently to make prediction, to reduce the possibility of errors in decision making.
- **Algorithm for Regression**
  Algorithm creates a statistical relation among an image and the place of the object in that image. The statistical image model offers quick recognition of image and sampling expanded to other objects without having massive human modeling that can be used for self-driving cars by neural network regression, Bayesian regression, and decision forest regression, among others.

2.2 *Algorithm for Obstacle Avoidance*

Convolutional Neural Network (CNN) is a neural network that has one or more convolutional layers used for image processing, segmentation, and classification for auto correlating data that simulates multilayer structure that can be extracted from the feature of input data to improve classification or prediction of image for decision making to avoid obstacles. This module is accessed only when the obstacle is detected and layer is activated with the presence of obstacles in critical distances.

1. Input is taken in the form of image.
2. Then image is detected and classified based on various regions.
3. Each region is determined as separate image in pixels.
4. Pass all these image to the CNN and classify them into various classes.

5. The divided corresponding classes of all regions are detected to get the original image with the objects.

The approach is that the objects in the image can have different aspect ratios and spatial point cloud.

2.3 Algorithm for Object Mapping

SLAM-based algorithm reads from the environment with the help of the sensor that reduces the computational requirements of dataset extracted from the features of image, but SLAM cannot deal individual pixel. LIDAR calculates the distance to an object by illustrating the object with manifold transceivers. Each transceiver instantly radiates pulsated light and measures the reflected pulses to ascertain position. By determining obstacle detection and object tracking algorithm with LIDAR sensor, neural network takes decision and sends it to the driver module to tell which direction to move to avoid obstacle.

3 Additional Information Required from Authors

MATLAB was chosen to build and implement the obstacle avoidance and navigation interfacing with LIDAR tool box. The LIDAR Labeller tool box enables to load signals from multiple types of data sources with a file or folder containing one or more signals to label as shown in Fig. 3. Use this process to load the data for a point cloud sequence.

Simultaneous Localization And Mapping (SLAM) algorithm is a collection of various LIDAR scans based on pose graph optimization and records the environment using the LIDAR tool box that provides accurate path planning data rather than dangerous human maps. Targets detected and boundary are shown in Fig. 4.

Implementation of SLAM algorithm:

- First, load the scanned data image from file.
- And then run SLAM-based algorithm that provides optimized map and plot target.
- Observe process with initial 10 scans.
- Observed data provides the issues in closure of loop and the process of optimization.

The LIDAR tool box in MATLAB simulation provides image processing that supports binary, indexed, grayscale, and true color *image types*. Each image is stored as pixels. Acquired LIDAR data from the image creates a 2D map of the surrounding environment by using laser beams based on time of flight and distance measured with field of view. The color image mapping is represented in Fig. 5.

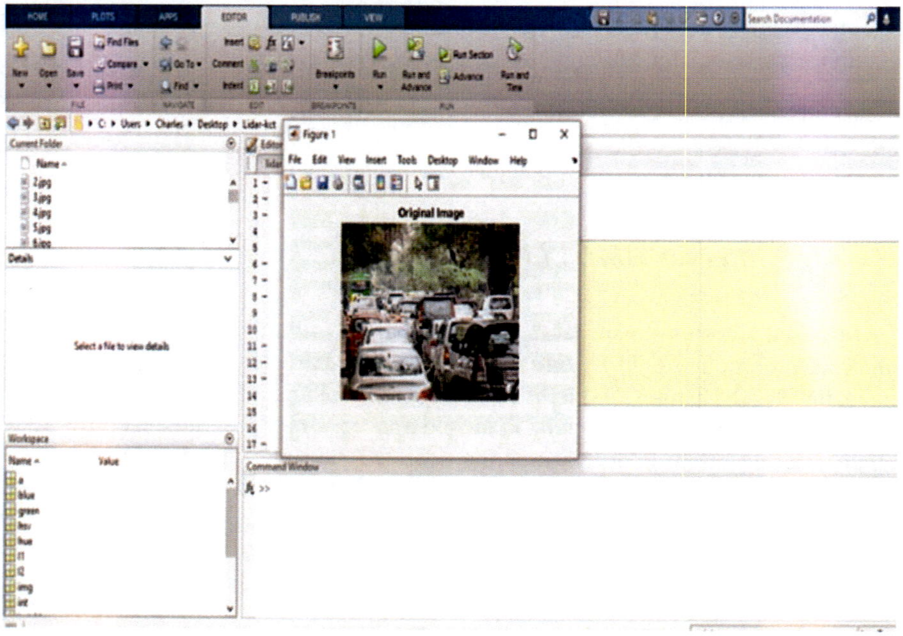

Fig. 3 LIDAR tool box

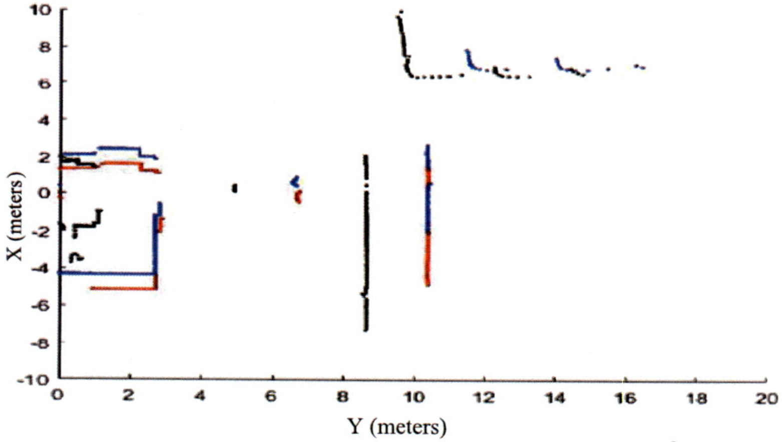

Fig. 4 Obstacle path of detection

The LIDAR sensor continuously detects the surroundings as the data frames on FOV with continuous buffered acquisition, which can be initiated using the start function. Finally, these data frames are stored in input buffer.

Fig. 5 RGB conversion

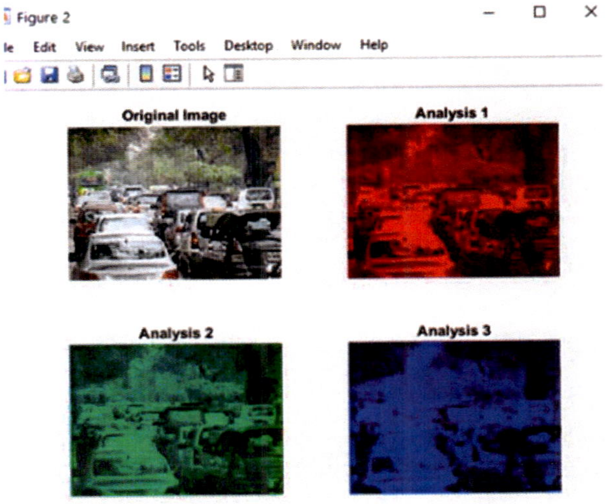

The histogram of the image after noise removal is shown in Fig. 6. Read one data frame from the LIDAR sensor from the point cloud represented in MATLAB. The data values represent a distance measurement, and the default units are meters (m).

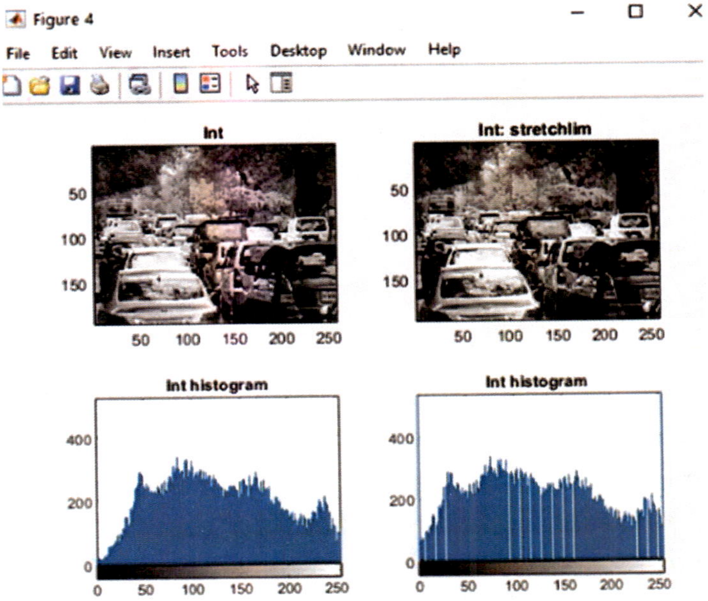

Fig. 6 Histogram from various conditions

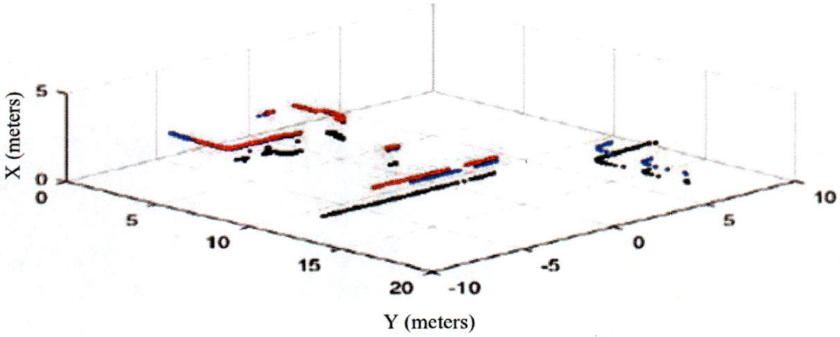

Fig. 7 Data frame visualization

Use the read function to read data frames into MATLAB workspace with different timestamp.

By using pc player, the data frame of point cloud is displayed with XY-limits (meters). The acquired data frame determines point cloud array, and the acquired timestamps determine number of frames per second (frame rate) based on field of view represented in Fig. 7.

Path Detection

The obstacle perception step produces an occupancy grid which contains all the perceived obstacle area within the grid to simplify the path. Once the grid has been completed it detects the shortest path from the current position as shown in Fig. 8.

4 Conclusion

This paper has provided the obstacle avoidance self-driving car with LIDAR sensor in order to identify different types of objects and to locate the distance of object. IT had estimated consecutive point clouds based on field of view (FOV) which is used for obstacle avoidance and ranging of object or vehicle nearby to avoid collision of data in crucial conditions. The histogram of images, data frame visualization and path planning are presented in this paper.

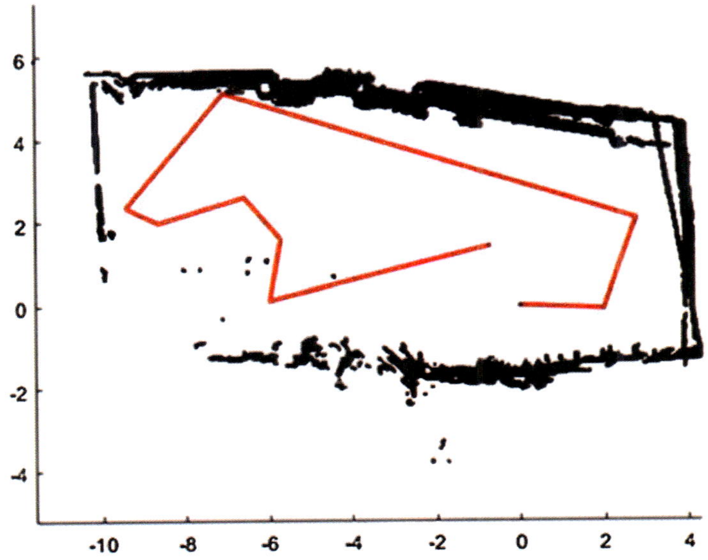

Fig. 8 Path planning result

References

1. A.J. McKnight, A.S. McKnight, Young novice drivers: careless or clueless. Accid. Anal Prev. **35**(6), 921–925 (2003)
2. N. Rhodes, K. Pivik, Age and gender differences in risky driving: the roles of positive affect and risk perception. Accid. Anal Prev. **43**(3), 923–931 (2011)
3. J.D. Lee, Technology and teen drivers. J. Safety Res. **38**(2), 203–213 (2007)
4. K. Jiang, D. Yang, C. Liu, T. Zhang, Z. Xiao, A flexible multi-layer map model designed for lane-level route planning in autonomous vehicles. Engineering (2019)
5. T. Litman, *Autonomous Vehicle Implementation Predictions: Implications for Transport Planning, ser. desLibris: Documents collection* (Victoria Transport Policy Institute, 2013)
6. S. Chae, T. Yoshida, Application of RFID technology to prevention of collision accident with heavy equipment. Autom. Constr. **19**(3), 368–374 (2010)
7. X. Zhao, P. Sun, Z. Xu, H. Min, H.Y. Asvadi, L. Garrote, C. Premebida, P. Peixoto, U.J. Nunes, Multimodal vehicle detection: fusing 3D-LIDAR and color camera data. Pattern Recogn. Lett. (2017)
8. Y. Zhang, J. Wang, X. Wang, J. M. Dolan, Road-segmentation-based curb detection method for self-driving via a 3D-LiDAR sensor. IEEE Trans. Intell. Transp. Syst. (2018)
9. K. Li, X. Wang, Y. Xu, J. Wang, Density enhancement-based long-range pedestrian detection using 3-D range data. IEEE Trans. Intell. Transp. Syst. **17**, 1368–1380 (2016)
10. J. Jia, Pyramid scene parsing network, in *IEEE Conference on Computer Vision and Pattern Recognition* (2017)
11. L. Chen, G. Papandreou, I. Kokkinos, K. Murphy, A. Yuille, DeepLab: semantic image segmentation with deep convolution nets, atrous convolution, and fully connected CRFs. IEEE Trans. Pattern Anal. Mach. Intell. (2017)
12. J. Long, E. Shelhamer, T. Darrell, Fully convolution network for semantic segmentation (2015). arXiv preprint arXiv:1411.4038v2

13. M. Cordts, M. Omran, S. Ramos, T. Rehfeld, M. Enzweiler, R. Benenson, U. Franke, R. Stefan, B. Schiele, The Cityscapes dataset for semantic urban scene understanding, in *IEEE Conference on Computer Vision and Pattern Recognition* (2016)
14. J. Ziegler, H. Lategahn, M. Schreiber, C.G. Keller, C. Knoppel, J.H.M. Haueis, C. Stiller, Video based localization for BERTHA, in *2014 IEEE Intelligent Vehicles Symposium Proceedings, IRRR*, pp. 1231–1238 (2014)
15. D. Frossard, R. Urtasun, End-to-end learning of multi-sensor 3d tracking by detection, in *ICRA* (2018), pp. 635–642
16. H.-N. Hu, Q. Cai, D. Wang, J. Lin, M. Sun, P. Krahenbuhl, T. Darrell, F. Yu, Joint monocular 3D vehicle detection and tracking, in *Proceedings of International Conference on Computer Vision* (2019), pp. 5390–5399
17. D. Fox, W. Burgard, S. Thrun, The dynamic window approach to collision avoidance. IEEE Robot. Autom. Mag. **4**(1), 23–33 (1997)
18. M. Mattei, L. Blasi, Smooth flight trajectory planning in the presence of no-fly zones and obstacles. J. Guid. Control Dyn. **33**(2), 454–462 (2010)

Admittance-Based Structural Health Monitoring of Pipeline for Damage Assessment

T. Jayachitra and Rashmi Priyadarshini

Abstract This paper reveals the experimental evaluation on the usage of electromechanical impedance techniques in pipeline structures. Pipelines are damaged due to its exposure to landslides, corrosion, overloading and earthquakes. To overcome this situation, the pipelines must be monitored continuously to avoid such damage. PZT (lead zirconate titanate) patch is used for sensing the damage in the structure. The most effective method in structural health monitoring for damage detection is electromechanical impedance technique. Impedance chip is used for structural health monitoring using this technique. The admittance measurements are measured using impedance chip. The results indicate that output obtained from the impedance chip is accurate. Damage detection in pipeline structure using impedance chip is the novel method in this proposed paper.

Keywords Structural health monitoring · Electromechanical impedance · Admittance · Pipeline · Lead zirconate titanate patches

1 Introduction

Structural health monitoring (SHM) is required for all structures in earth. Due to aging of the structures, SHM is being more focused in recent research [1]. Electromechanical impedance technique is one of the common techniques used by researchers for damage detection in the structures which was proposed first by Liang et al. [2]. The mechanical impedance of the structure is converted into electrical impedance, and vice versa is the basic property of piezoelectric which is involved in EMI technique [3].

T. Jayachitra (✉)
Electrical and Electronics Engineering Department, School of Engineering and Technology, Sharda University, Greater Noida, Uttar Pradesh 201310, India
e-mail: jayachitra.kishor@sharda.ac.in

R. Priyadarshini
Electronics and Communication Engineering Department, School of Engineering and Technology, Sharda University, Greater Noida, Uttar Pradesh 201310, India

S. Shakya et al. (eds.), *Mobile Computing and Sustainable Informatics*, Lecture Notes on Data Engineering and Communications Technologies 68,
https://doi.org/10.1007/978-981-16-1866-6_43

The piezosensor is bonded to the mechanical structure. The piezosensor is excited by the sinusoidal signal which is bonded to the structure, it generates a vibration on the structure, and in turn, the structure responses to the piezosensor by inducing an electrical signal [4]. The admittance signature is a function of tension, mass and structure dampness. Measurement of admittance provides the damage detection with respect to the base reading [5].

Impedance analyzers and LCR meters were commonly used instruments for the detection of damage. The impedance analyzer or LCR meter measurements was accurate and used for detection damage of various structures like concrete beam, aluminum plate, etc. Impedance analyzer or LCR meter is difficult to deploy. The disadvantage of these instruments is high cost and bulk size [6]. Quinn et al. were the first one to propose impedance chip in the curing stage of concrete [7]. Impedance chip is the best way for prototype applications [8]. The impedance chip is used for prototype structures like concrete beam, aluminum and steel for the detection of damage due to its less cost and small size [9, 10]. Corrosion was also detected using impedance chip [11].

Damage identification was performed in pipeline structure using different software. Abacus software is used to detect the damage in pipeline for guided mode propagation [12]. Imaging algorithm is used to detect the detection of small damages in pipeline structures [13]. Damage detection of cracks in pipelines is identified using analysis of wavelet packet. Piezoelectric transducers are used as sensor and actuators to detect the responses [14]. In signal processing, compressive sensing technique was used to detect the damage in pipeline [15]. Using signal processing techniques, the reflections due to the pressure wave detect the damage in the pipeline structures [16]. Kalman filter algorithm in wireless sensor network is utilized to detect the damage in pipeline of water [17].

Impedance chip is a impedance measurement chip which consists of analog to digital converter, frequency generator, digital signal processor and a sensor to measure the temperature. Frequency generator is used to excite, and processing takes place by DSP. The impedance data is acquired at every sweep frequencies. The user-defined software is used to predict the results. 200 Ω resistance is used for calibration. Repeatability is required for verification of data obtained [18].

The flow diagram for the proposed method is shown below in Fig. 1.

2 Experimental Analysis

The galvanized iron (GI) pipeline is chosen for experimentation. The size of the pipeline is 550 cm. Epoxy resin is used to bond the piezosensor with the structure. The joints are made for two GI pipelines using bolt. The bolt is tightened and the admittance signatures are measured with impedance chip. This is considered as base measurement. The bolt is removed in threads, and the corresponding admittance signatures are measured. The admittance value decreases as the removal of thread level is high. Figure 2 shows the connection for the proposed damage detection of

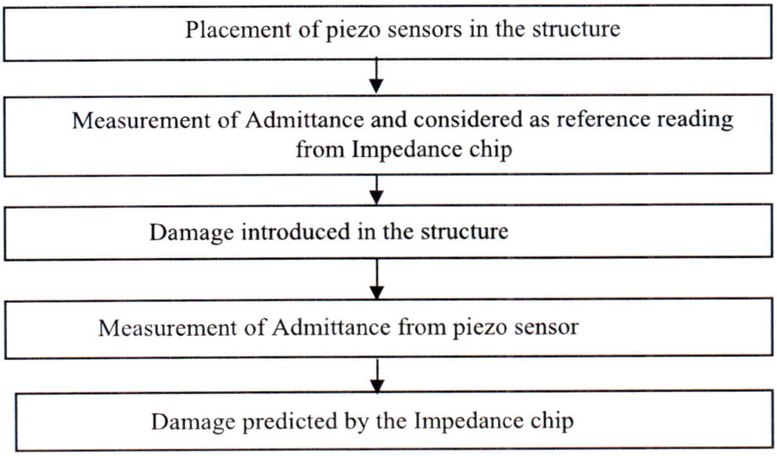

Fig. 1 Flow diagram of damage detection in pipeline

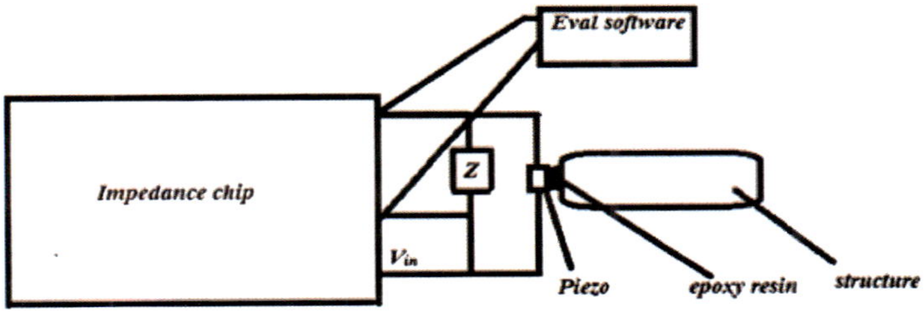

Fig. 2 Circuit connection of admittance measurement for pipeline structure

pipeline structure. The sinusoidal excitation is provided to the piezosensor which is bonded to the structure by impedance chip. The impedance chip is connected to the laptop through USB. Eval software is used for activation of impedance chip. The frequency range is provided from 30 to 80 kHz, and frequency sweep is performed. The output impedance is measured from impedance chip. Figure 3. shows the experimental setup of damage detection of pipeline.

3 Results

The admittance values are measured at each thread level.

Figure 4 shows the comparison curves of admittance with new bolt tighten and removal of bolt at the first thread level. The admittance value is decreased at 50 kHz which shows that there is damage indulged in pipeline.

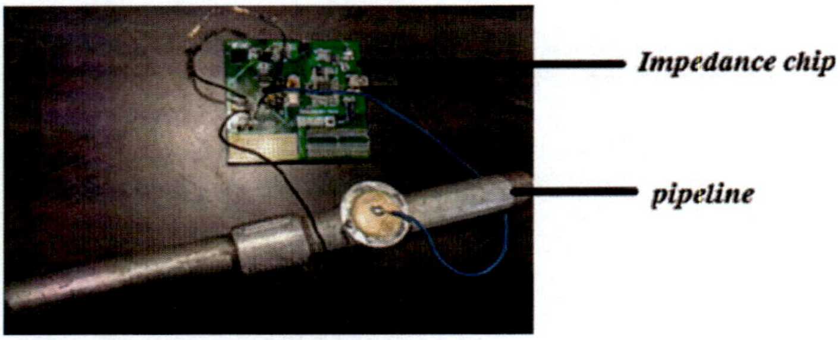

Fig. 3 Experimental setup for damage detection in GI pipeline

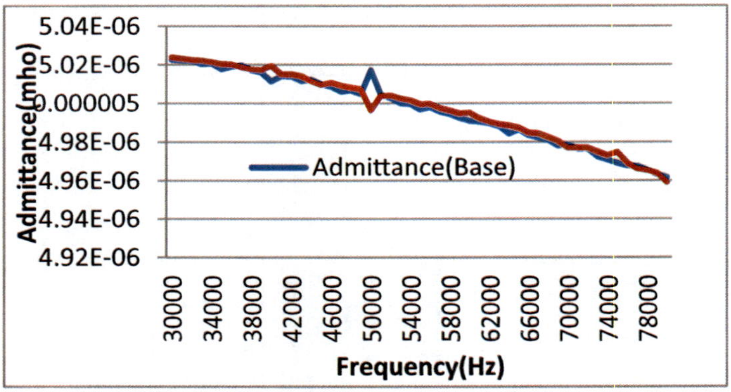

Fig. 4 Admittance signatures of pipeline after the removal of first thread

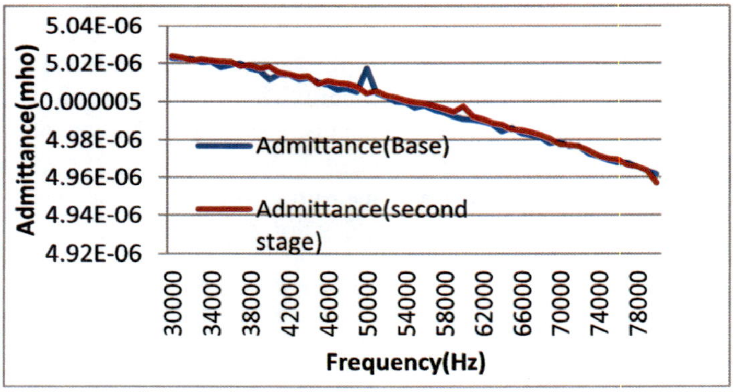

Fig. 5 Admittance signatures of pipeline after the removal of second thread

Figure 5 shows the admittance signatures of base which is fully tighten position and removal of thread at the second stage. The admittance value decreases as the thread level is removal at the second stage.

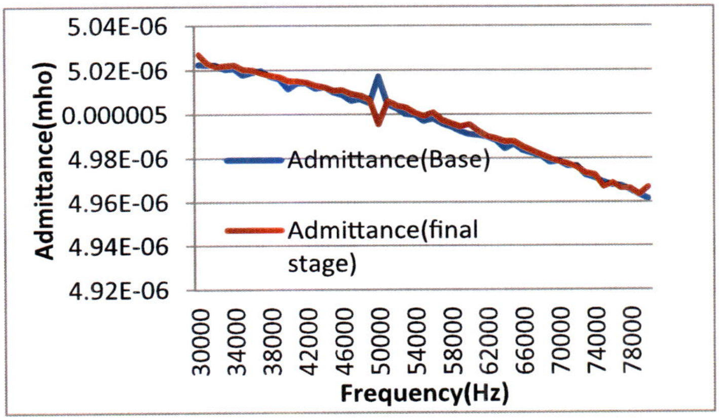

Fig. 6 Admittance signatures of pipeline at final stage

Figure 6 shows the admittance signatures of base which is fully tighten position and removal of thread at the final stage. The admittance value decreases to the maximum as the thread level is removed at the final stage. The detection of damage is clearly identified.

4 Conclusion

This paper clearly indicates that pipeline is damaged due to various reasons like aging, corrosion and overloading. Nowadays, water is more wasted due to leakage of pipes. Detection of damages is very important in the current scenario to avoid major accidents too. Damage can be detected to any kind of pipeline structures made of steel, iron, aluminum, plastic, etc. The cost-effective method to identify the damage in the pipeline is using impedance chip. Impedance chip is suitable for damage detection of smaller structures. Admittance value decreases at a particular frequency which indicates the detection of damage in the pipeline.

References

1. N. Kaur, S. Bhalla, S.C.G. Maddu, Damage and retrofitting monitoring in reinforced concrete structures along with long-term strength and fatigue monitoring using embedded Lead Zirconate Titanate patches. J. Intell. Mater. Syst. Struct. **9**, 100–115 (2019)
2. V.G.M. Annamdas, M.A. Radhika, Electromechanical impedance of piezoelectric transducers for monitoring metallic and non-metallic structures: a review of wired, wireless and energy-harvesting methods. J. Intell. Mater. Syst. Strucut. **24**(9), 1021–1042 (2013)
3. N.E. Cortez, J.V. Filho, F.G. Baptista, Design and implementation of wireless sensor networks for impedance-based structural health monitoring using ZigBee and global system for mobile

communications. J. Intell. Mater. Syst. Struct. **26**(10), 1207–1218 (2015)
4. D. Wang, H. Song, H. Zhu, Numerical and experimental studies on damage detection of a concrete beam based on PZT admittances and correlation coefficient. Constr. Build. Mater. **49**, 564–574 (2013)
5. N. Kaur, S. Bhalla, R. Shanker, R. Panigrahi, Experimental evaluation of miniature impedance chip for structural health monitoring of prototype steel/RC structures, in *Experimental Techniques, Society for Experimental Mechanics*, pp. 1–12 (2015)
6. C.B. Priya, A.L. Reddy, G.V.R. Rao, N. Gopalakrishnan, A.R.M. Rao, Low frequency and boundary condition effects on impedance based damage identification. Case Stud. Nondestr. Test. Eval. **2**(1), 9–13 (2014)
7. W. Quinn, G. Kelly, J. Barett, Development of an embedded wireless sensing system for the monitoring of concrete. Struct. Health Monitor. Int. J. **11**(4), 381–392 (2012)
8. B. Xu, V. Giurgiutiu, A low cost and field portable electromechanical (E/M) ımpedance analyzer for active structural health monitoring, in *Proceedings of the 5th International Workshop on Structural health Monitoring,stanford University, CA* (2005), pp. 15–17
9. D. Wang, H. Song, H. Zhu, Embedded 3D electromechanical impedance model for strength monitoring of concrete using a PZT transducer. Smart Mater. Struct. **23**(11) (2014)
10. S.K. Smantaray, S.K. Mittal, P. Mahapatra, S. Kumar, An impedance-based structural health monitoring approach for looseness identification in bolted joint structure. J. Civil Struct. Health Monit. **8**(5), 809–822 (2018)
11. S. Park, B.L. Grisso, D.J. Inamn, MFC-based structural health monitoring using a miniaturized ımpedance mesuring chip for corrosion detection. Res. Nondestr. Eval. **18**(2), 139–150 (2007)
12. S. Yan, Y. Li, S. Zhang, G. Song, P. Zhao, Pipeline damage detection using uysing piezoceramic transducers: numerical analyses with experimental validation. Sensors (Basel) **18**(7) (2018)
13. F. Deng, L.F. Texeria, Laocating small structural damages in pipes using space-frequency DORT processing. Results Phys. **7**, 1637–1643 (2017)
14. G. Du, Q. Kong, H. Zhou, H. Gu, Multiple crack Detection in pipeline using damage Index matrix method based on piezoelectric transducer–enabled stress wave propagation. Sensors (Basel) **17**(8) (2017)
15. Y. Wang, H. Hao, Damage detection scheme bades on compressive sensing. J. Comput. Civil Eng. **29**(2) (2015)
16. M. Golshan, A. Ghavamian, A.M.A. Adbul Shaheed, Pipeline onitoring sytem by using wireless sensor network. J. Mech. Civil Eng. **13**(3), 43–53 (2016)
17. F. Karray, A.G. Ortiz, M.W. Jaml, A.M. Obeid, M. Abid, Earnpipe: a testbed for smart water pipeline monitoring using wireless sensor network, in *20th International Conference on Knowledge Based and Intelligent Formation and Engineering Systems*, vol. 96, pp. 285–294 (2016)
18. Devices A (1 April) 1 MSPS, 12 Bit Impedance converter Network Analyzer. Available: https://www.analog.com/en/index.html

A Novel Replication Protocol Using Scalable Partition Prediction and Information Estimation Algorithm for Improving DCN Data Availability

B. Thiruvenkatam and M. B. Mukeshkrishnan

Abstract Data centre networks (DCN) is emerging as one of the striving research areas in the data communication and networking domain. The information access across DCN involves various issues due to the unexpected disconnection of networks. This means that the nodes access can be improved when the information presence on neighbouring nodes is aligned accordingly. This paper deals with a novel replication protocol by using an information presence algorithm and a cooperative caching to enhance the data availability across various nodes. The implementation shows that the proposed algorithm outperforms the existing data management algorithms available in DCN.

Keywords DCN · Replication · Information presence · Cooperative caching

1 Introduction

In general, DCN is considered as a collection of information hubs that are associated with both wired/wireless communication mode by utilizing an existing network framework or centralized organization [1]. The remote hubs are free to move arbitrarily and organize themselves in a subjective manner. A connection has been established between the transmitter hub a and collector hub b in case and as it were on the off chance that the control of the radio flag received by hub b is over a certain edge called the connecting edge. In this case, b is said to be a's neighbour. More often than not the joins between hubs would be bi-directional, but there may be cases when it contrasts in transmission control (or recipient affectability) grant rise to unidirectional joins. In case a cannot communicate specifically with v (single-hop), the hubs

B. Thiruvenkatam (✉)
Department of Computer Science and Engineering, SRM Institute of Science and Technology, Chennai, India

M. B. Mukeshkrishnan
Department of Information Technology, SRM Institute of Science and Technology, Chennai, India
e-mail: mukeshkm@srmist.edu.in

S. Shakya et al. (eds.), *Mobile Computing and Sustainable Informatics*, Lecture Notes on Data Engineering and Communications Technologies 68,
https://doi.org/10.1007/978-981-16-1866-6_44

communicate through multi-hop directing, in which one or more middle of the road hubs must act as a transfer between the communicating hubs nodes.

1. *Challenges in High-Speed DCN Environments with Respect Networking Environments*

The following are some of the challenges in high speed data network environment [2]

- Transmission errors
- Frequent disconnections with transmitting nodes
- Dynamically changing topologies
- Security issues
- Routing efficiency and quality of service
- Power efficiencies, energy consumption.

The transmission channel in a DCN is available to both genuine organize clients and noxious assailants. There is no standard security instrument in a DCN from the security plan point of view to address this issue [2]. Getting to farther administrations and information is one of the foremost critical applications in both settled and versatile networks. As a case, a war zone communication framework is utilized in military applications, protect operations, domestic organizing, authorization operations, sharing data in a conference, sensor systems, and so on.

2. *Data Replication in DCN*

Data items in one node are replicated or copied in one or more nodes to improve efficiency by decreasing the query delay and enhancing the data availability. The frequently accessed data or data item in the node that is critical is replicated [3]. So, if the DCN node requests for data item from the node that is far away from it, accessing delay could be reduced by retrieving it from the replica site.

Accessing from replicating sites will take less time when compared to accessing from the node that is actually holding the data. So, the data items that are being requested will be available very faster than normal retrieval. But as the memory of the nodes will be of lesser size, the memory of DCN nodes should be effectively used by replacing the unwanted data item with the most accessed data item.

3. *Network Partitioning in DCN*

The development in remote communication advances pulls in a significant sum of consideration in information systems. Since portable has in an advertisement hoc arrange as a rule move unreservedly, the topology of the arrange changes powerfully and detachment happens regularly [4]. These characteristics make it likely for an information arrangement that has to be isolated into a few disengaged allotments and then the information availability is subsequently diminished. A few plans are proposed to lighten the diminishment of information openness by reproducing information things. When the network connectivity among the nodes in the network is low, network partitioning occurs. The node disconnectivity may be due to the frequent topology changes in the ad hoc networks or due to the quick drain of battery power of the DCN nodes. This leads to serious consequences in the DCN environments.

Network partitioning will significantly diminish the information accessibility when the server that holds the information is not within the same parcel as the client hubs. By predicting the network partition time and replicating the data accordingly will improve the data availability [5].

The following are some of the factors that can be used to determine the network partitioning time in DCN [6].

(a) Link stability
(b) Critical node detection
(c) Mobility pattern

4. *Scalability in DCN*

Data networks include over tens of thousands of hubs, for illustration, in sensor systems, strategic systems. As the arrange measure increments, an inquiry sent by a client is ought to navigate a long way to reach the server hub, in this manner the inquiry is fetched and expanded the inactivity [7]. This may moreover lead to channel get to dispute which diminishes accessible transfer speed and increments channel get to delay. Ad hoc systems empower the unconstrained network among remote gadgets without requiring any specific foundation. At display, the expanding request has moved towards the applications like logical investigation, online gaming with numerous players, law requirement, individual communication, etc. In all these applications, bunch communication is more imperative. Multicasting gives fundamental administration for bunch communication in such sort of applications. With the expanding number of hubs, the adaptability issue in the multicasting of DCN is still remaining as an open problem.

The performance of DCN depends on how far it is scalable for larger networks. Scalability can be interpreted in two ways: asymptotic adaptability and in-practice adaptability. Asymptotic adaptability is the constrain available within the arrangement of development as a work of estimate and in-practice versatility is the number of hubs past with which the arrange will not be enough to work. A convention is said to be versatile on the off chance that it has the capacity to fulfil the requests in terms of arrange estimate without any extra overhead.

2 Problem Definition

The performance of DCN largely depends on the availability of data item and the access latency. Data replication algorithms will serve this purpose by making the data items cached in one or more DCN nodes by considering the different parameters [8]. The frequent topology changes lead to network partitioning in DCN environments. This will reduce data availability when the server that holds the desired data is not available in the same partition. Network partitioning may cause a DCN node to experience a set of neighbours according to the node's transmission range.

When the requesting data item is not within its communication range (data in availability), query delay and latency increases, which may affect the performance of DCN and cause serious consequences in the case of DCN applications like military and battle fields. Protocols are proposed for predicting the network partitioning time and possibly replicating the suitable data item based on above-mentioned parameters. The cache replication protocols so far proposed will either deal with network partitioning or scalability but not on both. Development of cache replication protocols will consider both the issues described above and it would be an alluring point to resolve the networking issues. This research work is primarily centred on creating a versatile segment forecast calculation to realize both the information accessibility and versatility.

3 Impact and Study

COACS [9] is not aware of what data items are actually cached, when the data item is replaced or when the node is disconnected from the network. If the server data update rate is greater than node request rate, the network traffic prevails and packet dropout rate increases with longer delays.

A server-based approach maintains a strategic distance from issues with a thrust-based cache consistency approach. It diminishes the remote activity by tuning the cache overhaul rate to the ask rate for cached information and maintains the cache consistency. In SSUM [10], the requesting node sends the request to its nearest query directory, which advances the ask rate for comparing the caching hub that returns the reaction to the RN. If not, it receives the response from the server if the query is not in any of the QD and now the RN becomes the CN and sends the server cache update packet (SCUP) to the server.

In group mobility and partition prediction in remote arrange, to foresee segment, gather-based developments of the DCN hubs are distinguished and characterized. Upgraded characterization of portability bunches based on models is utilized for the location of organized dividing. A straightforward information clustering calculation is utilized such that it recognizes the versatility of the bunch. Speed-based versatility demonstrate is proposed, long-term apportioning can be anticipated by the versatility parameters. In effective dispersed arrange wide broadcast calculation for tall speed systems, a strategy for organize wide broadcast (NWB) in DCN systems is proposed.

The weighted linear prediction is proposed with an efficient group-based partition prediction scheme for communication networks. It is based on the displacement of group rather than the velocity. Safe distance is used to predict the partition between adjacent groups. The clustering model uses distributed group mobility adaptive (DGMA) algorithm [11].

Some researchers have overviewed the existing information replication conventions and classified them into network ignorant and network mindful. Whereas most of the current arrangements are known to be either partition aware or versatile, there has no impressive exertion within the investigate of replication conventions that address all the DCN [12] issues, i.e. none of them is energy aware, parcel mindful as well as adaptable.

Data replication protocols are those that concentrate on data availability and performance of DCN and not on network partitioning, energy consumption and scalability.

Ad hoc aware data replication protocols concentrate on at least one of the following issues.

1. Network partitioning
2. Energy consumption
3. Scalability.

Selective replication for content management environments [13] presents an arrangement that addresses and coordinates the particular database and record replication within the setting of an undertaking voice entry. It is straightforward to existing applications with a negligible capacity overhead. Highly localized substance will not be reproduced. They utilize database tables to store the metadata. This does not consider the parcel arrangement and versatility issues. Caching in substance administration situations will altogether progress the proficiency of the data present in remote situations by decreasing the get to inactivity and utilization of transfer speed.

The consistency management strategies for data replication in DCN solves the issues concerning the consistency management of data operations on the replica. It attempts to classify consistency levels according to the requirements from the applications and provides protocols. The previous replication strategies focused only on service availability.

Chen et al. [14] use data lookup service. Each hub uses location data given by GPS and those gotten by other hubs of the gather to calculate development of each part within the gather to foresee segment but does not consider scalability issues. Some researchers consider replication procedure when arrange is divided. Three-copy assignment method assuming information item is not upgraded and after that amplified the framework where an occasional upgrades happen. Agreeable caching moves forward framework execution since it allows sharing and coordination of cached information among numerous DCN users within the arrange. By agreeably caching regularly gotten to data, DCN gadgets do not continuously got to send demands to the information source [15].

Clusters are formed by the election algorithm. Each cluster has a replica manager that holds the information about the replicas. The cluster head stores the replica information in Location Replica Cache (LRC). There is a replica keeper for every object. And the replica location information is stored in Global Replica Cache (GRC) [16].

In that, the clusters are composed utilizing steady joins. A way is steady on the off chance that the item of network likelihood of the joins that compose the way is additionally than a given esteem. Untrustworthy communication and the user's mobility make it troublesome to preserve cache consistency. It employs RPGM to recognize versatility bunches. The data convention is executed at each migration period. K-NN localized calculation upholds localized crossover information accessibility plot that combines thrust and drag-based administrations. The concept of replication affects the framework vitality consumption.

The major objective of cache replication is to improve the response time by reducing the query delay rate. Data replication concerns with generating data objects one time and copying it to multiple nodes in the DCN cluster. Network partitioning reduces data accessibility and increases query response time. Query cost and latency increases as the network size increases (scalability). The novel replication protocols so far proposed either deals with network partitioning or scalability but not on both. Replicating data in cluster nodes before the partition occurs would reasonably improve data availability and decrease the query delay time. The data replication algorithm must be also scalable to larger networks.

4 System Design for High Availability DCN

The design consists of the phases namely data replication, partition prediction, and scalability. The user requests the data from the cluster heads which maintains the information about the nodes in its cluster. Network partition is predicted using prediction algorithm. Once the partition is predicted the data replication algorithm is applied to the network. The algorithm is then extended to a scalable network. When the data item is modified, the changes in the hosts are noticed by the cluster head and reflected in appropriate replicating sites (Fig. 1).

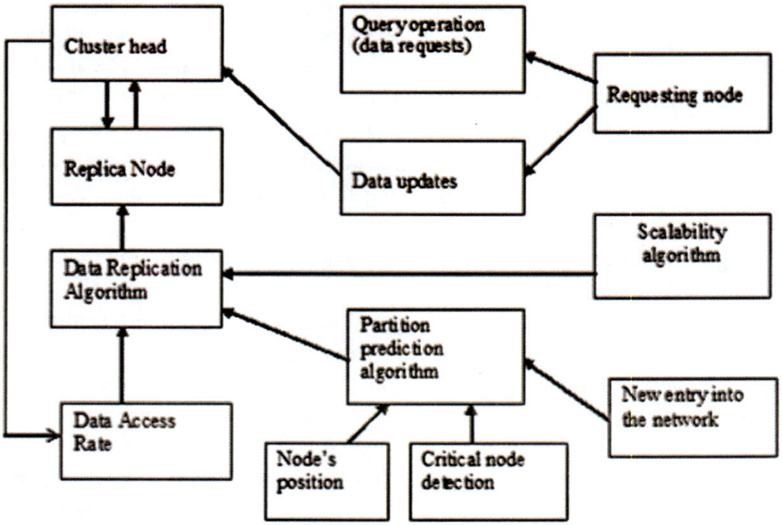

Fig. 1 Data flow architecture diagram

Performance Metrics

Replication cost—The number of replica servers is increased in the networks, higher is the cost incurred. Whenever network partition is predicted data items are chosen to replicate in the appropriate replica site which is based on the remaining battery power.

Update cost—If a copy of a data item residing in multiple nodes, the nodes need to perform larger number of hops for the update operation. Data nodes may update the temporary data values frequently which must be updated at regular interval not only in the actual data node that holds the data item but also in the replica node in order to maintain consistency.

Data availability is the probability value for accessing a data item when it is queried by the requesting node. Data availability can be ensured by making the data accessible and survivable. Predicting the network disconnection and replicating the data in the replica nodes beforehand ensures data accessibility. Replicating the data among the replicas with higher remaining energy would ensure data survivability.

Data consistency is accessing the most recently updated value of the data item. As many of the data are replicated in nodes that are accessing the data from the replica nodes and may sometimes cause data inconsistency if consistency strategies are used. In designing a data replication protocol, optimizing all the above metrics is hard to achieve. One must be optimized at the expense of others.

The proposed protocol aims at optimizing the following metrics:

Data availability at the expense of replication cost. Creating large number of replicas improves data availability greatly, but at the same time, the replication cost will be higher since each replica node receives changes in order to maintain up-to-date information.

Query cost at the expense of replication cost and update cost. When there are a large number of replica nodes updating data in each of these nodes incurs high update cost. On the other hand, the requesting nodes may query the data item from nearby servers which incurs lower query cost.

1. *Prediction Factors*

The existing proposals that detect that any network partitioning/switch-off include solutions like centralized servers, worldwide organize innovation at each hub, flooding, etc. The parameters that are considered here are

2. *Regular Updates*

Each hub is broadcasting customary upgrade of its area and speed to all quick neighbours. Recurrence of customary overhauls is parameter of the organize. Upgrades ought to be more frequent if the proportion of most extreme conceivable speed of hubs within the organize and their most extreme transmission run is huge and can be less visit in case it is moo. Subsequently, the recurrence of customary upgrades ought

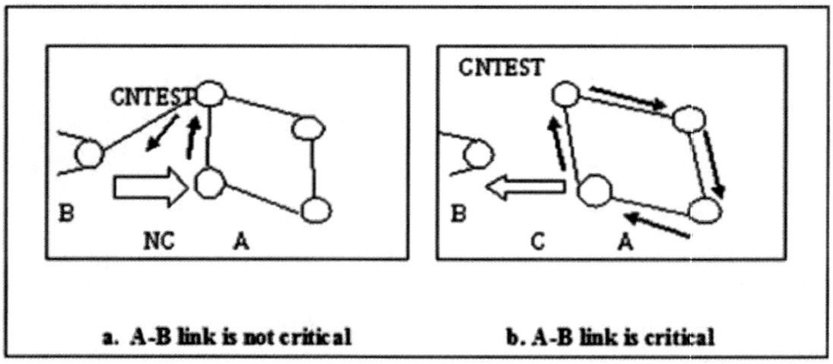

Fig. 2 Critical link test

to be the compromise between transfer speed utilization and required precision of information. The upgrade in which a case that essentially changes speed vector (v) or transmission run $r(t)$, it will broadcast locally the data. For each neighbour, the following information are stored: Transmission range $r(t)$, Node id, Location(x, y), and Current speed (v).

5 Link Sustainability and Partition Prediction

Critically Safe and Unsafe Links

The link breakage is calculated between any two nodes. If any two nodes come to know that the link between them is about to break, then the algorithm is initiated to know whether the link failure leads to network partitioning or not. In order to determine the link failure between the nodes, a node is taken to initiate, generally, the node with lowest node id is chosen (delegated node).

Figure 2 depicts the critical link test algorithm which predicts if there is connection between two data nodes even if a particular link between the two fails. Once it is decided, the CNTEST message will be sent by the delegated node to its neighbouring node. If CNTEST message reaches the destination node, if the delegated node receives not critical (NC) from the destination node, then it is considered that the link between the two nodes is stable, i.e. the network is still connected. If the delegated node receives critical (C) from the destination node, then it is considered that the link between the two nodes is critical, i.e. the network is partitioned.

Partition Detection-Path Break-Network Still Connected

Any node among the DCN network is chosen after certain relocation and made to deliver the test packet to another target node. If there exists path to the destination node the test packet reaches the destination node. If the shortest path between the sender and the receiver fails or if any intermediate node moves out of transmission range, it takes alternative path if it exists.

Fig. 3 Partition prediction

```
CLP ( )
 Begin
  If agent ( )
    Invoke CNTEST
  If CNTEST = destination node
    Stable Link
  Else
    Unstable link
 End
```

1. *Path Break-Network Partition*

Predicting the occurrence of network partition and replicating the data items (Fig. 3).

An intermediate node is assumed to be out of range and the test packet is made to send to a destination node. If the packet reaches the destination node directly or through alternative path, then the path is not critical. If the test packet reaches the source node in return, then based on the calculation of node's position and energy decision to replication is made in the appropriate node. Partitioning causes genuine issues within the arrange. Hubs from diverse allotments cannot communicate with each other, decreasing the quality of all administrations the networks, or making administrations inaccessible. In case, a density of hubs within the watched region is moo, it is conceivable that organize will never recapture its full network. The most exceedingly bad case is that a organize can go through a arrangement of dividing finishing up as an detached bunch of one-node subnets.

Prediction of Node's Position and Energy

Routing protocol is made as Border Gateway Protocol (BGP) and the energy value is set to 1000. In the node configuration set, the energy model, initial energy value, power spent in receiving mode, transmit mode, idle mode, and sleep mode. The link stability has to be considered, for making the decision of allocating replica to the appropriate server, disengagement time between two servers can be calculated, and the time in which they would be detached and evaluated. If the time of two servers is higher than their area time period, these two servers can share information dependably until another migration time period. Next thing to be considered is the remaining battery control. Hubs might not be able to communicate with each other in the event that they are out of control. Whenever the network disconnection is predicted, then the replication units are considered for replicating the data item in the appropriate replica sites.

Replication and System Assumptions

Information is dealt with as an information thing, which may be a collection of information. We dole out a one of a kind data identifier to each information thing found within the framework. The set of all data items is indicated by $D = \{D1, D2,$

…, $Dn\}$, where n is the whole number of information things and Dj $(1 \leq j \leq n)$ may be an information identifier. All information things are of the same measure and each information thing is held by a specific DCN have as the initial.

A unique host identifier to each DCN have within the framework is allotted. The set of all DCN has within the framework is indicated by $N = \{N1, N2, \ldots, Nk\}$, where k is the entire number of DCN has and $Nj(1 \leq j \leq k)$ may be a host identifier. Each DCN have, Ni, features a memory space of Ci data things for making copies barring the space for the initial information things that they have holds. The get to frequencies to information things from each hub are known and do not alter. For the reason of straightforwardness, all information things are accepted to be of the same measure.

Data items can be classified into

- Read only
- Read write data items.

The read write data items can be of either transitory or tireless. The worldly information things are those which are substantial as it were certain period of time and may alter as often as possible. For illustration in military applications, the area of trooper and the area of adversary are the worldly information things as they move as often as possible in war zone. The diligent information things are those which are substantial all through their presence in database. Area of restorative camps and list of security officers are the illustrations of determined information things.

Round-Trip Time

Once the partitioning time is predicted, based on the round-trip time the node, which has to be replicated is detected and replicated accordingly. The node with highest round-trip time is chosen for replication. Similarly, the querying node retrieves the response from the one with least round-trip time.

Relocation Interval and Access Rate Calculation

The replication decision at the appropriate nodes and the appropriate data items are obtained during certain time period and that the duration is relocation interval. The ratio between the remaining valid time interval and the relocation period is called the revised access rate. The remaining valid time interval is the remaining time up to which the data item will survive in that particular node. A data item that has the higher age relocation ratio is replicated before the one with a lower relocation ratio.

Absolute time interval is the absolute validity interval for the data item. Current timestamp value in the replica node is the time when the current value of the temporary data item d is obtained in replica node. For the temporary data item, the valid time period is calculated based on the current timestamp and the absolute time interval. Based on this valid time period, revised access rate is calculated.

Algorithm for Elimination of Replica Duplication

```
Begin
 If (D → temporary D)
  Valid Time period= Current timestamp
+Absolute time interval
  If (valid Time period >= relocation Interval)
   Revised access rate=1
  Else
   Revised access rate= valid time period/
relocation interval
   Access rate=access rate *revised access rate
 For each Node n_i
 For each Node n_j adjacent to n_i
 For all data items d_i which are in both nodes n_i,n_j
  If (access rate of D in n_i> access rate if D in n_j)
  Replace D in n_j with next highest access rate D
 Else
  Replace D in n_i with next highest access rate D
End
```

In order to avoid duplication, access frequency of data in a node is compared with its neighbouring node and the data that is in both the nodes are noted. The node that has least access frequency to that data item is copied with some other data with higher frequency.

Data Replication

Calculate the access frequencies of nodes to data items and arrange them in descending order of their frequencies. Store the data items with higher access frequencies in the nodes to fill its memory size. Every node broadcasts a message with its ID and the data ID that it holds to their neighbouring nodes. If the neighbouring nodes hold the replica of the same data item, the access rates of the nodes to that particular data item are compared and the one which has highest access frequency to the data is retained with that data.

An example of executing replication algorithm

Assume the following access rates for each of the DCN nodes. During a relocation interval, each DCN host preliminarily determines the allocation of replicas based on the above algorithm. Then, each node broadcasts its host identifier and information on access rates to data items in the entire network. After all data hosts complete the broadcasts, from the received host identifiers, every host will know about its connected DCN hosts. The network may be partitioned into several sets of connected DCN hosts.

Figure 4 clearly depicts the procedure of executing replication algorithm. When there is duplication of a data item either original or replica between the two neighbouring data hosts (the cluster head and its neighbour), and if one of them is the original, the host which holds the replica changes it to another replica. If both of them are replicas, the host whose access frequency to the data item is lower than the other one changes the replica to another replica.

Procedure for Eliminating Replica

Figure 5 depicts eliminating replica duplication by comparing its adjacent nodes. When data from one node is replicated in other nodes there may be a chance of

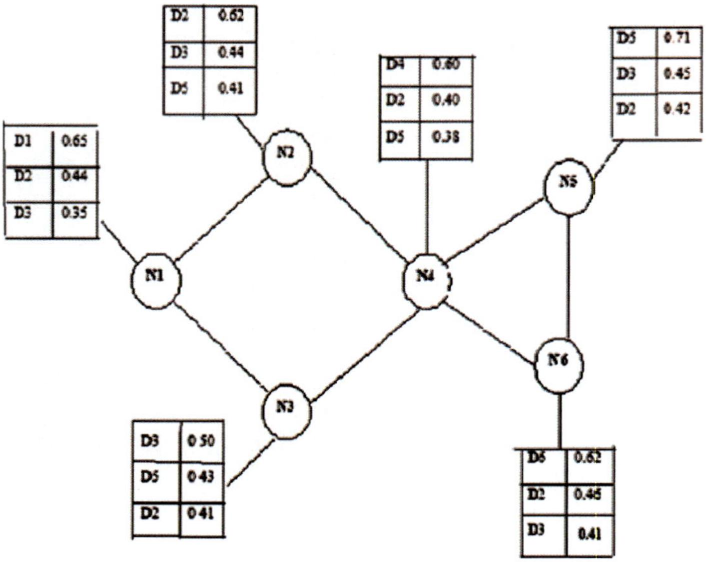

Fig. 4 Replication based on access frequencies

Fig. 5 Eliminating replication duplication

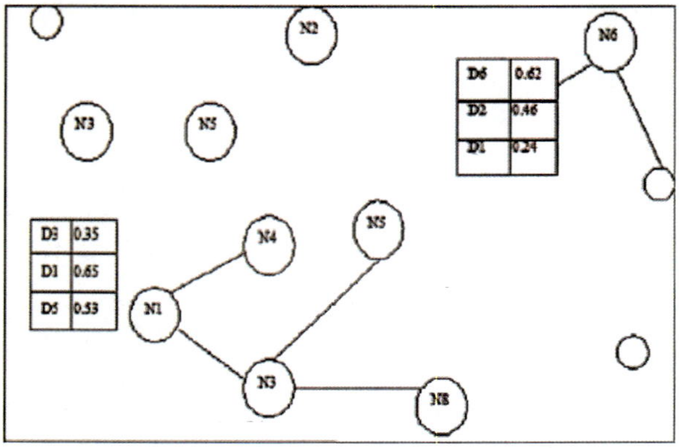

duplication of data item within the adjacent nodes. In order to avoid that data in one is compared with the neighbouring nodes and if two adjacent nodes have the copy of same data, the one with highest frequency is retained and the data in other node is replaced with some other data item by following the same procedure.

6 Assumptions and Algorithms in Using Replication to Improve High Data Availability DCN

The system consists of a set of *n* DCN nodes. The database of interest, denoted by *DB*, is either centralized or distributed. In the centralized database systems, a single node called server holds the whole database. The database in this case may also be a set of configuration items used by a given service. In the distributed database systems, each node *i* has a data item *Di* associated with it (i.e. *Di* is only updated by node *i*), and $DB = D1, \ldots, Dn$. A replicated object (i.e. *replica*) is a data item that is stored redundantly at multiple DCN nodes, called *replica servers*. We define data replication as a technique of creating and managing duplicate versions of data items. Each node can perform two types of operations: *update* and *query*. When a data item *Di* is updated, node *I* sends an *update* message to one or multiple replica servers, to modify the value of its replicas. When a node wants to query the value of a given data item, it sends a *query* message to one or multiple replica servers that hold a copy of this data.

Finding the Relevant Node

The node from where the data is to be replicated should be calculated first as shown in Fig. 6. The node can be identified by the technique of information presence system. The node should be decided based upon the resource constraint of it. If it is low in the case of communication or power, then it can be replicated. Here, the relevant node is found by using appropriate transit and relay counters and updating them to find the information presence so as to find the relevant DCN nodes at a distance to find where the replication should be done. The information of this relevant node will be

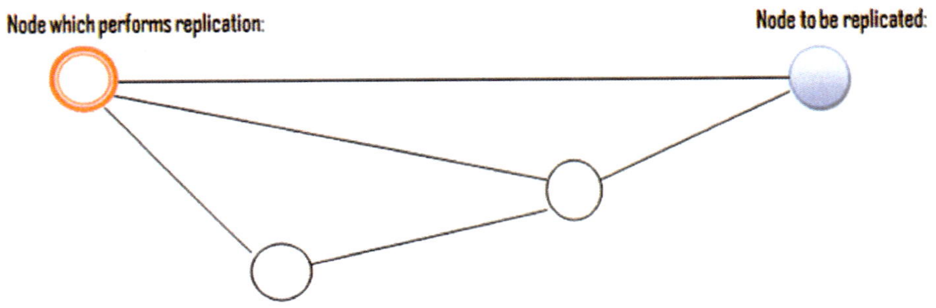

Fig. 6 Finding the relevant node

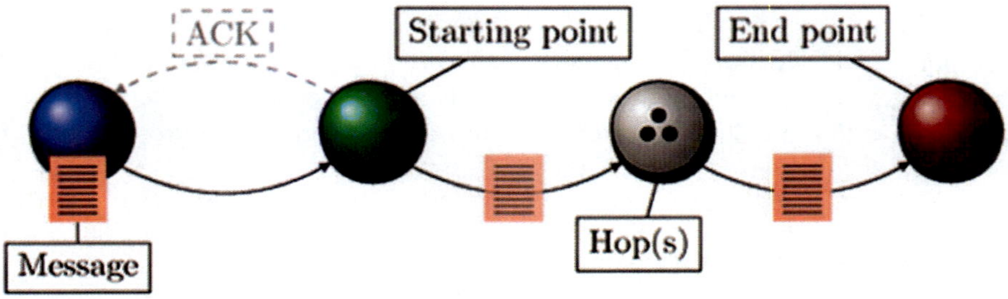

Fig. 7 Hop count estimation

thus used to do the actual replication so that there are more cases of data availability and more query hit ratio. Any queries served thereafter will have less average latency time.

Using Hop Count Information

Hop count calculation is done to find the farther node in the network as shown in Fig. 7. We can calculate the distance of every node in the network by using this technique. We locate the DCN nodes by assigning them relevant IDs and information is passed every node. Based upon the number of DCN nodes passed, a count is maintained and it is stored in the node. The hop count information thus computed could be used to select which node to replicate and what information to replicate so that the average query hit ratio is increased.

The data available in the neighbour DCN nodes are replicated and stored in node. Initially, the node is chosen based upon the valid parameters and then the data from the node is copied or replicated as shown in Fig. 8.

7 Information Presence Estimation

Queries between DCN nodes can be generated and statistics about information presence could be collected. After such information is collected, an appropriate replication protocol can be used. The information presence is done using provider and transit counters. The hop count information is collected using information replies got from various DCN nodes and correspondingly provider and transit counters are updated to reflect the information presence index. Thus, information presence estimation is done among the DCN node as shown in Fig. 9.

Choosing an Appropriate Replication Protocol

The information statistics collected is used to choose an appropriate replication scheme. This replication scheme should be energy aware, partition aware, and scalable. This could be achieved by doing replication such that no two similar data are cached in close proximity. Dissimilar data could be stored in close proximity so

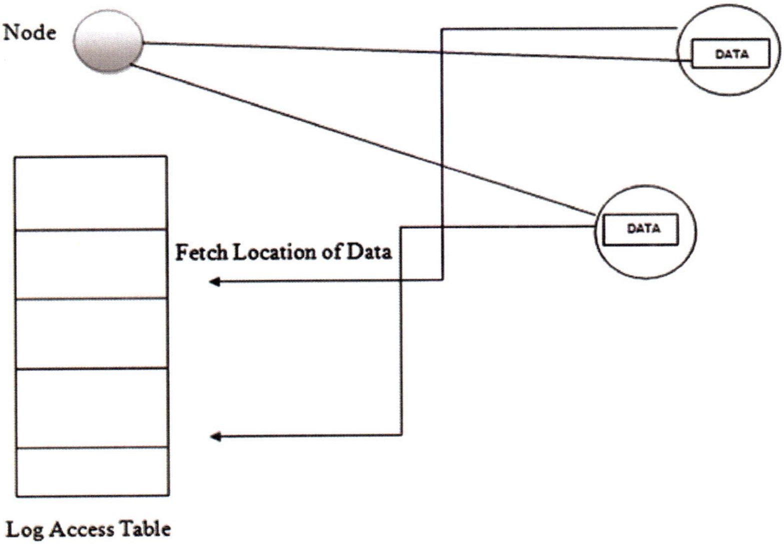

Fig. 8 Replicating data

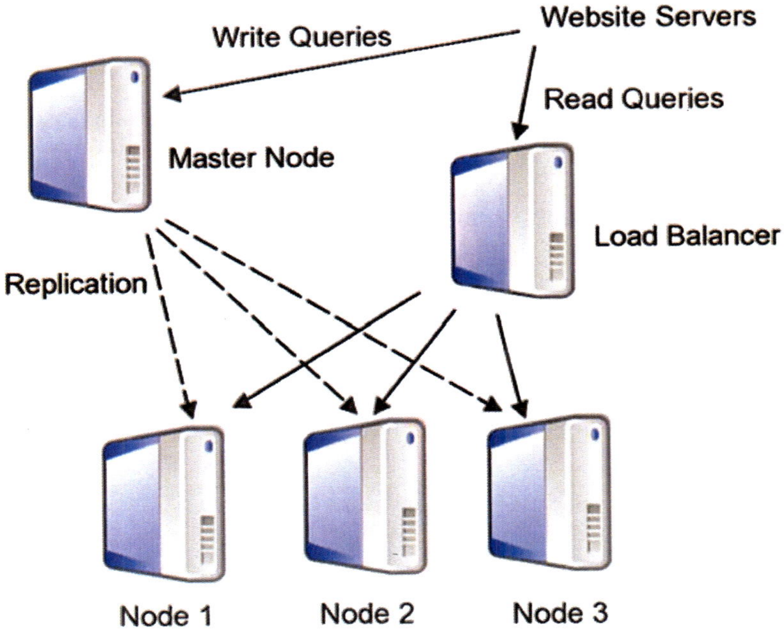

Fig. 9 Replication using information presence

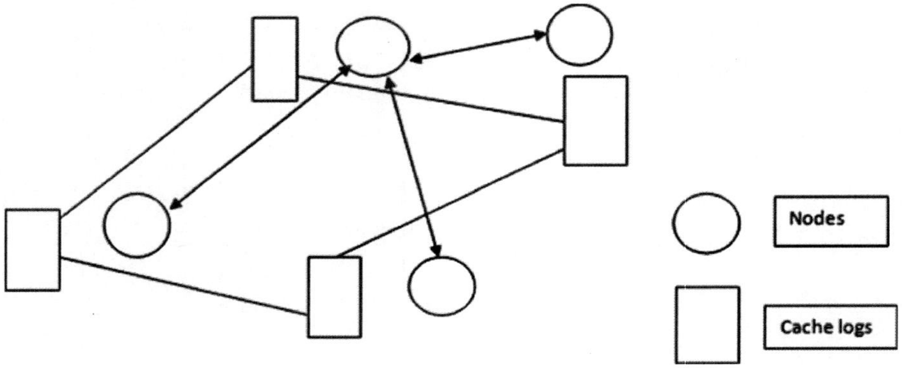

Fig. 10 Cooperative caching

that cache hits could be maximum, when queried among neighbouring DCN nodes. Thus, choosing a replication scheme that produces better cache hits and remaining aware of partitions have gained prime importance. Thus, if replication is done even before caching is done due to queries from DCN nodes, there is a better case of data availability and delivers good performance without traffic and guarantees query optimization.

Incorporating Cooperative Caching When Responding to Queries
In order that caching is done more efficiently, cooperative caching should be included so that DCN nodes collaboratively do caching to achieve a better performance. The caching techniques, Cache Path, Cache Data, and Hybrid Cache try to overcome the resource constraints and further improve performance by caching the data locally or caching the path to the data to save space. Such caching strategies as shown in Fig. 10 could also be included to reduce query delays and improve data accessibility.

Maintaining Consistency
Cache invalidation methods that reduce the number of accesses to old replicas whose original has been updated can be used. They are the update broadcast and connection rebroadcast methods. In the update broadcast and connection rebroadcast methods, a host holding an original data item broadcasts an invalidation report to connected DCN hosts every time it updates the data item. The invalidation report includes a data identifier and update time (time stamp). Such consistency schemes can be incorporated. A log-based consistency can be chosen where a data host records all the operations that it performs into a log with the order of the operations performed. This log of the operations performed is then sent to the other DCN hosts so that the hosts update their own copy of the replica. However, traffic would be more here. Choosing a cache invalidation method, log-based or state-based updations are of prime importance. Two cache invalidation methods were proposed that reduce the number of accesses to old replicas whose original has been updated. We call them the update broadcast and connection rebroadcast methods, respectively. In the update

broadcast and connection rebroadcast methods, a DCN host holding an original data item broadcasts an invalidation report to connected DCN hosts every time it updates the data item. The invalidation report includes a data identifier and update time (time stamp).

Information Estimation and Replication Algorithm

1. Create a random query for request of data items
2. di where *i*=0,1,2,3…… // di- data item
3. Get Information replies from the DCN nodes
4. Get hop counts from header of information messages
5. Increment transit and relay counters
6. transit++; // Ci- chunks of the data item, transit counter=0, relay counter =0;
7. relay++;
8. until value = hop count value;
9. Create groups G1,G2,G3,G4…. Of Gn DCN nodes each
10 Choose a node in each group
11. where, hopcount = high;
12. Calculate the distance:
13. Distance = hopcount/2
14. Choose the node at a distance "distance" from the previous node
15. Copy the data item di in the local cache of that node
16. repeat

Queries between DCN nodes can be generated and statistics about information presence could be collected. After such information is collected, an appropriate replication protocol can be used. The information presence is done using provider and transit counters. The hop count information is collected using information replies got from various nodes and correspondingly provider and transit counters are updated to reflect the information presence index. Thus, information presence estimation is done among the DCN node.

The information statistics collected is used to choose an appropriate replication scheme. This replication scheme should be energy aware, partition aware, and scalable. This could be achieved by doing replication such that no two similar data are cached in close proximity. Dissimilar data could be stored in close proximity so that cache hits could be at a maximum level when queried among the neighbouring DCN nodes.

Thus, choosing a replication scheme that produces better cache hits and aware of partitions has given prime importance. Thus, if replication is done even before caching is done due to queries from DCN nodes, there is a better case for data availability and good performance without traffic and also guarantees the query optimization.

The query processing with replication and caching is shown in Fig. 11.

8 Cooperative Caching Algorithm

When a data item *di* arrives:

Check if *di* = requested data by the current node

then

cache *di*

else /* Data passing by */

Check if copy of *di* = in the cache

update the cached copy if necessary

else if *si* < *Ts*

then

cache data item *di* and the path;

else if there is a cached path for *di*,

then

cache data item *di*;

update the path for *di*;

else

cache the path of *di*;

When a request for data item *di* arrives:

Check if there is a valid copy in the cache

then

send *di* to the requester;

else if there is a valid path for *di* in the cache

then

forward the request to the caching node

else

forward the request to the data center.

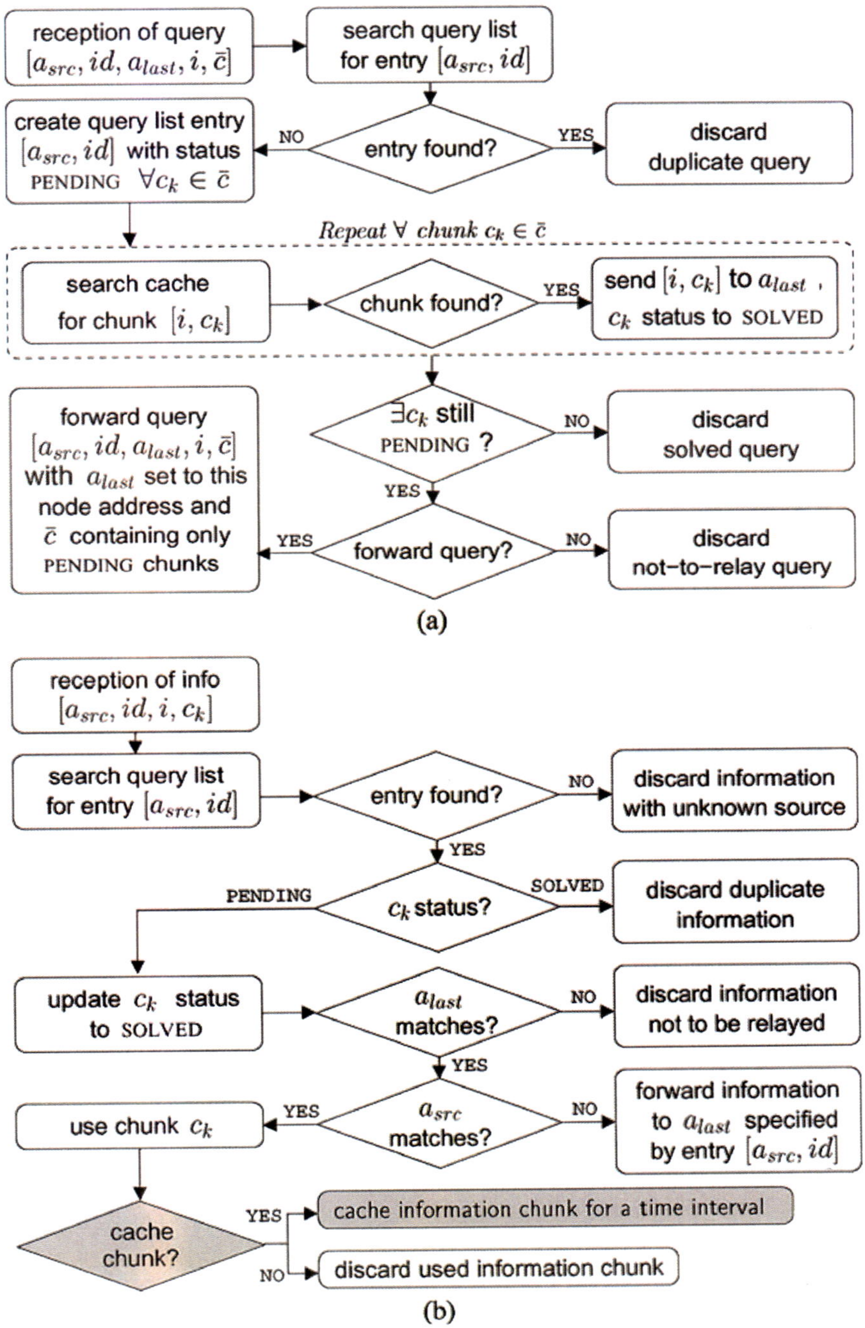

Fig. 11 Query processing with replication and caching

Replication decision is made based on the position of node and the replica site is identified based on node's energy. Node's address is retrieved with the help of its index through get_node_by_address and its position is frequently with the help of update position and its dimension in x is retrieved through xpos and ypos direction through ypos. Based on this, xpos and ypos time interval to replicate is decided. A threshold is set and when it exceeds, replica site in which the data has to be replicated and the data item to be replicated is chosen.

Ping agents are created and attached with respective nodes. 'recv' function is defined in the class agent/ping. The round-trip time calculated in recv function in order of milliseconds. The IP header and ping header are accessed for the received packet. A new packet is created and ping header is accessed for that packet. The 'ret' and send_time field are set to correct value. The round-trip time is calculated using address instance, source address, scheduler instance, and clock time.

Packets are made to forward packets among them. Data item in two nodes are compared and those data item that are in node $N1$ and not in the other node are found and replicated until its capacity to store the data item is full. Once data in one node is copied into the other after predicting network partition, the replica node is then compared with the neighbouring nodes to check if there is any duplication.

In order to improve scalability, setdest is used to generate a network topology with large number of nodes and at certain period of time a node is added into the network. It is then made to broadcast its ID to all the other data nodes within its transmission range. The node that receives this broadcast message acknowledges with its own ID. In this way, a newly added data node maintains the neighbouring table information and this node is also added in other's neighbouring table. Data access frequencies are also calculated based on how frequently it accesses data items.

9 Result Analysis

Group Information and Average Buffer Access

Figure 12 clearly states that the proposed protocol significantly improves group information access. This has been calculated using the presence of k-means algorithm included pro-ACK control proposed in Chapter 3. The eventual number of members is also constantly maintained across clusters. This can be witness from Fig. 12 the centroids and the relevant group members.

Figure 13 explains that the average buffer size is also utilized distributed across all the groups after introducing the cooperative caching in different DCN nodes. This also significantly improves the information access across all the clusters. The eminent improvement is utilizing the information access at optimal time t.

It is profound from Fig. 13 that the measured average time to access all the data buffers for the clusters has a close proximity among the groups, when there is a explosion in data access the proposed cooperative caching supports to utilize the

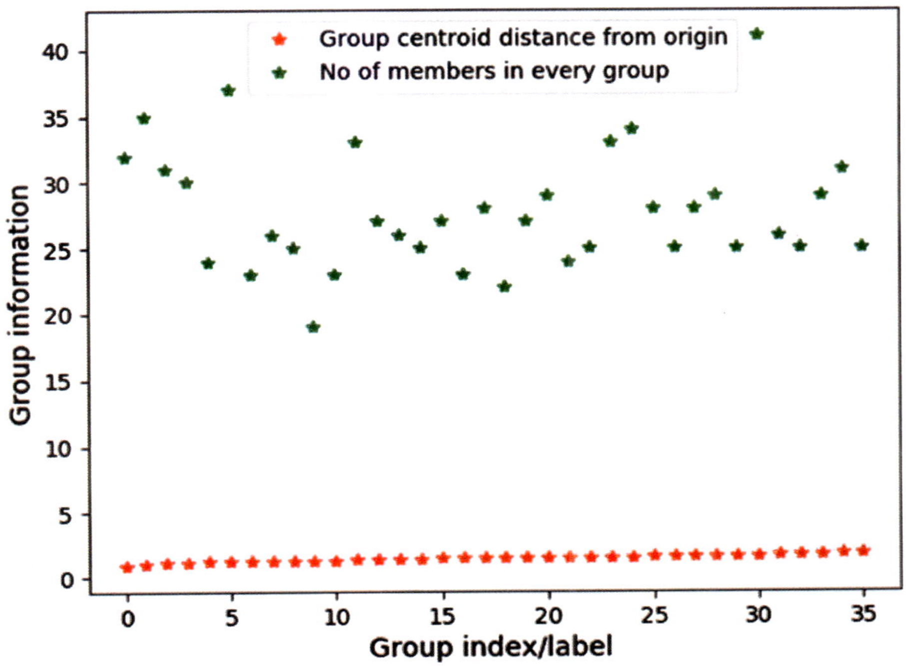

Fig. 12 Increase in group information access

same average time across all the groups. The scale between 0.0 and 0.3 in time reflects the peak information access utilization.

Group Priority

The proposed algorithm also witnesses the clusters access information based on the priority and this leads to organize and develop collision free information across the clusters. From Fig. 14, the involvement of replication and information estimation-based nearest neighbour caching improves various information access to the users.

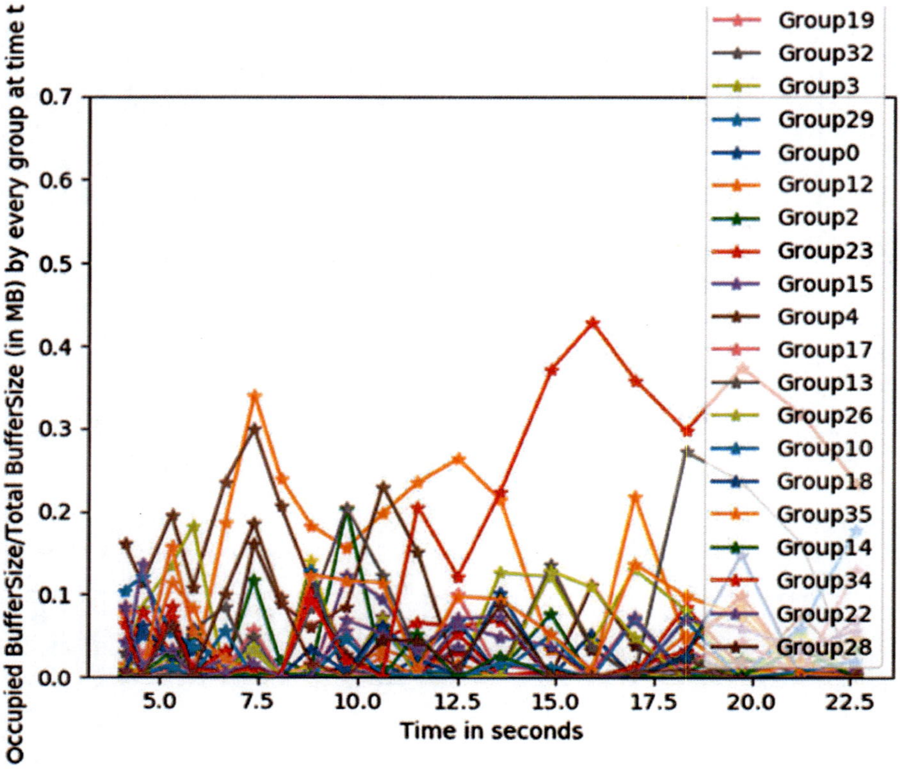

Fig. 13 Increase in information access after cooperative caching

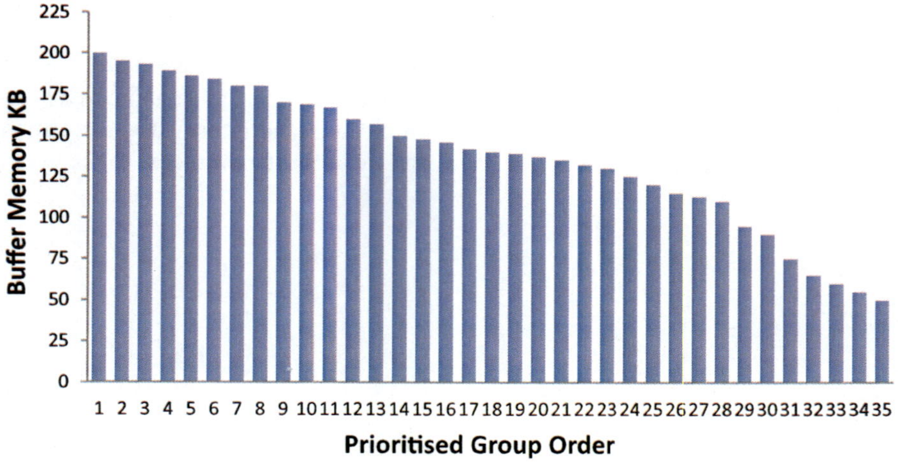

Fig. 14 Information presence and group segregation analysis

References

1. A.L. Edmonds, Deformation of maps to branched coverings in dimension two. Ann. Math. **110**(1), 113–125 (1979)
2. M.A. Sharkh, M. Jammal, A. Shami, A. Ouda, Resource allocation in a network-based cloud computing environment: design challenges. IEEE Commun. Mag. **51**(11), 46–52 (2013)
3. F.E. Abii, D.C. Ogula, J.M. Rose, Effects of individual and organizational factors on the turnover intentions of information technology professionals. Int. J. Manage. **30**(2), 740 (2013)
4. Q. Lv, P. Cao, E. Cohen, K. Li, S. Shenker, Search and replication in unstructured peer-to-peer networks, in *Proceedings of the 16th International Conference on Supercomputing* (2002), pp. 84–95
5. W.J. Ke, S.D. Wang, Reliability evaluation for distributed computing networks with imperfect nodes. IEEE Trans. Reliab. **46**(3), 342–349 (1997)
6. T. Hara, S.K. Madria, Data replication for improving data accessibility in ad hoc networks. IEEE Trans. Mob. Comput. **5**(11), 1515–1532 (2006)
7. W. Xia, P. Zhao, Y. Wen, H. Xie, A survey on data center networking (DCN): Infrastructure and operations. IEEE Commun. Surv. Tutor. **19**(1), 640–656 (2016)
8. J. Perelló, S. Spadaro, S. Ricciardi, D. Careglio, S. Peng, R. Nejabati, G. Zervas, D. Simeonidou, A. Predieri, M. Biancani, H.J. Dorren, All-optical packet/circuit switching-based data center network for enhanced scalability, latency, and throughput. IEEE Network **27**(6), 14–22 (2013)
9. A. Ioannou, S. Weber, A survey of caching policies and forwarding mechanisms in information-centric networking. IEEE Commun. Surv. Tutor. **18**(4), 2847–2886 (2016)
10. H. Artail, H. Safa, K. Mershad, Z. Abou-Atme, N. Sulieman, COACS: a cooperative and adaptive caching system for MANETs. IEEE Trans. Mob. Comput. **7**(8), 961–977 (2008)
11. K. Mershad, H. Artail, SSUM: smart server update mechanism for maintaining cache consistency in mobile environments. IEEE Trans. Mob. Comput. **9**(6), 778–795 (2010)
12. Y. Zhang, J.M. Ng, C.P. Low, A distributed group mobility adaptive clustering algorithm for mobile ad hoc networks. Comput. Commun. **32**(1), 189–202 (2009)
13. H. Qi, M. Shiraz, J.Y. Liu, A. Gani, Z.A. Rahman, T.A. Altameem, Data center network architecture in cloud computing: review, taxonomy, and open research issues. J. Zhejiang Univ. Sci. C **15**(9), 776–793 (2014)
14. E. Panagos, A. Delis, Selective replication for content management environments. IEEE Internet Comput. **9**(3), 45–51 (2005)
15. K. Chen, S.H. Shah, K. Nahrstedt, Cross-layer design for data accessibility in mobile ad hoc networks. Wireless Pers. Commun. **21**(1), 49–76 (2002)
16. F.H. Allen, S.D.B.M. Bellard, M.D. Brice, B.A. Cartwright, A. Doubleday, H. Higgs, T. Hummelink, B.G. Hummelink-Peters, O. Kennard, W.D.S. Motherwell, J.T. Rodgers, The Cambridge Crystallographic Data Centre: computer-based search, retrieval, analysis and display of information. Acta Crystallogr. Sect. B: Struct. Crystallogr. Cryst. Chem. **35**(10), 2331–2339 (1979)

Framework for Digitally Managing Academic Records Using Blockchain Technology

Ramalingam Dharmalingam, Hassan Ugail, Arun Nagarle Shivasankarappa, and Vaishnavi Dharmalingam

Abstract Research studies report that there are a growing number of falsified educational certificates being produced by dishonest job seekers and higher education applicants across the world. Technological development in the image-processing domain makes editing the document so simple that any individual can perform this kind of forgery without having high-level skills in image editing. Most national governments have put in place stringent policies and procedures to verify and authenticate academic documents. However, due to the amount of human intervention in the process, the efficiency of such measures is debatable. Such systems leave open the possibility that unethical insiders may engage in forgery. In addition, the process of document verification and authentication consumes substantial amounts of time and money. Existing systems of document attestation do not provide a simple and instant way of verifying the authenticity of certificates from transcript level onward. In response to the above issues, this project proposes a prototype model for digitally managing and attesting the academic records using permissioned blockchain technology. By this method, the block-chaining of a student record begins from the time of admission to the Higher Education Institute (HEI) and continues to record the academic progress until graduation, having the graduation details stored as the last block in the chain. The whole blockchain of the student record will remain in the system with the participants enabling any indirect stakeholder to verify the details instantly based on the hash code or QR code. Additional privileges will be provided

R. Dharmalingam (✉)
Faculty of Information Technology, Majan University College, Muscat, Oman
e-mail: ramalingam.d@majancollege.edu.om

H. Ugail
Visual Computing, Department of Media, Design and Technology, Faculty of Engineering and Informatics, University of Bradford, Bradford, UK
e-mail: H.Ugail@bradford.ac.uk

A. N. Shivasankarappa
Middle East College, Muscat, Oman
e-mail: arun@mec.edu.om

V. Dharmalingam
Department of Computer Science and Engineering, SRC, SASTRA Deemed to be University, Tamil Nadu, India

S. Shakya et al. (eds.), *Mobile Computing and Sustainable Informatics*, Lecture Notes on Data Engineering and Communications Technologies 68,
https://doi.org/10.1007/978-981-16-1866-6_45

for direct stakeholders such as (here in Oman) the Ministry of Manpower and the Ministry of Higher Education to access further details of the certificate-holder. The system includes the student as a stakeholder and also as a participant to ensure transparency for her/his academic records.

Keywords Blockchain · Higher education · Academic record verification

1 Introduction

Recent technological developments have provided opportunities to produce, edit and modify any documents, and academic documents are not an exception. Even though forgery of academic certificates has been prevalent for decades, it has become even easier with the emergence of advanced information and communication technology tools. According to studies, there are two major motivational factors for academic fraud. One of them is to gain an advantage in employment, and the other is to gain admission to highly regarded academic institutions for higher studies. Some other factors that contribute to academic fraud are: to get promoted at work; pressure from parents and peers; and the desire to seem more impressive than one actually is.

To ensure the integrity of the academic transcripts and degree certificates, various mechanisms such as authentication through an apostille, attestation, degree equivalency and degree verification process have been adopted by several countries. However, due to the nature of these mechanisms, which include considerable human involvement, there is always an opportunity for dishonest employees to provide counterfeit certificates.

This project aims to develop a prototype to digitally store, verify and authenticate academic records using the permissioned distributed ledger of blockchain technology. The prototype is a comprehensive system that uses IBM hyperledger fabric to provide a transparent and trustworthy system, with a high level of integrity and security, through which all stakeholders such as students, colleges, degree awarding universities, regulatory authorities and employers can seamlessly participate in the verification process.

2 Statement of the Problem

Even though it is very difficult to estimate the exact number of forged certificates that have been produced, due to the sensitivity of the issue, a report published in 2014 states that the Saudi Council of Engineers has identified more than 30,000 fake certificates among the engineering community [1].

Perhaps, bogus universities and degree mills such as United Nations University, Delhi, produce authentic-looking degrees and associated documentation [2]. Academic institutions around the world face the long-standing problem of counterfeit academic certificates. It is a Herculean task to differentiate genuine academic certificates from a bogus credential endorsed by illegitimate authorities. Many academic transcripts and degree certificates produced by diploma mills seem perfectly credible to the naked eye.

The use of verification services takes time and involves a cost. Nevertheless, these services are necessarily taken up by the vast majority of colleges, universities and employers. However, effective services of this kind are limited to just a few countries; hence, the scale and nature of fraud continue to increase. When recruiting students from other countries, the admission office faces several challenges to verify the legitimacy of documents. The employers face a similar issue when recruiting foreign talent since the process is very long and costs a lot of money. There is also no transparent mechanism to track the progress of students from entry to graduation by sponsoring organizations who award scholarships.

Various ministries in the Sultanate of Oman and employers sponsor students to take up higher studies. They aspire to track the progress of sponsored students to make decisions on whether or not to continue the sponsorship, based on the progress made by the students. However, there is currently no transparent mechanism to track the progress of students from entry to graduation, even though this is essential to sponsoring organizations, in order to make evidence-based decisions. Currently, this process is handled by the registration department of the respective higher education institution, and it takes a considerable amount of time to verify the data and send it to sponsors.

Legitimate academic documents are altered through new additions or changes to get an undue advantage during employment or seek admission to higher education. The degree of alteration ranges from changes in the date of birth, graduation dates, attendance and grades. There may also be cases where grades are changed by internal staff members for monetary benefit. This kind of unethical behavior is obviously unacceptable, so measures must be put in place to prevent it from happening.

In summary, this project aims to answer the following problems.

1. How can counterfeit academic certificates be identified?
2. How can the authenticity of academic transcripts from registration to graduation at any given HEI be ensured?
3. How can transparency be achieved in the academic record management process of HEIs?
4. How can instant verification of the accuracy and authenticity of the academic records of any HEI be provided to all stakeholders?
5. How can academic records be stored in a completely secure way so that no one can tamper with the content?

3 Literature Review

3.1 Apostille and Attestation Process

The risk advisory group found that 44% of the CVs investigated had made false claims in relation to their educational qualifications and grades [2]. Higher Education Degree Datacheck (HEDD), UK, provides degree verification services to protect UK graduates, universities and employees. HEDD states in a report that it had found more than 210 bogus providers and Web sites endorsing fake degree certificates [3]. A degree mill falsely using the name of the University of Wolverhampton was established which used genuine images from the real University of Wolverhampton and sold fake certificates [4]. A Pakistani software company, Axact, operated a network of diplomas and bogus Web sites, according to the International New York Times [5].

Forged documents are defined as those that are modified by editing the original document using sophisticated tools that are easily available and simple to use. Fabricated documents are counterfeit documents that are created to look like the original documents awarded by legitimate institutions. Manufactured in-house documents are those produced by a few unethical employees for monetary benefit which are nearly impossible to detect. Imaginative translations are done inaccurately specifically for foreign consumption. Degree or diploma mills produce illegitimate certificates. Italiano (2009) as cited in Lipson and Karthikeyan (2016) states that Touro College in New York has prosecuted some of its staff members for the changing grades and selling degrees for bribes totaling hundreds or thousands of dollars [6]. Further, Krupnick (2007) states that the Diablo Valley Community College in California, USA, has found few of its staff members changed the grades of students from "F" to "A" for a bribe of nearly 600 USD [7].

The origin and authenticity of the issuing authority of the public documents such as educational certificates are verified through a process called apostille. An apostille is produced for public documents issued by one country and used in another country. Countries which are part of the apostille convention known as "The Hague Convention" can make use of the apostille certificate. However, if either of the countries where the public documents are issued or used is not part of this apostille convention, then the traditional attestation process is followed [8].

The normal process of apostille starts with the applicant submitting the respective original document to the public department of the issuing country/state. This is then forwarded to the issuing authority that verifies the authenticity of the document, which stamps it and sends it back to the public department. Then, the applicant is required to submit the same to the Ministry of External Affairs of the issuing country for the apostille certificate. However, the apostille certificate only guarantees the origin of the document, verifies the designated signatory and does not take responsibility for the content of the certificate [9].

3.2 Blockchain Technology

Blockchain is a distributed database which can be accessed by multiple entities, and there is no single owner or controller of the data. The data entries made to the blockchain cannot be altered without consensus from all participants [10]. Participants are the key stakeholders who hold a copy of the database where the block-chained records are stored.

Blockchain technology is a recent advancement in digital communication technology which uses distributed computing technology and cryptography to digitally authenticate and authorize documents and transactions in relation to the documents with increased integrity and authenticity [11]. One of the most popular uses of blockchain technology is the Bitcoin cryptocurrency which is a buzz word in the information and communication technology industry [12]. The benefits of blockchain technology such as security, integrity, authenticity and scalability make it suitable for digital authentication of any documents and related transactions by the respective parties. Since the transactions are deposited as records in distributed systems, it becomes almost impossible to lose the contents [13]. The cryptography mechanisms used in this process ensure the integrity of the document transactions in such a way that unauthorized alterations are impossible [14].

Other key features of blockchain technology are explained below.

Transparency
Blockchain provides a platform for all stakeholders to act with integrity by allowing all interested parties to view and verify academic transcripts and certificates including the history of changes. This prevents fraudulent practices, because the academic information is verified publicly from the source and changes created by consensus among the chain's participants are immutable. In other words, no one can change, manipulate or remove any data once it is verified by the participants [15].

Cost
Blockchain technology eliminates third-party mediators who are the root cause of delay and possible academic frauds. In addition, the cost involved is virtually non-existent. Blockchain streamlines the exchange of information through a distributed hyperledger, thus minimizing the transaction fees charged by middlemen for the verification of academic documents [16].

Resilience
Blockchain is inherently a decentralized distributed ledger where data is recorded in multiple blocks with no single point of failure [17].

Trust
Blockchain technology can be trusted since it can retain a fully verifiable and immutable ledger which is not available in the academic community. A fault-tolerant consensus mechanism in blockchain systems provides trust in a non-biased way [18]. Blockchain technology is the backbone that represents a distributed ledger which is

tamperproof since it relies on cryptographic protocols and thus provides greater security. It also provides a layer of trust which did not exist earlier for verification and authentication purposes.

Permissionless or public blockchains such as Bitcoin and Ethereum are the ones where anybody can participate since they are more open to users. On the other hand, permissioned or private blockchains act as closed ecosystems where users are not freely able to join the network. These are preferred by centralized organizations where governance is decided by members of the business network [19].

3.3 Alternative Ways of Recognizing Learner Achievement

Open badge specification is a cloud-based learning recognition system developed by Mozilla that provides what are called "virtual badges" to recognize each learner's achievement. Even though it does not directly use blockchain technology, it consolidates all the badges of a particular learner so that it becomes easy for anyone to view learners' achievements [20]. However, this method is used simply to recognize the learner's achievement but does not provide a robust mechanism to verify the authenticity of the academic documents.

The popular blockchain-based certificate verification standard called "blockcerts" has been released by Massachusetts Institute of Technology under open-source license. This standard provides a blockchain-based system for verifying the certificate issued by any institution [21]. Huynh et al. (2018) state that blockcerts are unsuitable since the user interface is not fully developed [22].

3.4 Blockchain Platform

IBM hyperledger is the outcome of the Linux Foundation's open-source collaborative effort that created a distributed ledger platform with modular architecture. It runs programs called smart contracts and supports pluggable consensus protocols with identity, access control and strong security [23].

4 Aim and Objectives

Given the issues outlined above, this project aims to develop a prototype to digitally store, verify and authenticate the academic documents using the permissioned distributed ledger using blockchain technology. The proposed prototype will be a comprehensive system that is built on IBM hyperledger fabric to provide a transparent and trustable system with a high level of integrity and security, allowing all the stakeholders such as students, colleges, degree awarding universities and regulatory authorities to participate seamlessly in the document verification management process. It also provides a secure way for any interested third parties to verify instantly the authenticity of an academic document.

5 Research Methodology

Research methodology plays an important role in the success of the project. Even though there exists a variety of research methods, action research is chosen as the preferred methodology for this research, since it is an iterative process of deep enquiry consisting of 4 stages such as planning, acting, observing and reflecting, respectively.

Meyer (2000) states that the action research's strength lies in the focus of solution generation to the practical problems [24]. The article written by Koshy (2009) also states that the action research method involves action, evaluation and critical reflection before implementation. It further states that action research improves the knowledge of the researcher during the action, which in turn helps the researcher to fine-tune the solution and provide the best possible solution [25]. This is illustrated in Fig. 1 in the action research model proposed by Kemmis and McTaggart [26].

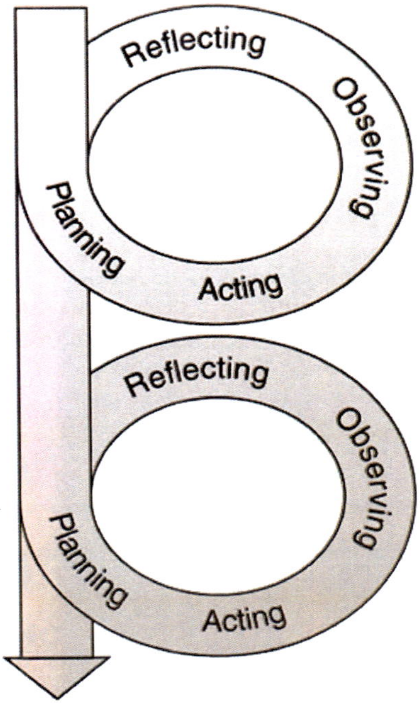

Fig. 1 Action research spiral model [26]

6 Prototype Design and Implementation

Model–view–controller (MVC) architecture is a popular design pattern for distributed software development. This architecture has three distinct layers as shown in Fig. 2 which abstracts its functions within the layer; thus, the changes in one layer are not affecting the other layers. For example, when the user interface (view layer) needs to be changed, then the business logic (control layer) is not affected. Similarly, when the business logic needs to be changed, then the database (model layer) shall not be affected. This architecture provides isolation between layers so that the changes in any layer can be done quickly and efficiently without affecting the other layer [27]. The view layer is built using AngularJS which supports MVC architecture and provides bidirectional data binding [28]. NodeJS is a cross-platform runtime environment which uses JavaScript across all components of MVC to build scalable Web applications.

Hyperledger is an open-source software for people to build custom blockchains as per their business needs and maintain confidentiality of transactions with private and permissioned network. It supports general-purpose programming languages without depending on native cryptocurrency, and users can develop pluggable components delivering high degrees of resilience and scalability. In contrast, other platforms like Ethereum provide transparency but with public/private and permissionless network, require mining and depend on built-in cryptocurrency called Ether. The authors have adopted a permissioned blockchain since it provides secure transactions among specific stakeholders to ensure transparency in verifying academic certificates. Chain code is the main program which contains the business logic and interacts between the control and model layers [29].

6.1 Stakeholders Involved in the Prototype

Figure 3 illustrates the process flow of the proposed system. The proposed system is having six stakeholders as follows:

Fig. 2 MVC architecture

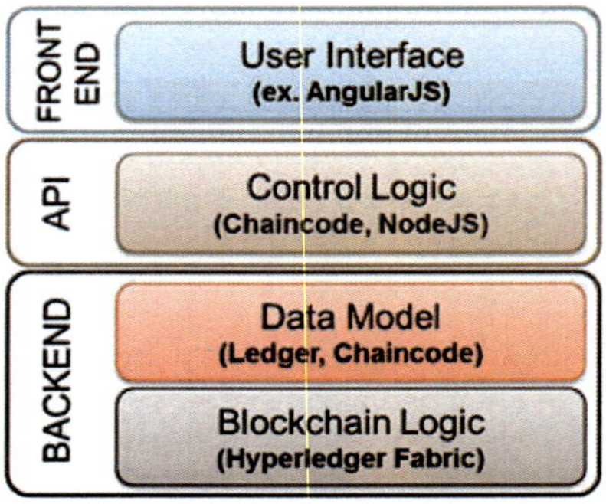

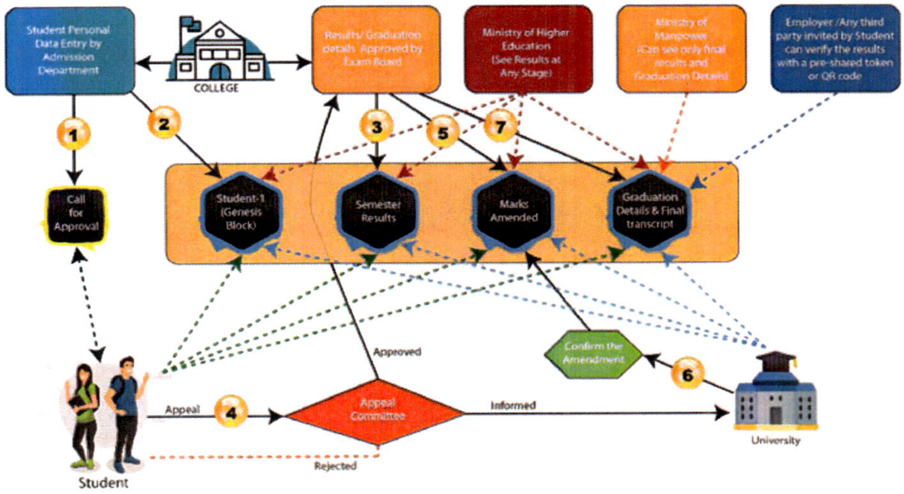

Fig. 3 Process flow diagram

1. The Higher Education Institution (HEI) where the student enrolls is called as the college.
2. The HEI is affiliated to the university which is the degree awarding body.
3. The candidate enrolling in the college is called as a student.
4. The Ministry of Higher Education (MOHE) provides the license for the HEI to operate and is also the regulatory authority. The MOHE normally verifies and attests the educational certificate for its authenticity and for ensuring qualification equivalency.
5. The Ministry of Manpower (MOMP) is the authority which registers the graduates and tracks the graduate destination.
6. The potential employer who intends to hire the candidate verifies the academic records of the candidate for its authenticity and accuracy.

6.2 *Process Flow*

Figure 3 illustrates the process flow of the proposed system.

1. The proposed process starts with the student enrollment into the HEI. The first step is to collect their personal data from the application form duly filled and signed by the applicant. The details from the form are entered into the student management system of the respective HEI. The entered details are to be verified by the applicant in the system.
2. The genesis block is created at this stage with the verified student's personal data. The applicant needs to login to the system and can see her/his personal information. In the future, this process may not be required if the student personal

information is fetched electronically from government-provided ID card such as civil resident card or passport.
3. The student appears for the examination, and the results are verified and confirmed by the examination board of the HEI and university outside the blockchain. The confirmed results are uploaded in the blockchain as the second block and are accessible by the HEI, student, university and MOHE.
4. The student can verify the result and if required can initiate the appeal process in the HEI.
5. Once the appeal is considered and approved and if there is any change in the grades or marks, the same will be entered into the next block in the blockchain system by the HEI.
6. The amendment due to the outcome of the appeal has to be approved by the respective university and informed to the student. This ensures transparency, and all stakeholders involved can track the reason for change in grades or marks.
7. Above steps 3–6 continue each semester until the student completes the graduation requirements.
8. Once the student meets the graduation requirement, the credit points are calculated and the classification of the award is generated by the student management system and the same is uploaded in the blockchain as the next block.

The above system allows the MOHE to verify the process at any stage which ensures transparency and integrity in the assessment process. The MOMP and potential employer can verify the graduation details and the final transcript at any time using the hash code or the QR code generated by the system. This hash code or QR code appears in the final transcript and in the graduate certificate.

This main benefit of the system is that it provides transparency in the graduate certificate verification process, also provides tamperproof security and overcomes the burden of attestation process done by MOHE. Any interested third party can also compare and verify the hard copy details submitted by the graduate for its accuracy in real time using the above-said hash code or QR code. The additional feature of this system is that there is no time or money spent by the student or by the employer in the verification of the academic documents. The HEI and the university who are the primary stakeholders increase their brand value by gaining the trust of students, employers and regulatory authority by providing a transparent and highly secure academic record verification system. Even though the research studies state that the vulnerabilities are emerging in the blockchain implementations, the proposed framework can be securely implemented as private blockchain [30, 31].

7 Conclusion

The authors were able to authenticate academic records submitted by candidates by providing an innovative and transparent system that maintains the integrity and security of the academic artifacts forever. This was possible by applying action research methodology and reflecting on the analysis to solve a genuine research problem of attestation and apostille and thereby contribute to the development of a trusted system by all stakeholders in the academic community.

8 Future Work and Link for Software Code

One of the HEIs in Oman is taken for case study to develop the prototype. This prototype can be further customized and used by any HEI as the source code of the prototype is made available to the public via the GNU General Public License in https://sourceforge.net/projects/academic-records-blockchain/files/latest/download.

Interested HEIs can explore further, and customize and adapt the proposed design to ensure transparency, integrity and the cyber resiliency for the academic records in a cost-effective way.

The additional benefit of this system is that it provides a way for any third party to verify the authenticity of the academic document instantly. This innovative mechanism will provide flexibility and ease of use for any stakeholder, avoiding the cumbersome process of document attestation and authentication which is currently practiced.

References

1. J.S. Al Suwaidi, Fake academic degrees: a crime against the present and the future. Op-eds-Gulf News. Spec. to Gulf News
2. CV Fraud Statistics 2017. Adv. Secur. Technol. https://www.advancedsecure.co.uk/wp-content/uploads/2018/01/CV-FRAUD-2017.pdf (2018). Accessed 11 Apr 2020
3. J. Rowley, Advice and guidance on degree fraud a toolkit for employers (2016)
4. Website selling "Wolverhamton University" degrees shut down. Express & Star. https://www.expressandstar.com/news/education/2017/01/04/website-selling-wolverhamton-university-degrees-shut-down/. Accessed 11 Apr 2020
5. W. Declan, Fake Diplomas, Real Cash: Pakistani Company Axact Reaps Millions—The New York Times. New York Times. https://www.nytimes.com/2015/05/18/world/asia/fake-diplomas-real-cash-pakistani-company-axact-reaps-millions-columbiana-barkley.html (2015). Accessed 11 Apr 2020
6. S.M. Lipson, S.F. College, L. Karthikeyan, The Art of Cheating in the 21st Millennium: Innovative The Art of Cheating in the 21st Millennium: Innovative Mechanisms and Insidious Ploys in Academic Deceit Mechanisms and Insidious Ploys in Academic Deceit The Art of Cheating in the 21st Millennium: Innovative Mechanisms and Insidious Ploys in Academic Deceit. 8:1948–5476 (2016). https://doi.org/10.5296/ije.v8i2.9117
7. M. Krupnick, Dozens implicated in Diablo Valley College grade scandal—The Mercury News. In: Mercur. News. https://www.mercurynews.com/2007/05/03/dozens-implicated-in-diablo-valley-college-grade-scandal/ (2007). Accessed 26 Oct 2020
8. A. Arazmuradov, Recalling benefits of international convention for modernization. SSRN Electron. J. (2016). https://doi.org/10.2139/ssrn.2011640

9. J.W. Adams Jr., The apostille in the 21st century international document certification and verification. Hous. J. Int. Law **34**, 519 (2011)
10. M. Nofer, P. Gomber, O. Hinz, D. Schiereck, Blockchain. Bus. Inf. Syst. Eng. **59**, 183–187 (2017). https://doi.org/10.1007/s12599-017-0467-3
11. M. Swan, Blockchain: Blueprint for a New Economy—Melanie Swan—Google Books, 1st edn. (O'Reilly Media, Inc., 2015)
12. A. Greenspan, Bitcoin and Cryptocurrency Technologies: A Comprehensive Introduction. By Arvind Narayanan, Joseph Bonneau, Edward Felten, Andrew Miller, and Steven Goldfeder. Princeton and Oxford: Princeton University. In: J. Econ. Lit. https://books.google.co.uk/books?hl=en&lr=&id=LchFDAAAQBAJ&oi=fnd&pg=PP1&dq=cryptocurrency&ots=ArsGeZ-LhJ&sig=DogCpSJd1nLRv_eLZCU6LtH8Yts&redir_esc=y#v=onepage&q=cryptocurrency&f=true (2017). Accessed 12 Apr 2020
13. R. Beck, Beyond bitcoin: the rise of blockchain world. Comput. (Long Beach Calif) **51**, 54–58 (2018). https://doi.org/10.1109/MC.2018.1451660
14. D. Arora, S. Gautham, H. Gupta, B. Bhushan, Blockchain-based Security Solutions to Preserve Data Privacy And Integrity. Institute of Electrical and Electronics Engineers (IEEE) (2020), pp. 468–472
15. Y. Zhang, C. Xu, X. Lin, X.S. Shen, Blockchain-based public integrity verification for cloud storage against procrastinating auditors. IEEE Trans. Cloud Comput. 1–1 (2019). https://doi.org/10.1109/tcc.2019.2908400
16. S. Ahluwalia, R.V. Mahto, M. Guerrero, Blockchain technology and startup financing: a transaction cost economics perspective. Technol. Forecast Soc. Change **151**, (2020). https://doi.org/10.1016/j.techfore.2019.119854
17. T. Harman, P. Mahadevan, K. Mukherjee, et al., Cyber resiliency automation using blockchain, in *2019 IEEE International Conference on Cloud Computing in Emerging Markets (CCEM)* (IEEE, 2019), pp. 51–54
18. M. Fleischmann, B. Ivens, Exploring the role of trust in blockchain adoption: an inductive approach, in *Proceedings of the 52nd Hawaii International Conference on System Sciences.* Hawaii International Conference on System Sciences (2019)
19. M. Iqbal, R. Matulevičius, Blockchain-based application security risks: a systematic literature review, in *Lecture Notes in Business Information Processing* (Springer, Berlin, 2019), pp. 176–188
20. D. Gibson, N. Ostashewski, K. Flintoff et al., Digital badges in education. Educ. Inf. Technol. **20**, 403–410 (2015). https://doi.org/10.1007/s10639-013-9291-7
21. (24/10/2016) PS-MIl, 2016 undefined Blockcerts—An open infrastructure for academic credentials on the blockchain
22. T.T. Huynh, T. Tru Huynh, D.K. Pham, A. Khoa Ngo, Issuing and verifying digital certificates with blockchain, *International Conference on Advanced Technologies for Communications* (IEEE Computer Society, 2018), pp. 332–336
23. P. Thakkar, S. Nathan, B. Viswanathan, Performance benchmarking and optimizing hyperledger fabric blockchain platform, in *Proceedings—26th IEEE International Symposium on Modeling, Analysis and Simulation of Computer and Telecommunication Systems, MASCOTS 2018* (Institute of Electrical and Electronics Engineers Inc., 2018), pp. 264–276
24. J. Meyer, Using qualitative methods in health related action research. BMJ **320**, 178–181 (2000)
25. V. Koshy, *Action Research for Improving Educational Practice: A Step-by-Step Guide* (Sage, 2009)
26. S. Kemmis, R. McTaggart, R. Nixon, *The Action Research Planner: Doing Critical Participatory Action Research* (Springer, Singapore, 2013)
27. J. Deacon, Model-View-Controller (MVC) Architecture (2009)
28. #1 MAJ, Sawant BR, Deshmukh A Single Page Application using AngularJS

29. E. Androulaki, A. Barger, V. Bortnikov, et al., Hyperledger fabric: a distributed operating system for permissioned blockchains, in *Proceedings of the 13th EuroSys Conference, EuroSys 2018* (Association for Computing Machinery, Inc., New York, 2018), pp. 1–15
30. Public Vs. Private Blockchain: A Comprehensive Comparison. https://www.blockchain-council.org/blockchain/public-vs-private-blockchain-a-comprehensive-comparison/. Accessed 25 Jan 2021
31. J.H. Lee, Systematic approach to analyzing security and vulnerabilities of blockchain systems (2019)

Development of Diameter Call Simulations Using IPSL Tool for IMS-LTE

R. Bharathi, T. Satish Kumar, G. Mahesh, and R. Bhagya

Abstract With the evolution of mobile communication, every country across the world are moving toward all-IP network for both wire line and wireless communication. The life has become unimaginable without the Internet. It is very significant to understand that who can have access to a particular service and for how long the services are utilized. The AAA protocols were created to fulfill these requirements. Diameter protocol is an advanced AAA protocol and very flexible protocol. It is believed to be next generation protocol for IMS-based architecture, 4G/LTE and 5G networks. Most of the cellular operators use customized tools for analyzing the performance and functionality of IMS-LTE networks. Currently, one such tool for call simulation, performance and functionality analysis is being developed, and call simulations are implemented and tested. This paper briefs how the diameter protocol is involved during the exchange of messages between the node and to check the performance and functionality of the IMS-LTE network elements and to analyze the call simulation.

Keywords Diameter protocol · Interface · HSS · TAS · Wireshark

R. Bharathi · T. Satish Kumar (✉) · G. Mahesh
Department of CSE, BMS Institute of Technology and Management, Bangalore, India
e-mail: satisht@bmsit.in

R. Bharathi
e-mail: bharavi_kumar@bmsit.in

G. Mahesh
e-mail: maheshg@bmsit.in

R. Bhagya
Department of ETE, R V College of Engineering, Bangalore, India
e-mail: bhagyar@rvce.edu.in

S. Shakya et al. (eds.), *Mobile Computing and Sustainable Informatics*, Lecture Notes on Data Engineering and Communications Technologies 68,
https://doi.org/10.1007/978-981-16-1866-6_46

1 Introduction

Mobile wireless communication has become popular and made remarkable changes in the last few decades. The demand for telecommunication services is increasing day by day with growth of the subscribers. The legacy networks like circuit switched networks were expensive and needed long time for deployment when compared to an all-IP. IP multimedia subsystem (IMS) is an all-IP network which provides the standard framework for delivering the mobile and fixed multimedia [1, 2].

The wireless standard body 3rd Generation Partnership Project (3GPP) has designed IMS network. It has its existence from 3GPP release 5 over a decade ago but is currently being implemented in next generation networks like 4G-LTE and IMS bases LTE networks [3]. Home subscriber server (HSS), mobility management entity (MME), application server (AS), call session control function (CSCF), policy and charging rules function (PCRF) and serving GPRS support node (SGSN) are the network elements in IMS-LTE network. HSS is the central function unit in IMS architecture. All other entities communicate with HSS through diameter interface. HSS enables implementation of the diameter signaling.

The life has become unimaginable without the Internet. Internet services are offered through mobile networks, wireless networks and other types of wire line networks. The authentication, authorization and accounting (AAA) protocols were created to understand who can have access to a services, for how long services are utilized and to keep track of all services [4]. Formerly, remote access dial-in user service (RADIUS) protocol was used to provide AAA services [5]. The evolution in the network required a better protocol for providing the authentication, authorization and accounting.

Diameter protocol is an advanced AAA protocol evolved from the RADIUS and is very flexible protocol. It was formulated to eradicate the drawbacks of RADIUS. It mainly operates on the application layer. The main function of the diameter is authentication, authorization, accounting and assist mobility management in IMS-LTE network. The 3GPP has adopted the diameter protocol for LTE, 5G and other upcoming services. It is being widely used across Internet protocol (all-IP)-based networks. Diameter protocol is believed to be next generation protocol for IMS-based architecture, 4G/LTE and 5G networks [6]. It is mainly involved in the exchange of the messages between the nodes.

IMS-LTE network element interacts with HSS, which is the central repository containing the subscriber information thorough diameter interfaces. The MME which is the functional unit of LTE network communicates with HSS through S6a Interface. The application server which the brain of the IMS architecture interacts with HSS through Sh interface and CSCF the key unit of IMS communicates to HSS through Cx interface.

The purpose of this paper is to check the performance and functionality of IMS-LTE network in real-time scenario. The main objective of this paper is to handle the location management procedure of LTE network and to implement SIP registration or third-party registration of IMS network. The location management procedure of LTE network is being implemented to update the location information of the subscriber. Here, basically, one end communication takes place where the request is sent to the HSS, and it responds back with answer. The third-party registration is implemented to let the TAS know that user is connected to the network and ready to communicate. The end-to-end communication is handled.

The remaining sections of this paper are described as follows. Section 2 introduces the diameter protocol and interface. Section 3 describes about diameter call flow. Section 4 proposes the design and implementation. Section 5 presents the results, and Sect. 6 presents the conclusions of this research.

2 Diameter Protocol and Interface

2.1 *Role of Diameter in IMS-LTE*

VoLTE network has three main network domains: The first one is LTE network, second one is IMS network and followed by PSTN network. All these domains are connected to each other by interfaces, links and protocols. The interface is the connecting point between the two network entities. Figure 1 shows the VoLTE network architecture and interfaces present between the IMS-LTE network entities [7, 8]. The network entities are connected to the HSS which is the main functional unit of IMS-LTE network through diameter protocol interface.

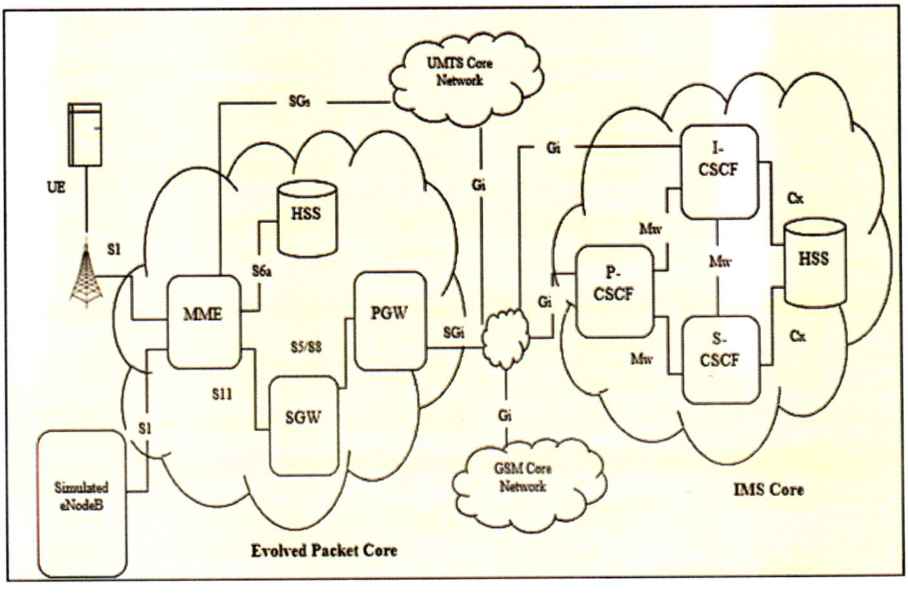

Fig. 1 VoLTE network architecture

2.2 Diameter Protocol

Diameter protocol is the most emerging protocol developed by 3GPP standard which plays a very vital role in the deployment of IMS-VoLTE network. It is the primary signaling protocol to provide the AAA services and mobility management. It is evolved from the RADIUS protocol and operates in application layer. The evolution of the next generation mobile network like IMS and LTE uses diameter protocol in various interfaces. It is designed as peer-to-peer architecture which is involved in exchange of the message between nodes of the network and in turn receiving the acknowledgment which may be either positive or negative.

2.3 Diameter Interfaces

The network entities of the IMS-VoLTE network communicate with HSS through diameter interface. The diameter protocol defined in IETF has been extended with the additional field like AVP by 3GPP standard to support the interfaces which exist between the IMS, LTE network entities [7]. The diameter interfaces in IMS-VoLTE network are S6a, Sh and Cx interfaces.

2.3.1 Diameter S6a Interface

The interface which is used to communicate between the HSS and MME nodes of the LTE network is S6a with and application Id 16777251 [9]. This interface is used to provide the services to the subscriber, to authenticate the user and to store the location related information of the subscriber that is sent by MME. There are totally eight messages exchanged between HSS and MME of LTE network.

2.3.2 Diameter Sh Interface

The telecom application server (TAS) is the brain of IMS network which provides a specific service to the end users. Sh is a diameter interface with an application Id 16777217 that exists between the application server and HSS. It helps to transport the user profile-related information, service-related information and user-related information. There are four messages transformed between AS and HSS.

3 Diameter Call Flow

The diameter message flows take place initially with the establishment of the transport layer connection, and then, the messages are being transferred [5]. The message will communicate if the diameter connection is successfully established. Once the transport layer connection is established, the initiator will send the capabilities-exchange-request (CER) to the other end. It will respond with a capabilities-exchange-answer (CEA). After the exchange of CER and CEA, now the connection will be ready to exchange the application messages like ULR, UDR, CLR, and SNR. If there is delay in message, then the device watchdog request and answer (DWR/DWA) is triggered. Finally, the transport layer is disconnected by triggering the disconnect peer request and answer (DPR/DPA). Figure 2 shows the flow chart of diameter call flow between the entities.

3.1 Call Simulation Procedure of S6a Interface

It is used for performing authentication procedure, location management procedure, subscriber data handling procedure, notification procedure and fault recovery procedure. The main objective of the project is to perform the location management procedure. The five main procedures are listed below.

3.1.1 Authentication Procedure

Initially, the authentication procedure takes place where the authentication-information-request (AIR) is sent from MME to HSS. MME takes the information from HSS to authenticate subscriber. If the data present, then HSS shall returns authentication-information answer (AIA).

3.1.2 Location Management Procedure

The update location request (ULR) is sent between MME to HSS, to inform the HSS about the location information of the MME currently serving the user where HSS responds with update location answer (ULA). If the user changes the location and moves to different or new MME, the cancel location request (CLR) is sent from HSS to MME, and in return, it get backs the cancel location answer (CLA).

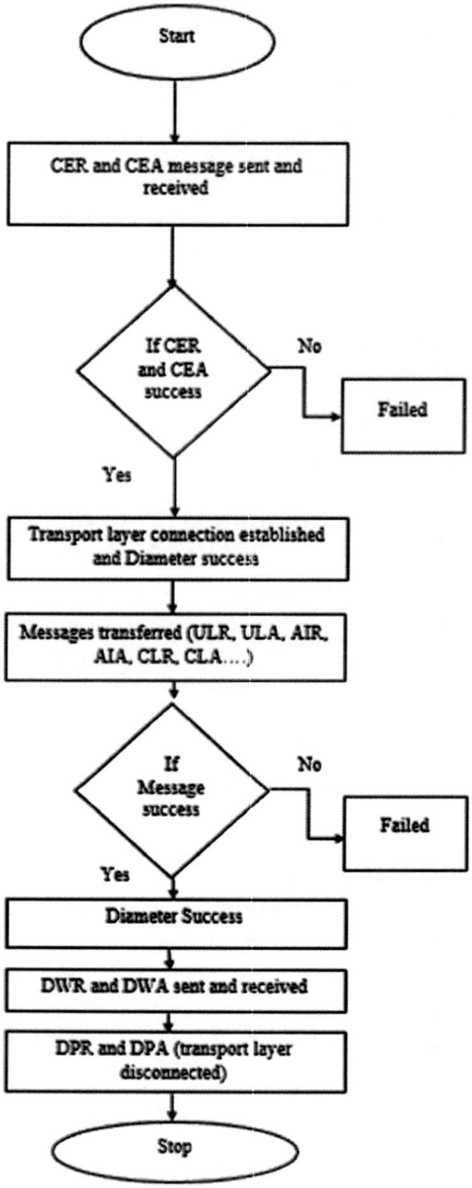

Fig. 2 Flowchart of diameter call flow

3.1.3 Subscriber Data Handling

The insert subscriber data (IDR) procedure takes place between the HSS and MME to update certain subscriber data in MME due to some changes made in the user data in HSS. The insert subscriber data answer (IDA) is sent back. To remove the data of subscriber information from MME, the delete subscriber data request (DSR) is sent. After the successful deletion, the delete subscriber answer (DSA) is sent back.

3.1.4 Notification Procedure

Whenever the updation in an inter-MME location does not occur but the notification has to be sent to HSS, hence, the notification request (NOR) has to be sent from MME to HSS. And in return, the HSS will send back the notification answer (NOA).

3.1.5 Reset Procedure

After the restart in order to indicate the MME that some failure has occurred, the reset request (RSR) is sent from HSS. In return, the MME sends back the reset answer (RSA).

3.2 *SIP Registration/Third-Party Registration*

Another objective of the project is SIP registration to update TAS. The SIP is the basis of the IMS and used as a signaling mechanism. The major functionality of the SIP in IMS network architecture is to register the subscriber, determine location of the user and to establish the VoLTE call. Whenever the UE gets activated, it has to register to the network. The main role of TAS is call diverting, forwarding, barring, etc. The procedure to update user profile in TAS is, initially the S-CSCF send the SIP register to register the mobile user with TAS. The TAS sends the user data request (UDR) to HSS. UDR mainly requests for the repository data from HSS for which it returns back with user data answer (UDA) message with appropriate information. Finally, the TAS completes the SIP registration by sending the SIP 200 OK message which means registration successful to S-CSCF.

4 Design and Implementation

The main concern of the project is to simulate the diameter call flows. Here, it is done by taking two scenarios: One is to handle the location management procedure of LTE network, and another is performing the third-party registration of IMS network. This section briefs about the design flow.

4.1 Location Management Procedure

When the subscriber moves from one MME to another MME, the location information of the subscriber has to be updated in the HSS which is the master database. The steps involved in executing this process are as follows: Initially, the subscriber with unique identifier and IMSI has to be created by using independent protocol simulator language (IPSL). IPSL is the protocol simulator language which belongs to Nokia networks [3, 10]. The flow chart of design flow is shown in Fig. 3.

The ULR message has to be initially registered to HSS with the current MME host address. If the registration is done successfully, HSS will respond back the ULA message with diameter success. If the user has moved from old MME to new MME, then the CLR message to cancel the previous request with old MME has to be sent from HSS to MME, and on successful cancelation, the CLA message is received back. Then, the new ULR message with new MME host is triggered. ULA is received back. The flow of ULR and CLR message exchanged between MME and HSS is shown in Fig. 4.

Fig. 3 Design flow

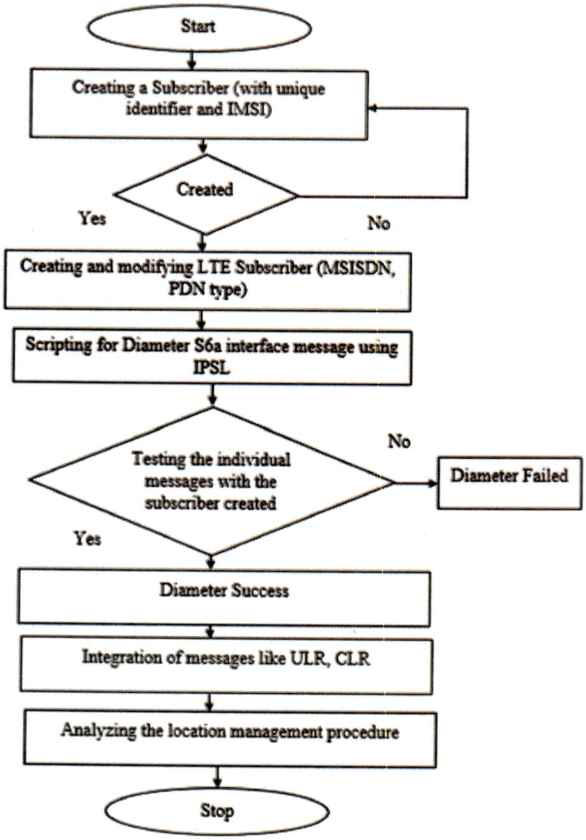

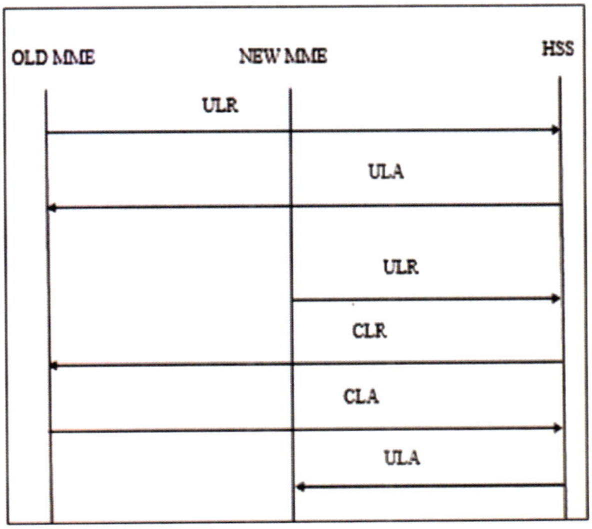

Fig. 4 Location management procedure

4.2 TAS Update in SIP Registration

In order to register the user profile to TAS, the SIP register is sent from S-CSCF to TAS core. The TAS will send the UDR which request for the repository data from HSS. The HSS will return back the UDA with all the basic subscriber user data. Once the registration is done, the TAS core will respond with the SIP 200 OK message to S-CSCF. Once the user is registered, the TAS will send subscriber notification request (SNR) to HSS for which it responds with subscriber notification answer (SNA). Once it is done, the TAS will send the UDR message for which HSS responds back with UDA, and finally, push notification request/answer (PUR/PUA) is exchanged. This is entire SIP registration or third-party registration procedure. The call flow is shown in Fig. 5.

5 Results and Discussion

This section presents the results of the diameter call simulation procedure. The call is simulated on IPSL tool, and traces are captured in the Wireshark environment.

5.1 Results of Location Management Procedure

The diameter connection takes place by establishment of the transport layer connection, and once it is established, the messages are transferred. The ULR/ULA and

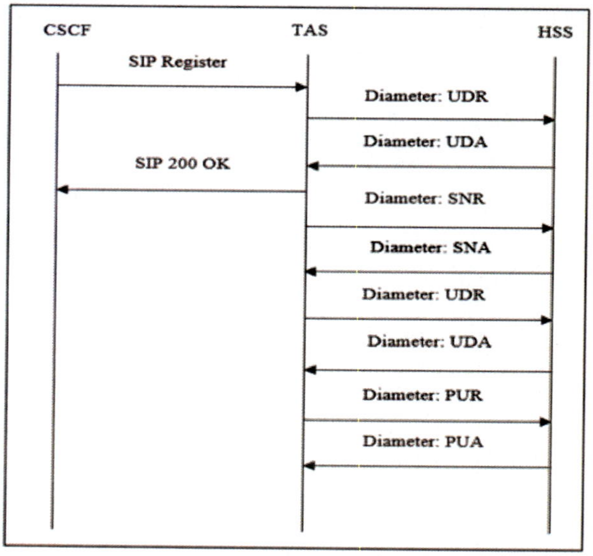

Fig. 5 SIP registration call flow

CLR/CLA messages are exchanged between MME and HSS during location management. Initially, CER/CEA is transferred to show that transport layer connection is established and ready to transfer the messages. Figure 6 shows the CER/CEA call flow.

Figure 7 shows the diameter message ULR that is sent to HSS to update the location of the user, and HSS replies the ULA showing that the user information is updated.

Figure 8 shows that the user has moved to new MME, and new location has to be updated so again the ULR message with new location information is sent to HSS.

Figure 9 shows call flow where that CLR is sent to old MME to cancel the previous location. Once it is done, CLA is received back, and once the new location is updated in HSS, the ULA message is received.

Figure 10 shows the traces captured in Wireshark during the call simulation.

5.2 *Results of SIP Registration*

Third-party or SIP registration is very important in the IMS network to make sure that user is registered to the network and ready to make the call. The output of the SIP registration success is shown in Fig. 11, and the message traces are also captured in the Wireshark environment as shown in Fig. 12. Initially, SIP register is sent after the successful registration of the user to the TAS core the 200 OK message is returned.

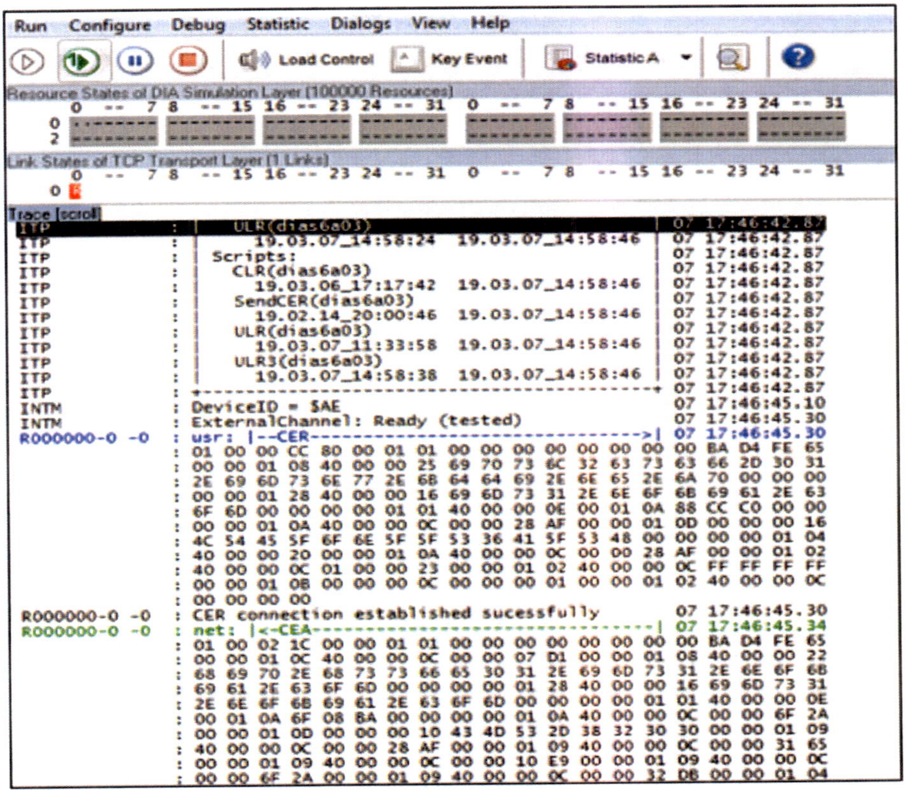

Fig. 6 CER/CEA call flow

6 Conclusion and Future Work

With the increase in subscribers, the telecom industry is growing drastically. It is very crucial to provide the AAA services to the users which are done through the diameter protocol. Diameter plays vital role in IMS-LTE network. It is very difficult to analyze the performance and functionality of the network elements in real-time scenario. For this purpose, call simulation is carried out on a tool to analyze the performance and functionality of the network elements. As a part of the diameter call flow simulation, the location management procedure of the LTE network and the SIP registration procedure by handling the diameter Sh interface message is being implemented, and the traces are captured in Wireshark to analyze the parameters. Similarly, many subscribers registration and VoLTE to VoLTE call flow can be taken as future work.

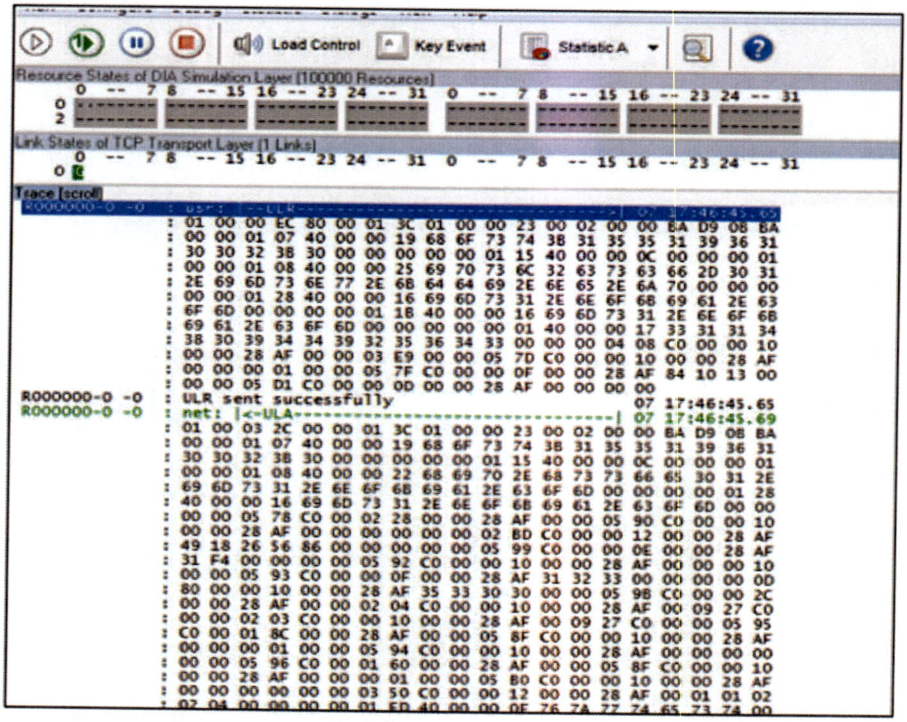

Fig. 7 ULR/ULA call flow

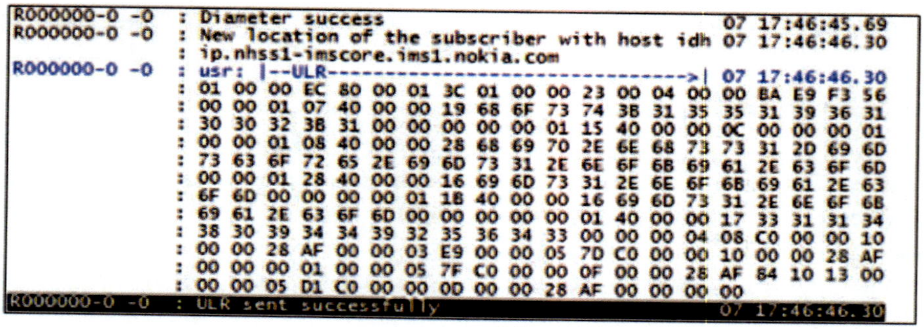

Fig. 8 New ULR sent

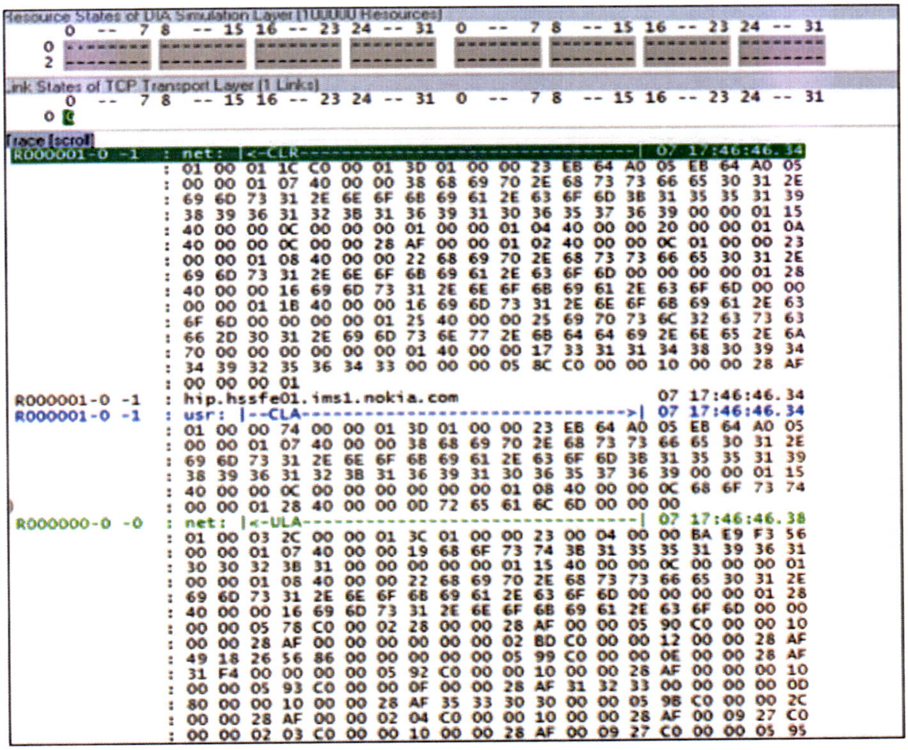

Fig. 9 CLR/CLA call flow

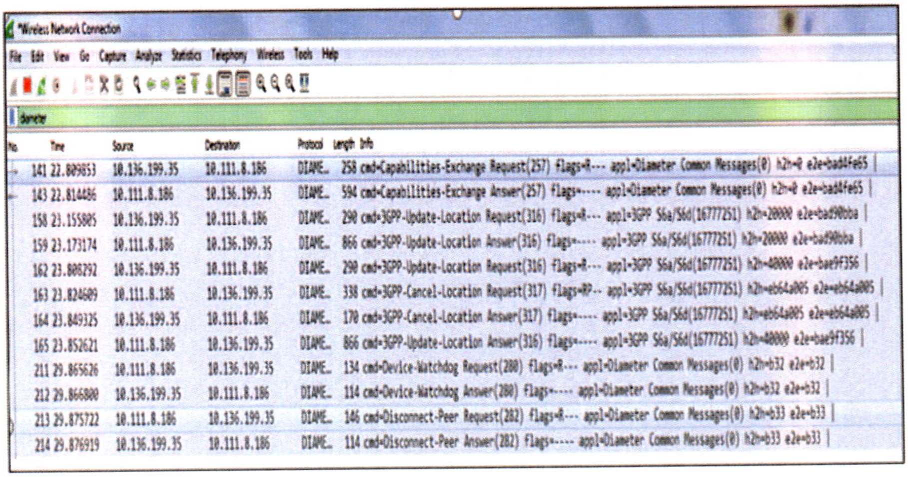

Fig. 10 Wireshark trace for Location management

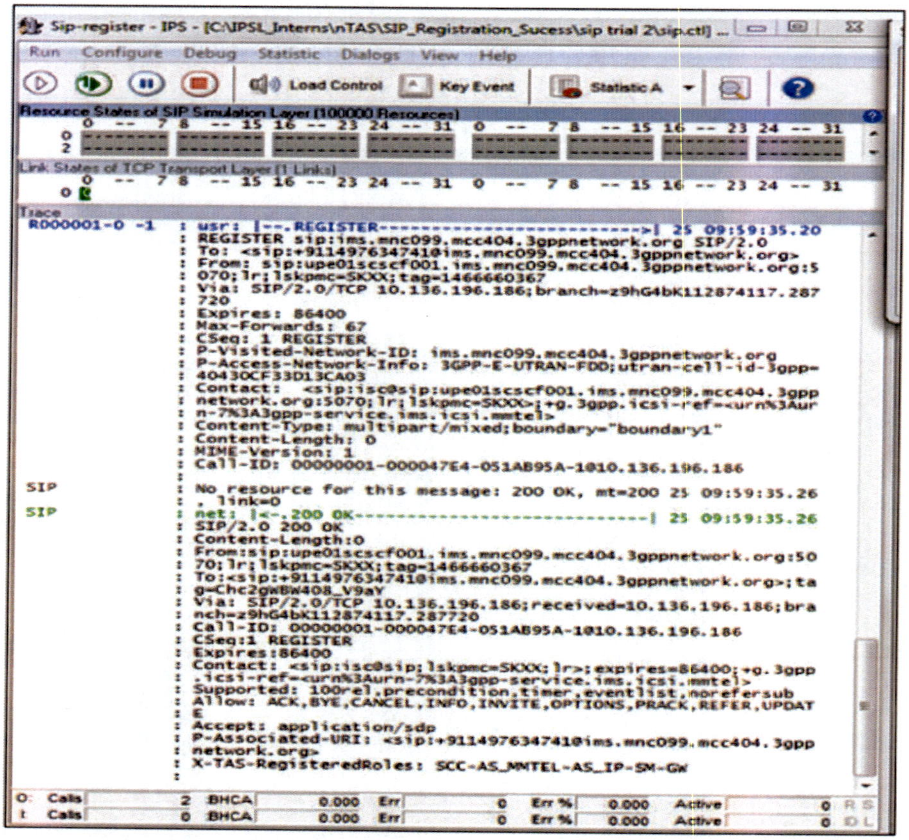

Fig. 11 SIP register with 200 OK

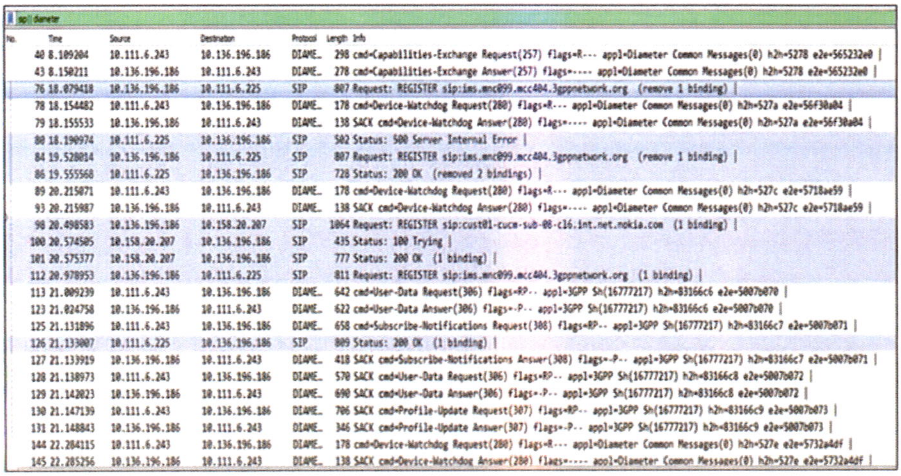

Fig. 12 Wireshark trace for SIP registration

References

1. P. Swetha, G.S. Nagaraja, P. Vasa, Testing mobile application part (MAP) protocol using MAP-simulator tool in a mobile communication testing laboratory (2019)
2. J. Jelinek, P. Satrapa and J. Fiser, "Experimental issues of the model of the enhanced RADIUS protocol," *2015 IEEE International Workshop of Electronics, Control, Measurement, Signals and their Application to Mechatronics (ECMSM)*, Liberec, 2015, pp. 1–6
3. G. Yang, Z. Lei, H. Wang, Y. Dou, Y. Xie, Analysis of the RADIUS signalling based on CDMA Mobile Network, in *2014 Sixth International Conference on Intelligent Human-Machine Systems and Cybernetics*, Hangzhou, 2014, pp. 270–274. https://doi.org/10.1109/ihmsc.2014.167
4. T. Jelassi, F.J. Martínez-López, Digital business transformation in silicon savannah: how M-PESA Changed Safaricom (Kenya), in *Strategies for e-Business. Classroom Companion: Business* (Springer, Cham, 2020). https://doi.org/10.1007/978-3-030-48950-2_23
5. M.M. Rahman, S.S. Heydari, A self-healing approach for LTE Evolved Packet Core, in *2012 25th IEEE Canadian Conference on Electrical and Computer Engineering (CCECE)*, Montreal, QC, 2012, pp. 1–5. https://doi.org/10.1109/ccece.2012.6335007
6. Z. Sun, W. Xing, X. Cui, A security test message generation algorithm of diameter protocol in the IMS network, in *2015 2nd International Symposium on Dependable Computing and Internet of Things (DCIT)*, Wuhan, 2015, pp. 72–79. https://doi.org/10.1109/dcit.2015.15
7. M. Matuszewski, M.A. Garcia-Martin, A distributed IP multimedia subsystem (IMS), in *2007 IEEE International Symposium on a World of Wireless, Mobile and Multimedia Networks*, Espoo, Finland, 2007, pp. 1–8. https://doi.org/10.1109/wowmom.2007.4351728
8. M.-S. Hwang, Y.-L. Tang, C.-C. Lee, An efficient authentication protocol for GSM networks, in *IEEE/AFCEA EUROCOMM 2000. Information Systems for Enhanced Public Safety and Security (Cat. No.00EX405)*, Munich, Germany, 2000, pp. 326–329. https://doi.org/10.1109/eurcom.2000.874826
9. S.B. Vinay Kumar, M.N. Harihar, Diameter-based Protocol in the IP Multimedia Subsystem. Int. J. Soft Comput. Eng. (IJSCE) 2231–2307 (2012)
10. T.Q. Thanh, Y. Rebahi, T. Magedanz, An open testbed for diameter networks, in *SoftCOM 2012, 20th International Conference on Software, Telecommunications and Computer Networks*, Split, 2012, pp. 1–5

Automated Classification of Alzheimer's Disease Using MRI and Transfer Learning

S. Sambath Kumar and M. Nandhini

Abstract Alzheimer's disease (AD) is a severe brain disease which adversely impacts thinking and memory. As a result, the collective brain size shrinks and prompts death. Therefore, the traditional diagnosis of AD is a crucial factor to decide on the therapeutics. Machine learning (ML) and deep learning (DL) techniques are applicable to diagnosing the AD disease in early stages with the use of brain "magnetic resonance imaging (MRI)". This paper is proposing a convolutional neural network model which utilizes transfer learning technique to classify the large amounts of MRI data with pre-trained AlexNet architecture. The proposed model is trained and tested with segmented images of Cerebral Spinal Fluid, Grey Matter, and White Matter for classification. While comparing the experimental results with the proposed model improves the performance of diagnosing disease. The presented method is trained, tested and validated on MRI image segmentation on ADNI dataset.

Keywords Alzheimer's disease · MRI · ImageNet · Deep learning · CNN · AlexNet · Transfer learning

1 Introduction

Dementia is not a disease technically, but rather a way for describing a set of symptoms such as difficulty in memorizing thing/events, difficulty in learning new information, which makes very difficult to survive day to day life themselves. Dementia is caused by some sort of brain cell damage [1]. The most prevalent cause of dementia

Data used in the preparation of this article were obtained from the Alzheimer's disease neuroimaging initiative (ADNI) database (www.loni.ucla.edu/ADNI). As such, the investigators within the ADNI contributed to the design and implementation of ADNI and/or provided data but did not participate in analysis or writing of this report. A complete listing of ADNI investigators is available at: http://adni.loni.usc.edu/wpcontent/upload/how_to_apply/ADNI_Acknowledgement_List.pdf.

S. Sambath Kumar (✉) · M. Nandhini
Department of Computer Science, Pondicherry University, Puducherry, India

M. Nandhini
e-mail: mnandhini.csc@pondiuni.edu.in

S. Shakya et al. (eds.), *Mobile Computing and Sustainable Informatics*, Lecture Notes on Data Engineering and Communications Technologies 68,
https://doi.org/10.1007/978-981-16-1866-6_47

is Alzheimer's disease (AD); it makes loss or degeneration of neurons in the brain especially in the cortex. The disease also damages the brain that can lead to loss and other cognitive abilities [2]. Keeping the individuals alive longer from their cancer, heart disease and causing AD epidemic. Therefore, in the next two to three decades, five million is anticipated to triple in the world especially in the USA [3]. In medical diagnosis, MRI is a powerful method used for precise visualization of the soft tissues. In the past decade, MRI images are used successfully to detect the brain tumours, multiple sclerosis and other diseases. Traditionally, radiologists produce diagnosis based on their practical, and learning and training. They identify and distinguish the concealed patterns in the MRIs and give diagnosis consistent with that informatics. The MRI images of soft tissue and the brain bone produced MRI instrument. This was the first radiological technology for brain imaging. The computer-assisted diagnosis (CAD) methods are needed to be designed to detect and identify the development of AD in clinical research. CAD utilizes machine learning and other computer algorithms to assist physicians to find the correct diseases. Radiology images are important in which individuals who were worried about being prosecuted tried to decrease the medical malpractice twenty years ago. Lots of individuals seeking methods to attempt and lower the legal cost. The researchers, therefore, found a ML and DL algorithm to assist doctors and create the correct diagnosis and healthcare workers. Various computer-assisted methods are developed to extract the important features from Volume of Interest (VoI) and Region of Interest (ROI) and patch-based method [4, 5]. The characteristic of features was acquired from the WM voxels, GM voxels and specific regions of hippocampus area. These are the methods are to detect the AD at early stages. Most of the researchers are focused only binary classification of AD, which proves only whether disease detected or not (0 or 1). The current image diagnosis is mainly diagnosed by a professional radiologist, and preliminary conclusions are drawn and submitted to the attending physician for in-depth examination and diagnosis. The number of patients increases and medical conditions develop rapidly, and the patient's medical image data also increases exponentially. It is increasingly difficult for professional radiologists to manually meet the needs of patients for timely diagnosis, which poses a huge challenge for subsequent treatment.

The majority of research work focuses on CAD system using MRI brain images, which are very useful in distinguishing the brain changes in diseases. Recently, adopting feature-based approach in machine learning is often replaced by different deep learning architecture [6, 7]. In this chapter, a proposal of one of the efficient models of CNN architecture to analyse the MRI brain images. Also, this model is expected to detect the AD based on pre-trained AlexNet CNN, a better classification model may be obtained for reducing the time complexity. The proposed technique is not only used for medical image data, but it can also be used for researcher and physicians for predicting new data. The 3D MRI brain image obtained from the ADNI organization. The proposed model has some important features.

- It is a modified Pre-trained CNN model. Since softmax is the major block in classifying new classes, modifying the softmax may directly help.

- The proposed model works based on the transfer learning model to classify the AD from normal control.
- For evaluation of the proposed model, different grey matter levels are used to obtain the various stages of AD.

The organization of this chapter consist, Sect. 2, describes the related work. Proposed methodology is demonstrated in Sect. 3. Experimental setup and results are explained in Sect. 4, and Sect. 5 accommodate conclusion of the chapter.

2 Related Works

Recently, various types of classification have been proposed in AD disease. To diagnosis the AD, various imaging modalities are available in ADNI organizations. The medical imaging modalities are functional MRI, structural MR scans, PET data, DTI and so on. This AD and MCI detect using CAD system. MRI biomarkers of dementia are a very active research field. In this section, some recent published papers using MRI biomarkers to discriminate dementia along with the level of neural network classification used for AD detection

2.1 Machine Leaning Model Classification Techniques

Machine learning (ML) is a subset of artificial intelligence which involves any technique trying to mimic human behaviour. Machine learning is concerned with the use of statistical methods and the development of algorithms that allow the machine to learn automatically from data and improve with experiences.

Raut et al. [8] extracted texture feature and shape feature from the hippocampal region of the brain using GLCM and machine learning technique of Moment Invariants (ML). Also, classified using Error Back Propagation (EPB) algorithm. The performance of artificial neural network classifier approach reached an average accuracy value of 86.8%. The different feature extraction and various automatic classifications (PCA, LDA, ICA, SVM, KNN, Naïve Bayes, and ANN) of MRI brain images were evaluated by using ML approach [9] with high accuracy for AD classification. The primary focus of this model is to improve the accuracy of AD and help researchers and radiologists to detect the various stages of AD with improved accuracy.

Farhan et al. [10] were proposed the automated image-processing approach. The proposed approach to identify the early stages of AD through brain MRI images. A dataset of 85 age and gender-matched individuals was selected from the OASIS database and the features selected were the size of the hippocampus and the volumes of GM, WM and CSF for identification of disease from MRI brain images. In the classification step, three different machine learning classification models—SVM,

MLP, and J48 were used for distinguishing AD patients from normal levels. Additionally, the ten-fold cross-validation method is utilized for evaluating the efficiency of the model. The proposed model reaches the classification sensitivity of 87.5%, accuracy of 93.75%, and specificity of 100% using the size of left hippocampus features. Zhe Xiao et al. combined GM volume, GLCM, and GF from the structural brain of the MRI images [1]. Then proposed feature selection algorithm SVM-RFE (extends to Recursive Feature Elimination) and covariance are combined to achieve the classification accuracy of AD or MCI from normal control.

Researchers in [11] have extracted features from MRI volumes and GM segmentation volume to reach the early stages of AD through brain MRI Images. Principal Component Analysis (PCA) was employed, for reducing the dimension of the extracted features for the diagnosis of AD with accurate results. The proposed model exhibits high accuracy of 93.86% for the OASIS dataset. Taqi et al. [12] proposed Normalized Hu Moment Invariants (NHMI) to extract the features through MRI image scan. Finally, the model implemented for the classification process was done and using K-nearest neighbour (KNN) and SVM classifier. Their model achieved accuracy is 100% for SVM and 91.4% for KNN on ADNI dataset.

Zhang et al. [13] feature extracted from MRI image using digital wavelet transform (DWT), principal compound analysis (PCA) employed for feature reduction. Then the author also used KSVM with RBF kernel for optimization. The model achieved an accuracy of 91.33%. In Reference [14], the author proposed automatic classification technique from brain sMRI images for AD detection, mostly important visual features, which are extracted from the hippocampal area to differentiate AD and MCI from NC. The model reaches the accuracy level of 87% and 85% for AD/NC and 72.23% and 78.22% for AD/MCI and MCI/NC on 218 ADNI dataset.

The authors in [15] have proposed a model to differentiate AD and NC patients using machine learning techniques on brain images. DWT was utilized to extract the features, PCA is employed to reduce the dimensionality features, and logistic regression is used for classification performance to achieve the high accuracy in AD detection. The proposed model reaches the classification sensitivity of 98.82%, the accuracy of 97.80%, and specificity of 80.00% with the help of 5-Fold cross-validation performance metrics.

Kim et al. [16] proposed 3 frontotemporal dementia (FTD) syndrome patients are classified for excellent classification. Then the author combined 3 FTD subjects into 1 diagnostic group by using machine learning classification approach through 3D MRI brain images. Finally, PCA and LDA were employed for the diagnosis of AD and cortical thickness with accurate results.

2.2 AD Detection Using Deep Learning

CNNs have achieved excellent classification accuracy, one of the major shortcomings of these networks is the lack of transparency in decision making, particularly the limited understanding of the internal features extracted by each convolutional layer. During training process, convolutional filters are optimized in each layer and convolved output is recorded as feature maps. Initial Conv filters are expected to learn lower level features and deeper layers extract higher label features [17, 18] and eventually converge to the final classification. As the filters are formatted as a simple matrix to explain these filters directly. Thus, the inadequate network transparency restricts the further optimization possibility as well as evaluation robustness [18, 19].

In recent days, the deep learning models have been increased to analyse the AD detection from normal control with high performance to various conventional techniques. DL approaches such as CNN architectures, AlexNet, ResNet, Inception block, along with the LeNet model have been reported to have high accuracies with a huge number of training data and the use of medical imaging high-resolution data. Gupta et al. [20] used sparse autoencoder algorithm to extract the handcrafted features from the ADNI dataset. Further, the authors have employed the combination of convolutional layer to obtain the accuracy with high performance classification from normal images.

In literature [21] the author proposed TL method for the clinical diagnosis of AD from brain MRI images and segmented into GM tissue using spm-Dartel tool and registered with AAL to identify the relevant brain regions. The feature selection was based on information obtained and it reduced the features from 90 to 7, controlling the redundancy efficiently. Further, the authors have employed TrAdaboost algorithm to classify AD from normal controls on the ADNI database. The proposed model reaches the highest accuracy of 93.7% and specificity of 100%, respectively.

In literature [22], the author proposed different stages and early stages of AD using deep convolutional neural network (CNN) through MRI brain images. The proposed model uses small OASIS dataset and achieved high accuracy for AD detection. They train a deep learning model to achieve high performance. Bi et al. [23] used CNN for feature extraction and employed unsupervised predictor to reach good performance in AD detection. Finally, the author proposed 3 orthogonal panels and 1 slice using MRI images. The deep CNN model attained an accuracy level of 97.01% for AD/MCI, 92.6% for MCI/NC on 1075 subjects from ADNI 1 1.5T. Anitha et al. [24] put forward a model to detect the AD in brain MRI by utilizing ANN and DNN techniques to extract the best features from cortex thickness, corpus callosum and hippocampus shape. In the model, both deep neural networks were compared for the AD and NCS of MRI scans (Table 1).

In this literature, two well-known commonly available datasets named as ADNI and OASIS are employed for the classification task. Image pre-processing is carried out with the help of SPM and Gaussian kernel. The dataset undergoes data augmentation for resolving the class issues in order to get the good performance. Generally, SVM is defined as a well-known and effective model used by various developers for supervised classification. If the input and output are provided, then input details are applied as a training set for data classification and it is similar to output. Also, developer has employed SVM as a classifier.

Many ML and CNN architectures are developed to solve image-processing tasks such as feature extraction, feature selection, segmentation, along with detection. Moreover, in the traditional pipeline, the researches focused mainly on the binary

Table 1 Summary of reviewed AD detection and classification models

Authors	Objective	ML/DL	Methods	Class labels	Performance	Limitations
Raut [8]	The author extracted texture shape features to predict the AD disease in the early stages with high accuracy	ML	GLCM, error back propagation (EPB) algorithm	AD, NL, MCI	Accuracy: 86.8%	Focused texture features from the hippocampus regions only
Lohar [9]	Different feature extraction and automatic classification techniques are used to analyse the AD with the help of CAD	ML	PCA, LDA, ICA, SVM, KNN, Naïve Bayes	AD, MCI, and normal	–	–
Farhan [10]	Early stages of AD	ML	SVM, MLP, and J48	AD, NL, MCI	Accuracy: 93.75%	The author used small set of features with one ROI of left hippocampus and GM volume with MRI dataset alone
Khajehnejad [11]	To detect the AD from MRI brain images GM segmentation volumes are used	ML	PCA	AD, NL, MCI	Accuracy: 93.86%	Extracted VBM features alone and obtained the whole set of labels
Taqi [12]	The author proposed normalized HuMoment invariants to extract the features	ML	SVM, KNN	AD, NL, MCI	SVM and KNN accuracy is 100%, 91.4%	Two machine learning classifiers (KNN, SVM) with a smaller number of datasets are used. This approaches performed with binary classification

(continued)

Table 1 (continued)

Authors	Objective	ML/DL	Methods	Class labels	Performance	Limitations
Zhang [13]	Extracted features using digital wavelet transform (DWT)	ML	PCA, KSVM, RBF	AD, NL, MCI	Accuracy: 91.33%	The author tested with RBF kernel function alone and classified with 90 MRI images with 5-fold cross-validation
Ahmed [14]	The visual features are extracted from the hippocampal area	ML	–	AD, NL, MCI	Accuracy: 87.00%	Focused visual features from the hippocampus area regions only
Jha [15]	The author have proposed a model to differentiate AD from normal level	ML	DWT, PCA	AD, NL, MCI	Accuracy: 97.80%	Small ML technique to differentiate the normal images from AD and computing time is high. The experimentation was carried from the 90 MRI images with 5-fold cross-validation
Kim [16]	Combined 3 FTD clinical subjects into 1 diagnostic group	ML	PCA, LDA	AD, NL, MCI	The hierarchical classification tree: 75.8% and non-hierarchical classification is 73.0%	Cortical thickness region alone can classified FTD clinical subtypes with cortical atrophy
Gupta [21]	The author used sparse autoencoder algorithm to extract the handcrafted features from MRI brain images	DL	Autoencoder	AD, NL, MCI	Accuracy: 93.7%	Data pre-processing steps do not incorporate prior domain knowledge of GM

(continued)

Table 1 (continued)

Authors	Objective	ML/DL	Methods	Class labels	Performance	Limitations
Islam [22]	Different stages and superior performance for early stages using CNN	DL	CNN	AD, NL, MCI	–	The author tested on AD data alone with limited dataset
Bi [23]	The author used CNN for feature extraction and unsupervised predictor to reach the good performance in AD detection	DL	CNN	AD, MCI and Normal control	Accuracy: AD versus MCI: 95.2% MCI versus NC: 90.63% MCI versus NC: 92.6%	Not investigated the k-means clustering algorithm. Fine-tune the k-means algorithm or employing other clustering algorithms, Gaussian mixture models
Anitha [24]	The author extracted the best features from hippocampus, corpus callosum and cortex thickness to reach the high accuracy	DL	ANN, CNN	AD, NL, MCI	SVM and CNN accuracy are 84.4%, 96.00%	The author has to increase the MRI images to train and get the high accuracy. This paper focused only coronal view of MRI alone. Axial and sagittal view is not focused in this study

classification and ternary classification to differentiate the AD patients. However, the chapter aimed to build a deep neural network to identify AD detection at various stages based on the medical brain MR images with the drawbacks of pre-processing complexity and unreliability of findings. The proposed model overcomes these limitations by proposing the efficient model to identify the AD at various stages. Pre-trained AlexNet work which utilizes transfer learning concept is used for training the model on ADNI dataset for classifying AD with normal individuals. To achieve the measures of accuracy, reproducibility and reliability of the results, scanners, and neuroimaging protocols are utilized.

3 Proposed Methodology

In this chapter, the proposed methodology and goals are going to be clearly exposed. Then the dataset used for this chapter is going to be described.

3.1 3D CNN Architecture: Basic Composition

Convolutional neural networks generally consist of four parts. The first part is the convolution layer (Convolution), the second part is the activation function (ReLU), the next part is pooling, and the last part is full connection layer (fully connected layer) [25]. First, input picture is passed through the convolution kernel to make the feature extraction, then the feature is extracted by the pooling layer by drawn sampling to obtain the feature map. Finally, the fully connected layer is connected to combine all the features to calculate the output score of each class, and the input is mapped to the output to obtain the final classification result. Neurons in a CNN are 3D filters that activate depending on their inputs and connected just to a small region called receptive field of previous neuron activations. The network computes a convolution operation between the connected inputs and their internal parameters and get activated depending on their output and a non-linearity function. The advantage of CNNs over ANNs is that assume inputs are images (width, height, channel). To encode certain properties of MRI images into their architecture: neurons activate when particular features extracted at different locations in the MRI image. Each of the 3D inputs volume called feature map and transforms it to another by means of convolutions and a non-linearity. By stacking layers and downsampling their outputs, the architecture extracts more number of feature maps and complex.

3.2 Convolutional Layer

The convolutional layer operates on the 3-dimensional objects in contrast to the 1-dimensional layers introduced before. Simple comparison is presented in Fig. 1. At

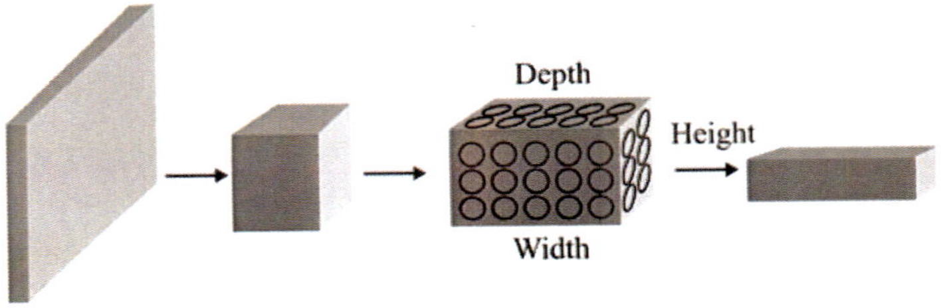

Fig. 1 Standard neural network

first glance, it can be misleading as CNN is supposed to be used for images that are 2-dimensional.

But in fact, most of the images are consisting of 2-dimensional maps of intensities of red, green and blue colour. That gives us the third-dimension depth, known in literature as channels. Other two dimensions are standard: width and height. For most cases, the depth is equal to 3 for colour images and to 1 for black-and-white images. The convolutional layers increase depth as going deeper in the network, as there are much more interesting abstract features. Low level features are edges, colours, blobs that are recognized by the first layers. Abstract concepts like humans, faces, animals can be distinguished in the upper layers, therefore there are more neurons.

In practice, there is even more dimensions as images are transferred through the network in batches, batch size forms 4th dimension. These multidimensional objects are called tensors.

The convolutional layer calculates dot product of the 3-dimensional maps (kernels) sliding (convolving) across spatial (width and height) dimension of the image. These maps have depth of the input tensor and share weights across all positions where applied. Because of that they are invariant to translation in the input image. They also have small spatial size, which is controlled by a hyperparameter called receptive field and is often set to 3 or 5. In that way, every neuron in the map is only locally connected to the input tensor. It is worth noticing the asymmetry between depth dimension and spatial dimension (depth and width). The tensor of weights is 4 dimensional.

$$o[i, j, d] = \sum_{d'=0}^{D-1} \sum_{i'=0}^{kW-1} \sum_{j'=0}^{kH} w\left[i', j', d', d\right].x\left[i.sW - pW + i', j.sH - pH + j', d'\right] + b_d$$

The convolutional layer has the following parameters:

- Input depth D_{in}: number of channels in the input tensor.
- Output depth D_{out}: number of channels in the output tensor.
- Kernel (filter) size kW, kH: size of the 2-dimensional map that is convolved with the spatial dimension.
- Stride size sW, sH: the size of jump between the neighboring applications of the kernel.
- Padding pW, pH: additional zeros on the border of the spatial input.

The stride controls overlapping factor between receptive fields. Bigger strides reduce the size of the output tensor and preserved information. To slide with the kernel across spatial dimension all the sizes of the kernel, stride, padding and spatial dimension must follow the constraint.

$$W_{\text{out}} = \frac{(W + 2.pW - kW)}{sW} + 1 \in N \quad (1)$$

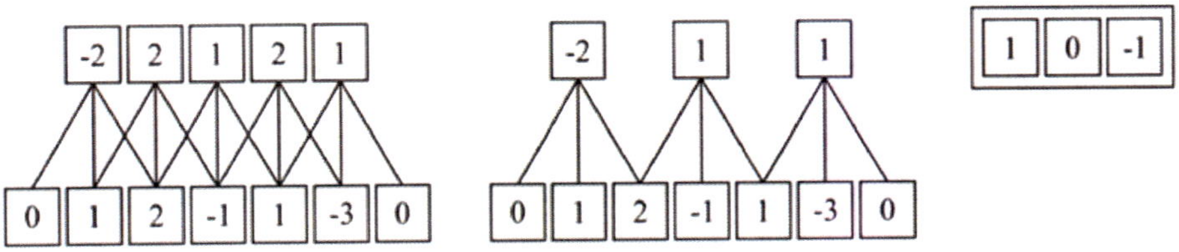

Fig. 2 One spatial dimension and one neuron with kernel size 3 and padding (Left stride is 1, right stride is 2, bias is 0) [26]

$$H_{\text{out}} = \frac{(H + 2.pH - kH)}{sH} + 1 \in N \tag{2}$$

The numbers above must divide without the rest.

Padding is especially convenient when set in such a way that preserve the spatial dimension for stride 1, i.e.

$$pW = (kW - 1)/2 \tag{3}$$

$$pH = (kH - 1)/2 \tag{4}$$

Then, $W_{\text{out}} = W_{\text{in}}$ and $H_{\text{out}} = H_{\text{in}}$. The example explaining stride and padding in the 1-dimensional case is presented on Fig. 2.

3.3 Fully Connected Layers

Fully connected layers have the same of a MLP, or one can see them as a convolutional layer but their filters are fully connected to a feature map input. Those layers perform a matrix multiplication with the feature map input, add a bias, and use nonlinearity function and get activated. The last fully connected layer must be replaced to corresponding categories. The weights are initialized $\sim N(0, 1_e - 4)$.

3.4 AlexNet

The objective of the chapter was to evaluate the features learned by a 3D ConvNet for AD detection with respect to the required layer structure. There were several networks that were trained and evaluated to achieve the higher accuracy with classification performance. The base structure of the 3D ConvNet is described which was modified to conduct various experiments. The base 3D ConvNet was comprised with three to four 3D Conv layers. Each Conv layer was followed by pooling layer and activation layer. After the last Conv layer, the input data was flattened connected with one or more fully connected layers with ReLU activation function. The last layer always was a fully connected layer with softmax activation function. Sometimes after last Conv layer, the output was connected directly to the classification layer. The number of filters, filter sizes, pool sizes were also changed for different structure (Fig. 3).

AlexNet has the following three advantages: (1) In terms of activation function, although the ReLU activation function has been proposed for a long time, it was not until AlexNet successfully using ReLU as the activation of CNN and verified that the outcome in the deep network is better than the sigmoid activation function, and the gradient diffusion problems of sigmoid are successfully avoided to able to continue training and learning. (2) In terms of overfitting problems, use dropout randomly ignores a part of the neurons to form a small neural network to avoid problems such as overfitting of the model. Using the data enhancement, the 224 × 224 size region is randomly intercepted from the original images of 192 × 192 × 192 and 256 × 256 × 256 and the basic processing of the image is performed by horizontal flipping. The large amount of data is added to effectively solve the CNN overfitting problem with many parameters. It greatly improves the generalization ability of the model. (3) Traditionally CNN generally uses average pooling to do downsampling operations, but AlexNet uses maximum pooling to do downsampling operations, which is better than the blurring effect produced by average pooling.

This 3D ConvNet was true with some slight modifications for supervised learning. Over the years, convolution neural networks showed themselves to be state-of-the-art for this challenge. Some of the networks are AlexNet, 2012 [17], VGGNet, 2014 [27], ResNet, 2015 [28] and DenseNet, 2016 [29]. These networks are pre-trained with CNN network on the ImageNet database are frequently used for transfer learning. They provide relatively good feature extractors like AlexNet architecture with the advantage of not being trained again from scratch, which could be highly time-consuming and resources expensive. Along with that AlexNet architecture is fine-tuned for AD detection. The imager size is 256 × 256 × 3, when the batch size is set larger than 16, the memory will be insufficient when it is read into the memory,

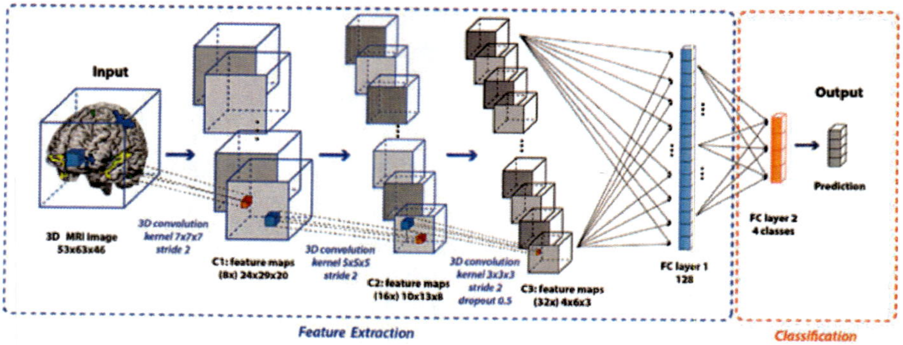

Fig. 3 3D ConvNet with 3 Conv blocks shows how input of size (53 × 63 × 46) transforms through the network. After each conv layer, max-pooling operation reduces the input space by half

Table 2 Experimental situation

Batch_size	16
Input_size	256
Epochs	20
Loss	Binary_crossentrophy
Optimizer	Adam

and it cannot be trained, when the learning rate used is too large, the loss result is difficult to coverage. Because in the end have to perform the two classification, so the loss function of binary class entropy is employed. In terms of the optimizer, used Adam with faster training and better results and adopted default parameters. The initialization parameters are shown in Table 2.

3.5 Transfer Learning

AlexNet is the entry to the ImageNet large-scale image recognition challenge and it was demonstrated good classification result in CNNs. Normally MRI brain image dataset has more than 100 s of sample to train the model and get the good accuracy with the help of transfer learning. The process of entire structure includes load pre-trained network, transfer learning, and train network. First level of the model is loaded with the pre-trained AlexNet architecture through the replacement of last three layers: fully connected layers, a softmax layer, and a classification layer. Finally, adding three layers (fully connected layer, softmax layer, classification layer) in AlexNet. Then fine-tuning the old layers and training the new layer of AlexNet architecture with the ADNI dataset. The pre-trained model already was trained with a huge database of ImageNet with millions of images and proved classification with the help of feature extraction technique. These pre-trained parameters need minor modification to adapt to the new MRI brain images. The modified parameters in the new transferred network are a very small portion of the entire network.

Transfer leaning is commonly used in ML technique. It consists of using the knowledge learnt from one task and applying it to a second one. A domain is a tuple of two elements, the feature space K and a marginal probability, $P(M)$, Where $M = \{m_1, m_2, \ldots, m_n\} \in K$. Let L be the output space. A task T is learnt from the training data consisting of pairs $(m_i, l_i) \in K_T X L_T$ and is definitely $(K_T, L_T, P_T(K, L))$. The main goal of transfer leaning is solving a specific target task $(K_T, L_T, P_T(K, L))$ by using the source data $(K_s, L_s, P_s(K, L))$. The condition has to at least be $K_s \neq K_T$, $L_s \neq L_T$, or $P_s \neq P_T$.

Transfer learning is a statistical technique and getting more attention to train the effective model in deep learning. Utilizing new modified parameters in pre-trained network can using the important features from MRI brain images. These models are mainly for feature extraction and classification with good convergence. Then transferred AlexNet requires high-end GPU to run the model. Adaptive Moment

Estimation (Adam) optimizer is utilized for this parameter learning. Furthermore, the model trained by Stochastic optimization methods for calculating individual learning rates by first and second moments of the gradients descent method to update parameters in training data. The starting learning rate was 1e−4 and the momentum was 0.9. The parameters are re-utilized for the classification of AD.

3.6 Modified Network Architecture

A pre-trained AlexNet architecture is incorporated, replace the final last three layers in a modified AlexNet structure with a 3-layer network. These 3 layers are the smallest "size" indicates the transferred network size is very small due to the layer of fc8 dimension, which is equal to the number of classes. The original AlexNet model is not suitable for brain detection disease. The training sets are similar in these models. Table 2 clearly states that training time is increasing and accuracy values are decreasing with the replacement of a greater number of layers. The activation function of (parameter) ReLU layers and dropout layers are predefined in the network. Only 6th and 9th fully connected layers are needed training because to improve the training accuracy in the model. The decreased accuracy value represents that, the predefined parameters in these layers cause the poor generalization of the trained networks. To overcome this issue, replaced the last three layers and reached the best accuracy. The original AlexNet model is not suitable for analysing the medical (sMRI) images. Hence, the TL concept is employed to retain and improve the AlexNet model.

4 Dataset Description

4.1 ADNI

For the validation of efficient performance of the introduced CNN-TL model, a dataset is collected from the ADNI (http://adni.loni.usc.edu) which consists of sMRI data along with clinical inputs. Researchers from around the world are encouraged by ADNI by providing unrestricted access to MRI data to potentially create techniques. The experiments have been carried out on a window operating system MATLAB R2018b with 8th Gen i7-7500 CPU and Google cloud. The dataset is gathered from a total of 812 baseline T_1 weighted MRIs subjects of different categories were acquired for the study and three levels namely severe, mild, and normal level subjects. Image acquisition standard used for all these images is ADNI harmonized protocols. These all-baseline MRI images are pre-processed by grad warping, B1 correcting and reduction of N3 non-uniformity. In the total number of 812 subjects, a total of 189 subjects comes under the AD category, a total of 396 subjects comes under the MCI category. Also, a total of 227 subjects comes under the NL category. This information is given

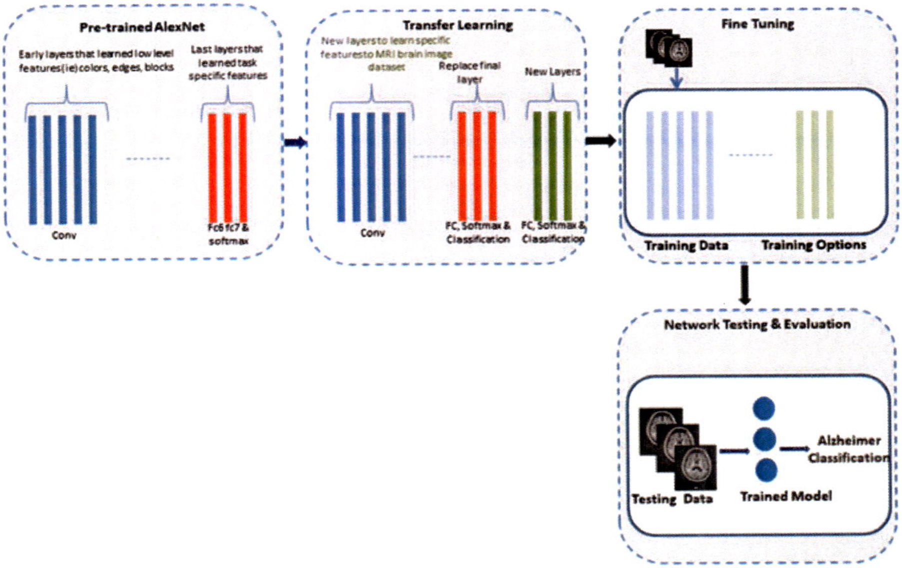

Fig. 4 Proposed methodology

in Table 1 and images are illustrated in Fig. 4. It is MRI structural image, Imaging the structure of each brain tissue with grey-scale 3-dimensional data which has so many slices. It has some labels Age, MMSE, and Education (Table 3).

4.2 *Pre-processing of MRI*

This image data downloaded from ADNI has already undergone some pre-processing steps. This was done in order to limit the difference in the MRI from the different scanning sites and to limit the number of pre-processing strategies in the research literature, which leads to better comparability between difference research method. The pre-processing steps from ADNI: Gradwarp, B1 non-uniformity and N3 algorithm.

The further pre-processing steps in this chapter are done using the SPM12 toolbox in MATLAB. This pre-processing step consists of: Normalization, segmentation, skull striping and smoothing.

Before normalization the images had varying resolution and voxel size, the resolution varied from the 192 × 192 × 160 to 256 × 256 × 166. The MRI scans were normalized to MNI152. The brain template of MNI152 was created from the 3D brain MRI. The normalization is done using the normalization procedure in SPM12, with default settings. After normalization processes, all the images have the resolution 79 × 95 × 79 with a voxel size of 2 × 2 × 2 mm. After normalization, the normalized MRI images are segmented into three tissues, namely GM, WM, and CSF, using

Table 3 Baseline characteristics of participants

Dataset	Group	Subjects	Male/Female	Age (SD) in y	MMSE (SD)	Education	CDR score
ADNI	NL	227	110/117	73.43 ± 7.42 (56.30–91.77)	27.38. ± 1.99(25–30)	14.7(3.1)	0.025(0.11)
	AD	189	112/77	75.30 ± 7.4 (55.72–91.83)	23.80 ± 2.42(18–27)	16.0(2.8)	4.304(1.60)
	MCI	396	165/231	73.38 ± 7.48 (55.21–91.75)	27.18 ± 1.99(24–30)	15.7(3.8)	1.255(0.77)

the standard segmentation procedure in SPM12. For skull stripping, the sum of the segmented GM and WM, with threshold value, is used as a mask to skull strip the MRI.

$$I_{SS} = I_{MNI} * ((I_{GM} + I_{WM}) > T \quad (5)$$

I_{SS} is the skull stripped brain image, I_{MNI} is the MNI152 normalized brain image, I_{WM} and I_{GM} are the segmented images, and T is the threshold value. Higher threshold value gives an more stringent skull stripping. In this work the threshold value is set to 0.2. Then smoothing is done suing a Gaussian smoothing kernel using a kernel size of $3 \times 3 \times 3$ mm.

Performance Measures

For analysing, the classification results obtained by the presented model are evaluated using three evaluation parameters, namely accuracy, specificity, and sensitivity are employed and are briefly defined below.

$$\text{Accuracy} = \frac{\text{TP} + \text{TN}}{\text{TP} + \text{TN} + \text{FP} + \text{FN}} \quad (6)$$

$$\text{Specificity} = \frac{\text{TN}}{\text{TN} + \text{FP}} \quad (7)$$

$$\text{Sensitivity} = \frac{\text{TP}}{\text{TP} + \text{FP}} \quad (8)$$

where FP, FN, TP, and TN means false positive, false negative, true positive, and true negative correspondingly.

5 Experimental Results and Discussion

5.1 Experimental Setting

The model tested in this chapter is implemented using MATLAB 2018b. The model was trained by splitting the training set into a new training set (80% of the original training set) and a validation set (15% of the original training set), with a single subject scan only being in either the training or validation set. The test set used for neural networks is equal to the test set that was used with the classification of features. Because of the excessive computational cost related to 3D CNN, there is a limited set of hyperparameters were tested. The network tested different number of convolutional layer and different dropout in this last fully connected layer.

5.2 Experiments Results

The validation accuracy and training loss of progress are presented in Fig. 5 and confusion matrices are constructed for visualizing the classification model on test or validation dataset. To assess the classification models, the performance metrics (specificity, accuracy, and sensitivity) calculated through confusion matrix. The proposed model attained 97.80% for accuracy, 86.00% for specificity and 98.80% sensitivity.

The accuracy over the training period in epochs represented in Fig. 7, The validation set denoted in orange colour and the training set indicated in Grey colour. The *X*-axis correspond to the epochs and the accuracy in *Y*-axis. Figure 5 shows that the training loss is decreasing with the increasing epochs together with validation loss so that it can certifies that the shared weigh parameters transfer learning can avoid overfitting problem and it works well both on training set and validation set. The less weight parameters, the less computation needs increase the speed our training time (Table 4 and Fig. 6).

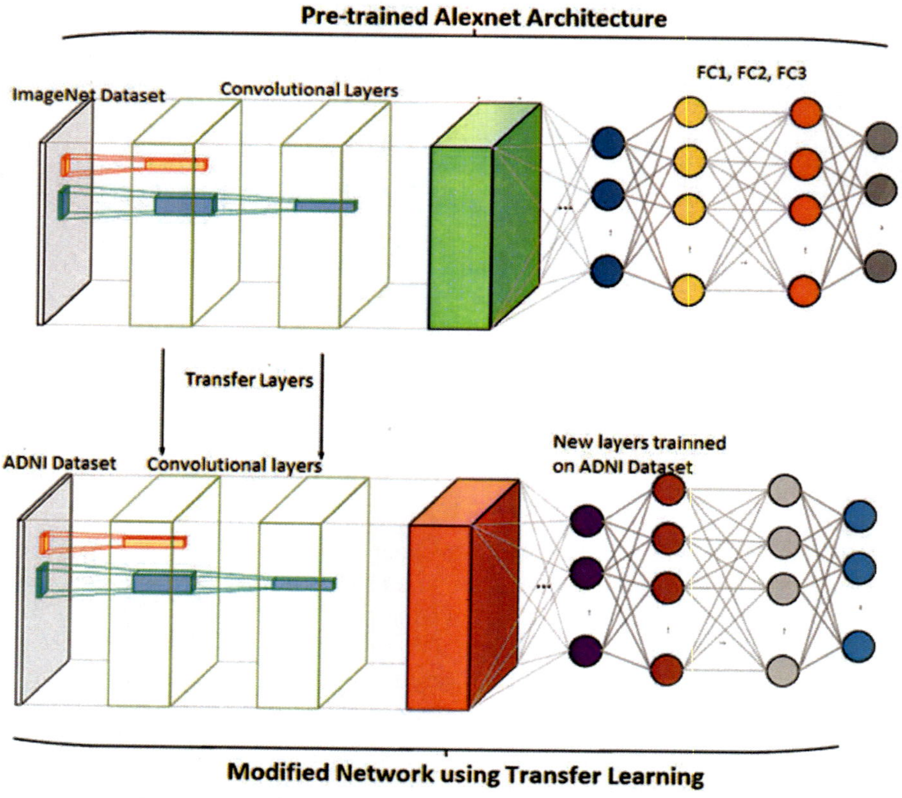

Fig. 5 Transferred CNN layers

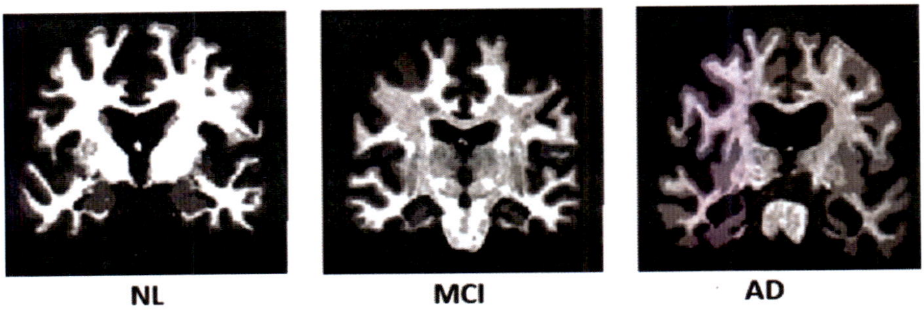

Fig. 6 Sample test images

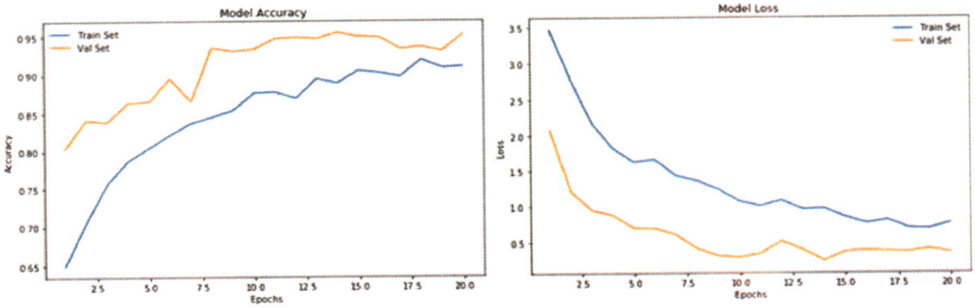

Fig. 7 Validation accuracy and training loss of 20 epochs

Table 4 Transfer learning and performance on ADNI dataset

Dataset	Classification	Number of layers	Accuracy (%)	Sensitivity (%)	Specificity (%)	Training time
ADNI	Binary	3	97.2	98.12	81	2 min 15 s
ADNI	Binary	6	96	97	84	3 min 49 s
ADNI	Binary	9	65.7	68.6	53.4	4 min 39 s

5.3 *Performance Matrix*

The method performed on the testing data one way to do this by creating a confusion matrix for our method. The confusion matrix represented in rows and columns corresponds to the predicted machine learning algorithm and known truth. Since there are only two categories to choose from the AD or does not have AD. The top-left corner contains true negatives (TN) and these are patients that did not have an AD that was correctly identified by the model. True positives (TP) are in the bottom right-hand corner. These are patients that had an AD that was correctly identified by the model. The bottom left-hand corner contains false negatives (FN) are when a patient has AD,

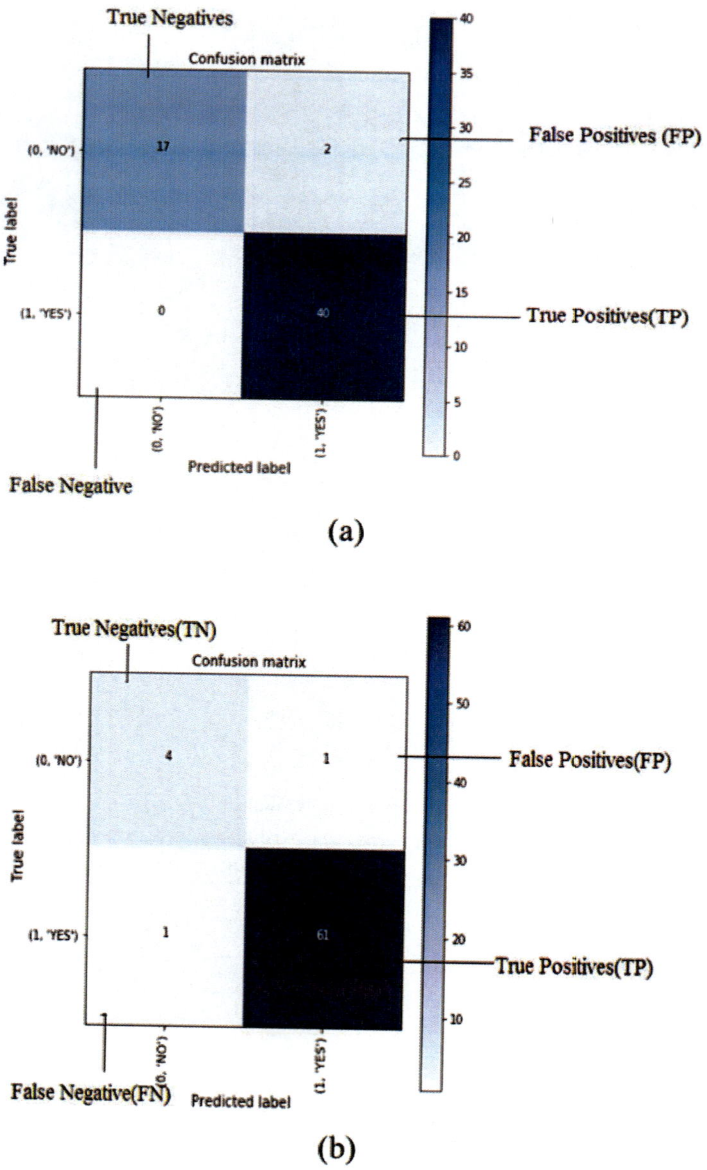

Fig. 8 Confusion matrix for neural network model: **a** MCI to NL, **b** MCI to AD

but the model said did not. Lastly, the top right-hand corner contains false positives (FP). False positive does not have AD, but the model said to do.

Table 5 Comparative results of different models

Author (year)	Imaging modality	Classification methods	Validation method	ACC (%)	SEN (%)	SPE (%)
Farhan et al. [30]	MRI	SVM + MLP + J48	10-Fold	93.75	87.5	100
Xiao et al. [31]	MRI	SVM-RFE	10-Fold	92.86	87.04	98.28
Taqi et al. [32]	MRI	KNN + SVM	–	91.4	83.3	–
Zhang et al. [6]	MRI	DWT + PCA + KSVM + RBF	5-Fold	91.33	92.47	72
Jha et al. [10]	MRI	DWT + PCA + LR	5-Fold	97.0	98.00	80
Proposed model	**MRI**	**AlexNet + TL**	**10-Fold**	**97.2**	**98.12**	**81**

Note The highest value shown in bold
ACC accuracy, *SPE* specificity, *SEN* sensitivity

5.4 Comparison of the Proposed Model Results with State-of-Arts Method

The experimental analysis of the proposed AlexNet with TL method is discussed with recent approaches. TL method is better than training from scratch method both in accuracy and speed. There are several reasons: there is a limited number of samples to train the model so that it is overfitting. The comparative results are displayed which is shown in Table 5. Many ML has developed with feature extraction and classification with the same MRI brain datasets for identification of AD.

The proposed model compares 5 methods and best performance model presented here.

It has been found that the accuracy and sensitivity hike in the proposed model, as shown in Table 5 and Fig. 9, is due to the transfer learning concept employed with revised AlexNet structure in a deep convolutional neural network. The low accuracy value found in sensitivity classification from the proposed model recommends that still some original parameters are needed to improve the generalization performance. The original AlexNet model is not suitable for analysing the medical (sMRI) images. Hence, the TL concept is employed to retain and improve AlexNet. The accuracy value compared with the result of different models as shown in Fig. 7. Early detection of AD proved to be complicated for the methods tested in this chapter. The differential diagnosis classification problem, with AD, MCI, and NL, the best achieved accuracy was 97.2% in Table 3. However, this had a much higher validation accuracy, which indicates poor generalizability (Fig. 8).

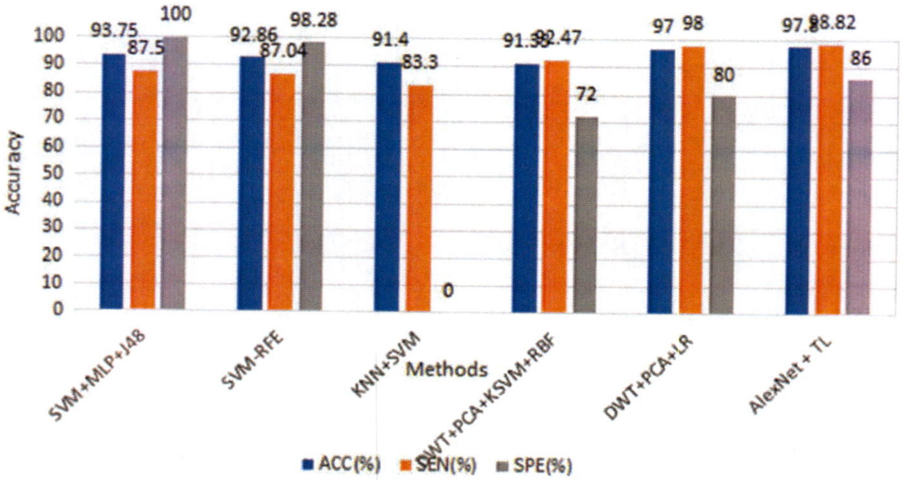

Fig. 9 Accuracy, specificity and sensitivity comparison with state-of-the-art methods

6 Conclusion

This chapter has discussed the detection of AD based on transfer learning classification model for binary classification and multiclass problems. The proposed model utilized the pre-trained CNN network (AlexNet) and fine-tuned with the segmented 3D MRI brain images for performance improvement. The conducted experiments showed that the pre-trained CNN architectures employed have shown better performance in terms of different measures such as accuracy, specificity and sensitivity. The research work can be extended in order to improvise the other CNN architectures and fine-tuning of all the convolutional layers.

References

1. Z. Xiao, Y. Ding, T. Lan, C. Zhang, C. Luo, Z. Qin, Brain MR image classification for Alzheimer's disease diagnosis based on multifeature fusion. Comput. Math. Methods Med. 1–13 (2017)
2. I. Beheshti, H. Demirel, Feature-ranking-based Alzheimer's disease classification from structural MRI. Magn. Reson. Imaging **34**(3), 252–263 (2016)
3. S. Belleville, C. Fouquet, S. Duchesne, D.L. Collins, C. Hudon, Detecting early preclinical alzheimer's disease via cognition, neuropsychiatry, and neuroimaging: qualitative review and recommendations for testing. J. Alzheimer's Dis. **42**, S375–S382 (2014)
4. Z. Lao, D. Shen, Z. Xue, B. Karacali, S.M. Resnick, C. Davatzikos, Morphological classification of brains via high-dimensional shape transformations and machine learning methods. Neuroimage **21**(1), 46–57 (2004)
5. G. Fung, J. Stoeckel, SVM feature selection for classification of SPECT images of Alzheimer's disease using spatial information. Knowl. Inf. Syst. **11**(2), 243–258 (2007)

6. M. Lim, D. Lee, H. Park, Y. Kang, J. Oh, J. Park, Convolutional neural network based audio event classification. KSII Trans. Internet Inf. Syst. **12**(6), 2748–2760 (2018)
7. J. Lee, D. Jang, K. Yoon, Automatic melody extraction algorithm using a convolutional neural network. KSII Trans. Internet Inf. Syst. **11**(12), 6038–6053 (2017)
8. A. Raut, V. Dalal, A machine learning based approach for early detection of Alzheimer's disease by extracting texture and shape features of the hippocampus region from MRI scans. IJARCCE **6**(6), 320–325 (2017)
9. R.P. Lohar, Mamata, A survey on classification methods of brain MRI for Alzheimer' s disease. Int. J. Eng. Res. Technol. **7**(05), 339–348
10. S. Farhan, M.A. Fahiem, H. Tauseef, An ensemble-of-classifiers based approach for early diagnosis of Alzheimer's disease: classification using structural features of brain images. Comput. Math. Methods Med. 1–11 (2014)
11. M. Khajehnejad, F.H. Saatlou, H. Mohammadzade, Alzheimer's disease early diagnosis using manifold-based semi-supervised learning. Brain Sci. **7**(8), 1–19 (2017)
12. A. Mohammed, F. Al-Azzo, M. Milanova, Classification of Alzheimer disease based on normalized Hu moment invariants and multiclassifier. Int. J. Adv. Comput. Sci. Appl. **8**(11), 10–18 (2017)
13. Y. Zhang, L. Wu, An MR brain images classifier via principal component analysis and kernel support vector machine. Prog. Electromagn. Res. **130**, 369–388 (2014)
14. O. Ben Ahmed, J. Benois-Pineau, M. Allard, C. Ben Amar, G. Catheline, Classification of Alzheimer's disease subjects from MRI using hippocampal visual features. Multimed. Tools Appl. **74**(4), 1249–1266 (2014)
15. D. Jha, G.-R. Kwon, Diagnosis of Alzheimer's disease using a machine learning technique. Alzheimer's Dement. **13**(7), P1538 (2017)
16. J.P. Kim, et al., Machine learning based hierarchical classification of frontotemporal dementia and Alzheimer's disease. NeuroImage Clin. **23** (2019)
17. A. Krizhevsky, L. Sutskever, G.E. Hinton, ImageNet classification with deep convolutional neural networks, in *NIPS'12 Proceedings of the 25th International Conference on Neural Information Processing Systems*, vol. 1, 2012, pp. 1097–1105
18. M.D. Zeiler, R. Fergus, Visualizing and understanding convolutional networks. Springer Int. Publ. Switz. **12**, 818–833 (2014)
19. S.J. Pan, Q. Yang, A survey on transfer learning. IEEE Trans. Knowl. Data Eng. **22**(10), 1345–1359 (2010)
20. A. Gupta, M.S. Ayhan, A.S. Maida, Natural image bases to represent neuroimaging data, in *Proceedings of 30th International Conference on Machine Learning (ICML 2013) will be held in Atlanta*, vol. 28, 2013, pp. 16–21
21. K. Zhou, W. He, Y. Xu, G. Xiong, J. Cai, Feature selection and transfer learning for Alzheimer's disease clinical diagnosis. Appl. Sci. **8**, 3–15 (2018)
22. J. Islam, Y. Zhang, Brain MRI analysis for Alzheimer's disease diagnosis using an ensemble system of deep convolutional neural networks. Brain Inform. **5**(2), 2–14 (2018)
23. X. Bi, S. Li, B. Xiao, Y. Li, G. Wang, X. Ma, Computer aided Alzheimer's disease diagnosis by an unsupervised deep learning technology. Neurocomputing (2019)
24. K.A.N.N.P. Gunawardena, R.N. Rajapakse, N.D. Kodikara, Applying convolutional neural networks for pre-detection of Alzheimer's disease from structural MRI data, in *2017 24th International Conference on Mechatronics and Machine Vision in Practice, M2VIP 2017*, pp. 1–7, 2017
25. CS231n: Convolutional neural networks for Visual Recognition. (Online). Available: http://cs231n.stanford.edu/2017/
26. F.-F. Li, J. Johnson, S. Yeung, Cs231N_2017_Lecture9, 2017
27. K. Simonyan, A. Zisserman, Very deep convolutional networks for large-scale image recognition, in *3rd International Conference on Learning Representations, ICLR 2015—Conference on Track Proceedings*, 2015, pp. 1–14
28. R.C. O'Reilly, D. Wyatte, S. Herd, B. Mingus, D.J. Jilk, Deep residual learning for image recognition kaiming. Front. Psychol. **4**, 770–778 (2013)

29. G. Huang, Z. Liu, L. Van Der Maaten, K.Q. Weinberger, Densely connected convolutional networks, in *Proceedings of 30th IEEE Conference on Computer Vision and Pattern Recognition, CVPR 2017*, 2017, pp. 2261–2269
30. H. Hanyu, T. Sato, K. Hirao, H. Kanetaka, T. Iwamoto, K. Koizumi, The progression of cognitive deterioration and regional cerebral blood fl ow patterns in Alzheimer' s disease: a longitudinal SPECT study. J. Neurol. Sci. **290**(1–2), 96–101 (2010)
31. D.H. Salat, J.A. Kaye, J.S. Janowsky, Prefrontal gray and white matter volumes in healthy aging and Alzheimer disease. Arch. Neurol. **56**, 338–344 (2017)
32. Y. Zhang, Z. Dong, P. Phillips, S. Wang, G. Ji, Detection of subjects and brain regions related to Alzheimer' s disease using 3D MRI scans based on eigenbrain and machine learning. Front. Comput. Neurosciecne **9**(June), 1–15 (2015)

Data Comparison and Study of Behavioural Pattern to Reduce the Latency Problem in Mobile Applications and Operating System

B. N. Lakshmi Narayan, Smriti Rai, and Prasad Naik Hamsavath

Abstract As the range of telecommunications applications is increasing unprecedentedly, mobile phones will most likely enhance the quality of communication among human lives. On the other hand, this constitutes to extreme social implications. The mobile phones are further progressed to become as instruments for leveraging instant correspondence and enhanced cooperation. Furthermore, it has been innovatively utilized for business purposes and among youngsters. Meanwhile, the number of sensors integrated in mobile phones and the applications running on them are also increased considerably. In this paper, a detailed comparative study and analysis have been performed on behavioural patterns in order to reduce the latency-related challenges in operating system and mobile applications.

Keywords Latency · Behavioural pattern · JIT

1 Introduction

The working structure of mobile phones, PDAs, tablets, etc., will be a portable working framework. For example, the framework of personal computers (PCs) is not considered versatile since they were initially developed for desktop usage, and it does not include any 'portable' components. This arrangement is vague in some of the recent frameworks, which are hybrid as it is developed for both the set-ups.

The portable framework will integrate the components of the framework available in PC along with various elements that remain suitable for delivering a versatile and handheld usage. It typically includes the currently available versatile frameworks that comprise a mobile phone with touchscreen, bluetooth, GPS portable route, Wi-Fi, discourse acknowledgment, camera, camcorder, voice recorder, music player, infrared, and close field correspondence.

B. N. Lakshmi Narayan (✉) · S. Rai · P. N. Hamsavath
Department of Master of Computer Applications, Nitte Meenakshi Institute of Technology, Yelahanka, Bengaluru, Karnataka, India

S. Shakya et al. (eds.), *Mobile Computing and Sustainable Informatics*, Lecture Notes on Data Engineering and Communications Technologies 68,
https://doi.org/10.1007/978-981-16-1866-6_49

Mobile phones with functional intersection capacities consist of a versatile working structure—the core client dealing with the programming phase and a low-level constant working construction that runs the radio and other equipment. Research shows that the low-level structure consists of an assessment on the security vulnerabilities by approving harmful base stations in order to build abnormal amounts of controller over the mobile phone.

In recent years, PDAs have begun to implement sensors like GPS, accelerometer, mouthpiece, video camera, and Wi-Fi technologies. They also provide the associated applications and cover other administration costs.

The widespread usage of mobile phones and the increasing quantity of data created from the applications and device sensors are contributing to the discovery of another field in data processing and sociology. Recently, the researchers are taking a special interest on exploring the behaviour and sociology from the perspective of big data. The extensive and versatile amount of data are utilized to observe and analyse certain characteristics and also the human portability, correspondence, and connection patterns.

The government and corporate regulations for surveillance and data insurance play a fundamental role in ensuring the security of every sensitive element present in the portable data. The examination shows that the portable data assets are uncommon and often it is not allowed to test the logical recommendations that are identified with valid exercises.

With the proliferation of end-clients' mobile phones, an extremely large development in the requests of diverse media portable substance is observed by the administrators. To deal with this huge increase in the data volume and number of customers, the administrators intend to consistently enhance their systems' limit and quickly build the system availability and coverage.

Different applications are used to maintain business, education, and services accessible as for business it is identified with client to business (C2B), business to client (B2C), business to business (B2B), and client to customer (C2C). With regards to instruction, there are online training, preparation, and improvement. Further, it manages the government-managed savings with the utilization of RFID. Moreover, these applications are accessible in portable interface, tablet and desktop.

Bergman et al. [1] proposed a method based on the user-subjective way to leverage personal data management systems layout, which gives the evidence and later applied on managing the personal data. Chau et al. [2] explain the search decline, which finds data by federation. Chen et al. [3] designed an algorithm iMecho: A close memory positioned desktop exploration system, which displays an adequate conclusion. Dumais [4] explained about the stuff I have seen: A system for personal data retrieval and re-use. Det al [5] characterize about the semantic file systems. Badis and Al Agha [6] approved a system on QOLSR, QoS routing for ad hoc wireless networks using OLSR. Hall and Anderson [7] explain operating systems for mobile computing. Freeman and Gelernter [8] describe the lifestreams, a storage model for personal data, which illustrates the distinct configuration for maintaining the personal information. Kamra et al. [9] expected an access on growth codes for maximizing sensor network data persistence. Karger and Quan [10] invented a method on Haystack: A user interface for creating, browsing, and organizing the arbitrary semistructured data. Al Agha [11] explained an approach on start and stop for Internet routers. Koetter and Médard [12] made an experiment on algebraic approach for network coding. Kadi and Al Agha [13] described a methodology on network coding-based flooding in ad hoc wireless networks under mobility conditions. Taivalsaari [14] explained a method on the event horizon user interface model for small devices. Teevan et al. [15] made a detailed report on the perfect search engine, which is a study of orienteering behaviour in the directed search. Terkourafi and Petrakis [16] explained an approach on a critical look at the desktop metaphor for 30 years. Marsden and Cairns [17] introduced an algorithm for developing the usability of the hierarchical file system.

2 Problem Statement

The concentration of the activities for data gathering will remain on the stream. A stream is a continuous flow of data, which can be considered to be virtual on the receiver device, where the data is stored by paying less attention towards the application side. The stream would be accessed at a period, where the recent data that enters the framework is placed at the start of the stream and the older data at the point will be pushed to the back.

In internetworking, it is the way moving towards a parcel of data from source to destination. Directing is regularly accomplished by a switch. Steering is a primary anecdote of the Internet, since it grants messages to move from one PC to another and finally reach the target machine [9, 10]. Each transitional PC is executed by directing the message towards the succeeding PC. Some portion of this system involves dissecting a directing table to finish up in the best way.

Business-related problem
Here, a Samsung mobile phone and different Web-based shopping brands like Amazon, Flipkart, and Quikr are taken into consideration. The concentration will be on the client to business (C2B) and business to business (B2B) platform, which analyses the offering and purchasing part of the items that are accessible in these shopping brands.

This research work will check the quick reaction time (QRT) and just in time (JIT) parameters, which fundamentally concentrate on the consumer loyalty, client administration & support, and also, the client mind, which symbolizes C3.

Computing related problems
The figure is computed with the three primary fundamental variables, which include time, space, and cost. Further, it utilizes the different portable OS like Android, IOS, and windows. Moreover, this research work will check the every single possible mix like time and space, space and cost, and time and cost.

3 Objectives

1. To enhance quick response time (QRS) and in the nick of time (JIT) by contrasting the data accessibility and Flipkart, Amazon, and Quikr and checking with C3 consumer loyalty, client benefit support, and client mind.
2. Details of behavioural example of versatile utilization and business administrator.
3. Evaluates the idleness issue regarding time and space by utilizing the multi-versatile working framework.
4. Preparing a model for upgrading calculation in the business handle.

4 Scope

The main goal is to discover the improved and disseminated calculations that can total the activity of the system to a base number of hubs to execute the most extreme of gadgets in the system. On the other side, it organizes coding, which is the another strategy that can enhance the execution of the system in specific circumstances. Those conditions rely on the open door, where the data gets together (e.g. sharing interfaces and multicast activity). At the point when the data gets together, the essential excess to interpret bundles could be built for a shared utilization, and it will consequently decrease the quantity of retransmissions.

5 Methodology

This research work considers the accessible 5000 records from Amazon, Flipkart, and Quikr of electronic and non-electronic items and checks the diagram by using the device.

By checking the records from 2000 to 2015, the softwares utilized in this period are NS2, MATLAB, SAS, and HADOOP.

Equipment for Mobile Set-up

1. **Samsung S6**
   1.5 GHz octa-centre Samsung Exynos 7420 processor, and it accompanies 3 GB of RAM.
2. **Lenovo vibe**
   Components 3G, 5.0″ IPS LCD capacitive touchscreen, 16 MP camera, Wi-Fi, GPS, and Bluetooth.
3. **Laptop: HP**
   Intel® Pentium® CPU N3540, 2.16 Ghz, 2.00 GB, 64-bit OS, x64-based processor.

6 Conclusion

The framework is acceptable for novice users. In addition, through the question-based associations, skilled users are more capable on utilizing the framework in quick and productive ways. While it excludes the complete functionality of a portable UI, the larger part of getting the data and the attached serialized frameworks are made available for use along with the UI.

The serialized framework will handle all data association and recovery and make it available whenever it is required. However, the actual access and data handling will be done by the suitable framework devices and any networking devices that are connected to Wi-Fi.

References

1. O. Bergman, et al., The user-subjective approach to personal data management systems design: evidence and implementations. J. Am. Soc. Data Sci. Technol. (2008)
2. D.H. Chau, et al., What to do when search fails: finding data by association, in *CHI '08 Proceeding of the Twenty-Sixth Annual SIGCHI Conference on Human Factors in Computing Systems* (ACM, New York, 2008), pp. 999–1008
3. J. Chen, et al., iMecho: an associative memory based desktop search system, in *CIKM '09 Proceeding of the 18th ACM Conference on Data and Knowledge Management* (ACM, New York, 2009), pp. 731–740
4. S. Dumais, Stuff I've seen: a system for personal data retrieval and re-use, in *SIGIR '03 Proceedings of the 26th Annual International ACM SIGIR Conference on Research and Development in Information Retrieval* (ACM, New York, 2003), pp. 72–79
5. D. Gifford, P. Jouvelot, M. Sheldon, J. Toole, Semantic file systems, in *Proceedings of the Thirteenth ACM Symposium on Operating System Principles* (ACM Press, New York, 1991)
6. H. Badis, K. Al Agha, QOLSR, QoS routing for ad hoc wireless networks using OLSR. Eur. Trans. Telecommun. J. **16** (2005)
7. S. Hall, E. Anderson, Operating systems for mobile computing. J. Comput. Sci. Colleges **25**(2), 64–71 (2009)
8. E. Freeman, E. Gelernter, Lifestreams: a storage model for personal data. SIGMOD Record **25** (1996)
9. A. Kamra, V. Misra, J. Feldman, D. Rubenstein, Growth codes: maximizing sensor network data persistence, in *SIGCOMM'06* (ACM, New York, 2006), pp. 255–266
10. D.R. Karger, D. Quan, *Haystack: A User Interface for Creating, Browsing, and Organizing Arbitrary Semistructured Data* (CHI Vienna, Austria, 2004)
11. K. Al Agha, Start and Stop for Internet Routers. Patent FR-INPI 09 58890, 2011
12. R. Koetter, M. Médard, An algebraic approach to network coding, in *INFOCOM 2002*
13. N. Kadi, K. Al Agha, Network coding based flooding in ad-hoc wireless networks under mobility conditions, in *Annals of Telecommunication*, vol. 65 (Springer, Berlin, 2010)
14. A. Taivalsaari, *The Event Horizon User Interface Model For Small Devices*, SML Technical Report, Sun Microsystems, 1999
15. Teevan, et al., The perfect search engine is not enough: a study of orienteering behavior in directed search, in *Proceedings of the SIGCHI Conference on Human Factors in Computing Systems*, 24–29 Apr 2004, Vienna, Austria, pp. 415–422

16. M. Terkourafi, S. Petrakis, A critical look at the desktop metaphor 30 years on, in *Researching and Applying Metaphor in the Real World*, Chapter 8
17. G. Marsden, E. Cairns, Improving the usability of the hierarchical file system, in *Proceedings of SAICSIT 2003*, pp. 111–113

Breast Cancer Detection Using Machine Learning

A. Sivasangari, P. Ajitha, Bevishjenila, J. S. Vimali, Jithina Jose, and S. Gowri

Abstract The most commonly diagnosed and second leading cause of cancer fatalities is breast cancer in women. AI and IoT integrated system that can help diagnose earlier breast cancer. The key tool for detecting breast cancer is mammograms. Yet cancer cells in breast tissue are difficult to identify. Since it has less fat and more muscle. To examine the irregular areas of density, mass and calcification that signify the presence of cancer, digitized mammography images are used. Several imaging techniques have been developed to detect and treat breast cancer early and to decrease the number of deaths, and many methods of diagnosis of breast cancer have been used to increase diagnostic accuracy. The current technique does not detect breast cancer reliably in the early stages, and most women have suffered from this.

Keywords Breast cancer · SVM · Mammogram · Accuracy · Machine learning

1 Introduction

Mammography is not widely used for screening because of ionizing radiation and associated health risks, whereas the techniques of magnetic resonance imaging (MRI) or positron emission tomography (PET) are expensive and not sufficiently available. Women who have come to undergo a regular mammography will detect this disease in time and thus have the chance to obtain a less aggressive disease. While this test was successful in early detection, there is still a high percentage of false positives and false negatives that cause patients to undergo unnecessary, more intrusive treatment and/or testing that leads to anxiety, higher costs, and long-term psychosocial harm. Artificial intelligence, as a scientist, maybe just as successful in predicting the spread of breast cancer. AI algorithms perform as well or better than pathologists in the detection of the spread of breast cancer to lymph nodes. Identifying cancer metastases in lymph node tissue is a complex, repeated, and time-consuming task.

A. Sivasangari (✉) · P. Ajitha · Bevishjenila · J. S. Vimali · J. Jose · S. Gowri
School of Computing, Sathyabama Institute of Science and Technology, Chennai, India

Bevishjenila
e-mail: bevishjinila.it@sathyabama.ac.in

S. Shakya et al. (eds.), *Mobile Computing and Sustainable Informatics*, Lecture Notes on Data Engineering and Communications Technologies 68,
https://doi.org/10.1007/978-981-16-1866-6_50

In reality, breast cancer is a malignant tumor formed by breast cells. A computer-aided detection (CAD) system that uses a machine learning technique that provide effective breast cancer diagnosis is required. These CAD systems can help at an early stage in the detection of breast cancer. The survival rate improves when breast cancer is diagnosed early enough, so better care can be given. Machine learning methodologies are commonly used in the detection of breast cancer through the development of computer-aided diagnosis. Deep learning has been highlighted in recent years, considering feature engineering, for the improved performance of the computer-aided system.

This paper presents a detailed study of current models of cancer detection and presents findings that are highly accurate and reliable. This paper is arranged to have four parts. In Sect. 2, the literature and current works are discussed. The suggested approach has been carried out in Sect. 3. Section 4 discusses the conclusions and discussions. In comparison with other models, the results presented have been shown to be reliable and effective.

2 Related Work

One of the most reliable methods used in hospitals and clinics for the early detection of breast cancer is mammography. Reducing mortality by as much as 30 percent has been shown to be successful. The primary aim of mammography screening is to identify and remove the cancerous tumor early before metastasis is developed. Masses and micro-calcification are the early symptoms of breast cancer, but lesions and normal breast tissues are also difficult to discern.

A clinical prototype for microwave breast cancer detection was proposed by Jeantide et al. [1]. This prototype will have a patient interface that is wearable, and no patient evaluation table is needed. More cost-effective is this wearable prototype. Using this wearable prototype, breast screening on a healthy volunteer is completed. With our table-based prototype, the resulting data is compared. Each signal and image analysis showed that the wearable prototype is an improvement over the prototype based on the table. In addition, as shown by the microwave prototype, the daily figures, crossing through a solitary menstrual cycle, gave a norm for safe tissue variation.

A compact instrument consisting of an expendable biochip for estimating electro-thermo-mechanical (ETM) properties of breast tissue has been designed and developed by Pandya et al. [2]. By integrating micro-engineered biochip and 3-D printing, they created a versatile tumor determination gadget. Using this gadget, we have demonstrated a measurably significant distinction between harmful, natural breast tissues in mechanical firmness, electrical resistivity, and thermal conductivity. For simultaneous ETM estimations of breast tissue examples, this gadget is ideal and can be a potential opportunity to represent ordinary and malignant breast tissue centers. It is also conceivable that the flexible tumor analysis approach will provide deterministic, quantifiable information on the characteristics of the bosom tissue as well as the initial and disease movement of the tissues. For other tissue-related diseases, the system may conceivably be used.

Late achievements in the improvement of SiNW-FET biosensors were suggested by Francesca Puppo et al. (2016) for unique, label-prone and extremely sensitive immunodetection. The feasibility of using SiNW-FET biosensors for early detection of disease markers in specific applications with tumor extricates was then demonstrated as a demonstration of this breakthrough in order to operate efficiently on the basis of genuine patients. The findings are seen in simple and highly sensitive antigen detection with *p*-type SiNW-FETs [3].

The classification algorithm was suggested by Mohammed et al. [4] and showed that the recursive feature selection algorithm selects the best subset of features, and that the SVM classifier achieved optimum classification efficiency for this best subset of features. To build a framework that allows for real-time and remote health monitoring based on IoT infrastructure and associated with cloud computing, the authors have applied various machine learning techniques and considered public health care datasets stored in the cloud. The framework will be capable of directing recommendations based on cloud-based historical and empirical data [5].

The authentication is carried out initially by the patient. Then, the sensed values are achieved from the sensors installed inside the innerwear. Later, the proposed algorithm securely uploads the collected data to the hospital public cloud server [6]. One-class support vector machines have recently been implemented extensively. It can be used in anomaly detection. It seeks to find an optimal hyperplane in high dimensional data that best separates the details from maximum-margin anomalies. The traditional one-class hinge failure, however, allows vector machines to be unbounded, resulting in greater losses caused by outliers affecting the efficiency of their anomaly detection. Established methods are computationally complex for larger data [7]. The identification of malignant cells is accomplished in the proposed method by extracting different shapes and textured-based characteristics, when the classification is performed out using three well-known classification algorithms. The most creative aspect of the proposed solution is the use of Evolutionary Algorithms (EA) to select optimal features, decrease computational complexity and speed up the cloud-based e-Health service classification process [8]. This paper presents the mammogram classification using the function extracted using Hough transformation. The mammogram classification using the function derived using the transformation of Hough is discussed in this paper. A two-dimensional transformation is the transformation of Hough. It is used in an image to differentiate features of a specific shape. The risk and mechanized identification of the two most important markers of miniaturized scale characterization and mass are extremely relevant for early breast cancer diagnosis. Since masses from the parenchymal position of the computerized mass containing them are always undefined [9]. Different types of machine learning techniques are used in predicting the disease in medical applications [10–14]. Efficient data mining algorithm and IoT devices are in medical applications [15–21].

3 Proposed Work

Calcification and anti-calcification, benign and malignant, thick and normal breasts, tumorous and non-tumorous are the types of data that need to be categorized. Signs of breast cancer at an early stage include:

- Chest or armpit area lumps
- A difference in the scale, shape or sensation of a breast
- Leaking fluid from the nipple
- Pain in the chest.

RF biosensors have been applied to identify biological tissues at particular frequencies, such as in the microwave spectrum, offering a promising new method for reliable, secure, sample-free and rapid detection of biomolecule and cancer cells. For biomolecule diagnostic systems, the RF biosensor provides low-cost, disposable, and highly sensitive solutions. While the electrode surface forms an antibody-antigen complex, changes in dielectric properties, scale, shape and charge distribution are monitored by electrochemical biosensors. Heat sensors are used to monitor the circadian temperature variations of the woman inside the breast cells and then transfer this information to the smartphone or PC of the wearer. Algorithms and neural networks which recognize and categorize abnormal temperatures and other cellular patterns will then analyze the data. Figure 1 illustrates the architecture of the proposed diagnostic device for breast cancer.

For patients with thick boobs, false positive and false negative mammography rates are very high. The precision of age-related mammography, medical history and diagnostic techniques. The wearable system proposed is composed of an infrared sensor, an electrochemical sensor and an optical sensor. When there is molecular recognition of particular components, it detects an electrical reaction. It is capable of detecting DNA which is damaged.

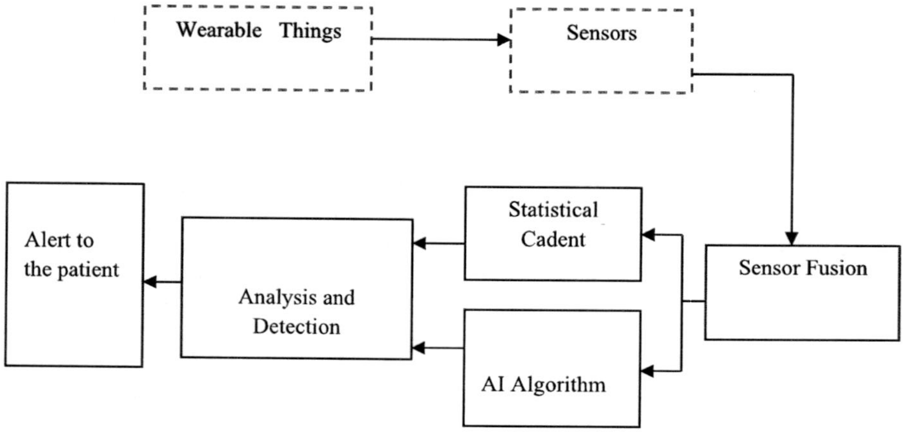

Fig. 1 Architecture of breast cancer diagnosis system

Sweat provides a wealth of knowledge that can be attached to wearable devices that are physiologically important. It absorbs molecular sweat components such as sodium ions, potassium ions, lactate and glucose sugar. The entire device was condensed into a Band-Aid-like wearable patch. During the wearable phase, the Bluetooth module transmits analyte readings to a smartphone for monitoring.

The electrochemical sweat sensor consists of two elements, such as the electrode sensing component and the electronic component for the processing and transmission to the smartphone of these signals. On the substrate, electrodes are produced and attached to the wireless communication modules. The micro-controller is attached to the sensors.

An significant secretory component in sweat is prolactin-induced protein (PIP). PIP has been reported in the human breast tumor range of 93%. Patch-style sweat sensor formats can be used. Within the wearable dress, a wearable electrochemical sensor is packed and fitted. The sensing portion is a flexible electrode plastic substrate and the sensing electrodes are surrounded by soft polymetric wells to absorb pressure variations. The micro-controller board is attached to six electrodes. The sensor tests the concentration of chemicals such as potassium and lactate in sweat. It sends data to a mobile phone via Bluetooth communication.

Related aspects of the surface temperature of the breasts are identified in order to diagnose breast cancer using IR thermography. To capture thermal images of the breast, a thermal-app camera is used. The Android phone suits the thermal camera. A mobile phone is transformed into a working thermal camera. Breast thermography works by discovering that the surface temperature of the breasts is rising. Sequences, including preprocessing, segmentation, feature extraction, classification and post-processing, further process the observed image.

Detection of breast cancer using thermography starts with breast screening and the study of thermal changes in thermography. To locate objects and boundaries in images, segmentation is used. The picture is split into $N * N$ block size classes. For edge detection, the Hough transform can be used. To remove the noise from the image, the Gaussian filter is used. Thermal gramme temperature variations can be defined by its texture. For further analysis, the segmented image is then fed into the classifiers. Data processing is required before implementing machine learning algorithms for classification problems. Standard scalar ensures that there is a mean of 0 and a variance of 1 for each function; thus, all features have the same coefficient. Due to the variety of nucleus morphologies, diverse morphological models are used. Characteristics such as area, perimeter, aspect ratio, strength, excentricity, signature of the form, perimeter, compactness, distance, length of the major axis, and length of the minor axis are extracted. The key activities used here are dilation and hole-filling, and we developed a binary edge map of the image with a reference using the gradient-dependent threshold technique.

The transformation of Hough is a tolerant edge gap that is not influenced by noise and a derivative of random transformation. From different angles, it gives projections. Before implementing the Hough transform, Canny edge detection is used in the preprocessed picture. It also utilizes the first Gaussian derivative to provide an effective edge detection filter for isolating phase edges. This edge detection procedure will decrease the preprocessing. Sample mammograms from the stated database are taken for all grades, such as malignant, benign and average. After gaining features from the mammogram image, the values are given to a classifier called the SVM-Support vector machine. It is a classifier which is used here to achieve greater effectiveness than other classifiers. For the determination of the maximum hyperplane margin between the two types, the values obtained from both cancerous and non-cancerous mammogram images are used.

4 Performance Analysis

This research uses the Wisconsin Breast Cancer (original) dataset 20 from the UCI Machine Learning Repository. There are 699 cases of breast cancer in Wisconsin (Benign: 500 Malignant: 20), 2 groups (55.5% Malignant: 20), and (45.5% benign), and 11 attributes with integer values.

The performance of the predictive model increases steadily as the number of characteristics increases in the feature set. Set the feature to 18 numbers for characteristics. High efficiency was achieved by the classifier, such as 96% accuracy, 96% specificity, 96% sensitivity, 97% F1 score, 96% MCC, 3% classification error and 0.002 s of execution time. The SVM polynomial model's output is high, and these characteristics are more appropriate for the diagnosis of breast cancer. The performance of the proposed system in terms of accuracy is high, as compared to previous methods.

The train SVM model calculates the performance metric values on the training dataset, such as accuracy, specificity, sensitivity, F1-score. When making the prediction on the testing dataset, determine the function is the least relevant and remove this feature from the feature set. The model proposed has now decreased its characteristics and chosen the feature set that offers the highest or lowest scoring metric. Attribute density details are described in Fig. 2.

Real valued features are calculated for each cell nucleus. It consists of the following below as

1. Radius—Mean of distances from center to points on the perimeter
2. Texture—Standard deviation of gray-scale values
3. Perimeter
4. Area
5. Smoothness
6. Compactness
7. Concavity
8. Concave points
9. Symmetry
10. Fractal dimension.

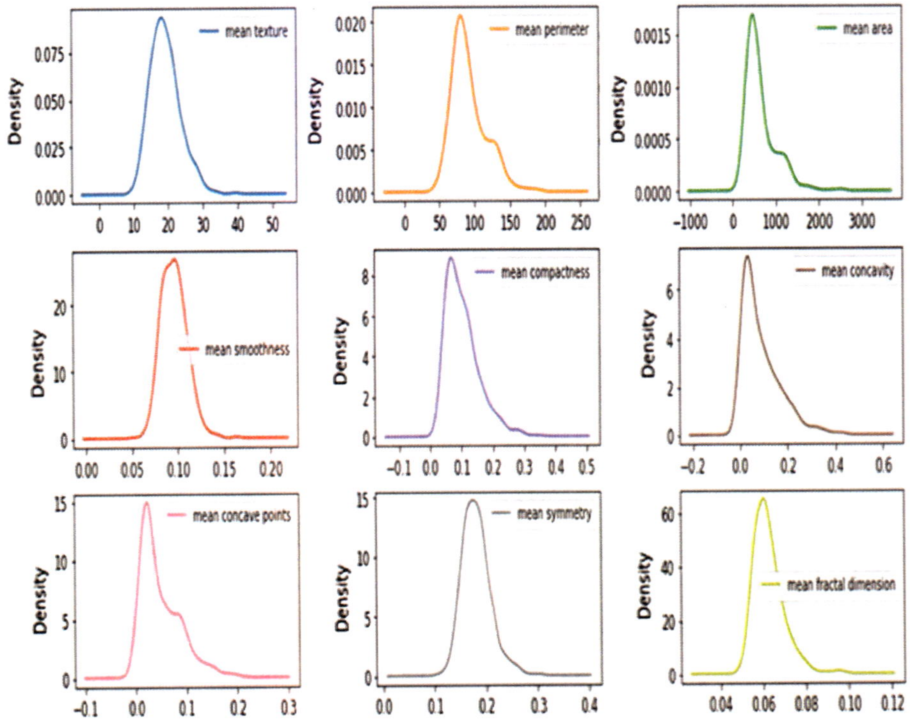

Fig. 2 Attributes density

As below, we calculate the following parameters of output measurement:

1. If Breast Cancer is identified as the subject, TPP (true positive)
2. TN (true negative) when it categorizes a positive as being negative: Healthy Healthy
3. FP (false positive) when it categorizes a healthy subject as being Cancer of the Mammary
4. If a BC is classified as safe, FNN shall specify that (false negative)

The SVM gained 84% accuracy, 100% precision, 3% sensitivity, 6 F1-score, and 16% classification error on one feature set, and model error. There was a computing time of 0.002 s. As the number of features increases in the feature set, the output of the predictive model increases gradually. Figure 3 shows the accuracy analysis with neighbors. Figure 4 shows the comparison result of proposed method with existing methods.

Table 1 shows the accuracy analysis of proposed method with existing algorithms.

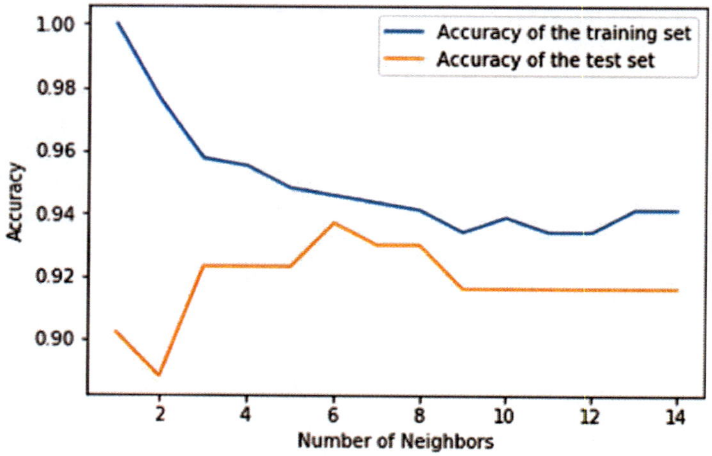

Fig. 3 Accuracy analysis

Fig. 4 Comparison analysis

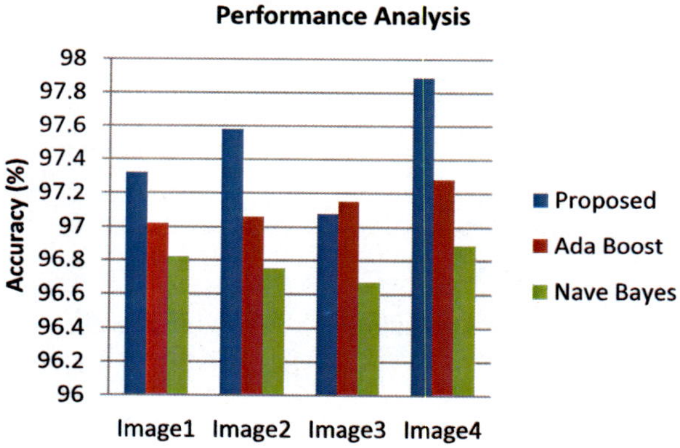

Table 1 Performance analysis

Methodology	Accuracy	Precision	Recall
Nave Bayes classifier	96.78	96.1	93.59
Ada boosting technique	97.8	97.34	94.67
RCNN classifier	91.6	91.78	92.3
Proposed methodology	97.8	97.67	97.10

5 Conclusion

All Detecting carcinogenic factors in the breast plays a significant role in SVM. In cancer detection, the procedure serves as a stepping stone. In several medical applications, SVM can be used. Automatic classification and segmentation were

SVM's key goals. Calcification and non-calcification, benign and malignant, thick and normal breasts, and tumorous and non-tumorous are the data categories to be categorized. For training purposes, various kinds of data are fed into a classifier. The analysis of the experimental findings suggests that malignant and benign individuals are effectively differentiated by the proposed technique.

References

1. T.F. Smith, M.S. Waterman, Identification of common molecular subsequences. J. Mol. Biol. **147**, 195–197 (1981)
2. J.S. Camilleri, L. Farrugia, Determining the concentration of red blood cells using dielectric properties, in *Conference: 2020 14th European Conference on Antennas and Propagation (EuCAP)*
3. H.J. Pandya, K. Park, W. Chen, M.A. Chekmareva, D.J. Foran, Simultaneous MEMS-based electro-mechanical phenotyping of breast cancer. Lab Chip. **15**(18), 3695–3706 (2015)
4. F. Puppo, F.L. Traversa, M. Di Ventra, G. De Micheli, S. Carrara, Surface trap mediated electronic transport in biofunctionalized silicon nanowires. Nanotechnology **27**, 345503 (2016)
5. M.H. Memon, J.P. Li, A. Ul Haq, M. Hunain emon, W. Zhou, Breast cancer detection in the IOT health environment using
6. Modified Recursive Feature Selection, Hindawi Wireless Communications and Mobile Computing Volume 2019, pp. 1–19
7. P. Kaur, R. Kumar, M. Kumar, A healthcare monitoring system using random forest and internet of things (IoT). Multim. Tools Appl. **78**, 19905–19916 (2019)
8. V. Savitha, N. Karthikeyan, S. Karthik, R. Sabitha, A distributed key authentication and OKMANFIS scheme based breast cancer prediction system in the IoT environment. J. Ambient Intell. Hum. Comput. (2020). https://doi.org/10.1007/s12652-020-02249-8
9. I. Razzak, K. Zafar, M. Imran, G. Xu, Randomized nonlinear one-class support vector machines with bounded loss function to detect of outliers for large scale IoT data. Future Gen. Comput. Syst. **112**, 715–723 (2020)
10. A. Sivasangari, P. Ajitha, I. Rajkumar, S. Poonguzhali, Emotion recognition system for autism disordered people. J. Ambient Intell. Hum. Comput. 1–7 (2019)
11. Sri, R. Jenitha, P. Ajitha, Survey of product reviews using sentiment analysis. Indian J. Sci. Technol. **9**, 21 (2016)
12. B.Y. Jinila, P.S. Shyry, Transmissibility and epidemicity of COVID-19 in India: a case study. Recent Patents Anti-Infect. Drug Discov. **15**, 1 (2020). https://doi.org/10.2174/1574891X15666200915140806
13. Madhukeerthana, Y. Bevish Jinila, Deepika, Enhanced rough set theory for denoising brain MR images using bilateral filter design. Res. J. Pharm., Biol. Chem. Sci. **7**(3). ISSN: 0975-8585
14. A. Sivasangari, D. Deepa, L. Lakshmanan, A. Jesudoss, M.S. Roobini, Lung nodule classification on computed tomography using neural networks. J. Comput. Theor. Nanosci. **17**(8), 3427–3431 (2020)
15. P. Ajitha, A. Sivasangari, G. Lakshmi Mounica, L. Prathyusha, Reduction of traffic on roads using big data applications, in *International conference on Computer Networks, Big data and IoT* (Springer, Cham, 2019), pp. 333–339
16. J.S. Vimali, S. Gupta, P. Srivastava, A novel approach for mining temporal pattern database using greedy algorithm, in *2017 International Conference on Innovations in Information, Embedded and Communication Systems (ICIIECS)* (IEEE, 2017), pp. 1–4
17. S. Vigneshwari, A. Mary Posonia, S. Gowri, An efficient framework for document retrieval in relationship strengthened personalized ontologies, in *Soft Computing in Data Analytics* (Springer, Singapore, 2019), pp. 735–742

18. R. Akshaya, N. Niroshma Raj, S. Gowri, Smart mirror-digital magazine for university implemented using Raspberry Pi, in *2018 International Conference on Emerging Trends and Innovations In Engineering And Technological Research (ICETIETR)* (IEEE, 2018), pp. 1–4
19. J.S. Vimali, Z. Sabiha Taj, FCM based CF: an efficient approach for consolidating big data applications, in *International Conference on Innovation Information in Computing Technologies* (IEEE, 2015), pp. 1–7
20. S.C. Mana, J. Jose, B. Keerthi Samhitha, Traffic violation detection using principal component analysis and viola Jones algorithms. Int. J. Recent Technol. Eng. (IJRTE) **8**(3) (2019). ISSN: 2277-3878
21. J. Jose, S.C. Mana, B. Keerthi Samhitha, An efficient system to predict and analyze stock data using Hadoop techniques. Int. J. Recent Technol. Eng. (IJRTE) **8**(2) (2019). ISSN: 2277-3878

Challenges and Security Issues of Online Social Networks (OSN)

A. Aldo Tenis and R. Santhosh

Abstract In the current scenario, social phishing is the dangerous security issue in Online Social Networks (OSN), in which the attacker tries to steal user's personal information, account details, passwords, and other login criterion and later which will be misused. It occurs due to the hacker, who misleads the client to open the fake web page and that particular malicious web page masquerading as a worthy link and the user may suppose to enter the personal information and other credentials which will be easily theft by the hacker. Malware installation is also possible in this technique which leads to ransom ware attack or the revealing of sensitive information. This phishing issue is broad, and so many phishing detection techniques are implemented to mitigate specific attacks. This paper surveys the literature on the existing anti-phishing techniques. Various categories and methodologies of anti-phishing techniques are also presented such us phishing detection techniques and recent anti-phishing frameworks

Keywords Online social network (OSN) · Malicious web page · Phishing · Anti-phishing · Ransom ware attack

1 Introduction

Online social media shrinks the world and changes the human lifestyle. Many researchers have made the detailed study and survey on this Online Social Network. The main goal of OSNs is sharing the message or information with maximum users. Even though there is lot of privileges on this Online Social Network, privacy issues and the security threats are the major barrier for the user. Privacy issues include identity cloning, malware, social phishing, and public shaming; these kinds of issues affect the privacy and threaten the Online Social Network (OSN) users. Phishing is an important issue in which the attacker hacks the user's valuable information

A. Aldo Tenis (✉) · R. Santhosh
Department of Computer Science and Engineering, Faculty of Engineering, Karpagam Academy of Higher Education, Coimbatore, Tamil Nadu, India

S. Shakya et al. (eds.), *Mobile Computing and Sustainable Informatics*, Lecture Notes on Data Engineering and Communications Technologies 68,
https://doi.org/10.1007/978-981-16-1866-6_53

which guides to leakage of sensitive information of the user, identity cloning, and destruction or property.

This paper focused the literature survey of the existing approaches to detect phishing attacks. This paper projects the comparative study and the limitations of existing anti-phishing tools.

By this study, the following anti-phishing methodologies were found.

- Heuristics-based strategy
- Blacklist strategy
- Similar text-based strategy
- Machine learning strategy
- Hybrid strategy
- Image-based strategy.

Section 2 projects the detailed study of the above said approaches. Section 3 describes the comparative study and features. Section 4 provides the observations, and Sect. 5 suggests new recommendations and conclusion.

2 Anti-phishing Approaches

We cannot imagine our lives without online communication. Phishing is privacy issue like online theft that steals user's personal details. Many researchers had already proposed huge volume of solutions against this phishing attack. Some of the existing approaches are examined here.

Heuristic-Based Strategy

Hackers design the phishing URLs which look like the authentic URLs. The users are tricked to trust and open the link and they may enter sensitive information. Sometimes, the phishing URLs trick the users for the malware installation [1–5]. Because of this, the sensitive data stored in the particular system can be stolen and forwarded to the attacker.

Heuristic-based approach is one of the techniques to identify such kind of phishing attacks. These techniques determine the internet credentials of phishing URLs. As a result of the detailed study of different research on this heuristic approach, the following techniques were identified.

- SVM
- Scoring and ranking
- Regression method
- Naïve Bayes classification.

These techniques make use of rule based or heuristics to segregate URLs.

Heuristics are nothing but the attributes to check URL or website. Some of the examples for heuristics are domain part of IP address, symbol '@' in URL, passwords of pop-up windows [6].

Blacklist-Based Approach
Blacklists are having malicious URLs. For every browsing, the entered URL is checked with the blacklist to identify whether it is a malicious link or not. If the currently entered URL is on the black list, then the necessary actions can be taken. If the visited URL is not present in the blacklist, then that page is identified as legitimate URL [7]. The browser or hosted at central server may be having the blacklists. The number of phishing pages on the Internet is measured by the blacklist coverage method [8]. The quality of the list is examined by analyzing the number of non-phishing sites which are incorrectly added on the blacklist. Because of such incorrectly added entry, the user may get a false warning even though they visit a legitimate URL.

There is some limitations in this blacklisting approach. The major drawback is that the blacklist may not contain the newly constructed phishing a URL which will not be detected as phishing URL.

Similarity Text-Based Approach
As the name of this approach, the websites will be compared to identify the resemblance of contents in the websites. For the comparison, the concepts of ranking method such as TF-IDF will be used. From which the phishing sites can be identified [9, 10]. The comparison of text using TF-IDF is as follows.

- TF-IDF scores will be calculated for each term in the particular web page.
- Next, the lexical signature will be created by specifying the terms with the top five TF-IDF scores.
- Then the identified lexical signature will be entered into the search engine.
- Compare the domain name and search results. If it is matched, we can consider it as the anti-phishy web page, or else consider it as phishing web page.

Machine Learning Approach
In this method, features will be classified based on the machine learning methodologies. Most of the methodologies deal with this phishing problem by using SVM. SVM is super vector machine method which helps to solve the classification problems [11]. It becomes popular because it produced accurate results for the unstructured problems such as text categorization.

Hybrid Approach
In this technique, more than one approach are used together to identify whether the particular website is legitimate or not. For example, both heuristic-based approach and blacklist-based approach may be used together for constructing a suitable anti-phishing framework.

Image-Based Approach

In this approach, visual similarity is used for comparing the phishy web pages and the non-phishy non-phishy web pages. In this method, based on the visual cues, the web pages are segregated into block regions. Based on the similarities such as block level, layout, and style [12] this visual resemblance is evaluated. If any of the above similarity having higher value than the predefined threshold, then that web page is considered as phishy.

3 Comparative Study and Feature Selection

Many researchers provide number of anti-phishing frameworks for detecting and classifying the phishing websites. Some of the browsers are using safe browsing technique which is known as Google Safe Browsing that may prevent the phishy URL. In this technique to detect the phishy link, the black list method is used. GSB technique cannot be used for detecting the phishy site if the black list is not updated. To identify or classify the phishing URL, following things need to be considered.

3.1 URL Collection

The particular web page link is given as the input for the phishing system. Phish Tank is one of the sites which having the suspicious URL. Alexa is the source from which most of the users collecting the legitimate links.

3.2 Feature Collection

The features of the particular URL are evaluated to recognize how phishing URL will be identified. By referring [13], the following features were identified in Table 1.

3.3 Feature Selection

Selection of features is a crucial thing for segregating the unwanted features which are not important in the process of classification.

Table 1 Features of phishing URL

Sr. No.	Feature	Description
1	Using IP address in domain name	Attackers may use numbers instead of name which also identified as phishy
2	Length of the link	The length of the link exceeded 1750 characters means that link is marked phishy
3	'-' symbol	If the URL has '-' symbol, it is considered as legitimate link
4	Subdomains	If the URL having more number of subdomains which is considered as phishing
5	HTTPS	For the secure connection, HTTPS is used
6	Request link	Entire text or images must be loaded
7	Anchor	<a> tags must have links which is similar to the domain name
8	Abnormal link	The URL is compared with the WHOIS record. If there is no record available or any mismatch, it will be considered as phishing website
9	"SFH"	Transferring data to some other domain may be avoided by Server Form Handler (SFH)
10	Redirection of another page	When the user clicks the link sometimes, it will be redirected to other pages. If there is more than four redirection, it also considered as phishy
11	Pop-up window	If pop-up window is used to enter password, it will be considered as phishy
12	Record of DNS	If there is no DNS record which is considered as phishy
13	Website traffic	If the particular website is frequently visited, it means that record is website traffic. For the particular website, there is no website traffic then that site is identified as phishy
14	Registered date of the domain	Website registered within one year is considered as phishing

3.4 Classification

Classification gives the required output. It is the final stage of the phishing system. There are lot of classification methods are available. Classification of the websites is important for the entire process. For making the effective approach, the proper algorithm should be chosen.

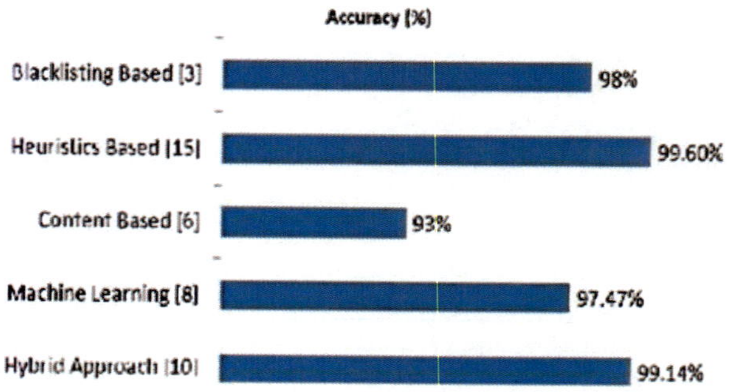

Fig. 1 Comparative analysis of accuracies for different approaches

4 Observations

By this study, it is concluded that hybrid method and heuristics method are better when compared with other approaches. Blacklist approach is used in [14] which detects the phishing URLs. The authors of [15] have derived many rules to examine the fuzzy model. In the heuristics approach even though the accuracy is high, it is not possible to identify the newly added spam features. Similarity of text-based method is used in [16] which depends on database of Google Image. The machine learning approach [17] provides a better accuracy than other methods (Fig. 1).

The effectiveness of the system will not be predicted based on the accuracy only. Also, it is mandatory to consider the number of features have analyzed. If any of the website is unnoticed, then there is possibility for considering it as legitimate. Hence, the number of features used for classification plays the vital role for designing the effective anti-phishing system.

5 Recommendation and Conclusion

Social phishing is becoming the dangerous security by which the attackers easily hack the user's sensitive information. Due to the highly impact on the usage of Online Social Network (OSN), it is crucial for detecting and classifying the phishing URL. This paper projects various existing anti-phishing frameworks for detecting and classifying the phishing site. From this study, it is observed that the best approach is heuristics approaches. In the existing approaches, there is lack of security in server side. It will be recovered in the future work. In the future, the accuracy rate of detection will also be increased. By providing some authentication, entering the sensitive information will also be avoided.

References

1. R.B. Basnet, A.H. Sung, Q. Liu, Feature selection for improved phishing detection, in *IEA/AIE 2012. LNCS (LNAI)*, ed. by H. Jiang, W. Ding, M. Ali, X. Wu, vol. 7345 (Springer, Heidelberg, 2012), pp. 252–261. https://doi.org/10.1007/978-3-642-31087-4_27
2. A.Y. Fu, L. Wenyin, X. Deng, Detecting phishing web pages with visual similarity assessment based on earth mover's distance (EMD). IEEE Trans. Dependable Secure Comput. **3**(4), 301–311 (2006)
3. M. Khonji, Y. Iraqi, A. Jones, Lexical URL analysis for discriminating phishing and legitimate websites, in *Proceedings of the 8th Annual Collaboration, Electronic Messaging, Anti-abuse and Spam Conference* (2011), pp. 109–115
4. S. Marchal, K. Saari, N. Singh, N. Asokan, Know your phish: novel techniques for detecting phishing sites and their targets, in *Proceedings—International Conference on Distributed Computing Systems 2016*, vol. 2016–August, no. Sect. V (2016), pp. 323–333
5. M. Khonji, Y. Iraqi, A. Jones, Phishing detection: a literature survey. IEEE Commun. Surv. Tutorials **15**(4), 2091–2121 (2013)
6. R.M. Mohammad, F. Thabtah, L. McCluskey, Intelligent rule-based phishing website classification **8**(3), 153–160 (2014) (published in IET Information Security)
7. G. Xiang, B.A. Pendleton, J.I. Hong, C.P. Rose, A hierarchical adaptive probabilistic approach for zero hour phish detection, in *Proceedings of the 15th European Symposium on Research in Computer Security (ESORICS'10)* (2010), pp. 268–285
8. P. Golle, B. Waters, J. Staddon, Secure conjunctive keyword search over encrypted data, in *Applied Cryptography and Network Security (ACNS'04)* (2004), pp. 31–45
9. Y. Zhang, J. Hong, L. Cranor, CANTINA: a content based approach to detecting phishing web sites, in *Proceedings of the 16th International Conference on World Wide Web*, Banff, Alberta, Canada, 8–12 May 2007
10. D. Miyamoto, H. Hazeyama, Y. Kadobayashi, A proposal of the AdaBoost-based detection of phishing sites, in *JWIS*, August 2007.
11. Y. Zhang, S. Egelman, L. Cranor, J. Hong, Phinding phish: evaluating anti-phishing tools, in *Proceedings of the 14th Annual Network & Distributed System Security Symposium (NDSS)* (2007)
12. T. Moore, R. Clayton, H. Stern, Temporal correlations between spam and phishing websites, in *Proceedings of the 2nd USENIX Conference on Large-Scale Exploits and Emergent Threats: Botnets, Spyware, Worms, and More (LEET'09)*, USENIX Association, Berkeley, CA, USA, (2009), p. 5
13. S. Patil, S. Dhage, A methodical overview on phishing detection along with an organized way to construct an anti-phishing framework, in *2019 5th International Conference on Advanced Computing & Communication Systems (ICACCS)*
14. A. Naga Venkata Sunil, A. Sardana, A PageRank based detection technique for phishing web sites, in *2012 IEEE Symposium on Computers & Informatics (ISCI)*, Penang (2012), pp. 58–63
15. P. Barraclough, G. Sexton, Phishing website detection fuzzy system modelling, in *2015 Science and Information Conference (SAI)*, London (2015), pp. 1384–1386
16. E.H. Chang, K.L. Chiew, S.N. Sze, W.K. Tiong, Phishing detection via identification of website identity, in *2013 International Conference on IT Convergence and Security (ICITCS)*, Macao (2013), pp. 1–4
17. P. Singh, N. Jain, A. Maini, Investigating the effect of feature selection and dimensionality reduction on phishing website classification problem, in *2015 1st International Conference on Next Generation Computing Technologies (NGCT)*, Dehradun (2015), pp. 388–393

Challenges and Issues of E-Health Applications in Cloud and Fog Computing Environment

N. Premkumar and R. Santhosh

Abstract Smart health care is the recent buzzword in the healthcare environment. Involvement of technology in the healthcare domain provides a platform for remote real-time monitoring and self-management of patients' health. The cloud computing is much utilized for developing technological solutions for an efficient smart healthcare system. Since cloud is a centralized environment, its response time becomes an issue in supporting latency-sensitive obligation in real-time applications. Most solutions in the e-health field depend on immediate decision making over real-time data where latency cannot be tolerated. Fog computing is a distributed infrastructure that overcomes this weakness of cloud computing by having the facility of local storage and processing in providing immediate response for decision making. Hence, thereby, it reduces the network latency and bandwidth usage by decreasing the amount of data sent to the cloud. This combination of fog computing and cloud computing can able to employ a reliable and efficient distributed e-health applications. This paper presents a literature review on the common use of cloud computing and fog computing for providing solutions in health care, along with the challenges that exist in providing solutions through fog computing.

Keywords Cloud computing · Health care · Internet of Things · Fog computing · E-health

1 Introduction

In the recent era, the healthcare sector identified the impact of using the Internet to enhance the medical data processing and analysis in real-time for speeding up as well as for offering a quality treatment to the patients. This thereby decreases the visit cost of doctor and improves the quality in patient care [1]. By the end of 2020, the usage IoT in the wearable tools and healthcare devices will be around 162 million [1].

N. Premkumar (✉) · R. Santhosh
Department of Computer Science and Engineering, Faculty of Engineering, Karpagam Academy of Higher Education, Coimbatore, Tamil Nadu, India
e-mail: premkumarn@kongunadu.ac.in

S. Shakya et al. (eds.), *Mobile Computing and Sustainable Informatics*, Lecture Notes on Data Engineering and Communications Technologies 68,
https://doi.org/10.1007/978-981-16-1866-6_54

The digital healthcare (i.e., e-health) applications are generally a collection of software and services dedicated on the gathering and communication of medical information which is used for delivering healthcare services [2, 3]. The data collected from the sensors in wearable tools and healthcare IoT devices allow us to track user habits, and this data needs to be processed in order to identify the criticality nature of the patients. Since huge amount of data are collected from the sensors, the concept of cloud computing gives an effective mechanism for storage and application processing in Internet [4].

The cloud-based solutions are effective for most IoT applications which are not sensitive to latency and not used in critical scenarios. But when the application deals with patients with critical needs, a traditional cloud-based solution creates a negative impression and leads to worst moments in emergency conditions [5]. Because any loss of network connectivity or latency deviations can create a negative impression and result in the worst-case scenario for critical patients in an emergency situation.

There arises a computing technology called fog computing that provides a solution in handling the limitation of cloud in healthcare systems. The fog computing enhances and enlarges the services presented by the traditional cloud computing to the network edges [6]. The fog computing provides low-latency infrastructure in which it allows processing of collected data from the sensors locally at the fog nodes instead of sending directly to the cloud for processing. Hence, thereby, it reduces the network latency and bandwidth usage by decreasing the amount of data sent to the cloud.

In the case of fog computing, the data generated by sensors are grouped, further processed and kept in a temporary database. This mechanism helps to escape from round-trip delays and network traffic, which may happen in the case of cloud computing. This advantage is specifically vital for e-health applications since it involves in data communication data for remote real-time processing over the Internet, such as remote ECG monitoring [7]. These applications help in decision making by providing crucial information on the health condition of the patients by continuous monitoring [8].

This study presents a review on cloud and fog computing-based IoT solutions for the healthcare system. This paper aims at providing a clear view on the stage-by-stage development on the solutions arrived in healthcare systems based on IoT which starts from initial solutions on patient monitoring to the current systems that involve the fog computing for smart health. This paper further discusses about the machine learning techniques and its models.

Recent e-health-related applications involve privileged levels of security services and also quality of service from deign level. Current studies about e-health applications do not express a complete classification of main and unique characteristics under fog computing infrastructure. In addition, it failed to show a detailed analysis about a comparison among cloud. The focus of this paper is to analyze and update the present state of the art on area of fog computing with cloud, focusing on the requirements of e-health applications, their issues, and open challenges that still need to be addressed. The organization of this paper is listed as follows:

1. A review of how e-health based real-time applications take advantage from fog computing environment in terms of infrastructure, communication pattern, deployment, and data security-related details.
2. A correlation between regarding the concept of e-health systems that employ fog computing versus application requirements.
3. An explanation about different issues and concerns regarding the integration of e-health applications into fog computing. In addition, discovering further research directions and developments associated with this context.

2 Reviewing Cloud and Fog Computing-Based IoT Solutions

2.1 Fog Computing

Fog computing also decreases the amount of data sent to the cloud since it provides the facility of local processing, storage, and analysis of data in giving immediate response for the frequent decision making activities in real-time. Hence, fog computing improves the response time and reduces the latency which supports the time-sensitive application in real-time with more number of nodes. The combination of fog and cloud computing can able to employ a reliable and efficient distributed e-health applications [9]. A fog can be placed between the cloud and IoT devices to carry out efficient data processing. The combination of fog and edge computing along with the cloud computing is able to provide a reliable and efficient distributed e-health applications. Generally, cloud computing has been used with IoT devices in most healthcare monitoring systems for storing and processing a large amount of data generated from the sensors [3]. Even cloud with IoT was used for healthcare systems, it is not efficient in handling latency-sensitive requirements of healthcare applications [5]. The growing interest toward fog computing gave an answer to handle this limitation by providing low-latency infrastructure. The architecture of fog computing enhances the storage and computational capability of cloud to the network edge. The fog computing supports applications which are sensitive to latency along with reduced amount of data traffic.

The architecture of fog computing extends the storage and computational capability of cloud to the network edge. Hence, by having the facility of data storage and processing locally, the fog computing overwhelms this weakness of cloud computing by having the facility of local processing in providing immediate response for decision making and also connects the IoT medical devices to the cloud. Thereby, the concept of fog computing also avoids the heavy network traffic that occurs in the cloud end.

The service of cloud computing is extended by the concept of fog computing. Fog computing decreases the volume of data sent to cloud by providing the facility of local processing, storage, and analysis of data in giving immediate response for the frequent decision making activities in real-time. Hence, fog computing improves the response time and reduces the latency which supports the time-sensitive application in real-time with more number of nodes. The fog computing has the facility of local processing in providing immediate response for decision making and also connects the IoT medical devices to the cloud.

Data collected by devices, like sensors, wearable devices, embedded devices, mobility prototype, and device usage permit to track a user's behavior, can be obtained and processed it to reveal vital conditions using either machine learning and deep learning-based approaches. Conventional cloud computing-based methods with big data investigation can able to offer reliable service and good performance when dealing decisive IoT-based applications [10, 11]. But when a patient needs time and critical medical service requirements, a more robust and higher degree of ease of access is needed. This is because of disconnection in the network bandwidth or latency variation and also produces serious consequences during emergency conditions [5].

2.2 *Review of IoT-Based Health Monitoring*

Deployment of a fog server in a cloud typically increases the overall performance of the underlying network and also reduces the requirement of bandwidth by supplying real-time data whenever required by mobile users those who reside to the boundary of the underlying network. Based on these perceptions, this paper focuses on two different but related areas: (1) usage of fog computing scheme in healthcare domain and (2) IoT-related remote health monitoring solutions.

2.3 *A Typical Structural Design of Internet of Medical Things (IoMT) System*

The integration of fog computing in the medical healthcare architecture typically makes use of the designing remote monitoring-based schemes that force sensor and wearable networks for realizing suitable responsive, preventive, and protective solutions. In this context, fog nodes act as a local server, processing and collecting health-related data in order to provide instant responses to a variety of services. There is several research communities have been inspecting previously for designing new solutions for supervising patients' vital health data and their present status with the objective of remotely affording medical report to physician or clinicians for further instant action.

Chen et al. [12] proposed a computer-based solution to monitor vital data of patient like ECG remotely in order to give instant advice about heart rate and taking/giving medical service if needed. Successive implementation might use microcontrollers and computers to collect data from patient's physiological behavior generated from specific sensors like e-health sensor which needs to be more analyzed using a computer-based setting [13, 14].

Recently, the development in IoT technology has cemented the manner to the growth of different sophisticated solutions that exploit both software and network architectures. Such methods focusing to deal with healthcare issues at various levels like aged people care, disease supervision and monitoring, fitness management, and more [15]. A general scheme employs various levels of architectures [16] as listed below.

- Edge level—In this level, portable devices like smartphones/watches and gateways are responsible for pre-processing operations and low-level processing on data obtained from body area network (BAN).
- Fog level—In this level, personal computers or servers and gateways collectively gather information from sensor networks, which is then responsible for local processing.
- Cloud level—In this level, tasks are computed by invoking cloud services. In addition, data is stored remotely and maintained.

All the above three levels need not necessarily implemented. For example, in a lively monitoring situation where a fog level seems to be not suitable, edge computing devices could directly work together with cloud-based services. Figure 1 shows multilevel architecture that encompasses of edge computing network, fog computing network, and cloud services.

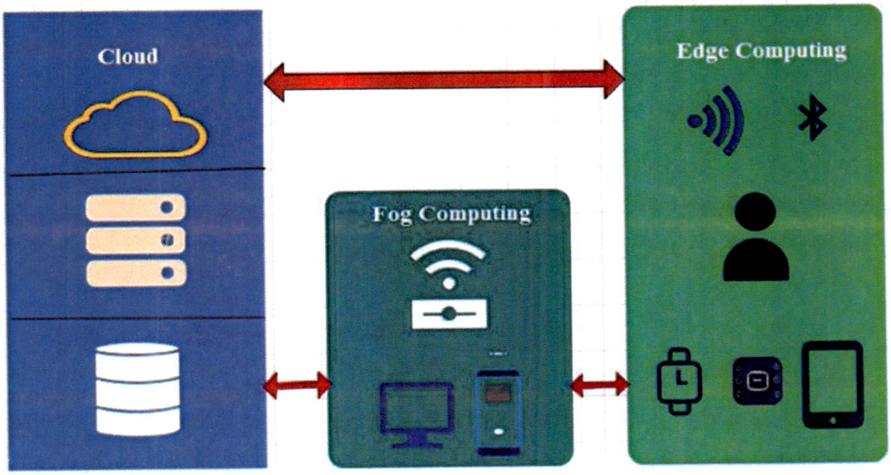

Fig. 1 Multilevel architecture for IoMT

3 A Review of IoMT Monitoring Solutions

Over last few years, many contributions have been proposed dealing IoT-based architectures suitable for remote health monitoring applications. The following review solutions cover diverse application areas of smart healthcare systems like monitoring patient's physiological parameters, behavior, and environmental condition and make use of various sensors and sensors. The following sub-sections discuss various issues of IoMT permitting to the issues they seek to deal with.

3.1 Analysis of Physiological Parameters

Magaña Espinoza et al. [17] proposed a solution in which WBSN is employed to observe heart rate of a patient and their activity rate when they reside in residence. The node located at the boundary of the network is linked with the Internet and raise alert messages on mobile phone to family relatives or physicians whenever sudden changes occur in vital physiological parameter values. Villarrubia et al. [18] aim to track patient's cardiac-related functions for the benefit of analyzing ECG-related data that have proposed a similar work. Through television-based user interface, patients can also permitted with the system. The work proposed in [19] explores the utilization of Bluemix type of cloud technology which is responsible to keep patient's vital physiological parameters, permitting physician to access the data remotely and can also visualize the final using IoT platform, IBM Watson.

In this connection, fog architectures can also be used for classifying activities of patients. In [20], the authors presented a solution for identifying various human activities through the use of wearable sensors. Neural network-based technique is also utilized which is running a fog server locally, while [16] utilizes extra sensors to track patient's movement and investigating them using random forest and support vector machines for classification to predict activities.

3.2 Systems of Analysis

Mathur et al. [21] proposed a scheme to forecast health issues of the limb that resides in residence by monitoring parameters like temperature. In this solution, a group of edge devices have been utilized for capturing data and transmit them to a fog server that finally applied machine learning-based algorithms for effective prediction. The schemes proposed in [22] and [23] employ fog technology to realize speech recognition and monitoring method for offering service to patient's who has been affected by Parkinson disease. Audio form of signals is collected that are generated from smart watch and transmitted the same to anode in a fog network for feature extraction. The classification process carried out in the cloud.

3.3 Techniques for Handling Diabetes Disease

An IoMT-based solutions are emerging in diabetes-based medical service where numerous portable and wearable monitoring devices like insulin pen and pump, blood glucose meter, and artificial pancreas device that are capable of relying on wireless technology with smartphone and offering diagnostic services without using cloud technology [24]. Numerous recent wireless devices in this area are used to forecast the inception of a disease. Abdel-Basset et al. [25] converse a new method for envisaging type-2-based diabetes, whereas [26] utilize a J48 decision tree classification algorithm which was incorporated on a mobile phone to realize risk levels of diabetic-affected patients.

4 Discussions and Research Directions

The aforementioned review studies have inveterate that a large class of advanced techniques mainly machine learning-based techniques which can be successfully employed for effective anomaly detection, decision making, analytical risk analysis monitoring, etc. Most solutions still rid deep investigation of tasks in cloud, due to restricted resources of low powered devices. However, an ever-increasing demand toward designing innovative solutions that incorporate techniques fog, edge, and cloud. Testing and training are very important when designing real-time applications for fog and cloud computing. However, many issues like security issue, advanced algorithms for intermediate analysis and prediction, power optimization, and latency are still needs to improved. Table 1 lists an overview of existing solutions.

5 Conclusions

Cloud and fog computing-based IoT solutions for the medical health care are still developing from effortless architecture to gather, broadcast, and envisage data in required formats and moving from complex sensor networks to smart systems which can able to identify activities automatically and infer decisions with no false positives. Most recent solutions related to e-health field depend on immediate decision making over real-time data where latency cannot be tolerated. In this context, fog computing is a suitable candidate for effective and instant response. It helps to reduce network latency and bandwidth usage. The enhancement of cloud with fog computing can able to employ reliable and efficient distributed e-health application. In addition, edge computing becomes a hot research area with improved capabilities for cloud computing. Edge data centers are typically utilized for storing device related data of the end users instead of frequently contacting the actual cloud. Thus, ensuring that

Table 1 Overview of the existing solutions

Reference paper	Aim of the paper	Sensors used	Computing device(s) used	Remarks
Abdellatif et al. [27]	Testing electroencephalogram-related data	Electroencephalogram sensors	Not mentioned	Autoencoders are applied for data compression, and feature extraction has been done using edge-based scheme; different classification rules applied against various machine learning techniques
Alwan and Rao [28]	Determining body temperature	Temperature sensor(s)	Arduino and Raspberry Pi	Collected information is exchanged between Raspberry Pi and Arduino device using ZigBee
Greco et al. [29]	Detecting variation in physiological data	Accelerometer, gyroscope, and magnetometer	Raspberry Pi 3	Anomaly detection has been done in distributed edge computing architecture using HTM-based algorithm
Muhammad et al. [19]	Recognizing voice pathology	Body temperature, ambient humidity, and electrocardiogram	Smartphone	Data collected from IoT devices has been directed to the app in smartphone using Bluetooth. Feature collection, removal, and categorization are done in the cloud. Local binary pattern applied on voice signal

(continued)

Table 1 (continued)

Reference paper	Aim of the paper	Sensors used	Computing device(s) used	Remarks
Pham et al. [24]	Remote health monitoring	Various environmental sensors, optitrack camera, and smartwatch sensors	Arduino Mega	Data collected from sensors are processed using home gateway device. Physiological information (along with contextual data) directed to the cloud for storage and user access
Priyadarshini et al. [30]	Predicting hypertension and diabetic attacks	Temperature sensor, SpO_2, and cholesterol sensors	PC server	Deep learning technique is applied for inferring a user's mental condition that can be used detecting type-2-based diabetes
Sareen et al. [31]	Detection and prevention of Zika virus	Mosquito sensors	Smartphone and fog server	Data collected from the sensors is processed in the cloud using fog servers
Sood and Mahajan [32]	Identify and control of chikungunya virus	Various environmental and wearable sensors	Not mentioned	Fog layer is used in real-time data analysis and processing. Cloud services then applied for deep analysis
Uddin [20]	Activity prediction	Accelerometer ECG sensor, gyroscope, and magnetometer	GPU-integrated fog server	Neural network methods is used for recognizing different activities
Yeh [33]	Secure communication using BSN-based architecture	Wearable sensors	Fog server and handheld device	Cryptographic operations are applied to ensure secure transmission. This offers authentication and confidentiality of smart objects

only legitimate EDCs can involve during the load balancing process really becomes an attractive research problem.

References

1. A.O. Akmandor, N.K. Jha, Smart health care: an edge-side computing perspective. IEEE Consum. Electron. Mag. **7**(1), 29–37 (2017)
2. I. Bisio, C. Estatico, A. Fedeli, F. Lavagetto, M. Pastorino, A. Randazzo, A. Sciarrone, Brain stroke microwave imaging by means of a newton-conjugate-gradient method in L^p Banach spaces. IEEE Trans. Microw. Theory Tech. **66**, 3668–3682 (2018)
3. P.G. Svensson, eHealth applications in health care management. Ehealth Int. **1**, 5 (2002)
4. F. Bonomi, R. Milito, P. Natarajan, J. Zhu, Fog computing: a platform for internet of things and analytics, in *Big Data and Internet of Things: A Roadmap for Smart Environments* (Springer, Cham, Switzerland, 2014), pp. 169–186
5. I. Azimi, A. Anzanpour, A.M. Rahmani, T. Pahikkala, M. Levorato, P. Liljeberg, N. Dutt, HiCH: hierarchical fog-assisted computing architecture for healthcare IoT. ACM Trans. Embed. Comput. Syst. (TECS) **16**(5s), 174 (2017)
6. I. Stojmenovic, S. Wen, The fog computing paradigm: scenarios and security issues, in *Proceedings of the 2014 Federated Conference on Computer Science and Information Systems (FedCSIS)*, Warsaw, Poland, 7–10 Sept 2014, pp. 1–8
7. H. Ozkan, O. Ozhan, Y. Karadana, M. Gulcu, S. Macit, F. Husain, A portable wearable tele-ECG monitoring system. IEEE Trans. Instrum. Meas. **69**, 173–182 (2020)
8. C. Habib, A. Makhoul, R. Darazi, R. Couturier, Health risk assessment and decision-making for patient monitoring and decision-support using wireless body sensor networks. Inf. Fusion **47**, 10–22 (2019)
9. K. Bierzynski, A. Escobar, M. Eberl, Cloud, fog and edge: cooperation for the future? in *2017 Second International Conference on Fog and Mobile Edge Computing (FMEC)* (IEEE, 2017)
10. B.B.P. Rao, P. Saluia, N. Sharma, A. Mittal, S.V. Sharma, Cloud computing for internet of things & sensing based applications, in *Sixth International Conference on Sensing Technology (ICST)*, Kolkata, 2012, pp. 374–380
11. P. Gaba, R.S. Raw, Vehicular cloud and fog computing architecture, applications, services, and challenges, in *IoT and Cloud Computing Advancements in Vehicular Ad-Hoc Networks* (IGI Global, 2020), pp. 268–296
12. C.M. Chen, H. Agrawal, M. Cochinwala, D. Rosenblut, Stream query processing for healthcare bio-sensor applications, in *20th International Conference on Data Engineering* (IEEE 2004)
13. I. Orha, S. Oniga, Automated system for evaluating health status, design and technology in electronic packaging (SIITME), in *IEEE 19th International Symposium for* (2013), pp. 219–222
14. O. Yakut, S. Solak, E.D. Bolat, Measuring ECG signal using e-health sensor platform, in *International Conference on Chemistry, Biomedical and Environment Engineering (ICCBEE'14)* (2014), pp. 65–69
15. S.M.R. Islam, D. Kwak, M.H. Kabir, M. Hossain, K.S. Kwak, The internet of things for health care: a comprehensive survey. IEEE Access **3**, 678–708 (2015)
16. S.S. Ram, B. Apduhan, N. Shiratori, A machine learning framework for edge computing to improve prediction accuracy in mobile health monitoring, in *International Conference on Computational Science and Its Applications* (Springer, Cham, July 2019), pp. 417–431
17. P. Magaña Espinoza, R. Aquino-Santos, N. Cárdenas-Benitez, J. Aguilar-Velasco, C. Buenrostro-Segura, A. Edwards-Block et al., WiSPH: a wireless sensor network-based home care monitoring system. Sensors **14**(4), 7096–7119 (2014)
18. G. Villarrubia, J. Bajo, D. Paz, F. Juan, J.M. Corchado, Monitoring and detection platform to prevent anomalous situations in home care. Sensors **14**(6), 9900–9921 (2014)

19. G. Muhammad, M.F. Alhamid, M. Alsulaiman, B. Gupta, Edge computing with cloud for voice disorder assessment and treatment. IEEE Commun. Mag. **56**(4), 60–65 (2018)
20. M.Z. Uddin, A wearable sensor-based activity prediction system to facilitate edge computing in smart healthcare system. J. Parallel Distrib. Comput. **123**, 46–53 (2019)
21. N. Mathur, G. Paul, J. Irvine, M. Abuhelala, A. Buis, I. Glesk, A practical design and implementation of a low cost platform for remote monitoring of lower limb health of amputees in the developing world. IEEE Access **4**, 7440–7451 (2016)
22. A. Monteiro, H. Dubey, L. Mahler, Q. Yang, K. Mankodiya, Fit: a fog computing device for speech tele-treatments, in *2016 IEEE International Conference on Smart Computing (SMARTCOMP)* (IEEE, May 2016), pp. 1–3
23. H. Dubey, J. Yang, N. Constant, A.M. Amiri, Q. Yang, K. Makodiya, Fog data: enhancing telehealth big data through fog computing, in *Proceedings of the ASE Bigdata & Socialinformatics 2015* (ACM, Oct 2015), p. 14
24. M. Pham, Y. Mengistu, H. Do, W. Sheng, Delivering home healthcare through a cloud-based smart home environment (coSHE). Future Gener. Comput. Syst. **81**, 129–140 (2018)
25. M. Abdel-Basset, G. Manogaran, A. Gamal, V. Chang, A novel intelligent medical decision support model based on soft computing and IoT. IEEE Internet Things J. 1–11 (2019)
26. M. Devarajan, V. Subramaniyaswamy, V. Vijayakumar, L. Ravi, Fog-assisted personalized healthcare-support system for remote patients with diabetes. J. Ambient Intell. Humanized Comput. **10**, 3747–3760 (2019)
27. A.A. Abdellatif, A. Mohamed, C.F. Chiasserini, M. Tlili, A. Erbad, Edge computing for smart health: context-aware approaches, opportunities, and challenges. IEEE Netw. **33**(3), 196–203 (2019)
28. O.S. Alwan, K.P. Rao, Dedicated real-time monitoring system for health care using ZigBee. Healthc. Technol. Lett. **4**(4), 142–144 (2017)
29. L. Greco, P. Ritrovato, F. Xhafa, An edge-stream computing infrastructure for real-time analysis of wearable sensors data. Future Gener. Comput. Syst. **93**, 515–528 (2019)
30. R. Priyadarshini, R. Barik, H. Dubey, Deep fog: fog computing-based deep neural architecture for prediction of stress types, diabetes and hypertension attacks. Computation **6**(4), 62 (2018)
31. S. Sareen, S.K. Gupta, S.K. Sood, An intelligent and secure system for predicting and preventing Zika virus outbreak using fog computing. Enterp. Inf. Syst. **11**(9), 1436–1456 (2017)
32. S.K. Sood, I. Mahajan, A fog-based healthcare framework for chikungunya. IEEE Internet Things J. **5**(2), 794–801 (2017)
33. K.H. Yeh, A secure IoT-based healthcare system with body sensor networks. IEEE Access **4**, 10288–10299 (2016)

Linguistic Steganography Based on Automatically Generated Paraphrases Using Recurrent Neural Networks

G. Deepthi, N. Vijaya SriLakshmi, P. Mounika, U. Govardhani, P. Lakshmi Prassanna, S. Kavitha, and A. Dinesh Kumar

Abstract Linguistic steganography is used for hiding information in the multimedia data in a traditional approach. For that, we are taking an image and converting the text into bits like 0's and 1's. In this paper, we are using both image and text dataset. Nowadays, providing secure data with confidentiality and integrity has become a big challenge in the present world. To improve the security of these existing parameters, we have proposed a methodology by referring through various reference papers and existing methodologies that can provide and attain confidentiality and integrity by utilizing various algorithms to provide a collective result. We have used algorithms like support vector machine (SVM), recurrent neural network (RNN), and convolutional neural network (CNN). For higher security, steganography algorithm is employed to mix with this algorithm. By applying the above algorithms, we will be doing encoding for providing more security. Image quality will not be loosed and there will be efficiency in decoding the information by doing encoding. If any users attack the code, due to the techniques that we used in our encoding, enhancing more security will be provided.

Keywords Linguistic steganography · Encryption · Decryption · Data hiding · Steganalysis

1 Introduction

Encryption methods or encryption algorithms encode message in a particular way that only officials' members can decode while the un-officials cannot decode. It highly secures that the data is hidden in the image. In some cases, the un-officials can access the important information because of that we use the privacy method so

G. Deepthi (✉) · N. Vijaya SriLakshmi · P. Mounika · U. Govardhani · P. Lakshmi Prassanna · A. Dinesh Kumar
Department of Computer Science and Engineering, Koneru Lakshmaiah Education Foundation, Vaddeswaram, AP, India

S. Kavitha
Department of ECE, Hindusthan Institute of Technology, Coimbatore, India

S. Shakya et al. (eds.), *Mobile Computing and Sustainable Informatics*, Lecture Notes on Data Engineering and Communications Technologies 68,
https://doi.org/10.1007/978-981-16-1866-6_55

that the only officials can access the information and the remaining persons cannot have that information by any means. While these two methods make sure that the data is security, they may also reveal the important information, making it more unsafe to get attacks, such as cracking the information.

Steganography is used to bury the information so that only the sender or officials know the existence of the information. If an individual or the person knows about the existence of the information, then that person may try to crack the existence of that hidden message or if that person does not know about the existence of the information, they will not do anything related to that message. Steganography structure is designed by using the concepts of "cover medium," "hidden message," "the key." The cover medium may be a painting or an image or music or video, etc. The cover medium is that thing or an item that carries the hidden message. The keys are used to crack the message in the decoding process. The generic description of the steganographic process is "medium + hidden message + key = stegno medium." In the cover medium, they used to hide the message, and using the key, they encrypt the hidden message present in the cover medium. The outcome is the stegno medium. The cover medium is actually an image or audio or video files. If a message is encrypted, it must be decrypted if discovered.

Here, we are hiding the message in the image. Firstly, there should not be any difference in original image and hidden image, or the message hidden in the hidden image should be unnoticeable. The message hidden within the image should not draw any attention to the hidden image so that no one would attempt to extract the hidden message in any way. Secondly, the hiding method should be reliable because it will not be difficult to extract the hidden message or it will be difficult to take out the hidden message if they do not have any extracting method or algorithm and proper key. Finally, the hidden message is of any length (long, short).

There are so many examples regarding linguistic steganography. Some examples of linguistic steganography are

1. Hiding the text in the image.
2. Hiding the image in an image.
3. Playing an audio track backward to reveal a secret message.
4. Playing a video within another video.
5. Playing the video faster to reveal a secret message.

Linguistic steganography can also be extracted the hidden message from a plain text. For ex:

"since everyone can read, encode text in neutral sentences is doubtfully effective"

The first letter in each word can produce a secret message or the secret information. Therefore, the secret message in the above sentence is "**SECRET INSIDE**".

Here, we are calculating the accuracy of three algorithms like support vector machine, convolutional neural network, and recurrent neural network. By using that accuracy, we can estimate or conclude which algorithm is best. In this, we can conclude that recurrent neural network is best because we got the highest accuracy for recurrent neural network.

Formula for calculating accuracy is

$$\text{Accuracy} = (\text{TP} + \text{TN})/(\text{TP} + \text{TN} + \text{FP} + \text{FN})$$

where

TP true positive value
TN true negative value
FP false positive value
FN false negative value.

By using this accuracy formula, we can calculate the accuracy of each algorithm.

2 Literature Survey

2.1 Safety of Linguistic Steganography

The main security service offered by steganography is data hiding. Steganography is done by using a steganographic key which is only known to sender and receiver so there is no data leak in between. These are some of the securities offered by using steganography.

2.2 Problems Through Steganography

If the attackers have any steganographic image, they can perform steganalysis. The term steganalysis means it is used by the intruders to extract the data forcefully from the steganographic images. This can result in data leaks and a lack of safety to the data.

3 Types of Steganography

There are two types of steganography as mentioned in Fig. 1. The two types of steganography are technical steganography and linguistic steganography.

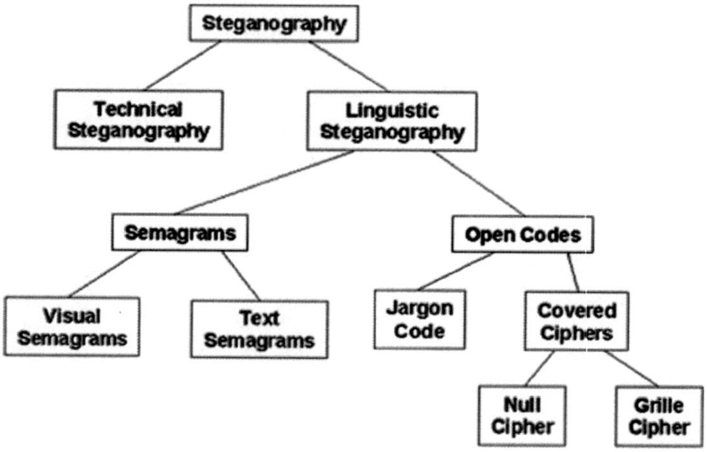

Fig. 1 Types of steganography

3.1 Technical Steganography

Technical steganography will communicate the information, but it does not deal with the written words. Technical steganography is one of the methods in steganography. In this method, we will use the tools or devices, etc., to extract the secret message. Some of the technical steganography methods are:

3.1.1 Invisible Ink

These invisible inks can be used in pubs or clubs, amusement parks as hand stamp so that these inks will be visible in black light.

3.2 Linguistic Steganography

Linguistic steganography is used for hiding information in the multimedia data in a traditional approach. For that, we are taking an image and converting the text into bits like 0's and 1's.

4 Methodology

While the improvement of this project, there were several various difficulties. Somehow, this project is split into three major sections as these three parts of this project are the most decisive parts toward the foundation of this project. In this project mainly, we have previously owned three algorithms. The three algorithms are support vector machine (SVM), convolutional neural network (CNN), and recurrent neural networks (RNN). We will calculate the accuracy of each algorithm and check which algorithm is best. Basically, text steganalysis models are worked to aim the examiner and thus extract some text attributes, and then they examine the comparison between these facets before and after steganography to regulate either the text contains unpublished data or not. Traditional character or string steganalysis paradigm largely believes a majority of uncomplicated char characteristics grabbed virtually, like alphabet density. The attributes they extracted as well as examined were very smooth and linear that created them a toughness to deal with the trending steganography sentence causation methods stationed on neural networks. Freshly, a few investigations have tried to inspect the aristocratic connotation connection allying appellation in narrations to regulate even if the subject matter carries camouflaged counsel or not.

4.1 Support Vector Machine (SVM)

Support vector machines are expert system design with the correlated learning algorithms which are used for the verification of the info used for differentiation and for any analysis. SVM applies a method known as kernel trick in which kernel gets hold of low spatial input space and converts it into a higher spatial space. Train and test taken because now we must do classification. For doing classification, we have spitted the data into train and test. As we created the dataset into database, the file is loaded into x and y values. When we are splitting into training and testing 75% of the data, we are spitted into training and rest of the data is spitted into testing and we are finding the accuracy for this algorithm.

4.2 Recurrent Neural Network (RNN)

There are many neural networks and recurrent neural network (RNN) is one of the unique neural networks. Recurrent neural network is not deep just it has hidden layers compared to another neural networks. The concept of the recurrent neural networks is the output of previous one is the input of next step, and it is connected to every step.

RNN is the neural network designed which has the hidden units, and that hidden units are used for performing or analyzing of the data within the applications like DNA sequence, text processing, etc., in which the next step depends upon the output of previous one. More text steganography methods have developed in the recent years. They will automatically generate the steganography text which supported to the learn the statistic pattern. For recurrent neural network, it has only one hidden

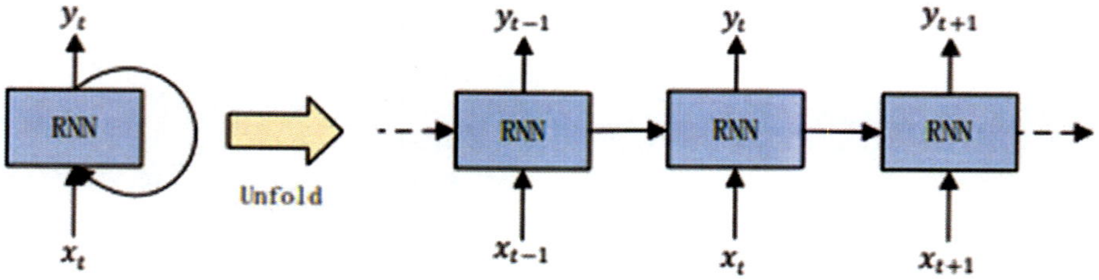

Fig. 2 Fundamental feature of recurrent neural network

layer and the formula is (Fig. 2)

$$m_t = f_h(w_h.x_t + u_t.h_{t-1} + b_h)$$
$$n_t = f_o(w_o.h_t + b_o)$$

where

x_t is the input and
y_t is the output.

RNN is widely used for modeling sequential signals. The above two equations are the simplest RNN model. In this paper, we have used long short-term memory and this long short-term memory can be used in the hidden layers.

4.3 Data Preprocessing

Data preprocessing is a technique that is wanting to change the data into a clean dataset. The information is gathered from different resources, and the gathered information is in the format of raw data which is not feasible for the analysis. The focus of this paper is to show few techniques that have been used in data science project. The techniques that we are getting to use are

- **Data Cleaning**
- **Data Integration**

4.3.1 Data Cleaning

It is the process which contains methods like detection and correcting and refers to identifying the incomplete data or irrelevant data parts and then replacing or modifying that incomplete or irrelevant data to a complete data.

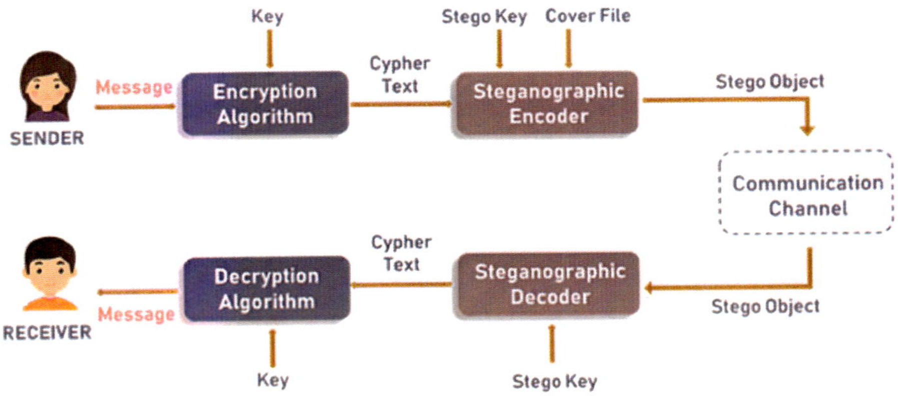

Fig. 3 Process of linguistic steganography

4.3.2 Data Integration

It is a module in the preprocessing that involves combining the data from multiple resources into a single resource and to provide the clear view of the dataset (Fig. 3).

5 Proposed System

The linguistic steganography is the handout address allows the user to slot their secret expurgated message in image in a way that cannot be microscopic and does not influence the constitution of the original image. The user who wants to secure their data or protect their data from others in any way can hide their message in the image using the concept of the linguistic steganography. It provides a well-organized way for secure convey of data of the information. This procedure can control different file formats like jpg, png, etc. The above-mentioned file formed can be used in our system and hide secret messages of text into the image.

6 Results

The results are the hidden message is stored in the image.

Figure 4a, b shows the input and output. The original image and the hidden image will look same so the un-officials cannot think, or they will not have any knowledge that there is any hidden message in the hidden image. But the original image and hidden image are not same because in the original image, there will not be any message or information but in the hidden message, there will a secret message or information.

Fig. 4 **a** Original image. **b** Hidden image

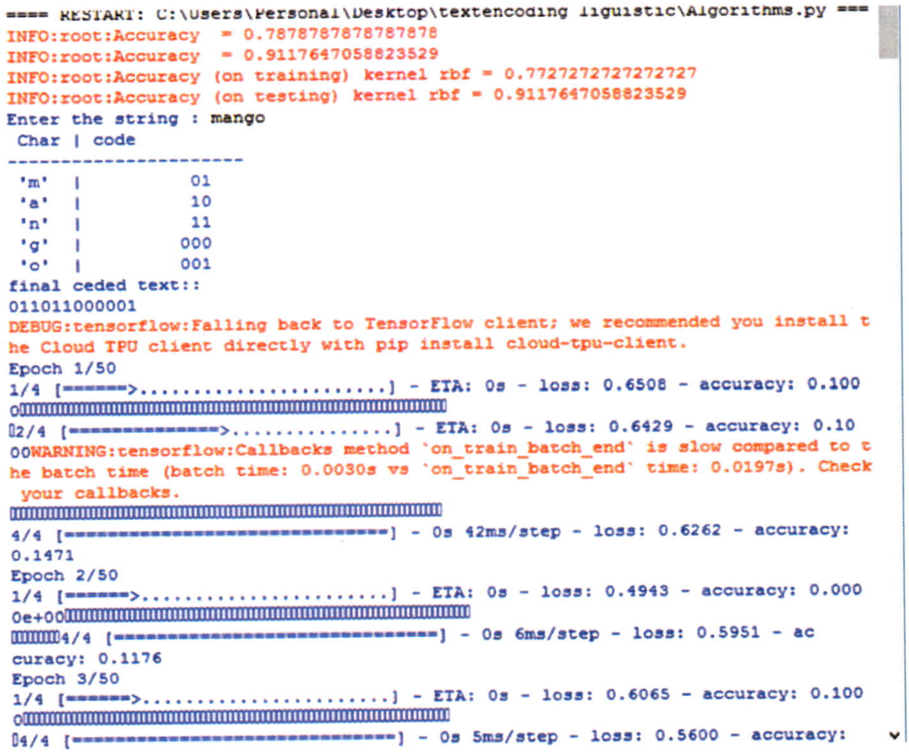

Fig. 5 Using text dataset

Figure 5 is the output using the text dataset. In this, we calculated the accuracy for each algorithm, and we got the highest accuracy for recurrent neural network. We are taking the string for example and that string is converted into 0's and 1's and it shows the binary value of each letter in the given word.

7 Conclusion

In this paper, we are proposing the linguistic steganography based on automatically generated paraphrases and recurrent neural networks which can implement finest text on the support of a set data in the binary form that needs to be converted. We deployed several investigations to test our recommended model from the availability of information hidden stamina or storage to equip it in the image and converting the text into bits. The result of our observations shows that the proposed model outperforms all the previous related work.

References

1. C. Zhi-Li, H. Liu-Sheng, Y.Z. Shan, Z. Xin-Xin, Z. Xue-Ling, Effective linguistic steganography detection, in *IEEE 8th International Conference on Computer and Information Technology Workshops*
2. D. Rawat, V. Bhandari, A steganography for hiding image in an image using LSB method for 24 bit color image. Int. J. Comput. Appl. **64**(20) (2013)
3. R. Yang, Z.-H. Ling, Linguistic steganography by sampling-based language generation, in *Proceeding of APSIPA Annual Summit and Conference* (2019)
4. P. Meng, L. Huang, Z. Chen, W. Yang, D. Li, Linguistic steganography detection based on perplexity, in *2008 International Conference on MultiMedia and Information Technology*
5. P. Meng, L. Hang, W. Yang, Z. Chen, H. Zheng, Linguistic steganography detection based on statistical language model, in *2009 International Conference on Information Technology and Computer Science*
6. Z.M. Ziegler, Y. Deng, A.M. Rush, Neural linguistic steganography, in *Proceedings of the 2019 Conference on Empirical Methods in Natural Language Processing and the 9th International Joint Conference on Natural Language Processing*, Hong Kong, China, 3–7 Nov 2019, pp. 1210–1215
7. C.-Y. Chang, S. Clark, Practical linguistic steganography using contextual synonym substitution and a novel vertex coding method, in *2014 Association for Computational Linguistics*
8. L. Xiang, W. Wu, X.L.-C. Yang, A linguistic steganography based on word indexing compression and candidate selection. Springer Publication on 11 May 2018
9. S.M. Masud Karim, Md. Saifur Rahman, Md. Ismail Hossain, A new approach for LSB based image steganography using secret key, in *International Conference on Computer and Information Technology (ICCIT)* (2011)
10. T. Fang, M. Jaggi, K. Argyraki, Generating steganographic text with LSTMs. arXiv preprint arXiv:1705.10742 (2017)
11. W. Bender, D. Gruhl, N. Morimoto, A. Lu, Techniques for data hiding. IBM Syst. J. (1996)
12. X. Ge, R. Jiao, H. Tian, J. Wang, Research on information hiding. US-China Educ. Rev. (2006)
13. Z. Zhou, Y. Mu, N. Zhao, Q.M.J. Wu, C.N. Yang, Coverless information hiding method based on multi-keywords (Springer International Publishing, 2016)
14. H.H. Moraldo, An approach for text steganography based on Markov chains. arXiv preprint arXiv:1409.0915 (2014)
15. H.Z. Muhammad, S.M.A.A. Rahman, A. Shakil, Synonym based Malay linguistic text steganography (IEEE, 2009)

Critical Analysis of Virtual LAN and Its Advantages for the Campus Networks

Saampatii Vakharkar and Nitin Sakhare

Abstract One of the sizzling regions of the networking systems is the VLAN technology. A VLAN permits network devices to be combined as virtual LANs in logical association rather than a physical association. In this paper, we have done critical analysis of VLAN. We have also seen the advantages of using VLAN as it helps to create multiple networks with one class of IP address, and by restricting inter-VLAN communication we can allow or deny the users from accessing a particular network. The basic need of implementing VLAN is the breaking of networks, and to better understand this, we have shown VLAN configuration on Cisco Packet Tracer. VLANs have various advantages, but in our study, we have found out that VLANs are used for many things which were not originally meant for it.

Keywords VLAN · IP address · Router · Ports · Segmentation · Troubleshooting

1 Introduction

A local area network (LAN) interconnects a set of computer networks inside a finite set of area like school, society, etc. It works on a single broadcasted domain by communicating between the connected devices. A virtual local area network (VLAN) is a type of LAN which enhances the capabilities of a LAN and makes it more efficient. VLAN is basically logical assembling of networking devices, and when a VLAN is created, a large broadcasted domain is broken into smaller broadcasted domains, i.e. virtual LANs (VLANs) allowing a single LAN to be partitioned into several separate LANs. VLAN provides segmented and organized flexibility in a switched network. VLAN is just like a subnet, i.e. two different subnets require a router to communicate, and two different VLANs also require a router to communicate [1, 2].

VLANs are built in such a way that they make partitioning a single switched network or fixing the security requirements of the system so effortless for the network

S. Vakharkar (✉) · N. Sakhare
Vishwakarma Institute of Information Technology, Pune, India

N. Sakhare
e-mail: nitin.sakhare@viit.ac.in

S. Shakya et al. (eds.), *Mobile Computing and Sustainable Informatics*, Lecture Notes on Data Engineering and Communications Technologies 68,
https://doi.org/10.1007/978-981-16-1866-6_56

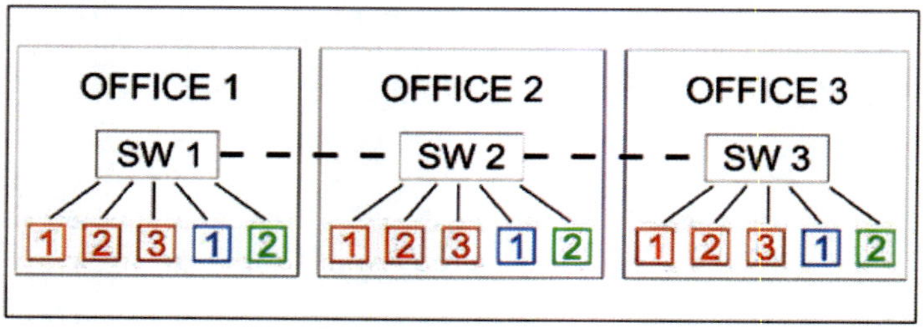

Fig. 1 Company's offices and its department (before using VLAN)

administrator, i.e. without making huge changes in the infrastructure or changes in the cables [3]. In VLAN, logically all the department's computers can be connected to the same switch but they are segmented to behave that they are different. VLANs are usually set up by large businesses to re-partition devices for better traffic management. The basic need of implementing VLAN is the breaking of networks. The networks are divided into workstations within a LAN to eliminate the blockage and load. Previously, basic LAN is limited to its capabilities and induces blockage in the network. Virtual LAN can only be made using switches or bridges whereas in LAN hubs, switches and routers are used [4, 1, 5].

Let us take an example to understand the need of VLAN better. A company Z has three offices located in the three corners of the city. As shown in Fig. 1, all of its offices are inter-connected with the dotted black links. The company Z has three important departments:

1. IT support (Red),
2. Accounting (Green),
3. Administration (Blue).

The IT support department has nine computers. Accounting and administration department has three computers each. Every office of this company has three computers from IT support department, one computer from accounting and administration department each. As the company is on a huge level, the administration and accounting department has very secretive information in them and needs to be protected from the development department. Now, all the computers share same broadcast domain because of default configuration, cause of which the IT support department will be able to access the administration or accounting department's resources, which might be very problematic for the company. Now, some logical boundaries over the physical network could be made with a VLAN; we could create which would indeed help us. Assume that we created three VLANs for the company's network and assigned those VLANs to the related computers, like,

1. Admin VLAN for administration department
2. IT VLAN for IT support department
3. Acc VLAN for accounting department.

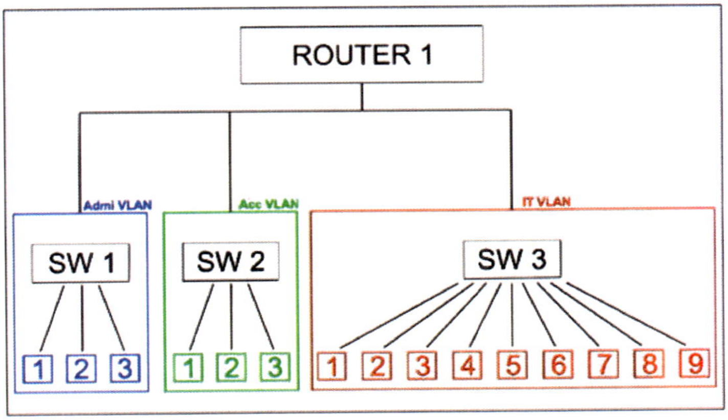

Fig. 2 Company's offices and its department (after using VLAN)

Therefore, physically we did not change anything but instead we changed logically, by grouping the computers according to their functionalities. These groups, i.e. VLANs, will communicate with each other using a router. After logically assigning the VLANs, the company's network will look like Fig. 2.

By doing this, it seems like the group of similar domain's computers within the VLAN are communicating to each other as if each computer is attached to the same cable. We solved the broadcast problem by reducing the size of the broadcast domain. It also secured our important resources, and most importantly instead of location, it logically allowed us the grouping of devices by the function. There are two methods by which we can assign VLAN to a device: static and dynamic [6, 1, 7].

The main reason of why any administrator would use VLANs is to:

1. Limit the scope of broadcast traffic.
2. Simplify access management policies.
3. Enable sleek host quality for wireless users.
4. Support decentralized network management.

In this paper, critical analysis of VLAN is done. We will also be looking at the advantages of using VLAN for the campus networks as VLAN helps to create multiple networks with one class of IP address, and by restricting inter-VLAN communication, we can allow or deny the users from accessing a particular network [2].

2 Literature Survey

A lot of research has been done in the area of VLAN and its implementation in campus networks to study the benefits of VLAN.

Yu et al. [8] have examined how VLANs can be effectively used for four campus networks. VLANs are not only used for only a certain purpose rather they are used for many different aspects. Although use of VLANs complicates the network configuration management, organizations must look at the VLAN with aim to make the management easier for campus and enterprise managers.

Zhu et al. [9] designed Secure VLAN (S-VLAN) architecture-based application and implemented a prototype system for the same. To prevent the security problems from interrupting, the proposed S-VLAN is acceptable for universities and enterprises. Further, S-VLAN is cost beneficial for real-time applications like VoIP, and it is easy to use and more efficient.

Yamasaki et al. [4] introduced a flexible campus VLAN system established on OpenFlow. A virtual group ID (GID) was responsible for running the communication access. Accessing the management functions based on NEC's OpenFlow controller and authentication was the purpose.

Odi et al. [10] implemented inter-VLAN routing in Ebonyi State University for effective distribution of network services. The VLAN architecture when deployed in Ebonyi State University showed immense benefit to the network users. This work indeed showed how VLAN and inter-VLAN routing in managing and maintaining Ebonyi State University Networks benefited.

Mathew and Prabhu [3] configured four layer two switches and two layer three switches to explain the concept of VLAN and routing in VLAN.

Prasad et al. [11] allowed only one specific VLAN to pass through from one network and reach the other specific VLAN of the different network using the access list to prevent communicating the other VLAN inside the network. A routing configuration to connect different VLANs in a network called inter-VLAN was proposed. The main purpose of this paper is to provide a safe and secure connection for an organization where higher-level information is protected at all times for others in a single network.

To study dependencies induced by VLANs on a campus network and study the layer 2 topology for the Gergia Tech campus network, Mansy et al. [12] developed a EtherTrace, a passive layer 2 topology discovery tool. Data from 94 routers, 114 switches, and almost 90,000 unique MAC addresses was taken in a day. Much work remains to fully understand the nature of interactions between VLANs and layer 3 topologies in campus [12].

Mario Ernesto Gomez-Romero, Mario Reyes-Ayala, Edgar Alejandro Andrade-González, Jose Alfredo Tirado-Mendez, designed and implemented a policy-based VLAN. The main motive behind creating this policy-based VLAN was to maximize the security levels so as to decrease undesirables' sites from being accessed and to avoid the attack of hackers and also wanted to avoid time wasting by increasing the network. Lots of plots that showed the user navigation and the information transferred by them were seen in the outcomes. Hence, the implemented solution could be changed according to current organization specification [13].

Aziz [6] introduced a technique to differentiate the areas of same classified networks into VLAN-based-independent subnetworks.

Krothapalli et al. [14] presented a surfeit system of virtual MAN which addresses the network operators face in managing VLANs in an enterprise. The most performed tasks while managing VLANs have been automated by a system, even though the management process of VLANs is error-prone. The beta version was tested on Purdue University campus network and was successful in understanding VLANs design.

Sun et al. [1] take a step towards systematic approach at network designs in the VLANs. While giving privilege and feasibility constraints on the design of VLANs, the algorithms trade multiple aspects like broadcast traffic costs, or maintaining spanning trees of each VLAN in the network and they also took care of the VLAN dependencies hence enables the automatic detection of network. The results when shared showcased that better designs can be made with the algorithms than the current practices without error. To better understand how networks reacted in different situations, the simulations were analysed. The results showcased that by using the work which was introduced, the response time and security are improved.

Shuizhen [15] to achieve a higher overall execution did detailed inspection and description on the campus network, which included everything from the method of planning and designing to network planning, and service systems of the campus network. The study gave a better solution which would meet the future needs.

Ingole and Bhise [2], to understand how VLAN works examined four campus networks, and the study showed that VLANs are used for various purposes which are not suitable at first place, and usage of VLANs complicates the network management.

Al-Khraishi and Quwaider [7] proposed the use of virtual local area network (VLAN) with the file transfer protocol (FTP) application and hyper-text transfer protocol (HTTP) application traffic using the OPNET 14.5 simulator over a wireless LAN network, which gave a better understanding that wireless is better than VLAN as VLAN resulted in reduction of both delay and throughput. According to the results, the OLSR protocol performed better as it does not need to find routes to the destination and is a proactive routing protocol regardless of its size.

Wang et al. [5] introduced a few methods to implement VLAN and studied the drawbacks in Ethernet environment. On that basis, a new strategy based on a new service was proposed to implement VASS architecture.

Zhou and Ma [16] put forward an algorithm which with incomplete address forwarding table, discovered a physical layout topology structure of Ethernet networks. This algorithm implemented itself in some universities where it worked well in its VLAN networks as it can handle both layer 2 switches with VLAN and layer 3 switches.

3 Methodology

VLAN is created from one or more LANs. It allows grouping of the available devices into logical networks which is in turn a virtual LAN that is operated like a physical LAN. A VLAN is made at the switch. For example, the consulting staff might be located anywhere in a building, but if they are assigned to a single VLAN, they can share and communicate resources if connected to the same segment, and other departments will not be able to access their resources and other departments' resources will be invisible to them. Virtual local area network is the full form of VLAN. A VLAN divides the devices on the network logically on layer 2 (data link layer), and usually, layer 3 devices divide the broadcast domain, but broadcast domain can be split by switches using the concept of VLAN [13, 5]. A broadcast domain is a network component where if the broadcasted device is a packet, then all the devices will receive it.

3.1 *Inter-VLAN Routing*

Forwarding the network traffic from one VLAN to other VLAN is inter-VLAN routing. It allows communicating between various different VLANs in the same switch or other with the help of layer 3 device router. Inter-VLAN can be configured in three different ways:

1. Traditional method
2. Router on stick
3. Inter VLAN in layer 3 switch.

3.2 *Types of VLAN*

There are five main types of VLANs. They are defined by the type of network traffic they carry for a specific VLAN and the kind of function they perform.

Default VLAN
It is VLAN 1, a default setting on Cisco switches and other vendors. Unless we specifically assign an access port to a particular VLAN, like VLAN 10 or VLAN 20, the access port belongs to VLAN 1. We cannot modify the default VLAN or delete it, it is the default setting, and it is VLAN 1. VLAN 1 is never intended to be used as a standard data VLAN [6].

Data VLAN
It can also be called as a user VLAN, as it is configured to carry only user-generated traffic.

Voice VLAN
It is configured to carry voice traffic and has the permission to transmit priority over other types of network traffic, and the phone calls are required as without it, the communication is not complete.

Management VLAN
Management VLAN is configured to access the management capabilities of a switch. Management VLAN is to be assigned with an IP address and subnet mask is the management VLAN. Any switch VLAN could be configured or define a unique VLAN which provides the services to the management VLAN.

Native VLAN
A native VLANs traffic traverses on the 802.1q trunk without a VLAN tag. A management VLAN and native VLAN could be the same, but for a good security practice, they are not. By default, the native VLAN is 1, but we can change it to any number like VLAN 2 or 20, etc. It can be configured on the trunk port. It is a er trunk, per switch configuration. The same native VLAN should be on both ends of the trunk, else the trunk will not operate properly. Native VLAN is defined in 802.1q standard, and older devices that do not support VLANs were used by native VLAN.

It is used to control and manage protocol traffic like Cisco Discovery Protocol (CDP), VLAN Trunking Protocol (VTP), Spanning Tree Protocols (STP) or some other network management traffic by the switch. It is also useful when we deal with the voice over IP or VoIP [6].

3.3 *Methods of Assigning VLAN*

There are two methods of assigning VLAN: Static and dynamic.

Static VLAN
Static VLAN is based on configuration of the ports. In static VLAN, ports on the switch are physically allocated to virtual network. Once ports are allocated, their pre-assigned VLANs are always associated [11].

Let us take an example, as shown in Fig. 3, ports 1, 2, 3, 4 are assigned to IT, ports 5, 6, 7 to administration and ports 8 and 9 to accounts. Static device has nothing to do with the devices. When we insert one device into port 9, for example, the device will become a member of the administration VLAN, and when we plug the same device into port 1, the device will be on the IT VLAN.

Dynamic VLAN
It is not based on a port location but instead on a device. It is mostly MAC-based, i.e. when VLAN membership is defined by device's MAC address. This membership is formed on the IP address of a device. VLAN Member Policy Server (VMPS) is needed for a dynamic VLAN [11].

For example, as shown in Fig. 4, we use AA, BB or CC to represent MAC addresses just for our understanding. As we know an actual MAC address is a 48-bit-long hexadecimal number. When a computer with the MAC address DD is plugged into port 5, port 5 will communicate with the server and check against the database for its membership. In a database, computer DD is pre-assigned to VLAN30. Suppose

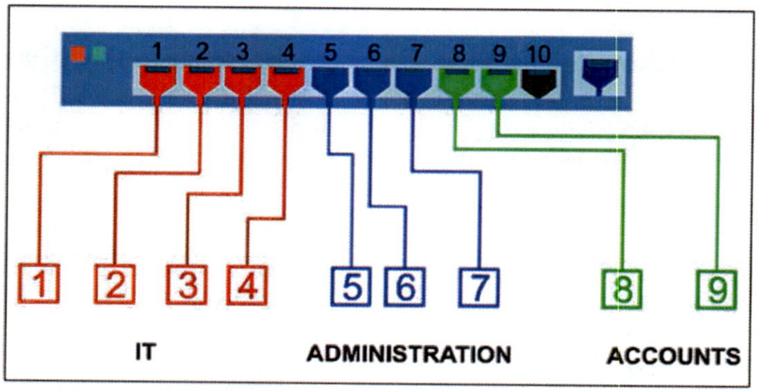

Fig. 3 Static VLAN (port-based VLAN)

computer DD is plugged into port 1, the result would be same, because port 1 will communicate with the server and find the pre-assigned VLAN to computer DD. Therefore, even if the port location is changed, it does not change the VLAN membership of a device.

Hence, dynamic VLAN has two advantages over static VLAN,

1. Flexibility: Instead of configuring every individual port of all switches, it controls the devices on a central server. In an organization, it simplifies the network management.
2. Security: The server's database checks for its membership with every device connected to a switch. Therefore, no outside device can plug in and +get access to the network [11].

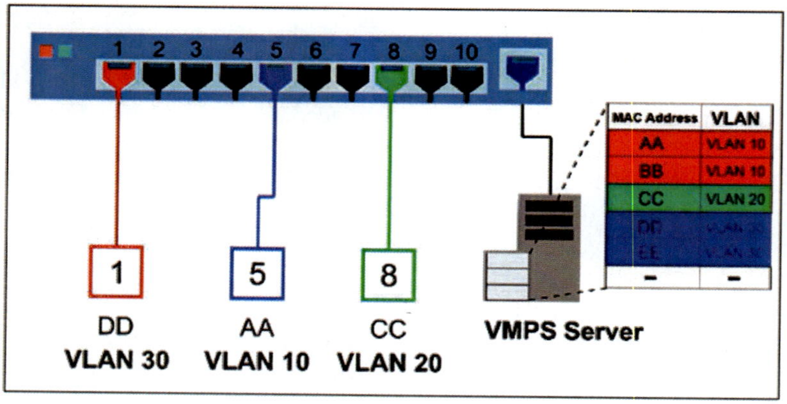

Fig. 4 Dynamic VLAN (MAC-based)

3.4 Switch Ports in VLAN

A switch port has two modes: access and trunk, and the switch port can settle down in two modes: static and dynamic. A switch port can be manually configured for access or trunk mode using static method. Dynamic Trunking Protocol can be used for an interface for enabling the trunking in dynamic method.

Access Port
Untagged Ethernet frames from one particular VLAN can be exchanged by a switch port in access mode. The switch interfaces attached to devices like printers ARE configured as access ports. Using the command switch port access vlan vlan-id, in interface configuration mode, the VLAN that a specific switch port is assigned to can be altered [3].

Voice Access Port
Voice access ports are well suited for connecting IP phones. Corporate users are using voice access ports over two networking devices—data traffic from the computer and voice traffic from IP phone. Former is known as data VLAN, whereas later is known as voice VLAN.

Trunk Port
Trunk ports carry traffic at the same time from multiple VLANs. They can be configured between a switch and router or server and switch, but mostly it does between two switches. Trunking is indeed an excellent feature because it shares one single physical link to various VLANs even when traffic is isolated between different VLANs. 1 to 4094 is by default the full range of VLAN IDs and is allowed to on a trunk port. VLAN spans across multiple switches with various access ports which belong to VLANs spread across various switches in the different parts of the network cause of trunking which in turn provides excellent flexibility to VLANs and assigning a host to a VLAN nevertheless knowing where exactly it located physically on a switched network is [3, 11].

Tunnel Port
To set apart the traffic of consumers in a network provider from other consumers, tunnel ports are used in 802.1Q. Tunnelling is the transferal of data which is used only inside a private corporate network via public network, such that the routing nodes in the public network are unaware that it is included in a private network (Table 1).

4 Experimentation

Cisco Systems designed platform visual simulation tool known as the Packet Tracer. It permits users to imitate modern computer networks and form network topologies. It is useful for VLAN simulations as it permits to make complex and large

Table 1 VLAN ranges

Sr. No.	VLAN type, range	Description
1	VLAN 0, 4095	Reserved VLAN
2	VLAN 1	Default VLAN of switches—can be used but cannot be deleted or edited
3	VLAN 2-1001	Normal VLAN range—can create, edit or delete
4	VLAN 1002-1005	For FDDI and token rings
5	VLAN 1006-4094	Extended range

networks, indeed not possible cause of huge costs in the physical hardware. It allows to reproduce the configuration of Cisco products using a command line interface [6].

4.1 VLAN Configuration on Cisco's Packet Tracer

Let us consider the following scenario: Company X has four important departments (Table 2).

Every department's PC is located on a different floor, and as the company is on a huge level, every department has very sensitive information in their PCs and needs to be protected from other departments. So, with the VLAN, logical boundaries are created rather than physical network, which would indeed help us.

Let us do it with the help of Packet Tracer.

1. Open Packet Tracer.
2. Click on network devices and select switches.
3. Select a Pt-Switch and drag and drop it.
4. Click on end devices and select 14 PCs and drag and drop it.
5. Now, make four groups:
   PC0 to PC4 as VLAN 2 for accounts
   PC5 to PC8 as VLAN 3 for IT
   PC9 and PC10 as VLAN 4 for HR
   PC11 to PC13 as VLAN 5 for sales
6. Now let us make the connections. Go on connections and select copper straight through wires, and connect every PC to Switch 0.
   Like, Connect PC0 (Fast Ethernet 0) to Switch 0 (Fast Ethernet 0/1) (Table 3).

Table 2 Company X's department and its Pcs

Sr. No.	Department	No. of PCs
1	Accounts	5
2	IT	4
3	HR	2
4	Sales	3

Table 3 PC to switch connections

PC No.	Switch 0
PC1 (Fast Ethernet0)	Fast Ethernet 0/2
PC2 (Fast Ethernet0)	Fast Ethernet 0/3
PC3 (Fast Ethernet0)	Fast Ethernet 0/4
PC4 (Fast Ethernet0)	Fast Ethernet 0/5
PC5 (Fast Ethernet0)	Fast Ethernet 0/6
PC6 (Fast Ethernet0)	Fast Ethernet 0/7
PC7 (Fast Ethernet0)	Fast Ethernet 0/8
PC8 (Fast Ethernet0)	Fast Ethernet 0/9
PC9 (Fast Ethernet0)	Fast Ethernet 0/10
PC10 (Fast Ethernet0)	Fast Ethernet 0/11
PC11 (Fast Ethernet0)	Fast Ethernet 0/12
PC12 (Fast Ethernet0)	Fast Ethernet 0/13
PC13 (Fast Ethernet0)	Fast Ethernet 0/14

Now, all the connections are made.

7. Let us do IP Configuration of all the 14 PCs now,
   Click on PC0 → Desktop → IP Configuration
   Set the IP address as: 192.168.10.1 then close it.
   Now, do the same for rest of the PCs (Table 4).
8. Now go to PC0 → Command Prompt
   Type
   →ping 192.168.10.10
   →ping 192.168.10.9

Table 4 IP configuration of PCs

PC No.	IP address
PC1	192.168.10.2
PC2	192.168.10.3
PC3	192.168.10.4
PC4	192.168.10.5
PC5	192.168.10.6
PC6	192.168.10.7
PC7	192.168.10.8
PC8	192.168.10.9
PC9	192.168.10.10
PC10	192.168.10.11
PC11	192.168.10.12
PC12	192.168.10.13
PC13	192.168.10.14

All the computers share same broadcast domain because of the default configuration; therefore, every department can access each other's department's resources, which might be very problematic for the company. Hence, we will the group similar domain's computers within the VLAN.

9. Now click on Switch → CLI → Press Enter
Type these commands and press enter after every command:
→En
→confi
→vlan 2
→name Accounts
→exit
→vlan 3
→name IT
→exit
→vlan 4
→name HR
→exit
→vlan 5
→name Sales
→exit
→exit
→show vlan
Press Enter and type these commands for configuration:
→confi
→interface f0/1
→switchport access vlan 2
→exit
→interface f0/2
→switchport access vlan 2
→exit
→interface f0/3
→switchport access vlan 2
→exit
→interface f0/4
→switchport access vlan 2
→interface f0/5
→switchport access vlan 2
→exit
→interface f0/6
→switchport access vlan 3
→exit
→interface f0/7
→switchport access vlan 3
→exit
→interface f0/8

→switchport access vlan 3
→interface f0/9
→switchport access vlan 3
→exit
→interface f0/10
→switchport access vlan 4
→exit
→interface f0/11
→switchport access vlan 4
→exit
→interface f0/12
→switchport access vlan 5
→interface f0/13
→switchport access vlan 5
→exit
→interface f0/14
→switchport access vlan 5
→exit
→exit
→copy running-config startup-config
Now, the VLANs are configured successfully.

10. Now go to PC0 → Command Prompt
Type
→ping 192.168.10.10
→ping 192.168.10.9
→ping 192.168.10.2
After grouping the VLANs, we can see that when we are pinging from accounts to HR or any other department, it is not successful but when we ping within the department it is successful.
11. Now let us check with the protocol data unit (PDU) message.
Click on PDU → click on PC0 and PC11
Click on PDU → click on PC5 and PC8
Click on PDU → click on PC9 and PC10
PC0–PC11 fails and PC5–PC8 and PC9–PC10 is a success because the VLANs are now grouped together.
12. Now click on Switch0 → CLI, type → show vlan brief. We can see the group of similar domain's computers within the VLAN are communicating to each other as if each computer is attached to the same cable. Therefore, physically we did not change anything but instead we changed logically, by grouping the computers according to their functionalities.

5 VLAN Advantages

5.1 Segmentation

We have two departments, finance and sales, and both belong to one broadcast domain or network cause of which both are connected to same switch. In a single-broadcast domain, everyone else can hear, when one computer broadcasts. This type of network design could cause us two potential problems, (1) traffic congestions and poor performance if there are too many users in one broadcast domain and (2) security concerns because different departments are sharing a single network.

To address these two problems, there are three possible solutions.

1. we can create two VLANs with one switch. Therefore, physically, these two departments share one physical switch, but virtually, they are in two different independent VLANs. Here, we can also add one router so that the two VLANs can communicate using a technique called trunking. Trunking carries different data over one physical line simultaneously.
2. we can use two independent VLANs and no need of a physical router, and instead, the two VLANs can talk to each other via inter-VLAN routing techniques [10, 5].

5.2 The Simplicity of Network Design and Deployment

With VLANs, we are no longer confined to physical locations which means that the similar departments computers could be on different floors and the devices having various functions could be connected to a same switch without VLANs reorganizing huge budget and labour for anyone [8].

5.3 Easier Troubleshooting and Management

When various groups are isolated and segmented from each other, the VLAN troubleshooting and management becomes faster, and VLANs make sure that the critical data keeps flowing by making traffic flow as a priority [8].

5.4 Enhance Network Security

Virtual boundaries are made by VLANs which could be crossed over only through a router. Hence, one can restrict access to a VLAN by using standard, router-based security measures [8, 2].

5.5 *Flexibility*

Flexibility to add or subtract the number of hosts is provided by the VLAN [10].

5.6 *Easy to Manage*

Adding or changing of networks and making other network changes can be done through switch web management interface instead of wiring the closet in a VLAN [10].

6 Conclusion and Future Scope

VLANs are classified into layers 1, 2, 3 and more by logical segmentation of a network, and only layers 1 and 2 are in the draft standard 802.1Q. To determine the origin and destination VLAN for collected data, tagging and filtering database is bridged. There are various advances done in the field of networks, and VLANs allow the evolution of virtual groups, better security than before, better performances, and low costs. If VLANs are implemented successfully, they will show sizeable promise in future. We will see different styles of VLAN in future due to the fastest innovations of new hardware, software, technologies, etc. established VLAN yet. But the key aim would be designing those VLANs that could have the full mechanism of secure access instead of unsecure VLANs.

References

1. X. Sun, Y.W.E. Sung, S.D. Krothapalli, S.G. Rao, A systematic approach for evolving VLAN designs, in *2010 Proceedings IEEE INFOCOM*, Diego, CA, USA (2010)
2. J.N. Ingole, K.S. Bhise, Study on virtual LAN usage in campus networks. Int. J. Res. Comput. Inf. Technol. (IJRCIT) **2**(4) (2017)
3. A.I. Mathew, S.R.B. Prabhu, A study on virtual local area network (VLAN) and inter-VLAN routing. Int. J. Curr. Eng. Sci. Res. (IJCESR) **4**(10) (2017)
4. Y. Yamasaki, Y. Miyamoto, J. Yamato, H. Goto, H. Sone, Flexible access management system for campus VLAN based on OpenFlow, in *IEEE/IPSJ International Symposium on Applications and the Internet*, Bavaria, Germany (2011)
5. X. Wang, H. Zhao, M. Guan, C. Guo, J. Wang, Research and implementation of VLAN based on service, in *GLOBECOM'03. IEEE Global Telecommunications Conference*, San Francisco, CA, USA (2003)
6. D.A. Aziz, The importance of VLANs and trunk links in network communication areas. Int. J. Sci. Eng. Res. **9**(9) (2018)
7. T. Al-Khraishi, M. Quwaider, Performance evaluation and enhancement of VLAN via wireless networks using OPNET modeler. Int. J. Wirel. Mobile Netw. (IJWMN) **12**(3) (2020)

8. M. Yu, M. Rexford, X. Sun, S. Rao, N. Feamster, A survey of virtual LAN usage in campus networks. IEEE Commun. Mag. **49**(7), 98–103 (2011)
9. M. Zhu, M. Molle, B. Brahmam, Design and implementation of application-based secure VLAN, in *29th Annual IEEE International Conference on Local Computer Networks*, Tampa, FL, USA
10. A.C. Odi, N.E. Nwogbaga, O.N. Chukwuka, The proposed roles of VLAN and inter-VLAN routing in effective distribution of network services in Ebonyi State University. Int. J. Sci. Res. (IJSR) **4**(7), 2608–2615 (2015)
11. N.H. Prasad, B.K. Reddy, B. Amarnath, M. Puthanial, Intervlan routing and various configurations on Vlan in a network using Cisco Packet Tracer 6.2. Int. J. Innov. Res. Sci. Technol. **2**(11) (2016)
12. A. Mansy, M.B. Tariq, N. Feamster, M. Ammar, Measuring VLAN-induced dependencies on a campus network. Res. Gate (2009)
13. M.E. Gomez-Romero, M. Reyes-Ayala, E.A. Andrade-González, J.A. Tirado-Mendez, Design and implementation of a VLAN
14. S.D. Krothapalli, S.A. Yeo, Y.W.E. Sung, S.G. Rao, Virtual MAN: a VLAN management system for enterprise networks
15. X. Shuizhen, Planning, designing and building large-scale network at campus, in *3rd International Conference on Computer Research and Development*, Shanghai, China (2011)
16. J. Zhou, Y. Ma, Topology discovery algorithm for ethernet networks with incomplete information based on VLAN, in *IEEE International Conference on Network Infrastructure and Digital Content (IC-NIDC)*, Beijing, China (2016)

A Memory-Efficient Adaptive Optimal Binary Search Tree Architecture for IPV6 Lookup Address

M. M. Vijay and D. Shalini Punithavathani

Abstract The Internet protocol version 6 which has an increase in address length with prefix length poses challenges in memory efficiency, incremental updates. So this work proposes adaptive optimal binary search tree (AOBT) based IPv6 lookup (AOBT-IL) architecture. An adaptive optimal binary search tree (AOBT) structure is introduced for the minimization of memory utilization. An Altera Quartus Stratix II device with Verilog HDL implements the IP lookup design. The proposed method performance is validated using different lookup table sizes with comparative analysis. The proposed method accomplished better outcomes in the case of maximum frequency, memory, SRAM, and logic elements results when compared to existing methods such as balanced parallelized frugal lookup (BPFL), linear pipelined IPv6 lookup architecture (IPILA) and parallel optimized linear pipeline (POLP) methods.

Keywords IPv6 architecture · Lookup tables · Adaptive optimal binary search tree · Memory consumption · Throughput

1 Introduction

The next-generation Internet protocol standard which replaces the currently used IP standard IPv4 is IPv6. This IPv6 is originally urbanized with a Internet Engineering Task Force (IETF) to enhance the services provided by IPv4 [1]. This is exploited to serve the spaces where there is a limitation of IP addresses created by the previous IPv4. This IPv6 includes a 128-bit address, allows 3.4 × 1038 IP spaces, whereas, IPv4 has only a 32-bit address and allows 4.3 × 109 address spaces. Moreover, it enhances the Quality of Services (QoS) and swift the routing and simplicity in routing tables, and end-to-end communication. Although it possesses enormous

M. M. Vijay (✉)
Electronics and Communication Engineering, V V College of Engineering, Tisaiyanvilai, Tirunelveli, India

D. Shalini Punithavathani
Computer Science and Engineering, Government College of Engineering, Tirunelveli, India

S. Shakya et al. (eds.), *Mobile Computing and Sustainable Informatics*, Lecture Notes on Data Engineering and Communications Technologies 68,
https://doi.org/10.1007/978-981-16-1866-6_57

merits, network security sustains as a predominantly important issue due to the security issues [2]. The IPV6 address lookup approaches with trie, range and hash have been discussed [3]. It states that the incremental update of look-up tables and its associated memory states that lookup performance should significantly match with update time [3]. For multiple customers on a shared infrastructure, the recent public clouds accomplish storage and different kinds of services. All tenants are organized in a virtualized network environment to ensure performance and security isolation. In the Service Level Agreements (SLAs), the data center operators are required to execute NFs to enforce tenant isolation. An extensively deliberate issue in computer networking is Internet protocol (IP) lookup. The routing table or databases of IP routes or prefixes to particular network destinations are maintained via an IP router. The result of IP lookup considers the longest corresponding prefix between each prefixes associate with the port number [4]. The developing amount of Internet-connected devices and available addresses run out 32-bit address length of IPv4. The long-anticipated IPv4 address tiredness is enhanced with a developing IPv6 in which the address lengths are extended into 128 bits. At 100 Gbps, the recent core routers and data center switches are required to handle data traffic [5]. Nowadays, a classic problem in recent research is Internet Protocol address lookup (IP lookup). Several sophisticated solutions were introduced in recent years [6]. Because of the longer prefix length, the robust and practical design of the IP lookup engine is a challenging one. Both basic matching and IP lookup problems regarding the prefixes as ranges are tackled via dynamic segment tree (DST). Higher cost and power were obtained by using ternary address content mapping (TCAM). The dynamic range additions and deletions are performed via the dynamic multi-way segment tree (DMST), which was determined better memory utilization, update speed, and search speed but the fewer keys with smaller nodes lead to time complexity as well as computational complexity. Finally, the packet arrival time, assignment of time slots, and the infinite ability routing buffer design reducing the throughput. In this paper, we have proposed an adaptive optimal binary search tree (AOBT) based IPv6 lookup (AOBT-IL) architecture to tackle these problems. The major contribution of this paper is summarized as below:

- One lookup procedure is generated using pipelining thereby enhancing the throughput.
- An adaptive optimal binary search tree (AOBT) structure is introduced to minimize memory utilization.
- The FPGA implementation is performed using Altera Quartus Stratix II EP2S180F1020C5.

Remaining part of the paper is organized as follows: Sect. 2 depicts the existing works related to the implementation of IPv6 lookup. The proposed methodology is delineated in Sect. 3 and the experimental outcomes are discussed in Sect. 4. Ultimately, the paper is concluded in Sect. 5.

2 Related Works

The discussion in [6], DrawerPipe of reconfigurable network processing pipeline in which the same interface connects the packet processing into multiple "drawers". The "drawers" in any logical order are connected by using the programmable module indexing scheme called PMI that executes various network functions. Based on the experimental results, the DrawerPipe demonstrated ultra-low latency (< 10 μs) and high performance up to 100 Mpps with simply offload modified packet dispensation to FPGA. DrawerPipe minimizes four NFs line of code (LoC) compared with individual NF development. But, a scalability performance is poor and the packet buffer based on SRAM consumes more on-chip memory resources. The programmable packet processor architecture (PPPA) [7], provides programmability by elevated flexibility at the network data plane. The three major operations such as processing of packet data, classification, and parsing required in switches are supported using PPPA. The register-transfer level (RTL) description on the FPGA is allowed by applying the high-level P4 language to implement this architecture. At the SDN data plane, the pipeline model within the architecture is deliberated in order to increase the processing speed. The 320 MHz clock speed is operated by PPPA compared to ML605, NetFPGA-SUME, and NetFPGA-10G peer architectures. The PPPA consumes 1.3% of memory blocks, 1.9% of flip-flops and 4.3% of lookup tables, which is implemented in the Virtex-7 FPGA VC709 stage PPPA provides improved resource consumption and resource consumption with higher processing delay.

The parallel bloom filter was discussed in [8] for IPv6 and the novel GPU-accelerated software router has been developed by implementing Graphics Processing Unit. The NVIDIA GeForce GTX 580 with its implementation evaluates the performance of GRv6. The engine IP lookup accomplishes 60 Gbps to static routing tables by utilizing five real-life IPv6 routing tables. For static routing tables, the IP lookup speed accomplishes 85MLPS with lower throughput. The Blind Packet Forwarding (BPF) architecture was discussed in [9]. The dynamic blindness, flexible and fine-grained has been introduced via BPF design that offering diverse levels and multiple Network Address Confidentiality (NAC). Superior the masking grade is fed to a concentrate on in the first taxonomy. The endpoint masks its approach within domains higher network by applying for superior masking positions in the secondary blindness taxonomy. Furthermore, the BPF implementation was achieved by adapting OpenFlow, which demonstrates superior performances with higher sending rates but the resource consumption is poor. Unique parallel hardware architecture was introduced in [10] for the classification of hash-based exact matches in each clock cycle. The central principle of their research is the general structure of the memory organization available in all modern FPGAs. The similar capacity and throughput at the cost of 882 Block RAMs implement the feasible IPv6 5-tuple matching. The experimental results demonstrated increased capacities of match tables, more throughput, and very effective on-chip memory resources.

3 Proposed Methodology

The throughput is improved in which the pipeline is used to create one lookup procedure. The search tree height controls the number of pipeline levels. Each individual memory and each tree level maps the pipeline stages [11]. Figure 1 depicts the basic structural design of the proposed work. The routing tables associated with the prefixes determine the pipeline stages in which each group compresses with N pipelines. The IP address destination is separated and sent in the direction of each branch during the entry of a packet. The largest prefix match is the next-hop address. The delay is produced by pipelining in which the variation in a number of stages. Significantly, every smaller pipeline to match the best pipeline delay adds the delay block.

Fig. 1 Proposed workflow diagram

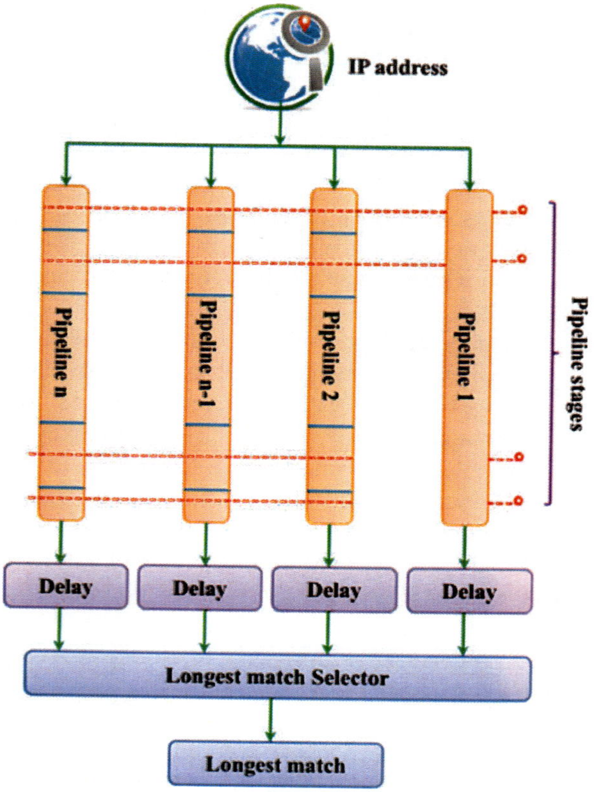

3.1 *Rapid Prefix Partitioning*

The smaller number of memory utilization with very high-speed backbone routers provides excessive Internet data. Consider the pipeline hardware configuration to accelerate the search speed. The FPGA hardware implements a dynamic multi-way segment tree and the greater delay accesses in memory [12]. The data structure is saved inside on-chip memory to circumvent the memory access delay thereby improving the search engine associates the throughput [13]. The active routing tables are introduced with the help of an adaptive optimal binary search tree (AOBT) structure, which is delineated in the following section.

3.2 *Adaptive Optimal Binary Tree (AOBT) Structure*

The prefix density variation is leveraged by designing the Adaptive Optimal Binary Tree (AOBT) structure. When the density adaptive trie covers the number of prefixes, the density adaptive trie is utilized by a hybrid data structure. Formulate the step to construct the hybrid trie-tree and the two data structures are presented and introduce lookup procedure at last.

3.2.1 Density Adaptive Trie

The prefix density distribution is constructed through the data structure called density adaptive trie (DAT). Further, the DAT merges the selective node merge (SNM) with the trie data structure. The method of SNM accepts the equi-sized section size to the prefix density while the single or multi-bit trie generates equi-sized sections, which are self-governing of the prefix density distribution [14]. The variable region sizes with the SNM method merge the equi-sized sections of lower density generated by the trie data structure (TDS). If the cumulative amount of prefixes subsequent to combining is equal to the largest number of prefixes, merge both equi-sized sections that retain them. An equi-sized region is merged using the SNM method from highest indices to lowest and lowest indices to highest ones. Initially, the lowest and highest index equi-sized sections are selected via the SNM method. After that, each chosen section is combined by evaluating the next contiguous equi-sized section. The numbers of merged sections with two constraints are present in the SNM method. (i) The number of equi-sized sections is covered by each merged section in which the prefix notation describes the space covered with the merged section. (ii) An adaptive trie of node header field size bounds the total number of merged sections. SNM technique minimizes the number of sections by combining the equi-sized section together. The memory efficiency of a data structure is enhanced with the SNM method.

Figure 2 shows the SNM impact on the duplication factor at the initial stage of the trie data organization. Figure 2a presents the SNM method benefit in terms

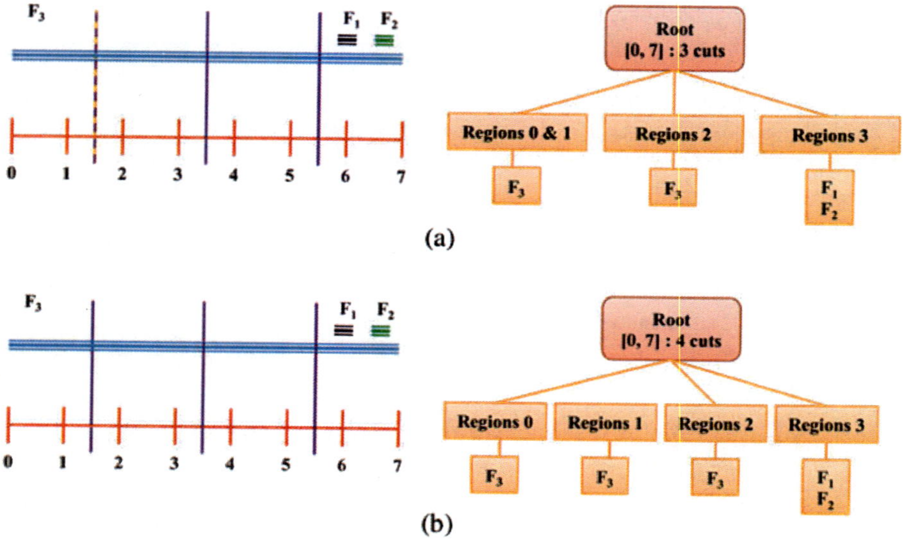

Fig. 2 SNM impacts on duplication factor to the initial stage of TDS, **a** with selective node merge and **b** without selective node merge

of memory efficiency. Figure 2b presents the initial stage of multi-bit trie without using the SNM method. Three-bit IP address is defined in which the prefix set is $F_1 = 110/3$, $F_2 = 111/3$, and $F_3 = 0/0$, respectively. Four equi-sized sections are classified from the section that corresponds to the combination of the different bit as zero to three [15]. Two leftmost equi-sized sections zero and one are merged with the SNM method in Fig. 2a in which dash lines separate it. When contrasted to the multi-bit trie described in Fig. 2b, the SNM method in Fig. 2a minimizes the number of nodes that hold the memory by 25%. The prefix replication factor is reduced by 33% but the prefix F_3 is replicated twice. Finally, the multi-bit trie data structure memory efficiency is increased.

The SNM model traverses the regions and saves them to the filed SNM, then categorized into the list of H to L and L to H. The indices of sections traversed from higher to lower indices and lower to higher indices are held with H to L and L to H array [16]. A single index is saved either in the H to L and L to H array for every region traversed with the SNM model. The initial indices describe the merged region as the merged region holds two or more multiple contiguous equi-sized sections. The SNM traverse an index of the next section determines the index of the last equi-sized region held in the merged section in which the un-merged section is enough to present this.

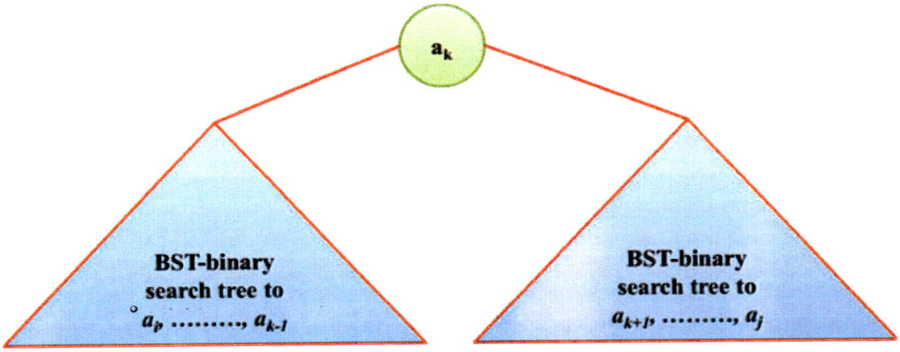

Fig. 3 Optimal binary tree structure

3.2.2 Adaptive Optimal Binary Search Tree

An information storage and retrieval are carried out by a widely used data structure called a binary search tree. Where T is a binary search tree to st keys as of a total order. Every node contains a key value in which every right subtree keys and left subtree keys are greater and less than the key at the root [17]. Let us consider m-keys and each key accesses the probabilities. A binary search tree on these m keys is constructed with the most favorable binary tree structure trouble in which the expected access time is minimized. The best alphabetic tree difficulty is a gap that contains non-zero access probabilities [18]. From the root, the minimum expected weighted way length with the binary tree is built and it is named as the problem of the Huffman tree. As a result, an optimal binary search tree comprises better memory efficiency. Figure 3 explains the structure of the optimal binary tree.

3.2.3 Reduced OBT

The proposed tree architecture is based on an optimal binary search tree which is enhanced using Leaf Size Reduction (LSR) technique. An adaptive trie reduces the memory consumption using prefix replication. No prefixes replications are used in OBT so it slightly increases the time complexity [19]. To reduce it an LSR technique is included in the OBT technique by minimizing the amount of information stored in OBT.

3.2.4 Procedure to Built AOBT

Algorithm 1 presents an adaptive optimal binary search tree procedure. All prefixes keep the number of prefixes with the root section and the root section below the threshold and then create an optimal adaptive search tree [20]. Otherwise, the equi-sized region is separated iteratively. Apply the equi-sized section using the SNM method [21]. An adaptive optimal search tree is constructed when the number of prefixes less than a threshold worth otherwise to built density adaptive trie and the section is separated. The number of separation in the section is the power of 2. The amount of bits from the IP address is represented by the bottom logarithm of the numeral of separations (Fig. 4).

Algorithm 2 illustrates the reduced OBT lookup procedure [22]. Initially, parse the header leaf and read it. Then, record each matched prefixes against the IP address destination with the prefix length. Return them to the longest prefix matches if all the prefixes are matched (Fig. 5).

The tree structure is the common way to represent prefixes. Finally, the search procedure included in the tree is managed to the address of destination. Adaptive

```
Algorithm 1: Adaptive Optimal Binary search Tree (AOBT)
Input: Stack S and prefix table
Generate each prefix that covers the root region
if
  The number of prefixes that a section maintains is less than the threshold value of a
  then
  Generate an adaptive optimal search tree to those prefixes
  else
  The root section is pushed into S
end if
while
  S is un-empty do
  The top node in S is removed and utilize as a reference section
  The number of partitioned is calculated and utilize as a reference section
  Based on the previous step to separate the reference section
  Apply selective node merge method to separate reference section
 for
    Every separated reference section do
   if
    The number of prefixes less than the threshold value is held
   then
    Generate an adaptive optimal search tree to those prefixes
   else
    Construct an ADT node to those prefixes
    This region is pushed into S
   end if
 end for
end while
Return the AOBT tree constructed
```

Fig. 4 Algorithm for adaptive optimal binary search tree (AOBT)

Algorithm 2: Reduced OBT lookup procedure
Input: IP address destination and reduced OBT
Parse the leaf header
Examine the prefixes held in the leaf
for
IP address destination is matched next to elected prefix
if
Positive match
then
The matched prefix length is recorded
end if
Recognize the best prefixes from each positive matches
return the next-hop data with the best prefix match

Fig. 5 Reduced OBT lookup procedure

optimal binary search tree (AOBT) structure effectively reduces memory utilization thereby reducing search complexity.

4 Experimental Results

The proposed methodology performance is evaluated in this section. The real and synthetic prefixes are used for both memory consumption and memory access evaluation [23]. Different state-of-the-art results are discussed to show the effectiveness of the proposed method. The experimental examinations are delineated as follows:

4.1 Implementation of FPGA

An anti-fuse-based and SRAM-based arrays are two major classes of Field Programmable Gate Array (FPGA). Mainly, Altera Quartus Stratix II EP2S180F1020C5 is utilized as an SRAM-based FPGA. An adaptive logic module (ALM) provided the importance of Stratix II that consists of two adaptive lookup tables (ALUTs).

One of the difficult tasks is FPGA implementation perhaps. Limit the sustained routing table size if the restriction of FPGA devices to few number of on-chip memory. An Altera Quartus Stratix II EP2S180F1020C5 with Verilog HDL implements linear pipelined IP lookup design. A serial connectivity of highest bandwidth is FPGA [24]. The chip consists of 930 blocks of M512 RAM, 9, 383,040 bits of on-chip RAM and 179,400 logical elements.

The proposed AOBT-IL implementation performance is evaluated with existing methods such as LPILA [25], BPFL [26], and POPL [27] thereby using 3 various sizes

Table 1 Resource utilization with respect to state-of-the-art methods

Methods	Lookup table size (K)	f_{max} (MHz)	Memory (MB)	SRAM (MB)	Logic elements
AOBT-IL	72K	169.2	2.3	0.20	25.1K
	144K	152.1	4.01	0.62	35.7K
	209K	149.7	6.03	0.76	40.9K
LPILA	72K	168.1	2.8	0.28	26.1K
	143K	150.2	4.76	0.71	35.2K
	209K	148.9	7.01	0.86	41.5K
BPFL	72K	110.5	4.03	0.32	52.1K
	143K	106.2	6.33	0.76	61.1K
	209K	99.9	8.86	0.95	89.7K
POPL	72K	146.5	3.4	0.43	29.3K
	143K	123.2	9.08	1.9	10.4K
	209K	123.2	9.08	3.08	10.4K

of lookup tables such as 72K, 143K, and 209K [28]. The state-of-the-art comparison results are depicted in Table 1.

Maximum frequency f_{max}: The maximum frequency for LPILA is 168.1 MHz, BPFL is 110.5 MHz and the proposed method is 169.2 for a lookup table size of 72K. The maximum frequency for POPL is 123.2 MHz, BPFL is 106.2 MHz, LPILA is 150.2 MHz and the proposed method is 152.1for a lookup table size of 143K. For the 209K lookup table size, we have attained a maximum frequency for BPFL is 99.9 MHz, POPL is 123.2 MHz, LPILA is 148.9 MHz and the proposed method is 149.7.

Memory: The proposed, POPL, BPFL, and LPILA requires 2.3 MB, 3.4 MB, 4.03 MB, and 2.8 MB memory for the lookup table size of 72K. For the lookup table size of 143K, we have achieved 4.01 MB, 9.08 MB, 6.33 MB, and 4.76 MB memory in the case of proposed, POPL, BPFL, and LPILA methods. The POLP needs 9.08 MB, BPFL needs 8.86 MB, LPILA requires 7.01 MB, and proposed AOBT-IL needs 6.03 MB memory in case of 209K lookup table size.

SRAM: The POPL, BPFL, LPILA, and proposed method stores 0.43 MB, 0.32 MB, 0.28 MB, and 0.20 MB for a lookup table size of 72K. For a 143K lookup table size, the POPL, BPFL, LPILA, and proposed method accumulate 1.9 MB, 0.76 MB, 0.71 MB, and 0.62 MB, respectively. Finally, the POPL, BPFL, LPILA, and proposed method store 3.08 MB, 0.95 MB, 0.86 MB, and 0.76 MB to 209K lookup table size.

Logic elements: The POPL, BPFL, LPILA, and proposed method occupies 29.3K, 52.1K, 26.1K, and 25.1K to a lookup table size of 72K. While the lookup table size is 143K, we have obtained 10.4K, 61.1K, 35.2K, and 35.7K logic element values for POPL, BPFL, LPILA, and the proposed method. Finally, the POPL, BPFL, LPILA, and proposed method occupies 10.4K, 89.7K, 41.5K, and 40.9K for a lookup table size of 209K.

A very rare case of worst-case scenario has occurred. As per ISO 3166, the statistic to allocate used oriented regional Internet registries and geographical area. The regression performs statistical analysis of the worst-case updates and table of side effects.

4.2 *Real Prefixes*

The memory access and memory consumption of the hash table is delineated in Table 2. For the real prefix table, the values between −0.7 and 0.9 bytes per-prefix bytes are evaluated. In all scenarios, the memory consumption is equivalent and tested as prefixes which mainly distributes 23 MSBs.

The performance of an adaptive optimal binary search tree structure with respect to memory consumption and memory access are illustrated in Figs. 6 and 7. For two-level prefixes classes, one to six classes are used in both figures. Also, the adaptive optimal binary search tree structure performance with zero class is presented. For all scenarios, the memory consumption ranges from 1.37 to 1.62 bytes per-prefix as shown in Fig. 6. For a single adaptive optimal binary search tree, it varied among 1.33 up to 3.16 bytes per-prefix bytes. AOBT overhead varies from 0.37 to 0.06-byte per-prefix byte using a two-level prefix class. For scenario SC0013, the overhead of a single-adaptive optimal binary search tree ranges from 0.85 up to 3.15, respectively. Thus, the AOBT overhead and memory consumption are reduced via a two-level class.

An influence of different classes for real prefix tables for memory accesses of AOBT is shown in Fig. 7. The number of *K* classes increase up to three minimizes the memory consumption. The memory consumption is never improved by using

Table 2 Memory consumption of AOBT technique for real prefix table

Scenario	Memory accesses	Hashing table sizes (KB)
SC001	2	20
SC002	2	19
SC003	2	24
SC004	2	19
SC005	2	19
SC006	2	20
SC007	2	21
SC008	2	20
SC009	2	20
SC0010	2	22
SC0011	2	21

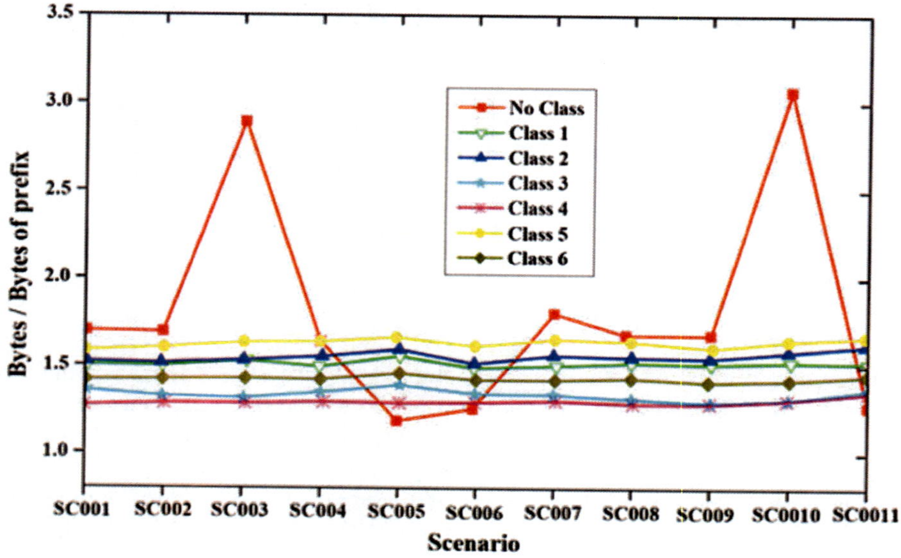

Fig. 6 Influence of different classes for real prefix tables for memory consumption of AOBT

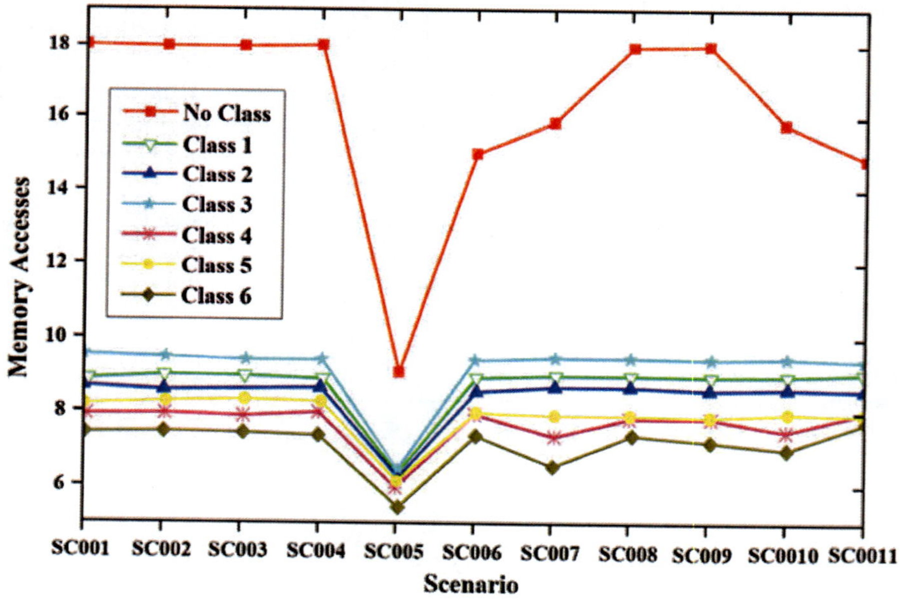

Fig. 7 Influence of different classes for real prefix tables for memory accesses of AOBT

more classes. Very low prefixes hold the increasing K value of most classes. The memory consumption is maximized by using too many groups. This varies from six to nine with 2-level prefix classes for memory access to traverse the AOBT. While a single AOBT varies between 9 and 18, respectively. Approximately, an average amount of memory access is minimized by a factor of 2. For all scenarios, eight memory access are required in the worst-case update using two or more classes. For those scenarios, two-level grouping creates few variations exist between the prefix classes and the performance equivalent to all scenarios calculated.

4.3 *Synthetic Prefixes*

The synthetic prefix table with cost initialization on the first 25 bits address the complexity of the perfect hash table as shown in Table 3. For five scenarios tested, an average of 2.7 bytes per-prefix byte is required. Linear memory consumption is shown with the help of an ideal hash table. The value is independent of the prefix table for the amount of memory accesses.

For the amount of memory consumption, memory access, and scaling, the synthetic prefix tables with the performance of AOBT are evaluated and depicted in Figs. 8, 9, and 10. For a two-level prefix class, one to six classes are used for each of the three figures. Also, the three figures present AOBT performance with no class.

Table 3 Synthetic prefix table with cost initialization on the first 25 bits

Parameters	Prefix table size				
Memory accesses	580K	290K	110K	50K	29K
Hashing table sizes (KB)	2	2	2	2	2

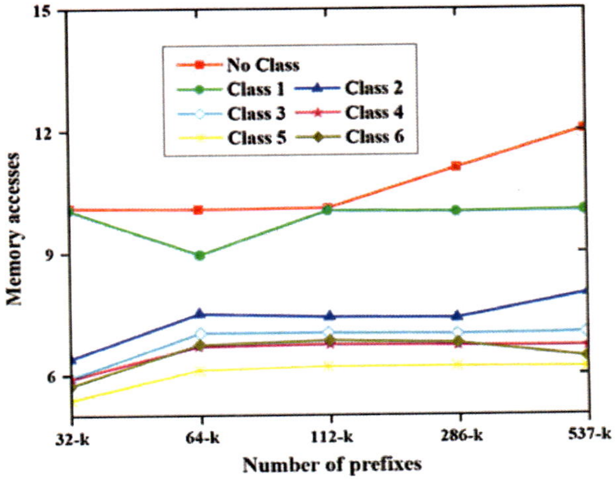

Fig. 8 Influence of different classes for synthetic prefix tables for memory consumption of AOBT

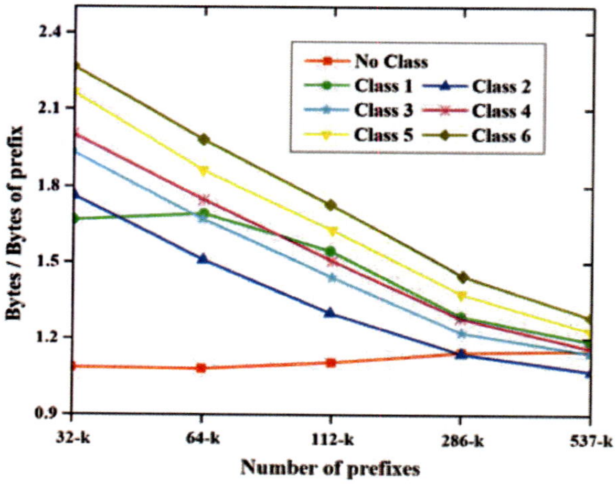

Fig. 9 Influence of different classes for synthetic prefix tables for memory accesses of AOBT

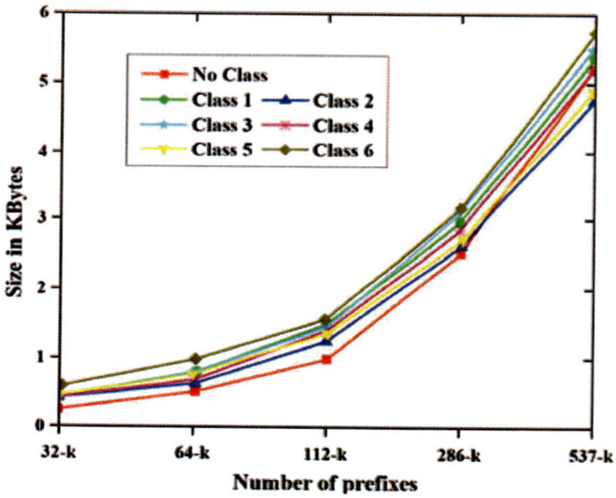

Fig. 10 Influence of different classes for synthetic prefix tables for scaling of AOBT

Figure 10 observes two characteristics for memory consumption. Multi-level prefix class utilized with two classes faintly minimizes the memory consumption in excess of a single AOBT for 290K prefixes. An AOBT consumes 1.20 and 1.10 byte per prefixes byte using this method with two classes. Finally, lower memory consumption is achieved.

The influence of different classes for synthetic prefix tables for memory consumption scaling of AOBT is delineated in Fig. 10. The two-level prefix class and *k* number of classes range from 1 to 6 using synthetic prefix tables. For prefix tables larger than 116K, an AOBT memory consumption with and without a two-stage prefix class developed exponentially. The scaling memory consumption of AOBT in case of with and with no class prefix class because the abscissa utilizes a logarithmic scale.

For the largest scenario with 580K prefixes, the memory consumption of AOBT is 4638 KB.

5 Conclusion

This paper proposed an adaptive optimal binary search tree (AOBT) based IPv6 lookup (AOBT-IL) structure with prefixes properties in IPv6. Experimentally, a device of Altera Quartus Stratix II with Verilog HDL implements the IP lookup design. When compared to the existing methods such as IPILA, BPFL, and POLP, the proposed AOBT-IL method requires less memory allocation. A AOBT shows the effectiveness of the proposed work. For all lookup table sizes, the frequencies with its memory bits and SARM of the AOBT-IL model accomplished higher than the existing methods such as IPILA, BPFL, and POLP. Different test scenarios with real and synthetic table performances are validated. In the future, work will emphasize on implement of throughput enhancement with network virtualization in IPV6.

References

1. J. Iurman, B. Donnet, F. Brockners,Implementation of IPV6 IOAM in Linux Kernel, in *Proceedings of Technical Conference on Linux Networking (Netdev 0x14)* (2020)
2. O. Erdem, A. Carus, H. Le, Value-coded trie structure for high-performance IPv6 lookup. Comput. J. **58**(2), 204–214 (2015)
3. Y.K. Li, D. Pao, Address lookup algorithms for IPv6. IEE Proc. Commun. **153**(6), 909–918 (2006)
4. Y. Qu, V.K. Prasanna, High-performance pipelined architecture for tree-based IP lookup engine on FPGA, in *2013 IEEE International Symposium on Parallel & Distributed Processing, Workshops and Ph.D. Forum* (IEEE, May 2013), pp. 114–123
5. H. Le, V.K. Prasanna, Scalable tree-based architectures for IPv4/v6 lookup using prefix partitioning. IEEE Trans. Comput. **61**(7), 1026–1039 (2011)
6. J. Li, Z. Sun, J. Yan, X. Yang, Y. Jiang, W. Quan, DrawerPipe: a reconfigurable pipeline for network processing on FPGA-based SmartNIC. Electronics **9**(1), 59 (2020)
7. A. Yazdinejad, R.M. Parizi, A. Bohlooli, A. Dehghantanha, K.-K.R. Choo, A high-performance framework for a network programmable packet processor using P4 and FPGA. J. Netw. Comput. Appl. **156**, 102564 (2020)
8. F. Lin, G. Wang, J. Zhou, S. Zhang, X. Yao, High-performance IPv6 address lookup in GPU-accelerated software routers. J. Netw. Comput. Appl. **74**, 1–10 (2016)
9. I. Simsek, Blind packet forwarding in a hierarchical level-based locator/identifier split. Comput. Commun. **150**, 286–303 (2020)
10. M. Kekely, L. Kekely, J. Kořenek, General memory efficient packet matching FPGA architecture for future high-speed networks. Microprocess. Microsyst. **73**, 102950 (2020)
11. Z. Chicha, L. Milinkovic, A. Smiljanic, FPGA implementation of lookup algorithms, in *2011 IEEE 12th International Conference on High Performance Switching and Routing* (IEEE, 2011), pp. 270–275
12. Y.K. Chang, Y.C. Lin, C.C. Su, Dynamic multiway segment tree for IP lookups and the fast pipelined search engine. IEEE Trans. Comput. **59**(4), 492–506 (2009)

13. D. Pao, Lu. Ziyan, A multi-pipeline architecture for high-speed packet classification. Comput. Commun. **54**, 84–96 (2014)
14. D. Xin, J. Han, K.C. Chang, Progressive and selective merge: computing top-k with ad-hoc ranking functions, in *Proceedings of the 2007 ACM SIGMOD International Conference on Management of Data*, June 2007, pp. 103–114
15. J.C. Vega, M.A. Merlini, P. Chow, FFShark: a 100G FPGA implementation of BPF filtering for Wireshark, in *2020 IEEE 28th Annual International Symposium on Field-Programmable Custom Computing Machines (FCCM)* (IEEE, 2020), pp. 47–55
16. T. Stimpfling, N. Bélanger, J.P. Langlois, Y. Savaria, SHIP: a scalable high-performance IPv6 lookup algorithm that exploits prefix characteristics. IEEE/ACM Trans. Netw. **27**(4), 1529–1542 (2019)
17. R.K. Sevakula, N.K. Verma, Balanced binary search tree multiclass decomposition method with possible non-outliers. SN Appl. Sci. **2**, 1–15 (2020)
18. C. Luo, Internet enterprise organization strategy based on FPGA and machine learning. Microprocess. Microsyst. 103714 (2020)
19. C. Huang, Modern art network communication based on FPGA and convolutional neural network. Microprocess. Microsyst. 103498 (2020)
20. P. Alapati, V.K. Tavva, M. Mutyam, A scalable and energy-efficient concurrent binary search tree with fatnodes. IEEE Trans. Sustain. Comput. **5**(4), 468–484 (2020)
21. T. Beneš, M. Kekely, K. Hynek, T. Čejka, Pipelined ALU for effective external memory access in FPGA, in *2020 23rd Euromicro Conference on Digital System Design (DSD)* (IEEE, 2020), pp. 97–100
22. A. Kushwaha, S. Sharma, N. Bazard, A. Gumaste, B. Mukherjee, Design, analysis, and a terabit implementation of a source-routing-based SDN data plane. IEEE Syst. J. (2020)
23. L. Kekely, J. Cabal, V. Puš, J. Kořenek, Multi buses: theory and practical considerations of data bus width scaling in FPGAs, in *2020 23rd Euromicro Conference on Digital System Design (DSD)* (IEEE, 2020), pp. 49–56
24. Y. Hu, G. Cheng, Y. Tang, F. Wang, A practical design of hash functions for IPv6 using multi-objective genetic programming. Comput. Commun. **162**, 160–168 (2020)
25. M.M. Vijay, D. Shalini Punithavathani, Implementation of memory-efficient linear pipelined IPv6 lookup and its significance in smart cities. Comput. Electr. Eng. **67**, 1–14 (2018)
26. D. Pao, Z. Lu, A multi-pipeline architecture for high-speed packet classification. Comput. Commun. **54**, 84–96 (2014)
27. J.J. Kester, Comparing the accuracy of IPv4 and IPv6 geolocation databases. Methodology **10**(11), 12–17 (2016)
28. M. Hemalatha, S. Rukmanidevi, N.R. Shanker, Searching time operation reduced IPV6 matching through dynamic DNA routing table for less memory and fast IP processing. Soft Comput. 1–14 (2020)

Congestion Control Mechanism for Unresponsive Flows in Internet Through Active Queue Management System (AQM)

Jean George and R. Santhosh

Abstract In this paper, we would like to analyze different type of congestion control methods which are using active queue management (AQM) system. Nowadays, we need fast and secure transmission of large amount of data, through network resources. So we need congestion control, also it has an important role to play in it. Best AQM method can provide an environment, which will provide high throughput, minimum response time and maximum resources utilization. Unresponsive flows are one of the factors, which are reducing quality of service (QoS). If we can control or remove unresponsive flows, it will considerably reduce congestion also increase the resources utilization as well as output. Many CHOKe series AQM and RED series AQM are using for congestion control in the network. Active queue management has an important role in congestion control; because they are handling information flow at ingress router. This paper reviews a detailed about major AQM-based congestion control methods.

Keywords AQM · Congestion control · CHOKe · RED · Unresponsive flow

1 Introduction

At the beginning of network communication onwards, active queue management has an important role in IP networks. It is complement the work of end-system protocols such as the transmission control protocol. In congestion regulation, it improves network efficiency and decreases packet drop and queuing delay [1]. Congestion control and queue administration have a central role to offer the robustness and fairness of the network communication. The unexpected rise of the online users and web applications, especially in this pandemic, make a rapid change on Internet usage. So congestion control has become focus to network analysis for improving efficient use of resource utilization.

J. George (✉) · R. Santhosh
Department of Computer Science and Engineering, Faculty of Engineering, Karpagam Academy of Higher Education, Coimbatore, Tamil Nadu, India

S. Shakya et al. (eds.), *Mobile Computing and Sustainable Informatics*, Lecture Notes on Data Engineering and Communications Technologies 68,
https://doi.org/10.1007/978-981-16-1866-6_58

In Internet, congestion control has an important role, it depends on different types of real-time applications and measures the network efficiency. Different types of AQM method have some limitations, so it is a highly demanding research area [2]. In networking, traffic development section seeks to effectively balance traffic load throughout current networks, so it has achieving the required quality of service and reducing typical costs of adding hardware and software implementations, common to the network engineering [3]. When one packet is moving through the Internet, it is traveling through many routers. Each router is using an AQM and buffer space for handling the inbound packets. In current situation, Internet traffic is fluctuating unexpectedly. In the recent years, research scholars have come out with various congestion avoidance algorithms and methods. But all these methods are hopeless especially in heavy traffic network. Because of the Congestion, large amount of packet loss occurs in network that leads to more expense. This is reduced by the different types of existing AQM methods. AQM schemes are divided based on congestion metrics with and without flow information as shown in Fig. 1.

Responsive flows are the information flows that are identical and unique to each other and ready to get serviced in a communication link. Unresponsive flows are other types of flows, and mainly the duplicate and similar information flows that will intersect each other [4]. Unresponsive flow causes of congestion. Unresponsive flows are generated by some virus attack, some types of repeated useless request and different types of hacking attempts. So to avoid them, we should use a perfect algorithm for identifying and destroying them. To solve the issues of unresponsive data streams, multiple algorithms and methods are implemented in routers [5, 6].

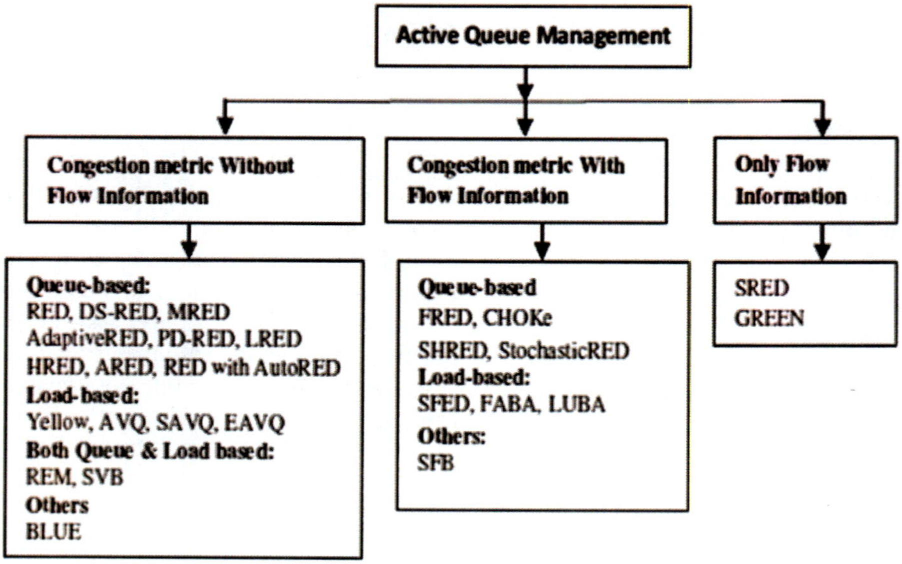

Fig. 1 Classification of AQM methods

The idea is used to detect unresponsive flows and control their rates, so other flows can utilize the network efficient way.

2 Related Work

In congestion control for Internet, sender computer controls packet transfer rate in the congestion portion of the network. A Packet Schedule is used to execute active queue management in an Internet network.

2.1 RED

Sally Floyd and Van Jacobson proposed Random Early Detection algorithm (RED). It is one of the active queue management methods, which helps in avoiding overcrowding in Internet. Commonly router drop packets, at the end of queues. But RED uses statistical methods in a "probabilistic" way to remove packets before queues surplus. In this way reduces packet transfer rate of a sender system down enough to control the queue steady state and reduces the number of packets in the network. This is what happens when a queue gets more packets more than its capacity or a host is transmitting in the high packet rate. Random Early Detection algorithm makes two choices, in that first one is packet dropping time and second one is select the packet which we need to drop. It keeps average size of queue in the network in a comfortable way to send packet without any loss. If maximum threshold more than the average size of queue, then this algorithm drops one by one each packet. The average queue size is calculated by count of each arrival of packet shown in Fig. 2.

RED uses average time calculation, if the queue size is rising and falling irregularly in number of packets. If any big congestion happened suddenly, then also RED will not take any action. At the same time queue is continuing a partially filled state, then RED will imagine congestion happened in the network, also drop large volume of packets [7]. Random early drop, also called as random early discard or Random Early Detection is a discipline for a network scheduler for congestion avoidance.

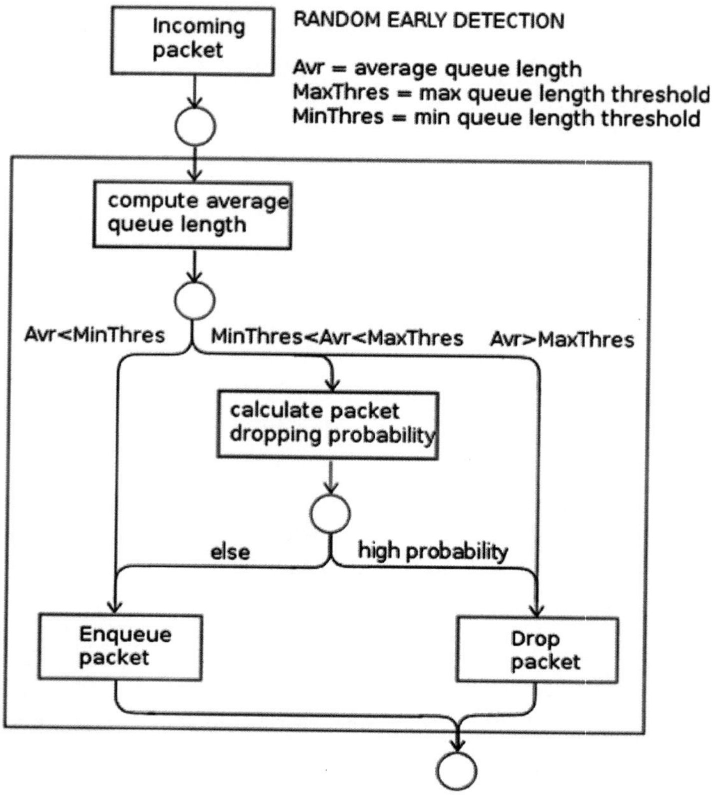

Fig. 2 RED

Pseudo Code for RED Algorithm

```
For each incoming packet {

        Calculate RQave
        if (RQave≥ maxth) {
        Remove the packet
        }
        else if (RQave>minth) {
        Find out the probability percentage to drop Rpa
        Remove the packet with respect to the value of Rpa,
        Else add that packet
        }
        else {
        Forward the packet
        }
}
```

Variables

RQ_{ave}: average size of queue
R_{pa}: probability percentage of incoming packet
$\max_{\text{th}}$: Maximum threshold
$\min_{\text{th}}$: Minimum threshold.

2.2 Flow-Based Random Early Detection

Lin and Morris created FRED algorithm. It is a flow-based RED algorithm. The specialty of flow-based Random Early Detection is altering non-adaptive streams by arranging highest level of temporary memory space and surpassing a part of it to average packet rate in stream holding area usage. This algorithm is caring easily broken information streams by predetermined streams of less transmission capacity channel, giving reasonable distribution for more sequence of information streams by consuming multiple buffers. When buffer is using, this will solve many problems of Random Early Detection algorithm. In this, it measures average length of queue at incoming buffer as well as outgoing buffer.

2.3 BLUE

BLUE [2] developed by Wu-Chang and Feng. It is another extension of RED, which uses packet loss and link utilization as control variables to measure the network congestion. In the place of average length of current queue, this algorithm using channel usage percentage and removed data packet percentage, through that it tries to improve efficiency. BLUE finds out probability to drop a packet and write on each incoming packet. If the queue is in the state of removing incoming packets then probability to drop a packet will increase else decrease. Therefore in case of larger queue, delay in queue organizes meritoriously. So how to arrange the value of different factors and unexpected problems is still there, but packet falling method is accepted as a solution for adaptive data flow [8]. RED has several variants like FRED, SRED, ARED, RRED, DRED, WRED, BLUE, etc. These methods are good enough for queue management but they possessed drawbacks in modern networks [4]. The main problems are they fail to recognize TCP and UDP flows and have not handle traffic fluctuations. So these methods face problems like Lockout and global synchronization, etc., when the fine-tuning of parameters are absent in the network. They cannot process short lived Internet packets. They fail to recognize or remove unresponsive flows. They cannot promise QoS in Internet.

2.4 CHOKe

To overcome these limitations, new active queue management mechanism is implemented, called CHOKe. "Choose and Keep for responsive information flows, Choose and Kill for unresponsive information flows" is the expansion of CHOKe [4]. Psounis et al. proposed CHOKe algorithm [9, 10]. When a fresh information packet comes to an already overcrowded router, then one information packet throws down randomly from the queue and evaluate with the incoming data packet. If two of them have the similar flow ID, then both is destroyed, else the packet that was selected randomly is retained in to buffer also admit new information packet into the queue with a Precedence with respect to the weight of congestion. Evaluation of precedence selection is like as Random Early Detection algorithm.

CHOKe is a modest and stateless algorithm where no extra data structure is required. CHOKe marks two thresholds on the buffer, one is lowest threshold min_{thd}, and other is largest threshold max_{thd}. If the size of average queue is less than min_{thd}, all incoming packet is queued into the first in first out buffer [11]. If Q_{avg} is lesser than min_{thd} then the packets are not dropped and arriving packet is allowed to occupy in buffer. If the size of average queue more than max_{thd}, then every arriving packet is dropped immediately like as in Fig. 3. This moves the queue status back to below max_{thd}. CHOKe provides better quality of service in burst traffic.

Algorithm for CHOKe

```
For each incoming packet {
        EvaluateCQavg
        If (CQavg<minthd) {
                Addarrived packet
                }
                else {
        Remove a packet randomly from the queue
                If both packets have same flow {
                        Remove that two packets
                }
                else{
                if (CQavg<maxthd){
                Add packet with priority p
                }
                else {
                Remove the packet
                }
        }
          }
}
```

Through this way, CHOKe can identify and remove unresponsive information flow, also decrease the congestion and increase the response time. There are many differences in first CHOKe system, where the first one is the original CHOKe. In

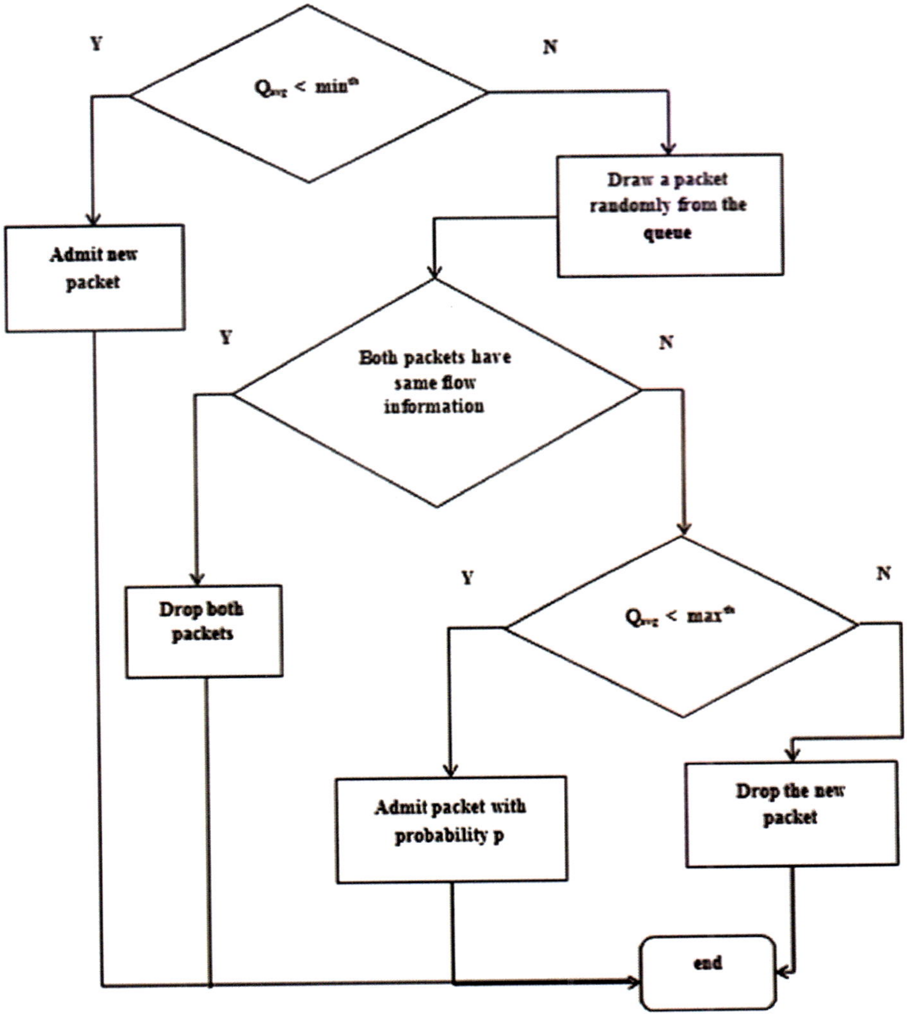

Fig. 3 Flowchart for CHOKe

the original CHOKe, the drop packet is selected randomly from the queue [9, 12]. The descendants of CHOKe listed below. Different type CHOKe algorithms are there, each one improves the QoS more than theirs previous one. This way they are improving the quality step by step. A CHOKe algorithm selects the packet from the head of the queue which is called front CHOKe, and a CHOKe algorithm drops the packet from the end of the queue is called back CHOKe, CHOKe algorithm, which drop multiple no: of packets from the buffer and compare with arrived packet is called M-CHOKe.

In the case of X-CHOKe algorithm, it uses data structure to save information and maintain a table to hold flow's hit counter *N*, this is called X-CHOKe [13]. CSa-X-CHOKe improves X-CHOKe by calculating packet drop probability based on congestion in network and link load. Congestion level is evaluating from link load and average queue length [14]. Self-Adjustable CHOKe, process TCP flows and UDP flows in a different way, and can adaptively change its parameters according to the existing traffic status [15].

2.5 A-CHOKe

CHOKe algorithm solves many drawbacks of RED type AQMs mainly lockout and global Synchronization problems. But sometime CHOKe is not suitable in case of Small number of packets from unresponsive flows in the queue. For overcome this problem, a new method Adaptive CHOKe (A-CHOKe) is proposed. It offers a stable functioning point for the queue size and better bandwidth allocation, irrespective of the variable congestion behavior of the information flows [16]. Also it gives high throughput, reduces queuing delay and information loss by fine-tuning of different parameters. The varying value of parameter adapts itself to the changing nature of the congestion and traffic. A-CHOKe is a more stylish way to do M-CHOKe such that the algorithm automatically selects the proper number of packets selected from the buffer, where the buffer is divided into a number of regions [7].

2.6 P-CHOKe

Adaptive CHOKe algorithm is the base of Piggybacking CHOKe (P-CHOKe) algorithm. P-CHOKe algorithm differentiates responsive information flow and unresponsive information flow and then selects only responsive one. Through this way, it avoids non-adaptive information flows [17]. In this algorithm, gateway receives data packet from M number of systems. Router collects all incoming packets and also reroutes it to the destination system. So destination system will send acknowledgment for all received packets one by one to the router. Then router collect these acknowledgments and send to the corresponding sender by its flow ID. When P-CHOKE compare with other types of CHOKe, It can send more percentage of packets and less delay in data packet transferring in the network.

Pseudo code for P-CHOKe

Step 1: Start
Step 2: Collect each incoming packet
Step 3: Packet redirect to router or switch
Step 4: Select random packet and get its flow ID
Step 5: The router compares new packet and chosen packet
Step 5: If the flow ID of two packets is same, then that two packet will be removed
Step 6: Else process it and send it to destination

Step 7: Receiver returns the ACK
Step 8: Router accumulates the ACKs
Step 9: Router forwards the ACK to the sender based on the flow ID
Step 10: Stop.

2.7 GCHOKe

Geometric CHOKe (gCHOKe) is an active queue management system, use to select adaptive streams from other streams. This is a stateless AQM. This algorithm is an overview of the original CHOKe. gCHOKe offers an extra protection layer, created by placing an additional flow matching trial after each successful flow evaluation of packets. gCHOKe offers an extra protection layer, after getting an effective stream evaluation of data, gCHOKe checking the flow of data if it is same then increase the data flow. Total evaluation allowed at packet entry is decided by max_{eval}. It ensures better data stream rate and maintains it. When we consider with original CHOKe algorithm, our observation and different types of modeling methods confirm that this algorithm provides more than 20% betterment in responsive information flow with channel availability and temporary memory space. Moreover, up to 14% of the total link capability can be saved from the unresponsive information stream. So this method is avoiding unresponsive or non-adaptive flows to maintain an improved utilization of network properties at the gateway [13].

gCHOKe provides an advanced flow protection, if any successful identity check occurs, which is giving one more stream equality checking trail [18]. Number of testing varies in gCHOKe compare with others. CHOKe algorithm allows the responsive information stream to use more temporary memory space, which is possibly controlling and the linking in individual identity checking of each incoming data unit. The series of this special equality of this stream checking provides a better safeguard of data transfer speed of information stream from non-adaptive ones [7]. Using these trail variation, a preferred level of security shield can be attained, so congestion control will be more adaptable than the basic CHOKe.

2.8 FAVQCHOKe

Consideration of length of queue as a measurement of congestion factor in many AQM, so this is an inherent problem of all previous AQM. For solving the problems of previous methods, a new method is proposed, that is called FAVQCHOKe. Incoming data flow per unit time in the network is evaluated as major measurement method of congestion to make better momentary action of the system. The total space usage of channel capacity confirms by this network. FAVQCHOKe algorithm suitable for real-time software applications also achieves very less information flow drop and maintain higher channel usage [19]. In FAVQCHOKe, virtual queues are used at the packet receiving part of router, so the real queue buffers are updated using the status of virtual queue. This method keeps internal buffers from the risk of harmful unresponsive information flows.

2.9 CHOKeW

Bandwidth variation and TCP protection are important for fulfilling quality of services in Internet. CHOKeW is a stateless active queue management system, which implements identical packet removal method generated by CHOKe. So it can manage the transmission capacity distribution, also avoid Random Early Detection method portion. With respect to the congestion, also precedence of incoming packet, CHOKeW varies number of packet droplet. In the valuation of different virtual model, this algorithm is proficient to manage different types of overcrowded situations [20]. By giving high priority flows, CHOKeW can support different bandwidth priority levels and provides high throughput. It restricts the effect of non-adaptive information stream, through that safeguard TCP flows. CHOKeW solve information flow starvation.

2.10 CHOKeR

CHOKeR is an improved version of CHOKeW algorithm which takes precedence over the problems of bandwidth differentiation in different priority levels and poor performance on burst traffic in large congested network which are faced by CHOKeW. CHOKeR implements innovative AQM concepts. It compares incoming information stream with the multiple information packet removed from the exit buffer, in this algorithm R representing ratio of Channel allocation. Like MISD, in this algorithm packet removing factor planned as multiple step data unit increment and uni-step data unit decrement, through that increase the efficiency of network. The efficiency of this algorithm is compared with investigative model as well as advanced virtual models. From the result of CHOKeR realize that, it can give constant transmission capacity provision and manage separated transmission capacity ratio for different types of information streams. CHOKeR retains fair mindedness between streams of the identical priority. CHOKe family have not yet proved better reliability and stability in modern social IP networks because they fail to recognize and isolate link prediction problems raised during the on-need formation of dynamic links [21].

3 Comparison of Different Active Queue Management Algorithms

See Table 1.

Table 1 Advantage and disadvantage of different AQMs

Sl. No.	AQM methods	Benefits	Drawbacks
1	RED	Quick congestion discovery Not suitable for burst traffic No global synchronization	Difficult to set parameters Insensitive to traffic fluctuation in network
2	FRED	It will help to avoid wrong data flow	Difficult to set parameters Insensitivity to traffic fluctuation rates
3	BLUE	Simple Provide high throughput	No early congestion detection Slow response
4	CHOKe	Identifies responsive and unresponsive flows	Single trail and helpless by burst Flows
5	M-CHOKe	Multiple trails for comparison	Global synchronization and lockout Problems
6	XCHOKe/RECHOKe/CSa-XCHOKe	Uses historical data for flow Protection	Lacks stateless property
7	A-CHOKe	Protects well-adaptive flows from non-adaptive flows	Fluctuation in heavy loads
8	P-CHOKe	It provides better packet delivery ratio and low queue delay	No fair bandwidth allocation
9	GCHOKe	Extra shield of protection by using bonus trial on successful matching	Does not work well in long lived and short lived TCP traffic
10	FAVQCHOKe	Uses virtual queue to guaranty more security for internal queue	Imperfection in differential service since load factor-based
11	CHOKeW	Supports differentiated services	It will not ensure, the availability of channel capacity for a High priority data
12	CHOKeR	Uses MISD scheme	No sustainability in heterogeneous real-time network

4 Conclusion

This paper detail explains about different active queue management-based congestion control algorithms. Many disadvantages of Random Early Detection series AQMs are solved by CHOKe series. But they are not efficient to provide desired stability for load balancing when the heterogeneous network is dealing with long and short lived TCP traffic and large UDP flows. The AQM algorithm is categorized with respect to the Metrics, which use to measure the data transfer delay and data transfer occur in unit time. Many algorithms are concentrated on the values of factors that effecting congestion, and some of the algorithms need only stream information, but some other algorithms need these two information for detecting congestion in the network. While we consider different type of data transfer etiquette, demand of congestion management is increased. Now a day's ICT is enhanced, also increase the network traffic and different type of information flows like data, audio, video, etc., are transferring through Internet. But many resources like transmission capacity provision, temporary memory space and efficiency of processor are limited. So the TCP protection in heterogeneous environment, investigation and enhancement on AQM for identifying unresponsive flow and accruing more throughput using available system. Use a method to control congestion for TCP-friendly transmission.

References

1. J. Pandey, A. Hiradhar, A survey on AQM control mechanism for TCP/IP flow. IJARCSSE (2014). ISSN: 2277 128x
2. W. Feng, K.G. Shin, D.D. Knadlur, D. Saha, The BLUE active queue management algorithms. IEEE/ACM Trans. Netw. **10**(4), 513–528 (2002)
3. M. Azath, R.S.D. Wahidabanu, A study of various load balancing techniques in internet. Int. J. Comput. Sci. Inf. Secur. **8**(3) (2010)
4. C. Vijith, M. Azath, Survey on CHOKe AQM family. Int. J. Comput. Sci. Eng. **2**(11), 81–85 (2014)
5. K. Fall, S. Floyd, Router mechanisms to support end-to-end congestion control (2002). ftp://ftp.ee.lbl.gov/papers/collapse.ps
6. V. Jacobson, Congestion avoidance and control, in *Proceedings of ACM SIGCOMM* (2001), pp. 314–329
7. B. Kiruthiga, E. George Dharma Prakash Raj, Survey on AQM congestion control algorithms. IJCSMC **2**(2), 38–44 (2014)
8. K. Hao, Z. Jin, P. Hao, Study on internet congestion control algorithm, in *Fifth International Conference on Intelligent Networks and Intelligent Systems* (2012)
9. R. Pan, B. Prabhakar, K. Psounis, CHOKe, a stateless active queue management scheme for approximating fair bandwidth allocation, in *IEEE INFOCOM*, Mar 2000
10. P. Young, *Recursive Estimation and Time-Series Analysis* (Springer, Berlin, 2000)
11. G.F. Ali Ahammed, R. Banu, Analysing the performance of active queue management algorithms. IJCNC **2**, 19 (2010)
12. Ao. Tang, J. Wang, S.H. Low, Understanding CHOKe: throughput and spatial characteristics. IEEE/ACM Trans. Netw. **12**(4), 694–707 (2004)
13. A. Eshete, Y. Jiang, Generalizing the CHOKe flow protection. Comput. Netw. **57**(1), 147–161 (2013)

14. M. Jiang, C. Wu, M. Zhang, H. Bian, CSa-XCHOKe: a congestion adaptive CHOKe algorithm. Chin. J. Electron. **19**(4) (2010)
15. Y. Jiang, J. Liu, Self adjustable CHOKe: an active queue management algorithm for congestion control and fair bandwidth allocation. IEEE Comput. Commun. **2**(4), 1018–1024 (2013)
16. K. Chitra, G. Padamavathi, Adaptive CHOKe: an algorithm to increase the fairness in internet routers. IJANA **01**(06), 382–386 (2010)
17. G. Sasikala, E. George Dharma Prakash Raj, P-CHOKe: a piggybacking-CHOKe AQM congestion control method. IJCSMC **2**(8), 136–144 (2013)
18. A. Eshete, Y. Jiang, Protection from unresponsive flows with geometric CHOKe. Centre for Quantifiable Quality of Service in Communication Systems, Feb 2012
19. K. Chitra, G. Padmavathi, FAVQCHOKE: to allocate fair buffer to a dynamically varying traffic in an IpNetwork. IJDPS **2**(1), 73–82 (2011)
20. S. Wen, Y. Fang, H. Sun, CHOKeW: bandwidth differentiation and TCP protection in core networks. IEEE Trans. Parallel Distrib. Syst. **20**(1), 34–47 (2009)
21. S. Chahar, Social networking analysis. Int. J. Comput. Sci. Eng. **02**(5), 159–163 (2014)

Anomaly Detection on System Generated Logs—A Survey Study

Jisha M. Jose and S. R. Reeja

Abstract Recently, the application and systems are generating huge amounts of log data. With the rise of different applications in different domains that are deployed in every environment, it has become inevitable that a system may run into either functional or performance faults. System analyst and administrators have to look at system log data at the time of fault, and analysing millions of lines of logs manually is an impossible task to find the root cause for any faults. Timely detection of any abnormal behaviour is very important to prevent negative impact on the service, thereby helping to build a secure and trustworthy system. The aim of this survey is to provide a comprehensive overview of recent research in supervised and unsupervised anomaly detection methods. A detailed review on the various adoption methods and data sets on which it has been applied across various application domains are discussed in this paper.

Keywords Anomaly detection · System logs · Anomaly types · Log analysis · Log analysis

1 Introduction

Today, every device in the world generates some amount of data. With the massive use of social media, smart phones, IoT devices, e-commerce, etc., the data are generated in a huge size on every second. Big data system is widely used as effective solution in various domains like education, e-commerce, health care, manufacturing, financial services and media. The process of storing large-scale records and information for audit or for solving any security related problems is called logging [1]. Every log differs based on the application that generates it. A logging statement is comprised of different log messages generated by the tasks specific to an application. Log data has become the most valuable resource to scrutinize the performance issues in big data

J. M. Jose (✉) · S. R. Reeja
Dayananda Sagar University, Bangalore, India

S. R. Reeja
e-mail: reeja-cse@dsu.edu.in

S. Shakya et al. (eds.), *Mobile Computing and Sustainable Informatics*, Lecture Notes on Data Engineering and Communications Technologies 68,
https://doi.org/10.1007/978-981-16-1866-6_59

and cloud systems and thereby it may be useful for monitoring online and thereby detection of anomaly for improving system performance [2]. These unstructured or semistructured logs contain many hidden information, which can be mined for performance tuning and anomaly detection [3].

The logs are generated in human-readable form that contains text messages, parameters or other important values of running processes. The structure of the logs is not fixed and their content depends on the type of application and other configuration settings like kind of messages to be entered (e.g. informative messages, errors and debug output), word length in history of logs, the time defined for writing logs by various components of the system [4]. And usually, assembly of logs from different sources is dumped to one file as a bulk job or as a single, that generate messages dynamically on the fly.

Most of the monitoring system focus on observing various software and metrics of hardware like utilization of CPU, memory consumed, web traffic, extent of live sessions and others. But currently many cloud applications do monitor the operational behaviour of the system from cloud logs. It provides very valuable information of all running operations in real time. But there are many challenges in analysing these dynamic logs as they provide noisy and low-level information. Also, they do not provide much information on change of state of system [5].

Along with huge volume and complexity of log data, being the major challenges in anomaly detection, administrators and operators also face difficulty in diagnosing the problem to analyse the actual root cause for the anomaly. Here, the challenge is to provide them the necessary tools and techniques to narrow down the diagnosis to the specific parts in logged data that can find out the root cause of the anomaly [6]. Logs occur is large volume and it is very difficult to track which line of log or set of logs is associated with a particular change of action of a system.

In many large-scale applications, which has lots of cloud operations, hundreds of shared resources keep changing due to multiple concurrent operations. In such cases, it becomes difficult to monitor and detect the abnormality of the execution behaviour of an operation solely by log contents. For these, it is important to consider additional information along with log information for validating such operations.

2 Type of Anomalies

There are three major classification anomalies: point anomaly, contextual (conditional) anomaly and collective anomalies. While designing an anomaly detection system, one needs to understand the type or characteristic of anomalies that are to be targeted, which may lead to development of the right approach.

- Point anomaly: It is the simplest and most common type of anomalies. Usually, when deviation from normal data set pattern is observed from a particular instance of data, it is considered as point anomaly, e.g. a sudden rise in blood pressure level on a normal day.

- Conditional (contextual) anomalies: It refers to anomalous behaviour of a data instance that can be considered as anomaly in only certain condition or particular context and not in others, e.g. receiving more phone calls during festive season.

  During Christmas and New Year, one may receive more than 50 phone calls when compared to an average of 15 calls on usual day. So, this may not be considered as anomalous but receiving same count during normal working day may be considered as conditional anomaly.

- Collective anomalies: If group of alike data cases, performing anomalous altogether, when compared with entire data set, it is considered as collective anomalies. Here, individual data instance may not be an anomaly by itself but its presence in a collection is identified as an anomaly, e.g. one low value in human electrocardiogram (ECG) output report may not represent an anomaly but occurrence of low values in ECG output continuously in consecutive manner over collection samples correspond to collective anomalies.

3 Approaches for Anomaly Analysis

The various techniques and algorithms adopted for anomaly analysis on different datasets can be broadly classified into three approaches - Statistical approach, Supervised approach, Unsupervised approach.

3.1 *Statistical/Distribution-Based Approach*

In statistical approach, in terms of some statistical distribution, log data set is organized. Data that do not match this distribution are removed or examined. This approach is easy to implement for a particular use case in a single domain but it becomes difficult when anomaly definition keeps changes across various domains. In other words, an anomaly detected in one application may not be a problem for another application. Detection of contextual or collective anomalies, especially for application in log data, which varies from application to application may be really challenging. So, such approaches require prior knowledge of data set [7].

3.2 *Supervised Approach*

Like any other classification problem, here also the entire log data set is labelled as anomalous and non-anomalous data. At initial step a subset of data is considered for training the model and then the trained model is used for testing the balance data. There is a high chance for overfitting the model in case of supervised approach, since log data is the collection of texts of different types. To generalize and to get accurate classification result, the data have to be well balanced [7].

Supports like vector machines, Bayesian networks and classification techniques are widely used for anomaly detection. The advantage of supervised approach is that these are fast and efficient as it learns from labelling and feedbacks given to the model. Hence, it gives better results to known problems as it would have learned while training the system. But it is hard to get well-balanced labelled data for training. Also, in case of anomaly detection, the labels to be given as normal data keep changing and it cannot be fixed. Moreover, for dynamically changing live stream data, the chances of anomalous data getting labelled as normal data are very high, which may lead to false positive errors. In other words, a simple supervised learning system can never be superior to human experts as its goal is to mimic the experts. And for the experts, to detect every anomaly in large live data will be really a difficult task.

3.3 *Unsupervised Approach*

Supervised learning approach requires lots of manual efforts from domain exerts to create a labelled training set. But such training set cannot be used permanently for a long time as any system-generated log line keeps changing over a period of time. So, it does not help much in detecting unknown attacks. Any method that classifies the logs, based on labelled fixed training samples, provides a static view of the data. Hence, a self-learning approach that can dynamically get adapted for triggering alarms on its own will be very helpful [4].

4 Related Works

Since past few years, detection of anomaly has been under study and many research and experimental works were carried in various domain like social networks, computer network and security, distributed system, cloud system, surveillance system, etc. Also, the purpose of detection has not just been for the security purpose but also for checking system performance and also reliability [8].

Usually, system symptoms and failure are mapped and various techniques are used for categorizing them. There are various automated approaches like machine learning, rule-based, statistical, model-based, count and threshold and visualization methods [9]. But there are always some limitations in using these approaches like requirement of detailed understanding of the system in model-based methods and maintenance of large knowledge bases for rule-based methods.

4.1 Statistical Approach Multiregression Technique Ordinary Least Squares (OLS) and Clustering

In this [5] paper, regression-based statistical approach is used to find out a correlation between operational behaviour (event logs) and cloud metric changes. At first, low-granular meaningful event logs are extracted by using regular expression and cloud metrics, using cloud watch. Later, these event logs and metric data are mapped by using weight timings and clustered using correlation coefficients. In this paper, multiregression technique and ordinary least squares (OLS) regression were used to find out link between event logs and metrics changes. To detect anomaly in running operations, VM failure events along with rollback upgrade operation on Amazon Web Services were introduced. For prediction, equations are derived from multiple regression models and if these predicted values do not match the actual value, alarms were raised. Deriving assertion between event logs and cloud metrics for anomaly detection is achieved in this paper, which can be extended for designing self-adaptive operations that can be used for healing the system on occurrence of an anomaly.

4.2 Isolation Forest and ANN

Farzad and Gulliver [10] have performed experimentation using unsupervised approach for detecting anomalies. For this, they used the concept of isolation forest to predict positive data by employing a threshold also unlike the usual method, where isolation forest is used for predicting negative data. Two deep autoencoder networks for feature extraction are used along with this. A feed-forward artificial neural network (ANN) which can learn data features is called autoencoder. Size of code layer, which is a hidden one, is typically small by its usage for feature extraction and dimensionality reduction in an autoencoder network. An autoencoder which has more than one hidden layer is called deep autoencoder. They have evaluated their proposed work by using popular data sets like OpenStack, BGL and Thunderbird for log communications. To predict positive data, isolation forest is fed with first autoencoder's result. Further, usage of second autoencoder improvised the separation of anomaly as second one required only fewer epochs to get trained from already extracted feature set [10].

4.3 Log Vector and Three 1 layer-CNN

In [11], authors have used convolutional neural network (CNN) to analyse the complex relationship in Hadoop logs. At first, to find the frequent log constants called as log key, a log template parser is used. These log keys were passed to another parser to get filtered out into structured useful log keys. After that, these log keys were encoded with a unique number and regrouped to different session groups as a matrix. This implanting process is called logkey2vec. For the next step, three one-layer convolutions (filters) of CNN are used on the embedded log vectors simultaneously. The output of the convolution layers is concatenated by a max-pooling layer that filters out weakly related features and thus passing only strongly related features for the next layer. In the second last layer, a dropout function is used for the regularization to avoid overfitting. And at last, to produce the probability distribution results, a function called softmax is added in the output layer.

To compare efficiency of their proposed CNN model, build a simple a multilayer perceptron (MLP) with easily adjustable parameters as a standard one and then work out it on logkey2vec pertaining to HDFS logs. As per their analysis, authors found that approach based on CNN achieved a superior precision on detection in comparison to MLP using big data system HDFS logs.

4.4 UnSupervised Clustering (Dynamic Cluster Evolution and Time Series Model)

In [4], the authors have experimented their dynamic learning clustering evolution technique over semisynthesized data. The logs were taken note out of web server, SQL database and reverse proxy components by synthesizing it from five virtual users. Three of them generated normal logs. One among them mimicked logs generated from automated software at every first 30 min in every hour. And another one injected abnormal behaviour by performing various actions like missing periodic event, sudden frequency peak and also increase in frequency for long term as well as gradual.

At first, log stream or log files were used to read log lines from them and preprocessing techniques were used to remove special characters or reduce consecutive spaces, etc. Dynamic cluster evolution consists of two phases: construction as well as allocation phase [4]. In the first phase, i.e. construction phase, each log line in a stream is allocated to the cluster to which its similarity index falls within the threshold that is being set. Here, various smaller maps of cluster are produced in a duration of delimited time frame. In allocation phase, connection is established among the cluster maps of various time window by earmarking the log lines of previous and subsequent ones. This allocated log lines are used to establish transition of cluster from preceding to present time window. Computation of evolution metrics of cluster is done based on the known transitions. These cluster metrics are estimated with models of time series, for which periodic effects as well as trends are considered. Forecast of one step in advance for the metric in the succeeding time window is made by arriving reasonable conclusions of these models. Later anomaly detections are computed by comparing the previous prediction with actual value. Along with anomaly detection, a correlation between log lines by clustering time series data is also monitored. Here also, the amount of false positive cases could not be resolved.

4.5 Semi-Supervised Approach

Real-time anomaly detection system (RADS) for cloud data centres was proposed by the authors Barbhuiya et al. [12]. It was mainly focused to detect CM-level anomalies occurring due to unknown DDoS and cryptomining attacks. One class classification (OCC) algorithm was used to learn normal pattern of CPU and network usage of each of the hosted VMs. This linear classification model was built to separate negative and positive classes, where positive class means combination of known normal and unknown spike samples whereas negative class indicates unknown anomaly samples. So, in this case, whenever CPU or network usage pattern deviated from the usual pattern, the algorithm used to flag it as an anomaly. To overcome false positives, authors combined average and standard deviation of the raw data by using a window-based time series analysis.

4.6 Graph Analytics and Deep Learning

In [8], authors identified a reliable party called PredictDeep, which has analytical provider for the detection of advance anomaly and security feature prediction. Apart from the usual statistical approach and machine learning techniques, they proposed a combination of deep learning mechanisms along with graph analytics. This combination added intelligent information, which is required to identify and envisage hidden and advanced pattern of anomalies pertaining to security. This model consists of graph model designer, feature extractor and anomaly predictor. The log information and isolated profile are modelled by graph model designer to a graph that depicted the various properties of nodes in a cluster with his interdependencies and linkage relationships.

This graph model is fed as input to feature extractor to extract rich features like graph structure, node properties and semantic information by using graph analytics. Then, these features are used by the anomaly predictor component that uses graph convolutional networking practices using unsupervised approach to learn node embeddings. These become quick initiation for the techniques of machine learning to forecast and identify nodes, which exhibit mischievous and abnormal activities. For this, they chose innovative profound techniques of graph convolutional neural network (GCN) and the task of acquiring knowledge was executed in an unsupervised method by using the technique of batch-training GCN [8]. They have conducted experiments on non-proprietary benchmark Hadoop log data set for detection of anomaly in HDFS. Also, authors have compared their PredictDeep with other existing models like PCA, CNN, IM, DeepLog.

4.7 Deep Learning LSTM Model and Clustering

In [2] paper, authors proposed a framework of general purpose called as DeepLog for detection of anomaly online, over system logs with LSTM neural network. DeepLog has considered various parameters like timestamp and log key for taking values from the log message unlike other systems, where only log key is used. Therefore, it is capable to pick up anomalies of distinctive types and perform detection of anomaly at log beginning. As part of training phase, it uses a small training data set consisting of normal log entries. It does detection of anomaly on every incoming log beginning in a streaming fashion. There are two stages of training for normal data and detection of abnormality. If a log entry is labelled as abnormal, then DeepLog's workflow provides more detail semantics to understand and diagnose the anomaly. It also considered user feedback and thereby, able allow to incrementally update the LSTM model to get adapted to new patterns. The threat model considered two types of attacks. The one that caused sluggish execution is called denial of service (DoS), which may lead to timestamp difference in log. And another attack that caused restart of server repeatedly called blind return-oriented programming (BROP) attack led to abortion of task by ending the log sequence early. Authors have performed an extensive evaluation on both HDFS log and OpenStack log to demonstrate the effectiveness of the system.

4.8 Deep CNN and Classification

The main highlight of this paper is an effective log pre-processing method which comprises various stages [13]. At first, a log collection tool called Rsyslog is used for data collection that provides timestamp, severity, facility, node and msg as a part of output format. The msg is further used for classifying event groups derived from its semantics, called as event category labelling stage. Each of these event categories is used for building dictionary libraries to convert text log into numerical data using TF-IDF method. Later to remove noise, a filtering is implemented to remove redundant even information like stopwords and punctuations. And in the last step of pre-processing step, authors have generated numeric semantic vector feature that correspond to semantics of every information from msg. In this, Levenshtein distance is used to measure semantic similarity and top ten similarity values of semantic for every category of event is selected to form a combine feature vector. And at last, a min-max normalization is used to normalize every feature into a [0, 1] interval. This numerical feature matrix is implemented on system for log classification, based upon deep CNN model. While experimenting, authors have tried to adjust various hyper-parameter configurations to find best learning model. In addition to these, authors tried to achieve decent precision for category prediction by applying their trained CNN model to CMRI-Hadoop and bluegene/L logs.

4.9 Cloud Monitoring of OpenStack Logs Using Elasticsearch and Kibana

Authors [14] have developed a monitoring system to help the administrators to easily identify the error logs from the dashboard and generator alerts accordingly. The logs are fetched, parsed and stored in schemaless database and Elasticsearch are later displayed on user-friendly dashboard. From OpenStack modules Nova—that host and manage cloud computing services, Neutron—that provide network connectivity as a service and Cinder—that offer continual storage space, logs are gathered to instances, which are managed by OpenStack EC2.

These logs are parsed into JSON format and saved it in the NoSQL database. The log generated from Nova, Neutron and Cinder are classified into to computation, network trace and storage logs. Later, they are transformed into different severity level of logs like warning, error, info, trace and debug by Logstash. Every day, it is also indexed, so that they can be queried by Elasticsearch. These logs are displayed on dashboard as two tables, one for monitoring real-time data and other for warning, trace and error logs [14]. For visual representation, Kibana is used, which works well with bot Logstash and Elasticsearch.

4.10 Autoencoders and LSTM in Deep Learning

The main outcome of this model [15] is that it does not demand pre-processing and feature extraction system of log that is competent to detect both abnormal message contents and time-based progression of log communications. For this, authors have utilized two numbers of autoencoders with units of LSTM. One of this is used for training of log lines lacking timestamp to study predetermined dimensional embedding of log memos. Other one is used for anomaly detection, which takes the text embedding messages and a numerical timestamp of the message as input. To find whether an input is anomalous or not, a distance measure is calculated between input and output. If the distance is above the chosen threshold, then it is considered as anomalous. They have carried out the experiment by using 1.1 crores links of HDFS logs by dividing 50% of data set for training and rest 50% for testing.

4.11 LogEvent2vec Along with Naive Bayes or Random Forests or Neural Networks Models

Authors [16] have experimented anomaly detection in three phases as log parsing followed by feature extraction and then applying anomaly detection models. For this, they used three well-known supervised model like random forests, Naive Bayes and neural networks on BGL data set. By log parsing, the semistructured information is converted to structured information. Here, every log is plotted to a log event. In the step for feature extraction, logs are apportioned into log sequence, based on a fixed window size using logEvent2vec. Here, for direct vectorization of log event vector, log events are given as input and thereby sequence vector of log could be obtained from log event vector with only one coordinate transformation. Word and LogEvent schemes are used for feature extraction. In word scheme, word2vec model accepts words as input, after eliminating the characters, which are non-alphanumeric. To pick up log file vector, the coordinate transformation is performed twice on the output words.

In the scheme for LogEvent, the word2vec model takes events of log as input. Here, coordinate transformation is performed once on event vector of log, to get the sequence vector of log with two coordinate transformation mechanisms known as bary and TF-IDF [16]. A combination of these is used on each of the three supervised model neural networks, random forests and Naive Bayes. Input to these anomaly detection models is the sequence matrix of log. The computational time for both the scheme with both bary and TF-IDF mechanism has been compared for all the three models. Also, authors have compared the result of these combination using $F1$-score on various dimensionality data. The result has shown that as classifier working with LogEvent2vec, functioning of random forest is superior to neural network and Naive Bayes models.

4.12 *Template2vec with LSTM Model*

Authors have used both HDFS and BGL data set to compare their log anomaly method with other unsupervised methods like LogCluster, DeepLog and invariant mining [17].

They have used a new approach of offline learning using deep learning to learn both sequential and quantitative patterns of the logs. For this, they have used template2Vec mechanism that converts the words in templates into word embedding vectors. For this, antonyms and synonyms of words are also considered. Then a weighted average of the word vectors of the words in the template is calculated to find template vector. Also, a log sequence is converted to a template vector sequence. This template vector sequence is given as input to LSTM model for training phase. The offline training is scheduled periodically so that new log templates can be updated into newly learned model.

In the online detection stage, the logs appear in real time and are matched with templates that are learned in the offline stage. If a new log does not match any existing template, then a temporary template is extracted using FT-tree and its template vector is calculated. Later, this temporary template vector is matched with existing one, to find similarity among the template vectors. The real-time logs that do not match any of the learned patterns are considered anomalous.

4.13 K-prototype and KNN Clustering Algorithm

Zhaoli Liu et al. collected logs from their own system that had more than fifty actual running server and a target server with labelled logs. Secure log, message log and audit log were mainly considered and they were massively collected in an efficient manner by using Flume. These logs were stored on HDFS, and processing was carried out using Spark. As a part of feature extraction, they tried to capture user behaviour from ten features on logging activities and session statistics. Logging activities are 3 W (who, where, when aspects) like username, host, physical location, time of connection and session lifetime. Statistics feature covers, what aspects like operating command sequence, password failed time, frequency of read, write, delete operations, etc. For clustering, they used K-prototype algorithm that can be used to cluster both categorical and numerical extracted feature attributes. For this, based upon coexistence of values, they designed a latest cost function and also distance measure. Further, to find the K value (sum of clusters), EM clustering algorithm was employed by them and later, grouped the data set into clusters using their K-prototype algorithm.

Later, these clusters were grouped into normal and anomalous, based on filtering threshold value. Normal clusters were the one that contained events greater than the threshold value. For refinement of anomalous clusters, two new distance-based features (first as distance between data point and cluster centres and second as distance between data points and its nearest neighbours) were used to measure their local and global anomaly degrees. And finally, KNN classifier with new distance features was used to find correct anomalies. They compared the performance of this integrated method with only K-prototype clustering algorithm, where refinement of clusters was not used. KNN algorithm and the results showed that integrated method outperformed when compared with other two methods [18].

In [19], authors have experimented with six models and a detailed comparison result has been prepared. At first log collection is followed by log parsing and feature extraction and at last applied appropriate anomaly detection models are the main four steps involved in their anomaly detection framework. For their work, they have taken publicly available production data set: HDFS data set and BGL data set. According to their studies, log parsing can be classified in two methods: clustering-based and heuristic-based. For their comparison work, they have used their own open-source log parsing toolkit. For feature extraction, authors used different window mechanism like fixed window, sliding window, session window and they performed analysis using these mechanisms on different models.

Finally, they have experimented and compared results among three supervised models like PCA, logistic regression and decision tree and another three models like PCA, log clustering and invariants mining, which are unsupervised [20]. They could insight some interesting findings like supervised learning achieved with high precision but their recall varied along with different data set and window settings, anomaly detection with sliding window achieved high accuracy than fixed window, invariant mining showed stable and high performance among unsupervised learnings, etc. Data sets and techniques used across various domains for anomaly detection and analysis are summarised in the below Table 1.

Table 1 Data sets and techniques used across various domains for anomaly detection

Year	Data sets	Technique used
2020 [8]	Authors have detected and predicted unseen and evolved pattern of security anomalies on Hadoop log data set	Combination of graph analytics and deep learning techniques as an unsupervised batch-training graph convolutional neural network technique
2020 [16]	BGL data sets from supercomputer system of BlueGene/L at National Labs of Lawrence Livermore. Log messages of 4, 747 and 963 and anomalous log messages of 348 and 460 in the BGL data set were used	Comparison among word2vec and logEvent2vec using both b-ary and TF-IDF coordinate transformation on models of random forests, Naive Bayes and neural networks
2020 [10]	Evaluated on well-known data sets like BGL, OpenStack, and Thunderbird	Isolation forest and two deep autoencoder networks
2019 [15]	For detecting anomalous message contents and of time evolution of log messages, HDFS logs are used	Two numbers of autoencoders with LSTM units are used. One of this to train text logs and other for anomalous detection
2019 [17]	HDFS and BGL data sets generated by the Blue Gene/L supercomputer, which consisted of 128 K processors and was deployed at Lawrence Livermore National Laboratory. 348, 460 anomalous and 4, 747, 963 normal log messages in the BGL data set were used	Template2vec for feature extraction and training model built using LSTM
2018 [5]	Rollback upgrade operations along with events of VM failure on Amazon Web Services	Multiregression technique (OLS) and cloud metrics using cloud watch
2018 [11]	HDFS log data set, which was generated from experiments conducted over 200 days in Amazon EC2	Implementation of CNN-based approach was done in TensorFlow, a baseline model on MLP for efficiency comparison
2018 [4]	Semisynthesized log data were collected from web server, the SQL database and the reverse proxy server	Dynamic clustering technique with evolution mechanism and time series analysis using ARIMA model
2018 [12]	Data sets were collected from OpenStack-based cloud data centre that hosts two characteristic cloud applications (graph analytics and media streaming), which are gathered from the workload collections of CloudSuit. Also, data sets collected from cloud date centre, called Bitbrains was used for carrying out real-world experiments with the real-world workload traces from it	Algorithm of one class classification (OCC) type and a time series analysis of window based were used

(continued)

Table 1 (continued)

Year	Data sets	Technique used
2018 [13]	Log events collected from distributed cluster system for a span of 5 months using Rsyslog tool	Labelling event categories, building dictionary libraries, feature engineering to build normalized feature vector and to train unique set of features, deep CNN based upon output of the previous layer was used
2018 [14]	Logs collected from OpenStack components Nova, Neutron and Cinder are monitored on real time on dashboard and alerts administrator timely	For transformation, Logstash is used whereas querying is done with Elasticsearch and Kibana to present data on dashboard in table form
2018 [18]	Live system with actual running servers of more than 50 and with single target server for labelled logs	Integrated *K*-prototype and KNN clustering algorithm with two new distance measures
2017 [2]	Anomaly detection on log data sets of HDFS and OpenStack	Workflow model—for every task, that used approaches of both deep learning (LSTM) and classic mining (density clustering)
2017 [19]	BGL and HDFS data set	Supervised learning: PCA, logistic regression and decision tree Unsupervised learning: PCA, log clustering and invariant mining along with different window setting for feature extraction

5 Conclusion

This research work has discussed various methods for anomaly detection and analysis by using system logs from various domains. A study on various solutions that is carried out in recent papers like statistical approach, combination of supervised, unsupervised and deep learning methods is presented in this paper.

Logs are the key source for any application for detecting the anomaly. A timely detection of the root cause of anomaly will help the enterprise for the development of counter measures besides appropriate framework for reducing the identified risk. The type of logs, its size, the source of generation is to be considered for proposing right anomaly detection mechanism. One cannot state a particular method is the best in generic for detection of anomaly [21]. For every application, one can find a better solution than an existing solution. With machine learning and deep learning methods, various combination can be proposed in different stages of anomaly detection like log parsing, feature extraction for better model development and training the model for finding the best solution for an application.

References

1. J. Zhu, S. He, J. Liu, P. He, Q. Xie, Z. Zheng, M.R. Lyu, Tools and benchmarks for automated log parsing, in *International Conference on Software Engineering: Software Engineering in Practice*, 2019, pp. 121–130. https://doi.org/10.1109/ICSE-SEIP.2019.00021
2. M. Du, F. Li, G. Zheng, V. Srikumar, DeepLog: anomaly detection and diagnosis from system logs through deep learning, in *Proceedings of 24th ACM Conference on Computer and Communications Security*, 2017, pp. 1285–1298
3. S. Lu, X. Wei, Y. Li, L. Wang, Detecting anomaly in big data system logs using convolutional neural network, in *International Conference on Big Data Intelligence and Computing and Cyber Science and Technology Congress*, vol. 1, 2018, pp. 15–158. https://dx.doi.org/doi:%2010.1109/DASC/PiCom/DataCom/CyberSciTec.2018.00037
4. M. Landauer, M. Wurzenberger, F. Skopik, G. Settanni, P. Filzmoser, Dynamic log file analysis: An unsupervised cluster evolution approach for anomaly detection. Comput. Secur. **79**, 94–116 (2018). https://doi.org/10.1016/j.cose.2018.08.009
5. M. Farshchi, J.G. Schneider, I. Weber, J. Grundy, Metric selection and anomaly detection for cloud operations using log and metric correlation analysis. J. Syst. Softw. **137**, 531–549 (2018). https://doi.org/10.1016/j.jss.2017.03.012
6. H. Zawawy, K. Kontogiannis, J. Mylopoulos, Log filtering and interpretation for root cause analysis, in *2010 IEEE International Conference on Software Maintenance (ICSM)*, 2010, pp. 1–5
7. L. Martí, N. Sánchez-Pi, J. Molina, A.C. Garcia, Anomaly detection based on sensor data in petroleum industry applications. Sensors **15**(2), 2774–2797 (2015). https://doi.org/10.3390/s150202774
8. M.A. Elsayed, M. Zulkernine, PredictDeep: security analytics as a service for anomaly detection and prediction. IEEE Access **8**, 45184–45197 (2020). https://doi.org/10.1109/ACCESS.2020.2977325
9. S. Kavulya, K.R. Joshi, F.D. Giandomenico, P. Narasimhan, Failure diagnosis of complex systems, in *Resilience Assessment and Evaluation of Computing Systems*, eds. by K. Wolter, A. Avritzer, M. Vieira, A.P.A. van Moorsel (Springer, 2012), pp. 239–261
10. A. Farzad, T.A. Gulliver, Unsupervised log message anomaly detection. ICT Express **6**(3), 229–237 (2020). https://doi.org/10.1016/j.icte.2020.06.003
11. S. Lu, X. Wei, Y. Li, L. Wang, Detecting anomaly in big data system logs using convolutional neural network, in *The Proceedings of 4th International Conference on Big Data Intelligence and Computing and Cyber Science and Technology Congress*, 2018, https://doi.org/10.1109/DASC/PiCom/DataCom/CyberSciTec.2018.00037
12. S. Barbhuiya, C. Zafeirios, K.P. Papazachos, D.S. Nikolopoulos, *RADS: Real-time Anomaly Detection System for Cloud Data Centres*, 2018, CoRR abs/1811.04481
13. R. Ren, J. Cheng, Y. Yin, J. Zhan, L.Wang, J. Li, C. Luo, Deep convolutional neural networks for log event classification on distributed cluster systems, in *International Conference on Big Data*, 2018, pp. 1639–1646. https://doi.org/10.1109/BigData.2018.8622611
14. V. Agrawal, D. Kotia, K. Moshirian, M. Kim, Log-based cloud monitoring system for OpenStack, in *2018 IEEE Fourth International Conference on Big Data Computing Service and Applications (BigDataService)*, 2018, pp. 276–281. https://doi.org/10.1109/BigDataService.2018.00049
15. B. Bursic, V. Cuculo, A. D'Amelio, Anomaly detection from log files using unsupervised deep learning. Formal methods, in *FM 2019 International Workshops*, 2020, pp. 200–207. https://doi.org/10.1007/978-3-030-54994-7_15
16. J. Wang, Y. Tang, S. He, C. Zhao, P.K. Sharma, O. Alfarraj, A. Tolba, LogEvent2vec: LogEvent-to-vector based anomaly detection for large-scale logs in internet of things. Sensors (Basel) **20**(9), 2451 (2020). https://doi.org/10.3390/s20092451.PMID:32357404;PMCID:PMC7249657

17. W. Meng, Y.Liu, Y. Zhu, S. Zhang et al., Log anomaly: unsupervised detection of sequential and quantitative anomalies in unstructured logs, in *Proceedings of the Twenty-Eighth International Joint Conference on Artificial Intelligence, IJCAI-19*, vol. 7, 2019, pp. 4739–4745
18. Z. Liu, T. Qin, X. Guan, H. Jiang, C. Wang, An integrated method for anomaly detection from massive system logs. IEEE Access **2018**(6), 30602–30611 (2018)
19. S. He, J. Zhu, P. He, M.R. Lyu, Experience report: system log analysis for anomaly detection, in*2016 IEEE 27th International Symposium on Software Reliability Engineering (ISSRE)*, Ottawa, ON, 2016, pp. 207–218. https://doi.org/10.1109/ISSRE.2016.21
20. S. Soja Rani, S.R. Reeja, in *A survey on different approaches for malware detection using machine learning techniques*, eds. by P. Karrupusamy, J. Chen, Y. Shi. Sustainable Communication Networks and Application. ICSCN 2019. Lecture Notes on Data Engineering and Communications Technologies, vol. 39 (Springer, Cham, 2019). https://doi.org/10.1007/978-3-030-34515-0_42
21. N.S. Uma, C.S.R. Prashanth, A detailed analysis of the various frequent itemset mining algorithms. J. Adv. Res. Dyn. Control Syst. **12**(2), 448–454 (2020)

Analysis to Increase the Term Deposits in the Banking Sector Using Decision Tree

Modem Asritha, D. Hemavathi, and G. Sujatha

Abstract At present, customer retention is one of the major challenges faced by the banking sector. In order to overcome the frequent shift of customers, customer loyalty is important. In this paper, marketing targets dataset from Kaggle was considered for the analysis. Logistic regression was applied for the dataset, and it was found that only four attributes are significant among all the attributes. And, in order to get the clear understanding on how the independent variables are affecting the dependent variable, decision tree classifier is used. The decision tree will provide a clear strategy to improve the customer satisfaction in the banking sector.

Keywords Customer retention · Logistic regression · Decision tree

1 Introduction

Big data analytics helps the banks in analyzing the behavior of customers based on various inputs like investment patterns and interest toward investment in the financial backgrounds. The data plays a vital role in getting the customer loyalty by providing the personalized banking solutions for them. Nowadays, the volume of data getting generated in banking sector is very high. In order to handle that high amount of data, big data analytics is needed [1]. It also helps to understand the market trends and gives an idea whether to increase or decrease the interest rates for various individuals. Data analytics helps to identify the customers with less credit scores, so banks do not provide any loans to them.

Banks use big data analytics to improve the level of services to the customers, and it also plays a vital role in creating the effective strategies through which they will attract and retain the customers [2]. In this paper, marketing targets dataset was

M. Asritha · D. Hemavathi (✉) · G. Sujatha
Department of Information Technology, School of Computing, College of Engineering and Technology, Faculty of Engineering and Technology, SRM Institute of Science and Technology, SRM Nagar, Kattankulathur, Kanchipuram, Chennai, Tamil Nadu 603203, India
e-mail: hemavatd@srmist.edu.in

G. Sujatha
e-mail: sujathag@srmist.edu.in

S. Shakya et al. (eds.), *Mobile Computing and Sustainable Informatics*, Lecture Notes on Data Engineering and Communications Technologies 68,
https://doi.org/10.1007/978-981-16-1866-6_60

considered for analysis. The marketing targets dataset consists of 31,647 records, out of which only 3715 customers have opted for the term deposit. Therefore, it is very clear that most of the customers have not opted for term deposit which indicates customer dissatisfaction. The aim of this paper is to find out the reasons that why most of the customers have not opted the term deposit. The dataset consists of attributes like subscribed, balance, age, campaign, duration, etc. The variable subscribed provides the information about the customers that has subscribed to term deposit or not [3].

Logistic regression was applied on the dataset to establish a relationship between target variable and predictor variables. From the results, p-value and coefficients were obtained. The p-value is used to identify whether there is a significant relationship or not between the variables. The lower the p-value (less than 0.05), the higher the level of significance. Attribute subset selection is a data reduction method applied on the dataset to find the significant attributes based on the *p*-value, and it is found that only four attributes are significant. In order to get a clear understanding that how independent variables are affecting the dependent variable, decision tree classifier is used. Decision trees are used for interpreting the data and constructed through an algorithmic approach that identifies the ways to divide a dataset based on various conditions. It is one of the widely used methods for the supervised learning.

2 Literature Review

The paper [4] deals with the development in the banking sector which helps many of the people to apply for bank loans. The problem is that the banks cannot provide loans to all the customers whoever has applied. For safety purpose, banks have to check their credit score and should provide loans to the eligible customers. At present, banks are providing loans after the verification and validation process but still there is no guarantee of selecting the right customer out of all customers. Therefore, this paper presents a decision tree algorithm which helps to identify the right customer. Predicting loans are very helpful for bank employees as well as for the customers. The main objective of this paper is to identify the right customers in simple and easy way. This paper is mainly for the management authority of banks, such that investors will not be able to alter anything in the complete process of prediction. Report will be sent to the various departments of banks for specific loan ID such that they can take proper action on the request.

Banking sector has been developing continuously beneath indicating influences in the big data era. This paper [5] deals with big data mining and digitalization in the banking sector. The era of big data came with many opportunities as well as challenges. The data mining skills plays a crucial role in the information science revolution. In addition to this, the improvement and popularization of mobile banking as well as e-banking serve as exponential growth in the real-time banking information. Mainly, the data analytics in the banking sector has been utilized to satisfy the needs of clients or customers, enhancing the strategic management and marketing. In this paper, it was concluded that banking domain has acquired data mining particularly for the fraud detection, to mitigate the risks and CRM.

This paper [6] deals with different areas where data mining can be used in financial domains like segmentation of customers, credit analysis, fraud transactions, etc. Generally, there are various types of loans that you have to take into consideration while you are borrowing money and also it is necessary to have an idea about available options. In this paper, it explains about different kinds of loans available for the customers like open-ended loans, security loans, student loans, etc. Open-ended loan is an augmentation of credit, where the money can be borrowed on the need basis and will be paid back on the continuous basis, this type of loans does not have any end date and secured loans are the one that depends on asset. In case, if there is any loan failure which has to be paid, the lender will make use of asset in order to cover up the loan. In this paper, three algorithms are used to create the predictive model which is used to identify and categorize the application of loans based on customer behaviors and previous credit scores.

The paper [7] deals with the multiple regression technique which is used to study the banking sector. In this paper, statistical tool called SPSS allows to identify how dependent variables are related to multiple independent variables. The value of the dependent variable can be predicted by using the equation which establishes relationship between dependent and independent variables, called as multiple regression technique. Analysis was carried out on the SBI and ICICI bank's data in which loyalty is a dependent variable and gender, education, occupation, age and income are considered as independent variables [8–10]. The results say that only two variables, income and age, are significant. For age, there is a negative coefficient which implies that increase of age of customer will decrease the loyalty. Therefore, to eliminate the insignificant variables from the regression technique, forward or backward stepwise regression can be used. From the results, it is clear that loyalty varies with respect to age and income in SBI bank and in the case of ICICI bank, loyalty varies gender-wise.

Many of the banks offer their services to the customers through Internet banking in free or low cost. Even after providing the Internet banking in low cost, still the usage of the Internet banking is very less. This paper [11] focuses on the reasons that are affecting the usage of Internet banking via logistic regression technique. So, a survey was conducted on 300 customers to know whether they are using Internet banking or not. Much effort has been carried out to find the causes that are affecting the usage of Internet banking through the implementation of logistic regression on data extracted via survey. In the end of analysis, it is clear that six factors are affecting the usage of Internet banking. They are level of income, accessibility to Internet over home and workplace, lack of security over Internet, difficulties during access and ease of communication. Therefore, it is concluded that banks should give high priority to the groups of high levels of income in order to improve the usage of Internet banking. In addition to this, there is necessity to improve the security systems associated with access to the Internet banking constantly in order to prevent the loosing of customers. Fake information of the nodes is verified by the technique used in [12] for the mobile nodes in the network.

3 Proposed Methodology

Logistic Regression

Logistic regression is one of the regression technique which consists of statistical process provides the relationship between target and predictor variables. Logistic regression analysis helps to analyze the changes in dependent variable with respect to change in one of the independent variables where other variables are set fixed. There are many regression techniques, and those techniques are differentiated based on various features like number of predictor variables, number of target variables and regression line shape. This technique is to predict the binary output (1/0) for the set of predictor variables, and this technique is mainly used when the response variable is categorical. In this paper, marketing targets dataset is considered for the analysis of customer satisfaction that consists of dependent variable subscribed (yes or no) which is a categorical value, so logistic regression is applied for the given dataset [13, 14]. In simple terms, it is used to predict the probability of occurring of an event. The main objective of the logistic regression is to find the probability of dependent variable (yes/no). The function of logistic regression is based on two things. Those are probability for success and failure, and it should satisfy the below conditions:

- Probability should be always positive ($p \geq 0$).
- Probability should be always less than or equal to one ($p \leq 1$).

Decision Tree

To get a clear understanding that how independent variables are affecting the dependent variable, decision tree classifier is used. Decision trees are used for interpreting the data and constructed through an algorithmic approach that identifies the ways to divide a dataset based on various conditions. It is one of the widely used methods for the supervised learning.

4 Implementation and Results

The marketing targets dataset consists of 31,647 records, out of that only 3715 customers have opted for the term deposit [15]. From this, it is very obvious that there is customer dissatisfaction. The main objective is to find out that why most of the customers have not opted for the term deposit. For the analysis purpose, marketing targets dataset is imported into the RStudio. The input dataset is split into two parts, namely training data (75%) and testing data (25%), and logistic regression is applied on the dataset to establish the relationship among the variables.

Marketing Targets Dataset Description

The dataset comprises eighteen attributes. They are subscribed, id, age, job, marital, education, default, balance, housing, loan, contact, day, month, duration, campaign, pdays, previous and poutcome. Among those attributes, only four attributes are considered as significant by applying feature selection technique.

4.1 Attribute Subset Selection

Attribute subset selection is one of the data reduction techniques, used to reduce the size of data and that helps to do the analysis more efficiently. The objective of the attribute subset selection is to get only the significant or relevant attributes, and this can be achieved with the help of statistical significance [16]. Statistical significance is based on *p*-value (Pr> |z|). *P*-value is used to identify whether there exists a significant relationship or not. The attributes with the *p*-value less than 0.05 are considered as significant, and the attributes with *p*-value greater than 0.05 are ignored. The attributes and *p*-values are given in Table 1.

Therefore, a model is created with the subset of attributes using the attribute subset selection, confusion matrix is created to calculate the accuracy and the values of confusion matrix are given in Table 2.

From the above table, accuracy obtained is 88%.

However, there is no clear idea from the results of logistic regression technique that how the dependent variable subscribed varies with respect to the independent variables, balance, duration and campaign. Therefore, decision tree an effective algorithm is used that considers all the possible outcomes and provides a conclusion.

Decision tree for subscribed as dependent variable and duration, campaign and balance as independent variables

The below decision tree has divided into many numbers of nodes. So, it is difficult to analyze the data properly and also to draw a conclusion. Therefore, the independent variables have to be analyzed individually or with set of two to make a proper decision (Figs. 1 and 2).

Table 1 Attributes with *p*-values and coefficients

S. No.	Attribute	Data type	*P*-value	Coefficients
1	Subscribed (intercept)	Categorical	2×10^{-16}	– 2.92
2	Balance	Numerical	2.06×10^{-11}	4.01×10^{-5}
3	Duration	Numerical	2×10^{-16}	3.45×10^{-3}
4	Campaign	Numerical	2×10^{-16}	-1.22×10^{-1}

Table 2 Results of confusion matrix

	False	True
No	20,658	291
Yes	2372	414

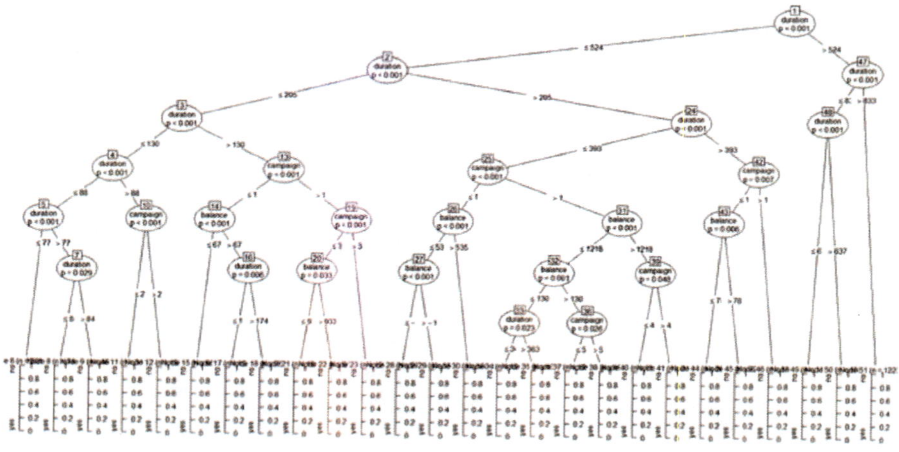

Fig. 1 Decision tree for duration, campaign and balance

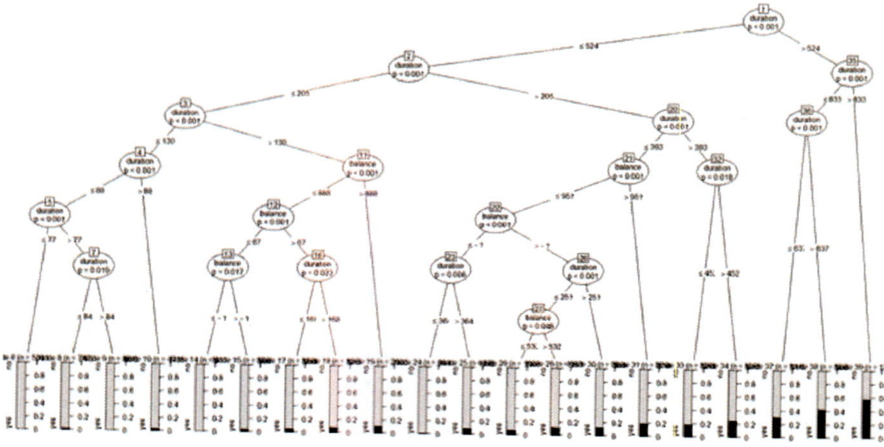

Fig. 2 Decision tree for balance and duration

Decision tree for subscribed as dependent variable and balance and duration as independent variables

The decision tree has divided the duration into two parts: One is greater than 524, and the other is less than or equal to 524. It is very clear that at Node 39 ($n = 1223$), mostly 60% (0.6) of the customers has opted for term deposit and it is inferred that as the duration increases, the number of customers for term deposit also increases. Along with the duration, balance greater than 951, there is slight increase in the number customers opted for term deposit.

Decision tree for subscribed as dependent variable and balance and campaign as independent variables

This decision tree has divided the campaign into two parts: One is greater than 1, and the other is less than or equal to one. From the below decision tree, it is inferred that there are more chances of getting subscribed for term deposit when balance is greater than 458 irrespective of campaign (at the time of campaign, the number of contacts performed) (Fig. 3).

Decision tree for subscribed as dependent variable and duration and campaign as independent variables

This decision tree has split the duration mainly into two segments: One is greater than 524, and the other is less than or equal to 524. Most of the customers have opted for term deposit when duration is more than 833. Therefore, it is obvious that increase in duration increases the number of customers for term deposit (Fig. 4).

Decision tree for subscribed as dependent variable and duration as independent variable

The decision tree has divided into two segments: One is greater than 524, and the other is less than or equal to 524. From the below decision tree, it is clear that as the duration increases, the number of customers subscribed for term deposit also increases (Fig. 5).

Decision tree for subscribed as dependent variable and campaign as independent variable

From the below decision tree, it is very clear that when the campaign is less than or equal to 1, the number of customers opted for term deposit is more when compared to the campaign greater than or equal to 1. And, when the campaign is greater than 17,

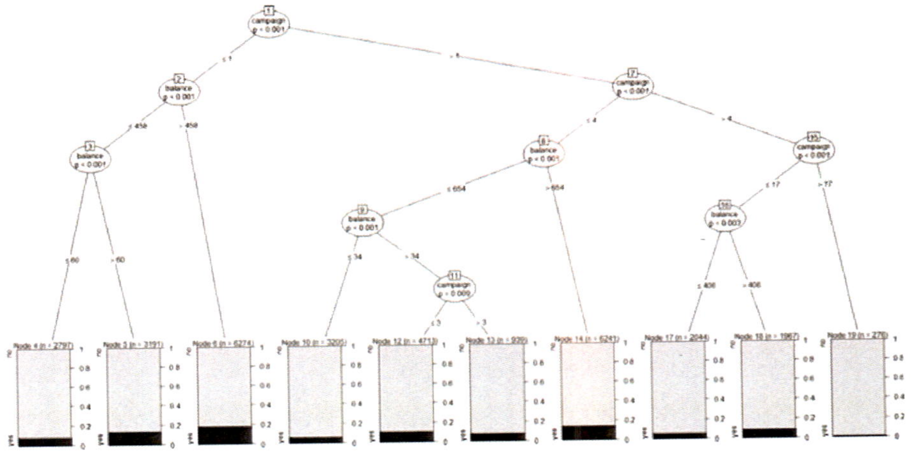

Fig. 3 Decision tree for balance and campaign

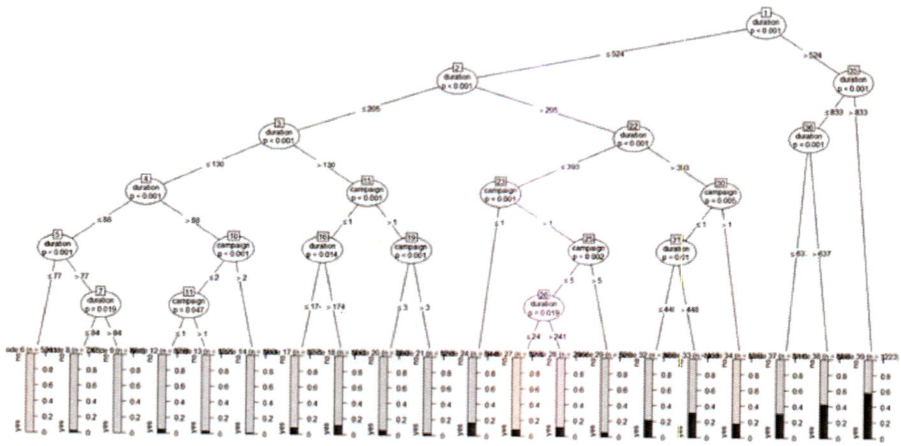

Fig. 4 Decision tree for duration and campaign

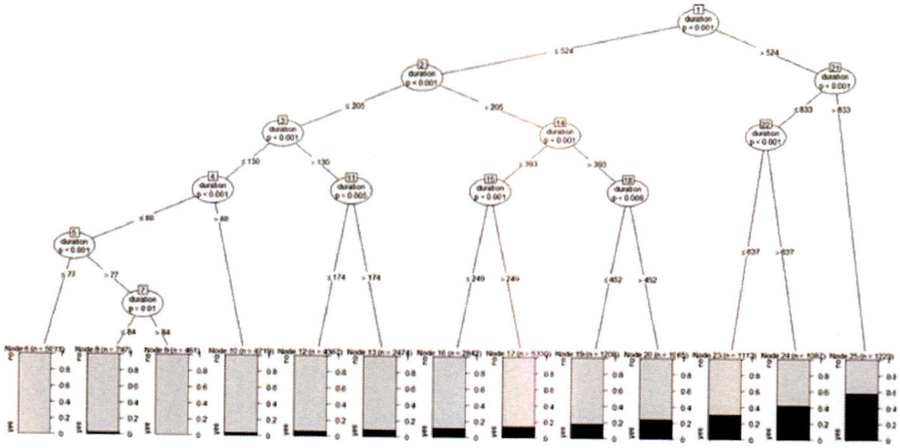

Fig. 5 Decision tree for duration

no customers have opted for the term deposit. So, it is good to maintain the campaign less than or equal to one [17] (Fig. 6).

Decision tree for subscribed as dependent variable and balance as independent variable

The decision tree shows that, as the balance increases, the number of customers subscribed for term deposit also increases (Fig. 7).

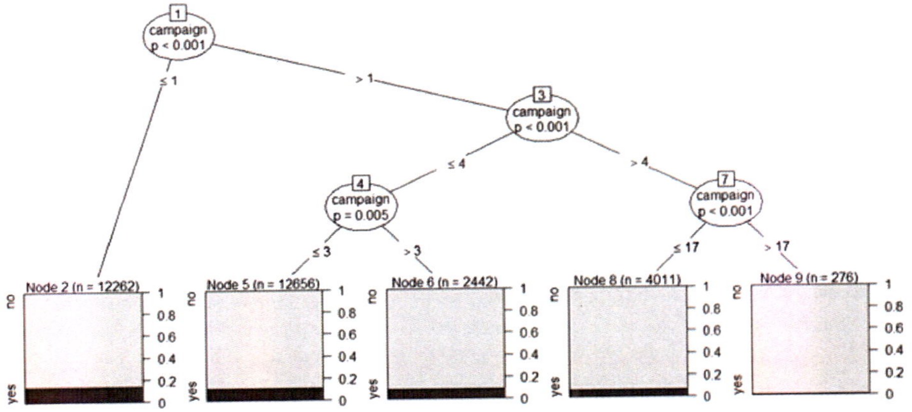

Fig. 6 Decision tree for campaign

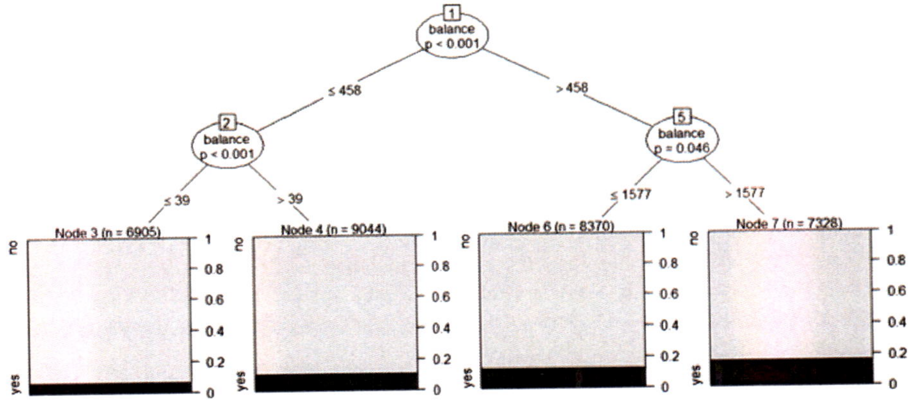

Fig. 7 Decision tree for balance

5 Conclusion

This paper focuses on identifying the reasons that why most of the customers have not opted for the term deposit. To get a clear understanding about the attributes in the dataset, decision tree classifier is used. From the results, it is very clear that as the balance and duration increase, the number of customers for the term deposit also increases. And, in the case of campaign attribute when the campaign increases, the number of customers opted for term deposit decreases. In order to avoid that, campaign should be less than or equal to 1. Therefore, this provides a clear idea to the banks about the attributes like balance and duration which need to improve to increase the number of customers for the term deposit.

References

1. U. Srivastavaa, S. Gopalkrishnanb, Impact of big data analytics on banking sector: learning for Indian banks. Proc. Comput. Sci. **50**, 643–652 (2015)
2. A. Arbore, B. Busacca, Customer satisfaction and dissatisfaction in retail banking: exploring the asymmetric impact of attribute performances. J. Retail. Consumer Serv. **16**, 271–280 (2009)
3. K. Agaliotis, M. Hadzic, Predicting retail banking consumer behavior using statistics. Euro. J. Appl. Econo. 2015
4. N. Madane, S. Nanda, Loan prediction analysis using decision tree. J. Gujarat Res. Soc. **21** (2019)
5. H. Hassani, X. Huang, E. Silva, Digitalisation and big data mining in banking. Big Data Cogn. Comput. 2018
6. A.J. Hamid, T.M. Ahmed, Developing prediction model of loan risk in banks using data mining. Mach. Learn. Appl. Int. J. (MLAIJ) **3**(1) (2016)
7. D. Upadhyay, C. Bhavsar, Multiple regression: a multivariate technique to study bank sector. Int. J. Sci. Eng. Technol. **3** (2015)
8. A. Srivastava, S.K. Singh, S. Tanwar, S. Tyag, *Suitability of Big Data Analytics in Indian Banking Sector to Increase Revenue and Profitability* (IEEE, 2017)
9. S.D. Kularathne, R.B. Dissanayake, N.D. Samarasinghe, L.P.G. Premalal, S.C. Premaratne, Customer behavior analysis for social media **3**(1) (2017)
10. S. Wang, I. Petrounias, Big data analysis on demographic characteristics of Chinese mobile banking users, in *19th Conference on Business Informatics* (IEEE, 2017)
11. A. Köse, Determination of reasons affecting the use of internet banking through logistic regression analysis. J. Glob. Strat. Manage. **3** (2009)
12. A. Rajaram, S. Palaniswami, A high certificate authority scheme for authentication in mobile ad hoc networks. Int. J. Comput. Sci. Issues (IJCSI) **7**(4), 37 (2010)
13. Md. Ahasan Haruna, G.P. Rokonuzzamanb, V.R. Prybutoka, How to influence consumer mindset: a perspective from service recovery. J. Retail. Consumer Serv. **42**, 65–77 (2018)
14. P. Foroudi, S. Gupta, U. Sivarajah, A. Broderick, Investigating the effects of smart technology on customer dynamics and customer experience. Comput. Hum. Behav. **80**, 271–282 (2018)
15. W. Mbaluka, *Big Data Management and Business Value in the Commercial Banking Sector in Kenya* (University of Nairobi, 2013)
16. D. Hemavathi, H. Srimathi, An efficient approach to identify an optimal feature selection method using improved principle component analysis in supervised learning process. J. Adv. Res. Dyn. Control Syst. **11**(07) (2019)
17. L.N.B. Srinivas, S. Ramasamy, An improvized missing data estimation algorithm for wireless sensor network applications. J. Adv. Res. Dyn. Control Syst. **9**(18), 913–918 (2017)

Server Data Auditing with Encryption Techniques

K. Venkatesh and L. N. B. Srinivas

Abstract The management of vast volumes of data with cloud service providers (CSPs) poses data privacy issues. Data confidentiality and privacy could be lost due to cloud administrator, malware, deceptive cloud providers, or other malicious users who may manipulate the data physically transferring the data from one location to another. Hence, corrections to saved data must be reviewed at regular intervals. Cryptography, remote (cloud) data verification, is carried out by third-party auditors (TPAs). Also ideal for public auditing are TPAs, which provide audit services with more efficient analytical and communication capabilities than standard users. Our work focuses on cryptographic algorithms for cloud data auditing and the problems surrounding the validity and privacy of these algorithms. Many approaches to preserving confidentiality and privacy have been suggested in the literature; they are usually categorized according to the different states of the data: static, dynamic, multi-owner, multi-user, etc.

Keywords Cryptography · Cloud Service Providers (CSPs) · Third - Party Auditors (TPAs) · Homomorphic authentication

1 Introduction

The proposed research work will create a confidentiality mechanism framework that facilitates public audits of shared cloud-based data. The public auditing and privacy preserving model is reviewed for improving security [1–3]. Message digest algorithm is studied in [4] and the business management model using unified modeling language is presented in [5]. The ring signatures are used to check the consistency of the shared data, particularly to measure metadata authentication. A general protect asset [6]

K. Venkatesh (✉) · L. N. B. Srinivas
Department of Information Technology, College of Engineering and Technology, Faculty of Engineering and Technology, SRM Institute of Science and Technology, SRM Nagar, Kattankulathur, Kanchipuram, Chennai, Tamil Nadu 603203, India
e-mail: venkatek2@srmist.edu.in

L. N. B. Srinivas
e-mail: srinival@srmist.edu.in

S. Shakya et al. (eds.), *Mobile Computing and Sustainable Informatics*, Lecture Notes on Data Engineering and Communications Technologies 68,
https://doi.org/10.1007/978-981-16-1866-6_61

and multipath routing model to improve QoS is discussed [7]. Security based data integrity model is created in [8].The identity of the owner is kept secret by public auditors, who are prepared to display shared data efficiently without accessing the entire file, on each block of common data. Still, instead of checking them one by one, our system will be able to perform multiple audit tasks simultaneously.

2 Related Work

1. Many have come up with an equivalent idea of auditing the info on the cloud or server. The authors "Cong Wang, Sherman S. M. Chow, Qian Wang, Kui Ren and Wenjing Lou have used techniques called homomorphism linear authenticator and random masking using MAC technique". This method makes sure that the TPA is unaware of the cloud server data in an effective auditing process that removes the cloud user's burden of repetitive and potentially expensive auditing, but that it also alleviates users' concern that their data is outsourced. And also, this technique is highly productive and stable.
2. The authors "Prasanna and Ramesh used ranking method and symmetric key encryption" to try an equivalent secure and efficient search of encrypted database models, where users of the database are protected from data protection violations. The system's performance is increased adequately with the usage of the symmetric key encryption technique instead of the public key encryption technique.
3. The authors "Boyang Wang, Baochun Li and Hui Li and Xi'an proposed privacy-preserving public auditing for shared data within the cloud." A modern method of Shared data in public auditing with successful cloud app revoking is presented. When a community member has been revocated, the semi-confidence cloud may be re-signed with proxy re-signatures for blocks signed by the revoked member. During user revocation, the community can save a great deal of time and money for computer resources.
4. Another method included is by Orruta who gave privacy-preserving public auditing using ring signatures. Measuring the authentication data required for the validity of shared data, ring signatures are used. A TPA will still be able to view the shared data without having to retrieve the file, and the privacy of the data will be at risk. They efficiently and responsibly exchange the data.
5. The authors "Bo Chen, Reza Curtmola and Giuseppe Ateniese proposed remote data checking for network coding-based distributed storage systems using remote data checking." A strong RDC architecture is based on a network application that relies on the insecure server for distributed storage systems. To ensure that the RDC-NC program, the information stays safe in the event of data manipulation, replay and contamination.
6. The authors "Dan Boneh1, Xavier Boyen2 and Hovav Shacham3 proposed short group signatures using RSA algorithm signatures." The signatures on our project are roughly the size with the same protection as a regular RSA signature. The

only disadvantage is that signature generation does not require any combined calculations and authentication requires one single combination.
7. The authors "K. Venkatesh, L. N. Srinivas, M. B. Mukesh Krishnan and A. Shanthini proposed QoS improvisation of delay-sensitive communication using SDN-based multipath routing for medical applications." The anonymous system of protection by the use of the protection mediator is isolated from the proofed data possession process. The system restricts the need for this mediator for measurement and bandwidth, but also the trust put in this mediator with regard to data and personal privacy.
8. The authors "Rajaram and Palanisami" enhanced the security scheme for the nodes in mobile nodes-based wireless sensor network.

3 Proposed Methodology

The system offers a public audit mechanism that protects the privacy of sharing cloud data. The ring signatures are used to check whether the original data is uploaded to the cloud or whether any changes have occurred to it without even having to retrieve the entire data.

The proposed structure is further broadened to facilitate batch auditing so that multiple audit activities are checked more effectively. In this case, when downloading the encryption code, the auditor did not have the owner data copy copied into another backup disk. Only the auditor could investigate whether this link to the file is breached or not. Even if the auditors are infected, the file can be restored without a local backup copy. During the audit, the data owner receives a mail regarding that his file is being checked.

This is the protocol used to email the site, which stands for the "quick mail transfer protocol." Your email shopper could send your message using SMTP to the server, and SMTP will be used to send the message to the correct mail server by the server. SMTP can merely be a sequence of commands which authenticate and direct electronic transfers of messages. Once you have set up your email system, you can customize the SMTP server to the requirements of your local network service provider. Consequently, the incoming mail server (IMAP or POP3), which could be divergent from the SMTP server, should be set to the Web server of your mail.

Advantages:

1. Any audit tasks that be performed concurrently by the proposed program.
2. They increase verification effectiveness for multiple audit tasks.
3. High file sharing protection offers.
4. Admin has control deleting users.
5. Users can send request to auditor.

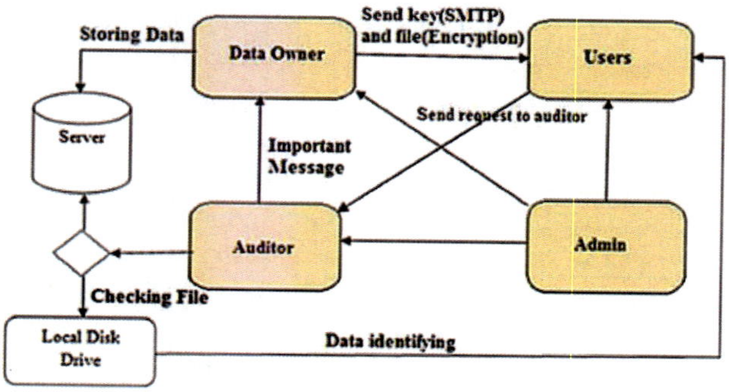

4 Algorithm Used

The ASE is a confidential but unclassified data protection encoding algorithm. The advanced encryption standards are used. As a possible result, government agencies will potentially become a private sector business communication encryption standard. In January 1997, a procedure was started by the NIST, a branch which is present in the USA. (See the Encryption of American and other sensitive communications for specific encrypted algorithms.) To consider a more reliable Data Encrypting Standard (DES), the weaker Triple DES substitution is added. The algorithm was to be world-wide free of royalties and to provide sufficient security for the safety of data over the next 20–30 years. The entire process of selection has been thoroughly public review and analysis, with the decision to ensure full exposure of the concepts. In 1998, the NIST chosen 16 AES candidates, who were at that time preliminary evaluated by, among them the national security agency, the world's cryptographical community. This led to the selection of five algorithms by NIST in August 1999 for further study.

A cryptographer named Joan Daemen and Vincent Rijmen both from Belgium RSA Security Bruce Schneier. The trusted cryptographer of Counterpane, were Mars, which is selected by an IBM research team of RC6 and it was submitted by RSA Security Rijndael. In all of the above, quick and reliable steps, such as encrypting and decrypting rates, setup key for time, and resistance to various attacks have been tested and verified with the code written in C and Java language in both software-centered and hardware-centric systems.

The elite community in cryptography once again received a comprehensive review. As a result, NIST declared on October 2, 2000, the option of the Rijndael standard. The Federal Information Processing Standard 197 has been formally approved by the Secretary of Commerce on December 6, 2001, specifying that all confidential unclassified information should be using Rijndael as an advanced encrypting system.

Steps of the Algorithm

1. KeyExpansion—round keys are derived from the cipher key.
2. Initial Round
   AddRoundKey—each byte of the state is combined with the round key using bitwise xor.
3. Rounds
   - SubBytes—a nonlinear substitution step where each byte is replaced with another according to a lookup table.
   - Shift Rows—a transposition step where each row of the state is shifted cyclically a certain number of steps.
   - Mix Columns—a mixing operation which operates on the columns of the state, combining the four bytes in each column.
   - AddRoundKey.
4. Final Round (no Mix Columns)
   - SubBytes
   - Shift Rows
   - AddRoundKey.

5 Modules

User Registration

The community manager automatically chooses a number when registering a user with an identification (ID). In the customer list for client tracing process, the company managing director adds. When registered, users will get a private key that they can use to create group signatures and decrypt files.

Public Auditing

Homomorphic authenticators are safely aggregated from individual data blocks to guarantee the person who checks the aggregate authenticator with clear combination of data blocks properly. A single combination of the homomorphic authenticator and a RMT is presented for public auditing in order to ensure the privacy and achieve privacy-conserving public auditing. In the proposed protocol, there are two phases, namely setup phase and audit phase.

Sharing Data

Information exchange is the canonical function. When a delegation is expected to be effective and versatile, the public audit property is especially useful. The schemes allow a service provider to exchange data in confidence and selectivity by providing each approved user with a single, small aggregate key, by means of a fixed and limited cipher text expansion.

Integrity Checking

It is therefore also of vital importance to promote data complexities to safeguard public risk audits. Now this research work demonstrates how the proposed key scheme can be adjusted to support data dynamics, including alteration, deletion and addition of block rates, based on existing plays. In the proposed design, this technique is used to perform a public risk audit of privacy with the help of dynamics of data. You do not download the entire file from the owner.

Homomorphic Authentication

Homomorphic authentication (HA) makes remote users like a CSP to store a data tag; the tag is made of {mi} data blocks in which $i = 1, 2, 3 \ldots n$ are secretly installed. The CSP then uses publicly accessible methods to use the corresponding succinct data tag to measure {mi} data points. HA enables anyone to certify the result of the complicated calculation done with the data tag A in authenticated datasets.

This can be done without the data for each department being disclosed. Two forms of HA exist: partially homomorphic and fully homomorphic encryption. A multiplication or a complementary homomorphism may be partially homomorphic encryption.

- File f is treated as an N-dimensional vector.
- Tag t is created for each block of file f.
- The client sends a random challenge vector c.
- The server returns proof of authentication: $\mu = \sum_i c_i, f_i$.

HA uses four algorithms: gen, encoding, proving, and *vrfy*.

Encode:

$$(f', st) \leftarrow Encode(f).$$

Vrfy:

$$a = Vrfy(pk, st, c, \mu).$$

It is inferred that the secret key is not required for verification using these algorithms. In addition, linear data block combinations provide the TPA with the appropriate information to recover the whole file.

How Message Authentication Works

An authenticated message code preserves a message's validity, validates the source's identity and does not republish the content. The MAC codes are generated using hash features with the hash value and message authentication. The recipient generates the message using a secure key with what the receiver and the sender are identified. While a MAC generally maintains message integrity, the privacy of the data is lost.

There were also reports of intruders able to change or share the message6 with others, which jeopardizes the data integrity.

The cloud auditing process is very simple: The cloud user loads the block of data together and sends hidden key (auditor) to the MAC and the corresponding file. The end user is allowed to check the data to avoid TPAs being curtailed. This requires that data blocks are released in the TPA. It is contradictory to the method of public audits. It can be used for cloud auditing following algorithm.

- key generator (k): $k \overset{s}{\leftarrow} k$
- code generator: Tag $\overset{s}{\leftarrow} \mathrm{MAC}_k(M)$
- verifier: $D \leftarrow \mathrm{VF}_k(M, \mathrm{Tag})$ where $D \in \{0, 1\}$.

How Privacy and Integrity are Preserved

The validated owning of a client's own data allows for verifying remote data, without uploading them. This is the technology which is focused on the probabilistic ownership of a random dataset from remote servers, through homomorphic linear authenticators. The data is confirmed through remote (integrity) data, apart from the users themselves. However, the consumer must access the entire data block to entire deterministic verification. Qian Wang and his collaborators have extended their work to provide proof for dynamically stored data with the use of tree data types for block-tag authentication.

These integrity schemes also faced problems with the verification of updates of small data. The Merkle hash tree, which maintains information integrity on the basis of the signing method, is one of the most common data integrity techniques and offers permission to conduct audits which prevents the presence of malicious users as TPA. This technique maintains data integrity based on the TPA.

The following equation is used for this tag generation scheme:

$$\sigma = \left(H(m_i \prod_{j=1}^{s_i} u_j m_{ij}) \right).$$

Analysis of the Architecture

CSPs have adapted their technical approach to diverse requirements due to the increasingly growing use of the cloud computing software and services. The diverse needs of elastic cloud computing are met by creating an effective infrastructure for customer-selected services. Non-monolithic cloud architecture can include multiple modules that are stored on various machines on the cloud infrastructure. Such machines or nodes have many complex tasks, and each action meets the demand. In addition, most nodes should be configured to provide services on a CSP basis. Based on the services offered, cloud architecture can vary. Service delivery is a key element of the cloud.

Using cloud service providers, dynamic resource allocation is typically provided to clients so that the CSP is neither over nor underfunded. Numerous CSPs are available on sale. Therefore, consumers must first question if the CSP offers the elasticity of its service and meets the defined service standard in the agreement before choosing for services of CSP. This auditing framework assigns the mission to all the auditing modules to work with responsible instructions. The backbone of the cloud is the cloud service architecture—a series of components or modules working together in order to perform those tasks. This group is loosely interconnected and varies according to the services offered by the CSP. CSPs need versatile resources in order to create cloud users' trust.

6 Conclusion

This paper has proposed a way in which auditing can be done without the auditor being able to view the data. The data stored by the data owner will be in the encrypted format and when a user would like to download the file and then receive an email regarding the key to download the file, then he can enter the key and download the original file. A request can be sent to the auditor regarding which file needs to be audited, and then with the help of hash values it will show whether the original uploaded file has been changed or not.

References

1. B. Wang, B. Li, and H. Li, Public auditing for shared data with efficient user revocation in the cloud. Proc. IEEE Infocom, 2904–2912 (2013)
2. C. Wang, S.S. Chow, Q. Wang, K. Ren, W. Lou, Privacy-preserving public auditing for secure cloud storage. IEEE Trans. Comput. **62**(2), 362–375 (2013)
3. B. Wang, B. Li, H. Li, Certificate less public auditing for data integrity in the cloud. Proc. IEEE Conf. Comm. Netw. Secur. (CNS'13), 276–284 (2013)
4. The MD5 Message-Digest Algorithm (RFC1321). https://tools.ietf.org/html/rfc1321 (2014)
5. C. Comes, L.O. Marian, N. Ghisoiu, I. Bircea, Business process management with unified modeling language. WSEAS Trans. Comput. **6**(2), 361–366 (2007)
6. C. Davis, M. Schiller, K. Wheeler, *IT AUDITING, Using Controls to Protect Information Assets*, ediţia a 2-a (The McGraw-Hill Companies, Osborne, New York, USA, 2011)
7. K. Venkatesh, L.N. Srinivas, M.B. Mukesh Krishnan, A. Shanthini, QoS improvisation of delay sensitive communication using SDN based multipath routing for medical applications. Future Gener. Comput. Syst. 256–265 (2019)
8. A. Rajaram, S. Palaniswami, The modified security scheme for data integrity in Manet. Int. J. Eng. Sci. Technol. **2**(7), 3111–3119 (2010)

Analysis of Cryptographic Hashing Algorithms for Image Identification in Deduplication Process

G. Sujatha, D. Hemavathi, K. Sornalakshmi, and S. Sindhu

Abstract Image deduplication is the process used to prevent the storing of duplicate images in the storage space. It helps to utilize the storage space efficiently by avoiding the duplicate copies of the intended image. Deduplication can be implemented by comparing the input image that has to be stored with the image existing in the storage space. But it is a time-consuming process. So instead of comparing the images directly, it is suggested to compare their hash values. In order to identify the duplication, initially, it is required to uniquely identify the image. This can be possible with the help of hash value. Hash value can be generated for each digital data, and it remains unique for all data. The main aim of this paper is to analyze various cryptographic hashing algorithms, which are being used to uniquely identify the image. Further, this research work is going to detail and compare the performance of various hashing algorithms by considering their operations and performance and finally concluded with a better solution to improve the process of achieving an image identification.

Keywords Digital image · Avalanche effect · Hash value · Storage as a service

1 Introduction

Cloud computing becomes more popular all around the world because of its variety of services to its users. The main advantage of using this cloud computing is transparency and user friendly. The users need not to be a master in using the services provided by these cloud computing [1]. One among the important services of cloud computing is cloud storage. Cloud storage service [1] provides storage facility to its users. They are the third-party servers, and even the users will not know where exactly their content gets stored. The cloud storage service provides the storage facility on demand. Whenever the user required storing their data, they can request

---

G. Sujatha (✉) · D. Hemavathi · K. Sornalakshmi · S. Sindhu
Department of Information Technology, College of Engineering and Technology, Faculty of Engineering and Technology, SRM Institute of Science and Technology, SRM Nagar, Kattankulathur, Kanchipuram, Chennai, Tamil Nadu 603203, India
e-mail: sujathag@srmist.edu.in

S. Shakya et al. (eds.), *Mobile Computing and Sustainable Informatics*, Lecture Notes on Data Engineering and Communications Technologies 68,
https://doi.org/10.1007/978-981-16-1866-6_62

and get serviced by the storage provider. The cloud computing models are classified as SaaS, PaaS and IaaS. IaaS offers computing resources such as storage and servers as shown in the Fig. 1.

The three important types of cloud computing [2] are private cloud storage, public cloud storage and hybrid cloud storage as shown in the Fig. 2.

In public cloud storage model, the resources can be accessed with the help of public Internet, and it is being shared by many users. Since many users are sharing the storage space, there is a possibility of having duplicate copies of digital data in the storage space. The storage space is simply wasted because of these duplicate copies. In order to prevent the duplication, the cloud storage provider employed the concept called deduplication, which [3] will prevent the duplicate copies that get stored in the storage space as shown in the Fig. 3. Especially, in case of image and

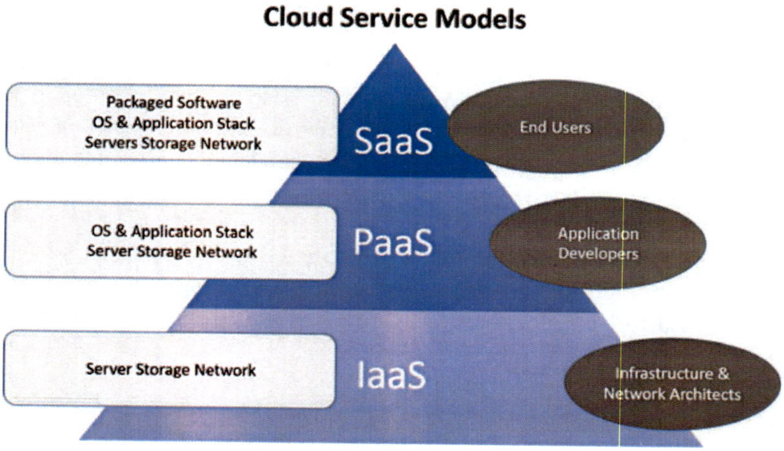

Fig. 1 Cloud service models

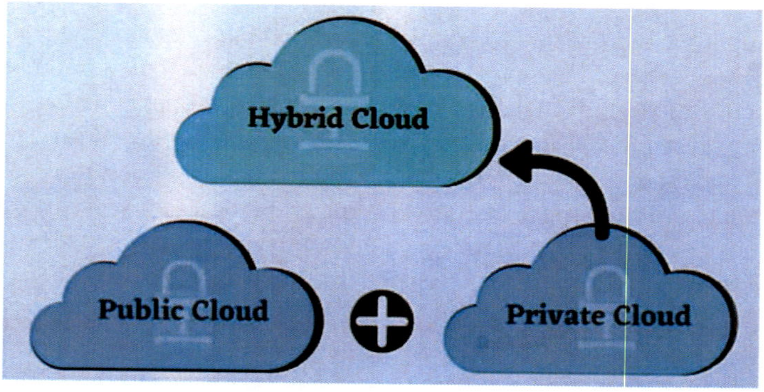

Fig. 2 Types of cloud

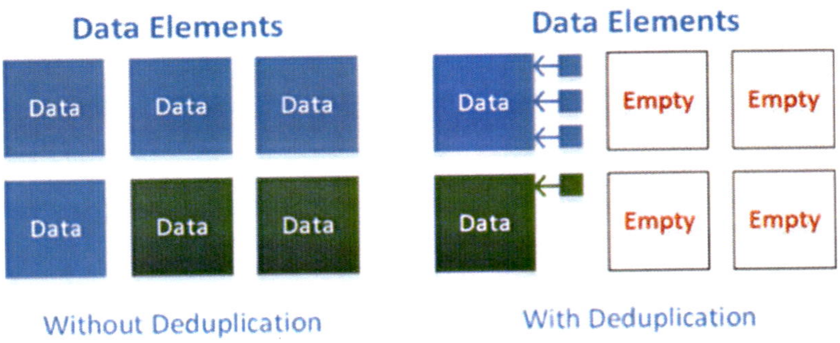

Fig. 3 Storage space with and without deduplication

video, because of this deduplication process, more than 50% of memory is prevented from the duplicate copies.

For implementing the image deduplication [4], initially, each individual image should be uniquely identified. For this unique identification, hash values are used. This is called hash-based deduplication. In hash-based deduplication method, hash value will be generated for each and every digital data. When a new image is to be uploaded in the cloud storage space, this hash value will be used to uniquely identify each image and check whether the incoming data is a duplicate copy or not. If it already exists in the storage space, then the new image will not be uploaded again in the storage space. There are many cryptographic algorithms that exist to calculate the hash value for digital data including text, image, audio and video. The proposed research work will analyze the performance of existing cryptographic hashing algorithms to achieve better image identification with respect to collision attack which is suitable for image identification in deduplication process.

Hashing is the technique which is used in the field of cryptography to uniquely identify the data. It is widely used for authentication and integrity check. There are many algorithms existing for hashing. The most widely used algorithms are family of MD5 and SHA algorithm. Piyush gupta [5] performed a comparative analysis of these algorithms based on their time and space complexity. He stated the differences and similarities in between MD5 and SHA512 with their pros and cons. He analyzed the operational steps of those algorithms in depth. He also analyzed its performance by various parameters like length of the key, speed, possibility of cryptanalysis and proved that the performance of SHA512 is better than MD5 in all aspects with an example.

To increase the reliability of hashing algorithm, NIST initiated a new algorithm called SHA3 which uses the concept Keccak algorithm. It is well suited for digital signatures and message authentication code. SHA3 is an advanced version of SHA2. There are many ways by which this SHA3 can be implemented. Nithin et al. [6] tried to implement this SHA3 in two ways. By step–by-step algorithm with 24 rounds and by using MUX to implement the same by reducing the area of operation, they proved that step-by-step algorithm has better performance since it has only less loops and has all direct assignments. It has less delay and also takes less time to generate the hash when compared to MUX implementation.

Kunhu et al. [7] proposed a blind watermarking technique in which they tried to prevent the content of the image as well as the authenticity of the image. They proposed robust watermarking to prevent the ownership protection and fragile to prove the authentication. They tested their technique in X-Rays and MRI scan report and found it sounds good.

SHA256 and SHA512 hashing algorithms play a major role in data integrity assurance. They are implemented in various ways. Ahmad et al. [8] proposed an alternative architecture for implementing this SHA256 and SHA512 algorithm using FPGA. To reduce the chip area, modulo-64 addition as modulo-32, modulo-16 and modulo-8 additions was used. And this is common architecture for implementing both SHA256 and SHA512.

Perceptual image [9] hashing plays a vital role in various image processing operations like image authentication, detecting image forgery, image deduplication and retrieval of image. Wang et al. [10] proposed an idea in which perceptual image hashing method using image alignment and hash-based image forgery detection and tampering localization method. They provide a framework through which they protect the image from various geometric distortions. The hash value is calculated from perceptual and DCT-based statistical features of the image. Throughput of the network which comprises of aggressive nodes is improved by the technique and mentioned by [11].

2 Image Identification

For the proposed experiment, different images are considered with varying sizes. This research work has employed various cryptographic hashing techniques like MD5, SHA2-512 and SHA3-512 [12] to calculate the hash values for those images. The online hash calculator is used to compute the hash values [13].

2.1 *Images*

From the Table 1, it is observed that the hash value is generated for each digital image as shown in the Fig. 4 by using different hashing algorithms. When a new image is to be uploaded in the cloud storage space, then its hash value will be calculated and checked against the hash values of all stored image in the storage space for its duplication. It will reduce the time of searching. If the user search the image for itself in the existing pool of memory, then it may take huge time to hit the image in the pool. Since the hash value is used for comparison, it reduces much of the searching time.

Table 1 Hash value of Digital images shown in Fig. 4 calculated using MD5, SHA-512 and SHA3-512

Images	MD5 hash value	SHA2-512 value	SHA3-512 value
Image1	ffd647603ce634f3 0f6c89fae96d2451	6f1f4efc45e594d48d736f7be1ad 5427993b084ca1971f164313eda 7ea7e8f678ad002cd53e9ed4725 30a5a6483d5ef9c3d26d1afdb58 a2971b4d5f2b905db00	248b2ef33534f447b9d06 ded1b3032f320220ab382 ecd8619327664e5f40991 47ac7aced1835fbc64284 bf4b9c4789fafe1bcacbc0 28ebcd71a5f69a872065a9
Image2	005e0cda0b6bc78e c6d707f5c8aea514	2e2422c6fc912fc538340bc0 3882c1d6d40e99a29c74266 3be7d4dd723c6d560e19f88 332ff2d6ceeb0c100ffcff94d4 ddeaf85574f1d41b75de258d 48b380ef	246225b6e7bfb1c8e32e9b 8534da5a619ef8187fee4e 56d781c734cbde8fa7bc23 039616a034ffb9b0e94742 848c6690c1f2d7e6ed8cf1 690508ba6a8817dbcd
Image3	1d4bd280f3385733 064aa12f5cb43950	8f38b46210a45421ba920d215 815b0c19926be85f7ea882a0d b622af1fd65f276ae971482433 9f777435f1386726510a32b620 ed64e02fd331ae456d6012b26a	cf0029e1f903c0eca92e6b cdd9f71e46af61e636030e 90a55ac1abd60c52bbf32 cd1a156053484b9faf9c66 0e4691b10e141f790f6c54 1063cbed5e8c067c53c
Image4	461ca491a7e01a6a 130c951265a52437	346dadb86e46799d68742606f 02c0f593de8603ca64f3ccfccf cb1423ad56f021d58faac3df48 058df248b3710e8997d126c65 4e2bcbd0dcd3f57624bebf4206	5c2504d6c2c7b9c2ce838 609ef4880cc511641c6258 0d8890e561d8238d71fb18 2cb919cf539ec5aaa267ef 19c49fac491eb931aedf88 87beac26132db0ccb65

Fig. 4 Images

3 Performance Analysis of various Cryptographic Hashing Algorithms

The term image identification is nothing but unique identification of image among countless images in the storage space. The commonly used method for identifying the image is "Hashing." The important property of hashing is that it should be computationally infeasible, which means that it will not be able to find two different data resultants with same hash value. This is the key property to identify the image uniquely. The cryptographic hash functions are very sensitive when compared to other image hashing algorithms like difference hash, average hash, wavelet hash and block hash algorithms. They are sensitive in the sense meant that if a single pixel in the image get modified, there will be a huge difference in the hash value. This is called "Avalanche effect." Since this dealing with deduplication, this property is very much needed. Many cryptographic hashing algorithms are currently existing. They are MD4, MD5, SHA0, SHA1, SHA224, SHA256, SHA384, SHA512, SHA3-224, SHA3-256, SHA3-384, SHA3-512 and few more.

3.1 Avalanche Effect

To prove the avalanche effect, visually similar images as shown in the Fig. 5 are considered and calculated the hash value for those two images using various hashing algorithms.

From the Table 2, it can be clearly understand that even for images, which are visibly same or small modification in original image leads to a large variation in the resultant hash value. This is called "Avalanche effect," and it is very much required to prove the image identity.

In 1991, Ronald Rivest designed the Message Digest Algorithm MD5 an advanced version of MD4 since it has many drawbacks. It was a widely used hash algorithm. It takes the input block size as 512 bits and produces 128 bit hash value. But it has a hash collision weakness and the attacker proved that by producing many inputs resulted with same hash. Then, NIST published a new hashing standard called SHA family which includes SHA0, SHA1, SHA224, SHA256, SHA384 and SHA512. SHA uses a

Fig. 5 Visually similar images

Table 2 Hash value of visually similar images shown in Fig. 5 calculated using various cryptographic hashing algorithms

S. No.	Hashing algorithm	Image1	Image2
1	MD5	e7d7d4a53487eb847c9 218e38872e59a	6adabb8f9dc3cbe 153ba4c545ffff555
2	SHA1	56492da763b5cdea2154 f2c0c185d461b2f13bce	290e5406b59165f45c6 7ba4068f027eb80114677
3	SHA2-224	16db3e651b38b342af24d 7e551f3dafa7387e7e6c6 a2a653291d814c	86ffc536be5479c4e908 ac429e75909576cc3 0c0e9fe5f9cfa6f5c6d
4	SHA2-256	938c7a1d06b195ca334c 63ca68bb2e67b2c78cea 03942a39764c33bce1bbed3d	c75e9d1b78261ba93458b8 8850fd59eb19cab756e 52d3e0cbd0af0befe8b6820
5	SHA2-384	0d71db6442ca0452c0e1b 9a6b612fc4a4e451a8bd5b 033902f6fe2cf5655daa0a6 00e7f2a5e5645355d33bd0 e23edada	5511ee5f6d46fa67107523 65a0b67ad13f4b54305b438c 0d245b2c9c9e838d456793 cc732743d2a8c78710b944 68804d
6	SHA2-512	de98fe7cc7ae489e12a8 dfb34f6256f66db34628e 7604927e25970aac5e1 d81bd8b9f2255c89b272 4f63285ad26cd542bb6e 2f1ba920df73a1fab16ec35b9b9b	66dbbacb731e33f9db2bf a3daa1311d7206268c46e2a 1f853ffec951d8e5ed05d 15d5be42dcf7a8095a3728c 816117c9578bc5a27f96 c96b60c52fbcb42f6b4e
7	SHA3-224	2cceed42ef2c8447868 dc14b51fdeed64da70f 479e843d03ce5d89a8	38119899a3e4839bd91e 87e7ea598e89e417b7e 838c24c8d79a3cfef
8	SHA3-256	433b90f2addcbceb5c4d 140069675dca85004bb 9485672a38058a41d3e b749f0	32e99d7d3efffaba3db133 37e5ac4dd2d6e17fb9e9a b7f906fb68962315bc294
9	SHA3-384	d8c31344526e083fc853f7e 8cd6685b4c738ccafb5d4bf 5c97952b31491b6d761843 46020ffb4da441a76c798cf4c405	5684fbc03d684904f0e 3a1a9983902f3b2faf56 b8bf7e8c7b43391d19e f1ed9462192f0bda851 af6aabfd7dd945bc2aa
10	SHA3-512	cd56b2d6fc4e7b5019beb37 93f608e8a021fb5fb3797ef9d 7f001ec80e934bc15979fb168 ad7bf78002face2bdba82d1e64c f3854e9bef15612e9879e522a8b9	c81c9023f439f4111ea 307e3ff6217ed1e3789c6 d09aeda6eafd313a8d3b8 cfd6a661bd51be94d09c5 9d9736ef30937da0b72dd 01b6c5b0212ab266333d642e0

key component called one-way function. It provides the three important properties of cryptographic security. They are pre-image resistance, second pre-image resistance and collision resistance.

4 Results and Discussion

Table 3 shows the comparison of various cryptographic hashing algorithms with respect to their input block size, output block size, number of rounds, internal operations and their resistance against collision attack. Table 4 and Fig. 6 shows the collistion resistance capacity of various cryptographic algorithms.

From the above analysis, SHA2-512 and SHA3-512 are equally good in performance. To have better image identification, SHA2-512 or SHA3-512 hash algorithm is utilized over digital images to calculate the hash in order to prove its identity, while it is stored in the public cloud storage space.

Table 3 Comparison of various cryptographic hash algorithms

Algorithm	Input block size (in bits)	Output size (in bits)	Number of rounds	Internal operations	Security against collision attack (in bits)
MD5	512	128	64	AND, XOR, ROT, ADD (mod 232), OR	Approximately less than or equal to 18
SHA0	512	128	80	AND, XOR, ROT, ADD (mod 232), OR	Approximately less than or equal to 34
SHA1	512	128	80	AND, XOR, ROT, ADD (mod 232), OR	Approximately less than or equal to 63
SHA2-224	512	224	64	AND, XOR, ROT, ADD (mod 232), OR, SHR	Approximately less than or equal to 112
SHA2-256	512	256	64	AND, XOR, ROT, ADD (mod 232), OR, SHR	Approximately less than or equal to 128

(continued)

Table 3 (continued)

Algorithm	Input block size (in bits)	Output size (in bits)	Number of rounds	Internal operations	Security against collision attack (in bits)
SHA2-384	1024	384	80	AND, XOR, ROT, ADD (mod 264), OR, SHR	Approximately less than or equal to 192
SHA2-512	1024	512	80	AND, XOR, ROT, ADD (mod 264), OR, SHR	Approximately less than or equal to 256
SHA3-224	1152	224	24 [14]	AND, XOR, ROT, NOT	Approximately less than or equal to 112
SHA3-256	1088	256	24 [14]	AND, XOR, ROT, NOT	Approximately less than or equal to 128
SHA3-384	832	384	24 [14]	AND, XOR, ROT, NOT	Approximately less than or equal to 192
SHA3-512	576	512	24 [14]	AND, XOR, ROT, NOT	Approximately less than or equal to 256

Table 4 Collision resistance of various cryptographic algorithms

Hash algorithm	Security against collision attack (in bits) (approximate)
MD5	18
SHA0	34
SHA1	63
SHA2-224	112
SHA2-256	128
SHA2-384	192
SHA2-512	256
SHA3-224	112
SHA3-256	128
SHA3-384	192
SHA3-512	256

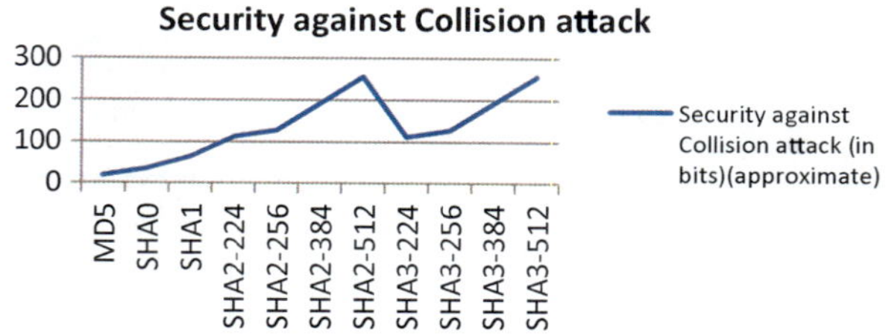

Fig. 6 Security against collision attack

5 Conclusion

The deduplication process plays a vital role in the better utilization of cloud storage space. For implementing deduplication, the images should be uniquely identified, and hashing technique is used to identify the images in the cloud storage space. The proposed research work has proved the avalanche effect of various cryptographic hashing techniques with the help of the experimental results obtained. The proposed research work has analyzed the operational details and performance of various cryptographic algorithms, and it is found that SHA512 is delivering a better performance over other hashing techniques, which are considered in the proposed analysis.

References

1. T. KamalaKannan, K. Sharmila, M.C. Shanthi, M.R. Devi, *Study on Cloud Storage and its Issues in Cloud Computing*
2. R. Arokia Paul Rajan, S. Shanmugapriyaa, Evolution of cloud storage as cloud computing infrastructure service. IOSR J. Comput. Eng. (IOSRJCE), 38–45 (2012)
3. A. Bhalerao, A. Pawar, A survey: on data deduplication for efficiently utilizing cloud storage for big data backups, in *2017 International Conference on Trends in Electronics and Informatics (ICEI)* (IEEE, 2017), pp. 933–938
4. M. Chen, Y. Wang, X. Zou, S. Wang, G. Wu, A duplicate image deduplication approach via Haar wavelet technology, vol. 2, in *2012 IEEE 2nd International Conference on Cloud Computing and Intelligence Systems* (IEEE, 2012), pp. 624–628
5. P. Gupta, S. Kumar, A comparative analysis of SHA and MD5 a comparative analysis of SHA and MD5. IJCSIT Int. J. Comput. Sci. Inf. Technol. **5**(3) (2014)
6. N.R. Chandrana, E.M. Manuelb, Performance analysis of modified SHA-3. Int. Conf. Emerg. Trends Eng. Sci. Technol. Proc. Technol. (2015)
7. A. Kunhu, H. Al-Ahmad, F. Taher, Medical images protection and authentication using hybrid DWT-DCT and SHA256-MD5 hash functions, in *2017 24th IEEE International Conference on Electronics, Circuits and Systems (ICECS)* (IEEE, 2017), pp. 397–400
8. I. Ahmad, A.S. Das, Hardware implementation analysis of SHA-256 and SHA-512 algorithms on FPGAs. Comput. Electr. Eng. **31**(6), 345–360 (2005)
9. L. Du, A.T.S. Ho, R. Cong, Perceptual hashing for *image authentication*: a survey. Signal Proc. Image Commun.
10. X. Wang, X. Zhou, Q. Zhang, B. Xu, J. Xue, Image alignment based perceptual image hash for content authentication. Signal Process. Image Commun. **80**, 115642 (2020)

11. G. Harikrishnan, A. Rajaram, Improved throughput based recognition connection denies for aggressive node in wireless sensor network. J. Comput. Theor. Nanosci. **14**(12), 5748–5755 (2017)
12. S.B. Sarmadi, M.M. Kermani, A. Aziz, Efficient and concurrent reliable realization of the secure cryptographic SHA-3 algorithm. IEEE Trans. Comput. Aided Des. Integr. Circ. Syst.
13. https://emn178.github.io/online-tools/sha512.html
14. The Keccak sponge function family. Retrieved 2016-01-27

A Comprehensive Analysis on Question Classification Using Machine Learning and Deep Learning Techniques

S. V. Kogilavani, S. Malliga, A. Preethi, L. Nandhini, and S. R. Praveen

Abstract The competence of any online Web site depends on the type of experience it gives to its users, which depends largely on the content they put up on their Web site. Hence, the content which is being put online should be really taken care. There are many Web sites that provide content to their user in terms of questions and answers, for example, the online Web site Quora, which has large scale of data in terms of questions and answers of users. Users are the one who put up questions and also provide answers to those questions. In this paper, a system is proposed by considering significant amount of data from Kaggle, where it is utilized for different approaches to predict the insincere questions. The goal of the proposed work is to develop a model that considers question as an input in English language and produces output as either 0 or 1. In that, 0 represents sincere type of questions and 1 represents insincere type of questions. Hence, the data are transformed into result including $F1$ score using preprocessing, TF-IDF and pre-trained word embedding. For each method, preprocessing tasks like tokenization, case normalization and punctuation removal are used. Each individual word after preprocessing is used as word vector in this work. These word vectors are then used for the insincere and sincere question classification which is given as input to the machine learning models like SVM, Naïve Bayes, logistic regression and deep learning model called RNN with word embedding. It is used to detect the toxic content in the reviews as 1 for insincere and 0 for sincere questions. The accuracy of these different methods is critically examined with the help of LSTM, word embedding and model ensembling.

Keywords Insincere · Sincere · SVM · Naïve Bayes · Logistic regression · RNN · Word embedding · Classification

S. V. Kogilavani (✉) · S. Malliga · A. Preethi · L. Nandhini · S. R. Praveen
Department of CSE, Kongu Engineering College, Perundurai, India

S. Shakya et al. (eds.), *Mobile Computing and Sustainable Informatics*, Lecture Notes on Data Engineering and Communications Technologies 68,
https://doi.org/10.1007/978-981-16-1866-6_63

1 Introduction

Machine learning is a subfield of artificial intelligence that made the system to automatically improve from the data history called past experience. Machine learning mainly concentrates on the creation of the computer programs that can obtain data from past experience, and it learns by itself. The main job is to build algorithms that can get input data and use various techniques to predict an output, and at the same time this output acts as new data for further processing. The objective of this research work is to classify given question as sincere or insincere category. A question intended not to seek helpful answers but to make a statement is termed as an insincere question. Mediratta et al., 2019, described that insincere questions are not meant to ask for useful answers on a topic of interest, and they can in fact be rhetorical or statements [1].

Given this peculiarity, people can distinguish them from usual and sincere questions by looking at some factors: They sometimes have sexual content to be more provocative and generate a wider response (whether disdain or shock) from the community of users. Basically, any question leads to create anger or offend and those questions are not used for information gathering purposes. The tone, which rather than being non-neutral, is exaggerated to draw attention to a point or a position. Their ultimate intent is to inflame, suggesting a discriminatory idea, seeking confirmation of a stereotype, insulting a person or a group, and often basing upon characteristics that are not measurable. They are generally not based on evidence, nor have a solid connection with reality. Various parameters like question tone, jargon usage, slang usage, abbreviations, phrasing and exclamations made the process of identifying a questioner's intentions as very difficult.

Machine learning and deep learning involve some common tasks such as feature selection and text classification. Machine learning and deep learning models become very easier by utilizing the available techniques, and it eases the job of researchers. These techniques also reduce the effect of the curse of dimensionality and overfitting problem by applying generalization.

2 Literature Review

The most important part of text classification is preprocessing. It is important to use appropriate methods for accurate results. Word2vec was used along with TF-IDF for automating text classification. Liu et al. [2] used Word2vec for vectorizing and TF-IDF for calculating the weight of the feature words [2]. The approach used multiplied the weight of the word to the word vector. Further, word vectors are grouped to give accurate results.

Aborisade et al. [3] used TF-IDF and bag of words for vectorization [3]. This was used by an author to classify tweets in the users on the basis of given corpora. The author trained models of logistic regression and Naive Bayes after preprocessing and extracting features using vectorization. The total number of tweets was 46,895 which gave an accuracy of 91% for logistic regression and 89.9% for Naive Bayes. The implemented model has a very small dataset, and hence a proper training is not provided for the models. The proposed project will implement models on a huge dataset, thus enabling more training.

Indra et al. [4] proposed a model for classifying tweets [4]. The author had implemented a logistic regression model of machine learning for classifying tweets according to the topic they belong. The system transformed the tweets into vector which is acceptable by the model. The confusion matrix showed an accuracy of around 92%. This system uses just one algorithm which does not give any evidence that this model has performed the best. The proposed system will implement 3 models which will enable to analyze which is performing better.

Haryanto et al. [5] used support vector machine model to classify BBC documents into five categories [5]. The model was enhanced by using chi-squared along with SVM. Both stemming (lancaster stemmer) and lemmatization (WordNet Lemmatizer) were used and fed separately to the model. The results stated that stemming provided better results and chi-square was an added advantage as it improved the model. The proposed model will also use both stemming and lemmatizer for different models for a comparison of the results.

Gurcan [6] implemented SVM, kNN, decision trees, multinomial Naive Bayes and Bernoulli Naive Bayes for classifying Turkish News Text [6]. The text was classified into five predefined classes and trained with different systems. The model which outperformed others was multinomial Naive Bayes with an accuracy of 90%.

Akshaya et al. [7] implemented several models are able to classify between sincere and insincere questions [7]. It used multiple models which follow supervised learning algorithm. In [7], the authors compared various machine learning models like Naive Bayes, Logistic Regression, Decision Tree, Random Forest all of which utilizes different preprocessing methods gives highest accuracy but their $F1$-score is less. An ensemble model is also implemented which combines the advantages of both Naive Bayes and logistic regression model. Using various preprocessing methods, model accuracy may be changed. According to various metrics, it can be stated that decision tree performs the best followed by random forest. Other models also perform considerably well, and there is no marginal difference between the metrics of different models.

Georgakopoulos et al. [8] and Mujadia et al. [9] used different machine learning and neural network-based models [8, 9]. The best performing model was an ensemble model of gradient boosting, random forest and 3-nearest neighbor classifiers with majority voting. Accuracy of the model is 62.37%.

Aldwairi et al. [10] and Chiroma et al. [11] determined the performance of machine classifiers in identifying suicide-related text from social media networks like Twitter (tweets) [10, 11]. They used four popular machine learning classifiers such as decision tree, Naive Bayes, random forest and support vector machine—measure of decision tree classifier ranging from 0.346 to 0.778 for suicide-related communication gives best results compared to other classifiers. Authenticating the behavior of the nodes in the mobile ad hoc network is proposed in [12].

3 Proposed System

3.1 Dataset

The dataset used in this project has been provided by Quora for the online competition on Kaggle. In order to classify the questions as sincere or insincere, this work utilizes training data as 1.3 million questions and features like question id, question text and question type as insincere or sincere with the value of 1 or 0.

There are more than one million rows of training data. The big dataset is expected to be quite challenging. Memory errors and excessive processing times may be challenging. There are several methods to attempt to counter the big size of the dataset, including using larger information samples to train and reduce dimensionality. The selection of features will be essential to optimize the model.

3.2 Preprocessing

Data preprocessing is nothing but the readying of data for experimentation—transforming raw data for further processing. For better improvement of the scores, text data cleaning like data preprocessing needs to be done using Html tag removal, punctuation removal process in which unnecessary punctuation and tags are removed, tokenization which converts sentences to words and lemmatization in which inflection is removed by identifying part of speech tag and contraction mapping. But for this database, we do not need to use standard preprocessing steps. Instead, pre-trained embedding can be used as preprocessing.

3.3 Pre-trained Embedding

The pre-trained word embedding, which uses billions of phrases, gives us the additional advantage over training our own embedding. These pre-trained models are the pretty easiest way to get started with embedding words. According to this, model is a collection of words embedded elsewhere which then merely load into our computer and memory. Pre-trained models are also available in other languages also. The benefit is that these models can exploit enormous datasets that were not possible to utilize, constructed using separate words, and meaning of these words can be identified in a statistical way. Examples of training datasets include the entire Wikipedia text corpora, the Web crawl database and Google News information. This method of pre-trained word embedding reduces the need for time (intensively) to acquire, clean and process such large sets of information.

3.4 Word Embeddings

Word embeddings represent words in a distributed vector format with the help of contextual information available using neural network architectures. It is a text depiction where words with the same significance are represented in a comparable way. In other words, in a coordinate system, it reflects words in which associated words are positioned nearer together, based on a corpus of interactions. In general, embedding represents geometric word encoding depending on how often a text corpus appears together. Various implementations of word embeddings described below differ in the way as how they are constructed. Word embedding is one of the document vocabularies' most common depictions. It captures a word's background in a text, similarity between semantic and syntactic representation, and relationship between one word and other word. They are vector representations of a particular word. The three-word embeddings are Word2vec, GloVe and fastText.

3.5 Word to Vector

In most of the situations, similar words occur more frequently with the same context. The functionality of Word2vec is to represent each word as a vector. The idea is if we train the model based on context, then similar words will have identical representation. In normal feed-forward densely connected neural network (NN), there is a set of independent variables and a target-dependent variable which is needed to predict. Like that first split the sentences into tokens and then develop pairs of tokens. So, one of the combinations could be a pair of words such as "cat" and "purr," where cat is the independent variable known as X and "purr" is the target-dependent variable known as Y that we are aiming to predict. We feed the "cat" into the NN through an embedding layer initialized with random weights and pass it through the softmax layer with ultimate aim of predicting "purr." The optimization method such as SGD minimizes the loss function "(target word|context words)" which seeks to minimize the loss in the process of predicting the target words provided the context words. If we do this with enough epochs, the weights in the embedding layer would eventually represent the vocabulary of word vectors, which is the "coordinates" of the words in this geometric vector space. The above example assumes the skip-gram model. For the continuous bag of words (CBOW), we would basically be predicting a word given in the context.

3.6 GloVe

GloVe is nearly like Word2vec, but it is a predictive model which is used to predict a particular context of words. GloVe learns by developing a co-occurrence matrix that basically created when a word is represented in a particular context. Because it will be a giant matrix, to achieve a diminished representation of dimensions, this matrix is factorized. GloVe has a lot of news, but that is the hard concept.

3.7 fastText

fastText is very distinct from the embedding of the above two. While each word is treated by Word2vec and GloVe as the lowest unit to train, fastText utilizes n-gram characters as the lowest unit. For instance, the word vector "apple" could be split into distinct units of word vectors as "ap," "app," "ple." The greatest advantage of using fastText is that it generates better word embedding for unusual words or even words that are not seen during practice as the character vectors of n-gram are shared with other words. This cannot be achieved by Word2vec and GloVe.

3.8 TF-IDF

The TF-IDF weight is a weighting scheme that is normally used in information retrieval and text mining tasks. It is a statistical measure used to identify the importance of word in a document collection. If the number of times a word occurs in the document, then that particular word's importance is increased.

3.9 Naïve Bayes Classifier

A Naive Bayes classifier is a machine learning algorithm that is based on Bayes theorem with the assumption that features are independent of each other. Lie et al. [13] denoted that Naïve Bayes lies on the assumption that the input features are independent of each other; i.e., it is not possible to know anything about other features even though additional variables are available [13].

3.10 Long Short-Term Memory

Long short-term memory networks known as "LSTM" are a type of recurrent neural network (RNN) which is used to learn long-term dependencies. Ranasinghe et al. [14] specified about deep learning models in general recurrent neural network which is a series of repeating modules of neural network [14]. RNN may consist of single which is monitored by gates that is used for information flow. It consists of sigmoid neural network layer and a point-wise multiplication operator.

4 Performance Evaluation

The dataset used 500 samples in csv format with the different questions in a three-column format. For machine learning models, text format is not compatible and has to be converted in a suitable format.

4.1 *Accuracy*

Accuracy helps to measure the closeness of the output to a certain value. The ratio of correct predictions over the total predictions is

$$\text{Accuracy} = \frac{\text{TP} + \text{TN}}{\text{TP} + \text{TN} + \text{FP} + \text{FN}}$$

It does not work well with imbalanced data. Hence, in this project other metrics are used for evaluation.

4.2 *Precision*

Precision is defined as the number of true positives divided by all positive predictions. It is a measure of a classifier's exactness. It tells us how repeatedly the classifier is correct when it forecasts positive. Low precision denotes that there are a high number of false positives.

$$\text{precision} = \frac{\text{TP}}{\text{TP} + \text{FP}}$$

4.3 *Recall*

It is the ratio of correctly predicted outcomes to the total outcomes. It is also known as the sensitivity of the model.

$$\text{recall} = \frac{\text{TP}}{\text{TP} + \text{FN}}$$

Table 1 Confusion matrix

	Predicted YES	Predicted NO
Actual YES	True positive	False negative
Actual NO	False positive	True negative

4.4 F1 Score

$F1$ score is the one that is calculated by combining the precision and recall measures. It is the harmonic mean of the two. It results nearly the same as the average of the two measures when they are closely related.

$$F1\ \text{score} = 2 * \frac{\text{precision} * \text{recall}}{\text{precision} + \text{recall}}$$

4.5 Confusion Matrix

Confusion matrix is a table that is used to describe the performance of a classifier on the test data for which the true values are known. Confusion matrix is represented in Table 1. Figures 1, 2, 3, 4, 5 and 6 represent the result of applying various machine learning and deep learning algorithms without preprocessing and with preprocessing. The result shows that with preprocessing all the algorithms produce better precision, recall, f-measure, support and accuracy. Figure 7 represents the accuracy of word embedding.

5 Conclusion

This paper focuses on the classification of questions as insincere or sincere using various machine learning and deep learning methods. After attempting different models and parameters' adjustment combinations, it was discovered that using word embeddings in the information processing era and using LSTM to suit the information produced the highest output. As the dataset was pretty huge, several iterations were required to run the algorithm. Embeddings turn out to be as a good boost for the dataset, as it performed better. For future work, add attention to the LSTM and suggest a long short-term memory network based on attention for classification of aspects-level sentiments. Convolutional neural network can be used as a recurring structure to capture context information as far as possible, resulting in significantly less noise compared to traditional neural networks based on windows.

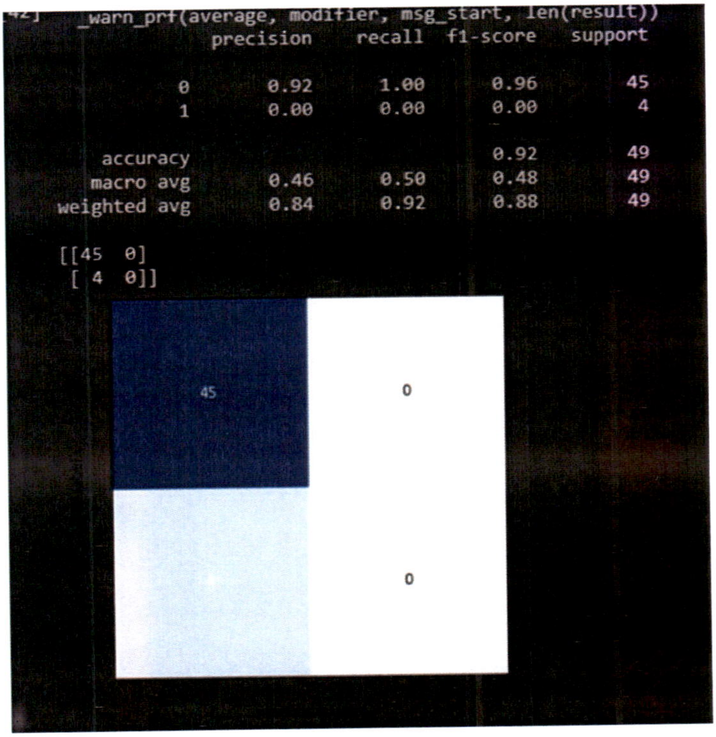

Fig. 1 Naïve Bayes before preprocessing

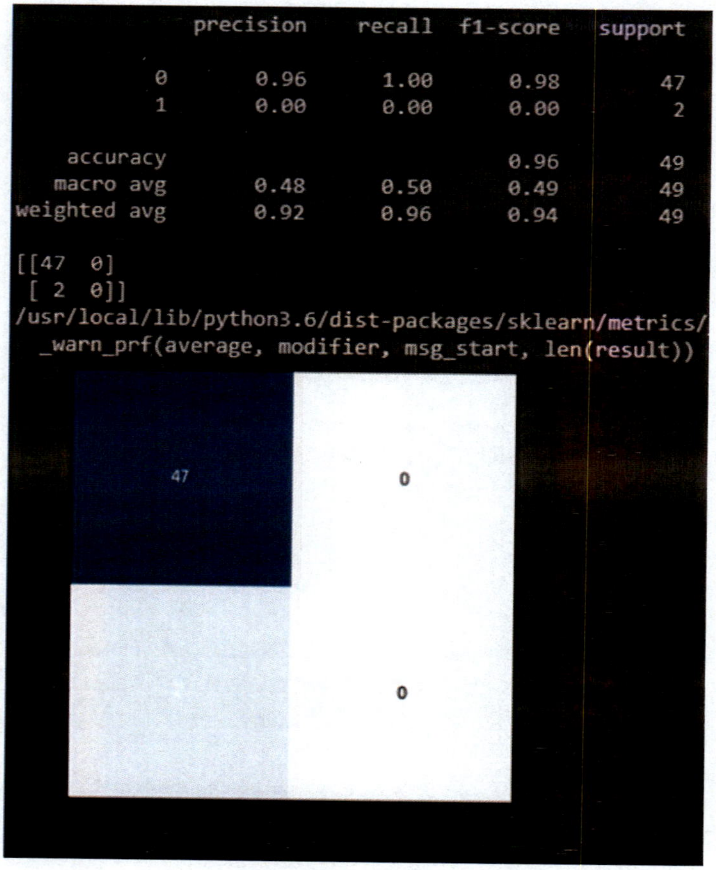

Fig. 2 Naïve Bayes after preprocessing

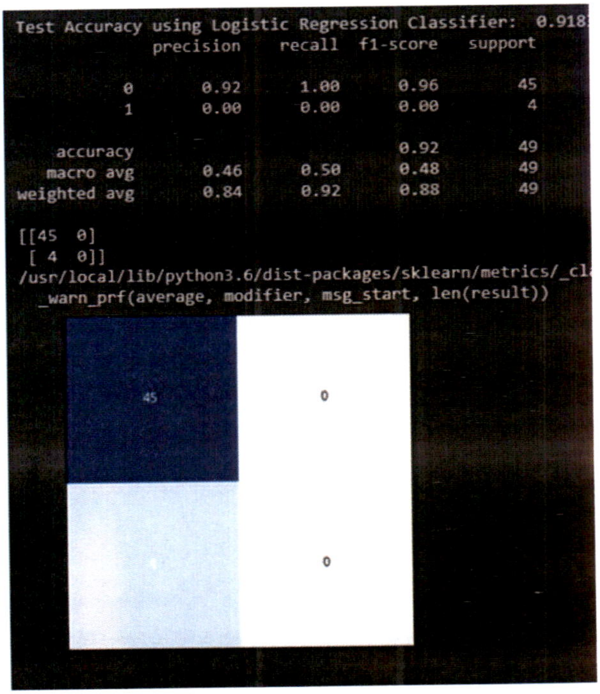

Fig. 3 Logistic regression before preprocessing

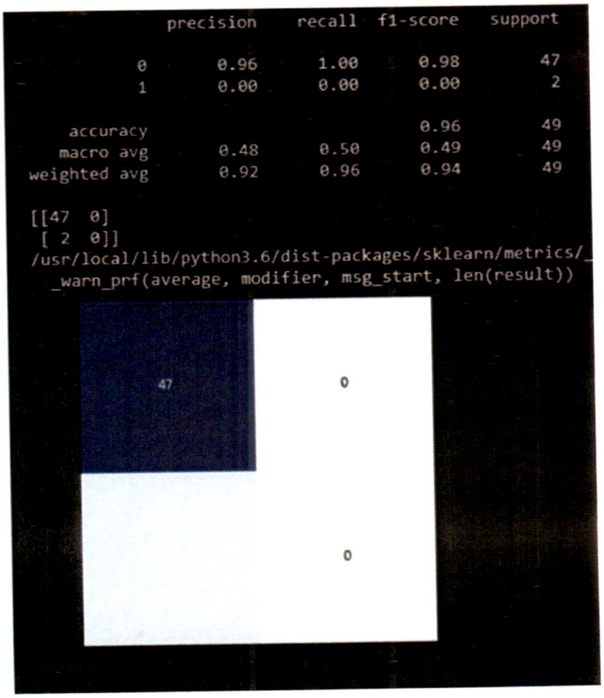

Fig. 4 Logistic regression after preprocessing

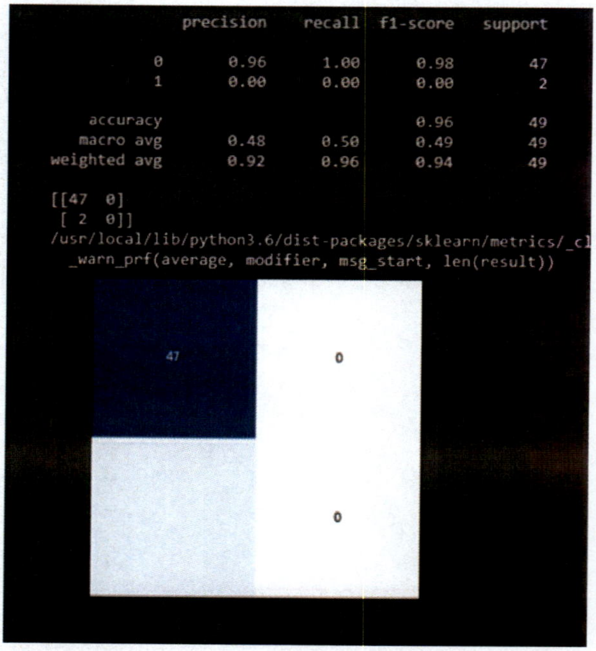

Fig. 5 SVM before preprocessing

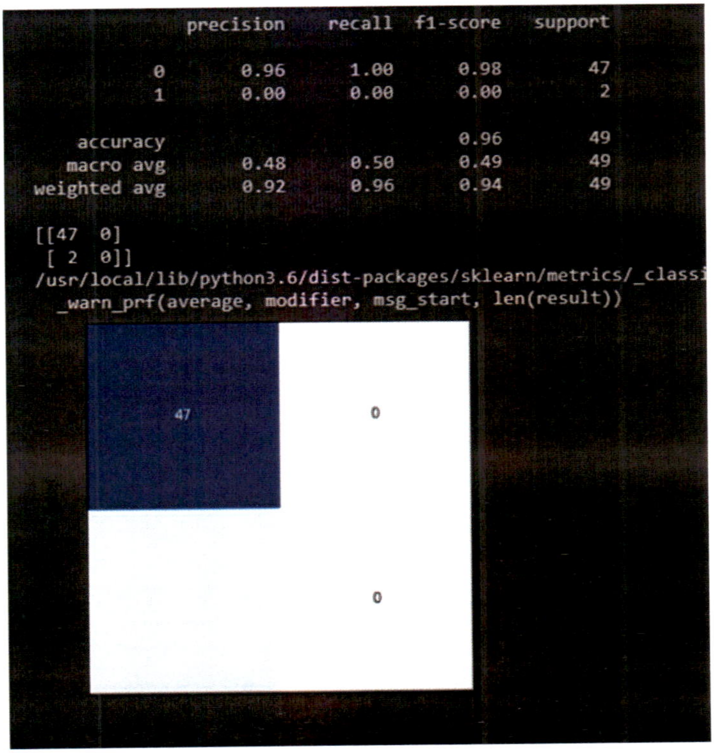

Fig. 6 SVM after preprocessing

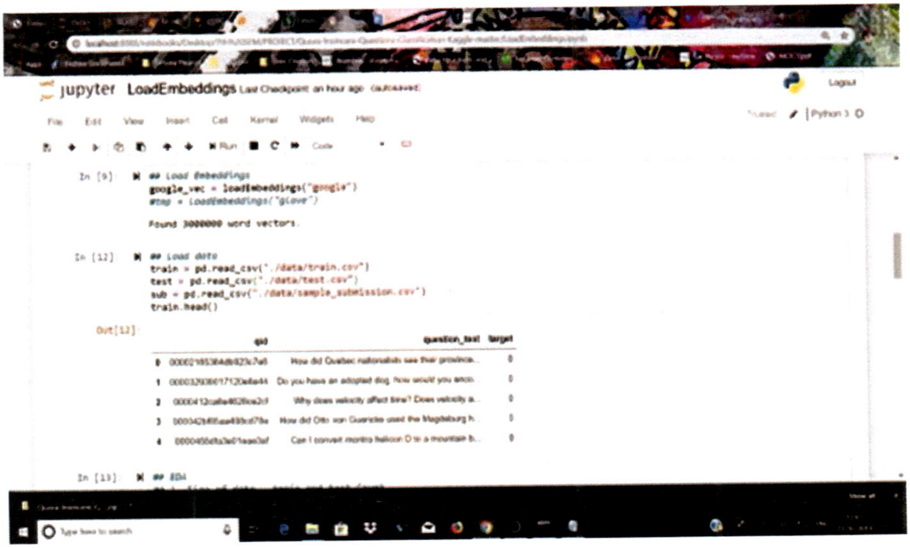

Fig. 7 Accuracy of word embedding

References

1. D. Mediratta, N. Oswal, *Detect Toxic Content to Improve Online Conversations*. arXiv:1911.01217 (2019)
2. C. Liu, Y. Sheng, Z. Wei, Y. Yang, Research of text classification based on improved tf-idf algorithm, in *2018 IEEE International Conference of Intelligent Robotic and Control Engineering (IRCE)*, Aug 2018
3. O. Aborisade, M. Anwar, Classification for authorship of tweets by comparing logistic regression and Naive Bayes classifiers, in *2018 IEEE International Conference on Information Reuse and Integration (IRI)*, July 2018
4. S.T. Indra, L. Wikarsa, R. Turang, Using logistic regression method to classify tweets into the selected topics, in *2016 International Conference on Advanced Computer Science and Information Systems (ICACSIS)*, Oct 2016
5. W. Haryanto, E.K. Mawardi, Influence of word normalization and chi-squared feature selection on support vector machine (svm) text classification, *in 2018 International Seminar on Application for Technology of Information and Communication*, Sept 2018
6. F. Gurcan, Multi-class classification of turkish texts with machine learning algorithms, in *2018 2nd International Symposium on Multidisciplinary Studies and Innovative Technologies (ISMSIT)*, Oct 2018
7. A. Ranganathan, H. Ananthakrishnan, D. Thenmozhi, C. Aravindan, Classification of insincere questions using SGD optimization and SVM classifiers. CEUR Proc. **2517** (2019)
8. S.V. Georgakopoulos, S.K. Tasoulis, A.G. Vrahatis, V.P. Plagianakos, Convolutional neural networks for toxic comment classification, in *Proceedings of the 10th Hellenic Conference on Artificial Intelligence*, 2019, pp. 1–6
9. V. Mujadia, P. Mishra, D.M. Sharma, Classification of insincere questions with ML and neural approaches, in *FIRE (Working Notes)*, 2019
10. M. Aldwairi, A. Alwahedi, Detecting fake news in social media networks. Proc. Comput. Sci. **141**, 215–222 (2018)
11. F. Chroma, H. Liu, M. Cocea, Text classification for suicide related tweets. Int. Conf. Mach. Learn. Cybern. (ICMLC) **2**, 587–592 (2018)
12. R. Ayyasamy, P. Subramani, An enhanced distributed certificate authority scheme for authentication in mobile ad-hoc networks. Int. Arab J. Inf. Technol. **9**(3), 291–298 (2012)
13. P. Liu, H. Yu, T. Xu, C. Lan, Research on archives text classification based on Naive Bayes, in *2017 IEEE 2nd Information Technology, Networking, Electronic and Automation Control Conference (ITNEC)*, 2017
14. T. Ranasinghe, M. Zampieri, H. Hettiarachchi, BRUMS at HASOC 2019: deep learning models for multilingual hate speech and offensive language identification, in *FIRE (Working Notes)*, 2019, pp. 199–207

SenticNet-Based Feature Weighting Scheme for Sentiment Classification

K. S. Kalaivani, M. Rakshana, K. Mounika, and D. Sindhu

Abstract Sentiment analysis aims to study the peoples' perceptions, sentiments and emotions in comments, blogs, review and so on. The performance of classifier can be improved by the use of efficient feature weighting schemes. The existing feature weighting schemes are based on frequency. They do not consider the meaning of the words for weighting. This leads to the degradation in the performance of machine learning algorithms. To address this issue, in the current study, SenticNet is used. SenticNet is a sentiment lexicon resource extracted from ConceptNet consisting of the polarity value in a range from − 1 to + 1. Classifiers like Naive Bayes and support vector machine are used to classify the reviews. *The performance of SenticNet and existing feature weighting schemes are evaluated for sentiment classification.* From the experiments conducted, it is found that SenticNet-based feature weighting scheme outperforms the existing schemes like binary, TF and TF-IDF.

Keywords SenticNet · Sentiment classification · Polarity detection · Feature weighting

1 Introduction

Increase in development of Web has been a very important factor in every part of our lives, and hence, there is a gradual increase in the user point analysis [1]. Sentiment classification is the process of separating reviews as either positive or negative. Human analysts agree that the results of automated test classification are reliable around 85% of time.

K. S. Kalaivani (✉) · M. Rakshana · K. Mounika · D. Sindhu
Department of CSE, Kongu Engineering College, Perundurai, India
e-mail: kalaivani@kongu.ac.in

S. Shakya et al. (eds.), *Mobile Computing and Sustainable Informatics*, Lecture Notes on Data Engineering and Communications Technologies 68,
https://doi.org/10.1007/978-981-16-1866-6_64

1.1 Levels of Sentiment Analysis

Sentiment analysis examines the people's perceptions like emotions, feelings and so on. It helps organizations to improve the product quality and develop new marketing strategies. In general, sentiment classification has been broadly classified into three categories:

- Sentence-level classification.
- Document-level classification.
- Feature-level classification.

1.1.1 Sentence-Level Classification

Sentence-level classification determines whether every sentences in the document express a negative, positive or neutral opinion for a product. Sentence-level classification is associated with subjectivity classification which makes difference between subjective and objective sentences. It has no opinion or judgment about the entity or object while subjective sentence has opinions. Consider the sentence, "India's economy is dependent on tourism and IT sector. It is a very good place to survive in." The first sentence is a factual one and does not convey any opinion toward India. Hence, this should not play any role in deciding the polarity of the sentence and should be removed out.

1.1.2 Document-Level Classification

To classify a whole opinion document into a positive or a negative statement, document-level classification is used. It is the process of labeling the documents into categories according to the content. It can be manual or automated, and it is used to sort and manage texts, images. This classification is not successful, if the entire document contains opinion of different classes.

1.1.3 Feature-Level Classification

Feature engineering is a basic and important task for sentiment classification. The core concept in this classification is to identify the text in a document which describes about the feature of any product. In the review, "The cover is of very high quality and has the right amount of cushion," and "very high quality" is the sentiment word. The basic steps for feature-based sentiment classification are as follows:

- Preprocessing.
- Feature Extraction.
- Word Extraction.
- Word Polarity Identification.

2 Literature Review

Kim (2008) proposed a semi-supervised dimensionality framework which not only reduces the cost and complexity but also prevents document from information loss. Experimental results proved that accuracy has improved since both label and structural information of data are taken into account.

Concept-level sentiment analysis (CLSA) model was proposed by Cambria (2015) to address the issue of opinions which does not occur at document level. CLSA model takes all the natural language processing tasks that are necessary for extracting assertive information from text, namely subjectivity detection, topic spotting, semantic parsing, aspect extraction, polarity detection and microtext analysis.

Dragoni (2015) introduced an approach by using graph that combines WordNet and SenticNet. Graph propagation algorithm is used to propagate sentiment information from datasets. The graph is plotted using sentiment information. The results demonstrate its viability in real-world cases.

Agarwal and Namita (2016) have proposed a novel model which is drawn out from ConceptNet, and it is based on commonsense knowledge. Domain-specific features are determined by ConceptNet-based ontology, and polarities are determined by contextual polarity lexicon. Results improve effectiveness since the feature's importance is evaluated by the feature's depth.

For achieving more accurate classification, generalized syntactic dependency features and negation handling were proposed by Kalaivani and Kuppuswami (2019). Further to reduce the dimensionality of feature space, weighted frequency and odds and information gain are used. Term frequency-inverse document frequency is also deployed to observe if there is any dissimilar distribution of features between two or more different classes. From the experiments conducted, it is found that for sentiment classification, generalized syntactic dependency features are more effective than standard features. Rajaram and Palaniswami [2] proposed an authentication approach for verifying the nodes in the mobile ad hoc network.

The existing weighting methods are based on word co-occurrence counts, and hence, the polarity of the words is not considered for weighting. SenticNet-based feature weighting scheme is introduced in this study to improve the performance of sentiment classification.

3 Proposed System

The flow diagram of the proposed work is given in Fig. 1. Firstly, reviews are preprocessed to remove punctuations, numerals and stop words. Then, the remaining words are extracted and used as features. Case normalization is done to change the reviews into lowercase. After preprocessing, for each word, the polarity value is found using SenticNet which is further used for document term matrix construction. Constructed document term matrix is given as input to SVM and Naive Bayes classifiers.

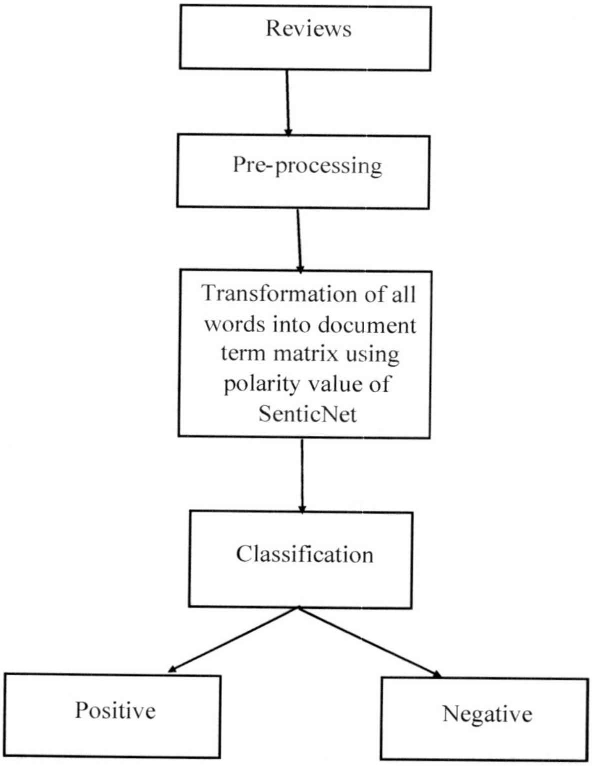

Fig. 1 Flow diagram of SenticNet-based feature weighting scheme

3.1 Dataset

To determine the performance of SenticNet-based feature weighting scheme, the popular Amazon product review is taken as the dataset. This dataset contains 2000 reviews. Out of them, 1000 reviews are positive labeled, and the remaining 1000 reviews are negative labeled. For training, 90% of reviews are used, and the remaining 10% of reviews are used for testing.

3.2 Preprocessing

Preprocessing is an initial step to prepare the data for further analysis. It can be applied to any preparatory processing stage. It corrects misspelled words, removes punctuation symbols and so on [3]. In this work, case normalization and stop word removal are done to prepare the data for processing.

3.2.1 Case Normalization

The reviews are in both uppercase and lowercase. For uniformity, case normalization is performed to convert all the reviews into lowercase.

3.2.2 Stop Words

Stop words refer to words which do not provide much meaning to the sentence. Therefore, stop words can be ignored in the sentence. By filtering stop words, size of the document decreases, and the model will be trained in less time without any impact on accuracy.

3.3 Existing Work

Feature weighting plays a major role in evaluating the performance of machine learning-based sentiment classification. The existing weighting schemes are based on the number of times that the feature occurs in a document.

3.3.1 Feature Weighting

In text categorization, to determine the numerical value for every feature present in the document, feature weighting is used [4]. By assigning distinct values for every feature, the importance and relationship between attributes can be identified, and hence, the problem space gets minimized. In sentiment analysis, there are three traditional feature weighting schemes [5].

Binary Weighting

Binary weighting is about representing the presence and absence of a feature in a document. The feature weight is 1 if the keyword is present in the overall document; otherwise, the feature weight is 0, e.g., a document containing 100 words has word "speed" 5 times. In this weighting scheme, how often the word occurs in the document will not be taken into account. It just denotes 1 for the keyword presence and 0 for the keyword absence (Table 1).

Term Frequency

Term frequency calculates the feature density in the document. Using TF, significance of every feature can be identified. In most cases, words appear more frequently

Table 1 Sample document term matrix using binary weighting

Docs	Book	Best	Good	Speed	Read	Bad
1	1	1	1	0	0	0
2	0	0	0	1	1	1
3	0	1	0	1	0	1

Table 2 Sample document term matrix using TF weighting

Docs	Book	Best	Good	Great	Read	Bad
1	10	1	1	1	2	0
2	2	0	0	0	0	12
3	6	2	3	4	0	1

Table 3 Sample document term matrix using TF-IDF weighting

Docs	Book	Best	Good	Great	Read
1	0.0146	0.0137	0.0126	0.0157	0.0169
2	0.0157	0.0166	0.0195	0.0169	0.0143
3	0.0186	0.0176	0.0189	0.0127	0.0178

in longer sentences than in shorter ones. To normalize it, the number of word occurrences is divided by overall count of the words present in the document. If a word "speed" occurs 8 times in a document which has 100 words, then the frequency for word speed will be 8/100 = 0.08 (Table 2).

Term Frequency—Inverse Document Frequency

TF-IDF calculates word's originality by comparing the number of word occurrences in a document with the number of documents the word appears in. IDF diminishes the weight of scarcely appeared words which has to be scaled up [6].

$$\mathrm{IDF} = \log_e \frac{\text{Total\_ number\_ of\_ documents}}{\text{Number\_ of\_ documents\_ with\_ term\_ t\_ in\_ it}} \tag{1}$$

For example, the word "speed" has 5 occurrences in a 100 words document, and then, term frequency is (5/100) = 0.05. Consider there are 5 million documents, and word "speed" has appeared in 500 of those that the IDF is log (5,000,000/500) = 4. TF-IDF for the above example is 0.2 (Table 3).

3.4 Proposed Work

Existing weighting schemes are frequency based, they do not consider the polarity of the word for identifying the feature weights. So in this work, SenticNet is used which is a lexicon resource to find the polarity of the word and use it for frequency weighting.

3.4.1 Senticnet

SenticNet, which is emotional analysis at the conceptual level, that is, through semantics and linguistics to detect polarity and to recognize emotion rather than relying solely on word co-occurrence frequency [7]. SenticNet can be regarded as one of the following:

- A conceptual knowledge base.
- A multi-disciplinary framework.

As a knowledge base, SenticNet contains a set of 100,000 concepts that are related to semantics, emotions (enjoyment, disappointment, misery, doubtful) and polarity. Polarity value ranges from − 1 to + 1 where − 1 is extremely negative and + 1 is extremely positive and polarity value is 0 if the term is not present in SenticNet lexicon resource [8]. As a framework, SenticNet includes a series of sentiment analysis tools and techniques that combine common sense reasoning, psychology, linguistics and machine learning. It is a multi-disciplinary paradigm that goes beyond the purely statistical sentiment analysis method by focusing on the semantic-preserving representation and sentence structure of natural language concepts. SenticNet covers the latest findings of conceptual-level sentiment analysis, in order to provide easy-to-use tools for large-scale social data analysis. It can be used in different fields, languages and automated tasks such as trend discovery and social media marketing.

3.5 Classification

Classification is the process of assigning the latest observed data to a group of attributes keeping the training dataset as base containing different attributes and whose test data class labels are unknown. The type of classifiers used in this study is as follows:

- Naïve Bayes.
- Support vector machine.

3.5.1 Naïve Bayes

Naive Bayes classifier is not only simple but also very efficient. It was developed under the assumption that the features present in the dataset are mutually independent [9]. This classifier does not need much training, and it is highly scalable.

3.5.2 Support Vector Machine

SVM is a group of supervised algorithms to classify the problems. SVM represents different classes in a hyperplane [10]. The main goal of SVM is to separate the datasets into classes to find a maximum marginal hyperplane which can be done in two following steps:

- SVM will generate hyperplane in a looped manner that segregates the classes in a great way.
- It chooses the hyperplane which separates the classes correctly.

4 Performance Metrics

To evaluate the machine learning algorithms, performance metrics are used [11]. The following equations are used to evaluate the performance of support vector machine and Naïve Bayes classifiers.

$$\text{PRECISION} = \frac{\text{Document\_ correctly\_ classified\_ to\_ category\_}\ c_i}{\text{Total\_ documents\_ classifiedto\_ category\_}\ c_i} \tag{2}$$

$$\text{RECALL} = \frac{\text{Document\_ correctly\_ classified\_ to\_ category\_}\ c_i}{\text{Total\_ documents\_ in\_ category\_}\ c_i} \tag{3}$$

$$\text{ACCURACY} = \frac{\text{Total\_ number\_ of\_ classified\_ documents}}{\text{total\_ number\_ of\_ documents}} \tag{4}$$

$$F1 = 2 * \frac{(\text{Precision} * \text{Recall})}{(\text{Precision} + \text{Recall})} \tag{5}$$

5 Results

Table 4 shows the performance of existing weighing scheme using Naïve Bayes classifier and SVM.

Table 4 Performance comparison of existing work

Classifier	Feature weighting	Precision	Recall	Accuracy	$F1$-score
Naïve Bayes	Binary	98.77	86.25	86.25	92.62
	TF	99.38	91.25	91.00	95.43
	TF-IDF	98.89	99.00	94.00	99.50
SVM	Binary	99.65	82.75	82.75	82.75
	TF	100.00	77.25	75.70	75.70
	TF-IDF	100.00	75.00	75.00	75.00

Table 5 Performance comparison of proposed work

Classifier	Precision	Recall	Accuracy	$F1$-score
Naïve Bayes	98.57	99	98.61	96.76
SVM	98.57	100	99.28	99.28

Table 5 shows the performance of proposed weighing scheme using Naïve Bayes classifier and SVM. It is found that SenticNet-based feature weighting scheme provides the best performance in comparison with the existing scheme. SVM outperforms Naïve Bayes classifier in all weighting schemes. Among the binary, TF, TF-IDF, binary weighting scheme outperforms the other two schemes. In the proposed weighting scheme, SVM performs better than Naïve Bayes.

6 Conclusion

Existing weighting schemes like binary, TF, TF-IDF are frequency based. Depending on the number of occurrences of word in the document, weights are assigned. They do not consider the polarity of the words for assigning weight. In this work, polarity is identified using SenticNet scheme, and thus, accuracy is improved.

References

1. P. Vasile-Daniel, et al., Opinion mining on social media data: sentiment analysis of user preferences. Sustainability **11**(16), 4459 (2019)
2. A. Rajaram, S. Palaniswami, A high certificate authority scheme for authentication in mobile ad hoc networks. Int. J. Comput. Sci. Issues (IJCSI) **7**(4), 37 (2010)
3. A. Giulio, et al., A comparison between preprocessing techniques for sentiment analysis in Twitter. KDWeb (2016)
4. X.L. Haddi, et al., The role of text pre-processing in sentiment analysis. Proc. Comput. Sci. **17**, 26–32 (2013)

5. Y. Wang, H.Y. Youn, et al., Feature weighting based on inter-category and intra category strength for Twitter sentiment analysis. Appl. Sci. **9**(1), 92 (2019)
6. J. Amir, et al., *Semantic Sensitive TF-IDF to Determine Word Relevance in Documents*. arXiv: 2001.09896 (2020)
7. E. Amir Hussain, et al., *SenticNet. Sentic Computing* (Springer, Cham, 2015), pp. 23–71
8. S. Ranjan et al., *SenticNet. Sentiment Analysis in the Bio-Medical Domain* (Springer, Cham, 2017), pp. 39–103
9. L.-C. Concha Bielza, et al., Directional Naive Bayes classifiers. Pattern Anal. Appl. **18**(2), 225–246 (2015)
10. S. Suthaharan, *Support Vector Machine. Machine Learning Models and Algorithms for Big Data Classification* (Springer, Boston, MA, 2016), pp. 207–235
11. S. Sugrim, et al., *Robust Performance Metrics for Authentication Systems. NDSS* (2019)

Classification of Intrusions in RPL-Based IoT Networks: A Comparison

P. S. Nandhini, S. Kuppuswami, and S. Malliga

Abstract Internet of things is a group of objects or devices that establish communication through Internet. Since, the communication is through Internet, more processing power is required in the devices. The IoT devices are heterogeneous and energy constrained. Owing to constrained processing power, the traditional security procedures are not applicable to be executed in the devices. This generates different attacks like user to root, DoS, remote to local and probe attacks. In the existing systems, the classification is carried out to classify different intrusions by using various machine learning algorithms based on KDD dataset. The results were analyzed with various parameters. ML algorithms led to false classifications with medium accuracy. So, it was unable to classify various attacks perfectly. In this paper, the two variants of recurrent neural network such as long short-term memory and bidirectional LSTM are used for the classification of different IoT attacks. The proposed algorithms reduced the false classification with good accuracy. The parameters are computed for the proposed algorithm and are compared with existing machine learning algorithm.

Keywords IoT · LSTM · BiLSTM · Intrusion · Classification

1 Introduction

IoT is a cluster of computing devices, mechanical and digital machines with unique identifiers (UIDs). The devices have the ability to transfer data over Internet. It does not require human-to-human or human-to-computer interactions. Due to scarcity of resources, the devices in the network are prone to various intrusions. An intrusion is

P. S. Nandhini (✉) · S. Malliga
Department of CSE, Kongu Engineering College, Perundurai, India
e-mail: nandhini.cse@kongu.edu

S. Malliga
e-mail: mallisenthil.cse@kongu.edu

S. Kuppuswami
Vivekandha Educational Institution, Tiruchengode, India

S. Shakya et al. (eds.), *Mobile Computing and Sustainable Informatics*, Lecture Notes on Data Engineering and Communications Technologies 68,
https://doi.org/10.1007/978-981-16-1866-6_65

an unauthorized activity that happens in the network. The intrusions involve stealing a valuable network resource and always threaten the security of networks or their data. There are applications that monitor the network for malicious activity. If any intrusions occur in the network, it may be reported directly to the admin or may be collected using security information. So, there is a need for an intrusion detection system. IDS is an application that monitors the malicious activity in the network [1]. Cyber-attacks are based on the activities and targets of the attackers. Novel attacks were developed by contemplating the weaknesses in the network.

IDS is a remarkable research achievement in the securing the information. It can identify an attack, which is prevailing or already occurred in the network. Intrusion detection is correspondent to a classification problem such as a binary or a multiclass classification problem [2]. It is used for identifying whether network traffic behavior is normal or anomalous. In this paper, the dataset consists of four types of attacks such as: DOS, U2R, Probe and R2L. The objective of this paper is to increase the performance of classifiers in effectively identifying the intrusive behavior. Now, deep learning is an emerging technique that plays a major role in the classification of attacks.

Deep learners have knowledge to extract the representations from the processed data and design a better model for classifying various attacks that occur in IoT. Due to increase in the usage of computational resources, RNN have achieved a significant development in the domain of deep learning. It is the suitable approach for the classification and analysis on the data. LSTM is a subset network of RNN. The objective of classifying the attack is based on enhancing the performance of the classifiers. The aim of this paper is to implement LSTM and BiLSTM algorithms for classification and to compare the parameters with already existing machine learning algorithms.

In this paper, the existing systems are explained in Sect. 2. Section 3 details about the various procedures in the proposed work. In Sect. 4, the proposed algorithm is explained. Evaluation metrics are discussed in Sect. 5. The results are discussed in Sect. 6. Conclusion is presented in Sect. 6. The algorithms discussed in this paper are proposed to increase the classification efficiency and decrease the energy consumption in the nodes.

2 Related Work

Nowadays, deep learning algorithms are gaining popularity. It is suitable for classifying various intrusions that occur in the IoT networks. The studies revealed that DL algorithms have completely replaced the traditional way of intrusion classification. Increase in data generation leads to more consumption of processing power. DL algorithms are gaining popularity in most of the fields.

Liu et al. [3] examined particular attacks in NSL-KDD dataset. 11 machine learning algorithms were considered, and the results were discussed. NSL-KDD dataset was used. ML methods were used for pre-training. From the result, it is inferred that tree-type methods are suitable. XGBoost is one of the supervised algorithms that had good accuracy in classification. The disadvantage is that classification of multiclass problem is not considered.

Sharma et al. [4] have simulated various attacks in this paper. They have also analyzed the features and proposed a ML model. The significance of the proposed model is feature reduction. The features were reduced to 63.7%. The processing power and energy for communication were reduced. The authors have also described the various parameters that are suitable for classifying various attacks. The selection of right parameters for classification reduces the training and testing time. It also improved the efficiency of classification. The disadvantage of the proposed system is that due to large amount of data generation, it cannot be applied for real-time dataset. To overcome that, dimension reduction technique is used. Principal component analysis is used for reducing the dimension.

Sheikhan et al. [5] have explained three-layer RNN architecture in this paper. The input consists of 41 features and four classes of intrusion as outputs. The nodes in the layers are connected partially. The reduced RNNs can be used for reducing the high-dimensional features. The disadvantage is that the authors have not studied the performance parameters of the model for binary classification.

Kim et al. [6] have deployed LSTM algorithm to the KDD Cup 1999 dataset. The authors have conducted two experiments. One is regarding hyper-parameter values calculation and the other is the calculation of performance parameters. The performance parameters are detection rate and false alarm rate.

Foley et al. [7] have analyzed the vulnerability of MRHOF and OF0 OFs. ML algorithms were used to classify combined attacks. In addition, the authors have initiated the automation of detection of the attacks. The dataset includes the network and power features. RF and MLP classifiers had good accuracy in the classification of attacks. Ayyasamy and Subramani [8] suggested an authentication scheme for the mobile nodes in the mobile ad hoc network.

3 Proposed Work

The suggested method for the classification of attacks lies on using the correct datasets that includes different attacks. The steps involved in preparing the data [4] are (1) collection of data, (2) pre-processing of collected data, (3) normalization of pre-processed data and (4) feature selection. These steps are detailed below.

3.1 Data Collection

Most of the intrusion experts trust that most of the upcoming new attacks are the flavors of the already known attacks. The signatures of the already known attacks are enough to identify the novel variants [9]. For accurate classification, it is necessary that the data used for testing should not have the identical likelihood as in case of training. Data used for testing can have attack types that are not same from the training set. In this paper, the dataset used for classification incorporates 24 types of attack [10] in the training set and 14 more attack categories in the testing data. The features in NSL-KDD can be combined into three groups.

3.1.1 Basic Features

This category consists of features that are derived from a TCP/IP connection. These features cause an indirect delay in classification of attacks.

3.1.2 Traffic Features

This class incorporates the attributes that are derived in correspondence to the interval in a window. It is grouped into two groups. These attributes are known as time-based [11].

(a) Same Host: In the last two seconds, the connections that are destined for the same host as that of the present connections are examined. The statistics data related to conduct of the protocol and services are calculated.
(b) Same Service: In the past two seconds, the connections that incorporate the same services as present connection are examined.

3.1.3 Content Features

Probing attacks and DoS involve more interconnections to certain host in a limited time span. R2L and U2R attacks are involved in the payload part of the packets. A single connection is sufficient for this attack. To classify the above attacks, certain features are essential to focus the mistrustful behavior. These features are also known as content features.

3.2 *Pre-Processing of Data*

Pre-processing of data is an important step in data mining. It is the time consuming process. Realistic data are derived from heterogeneous platforms which may be noisy, incomplete, redundant and inconsistent. So, it is necessary to transform raw data into a suitable format that can be used for analysis and discovery of knowledge. Removing redundant instances, outliers, transformation of raw data [12] and normalization are the pre-processing carried out in this paper.

3.3 Data Normalization

During the training, input variables with higher values are removed to overcome the persistence of the smaller ones. Simulating neurons may be saturated, if raw input variables are fed into the neural network directly. If the neurons in the network are soaked, the modification in the input will lead to a minimum or no output change. This leads to the disturbance in the training of the network considerably and it also causes some irrelevant errors. To minimize the impacts caused by the saturation of the neuron, the input data should be normalized earlier before giving it to the neural network.

3.4 Selection of Features

This process involves in deletion of the irrelevant, not required and repeated variables. Non-essential data has impact on performance parameters such as accuracy. The output of this is the best representation of data. It also allows learning a solution for the fundamental problem [13]. If the section of feature is skipped, it has destructive effect on the prediction model's accuracy.

3.5 Network Training and Testing

After identifying the correct structures of the system, the model should be trained to grab the connection between the input and the output with the help of the training dataset. After training, networks should be tested with test data. Testing of the developed network is done to evaluate the generalization capability.

3.6 Attack Class Detection

There are four types of attack classes in NSL-KDD dataset [3]. They are DoS, U2R, R2L and Probe attack. DoS attack floods the network with unnecessary traffic. In R2L, the user with remote capability attempts to gain the local accessibility. In U2R, the attacker tries to attempt to gain privileges of root. Probe attack is a program that is used for monitoring data about the activities of the network.

4 Methodology

The suggested methodologies for classification of intrusions in IoT are LSTM and BiLSTM. The performance is compared with machine learning classifiers. It is used for classification of intrusions that occur in IoT.

4.1 LSTM

LSTM consists of various memory blocks called as cells. The two states are cell and the hidden state. Two states will be transferred to the next cell. The memory block is responsible for remembering and manipulations through major mechanisms called gates [14] (Figs. 1 and 2).

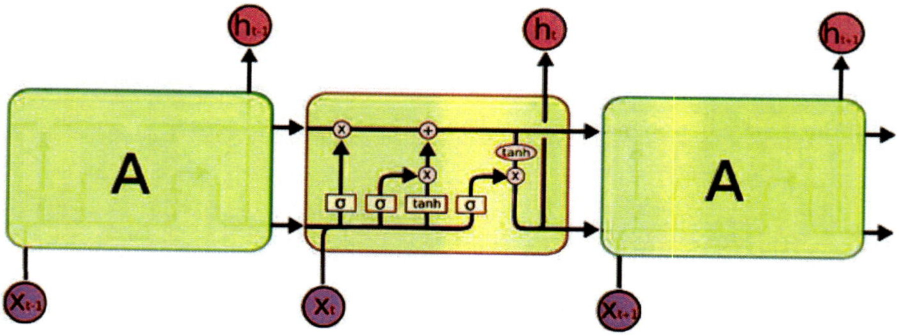

Fig. 1 LSTM architecture

Fig. 2 Forget gate

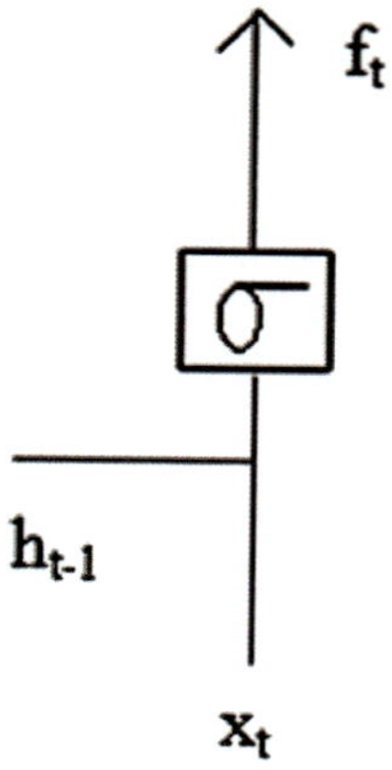

4.1.1 Forget Gate

Forget gate is required to optimize the performance of LSTM. It is subjected to delete the no longer information required for LSTM [15]. The less important information is deleted through filter multiplication. Two inputs to the gate are h_{t-1} and x_t. h_{t-1} is the previous cell output. x_t is the input at the specified time. The given inputs are produced with the weight matrices. Bias is added to the multiplied output. To the biased value, sigmoid function is applied. The sigmoid function output is a vector that ranges from 0 to 1 [3]. The values correspond to number in the cell state. Sigmoid function is responsible for deciding whether to have the values or to discard it. If zero is the output, then it means that the cell state has to forget that piece of information entirely. Similarly, if the output is one then the cell state has to remember that whole information. The output of the sigmoid is multiplied with the cell state.

4.1.2 Input Gate

It is accountable for adding the cell state information. Addition of information consists of three steps (Fig. 3).

- With the help of the sigmoid function, the values that are to be added to the cell are regulated. This correspondence to the forget gate and it will also be a filter for the values from h_{t-1} and x_t to the cell state. This is performed by using the tanh function and the output values ranges from − 1 to + 1
- Finding the product of the value from the sigmoid gate and tanh function and then adding the cell state information
- After this is completed, make sure that particular information is only given to the cell state. It is significant and it should not be redundant [16].

Fig. 3 Input gate

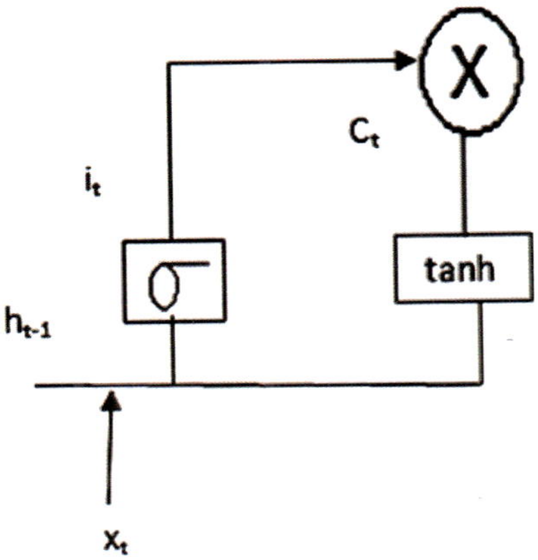

Fig. 4 Output gate

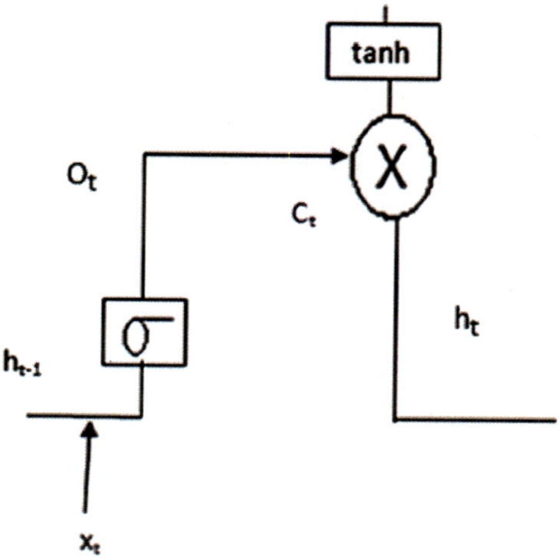

4.1.3 Output Gate

- The operation of an output gate can be split into three steps:
- After applying tanh function for the cell state, the vector is created. The values range from − 1 to + 1.
- Filter can use h_{t-1} and xt values, so that the values can be adjusted which can be the output of the above vector. The filter uses the sigmoid function again. Multiply regulatory filter value along with the vector value (created in step-1). It is sent as an output and also to the next cell hidden state (Fig. 4).

4.2 BiLSTM: Bidirectional Long Short-Term Memory

BiLSTM originated from bidirectional RNN. It processes the sequenced data in both the forward and the backward directions. It is done using two separate hidden layers [17]. BiLSTM joins the hidden layer to the corresponding output layer. The shortcoming of the conventional RNNs is that the ability of using the only previous context. Bidirectional RNN solves this by distributing the data in both the directions. BiLSTM network calculates the output of forward hidden layer, which is backward hidden layer output. Output layer is computed by repeating the forward layer for the value of t from 1 to T and the backward hidden layer repeated from the value of t from T to 1 and then the improvement of final output.

BiLSTM layer produces an output vector, y_t:

$$y_t = \sigma(\overrightarrow{ht}, \overleftrightarrow{ht})$$

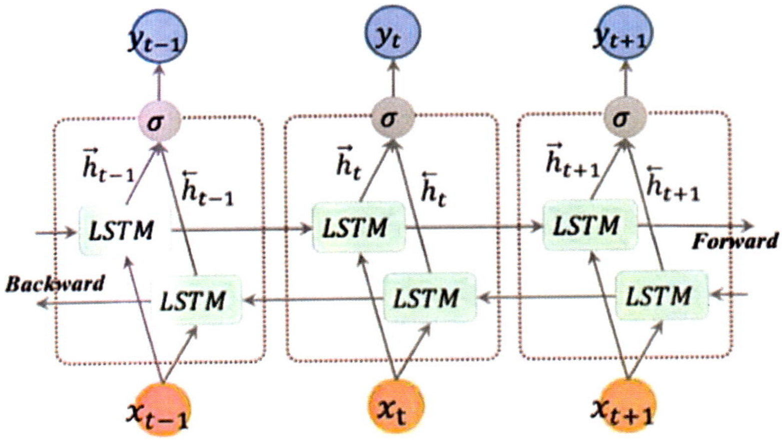

Fig. 5 BiLSTM architecture

The above function merges both the output sequences. The function used of four categories such as summation, concatenating, multiplication and average. Integrating BRNNs along with LSTM neurons brings out a BiLSTM RNN. It is a structure which has the capability of dealing through both forward and backward directions. It is also capable of retrieving long-term context data in both forward and backward directions. Blend of both LSTM layers is calculated as a single BiLSTM layer [18].

Figure 5 depicts a BiLSTM structure with three successive time-steps. In numerous domains such as classification and speech recognition, bidirectional models perform better than classical unidirectional approaches. The BiLSTM RNN architecture permits the combination of both previous and forthcoming context through bidirectional optimization process.

5 Parameter Evaluation

Selecting the appropriate parameters for RNN models is a big challenge because it causes the variation in performance parameters such as accuracy. By trial and error, the suitable parameters for obtaining the better performance are identified. The parameters that are considered for constructing the RNN models are learning rate, neurons in the hidden layers and number of hidden layers, and the parameters used for evaluation is presented in Table 1.

Table 1 Parameter evaluation

Parameter evaluation	Formula
Accuracy	Accuracy = (TP + TN)/TP + TN + FP + FN
Recall	Recall = TP/(TP + FN)
Precision	Precision = TP/(TP + FP)
$F1$ measure	$F1$ measure = 2TP/(2TP + FP = FN)

5.1 Learning Rate

It controls the adaptation that must be made to a given network weights corresponding to the gradient loss. If the learning or dropout rate is too large, it will prevent the model from learning constructively. To obtain satisfactory results, the learning rate should be between 0.01 and 0.1 [7].

5.2 Hidden Layers

The basic design of the RNN model consists of one hidden layer, followed by an output layer which is standard feed forward. Number of hidden layers to be included depends on the size and dimensions of the dataset.

5.3 Hidden Units

It is the count of the neuron in the hidden layer. If it is more, the network turns into more powerful. Number of parameters to be learnt is also increased which consumes more training time for predicting the model [19].

5.4 Time-Steps

It is the number of steps in time backpropagation. During training, it calculates the gradients for updating weight. The number of time-steps influences learning. For classification of intrusions, time step plays an important role. Number of time-steps that are required for backpropagation influences the identification of the correct patterns.

6 Results

In this paper, TensorFlow is used for implementing the architecture model. The models used in this work are two variants of RNN architecture. Experiments were conducted to assess the model performance by using NSL-KDD dataset. The dataset is divided into two sets such as training and testing. 80% of the dataset is used for training and the remaining 20% is used for testing. The prediction model was run on both training and testing classifiers. Accuracy and loss rate for each model are calculated (Table 2) [20].

The parameters of LSTM and BiLSTM considered for the classification were presented in Table 3. For each classification model, the suitable parameter setup was recognized [21]. Several runs were conducted with different values.

Learning rate is tuned for efficient training because of its effects is faster on the training model. It should not over fit the training model. Thus, the best learning

Table 2 Accuracy rate of machine learning classifier

Machine learning classifiers	Accuracy rate (%)
Naive Bayes	90.78
Logistic regression	91.39
K-neighbors	93.23
Decision tree	94.08
LSTM	95.43
BiLSTM	96.24

Table 3 Parameters of LSTM and BiLSTM

Parameters	LSTM	BiLSTM
Epoch	10	10
Input layer	1	1
Hidden layers	3	3
Output layer	1	1
Activation functions	Sigmoid	Sigmoid
Batch size	1	1
Optimizer	Nadam	Nadam
Classification engine	Binary_crossentropy	Binary_crossentropy
Time-steps	3 ms/step	2 ms/step
Learning rate	0.01	0.01
Dropout rate	–	0.5
One hot encoder	Yes	Yes
Weights	Random	Random
Biases	Random	Random

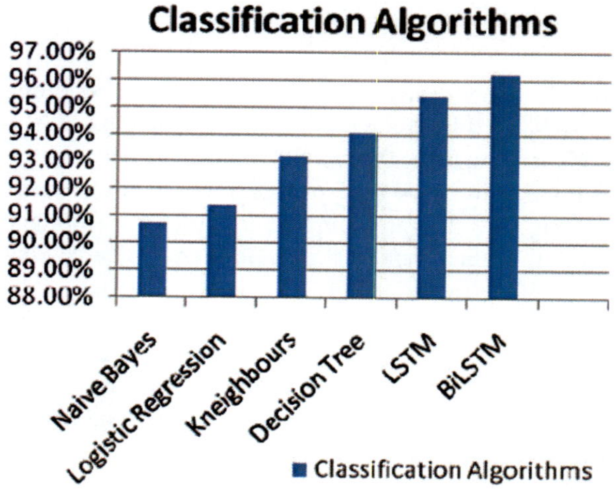

Fig. 6 Accuracy of classification algorithms

rate for NSL-KDD dataset is 0.01. LSTM had an accuracy of 95.43%. It is inferred that addition of more training cycles does not improve accuracy. Training time took 10,648.93 s for the LSTM 500 cycle. NSL-KDD dataset contains 498,431 samples with 4 attack classes. At different cycles, the algorithms were trained and tested. Loss rate for LSTM is 2.10% and BiLSTM is 1.10%. From Fig. 6, BiLSTM outperformed LSTM at 100 training cycles. BiLSTM required less time for training and also provided great accuracy [22].

7 Conclusion

In this paper, the proposed models were implemented and performance parameters such as accuracy and training time were calculated. Performance parameters of ML algorithms along with two RNN models (LSTM and BiLSTM) were implemented and compared. Thus, the performance obtained from LSTM and BiLSTM had a higher accuracy and the classification rate with a low false positive rate.

The results of this evaluation revealed that BiLSTM stood up and outperformed other architectures. LSTM showed better performance in terms of accuracy with high training cycles that resulted in a longer training time. BiLSTM offered a reduced training time with greater accuracy. BiLSTM not only has a strong case to be used in intrusion classification but has a high accuracy in classification. Thus, BiLSTM effectively improved both the accuracy of classification of intrusions and the ability to recognize the intrusion type.

References

1. N. Chaabouni, M. Mosbah, A. Zemmari, C. Sauvignac, P. Faruki, Network intrusion detection for IoT security based on learning techniques. IEEE Commun. Surv. Tutor. **21**(3), 2671–2701 (2019)
2. A. Mayzaud, R. Badonnel, I. Chrisment, *A Taxonomy of Attacks in RPL-Based Internet of Things* (2016)
3. J. Liu, B. Kantarci, C. Adams, Machine learning-driven intrusion detection for contiki-NG-based IoT networks exposed to NSL-KDD dataset, in *Proceedings of the 2nd ACM Workshop on Wireless Security and Machine Learning* (2020), pp. 25–30
4. M. Sharma, H. Elmiligi, F. Gebali, A. Verma, Simulating attacks for RPL and generating multi-class dataset for supervised machine learning, in *2019 IEEE 10th Annual Information Technology, Electronics and Mobile Communication Conference (IEMCON)* (IEEE, 2019), pp. 0020–0026
5. M. Sheikhan, Z. Jadidi, A. Farrokhi, Intrusion detection using reduced-size RNN based on feature grouping. Neural Comput Appl **21**(6), 1185–1190 (2012)
6. J. Kim, et al., Long short term memory recurrent neural network classifier for intrusion detection, in *2016 International Conference on Platform Technology and Service (PlatCon)* (IEEE, 2016)
7. J. Foley, N. Moradpoor, H. Ochen, Employing a machine learning approach to detect combined internet of things attacks against two objective functions using a novel dataset. Secur. Commun. Netw. **2020** (2020)
8. R. Ayyasamy, P. Subramani, An enhanced distributed certificate authority scheme for authentication in mobile ad-hoc networks. Int. Arab J. Inf. Technol. **9**(3), 291–298 (2012)
9. P.S. Nandhini, B.M. Mehtre, Intrusion detection system based RPL attack detection techniques and countermeasures in IoT: a comparison, in *2019 International Conference on Communication and Electronics Systems (ICCES)*. (IEEE, 2019)
10. P.S. Nandhini, B.M. Mehtre, Directed acyclic graph inherited attacks and mitigation methods in RPL: a review, in *International Conference on Sustainable Communication Networks and Application* (Springer, Cham, 2019)
11. M. Almulhim, N. Zaman, Proposing secure and lightweight authentication scheme for IoT based E-health applications, in *2018 20th International Conference on Advanced Communication Technology (ICACT)* (IEEE, 2018), pp 481–487
12. P. Ioulianou, V. Vasilakis, I. Moscholios, M. Logothetis, A signature-based intrusion detection system for the internet of things. Inf. Commun. Technol. Form (2018)
13. M. Conti, P. Kaliyar, C. Lal, REMI: a reliable and secure multicast routing protocol for IoT networks, in *Proceedings of the 12th International Conference on Availability, Reliability and Security* (2017), pp. 1–8
14. J. Ashraf, A.D. Bakhshi, N. Moustafa, H. Khurshid, A. Javed, A. Beheshti, Novel deep learning-enabled LSTM autoencoder architecture for discovering anomalous events from intelligent transportation systems. IEEE Trans. Intell. Transp. Syst.
15. A. Karaagac, E. De Poorter, J. Hoebeke, In-band network telemetry in industrial wireless sensor networks. IEEE Trans. Netw. Serv. Manage. **17**(1), 517–531 (2019)
16. S.M. Mostafa, I.M. Darwish, M.R. Saadi, Improved lightweight security approach routing protocol in internet of things. Internet of Things, 100208 (2020)
17. M. Jyothisree, S. Sreekanth, Design and implementation of RPL in Internet of Things, in *Advances in Decision Sciences, Image Processing, Security and Computer Vision* (Springer, 2020), pp. 707–718
18. N. Richerzhagen, *Transition in Monitoring and Network Offloading-Handling Dynamic Mobile Applications and Environments* (Technische Universität, 2019)
19. A. Yahyaoui, F. Yaakoubi, T. Abdellatif, Machine learning based rank attack detection for smart hospital infrastructure. Paper presented at the International Conference on Smart Homes and Health Telematics (2020)

20. F.Y. Yavuz, Ü. Devrim, G. Ensar, Deep learning for detection of routing attacks in the internet of things. Int. J. Comput. Intell. Syst. **12**(1), 39–58 (2018)
21. V.K. Asati, E.S. Pilli, S.K. Vipparthi, S. Garg, S. Singhal, S. Pancholi, *RMDD: Cross Layer Attack in Internet of Things. In: 2018 International Conference on Advances in Computing, Communications and Informatics (ICACCI)* (IEEE, 2018), pp. 172–178
22. D. Airehrour, J.A. Gutierrez, S.K. Ray, SecTrust-RPL: a secure trust-aware RPL routing protocol for Internet of Things. Future Gener. Comput. Syst. **93**, 860–876 (2019)

Intravenous Fluid Monitoring System Using IoT

K. Sangeetha, P. Vishnuraja, S. Dinesh, V. S. Gokul Anandh, and K. Hariprakash

Abstract Simple electrolyte bottles are used in hospital without any proper indication. This creates a trouble to patient as the blood will start flowing in reverse flow from body toward bottle. To overcome this critical situation, an IoT-based automatic alerting and indicating device is proposed where sensor is used to monitor the level. When intravenous fluid level is less, it alerts the nurse through the LCD display, Web page and buzzer. All the departments of hospital require a similar kind of automatic monitoring and indication system. Healthcare industries can also be one of the users. This arrangement can be helpful in small, medium and large-scale hospitals and also it is valuable during home care. This supervising system will decrease the chances of patients risk and enhances the accuracy of healthcare delivered in hospital. The regular need to check the level of bottles manually is avoided. This project is very useful to monitor the patients during night time. This research work helps to eradicate the serious and fatal risk on the air bubbles that can enter the patient's bloodstream. This application setup will generate assurance of delivering a safe situation to the patients.

Keywords Intravenous fluid · Flow sensor · Air bubble · Buzzer · IoT

1 Introduction

In existing framework, observing of patients is done by manual procedure, which may result in switching the stream of blood during the trickle process. At the point when bottle gets unfilled and if healthcare faculties have not made aware of it, it might end up in developing a reverse bloodstream. Observing the patients in late night is troublesome and the communication among specialist and patient will also be comparatively less. The answer for the previously mentioned issue is a flow sensor, which reads the amount of water discharge and the values that are displayed in the LCD display. The same values are also updated in the database. Central monitor will

K. Sangeetha (✉) · P. Vishnuraja · S. Dinesh · V. S. Gokul Anandh · K. Hariprakash
Computer Science and Engineering, Kongu Engineering College, Perundurai, India
e-mail: sangeetha_k@kongu.ac.in

S. Shakya et al. (eds.), *Mobile Computing and Sustainable Informatics*, Lecture Notes on Data Engineering and Communications Technologies 68,
https://doi.org/10.1007/978-981-16-1866-6_66

display the status of the patient's fluid bottle. Through the display staffs are able to know the level of the fluid in the bottle. If the level of the bottle reaches the fixed threshold value it will alert the staff and then the staff came to know that the bottle will get empty soon. This will prevent the reverse flow of bloodstream and so it will eliminate the serious consequences in patients.

2 Literature Review

Oros et al. [1] worked in intravenous infusion dosing system by detection, signaling and monitoring of liquid in an IV bottle at a remote location. This system alerts the medical staff to continue and timely changes of IV bottles which can have positive effects on increasing the success of IV therapy.

Anagha et al. [2] focused in automation of level monitoring using ultrasonic sensor which is placed at the top of the fluid bottle. They aimed to make patients comfortable with their system. They also worked for fan automation for patients ease of use. They place the flex sensor under the finger, when the patients need/no need of fan they can turn on/off by bending the finger. They also automated the stoppage of fluid using solenoid valve.

Mathew et al. [3] worked on the novel intravenous fluid level indicator for smart IV systems by checking the level of the fluid given to the patient and controlling the flow of the fluid automatically by using sensors.

Walsh et al. [4] worked in using various integrated sensor to measure the flow of the fluid and alerting the staff nurse when the fluid level reaches certain percentage.

Joseph et al. [5] worked in reducing the workload of staffs and overcome the critical situation in the area of an intravenous drip monitoring system using automated intravenous drip monitoring system.

Novak [6] worked in IV pump remote alarm system includes a modular unit detachably connected to a stationary unit, which is associated with the IV pump from which the nursing station can be alarmed when the status is critical.

Sampath Kumar et al. [7] worked with the fluid monitoring system with the help of infrared sensor. With the infrared sensor they calculated the height by fixing the infrared sensor at certain level. When the bottle is filled with fluid infrared receiver will not sense any signal as the signal gets refracted. If the bottle is with low amount of fluid that is lesser than the height of the infrared sensor it alerts. They displayed the result in the Web page.

Arulious Jora et al. [8] worked in completely different way. They used LDR sensor to sense the light intensity through the liquid. When the fluid is at higher level light spreads so the intensity will be minimum, resistance will be maximum and when it is at lower rate the intensity with be high, resistance will be minimum. Then, it alerts the nurse with radio frequency transmitter which transmits the data wirelessly to the nurse's room. The signal is received by radio frequency receiver which is connected with the buzzer.

Bhavasaar et al. [9] worked in automated intravenous fluid monitoring and alerting system by using load cell and heartbeat sensors and monitor the status of the patients and alert the staffs accordingly.

Rao et al. [10] worked by adopting IoT cloud with wireless sensor networks that is beneficial, especially when administering the condition of a greater number of patients and their resulting data storage are taken into account. Liquid level sensor, ultrasonic sensor and temperature sensor are used. Rajaram and Palaniswamy [11] described about the privacy methods to safeguard the information in network.

3 Methodology

Intravenous fluid monitoring system consists of Raspberry Pi 3, flow sensor, LED lights, buzzer and LCD display. The flow sensor calculates the discharge of amount of fluid from the bottle. It calculates the rate of flow of fluid continuously. These values calculated by the Raspberry are displayed in the LCD display and also the values are updated in the firebase. The Web page will retrieve the values from the firebase and display the values. LED lights and buzzer are to alert the hospital staffs. When the level of fluid in the bottle is lesser than the threshold level, then it alerts the staffs. This system consists of three states NONE which is the initial state before the start of the system, SAFE which indicates that the intravenous fluid bottle is not in critical state, and the other state is CRITICAL, in this state the level of the bottle is lesser than the threshold level. In the safe state, the green LED light will glow, buzzer will not alert, and in the critical state, the red LED light will glow and at the same time buzzer will also alert. The status will be displayed in the LCD display and the Web page also so the staffs came to know about the patient's IV fluid level and status (Figs. 1 and 2).

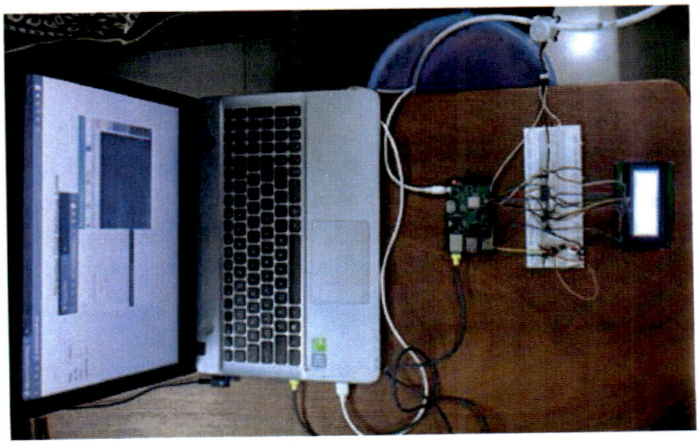

Fig. 1 System setup

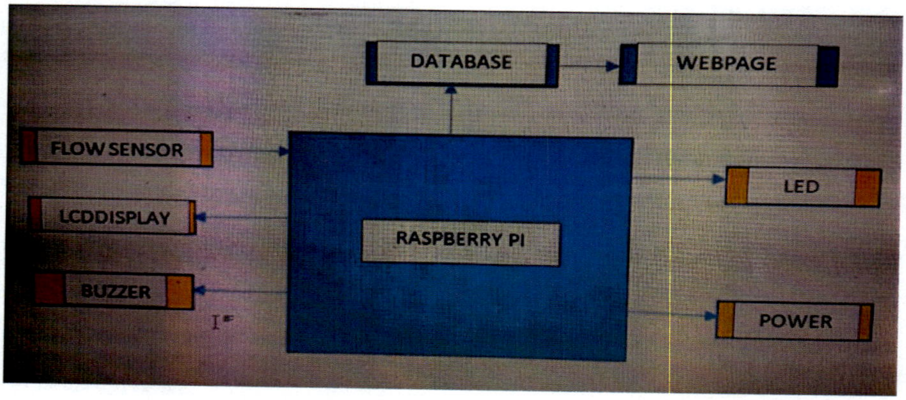

Fig. 2 Architectural diagram

Fig. 3 Raspberry Pi

4 Components Used

4.1 *Raspberry Pi 3 Model B*

Raspberry Pi 3 Model B is single-board computer with wireless LAN and Bluetooth connectivity. It requires USB power cable to give power supply to the Raspberry Pi and micro SD card with Raspberry Pi OS. In order to use Raspberry Pi as a desktop computer, it needs monitor, HDMI cable, keyboard and mouse. The flow sensor, LCD display, LED lights and buzzer are connected with the Raspberry Pi. This processes the data from the flow sensor and gives the values and status in the LCD display and updates the value in the firebase (Fig. 3).

4.2 *Flow Sensor*

1/8 inch water flow sensor YF-S401 consists of a plastic valve body, a water rotor and a Hall Effect sensor. When water flows through the sensor, the internal rotator will

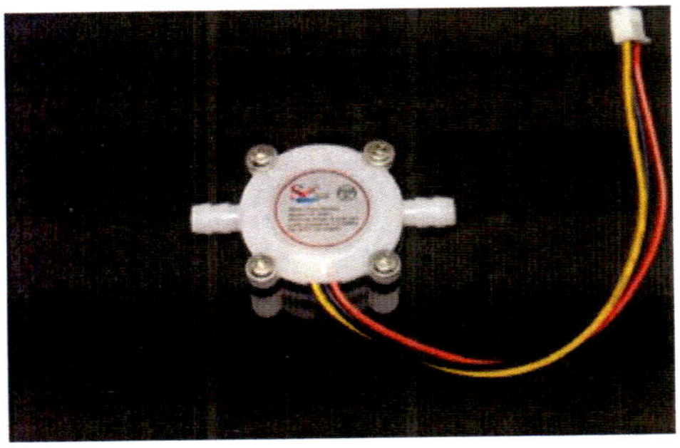

Fig. 4 Flow sensor YF-S401

alter its speed with various rate of flow and Hall Effect sensor outputs the equivalent pulse signal. Flow sensor has three wires, namely power-in (red), ground (black) and output (yellow). This is fixed in between the fluid pipe to monitor whether the fluid is flowing through the sensor (Fig. 4).

4.3 *LCD Display 20 × 4*

A liquid–crystal display (LCD) is a flat-panel display uses the light-modulating properties of liquid crystals combined with polarizers. Liquid crystals do not emit light directly, instead using a backlight or reflector to produce images in color or monochrome. In LCD display 20 × 4, 20 represents number of characters per row and 4 represents the number of rows. This uses blue backlight to view the letters. This gets the value from the Raspberry Pi and displays the values and status in the screen (Fig. 5).

Fig. 5 LCD display 20 × 4

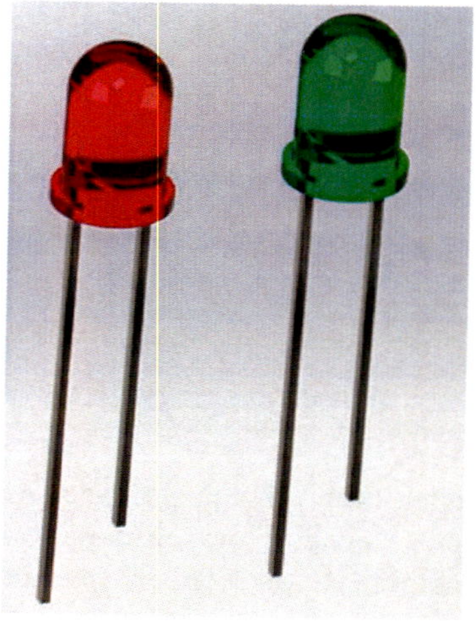
Fig. 6 LED lights

4.4 LED Lights

A light-emitting diode (LED) is a semiconductor mild supply that emits light when current flows through it. Electrons within the semiconductor recombine with electron holes, releasing energy inside the shape of photons. The system consists of two LED lights green and red for safe and critical state of the fluid bottle, respectively. According to the level of the fluid bottle, LED light will glow (Fig. 6).

4.5 Buzzer

5 Experimental Results

5.1 Status Displayed in the LCD Display as per the Data given by the Raspberry Pi

LCD display shows the condition of the bottle, amount of fluid flows per second, total discharge of fluid from the bottle and the time taken for the total discharge of fluid from the bottle (Fig. 8).

The buzzer is a sounding device that can convert audio signals into sound signals. It is usually powered by DC voltage. It is widely used in alarms, computers, printers and other electronic products as sound devices. This will alert the hospital staff when the level reaches the threshold point (Fig. 7).

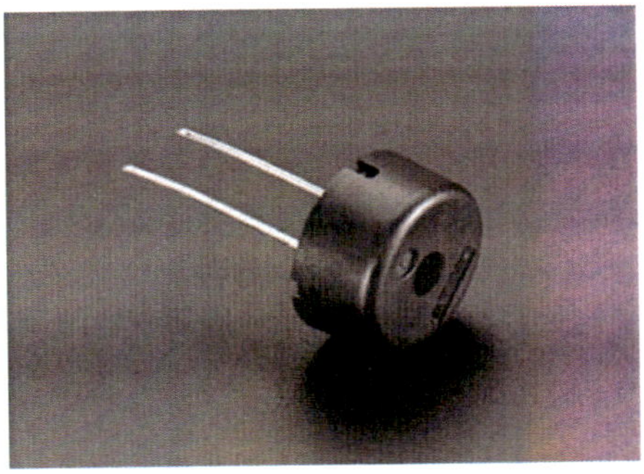

Fig. 7 Buzzer

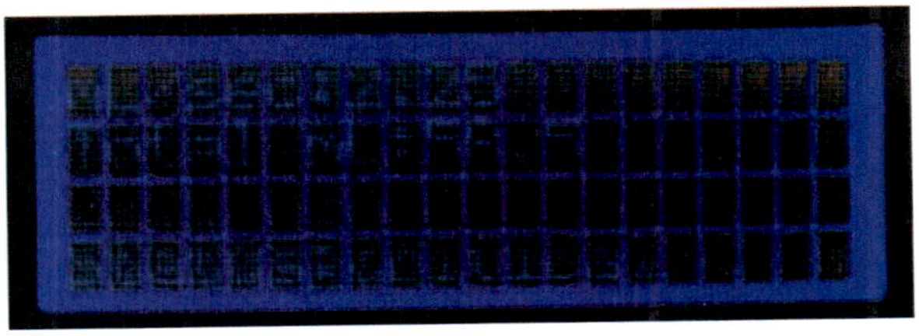

Fig. 8 LCD status display

5.2 *Data Updated in the Firebase*

If the total discharge is lesser than the threshold value, then the condition is safe, and the same will be updated in the database together with the total discharge and the rate of flow of fluid per second. If the total discharge is greater than the threshold value, then the condition is critical and it is updated in the database with the values of total discharge and the rate of flow of fluid from bottle (Fig. 9).

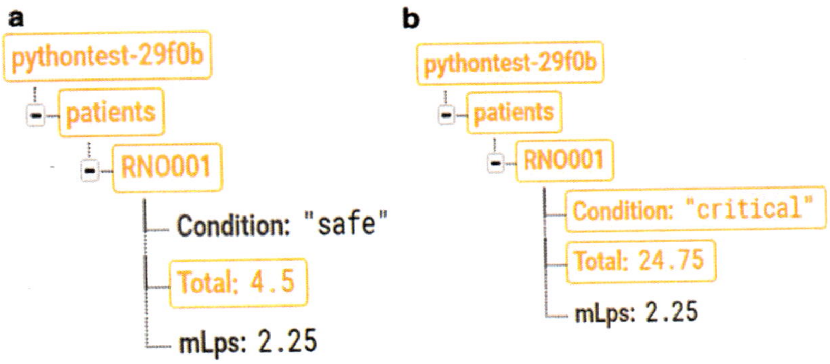

Fig. 9 **a** Firebase safe state, **b** firebase critical state

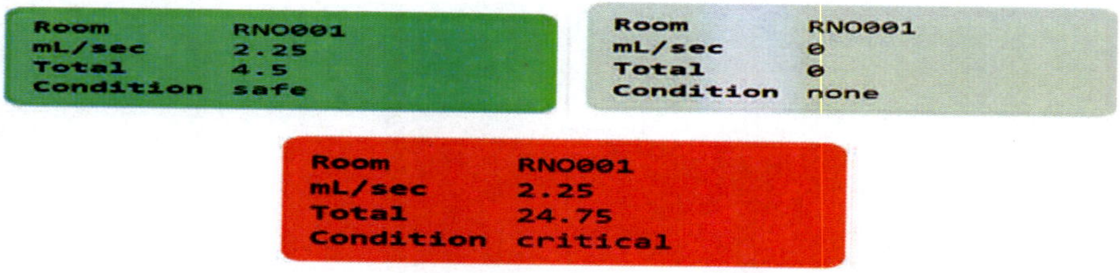

Fig. 10 Web page status

5.3 Data Retrieved from the Firebase

The Web page shows the condition, total discharge of fluid and the rate of flow of fluid. According to three different conditions none, safe and critical, the data is displayed with three different colors (Fig. 10).

6 Future Scope of Improvement

6.1 Mobile Application Monitoring

Mobile application is developed and the information of the fluid bottle is retrieved from the firebase and displays the same. To fix the threshold value, the capacity of the bottle is taken from the staff through the application before the start of the system.

6.2 Preventing the Reverse Flow

By fixing the solenoid valve, the system can operate a mechanism of opening and closing of fluid flow in a valve.

7 Conclusion

This project overcomes the consequences that occur due to negligence of monitoring the IV flow. Using proposed monitoring, one can monitor the level of bottle from a distant position which will aid in building smart healthcare system. This is an affordable, precise and efficient system that works in a smooth manner. Flow sensor will give data flow of fluid so that the discharge of water is monitored. This will be helpful especially at night times. Anyone can view the LCD display and report it to the staff if they did not see the monitor. This will prevent serious hazards of patients and will eliminate the manual monitoring of the patients by hospital staffs.

References

1. D. Oros, M. Penic, J. Sulc, M. Cavic, S. Stankovski, G. Ostojic, O. Ivanov, Smart intravenous infusion dosing system. Appl. Sci. **11**, 513 (2021)
2. R. Anagha, S. Ashwini, G. Keerthana, M. Monica, IoT based intravenous flow monitoring system. Int. Res. J. Eng. Technol. (IRJET) **07**(05) (2020)
3. E.L. Mathew, J.K. James, A. Radhakrishnan, B. Sebastian, H. Mathew, The novel intravenous fluid level indicator for smart IV systems. Int. Res. J. Eng. Technol. (IRJET) **07**(06) (2020)
4. T.A. Walsh, C. Sandmann, *Integrated Sensor to Monitor Fluid Delivery* (United States Patent Application Publication, 2020)
5. S. Joseph, N. Francis, A. John, B. Farha, Intravenous drip monitoring system for smart hospital using IoT, in *2nd International Conference on Intelligent Computing, Instrumentation and Control Technologies*, 2019
6. M.R. Novak, *Detachable Intravenous Alarm Monitoring System* (United States Patent Application Publication, 2019)
7. K. Sampath Kumar, S. Abinaya, M. Abirami, S. Kesavan, S. Sethu, Innovative intravenous fluid control and emergency monitoring system. Int. J. Eng. Manuf. Sci. **8**(3). ISSN 2249-3115
8. A. Arulious Jora, A. Divya Laveena, D. Earlina, S. Nirmala, Intravenous fluid level indicator. Int. Res. J. Eng. Technol. (IRJET) **05**(10) (2018)
9. M.K. Bhavasaar, M. Nithya, R. Praveena, N.S. Bhuvaneswari, T. Kalaiselvi, Automated intravenous fluid monitoring and alerting system, in *IEEE International Conference on Technological Innovations in ICT For Agriculture and Rural Development (TIAR 2016)*
10. K.R. Rao, K.E. Supriya, Design and development of IoT based intravenous infusion system, Emerging Trends in Electrical, Communications and Information Technologies, Lecture Notes in Electrical Engineering (2019)
11. A. Rajaram, S. Palaniswami, The modified security scheme for data integrity in Manet. Int. J. Eng. Sci. Technol. **2**(7), 3111–3119 (2010)

Author Index

S. Shakya et al. (eds.), *Mobile Computing and Sustainable Informatics*, Lecture Notes on
Data Engineering and Communications Technologies 68,
https://doi.org/10.1007/978-981-16-1866-6